T0299134

Stellar Evolution Physics
Volume 2: Advanced Evolution of Single Stars

This volume explains the microscopic physics operating in stars in advanced stages of their evolution and describes with many numerical examples and illustrations how they respond to this microphysics. Models of low and intermediate mass are evolved through the core helium-burning phase, the asymptotic giant branch phase (alternating shell hydrogen and helium burning), and through the final cooling white dwarf phase. A massive model is carried from the core helium-burning phase through core and shell carbon-burning phases. Gravothermal responses to nuclear reaction-induced transformations and energy loss from the surface are described in detail. Written for senior graduate students and researchers who have mastered the principles of stellar evolution, as developed in the first volume of *Stellar Evolution Physics*, sufficient attention is paid to how numerical solutions are obtained to enable the reader to engage in model construction on a professional level.

The processes in this volume build upon those in Volume 1 of *Stellar Evolution Physics: Physical Processes in Stellar Interiors* (ISBN 978-1-107-01656-9), which describes the microscopic physics operating in stars and demonstrates how stars respond from formation, through hydrogen-burning phases, up to the onset of helium burning. *Stellar Evolution Physics* is also available as a 2-volume set (ISBN 978-1-107-60253-3). Taken together, the two volumes will prepare a graduate student for professional-level research in this key area of astrophysics.

Icko Iben, Jr. is Emeritus Distinguished Professor of Astronomy and Physics at the University of Illinois at Urbana-Champaign, where he also gained his MS and PhD degrees in Physics and where a Distinguished Lectureship in his name was established in 1998. He initiated his teaching career at Williams College (1958–61), engaged in astrophysics research as a Senior Research Fellow at Cal Tech (1961–4), and continued his teaching career at MIT (1964–72) and Illinois (1972–99). He has held visiting Professorships at over a dozen institutions, including Harvard University, the University of California at Santa Cruz, the University of Bologna, Italy, and Niigata University, Japan. He was elected to the US National Academy of Sciences in 1985, and his awards include the Russell Lectureship of the American Astronomical Society (1989), the George Darwin Lectureship (1984) and the Eddington Medal (1990) of the Royal Astronomical Society, and the Eminent Scientist Award of the Japan Society for the Promotion of Science (2003–4).

Stellar Evolution Physics

Volume 2: Advanced Evolution of Single Stars

ICKO IBEN, JR.

University of Illinois at Urbana-Champaign

CAMBRIDGE
UNIVERSITY PRESS

Shaftesbury Road, Cambridge CB2 8EA, United Kingdom

One Liberty Plaza, 20th Floor, New York, NY 10006, USA

477 Williamstown Road, Port Melbourne, VIC 3207, Australia

314–321, 3rd Floor, Plot 3, Splendor Forum, Jasola District Centre, New Delhi – 110025, India

103 Penang Road, #05–06/07, Visioncrest Commercial, Singapore 238467

Cambridge University Press is part of Cambridge University Press & Assessment,
a department of the University of Cambridge.

We share the University's mission to contribute to society through the pursuit of
education, learning and research at the highest international levels of excellence.

www.cambridge.org
Information on this title: www.cambridge.org/9781107016576

First published 2013
Reprinted 2013

A catalogue record for this publication is available from the British Library

Library of Congress Cataloging-in-Publication data
Iben, Icko, 1931–
Stellar evolution physics / Icko Iben, Jr.
p. cm.
Includes bibliographical references and index.
ISBN 978-1-107-01657-6 (Hardback)
1. Stars–Evolution.   2. Stellar dynamics.   I. Title.
QB806.I24 2012
523.8′8–dc23
2012019504

ISBN   978-1-107-01657-6   Hardback

Also available as part of a two-volume set, *Stellar Evolution Physics* ISBN 978-1-107-60253-3

Contents

Preface

One might think that the most appropriate division of topics in a two volume book on stellar evolution physics would be the placement of all chapters describing the input physics required for the construction of stellar models in the first volume and the placement of all chapters describing stellar evolutionary models in the second volume. However, such a division disguises the fact that it is the operation of the input physics in stars that explains why they shine and evolve and that, therefore, both the input physics and the response of stars to the operation of this physics comprise the science of stellar evolution physics.

In preparing this book, after describing much of the input physics required for the construction of stellar models during early evolutionary stages, I constructed models in these early stages of evolution. Then, after describing some of the more complicated physical processes that play important roles during more advanced stages of evolution, I constructed models in these more advanced stages. The ordering of topics in the two volumes of this book reflects this chronological development.

After providing a general introduction to the observed properties of real stars and to the results of stellar evolution calculations, the first volume focusses on equations of state, energy generation by hydrogen-burning reactions, energy transport by radiation and convection, and on the elementary equations of stellar evolution and methods of solution. This is followed by a description of stellar models evolving during gravitationally contracting phases onto the main sequence, during the main sequence phase when core hydrogen burning is the primary source of surface luminosity, and during shell hydrogen-burning phases up to the onset of helium burning as the primary factor in controlling the evolutionary time scale and as a source of surface luminosity second only to hydrogen burning.

In the first part of this second volume, which is divided into three parts, input physics of a somewhat more subtle nature than addressed in the first volume is presented – diffusion, heat conduction by electrons, beta decay and electron capture at high densities, and weak interaction processes that are responsible for the production of neutrino–antineutrino pairs. The first part ends with a discussion of helium-burning nuclear reactions, which control the evolutionary time scale during the quiescent core helium-burning phase and play starring roles in the intricate dance between hydrogen burning and helium burning during the thermally pulsing asymptotic giant branch phase, leading to the formation of carbon and neutron-rich s-process elements that make their way from interior regions of production to the surface and thence into the interstellar medium.

All of the input physics developed in the first volume plays a role in the evolutionary models described in both volumes. Some of the input physics developed in this volume also plays a role in several evolutionary models presented in the first volume. For example, electron conduction and energy loss by neutrinos and antineutrinos play decisive roles in establishing the thermal structure of the hydrogen-exhausted, electron-degenerate cores of low mass post-main sequence stars as they ascend the red giant branch for the first time, with the consequence that helium is ignited off center and proceeds in a series of flashes until a quiescent core helium-burning phase is reached. Evolution to the off center ignition of helium is described in Chapter 11 at the end of the first volume, whereas the helium flashing phase and the subsequent quiescent core helium-burning phase are described in Chapter 17 in this volume.

Another example of the relevance of the input physics developed in this volume to models described in the first volume is the discussion in Chapter 15 of the interaction Hamiltonion for weak interactions. The interaction between electron neutrinos and electrons predicted by this Hamiltonion has been used in Chapter 10 in Volume 1 to demonstrate that some of the electron neutrinos generated in the Sun are converted into muon neutrinos on their passage outward through the Sun, a demonstration which contributed significantly to the resolution of the solar neutrino problem which plagued stellar astrophysics for many years.

Results of stellar model calculations during helium-burning phases for models of mass $1\,M_\odot$, $5\,M_\odot$, and $25\,M_\odot$ are described in the middle part of this second volume in Chapters 17, 18, and 20, respectively. In the case of the two lower mass models, special attention is given to the thermally pulsing AGB (TPAGB) phase when the main nuclear burning source of energy alternates between hydrogen burning and helium burning. In the case of the $5\,M_\odot$ model, special attention is given in Chapter 19 to the activation of s-process nucleosynthesis and to the dredge-up of freshly made carbon and s-process elements during the TPAGB phase. In the case of the $25\,M_\odot$ model, evolution is carried into the core and shell carbon-burning stages.

In the third and last part of this volume, consisting of Chapter 21, attention is focussed on the final stages of evolution of low and intermediate mass stars which become white dwarfs, with the evolution of a model of initial mass $1\,M_\odot$ being highlighted. Described in the first three sections of Chapter 21 are (1) the wind mass loss which a TPAGB star experiences in consequence of radiation pressure on grains in a shock-inflated atmosphere, the shocks being due to Mira-like acoustical pulsations, and (2) the resulting planetary nebula stage, when the contracting remnant emits radiation of sufficient energy to cause the ejected material to fluoresce. After a discussion in the next two sections of the equation of state and the specific heat of stellar matter in the solid and liquid states, the final evolution of the remnant as it cools as a white dwarf is described in the sixth section. In concluding sections, the formation of monoelemental surface layers due to diffusion, the dependence of final surface abundances on evolutionary history, and quantitative estimates of the birthrate of low mass stars in the Galaxy and of the age of the Galactic disk are presented.

PART IV

TRANSPORT PROCESSES, WEAK INTERACTION PROCESSES, AND HELIUM-BURNING REACTIONS

12 Particle diffusion and gravitational settling

In constructing models of evolving stars in volume 1 of this monograph, it has been assumed that particle diffusion can be neglected. However, the current abundance of Li at the solar surface is much smaller than predicted by models neglecting diffusion and this suggests that, during the gravitationally contracting phase which preceeds the main sequence, diffusion carries Li from the convective envelope into higher temperature regions below the base of the envelope where it can be destroyed. The abundance of Fe observed at the solar surface is smaller than the interior Fe abundance indicated by comparison between neutrino fluxes observed from the Sun and fluxes predicted by solar models, suggesting that Fe has diffused out of the convective envelope during the main sequence phase into regions below the base of the convective envelope. The fact that many low luminosity white dwarfs exhibit monoelemental surface abundances is a dramatic demonstration that, in regions where the gravitational acceleration is orders of magnitude larger than near the surface of the Sun, gravitationally induced diffusion is a first order effect. Thus, there is ample motivation for studying the physics of diffusion.

A description of the physics of particle diffusion can be given on many levels of sophistication. The description adopted in this chapter is based on an analysis of Boltzmann transport equations constructed on the assumptions that there exists an equilibrium distribution function for every species of particle and that, when a system is not in equilibrium, the time rate of change of each distribution function can be determined as the consequence of binary interactions between all particles.

In a star, three distinct types of particle diffusion occur simultaneously. They are induced, respectively, by the existence of a gravitational field, the existence of a temperature gradient, and the existence of concentration gradients. In regions of nuclear energy generation, concentration gradients are primarily a consequence of nuclear transformations. Elsewhere, they are a consequence of gravitationally and thermally-induced diffusion. All three types of diffusion are familiar in terrestial contexts. Gravitationally-induced diffusion is responsible for the fact that, in the Earth's atmosphere, the abundance of any given type of molecule decreases exponentially with distance above the Earth's surface over a scale height that is inversely proportional to the mass of the molecule. The second and third forms of diffusion compete: thermal diffusion moves faster-moving lighter particles into regions containing slower-moving heavier particles but concentration gradients set up by this diffusion act in the reverse direction.

In that it demands the existence of an electrical field of strength proportional to the gravitational acceleration, perhaps the most intellectually entertaining process is gravitationally induced diffusion in a region in which ions are at least partially ionized. The discussion of this type of diffusion can be broken into four parts: (1) analysis of the consequences of the equations when all processes are in equilibrium, (2) establishment of equations which describe how binary interactions affect the time rate of change of each distribution function, (3) estimation of the cross sections which characterize the interactions required to establish and maintain equilibrium, and (4) construction of algorithms for solving the diffusion equations. The discussion in this chapter focuses on parts (1), (2), and (4), with considerations involved in part (3) being described more thoroughly in Chapter 13.

The Boltzmann transport equations and their applications are discussed at length in many textbooks on statistical mechanics. A good example is the discussion by Kerson Huang in *Statistical Mechanics* (1963). Application to diffusion in the stellar context has been treated exhaustively by S. Chapman and T. G. Cowling in *Mathematical Theory of Nonuniform Gases* (1970). The discussion in this chapter leans heavily on the treatment by J. M. Burgers in *Flow Equations for Composite Gases* (1969).

In Section 12.1, it is shown that the lowest order moment of the Boltzmann transport equation for a single element species in an external force field vanishes when number conservation prevails and that the next order (linear momentum) moment vanishes when the pressure-gradient force exerted by the species is in exact balance with the external force on the species. In Section 12.2, it is shown that the distribution in space of an element which can be present in two ionization stages in a constant gravitational field at constant temperature can be found exactly but that a simple solution does not exist when more than two ionization states are present. In Section 12.3, the number conservation and linear momentum moments are found for a multicomponent system in which each species of particle is assumed to move, at any given point, with a well defined diffusion velocity with respect to the center of mass of the entire system; the number conservation moments for each species and for the entire system vanish; the overall linear momentum moment vanishes when overall balance between gravitational forces and pressure-gradient forces is assumed, but, in general, the linear momentum moment of an individual species involves an electrical field strength proportional to the gravitational acceleration and the moment does not vanish unless equilibrium with respect to diffusion has been achieved. The relationship between the electrical and gravitational field strengths when pressure balance prevails is described in Section 12.4.

In Section 12.5, it is shown that, for a system consisting of two species in prescribed stages of ionization, if it is assumed that the ratio of species abundances is independent of position, the linear momentum moment for each species can be found in closed form. The moments are identical in absolute value but opposite in sign, depending on species mass and ionic charge in such a way that the net gravitational force on the species with the larger mass to charge ratio is directed inward and is larger than the pressure-gradient force exerted by the more massive ions *plus the pressure-gradient force exerted by the electrons derived from the ions*. Exactly the opposite is true for the lighter ions plus attendant electrons. The forces are such that ions with the larger mass to charge ratio diffuse inward and those with the smaller mass to charge ratio diffuse outward, with the difference in linear momentum moments being proportional to the difference in diffusion velocities.

In Section 12.6, the concept of a resistence coefficient is introduced in the context of a three component system consisting of two ions and attendant electrons. The coefficient is identified as a consequence of scattering collisions between ions and an estimate of its dependence on ionic charges, on temperature and density, and on differences between diffusion velocities is obtained heuristically. By equating the resistence coefficient for every ion with its linear momentum moment, one establishes a set of relationships between diffusion velocities and local properties. In Section 12.7, it is shown that the constraints imposed by doing all calculations in the center of mass frame of reference and insisting on no electric current permit one to determine individual diffusion velocities. Resistence coefficients involving scattering interactions between electrons and ions are introduced and shown to be of secondary importance in estimating diffusion velocities. Generalization to a multicomponent gas is made in Section 12.8.

In Sections 12.9 and 12.10, gravitational diffusion velocities at the base of the convective envelope in solar models are estimated, indicating that, even when thermal diffusion is neglected, iron and other heavy elements diffuse inward rapidly enough to affect the estimate of the interior heavy element abundances based on observationally based estimates of surface abundances. Including the effects of thermal diffusion increases the rate at which heavy elements are depleted in the convective envelope (by over a factor of two in an early treatment, but perhaps by much less according to more recent estimates of the relative importance of thermal diffusion).

In Section 12.11, equations for the rates of change of abundances due to gravitational and concentration-gradient induced diffusion are presented and algorithms for calculating the changes in abundance due to diffusion in theoretical stellar models are constructed. Evidence that these algorithms lead to a monoelemental surface abundance in a cooling white dwarf is presented in Section 21.8.

12.1 Moments of the Boltzmann transport equation for a species under conditions of complete equilibrium

Consider a non-relativistic gas composed of particles of type s in thermal equilibrium in an environment where each particle of mass m_s and electrical charge e_s is acted upon by a gravitational field $\mathbf{g}(\mathbf{r}, t)$ and an electrical field $\mathbf{E}(\mathbf{r}, t)$, both of which may be functions of time t and position $\mathbf{r}$. The force on each particle is given by

$$\mathbf{F}_s(\mathbf{r}_s, t) = m_s \mathbf{g}(\mathbf{r}_s, t) + e_s \mathbf{E}(\mathbf{r}_s, t), \qquad (12.1.1)$$

where $\mathbf{E}$ is the strength of an electrostatic field of the sort introduced in Section 3.6 in Volume 1 as the necessary consequence of conditions in ionized matter in a gravitational field of strength $\mathbf{g}$.

Suppose that, at each point in a seven-dimensional phase space where the coordinates are position $\mathbf{x}$, velocity $\mathbf{v}$, and time t, there exists a distribution function

$$f_s = f_s(\mathbf{x}, \mathbf{v}, t) \qquad (12.1.2)$$

which has the properties that, at any time t and at any position $\mathbf{x}$, the number density of particles is given by

$$n_s(\mathbf{x}, t) = \int f_s(\mathbf{x}, \mathbf{v}, t) \, \mathrm{d}^3\mathbf{v} \qquad (12.1.3)$$

and the mean, or flow, velocity is given by

$$n_s \mathbf{u}_s(\mathbf{x}, t) = \int f_s(\mathbf{x}, \mathbf{v}, t) \, \mathbf{v} \, \mathrm{d}^3\mathbf{v}. \qquad (12.1.4)$$

In these expressions, $\mathrm{d}^3\mathbf{v}$ is a three dimensional volume element in velocity space and the integration is over all three dimensions in velocity space.

The rate at which the distribution function changes in phase space may be written as

$$\frac{\delta f_s}{\delta t} = \frac{\partial f_s}{\partial t} + \sum_i \frac{\mathrm{d}x_i}{\mathrm{d}t} \frac{\partial f_s}{\partial x_i} + \sum_i \frac{\mathrm{d}v_i}{\mathrm{d}t} \frac{\partial f_s}{\partial v_i} = \frac{\partial f_s}{\partial t} + \sum_i v_i \frac{\partial f_s}{\partial x_i} + \sum_i \frac{\mathrm{d}v_i}{\mathrm{d}t} \frac{\partial f_s}{\partial v_i}.$$

$$= \frac{\partial f_s}{\partial t} + \sum_i v_i \frac{\partial f_s}{\partial x_i} + \sum_i \frac{F_{si}}{m_s} \frac{\partial f_s}{\partial v_i}, \qquad (12.1.5)$$

where the subscript i specifies any of three orthogonal vector directions. The quantity $\delta f_s/\delta t$ at the extreme left in eq. (12.1.5) has been introduced at this point primarily to provide a label for identifying the sums to the right of the equal signs. Under very special circumstances, corresponding to various levels on which equilibrium has been achieved, either $\delta f_s/\delta t$ or the integral over $\mathrm{d}^3\mathbf{v}$ of some quantity times $\delta f_s/\delta t$ vanishes.

For example, multiplying the leftmost and rightmost sides of eq. (12.1.5) by $\mathrm{d}^3\mathbf{v}$ and integrating produces

$$\int \frac{\delta f_s}{\delta t} \, \mathrm{d}^3\mathbf{v} = \frac{\partial}{\partial t} \int f_s \mathrm{d}^3\mathbf{v} + \sum_i \frac{\partial}{\partial x_i} \int v_i \, f_s \mathrm{d}^3\mathbf{v} + \sum_i \frac{F_{si}}{m_s} \int \frac{\partial f_s}{\partial v_i} \mathrm{d}^3\mathbf{v} = 0. \quad (12.1.6)$$

The integral has been set equal to zero because, when it is translated into a form involving time and spatial derivatives of the familiar entities, number density and bulk velocity, the term between the equal signs is identical with eq. (8.1.9) in Volume 1, which follows when mass is conserved. Ultimately, $\delta f_s/\delta t$ may be viewed as a device for taking into account interactions between different species of particle, effects of which are absent in the zeroth order distribution functions constructed for isolated species.

Equations (12.1.3)–(12.1.6) are basic in the development of Boltzmann transport equations and the quantity f_s is known as a Boltzmann distribution function. If, in the frame of reference moving with the velocity $\mathbf{u}_s$, the distribution function is isotropic (depending on the velocity as $|\mathbf{v} - \mathbf{u}_s|$), then $f_s(\mathbf{x}, \mathbf{v}, t)$ is an even function of each $v_i - u_{si}$ and the third integral between equal signs in eq. (12.1.6) vanishes. Making use of eqs. (12.1.3) and (12.1.4), the final result in this case is

$$\frac{\partial n_s}{\partial t} + \sum_i \frac{\partial}{\partial x_i} (n_s u_{si}) = 0, \qquad (12.1.7)$$

which can be written in vector notation as

$$\frac{\partial n_s}{\partial t} + \nabla \cdot (n_s\, \mathbf{u}_s) = \left(\frac{dn_s}{dt}\right)_s + n_s\, \nabla \cdot \mathbf{u}_s = 0, \tag{12.1.8}$$

where

$$\left(\frac{dn_s}{dt}\right)_s = \frac{\partial n_s}{\partial t} + \mathbf{u}_s \cdot \nabla n_s \tag{12.1.9}$$

is the time derivative in the frame of reference moving with the flow velocity $\mathbf{u}_s$. Multiplying eqs. (12.1.7) and (12.1.8) by m_s and setting $\rho_s = m_s n_s$, the results are identical in form with the continuity, or mass conservation, equation for the gas as a whole, as expressed by eq. (8.1.9) in Volume 1. In Section 8.1 of Volume 1, the symbol $\mathbf{v}$ is defined as the flow velocity for the gas as a whole, in contrast with the notation in this section where $\mathbf{u}_s = \mathbf{u}_s(\mathbf{x}, t)$ is the flow velocity for a particular species and $\mathbf{v}$ is a variable parameter.

In the absence of nuclear transformations, individual particle species are conserved. When nuclear reactions occur, because the number of nucleons is conserved, mass is, to a good approximation, also conserved. Thus, the identity between eq. (8.1.9) and eqs. (12.1.7) and (12.1.8) means that the latter two equations are essentially rigorously true, even if full equilibrium has not been achieved. The fact that the integral of $\delta f_s/\delta t$ over $d^3\mathbf{v}$ must vanish also places a constraint on the manner in which interactions between particles of different species enter into the construction of $\delta f_s/\delta t$.

Multiplying the integrands in eq. (12.1.6) by any quantity Q and integrating over all velocity space produces what may be called the moment of Q. When $Q = m_s$, the resultant moment M_{s0} may be called the continuity moment, in recognition of the fact that the assumption that M_{s0} vanishes under equilibrium conditions leads to the same continuity equation which is obtained when particle or mass conservation is assumed.

Choosing $Q = m_s\mathbf{v}$ produces what may be called a linear momentum moment. In a reference frame moving with velocity $\mathbf{u}_s(\mathbf{x}, t)$, the velocity parameter $\mathbf{c}$ is related to the velocity parameter $\mathbf{v}$ in the fixed frame by

$$\mathbf{c} = \mathbf{v} - \mathbf{u}_s(\mathbf{x}, t). \tag{12.1.10}$$

If in the moving frame the distribution function has the Maxwell–Boltzmann form

$$f_s(\mathbf{x}, \mathbf{c}, t) = A_s(\mathbf{x}, t)\, e^{-m_s c^2/2kT}, \tag{12.1.11}$$

where

$$A_s(\mathbf{x}, t) = n_s(\mathbf{x}, t)\left(\frac{m_s}{2\pi kT}\right)^{3/2}, \tag{12.1.12}$$

in the stationary frame it has the form

$$f_s'(\mathbf{x}, \mathbf{v}, t) = A_s(\mathbf{x}, t)\, e^{-m_s|\mathbf{v}-\mathbf{u}_s|^2/2kT}. \tag{12.1.13}$$

It is evident that $f_s(\mathbf{x}, \mathbf{c}, t) = f_s'(\mathbf{x}, \mathbf{v}, t)$ and that $\partial f_s(\mathbf{x}, \mathbf{c}, t)/\partial c_i = \partial f_s'(\mathbf{x}, \mathbf{v}, t)/\partial v_i$. In both frames,

$$n_s = \int f_s(\mathbf{x}, \mathbf{c}, t)\, \mathrm{d}^3\mathbf{c} = \int f_s'(\mathbf{x}, \mathbf{v}, t)\, \mathrm{d}^3\mathbf{v}, \tag{12.1.14}$$

but

$$\bar{\mathbf{c}} = \int \mathbf{c}\, f_s(\mathbf{x}, \mathbf{c}, t)\, \mathrm{d}^3\mathbf{c} = 0 \tag{12.1.15}$$

and

$$\bar{\mathbf{v}} = \int \mathbf{v}\, f_s'(\mathbf{x}, \mathbf{v}, t)\, \mathrm{d}^3\mathbf{v} = \mathbf{u}_s. \tag{12.1.16}$$

Derivatives in the two frames are related by

$$\frac{\partial f_s'(\mathbf{x}, \mathbf{v}, t)}{\partial t} = \frac{\partial f_s(\mathbf{x}, \mathbf{c}, t)}{\partial t} + \sum_i \frac{\partial f_s(\mathbf{x}, \mathbf{c}, t)}{\partial c_i} \frac{\partial c_i}{\partial t}$$

$$= \frac{\partial f_s(\mathbf{x}, \mathbf{c}, t)}{\partial t} - \sum_i \frac{\partial f_s(\mathbf{x}, \mathbf{c}, t)}{\partial c_i} \frac{\partial u_{si}(\mathbf{x}, t)}{\partial t} \tag{12.1.17}$$

and

$$\frac{\partial f_s'(\mathbf{x}, \mathbf{v}, t)}{\partial x_j} = \frac{\partial f_s(\mathbf{x}, \mathbf{c}, t)}{\partial x_j} + \sum_i \frac{\partial f_s(\mathbf{x}, \mathbf{c}, t)}{\partial c_i} \frac{\partial c_i}{\partial x_j}$$

$$= \frac{\partial f_s(\mathbf{x}, \mathbf{c}, t)}{\partial x_i} - \sum_i \frac{\partial f_s(\mathbf{x}, \mathbf{c}, t)}{\partial c_i} \frac{\partial u_{si}(\mathbf{x}, t)}{\partial x_j}, \tag{12.1.18}$$

where, again, the index i specifies one of three orthogonal directions. The second line in each of the last two equations follows from the first line by differentiating c_i in eq. (12.1.10) with respect to x_j, recognizing that $\mathbf{v}$ in this equation is independent of $\mathbf{x}$.

Using eqs. (12.1.15)–(12.1.17) in the last form of eq. (12.1.5) gives

$$\frac{\delta f_s'}{\delta t} = \frac{\partial f_s}{\partial t} + \sum_i (u_{si} + c_i) \frac{\partial f_s}{\partial x_i}$$

$$- \sum_i \frac{\partial f_s}{\partial c_i} \left[\frac{\partial u_{si}}{\partial t} + \sum_i (u_{sj} + c_j) \frac{\partial u_{si}}{\partial x_j} \right] + \sum_i \frac{F_{si}}{m_s} \frac{\partial f_s}{\partial c_i}, \tag{12.1.19}$$

or

$$\frac{\delta f_s'}{\delta t} = \frac{\mathrm{d} f_s}{\mathrm{d} t} + \sum_i c_i \frac{\partial f_s}{\partial x_i} - \sum_i \frac{\partial f_s}{\partial c_i} \left[\frac{\mathrm{d} u_{si}}{\mathrm{d} t} + \sum_j c_j \frac{\partial u_{si}}{\partial x_j} \right] + \sum_i \frac{F_{si}}{m_s} \frac{\partial f_s}{\partial c_i}. \tag{12.1.20}$$

Multiplying by $Q \, d^3\mathbf{c}$ and integrating gives

$$\int \frac{\delta f_s}{\delta t} Q \, d^3c = \frac{d}{dt} \int f_s \, Q \, d^3c + \sum_i \frac{\partial}{\partial x_i} \int c_i \, f_s \, Q \, d^3c$$

$$- \sum_i \frac{du_{si}}{dt} \int \frac{\partial f_s}{\partial c_i} Q \, d^3c - \sum_i \sum_j \frac{\partial u_{si}}{\partial x_j} \int \frac{\partial f_s}{\partial c_i} c_j \, Q \, d^3c$$

$$+ \sum_i \frac{F_{si}}{m_s} \int \frac{\partial f_s}{\partial c_i} Q \, d^3c. \qquad (12.1.21)$$

The three terms containing $\partial f_s / \partial c_i$ can be integrated by parts and, since f_s must vanish at the limits, one has that

$$\int \frac{\delta f_s}{\delta t} Q \, d^3c = \frac{d}{dt} \int f_s Q \, d^3c + \sum_i \frac{\partial}{\partial x_i} \int c_i \, f_s \, Q \, d^3c$$

$$+ \sum_i \frac{du_{si}}{dt} \int f_s \frac{\partial Q}{\partial c_i} d^3c + \sum_i \sum_j \frac{\partial u_{si}}{\partial x_j} \int f_s \frac{\partial (c_j \, Q)}{\partial c_i} d^3c$$

$$- \sum_i \frac{F_{si}}{m_s} \int f_s \frac{\partial Q}{\partial c_i} d^3c. \qquad (12.1.22)$$

It follows from eq. (12.1.15) that, if $Q = m_s c_k$, the first and fourth terms on the right hand side of eq. (12.1.22) vanish. The integrals in the third and fifth terms are just $m_s n_s \delta_{ki}$, where δ_{ki} is a delta function, and the ith integral in the second term is $m_s n_s \overline{c_i^2} \, \delta_{ki}$, where the bar over c_i^2 denotes an average over the distribution function. Summing over i then gives for the kth component of the linear momentum moment:

$$M_{sk} = \frac{\partial}{\partial x_k} \left(\overline{c_k^2} \, m_s n_s \right) + \frac{du_{sk}}{dt} m_s n_s - F_{sk} \, n_s. \qquad (12.1.23)$$

If the particles obey Maxwell–Boltzmann statistics, the appropriate distribution function is given by eqs. (12.1.11) and (12.1.12), and, using eq. (12.1.3), one has that

$$\overline{c_k^2} \, m_s n_s = n_s \left(\frac{m_s}{2\pi kT} \right)^{3/2} \int c_k^2 \, e^{-m_s c^2 / 2kT} \, d^3c = n_s kT = P_s, \qquad (12.1.24)$$

where it has been recognized that the product $n_s kT$ is the isotropic pressure P_s exerted by the s-type particles in the frame of reference moving with the flow velocity $\mathbf{u}_s$. As long as particles do not move with relativistic velocities, the relationship between the isotropic pressure and the quantity $\overline{c_k^2} \, m_s n_s$ is independent of the statistics adopted in deriving the distribution function, as may be seen by setting $p_x = p_{sk} = m_s c_{sk}$ and $n = n_s$ in eq. (4.3.1) in Chapter 4 of Volume 1, which follows from the definition of pressure as the rate at which momentum is transferred to a specularly reflecting wall. On the other hand, the pressure calculated in this way is a function of the frame of reference in which the calculation is performed. For example, adopting the distribution function given by eq. (12.1.13), and choosing the k direction so that $\hat{\mathbf{k}} \cdot \mathbf{u}_s = u_s$, one has that

$$(P_s)_k' = \int f_s \, v_k^2 \, d^3\mathbf{v} = \int \left[(v_k - u_k)^2 + 2(v_k - u_k) + u_k^2 \right] f_s \, d^3\mathbf{v}$$

$$= n_s k T + n_s m_s u_s^2, \qquad (12.1.25)$$

which may be written as

$$(P_s)_k' = P_s + P_{s,\text{ram}} \, \hat{\mathbf{k}} \cdot \hat{\mathbf{u}}_s, \qquad (12.1.26)$$

where

$$P_{s,\text{ram}} = n_s m_s u_s^2 \qquad (12.1.27)$$

is called a ram pressure. Thus, in the stationary frame, the pressure is composed of a component which is identical with the isotropic gas pressure in the moving frame and of a vector component which, being the product of the bulk linear momentum density and the bulk velocity, is the flux of bulk linear momentum.

In any case, in the frame of reference moving with the flow velocity $\mathbf{u}_s$, the linear momentum moment, eq. (12.1.23), can be written as

$$M_{sk} = \frac{\partial P_s}{\partial x_k} + \frac{du_{sk}}{dt} m_s n_s - F_{sk} \, n_s, \qquad (12.1.28)$$

or, setting $\rho_s = m_s n_s$,

$$M_{sk} = \frac{\partial P_s}{\partial x_k} + \rho_s \left(\frac{du_{sk}}{dt} - \frac{F_{sk}}{m_s} \right). \qquad (12.1.29)$$

Setting $M_{sk} = 0$ is tantamount to stating that particles of type s are in exact pressure balance with themselves, whether or not other types of particle are present. If other types of particle are present, the assumption that $M_{sk} = 0$ places constraints on the nature of the interactions between particles of different types which will permit this condition to be achieved. In vector notation, the condition $M_{sk} = 0$ may be written as

$$\rho_s \frac{d\mathbf{u}_s}{dt} = -\nabla P_s + \frac{\mathbf{F}_s}{m_s}, \qquad (12.1.30)$$

or, adopting $\mathbf{F}_s$ given by eq. (12.1.1),

$$\rho_s \frac{d\mathbf{u}_s}{dt} = -\nabla P_s + \left(\mathbf{g} + \frac{e_s}{m_s} \mathbf{E} \right) \rho_s. \qquad (12.1.31)$$

If the particles of species s are not ionized, eq. (12.1.31) is identical with the overall linear momentum balance equation, eq. (8.1.14) in Volume 1, an identity which motivates the choice of the name linear momentum moment for M_{sk}.

12.2 A monoelemental gas in complete equilibrium at constant temperature in a constant gravitational field

Applying eq. (12.1.31) to a static, isothermal, and plane parallel atmosphere of electrically neutral particles in a constant gravitational field gives the familiar result that

$$n_s(h) = n_s(0) \, \exp\left(-\frac{gm_s}{kT} \, h\right) = n_s(0) \, \exp\left(-\frac{h}{l_s}\right), \qquad (12.2.1)$$

where $n_s(0)$ is the density of particles of type s at the base of the atmosphere, h is the height above the base of the atmosphere and

$$l_s = \frac{kT}{gm_s} \qquad (12.2.2)$$

is a scale height.

Equations (12.2.1)–(12.2.2) may be applied with no modification to describe the equilibrium distributions of different species of element in a gas consisting of an arbitrary number of neutral species in a plane parallel atmosphere at constant temperature in a constant gravitational field. In complete equilibrium, more massive particles, with smaller scale heights, are concentrated more toward the base of the atmosphere than are lighter particles. This is the end result of the gravitational settling process.

The simplicity and familiarity of this result disguises the fact that, on a molecular level, the situation is far from static. Particles are continually streaming downward because of acceleration by the gravitational field and, as required by number conservation, particles are continually streaming upward to match the downward flow. Thus, in any fixed volume of space, while the mean number of particles of any kind remains the same, the provenances of the particles in the volume are actually functions of the time.

One may think of the upward flow as being a consequence of the existence of number-abundance gradients. If the gravitational field were to suddenly disappear, across any plane there would be more particles moving upward from regions of higher density than would be flowing downward from regions of lower density above the plane. In terrestial laboratories, the net flux of particles of a given type is related to the abundance gradient by Fricke's law as expressed by eq. (8.9.5) in Volume 1, where the diffusion coefficient D_s is a consequence of collisional interactions, being smaller the stronger the interactions. Thus, one may write the upward flux of particles of type s as

$$\text{upward flux} = -D_s \, \frac{\partial n_s}{\partial h} = D_s \, \frac{n_s}{l_s} = D_s \, n_s \, \frac{m_s g}{kT}, \qquad (12.2.3)$$

where eq. (12.2.1) has been used to find $\partial n_s/\partial h$ in terms of n_s and l_s.

The downward motion of particles must also be limited by collisional interactions. In terrestial experiments, the force experienced by a macroscopic, spherical body passing through a viscous medium can, under appropriate conditions, be described by Stokes' law (George Gabriel Stokes, *Mathematical and Physical Papers*, 1880),

$$F = 6\pi \eta R V, \qquad (12.2.4)$$

where η is the viscosity of the medium, R is the radius of the macroscopic body, and V is the speed of the body with respect to the medium. This expression is applicable if R is large compared with a mean free path. Assuming that a qualitatively similar viscous force acts on a molecule moving through other molecules, one may envision that s-type particles accelerated downward by gravity reach an effective limiting velocity V_s determined by equating the gravitational force with the viscous force. Thus,

$$m_s g \sim a_s \eta_s V_s, \qquad (12.2.5)$$

where a_s is a length of the order of the linear dimension of an s-type particle. A qualitative estimate of the downward moving flux is then given by

$$\text{downward flux} \sim n_s V_s = n_s \frac{m_s g}{a_s \eta_s}. \qquad (12.2.6)$$

Equating estimates of upward and downward fluxes gives

$$D_s \sim \frac{kT}{a_s \eta_s}, \qquad (12.2.7)$$

supporting the intuitive expectation that the diffusivity and the viscosity are determined by the same physics, with the diffusivity decreasing and the viscosity increasing with an increase in the effectiveness of collisional forces. This exercise underlines the fact that, although a system may be in a state of equilibrium, the same dynamical forces that bring about equilibrium continue to operate in maintaining equilibrium.

When temperatures are large enough that the gas contains positively charged ions, the disparity between the gravitational force on the electrons freed in the ionization process and the pressure-gradient force exerted by these electrons means that a macroscopic electrostatic field must be present. As described briefly in Section 3.6 of Volume 1 and more extensively in Section 13.4 of this volume, for a monoelemental gas in a specific ionization stage,

$$e\mathbf{E} = -\frac{\left(m_s^{\mathrm{ion}} - m_e\right)}{1 + Z_s} \mathbf{g}, \qquad (12.2.8)$$

where Z_s is the number of free electrons per ion and m_s^{ion} is the mass of the ion. Inserting this relationship into eq. (12.1.31), one has that the motions of the electrically positive ions are governed by

$$\rho_s \frac{d\mathbf{u}_s}{dt} = -\nabla P_s + \mathbf{g} \left(1 - \frac{Z_s e}{m_s} \frac{(m_s - m_e)}{1 + Z_s}\right) \rho_s, \qquad (12.2.9)$$

where the superscript has been dropped from the mass m_s of the ion. Now, however, there are two types of particle present, ions and electrons. For the electrons, one can write

$$\rho_e \frac{d\mathbf{u}_e}{dt} = -\nabla P_e + \mathbf{g} \left(1 + \frac{e}{m_e} \frac{(m_s - m_e)}{1 + Z_s}\right) \rho_e, \qquad (12.2.10)$$

Adding eqs. (12.2.9) and (12.2.10) gives

$$\rho_s \frac{d\mathbf{u}_s}{dt} + \rho_e \frac{d\mathbf{u}_e}{dt} = -(\nabla P_s + \nabla P_e) + \mathbf{g} \left[(\rho_s + \rho_e) + e \frac{m_s - m_e}{1 + Z_s} \left(\frac{Z_s}{m_s} \rho_s - \frac{1}{m_e} \rho_e \right) \right].$$
(12.2.11)

Postulating that charge neutrality holds over regions of space containing many particles, one has that

$$n_e = \frac{\rho_e}{m_e} = Z_s n_s = Z_s \frac{\rho_s}{m_s}$$
(12.2.12)

and

$$\frac{d\mathbf{u}_e}{dt} = \frac{d\mathbf{u}_s}{dt},$$
(12.2.13)

so that

$$\rho \frac{d\mathbf{u}_s}{dt} = -\nabla P + \mathbf{g}\rho,$$
(12.2.14)

where

$$\rho = \rho_s + \rho_e = m_s n_s + m_e n_e = (m_s + Z_s m_e) n_s$$
(12.2.15)

and

$$P = P_s + P_e = (n_s + n_e) kT = (1 + Z_s) n_s kT.$$
(12.2.16)

The physics that culminates in eq. (12.2.14) is absolutely stunning. In order for charge neutrality to be maintained in a gravitational field, a gas becomes polarized. The degree of polarization is so tiny that the charge density associated with it is completely negligible compared with the charge density associated with either free electrons or ions (see Section 13.4), but it nevertheless sustains an electrostatic field which exerts a force on both free electrons and positive ions which is comparable in magnitude with the gravitational force exerted on the ions. Yet, the bulk acceleration of the fluid, as expressed by eq. (12.2.14), turns a completely blind eye to the absolutely essential existence of the electrostatic field.

In complete equilibrium at constant temperature, one has that $d\mathbf{u}_s/dt = 0$, and eqs. (12.2.14)–(12.2.16) give, in the direction of the steepest rate of change in n_s,

$$\frac{1}{n_s} \frac{dn_s}{dr} = \frac{1}{n_e} \frac{dn_e}{dr} = -\frac{g}{kT} \frac{m_s + Z_s m_e}{1 + Z_s} = -\frac{g}{kT} \frac{m_s^a}{1 + Z_s},$$
(12.2.17)

where

$$m_s^a = m_s + Z_s m_e$$
(12.2.18)

is the mass of the ion plus the mass of Z_s electrons and is larger than the mass of the neutral atom by the binding energy of the neutral atom minus the binding energy of the ion divided by c^2. Here and hereinafter, m_s^a is taken to be the mass of the neutral atom and $m_s + Z_s m_e$ and m_s^a are taken to be interchangeable.

In a plane parallel atmosphere, the solution of eq. (12.2.17) is given by eq. (12.2.1), where h is the distance above the base of the atmosphere and

$$l_s = (1 + Z_s) \frac{kT}{(m_s + Z_s m_e) g} = (1 + Z_s) \frac{kT}{m_s^a g}. \qquad (12.2.19)$$

Again, this result makes no acknowledgement of the existence of an electrostatic field; it simply reflects the fact that, when an element species is completely ionized, the number of particles contributing to the pressure-gradient force is larger by a factor of $1 + Z_s$ than when it is neutral.

As an aside, in a quasistatic stellar model composed of pure hydrogen, the fact that overall pressure balance is maintained automatically ensures that there is a also an exact balance between gravity-induced diffusion and diffusion due to a gradient in the density. Only three types of particle must be considered: the hydrogen atom at number density n_H, the proton at number density n_p, and the electron at number density n_e. In complete equilibrium at constant temperature in a constant gravitational field, eqs. (12.2.14)–(12.2.16) give

$$\frac{1}{\rho} \left(\frac{dP_H}{dr} + \frac{dP_p}{dr} + \frac{dP_e}{dr} \right) = -g, \qquad (12.2.20)$$

where

$$\rho = m_H n_H + m_p n_p + m_e n_e. \qquad (12.2.21)$$

If charge neutrality is assumed, then $n_e = n_p$ and $\rho = m_H n_H + (m_p + m_e) n_p \sim m_H (n_H + n_p)$. Assuming a non-relativistic perfect gas equation of state, one has that $P_e = P_p = n_p kT$. With these assumptions, eq. (12.2.20) can be written as

$$\frac{kT}{n_H + n_p} \left(\frac{dn_H}{dr} + 2 \frac{dn_p}{dr} \right) = -g m_H. \qquad (12.2.22)$$

The Saha equation for pure hydrogen is (eq. (4.13.19) in Volume 1)

$$\frac{n_p n_e}{n_H} = \frac{n_p^2}{n_H} = \left(\frac{2\pi m_e kT}{h^2} \right)^{3/2} e^{-\chi_H/kT}, \qquad (12.2.23)$$

where χ_H is the first ionization potential, so, at constant temperature,

$$2 \frac{1}{n_p} \frac{dn_p}{dr} = \frac{1}{n_H} \frac{dn_H}{dr}. \qquad (12.2.24)$$

Inserting eq. (12.2.24) into (12.2.22) gives

$$\frac{1}{n_H} \frac{dn_H}{dr} = -\frac{g m_H}{kT}, \qquad (12.2.25)$$

which, of course, has the solution

$$n_H(h) = n_H(0) \exp\left(-\frac{g m_H}{kT} h \right). \qquad (12.2.26)$$

The length h for atoms in eq. (12.2.26) is not to be confused with Planck's constant h in eq. (12.2.23). Combining eqs. (12.2.25) and (12.2.24) and solving gives

$$n_p(h) = n_p(0) \exp\left(-\frac{gm_H}{2kT} h\right). \tag{12.2.27}$$

Thus, although the scale height for protons and electrons is twice as large as the scale height for neutral hydrogen, ionization equilibrium as expressed by eq. (12.2.23) is maintained everywhere. As a corollary, the degree of ionization increases with height, approaching unity after several scale heights.

This provides a qualitative explanation for the existence of an ionosphere in the Earth's atmosphere that does not explicitly invoke sunlight for the energy necessary to achieve ionization. Of course, it is molecular nitrogen and oxygen rather than hydrogen that are the main constituents of the atmosphere, so the relevant scale heights are an order of magnitude smaller than the scale height for hydrogen, and it is the interaction between sunlight and molecules and atoms in the atmosphere that maintains temperatures in the atmosphere at relatively high levels.

These simple results tempt one to anticipate that, for a partially ionized, multicomponent gas, it is possible to link an electron number density n_{se} with each ion number density n_s in such a way that $Z_s n_s = n_{se}$ and to guess that the equilibrium distribution for each ion type may be approximated by eq. (12.2.1), with the scale height of an ion of charge Z_s being proportional to the inverse of $m_s^a = m_s + Z_s m_e$. However, even for the case of pure helium, with only four types of particle to consider, this approximation is incompatible with the requirement of ionization equilibrium as expressed by the relevant Saha equations.

In equilibrium, the sum of the linear momentum moments for the four particles is

$$\frac{dP_0}{dr} + \frac{dP_1}{dr} + \frac{dP_2}{dr} + \frac{dP_e}{dr} = -gm_{He}(n_0 + n_1 + n_2), \tag{12.2.28}$$

where n_0, n_1, and n_2 denote, respectively, the number abundances of neutral, singly ionized, and doubly ionized helium. Assuming that each partial pressure is given by the perfect gas law for non-relativistic particles and that the abundance of electrons is given by $n_e = 2n_2 + n_1$, one has that

$$\frac{dn_0}{dr} + 2\frac{dn_1}{dr} + 3\frac{dn_2}{dr} = -\frac{gm_{He}}{kT}(n_0 + n_1 + n_2). \tag{12.2.29}$$

Supposing that the abundance of each ion type may be approximated by

$$n_i = n_i(0) \exp\left(-\frac{h}{l_i}\right), \tag{12.2.30}$$

it follows that $l_1 = 2 l_0$ and $l_2 = 3 l_0$, where $l_0 = kT/m_{He}g$. The relevant Saha equations are

$$\frac{n_2 n_e}{n_1} = f_1(T) \tag{12.2.31}$$

and

$$\frac{n_1 n_e}{n_0} = f_0(T), \tag{12.2.32}$$

where the details of the functions $f_1(T)$ and $f_0(T)$ are irrelevant, and, on dividing one Saha equation by the other, it follows that, at constant temperature, the Saha equations can be satisfied everywhere simultaneously only if

$$\frac{1}{l_2} - \frac{1}{l_1} = \frac{1}{l_1} - \frac{1}{l_0},$$

or

$$\frac{2}{l_1} = \frac{1}{l_0} + \frac{1}{l_2}. \tag{12.2.33}$$

However, the prescription postulated by using eq. (12.2.30) in eq. (12.2.29) gives

$$\frac{2}{l_1} = \frac{1}{l_0} \text{ and } \frac{1}{l_0} + \frac{1}{l_2} = \frac{4}{3}\frac{1}{l_0}, \tag{12.2.34}$$

both relationships being incompatible with eq. (12.2.33). Thus, when more than two ionization stages are involved, a simple solution involving a single scale height for each ion does not exist.

12.3 Diffusion velocities and moments in a multicomponent gas in a gravitational field

In a gas which is composed of elements characterized by a variety of atomic masses and which is not in complete equilibrium with respect to diffusion, the mean velocity $\mathbf{u}_s$ of particles of a given species depends on the species particle mass m_s and differs from the center of mass velocity of the entire system, the latter velocity being defined by

$$\mathbf{u}(\mathbf{x}, t) = \frac{1}{\rho} \sum_s \rho_s \mathbf{u}_s(\mathbf{x}, t), \tag{12.3.1}$$

where

$$\rho = \sum_s \rho_s = \sum_s m_s n_s, \tag{12.3.2}$$

and the sums are over all types of particle, including free electrons.

Defining a velocity parameter relative to $\mathbf{u}(\mathbf{x}, t)$ by

$$\mathbf{c}^* = \mathbf{v} - \mathbf{u}(\mathbf{x}, t), \tag{12.3.3}$$

one has that the average velocity of s-type particles relative to the center of mass velocity is

$$\overline{\mathbf{c}_s^*}(\mathbf{x}, t) = \frac{1}{n_s} \int \mathbf{c}^* f_s(\mathbf{x}, \mathbf{v}, t) \, \mathrm{d}^3 c^* = \mathbf{u}_s(\mathbf{x}, t) - \mathbf{u}(\mathbf{x}, t). \tag{12.3.4}$$

The quantity

$$\mathbf{w}_s(\mathbf{x}, t) = \overline{\mathbf{c}_s^*}(\mathbf{x}, t) = \mathbf{u}_s(\mathbf{x}, t) - \mathbf{u}(\mathbf{x}, t) \tag{12.3.5}$$

is called the diffusion velocity for s-type particles. Note from eqs. (12.1.10) and (12.3.3) that

$$\mathbf{c}^* = \mathbf{c} + \mathbf{u}_s(\mathbf{x}, t) - \mathbf{u}(\mathbf{x}, t) = \mathbf{c} + \mathbf{w}_s(\mathbf{x}, t) \tag{12.3.6}$$

and that, in the center of mass reference frame,

$$P_s^* = m_s \int \left(c_k^*\right)^2 f_s(\mathbf{x}, \mathbf{c}^*, t)\mathrm{d}^3\mathbf{c}^* = m_s \int (c_k + w_{sk})^2 \, f_s(\mathbf{x}, \mathbf{c}, t) \, \mathrm{d}^3c$$

$$= P_s + \rho_s w_s^2 \, \hat{\mathbf{k}} \cdot \hat{\mathbf{u}}_s, \tag{12.3.7}$$

where P_s is the isotropic pressure in the reference frame moving with velocity $\mathbf{u}_s$. In light of the discussion centered on eqs. (12.1.25)–(12.1.27), a factor of $\hat{\mathbf{k}} \cdot \hat{\mathbf{u}}_s$ has been inserted after the ram pressure $\rho_s w_s^2$.

Multiplying each $\mathbf{w}_s$ in eq. (12.3.5) by ρ_s, summing over all types of particle, and using eq. (12.3.1) gives

$$\sum_s \rho_s \mathbf{w}_s(\mathbf{x}, t) = \sum_s \rho_s \mathbf{u}_s(\mathbf{x}, t) - \rho \, \mathbf{u}(\mathbf{x}, t) = 0. \tag{12.3.8}$$

Divergences in the reference frames moving with velocities $\mathbf{u}_s$ and $\mathbf{u}$ are related by

$$\nabla \cdot \mathbf{u}_s = \nabla \cdot \mathbf{u} + \nabla \cdot \mathbf{w}_s, \tag{12.3.9}$$

and the total time derivatives in the two frames are related by

$$\left(\frac{\mathrm{d}}{\mathrm{d}t}\right)_s = \frac{\partial}{\partial t} + \mathbf{u}_s \cdot \nabla = \frac{\partial}{\partial t} + \mathbf{u} \cdot \nabla + \mathbf{w}_s \cdot \nabla = \left(\frac{\mathrm{d}}{\mathrm{d}t}\right)_u + \mathbf{w}_s \cdot \nabla. \tag{12.3.10}$$

Multiplying eq. (12.1.8) by m_s gives the mass-conservation moment for s-type particles in the frame of reference moving with velocity $\mathbf{u}_s$:

$$M_{s0} = \left(\frac{\mathrm{d}\rho_s}{\mathrm{d}t}\right)_s + \rho_s \nabla \cdot \mathbf{u}_s = 0. \tag{12.3.11}$$

In the center of mass frame, this becomes

$$M_{s0} = \left[\left(\frac{\mathrm{d}\rho_s}{\mathrm{d}t}\right)_u + \mathbf{w}_s \cdot \nabla\rho_s\right] + \rho_s \nabla \cdot (\mathbf{u} + \mathbf{w}_s)$$

$$= \left(\frac{\mathrm{d}\rho_s}{\mathrm{d}t}\right)_u + \rho_s \nabla \cdot \mathbf{u} + \nabla \cdot (\rho_s \mathbf{w}_s) = 0. \tag{12.3.12}$$

Since, by definition, $\sum_s \rho_s \mathbf{w}_s = 0$, summing eq. (12.3.12) over all species produces

$$M_0 = \sum_s M_{s0} = \left(\frac{\mathrm{d}\rho}{\mathrm{d}t}\right)_u + \rho \nabla \cdot \mathbf{u} = 0, \tag{12.3.13}$$

which expresses overall mass conservation in the center of mass frame of reference.

Usings eqs. (12.3.7) and (12.3.10) in eq. (12.1.28), the kth component of the linear momentum moment for s-type particles is expressed in the center of mass frame by

$$M_{sk} = \frac{\partial P_s}{\partial x_k} + \rho_s \left[\frac{d}{dt} + \mathbf{w}_s \cdot \nabla \right] (u_k + w_{sk}) - F_{sk}\, n_s$$

$$= \frac{\partial P_s}{\partial x_k} + \rho_s \left[\frac{du_k}{dt} + \frac{dw_{sk}}{dt} + (\mathbf{w}_s \cdot \nabla)u_k \right] - F_{sk}\, n_s$$

$$+ \rho_s\, (\mathbf{w}_s \cdot \nabla)w_{sk} + \frac{\partial}{\partial x_k} \left(\rho_s w_s^2 \right), \tag{12.3.14}$$

where, in the second formulation, terms which are of second order in powers of the diffusion velocity are placed at the end of the equation.

Adding $M_{s0} = 0$ from eq. (12.3.12) to eq. (12.3.14) and rearranging gives

$$M_{sk} = \frac{\partial P_s}{\partial x_k} + \rho_s \frac{du_k}{dt} - F_{sk}\, n_s + \lambda_{sk} + \epsilon_{sk}, \tag{12.3.15}$$

where

$$\lambda_{sk} = \frac{d}{dt}(\rho_s w_{sk}) + (\rho_s \mathbf{w}_s) \cdot \nabla u_k + (\rho_s w_{sk})\nabla \cdot \mathbf{u} \tag{12.3.16}$$

and

$$\epsilon_{sk} = \rho_s(\mathbf{w}_s \cdot \nabla)w_{sk} + w_{sk}\rho_s \nabla \cdot \mathbf{w}_s + w_{sk}\mathbf{w}_s \cdot \nabla\rho_s + \frac{\partial}{\partial x_k} \left(\rho_s w_s^2 \right). \tag{12.3.17}$$

In spherical geometry, if all velocities are in the radial direction, one has, in this direction,

$$\lambda_s = \frac{d(\rho_s w_s)}{dt} + 2\, (\rho_s w_s) \left(\frac{du}{dr} + \frac{u}{r} \right) \tag{12.3.18}$$

and

$$\epsilon_s = 2 \left[\frac{d\left(w_s^2 \rho_s \right)}{dr} + \frac{w_s^2 \rho_s}{r} \right]. \tag{12.3.19}$$

Equation (12.3.15) differs from eq. (12.1.28) only by the addition of the terms λ_{sk} and ϵ_{sk}.

Summation of eq. (12.3.16) over all species of particle gives zero, and, on the supposition that the ϵ_{sk} term given by eq. (12.3.17) may be neglected, summation of eq. (12.3.15) over all species of particle, including electrons, gives

$$M_k = \sum_s M_{sk} = \frac{\partial P}{\partial x_k} + \rho \frac{du_k}{dt} - \sum_s F_{sk}\, n_s. \tag{12.3.20}$$

Setting

$$n_s F_{sk} = -\rho_s g_k + n_s e_s E_k, \tag{12.3.21}$$

and demanding that $\sum e_s n_s = 0$, eq. (12.3.20) becomes

$$M_k = \sum_s M_{sk} = \frac{\partial P}{\partial x_k} + \rho \frac{du_k}{dt} + \rho g_k. \tag{12.3.22}$$

When overall pressure balance prevails, the right hand side of eq. (12.3.22) vanishes. Thus, setting

$$M_k = \sum_s M_{sk} = 0 \tag{12.3.23}$$

is equivalent to assuming overall pressure balance.

Multiplying M_k as expressed by eq. (12.3.22) by ρ_s / ρ and subtracting the result from M_{sk} given by eq. (12.3.15) gives, when ϵ_{sk} is neglected,

$$M'_{sk} = M_{sk} - \frac{\rho_s}{\rho} M_k = \frac{\partial P_s}{\partial x_k} - \frac{\rho_s}{\rho} \frac{\partial P}{\partial x_k} - (F_{sk} n_s + \rho_s g_k) + \lambda_{sk}. \tag{12.3.24}$$

Using eq. (12.3.21) again, the explicit reference to gravity disappears from the term in parentheses on the right hand side of eq. (12.3.24), and one has

$$M'_{sk} = M_{sk} - \frac{\rho_s}{\rho} M_k = \frac{\partial P_s}{\partial x_k} - \frac{\rho_s}{\rho} \frac{\partial P}{\partial x_k} - n_s e_s E_k + \lambda_{sk}. \tag{12.3.25}$$

However, if it is assumed that overall pressure balance is maintained, gravity appears explicitly once again. As follows from eq. (12.3.22), when $M_k = 0$, $\partial P / \partial x_k = -\rho (du_k/dt + g_k)$, and, using this in the third expression in eq. (12.3.25), one obtains

$$M'_{sk} = M_{sk} - \frac{\rho_s}{\rho} M_k$$

$$= \frac{\partial P_s}{\partial x_k} + \left[m_s \left(g_k + \frac{du_k}{dt} \right) - e_s E_k \right] n_s + \lambda_{sk}. \tag{12.3.26}$$

Henceforth, M_{sk} (with no prime) refers to the renormalized expressions to the right of the second equal signs in eqs. (12.3.24)–(12.3.26).

12.4　The strength of the electrostatic field when equilibrium with respect to diffusion prevails

If conditions are such that there is no migration of any species of particle with respect to any other, all of the quantities w_{sk}, ϵ_{sk}, and λ_{sk} are zero and the quantity on the right hand side of eq. (12.3.26) vanishes. Assuming quasistatic equilibrium, so that the acceleration of the center of mass may be neglected, for positive ions of charge e_s one has that

$$m_s g - e_s E = -\frac{1}{n_s} \frac{dP_s}{dr}, \tag{12.4.1}$$

where the subscript k has been dropped and it has been assumed that all variations occur in the radial direction. For electrons, one has that

$$m_e g + eE = -\frac{1}{n_e}\frac{dP_e}{dr}. \tag{12.4.2}$$

Thus, in complete equilibrium with respect to diffusion, each particle species obeys its own pressure-balance equation in which the electrostatic field plays a prominent role.

When electrons are not degenerate, the right hand sides of eqs. (12.4.1) and (12.4.2) are essentially the same. For a monoelemental gas in which there are Z_s electrons for every positive ion, one has in good approximation that

$$m_s g - Z_s eE = m_e g + eE, \tag{12.4.3}$$

or

$$eE = \frac{(m_s + m_e)g}{(1 + Z_s)}. \tag{12.4.4}$$

Combining eqs. (12.4.2) and (12.4.4), one has

$$-\frac{1}{n_e}\frac{dP_e}{dr} = m_e g + eE = \frac{m_s + m_e}{1 + Z_s}g\left(1 + \frac{m_e(1 + Z_s)}{m_s + m_e}\right)$$

$$= eE\left(1 + (1 + Z_s)\frac{m_e}{m_s + m_e}\right), \tag{12.4.5}$$

which may be written as

$$eE = -\left(1 + (1 + Z_s)\frac{m_e}{m_s + m_e}\right)^{-1}\frac{1}{n_e}\frac{dP_e}{dr} \sim -\frac{1}{n_e}\frac{dP_e}{dr}, \tag{12.4.6}$$

where the last approximation follows from the fact that $m_e/m_s \ll 1$.

When electrons are degenerate, this approximation becomes even more precise. Given that $\sum_s Z_s n_s = n_e$ and that $dP/dr = -\sum_s n_s(m_s + Z_s m_e)\, g$, where Z_s is the number of free electrons contributed by each s-type ion, one may write eq. (12.4.2) as

$$eE \sim -\frac{1}{\sum_s Z_s n_s}\frac{dP_e}{dr}\left(\frac{-\sum_s n_s(m_s + Z_s m_e)g}{dP/dr}\right) - m_e g$$

$$= g\frac{\sum_s n_s(m_s + Z_s m_e)}{\sum_s Z_s n_s}\frac{dP_e}{dP} - m_e g$$

$$\sim g\frac{\bar{m}_{atoms}}{\bar{Z}_{ions}}\frac{dP_e}{dP}, \tag{12.4.7}$$

where $\bar{m}_{atoms}$ is the mean mass of the atoms from which ions are derived and $\bar{Z}_{ions}$ is the mean number of free electrons per ion contributed by ions. When electrons are sufficiently degenerate that $P_e \sim P$,

$$eE \sim \frac{\bar{m}_{atoms}}{\bar{Z}_{ions}}g. \tag{12.4.8}$$

A similar result follows by summing over s the term on the left hand side of eq. (12.4.1) and setting the result equal to zero, the only difference being that $\bar{m}_{\text{ions}}$, the mean mass of the ions, replaces $\bar{m}_{\text{atoms}}$ in eq. (12.4.7).

It is shown in Section 13.3 that, when electrons behave as a perfect gas,

$$eE \sim \frac{(\bar{m}_{\text{ions}} - m_e)g}{1 + \bar{Z}_{\text{ions}}} \left[1 - \left(\frac{kT}{(\bar{m}_{\text{ions}} - m_e)\, g\, r} \right) \frac{r}{\bar{Z}_{\text{ions}}} \frac{\mathrm{d}\bar{Z}_{\text{ions}}}{\mathrm{d}r} \right], \qquad (12.4.9)$$

which differs from eq. (13.3.20) only by the arrangement of terms and insertion of $r/r = 1$. Since diffusion in a gravitational field tends, in general, to force heavier ions to move in the direction of the gravitational field, one expects $(r/\bar{Z}_{\text{ions}})(\mathrm{d}\bar{Z}_{\text{ions}}/\mathrm{d}r)$ to be negative, ensuring that the term in square brackets in eq. (12.4.9) is positive. Thus, the electrostatic field is always directed opposite to the gravitational field.

In summary, when electrons behave as a perfect gas, the inward force on a free electron is overwhelmingly electrical and is proportional to $\bar{m}_{\text{ions}}\, g/(1 + \bar{Z}_{\text{ions}})$. The net average inward force on ions is made up of both a gravitational component and an electrical component and is proportional to $\bar{m}_{\text{ions}}\, g - \bar{Z}_{\text{ions}}\bar{m}_{\text{ions}}\, g/(1+\bar{Z}_{\text{ions}}) = \bar{m}_{\text{ions}}\, g/(1+\bar{Z}_{\text{ions}})$. The sum of the forces on electrons and ions combined is proportional to $\bar{m}_{\text{ions}}\, g$. When electrons are degenerate, the net force on an electron is entirely electrical and equal to $\bar{m}_{\text{atoms}}\, g/\bar{Z}_{\text{ions}}$ and the net force on an ion effectively vanishes! Thus, whatever the degree of electron degeneracy, in maintaining the balance between outward partial pressure-gradient forces and inward forces, an electrostatic field plays an absolutely essential role, despite the fact that it does not enter explictly into the overall pressure-balance equation.

12.5 Driving forces for diffusion in an initially homogeneous medium consisting of two ion species in a gravitational field

In order to construct an initial stellar model, one must specify, among other things, the relative distributions of all species of element to be included in the model. These distributions will, in general, not be consistent with diffusion equilibrium. In practice, a homogeneous initial composition is not only the simplest choice, but, in the case of pre-main sequence models, it is also the most reasonable one. It is of interest to examine linear momentum moments when this choice has been made.

Consider a model composed of just two species of element at temperatures and densities such that only one ionization state need be considered for each element. Let the model be spherically symmetric and in overall pressure balance, i.e., in quasistatic equilibrium. Furthermore, drop the terms λ_{sk} from the normalized moments in eq. (12.3.26). Adopting eE given by eq. (12.4.6) for the electrical field strength and using the convention that

$$m_s^a \sim m_s + Z_s m_e, \qquad (12.5.1)$$

the linear momentum moments in the radial direction for the two species of isotopes as given by eq. (12.3.26) are (with the prime being dropped from the renormalized moment furthest to the left)

$$M_{s1} = \frac{dP_s}{dr} + n_s m_s^a g + \frac{Z_s n_s}{n_e} \frac{dP_e}{dr} \qquad (12.5.2)$$

and

$$M_{t1} = \frac{dP_t}{dr} + n_t m_t^a g + \frac{Z_t n_t}{n_e} \frac{dP_e}{dr}. \qquad (12.5.3)$$

The sum of the two linear momentum moments is

$$M_{s1} + M_{t1} = \frac{dP_s}{dr} + \frac{dP_t}{dr} + \frac{n_s Z_s + n_t Z_t}{n_e} \frac{dP_e}{dr} + \left(m_s^a n_s + m_t^a n_t\right) g$$

$$= \frac{dP}{dr} + \left(m_s^a n_s + m_t^a n_t\right) g = \frac{dP}{dr} + \rho g = 0. \qquad (12.5.4)$$

The sum vanishes because the composite system is, by assumption, in overall pressure balance. If all constituents act as perfect gases, one can differentiate the partial pressures with respect to number densities and temperature, and the difference in the two moments can be written as

$$M_{s1} - M_{t1} = n_s kT \left(\frac{1}{T}\frac{dT}{dr} + \frac{1}{n_s}\frac{dn_s}{dr}\right) - n_t kT \left(\frac{1}{T}\frac{dT}{dr} + \frac{1}{n_t}\frac{dn_t}{dr}\right)$$

$$+ \left(n_s Z_s - n_t Z_t\right) kT \left(\frac{1}{T}\frac{dT}{dr} + \frac{1}{n_e}\frac{dn_e}{dr}\right) + \left(m_s^a n_s - m_t^a n_t\right) g. \quad (12.5.5)$$

If number-density ratios n_s/n_t are constant in space, then

$$\frac{1}{n_s}\frac{dn_s}{dr} = \frac{1}{n_t}\frac{dn_t}{dr} = \frac{1}{n_e}\frac{dn_e}{dr}, \qquad (12.5.6)$$

and eqs. (12.5.4) and (12.5.5) become, respectively,

$$M_{s1} + M_{t1} = (n_s + n_t + n_e) kT \left(\frac{1}{T}\frac{dT}{dr} + \frac{1}{n_e}\frac{dn_e}{dr}\right) + \left(m_s^a n_s + m_t^a n_t\right) g$$

$$= \left[n_s(1 + Z_s) + n_t(1 + Z_t)\right] kT \left(\frac{1}{T}\frac{dT}{dr} + \frac{1}{n_e}\frac{dn_e}{dr}\right)$$

$$+ \left(m_s^a n_s + m_t^a n_t\right) g = 0 \qquad (12.5.7)$$

and

$$M_{s1} - M_{t1} = \left[n_s(1 + Z_s) - n_t(1 + Z_t)\right] kT \left(\frac{1}{T}\frac{dT}{dr} + \frac{1}{n_e}\frac{dn_e}{dr}\right) + \left(m_s^a n_s - m_t^a n_t\right) g.$$

$$(12.5.8)$$

Solving eq. (12.5.7) for $kT\,((1/T)(dT/dr)+(1/n_e)(dn_e/dr))$, using the result to eliminate the same quantity in eq. (12.5.8), and rearranging produces

$$M_{s1} - M_{t1} = 2n_s n_t \, \frac{m_s^a(1+Z_t) - m_t^a(1+Z_s)}{n_s(1+Z_s) + n_t(1+Z_t)} \, g. \tag{12.5.9}$$

Since $M_{s1}+M_{t1} = 0$, it follows that $M_{s1} = -M_{t1} = \frac{1}{2}(M_{s1} - M_{t1})$. Equations (12.5.2) and (12.5.3) may then be written, respectively, as

$$M_{s1} = n_s m_s^a g - \left[-\left(\frac{dP_s}{dr} + \frac{Z_s n_s}{n_e} \frac{dP_e}{dr} \right) \right] = n_s n_t \, \frac{m_s^a(1+Z_t) - m_t^a(1+Z_s)}{n_s(1+Z_s) + n_t(1+Z_t)} \, g \tag{12.5.10}$$

and

$$-M_{t1} = \left[-\left(\frac{dP_t}{dr} + \frac{Z_t n_t}{n_e} \frac{dP_e}{dr} \right) \right] - n_t m_t^a g = n_s n_t \, \frac{m_s^a(1+Z_t) - m_t^a(1+Z_s)}{n_s(1+Z_s) + n_t(1+Z_t)} \, g. \tag{12.5.11}$$

The quantities in square brackets in eqs. (12.5.10) and (12.5.11) are the outwardly directed pressure-gradient forces exerted by ions of the designated type along with the free electrons associated with them. It is clear that, if masses and ionization degrees are such that $m_s^a(1 + Z_t) > m_t^a(1 + Z_s)$, the net inward gravitational force on ions of type s and their attendant electrons is greater than the outward pressure-gradient force exerted by these ions and electrons. Exactly the reverse is true for ions of type t and their attendant electrons.

One may infer that, when a multicomponent system of homogeneous composition in local thermodynamic equilibrium is in overall pressure balance, the net pressure-gradient force provided by any given constituent ion species and its attendant electrons does not, in general, equal the net gravitational and electrical force on that species. This means that there must be a momentum transfer between particles of different species. But, transfer of momentum between species implies a difference in the average velocities associated with each species, with the particles of the species giving up momentum moving outward relative to the particles of the species accepting momentum. In the example of two species with $m_s^a(1 + Z_t) > m_t^a(1 + Z_s)$, particles of type t must be drifting outward and particles of type s must be drifting inward with respect to the center of mass of the system.

One may formally write the net rate of transfer of momentum as

$$K_{st}(w_t - w_s) = M_{s1} = -M_{t1} = \frac{1}{2}(M_{s1} - M_{t1}) = n_s n_t \, \frac{m_s^a(1+Z_t) - m_t^a(1+Z_s)}{n_s(1+Z_s) + n_t(1+Z_t)} \, g, \tag{12.5.12}$$

where w_s and w_t are drift velocities with respect to the center of mass and K_{st}, which is called a resistance coefficient and has a value yet to be determined, has the dimensions of a number density times a rate of change of momentum divided by a velocity and is a consequence of collisional interactions between members of the two species.

Before examining the dependence of K_{st} on local physical conditions and on element species characteristics, it is of interest to reformulate the difference in linear momentum moments for a two element gas in such a way that gravity does not appear explicitly. Subtracting eq. (12.5.3) from eq. (12.5.2) and setting

$$\left(n_s m_s^a - n_t m_t^a\right) g = (\rho_s - \rho_t) g = \frac{\rho_s - \rho_t}{\rho} \rho g = -\frac{\rho_s - \rho_t}{\rho} \frac{dP}{dr}, \qquad (12.5.13)$$

one has

$$M_{s1} - M_{t1} = \frac{dP_s}{dr} - \frac{dP_t}{dr} - \frac{(\rho_s - \rho_t)}{\rho} \left(\frac{dP_s}{dr} + \frac{dP_t}{dr} + \frac{dP_e}{dr}\right) + \frac{Z_s n_s^a - Z_t n_t^a}{n_e} \frac{dP_e}{dr}$$

$$= 2\frac{\rho_t}{\rho}\frac{dP_s}{dr} - 2\frac{\rho_s}{\rho}\frac{dP_t}{dr} + 2\frac{m_t^a n_t Z_s n_s - m_s^a n_s Z_t n_t}{\rho\, n_e}\frac{dP_e}{dr}$$

$$= \frac{2\, n_s n_t}{\rho}\left[m_t^a \frac{1}{n_s}\frac{dP_s}{dr} - m_s^a \frac{1}{n_t}\frac{dP_t}{dr} + \left(m_t^a Z_s - m_s^a Z_t\right)\frac{1}{n_e}\frac{dP_e}{dr}\right]. \qquad (12.5.14)$$

If all particles obey a perfect gas equation of state and all abundance ratios are constant in space, eq. (12.5.14) is equivalent to eq. (12.5.9), for, under these circumstances, eq. (12.5.14) becomes

$$M_{s1} - M_{t1} = \frac{2\, n_s n_t}{\rho}\left(m_t^a\,(1 + Z_s) - m_s^a\,(1 + Z_t)\right) kT \left(\frac{1}{T}\frac{dT}{dr} + \frac{1}{n_e}\frac{dn_e}{dr}\right), \qquad (12.5.15)$$

and use of eq. (12.5.7) to eliminate the temperature dependent quantity in parentheses gives eq. (12.5.9) once again.

The formulation of eq. (12.5.14) is particularly instructive when electrons are quite degenerate. Then, $P_e \gg P_{\text{ions}}$ and the electron pressure-gradient force nearly balances the gravitational force. As long as the quantities A_s/Z_s and A_t/Z_t, where the As are atomic numbers, differ by an amount comparable to unity, the third term in square brackets in eq. (12.5.14) dominates the first two terms, so that

$$M_{s1} - M_{t1} \sim \frac{2\, n_s n_t}{\rho}\left(m_t^a Z_s - m_s^a Z_t\right)\frac{1}{n_e}\frac{dP_e}{dr}$$

$$\sim 2\frac{n_s n_t}{n_e}\left(m_s^a Z_t - m_t^a Z_s\right) g \sim 2\frac{n_s n_t}{n_e} m_H\,(A_s Z_t - A_t Z_s)\, g, \qquad (12.5.16)$$

where A_s and A_t are atomic numbers. Thus, under electron-degenerate conditions, the linear momentum moment for the ion with the larger mass to charge ratio is unambiguously larger than the linear momentum moment for the ion with the smaller mass to charge ratio and a separation of species types is inevitable. On the other hand, if $A_s/Z_s \approx A_t/Z_t$, as is the case in a mixture of ^{12}C and ^{16}O in the CO-core of a white dwarf formed by an intermediate mass star, the first two terms in square brackets in eq. (12.5.14) may well dominate the third, and the implied rate of species separation is considerably smaller than it would be if the atomic number to charge ratios were noticeably different.

12.6 On the determination of resistance coefficients for diffusion

The discussion thus far of two and three component systems in a gravitational field has established two main facts: (1) if it is assumed that $\delta f_s/\delta t = \delta f_t/\delta t = 0$, so that the ions and attendant electrons of each species type are separately in pressure balance, solutions may be derived in which the relative abundances of elements with large mass to charge ratios increase in the direction of the gravitational acceleration and the relative abundances of elements with a smaller mass to charge ratio increase in the opposite direction; (2) if it is assumed that all elements are distributed in such a way that relative abundances are independent of position, $\delta f_s/\delta t \neq 0$ and $\delta f_t/\delta t \neq 0$. These facts can be understood by recognizing that the distribution function for a single-component system in thermodynamic equilibrium takes no cognizance whatsover of the details of interactions between individual particles, despite the fact that these interactions are essential for maintaining equilibrium (see Sections 4.2 and 4.12 in Volume 1), whereas, in a multicomponent medium of homogeneous composition, the assumption of overall pressure balance has the consequence that every component is distributed in space in a manner inconsistent with the distribution it would have if influenced only by interactions between members of its own species. The inescapable conclusion is that, in a multicomponent system in a gravitational field, an assessment of the behavior of the system must take explicitly into account characteristics of the interactions between individual members of different species.

The determination of resistance coefficients involves an analysis of how collisions of every sort, including those between particles of the same species, affect the quantities $\delta f_s/\delta t$ in the Boltzmann transport equations and their moments. At the core of any determination is (1) an estimate of the rates at which close encounters occur between particles of different species, (2) an understanding that linear momentum is the primary quantity of interest which is being transferred between species, and (3) an evaluation of the linear momentum moments of the collision terms in the Boltzmann transport equations.

For the case of just two ion species s and t, a plausible first approximation to the resistance coefficients can be obtained heuristically. Because of mass conservation, one has

$$\int \frac{\delta f_s}{\delta t}\, d^3\mathbf{c} = \int \left(\frac{\delta f_s}{\delta t}\right)_{st} d^3\mathbf{c} + \int \left(\frac{\delta f_s}{\delta t}\right)_{ts} d^3\mathbf{c} = 0, \qquad (12.6.1)$$

where $(\delta f_s/\delta t)_{st}$ is the time rate of decrease of f_s due to scattering of s-type particles by t-type particles and $(\delta f_s/\delta t)_{ts}$ is the time rate of increase of f_s due to these same interactions. The same holds true for the t species. On the other hand, it must also be true that

$$\left| \int \left(\frac{\delta f_s}{\delta t}\right)_{st} d^3\mathbf{c} \right| = \left| \int \left(\frac{\delta f_s}{\delta t}\right)_{ts} d^3\mathbf{c} \right|$$

$$\propto \dot{n}_{st} = n_s n_t \langle \sigma_{st} v_{\text{rel}} \rangle, \qquad (12.6.2)$$

where σ_{st} is a cross section for close encounters between individual members of the two types of particle, v_{rel} is the relative velocity between individual members of the two types, the brackets denote an average over thermal velocities, and the rate $\dot{n}_{st}$ is the number of close encounters per second per unit volume between all members of the two species. In the stellar interior, most species are ionized to some degree and, assuming that cross sections are dominated by Coulomb forces, one has that

$$\left\langle \frac{1}{2} \mu_{st} M_H \, v_{st}^2 \right\rangle \propto kT \propto \frac{Z_s Z_t e^2}{r_{\text{close}}}, \tag{12.6.3}$$

where

$$\mu_{st} = \frac{A_s A_t}{A_s + A_t}, \tag{12.6.4}$$

v_{st} is a relative velocity, Z_s and Z_t are degrees of ionization, A_s and A_t are atomic numbers, and r_{close} is a distance of closest approach whose square gives a measure of the average cross section. Then,

$$\sigma_{st} \propto r_{\text{close}}^2 \propto \left(\frac{Z_s Z_t e^2}{kT} \right)^2 \tag{12.6.5}$$

and

$$\dot{n}_{st} \propto n_s n_t \left(\frac{Z_s Z_t e^2}{kT} \right)^2 \left(\frac{kT}{\mu_{st} M_H} \right)^{1/2}. \tag{12.6.6}$$

The complete Boltzmann transport equation for s-type particles may be written as

$$M_{s1} = R_{st}, \tag{12.6.7}$$

where

$$M_{s1} = \frac{dP_s}{dr} + n_s m_s g - n_s Z_s e E \tag{12.6.8}$$

is the linear momentum moment, and

$$R_{st} = \int m_s v_s \frac{\delta f_s}{\delta t} \, d^3 \mathbf{v}_s - \int m_t v_t \frac{\delta f_t}{\delta t} \, d^3 \mathbf{v}_t \tag{12.6.9}$$

is the rate at which momentum is transferred between the two types of particle as determined by taking collisions between the two types of particle explicitly into account.

From the discussion leading to eq. (12.5.12), R_{st} is proportional to the difference in the diffusion velocities of the two species and has the dimensions of a number density times a time rate of change of momentum per particle. In the center of mass frame of reference, the per particle difference in the bulk linear momentum of s-type and t-type particles is

$$m_t w_t - m_s w_s = 2\mu_{st} m_H (w_t - w_s). \tag{12.6.10}$$

Thus, one may guess that

$$R_{st} \propto 2\mu_{st}m_{\mathrm{H}}\,(w_t - w_s)\,\dot{n}_{st}$$

$$\propto 2\mu_{st}m_{\mathrm{H}}\,(w_t - w_s)\,n_s n_t\left(\frac{Z_s Z_t e^2}{kT}\right)^2\left(\frac{kT}{\mu_{st}m_{\mathrm{H}}}\right)^{1/2}$$

$$\propto 2\mu_{st}^{1/2}m_{\mathrm{H}}c\left(\frac{Z_s Z_t e^2}{kT}\right)^2\left(\frac{kT}{M_{\mathrm{H}}c^2}\right)^{1/2}n_s n_t\,(w_t - w_s). \qquad (12.6.11)$$

In a rigorous derivation, this heuristic guess becomes multiplied by a factor which takes geometric effects and electron shielding into account in estimating the effective Coulomb cross section (see, e.g., the discussions in Sections 13.3 and 13.5). Writing

$$R_{st} = K_{st}\,(w_t - w_s), \qquad (12.6.12)$$

the treatment by Chapman and Cowling in the *Mathematical Theory of Nonuniform Gases* (1970) gives

$$K_{st} \sim K_{st}^0 = \frac{2}{3}\,\sqrt{2\pi}\,\mu_{st}^{1/2}m_{\mathrm{H}}c\left(\frac{Z_s Z_t e^2}{kT}\right)^2\left(\frac{kT}{m_{\mathrm{H}}c^2}\right)^{1/2}\log_e\left(1 + x_{st}^2\right)n_s n_t,$$

$$(12.6.13)$$

where

$$x_{st}^2 = \left(\frac{4\,kT}{Z_s Z_t\,e^2}\right)^2\lambda^2, \qquad (12.6.14)$$

and λ is the consequence of placing a restriction on permissible impact parameters by using a screened Coulomb potential or applying a cutoff to the collision impact parameter equal to the average separation between adjacent ions. The canonical practice is to choose λ to be the larger of the Debye length,

$$\lambda_{\mathrm{D}} = \left(\frac{kT}{4\pi e^2\left(\sum_i n_i Z_i^2 + n_e\right)}\right)^{1/2}, \qquad (12.6.15)$$

where the sum is over all ions, and the mean ionic separation,

$$\lambda_{\mathrm{ion}} = \left(\frac{3}{4\pi\sum_i n_i}\right)^{1/3}. \qquad (12.6.16)$$

Rewriting eq. (12.6.13) as

$$K_{st}^0 = \frac{2}{3}\,\sqrt{2\pi}\,\mu_{st}^{1/2}\,Z_s^2 Z_t^2\left(\frac{m_{\mathrm{H}}}{a_0^3}\right)\left(\frac{c}{a_0}\right)\left(\frac{kT}{m_{\mathrm{H}}c^2}\right)^{1/2}$$

$$\times\left(\frac{e^2/a_0}{kT}\right)^2 a_0^6\,n_s n_t\,\log_e\left(1 + x_{st}^2\right), \qquad (12.6.17)$$

where a_0 is the Bohr radius of hydrogen, noting that

$$\left(\frac{m_H}{a_0^3}\right)\left(\frac{c}{a_0}\right)\left(\frac{kT}{m_H c^2}\right)^{1/2}\left(\frac{e^2/a_0}{kT}\right)^2$$

$$= 11.2935 \times (5.665\,26 \times 10^{18}) \times \left(3.029\,73 \times 10^{-4} T_6^{1/2}\right) \times \frac{0.099\,7137}{T_6^2}$$

$$= 1.932\,89 \times 10^{15} \frac{1}{T_6^{3/2}} \text{ g cm}^{-3} \text{ s}^{-1}, \tag{12.6.18}$$

and that

$$a_0^6 n_s n_t = \left(\frac{a_0^3}{m_H}\right)^2 \rho^2 Y_s Y_t = 7.840\,48 \times 10^{-3} \rho^2 Y_s Y_t, \tag{12.6.19}$$

one has that

$$K_{st}^0 (\text{g cm}^{-3} \text{ s}^{-1}) = 2.532\,49 \times 10^{13} \mu_{st}^{1/2} Z_s^2 Z_t^2 \log_e \left(1 + x_{st}^2\right) \frac{\rho^2}{T_6^{3/2}} Y_s Y_t. \tag{12.6.20}$$

In these equations ρ is in units of g cm^{-3} and the Ys are number abundance parameters defined by eq. (8.7.2). Equation (12.6.14) may be rewritten as

$$x_{st}^2 = \left(\frac{1}{Z_s Z_t} \frac{4kT}{e^2/a_0} \frac{\lambda}{a_0}\right)^2 = \frac{160.459}{Z_s^2 Z_t^2} \left(\frac{\lambda}{a_0}\right)^2 T_6^2. \tag{12.6.21}$$

For electrons interacting with ions, one must set

$$\mu_{se} \sim \frac{m_e}{m_H}, \tag{12.6.22}$$

and it is evident that resistence coefficients involving electrons and ions are typically over forty times smaller than those involving ions only.

Other estimates of resistance coefficients exist in the literature. These include numerical estimates by G. Fontaine and G. Michaud (1979) and estimates by D. Muchmore (1984). An analytical fit to the numerical results of Fontaine and Michaud by Iben and J. MacDonald (1985) is

$$K_{st} \sim 1.6249 \frac{\log_e \left(1 + 0.187\,69 \, x_{st}^{1.2}\right)}{\log_e \left(1 + x_{st}^2\right)} K_{st}^0, \tag{12.6.23}$$

which agrees with the Muchmore estimates for values of $x_{st} > 0.14$.

12.7 Inclusion of electron-flow properties and ion–electron interactions and determination of diffusion velocities

In order to determine individual diffusion velocities, it is necessary to invoke constraints which take the flow of electrons explicitly into account. Since there is no net mass flow in the center of mass frame, one has that, in the system consisting of two ions and their associated electrons,

$$m_s n_s w_s + m_t n_t w_t + m_e n_e w_e = 0, \tag{12.7.1}$$

where w_e is the flow velocity of electrons. Insisting on no electric current in the center of mass frame gives

$$(eZ_s) \, n_s w_s + (eZ_t) \, n_t w_t + (-e) \, n_e w_e = 0. \tag{12.7.2}$$

Eliminating w_e from eqs. (12.7.1) and (12.7.2) gives

$$(m_s + Z_s m_e)n_s w_s + (m_t + Z_t m_e)n_t w_t$$

$$= m_s^a n_s w_s + m_t^a n_t w_t = \rho_s w_s + \rho_t w_t = 0, \tag{12.7.3}$$

showing that, even though the electrons have been given a diffusion velocity of their own, one may think of fractions $Z_s n_s / n_e$ and $Z_t n_t / n_e$ of them flowing with the ions from which they have been derived. The quantities ρ_s and ρ_t in eq. (12.7.3) are the mass densities associated with the ions and their attendant electrons and it follows from this equation that

$$w_t = -\frac{n_s m_s^a}{n_t m_t^a} \, w_s = -\frac{\rho_s}{\rho_t} \, w_s. \tag{12.7.4}$$

Using this in eq. (12.7.3) gives

$$w_e = \frac{n_s}{n_e} \frac{m_t^a Z_s - m_s^a Z_t}{m_t^a} \, w_s. \tag{12.7.5}$$

Further,

$$w_t - w_s = -\frac{m_s^a n_s + m_t^a n_t}{m_t^a n_t} \, w_s = -\frac{\rho}{\rho_t} \, w_s, \tag{12.7.6}$$

$$w_s - w_e = \frac{n_t Z_t}{n_e} \frac{m_s^a n_s + m_t^a n_t}{m_t^a n_t} \, w_s = \frac{n_t Z_t}{n_e} \frac{\rho}{\rho_t} \, w_s, \tag{12.7.7}$$

and

$$w_t - w_e = -\frac{n_s Z_s}{n_e} \frac{\rho}{\rho_t} \, w_s = \frac{n_s Z_s}{n_e} \frac{\rho}{\rho_s} \, w_t. \tag{12.7.8}$$

In the two ion approximation, one has that, when the details of momentum transfer between ions and electrons are not taken into account,

$$M_{s1} = -M_{t1} = K_{st}(w_t - w_s) \text{ or } M_{s1} - M_{t1} = 2K_{st}(w_t - w_s). \tag{12.7.9}$$

Combining eqs. (12.7.6) and (12.7.9) gives

$$w_s \sim -(M_{s1} - M_{t1}) \frac{\rho_t}{\rho} \frac{1}{2K_{st}} = -M_{s1} \frac{\rho_t}{\rho} \frac{1}{K_{st}}. \tag{12.7.10}$$

Since the difference in linear momentum moments is finite in an initially homogeneous medium, eq. (12.7.10) establishes this difference as a driving force for the bulk motion of s-type particles relative to the motion of the center of mass of the system, with $1/K_{st}$ acting as a diffusion coefficient. Inserting into eq. (12.7.10) the estimate of $M_{s1} - M_{t1}$ given by eq. (12.5.9), one has that, in an initially homogeneous, fully ionized, mixture of two elements in which all particles behave as perfect gases,

$$w_s \sim -n_s n_t \frac{A_s(1+Z_t) - A_t(1+Z_s)}{n_s(1+Z_s) + n_t(1+Z_t)} \, m_H \, g \, \frac{\rho_t}{\rho} \frac{1}{K_{st}}, \qquad (12.7.11)$$

where A_s and A_t are atomic numbers and the approximations $m_s^a = m_H A_s$ and $m_s^a = m_H A_s$ have been made.

If $A_s(1+Z_t) > A_t(1+Z_s)$, $w_s < 0$ and, from eq. (12.7.4), $w_t > 0$. In a fully ionized mixture of hydrogen and helium, helium will sink and hydrogen will float. In a fully ionized medium of two intermediate mass elements for which $A_t/Z_t \sim$ constant, again the heavier ion sinks and the lighter one floats. Since A_t/Z_t tends to increase with increasing A_t, this tendency for separation increases with increasing atomic mass. Thus, in accord with intuition, in a homogeneous, fully ionized non-degenerate gas, a heavier ion is forced to move inward and a lighter ion is forced to move outward.

For an initially homogeneous mixture in which electrons are thoroughly degenerate and $|A_s/Z_s - A_t/Z_t| \neq 0$, eq. (12.5.16) may be used for the difference between moments in eq. (12.7.10), giving

$$w_s \sim -n_s n_t \frac{A_s Z_t - A_t Z_s}{n_s Z_s + n_t Z_t} \, m_H \, g \, \frac{\rho_t}{\rho} \frac{1}{K_{st}}. \qquad (12.7.12)$$

Under highly electron-degenerate conditions, most ions are bare nuclei and, since, for heavier nuclei, A_i/Z_i tends to increase with increasing A_i, there is no question but that heavier nuclei will preferentially diffuse inward. However, the tendency towards separation is significantly diminished for ions with identical charge to atomic number ratios.

Having learned that including the flow properties of electrons is essential for determining individual diffusion velocites, it is reasonable to explore the effect of including details of interactions between electrons and ions. In a multicomponent system, one has that

$$M_{s1} = R_{s1}, \qquad (12.7.13)$$

where

$$M_{s1} = \frac{dP_s}{dr} + n_s m_s g - n_s Z_s \, eE \qquad (12.7.14)$$

and

$$R_{s1} = \int \frac{\delta f_s}{\delta t} \, m_s v_s \, d^3\mathbf{v}$$

$$= \sum_t K_{st} \, (w_t - w_s) + K_{se} \, (w_e - w_s), \qquad (12.7.15)$$

where the sum is over all ion types other than type s. R_{s1} may be thought of as the system response to driving by the corresponding linear momentum moment M_{s1}. In dynamical equilibrium, the driving forces and response forces must balance, so $M_{s1} = R_{s1}$. For convenience, the response function is broken into two parts, a term expressing the interaction between ion s and all other ions and a term due to interactions between ion s and electrons. The quantities K_{st} are resistance coefficients which are proportional to elastic interaction cross sections between s- and t-type particles and are symmetric in the indices, so that $K_{ts} = K_{st}$. Similarly, $K_{es} = K_{se}$.

For electrons, one has

$$M_{e1} = \frac{dP_e}{dr} + n_e(m_e g + eE) = R_{e1} = \sum_s K_{es} (w_e - w_s), \qquad (12.7.16)$$

with the consequence that, when the system is not in equilibrium with respect to diffusion, the electrostatic field strength differs from the approximation employed up to this juncture.

The resistance coefficients are basically the inverse of diffusion coefficients. Since the latter are proportional to relative particle velocities, one expects the various K_{es}s to be of the order of forty times smaller than the various K_{st}s and that, therefore, the approximation to eE given by setting $M_{e1} \sim 0$ remains reasonable. Nevertheless, it is worth examining the degree to which the approximate estimate of the electric field strength adopted thus far differs from a more rigorous estimate which recognizes the finiteness of the resistance coefficients between electrons and ions. For electrons in the two element system, one has

$$\frac{dP_e}{dr} + n_e(m_e g + eE) = K_{es}(w_s - w_e) + K_{et}(w_t - w_e). \qquad (12.7.17)$$

Using eqs. (12.7.7) and (12.7.8), this becomes

$$\frac{dP_e}{dr} + n_e(m_e g + eE) = \bar{K}_{es} \, w_s, \qquad (12.7.18)$$

where

$$\bar{K}_{es} = \frac{\rho}{\rho_t} \frac{K_{es} \, n_t Z_t - K_{et} \, n_s Z_s}{n_e}. \qquad (12.7.19)$$

Thus,

$$eE = -m_e g - \frac{1}{n_e} \frac{dP_e}{dr} + \frac{1}{n_e} \bar{K}_{es} \, w_s. \qquad (12.7.20)$$

Similar manipulations yield $\bar{K}_{et} = -(\rho_t/\rho_s) \, \bar{K}_{es}$, so that $\bar{K}_{et} \, w_t = \bar{K}_{es} \, w_s$ and the value of eE given by eq. (12.7.20) is independent of the choice of ion indices.

Using eq. (12.7.20) in eq. (12.7.2), the linear momentum moment for s-type ions becomes

$$
\begin{aligned}
M_{s1}^{new} &= \frac{dP_s}{dr} + n_s(m_s g - Z_s \, eE) \\
&= \frac{dP_s}{dr} + n_s m_s g + n_s Z_s \, m_e g + \frac{n_s Z_s}{n_e} \frac{dP_e}{dr} - \frac{n_s Z_s}{n_e} \bar{K}_{es} \, w_s \\
&= \frac{dP_s}{dr} + \frac{n_s Z_s}{n_e} \frac{dP_e}{dr} + n_s m_s^a g - \frac{n_s Z_s}{n_e} \bar{K}_{es} \, w_s, \qquad (12.7.21)
\end{aligned}
$$

which differs from the linear momentum moment given by eq. (12.5.2) only by the addition of a term expressing the effect of interactions between electrons and the ion s. From eq. (12.7.17), with the help of eqs. (12.7.7) and (12.7.8), this moment drives diffusion velocities according to

$$R_{s1} = K_{st}(w_t - w_s) + K_{se}(w_e - w_s) = -\frac{\rho}{\rho_t}\left(K_{st} + K_{se}\frac{n_t Z_t}{n_e}\right)w_s. \qquad (12.7.22)$$

Equating the driving force given by eq. (12.7.21) with the response force given by eq. (12.7.22) yields

$$M_{s1}^{\text{old}} = \frac{dP_s}{dr} + \frac{n_s Z_s}{n_e}\frac{dP_e}{dr} + n_s m_s^a g = -\frac{\rho}{\rho_t}\left(K_{st} + \tilde{K}_e\right)w_s, \qquad (12.7.23)$$

where M_{s1}^{old} is defined, as originally, by eq. (12.5.2) and

$$\tilde{K}_e = \frac{1}{n_e}\left(n_t Z_t K_{se} - \frac{\rho_t}{\rho}n_s Z_s \tilde{K}_{es}\right). \qquad (12.7.24)$$

Using eq. (12.7.19) to eliminate $\tilde{K}_{es}$, eq. (12.7.24) becomes

$$\tilde{K}_e = \frac{1}{n_e^2}\left[(n_t Z_t)^2 K_{se} + (n_s Z_s)^2 K_{te}\right]. \qquad (12.7.25)$$

By symmetry,

$$M_{t1}^{\text{old}} = \frac{dP_t}{dr} + \frac{n_t Z_t}{n_e}\frac{dP_e}{dr} + n_t m_t^a g = -\frac{\rho}{\rho_s}\left(K_{ts} + \tilde{K}_e\right)w_t, \qquad (12.7.26)$$

where M_{t1}^{old} is defined, as originally, by eq. (12.5.3). Thus, w_s is still given by eq. (12.7.10), the only difference being that the ion–ion resistance coefficient $K_{ts} = K_{st}$ is replaced by $K_{ts} + \tilde{K}_e$.

12.8 Generalization to a multicomponent gas

Generalization to the case when there are an arbitrary number of components is straightforward. From eq. (12.7.17), one has for the electric field

$$eE = \frac{1}{n_e}\left(-n_e m_e g - \frac{dP_e}{dr} + \sum_s K_{es}(w_s - w_e)\right). \qquad (12.8.1)$$

Insistence on no net mass flow with respect to the center of mass and on zero electric current gives two relationships for w_e in terms of the ionic diffusion velocities. The two relationships are

$$w_e = -\frac{\sum_s m_s n_s w_s}{m_e n_e} = \frac{\sum_s Z_s n_s w_s}{n_e}, \qquad (12.8.2)$$

from which it follows that

$$\sum_s (m_s + Z_s m_e) n_s w_s = \sum_s m_s^a n_s w_s = \sum_s \rho_s w_s = 0. \qquad (12.8.3)$$

Using eq. (12.8.1) in eq. (12.7.14) gives

$$M_{s1} = \frac{dP_s}{dr} + n_s m_s g - n_s Z_s \frac{1}{n_e} \left(-m_e n_e g - \frac{dP_e}{dr} + \sum_t K_{et}(w_t - w_e) \right)$$

$$= \frac{dP_s}{dr} + n_s m_s^a g - n_s Z_s \frac{1}{n_e} \left(-\frac{dP_e}{dr} + \sum_t K_{et}(w_t - w_e) \right). \qquad (12.8.4)$$

Combining eqs. (12.7.13), (12.7.15), and (12.8.4) gives

$$M_{s1}^{\text{old}} = \frac{dP_s}{dr} + \frac{n_s Z_s}{n_e} \frac{dP_e}{dr} + n_s m_s^a g$$

$$= R_{s1}^{\text{new}} = \sum_t K_{st}(w_t - w_s) + K_{se}(w_e - w_s) + \frac{n_s Z_s}{n_e} \sum_t K_{et}(w_t - w_e),$$

$$(12.8.5)$$

where M_{s1}^{old} is given by eq. (12.5.2), which may still be considered to be the forcing term for diffusion, and R_{s1}^{new} differs from R_{s1}^{old} given by eq. (12.7.15) by the addition of the third term on the far right of eq. (12.8.5). Using eq. (12.8.2) to eliminate w_e from eq. (12.8.5) yields

$$M_{s1}^{\text{old}} = R_{s1}^{\text{new}} = \sum_t K_{st}(w_t - w_s) - K_{se} w_s + \sum_t K_{tse} w_t, \qquad (12.8.6)$$

where

$$K_{tse} = \frac{n_s Z_s K_{et} + n_t Z_t \bar{K}_{se}}{n_e} \qquad (12.8.7)$$

and

$$\bar{K}_{se} = K_{se} - \frac{n_s Z_s}{n_e} \sum_t K_{et}. \qquad (12.8.8)$$

Given the uncertainties in the evaluation of the resistance coefficients, as well as the incompleteness of the formulation which has here ignored niceties such as thermal diffusion, neglect of all terms which involve electron–ion collisions is a reasonable first approximation. For a system of N ions, there are N equations like eq. (12.8.6) which are to be solved for the w_ss in terms of stellar model characteristics as described by the forcing terms M_{s1}. However, the existence of the constraint expressed by eq. (12.8.3) means that the N equations are not entirely independent of one another. This is immediately evident when electron–ion interactions are neglected. Then, all of the ionic w_ss appear as differences, of which there are only $N - 1$. Thus, in the general case, one must use eq. (12.8.3) to eliminate one of the w_ss in terms of the others, leaving $N - 1$ equations to find the $N - 1$ w_ss selected for explicit solution in terms of stellar model characteristics. Alternatively, one may substitute eq. (12.8.3) for one of the N equations described by eq. (12.8.6) and solve for all of the N w_ss simultaneously.

By way of illustration, consider a two-ion system and neglect ion–electron interactions. Equations (12.7.9) and (12.7.3) may be written in matrix form as

$$\begin{pmatrix} \Delta M_{st} \\ 0 \end{pmatrix} = \begin{pmatrix} -2K_{st} & 2K_{st} \\ \rho_s & \rho_t \end{pmatrix} \begin{pmatrix} w_s \\ w_t \end{pmatrix}, \tag{12.8.9}$$

where, from eq. (12.8.5),

$$\Delta M_{st} = M_{s1} - M_{t1} = \left(\frac{dP_s}{dr} - \frac{dP_t}{dr} \right) + \left(\frac{Z_s n_s}{n_e} - \frac{Z_t n_t}{n_e} \right) \frac{dP_e}{dr} + (\rho_s - \rho_t)\, g. \tag{12.8.10}$$

The solution is, by Cramer's rule,

$$w_s = \begin{vmatrix} \Delta M_{st} & 2K_{st} \\ 0 & \rho_t \end{vmatrix} \div \begin{vmatrix} -2K_{st} & 2K_{st} \\ \rho_s & \rho_t \end{vmatrix}$$

$$= \frac{\Delta M_{st}\rho_t}{-2K_{st}(\rho_t + \rho_s)} = -\frac{1}{2K_{st}} \frac{\rho_t}{\rho} \Delta M_{st}, \tag{12.8.11}$$

in agreement with eq. (12.7.10). Similarly,

$$w_t = \begin{vmatrix} -2K_{st} & \Delta M_{st} \\ \rho_s & 0 \end{vmatrix} \div (-2K_{st}\, \rho) = \frac{1}{2K_{st}} \frac{\rho_s}{\rho} \Delta M_{st}. \tag{12.8.12}$$

Given that $M_{s1} = -M_{t1}$, one can also write

$$\begin{pmatrix} M_{st} \\ 0 \end{pmatrix} = \begin{pmatrix} -K_{st} & K_{st} \\ \rho_s & \rho_t \end{pmatrix} \begin{pmatrix} w_s \\ w_t \end{pmatrix}, \tag{12.8.13}$$

which results in

$$w_s = -\frac{\rho_t}{\rho} \frac{M_{st}}{K_{st}}, \tag{12.8.14}$$

which is identical with eq. (12.8.11).

For a system with three ions and attendant electrons, the most straightforward formulation is

$$\begin{pmatrix} M_{11} \\ M_{21} \\ 0 \end{pmatrix} = \begin{pmatrix} -(K_{12} + K_{13}) & K_{12} & K_{13} \\ K_{21} & -(K_{21} + K_{23}) & K_{23} \\ \rho_1 & \rho_2 & \rho_3 \end{pmatrix} \begin{pmatrix} w_1 \\ w_2 \\ w_3 \end{pmatrix}. \tag{12.8.15}$$

where the w_is are the diffusion velocities for the three ions and the M_{i1}s are the linear momentum moments for these ions. This result follows from eqs. (12.8.6)–(12.8.8) with $K_{se} = 0$ for all s and from the furthest equality to the right in eq. (12.8.3).

On taking into account the symmetry of the resistance coefficients, the determinant of the 3×3 matrix in eq. (12.8.15) reduces to

$$D = (K_{12}K_{23} + K_{21}K_{13} + K_{13}K_{32})\,\rho. \tag{12.8.16}$$

Using Cramer's rule and the fact that $M_{11} + M_{21} + M_{31} = 0$ (see eqs. (12.3.22) and (12.3.23)), the ws given by eqs. (12.8.15) and (12.8.16) are

$$D\,w_1 = K_{12}\,M_{31}\,\rho_3 + K_{13}\,M_{21}\,\rho_2 - K_{23}\,M_{11}\,(\rho_2 + \rho_3), \tag{12.8.17}$$

$$D\,w_2 = K_{12}\,M_{31}\,\rho_3 - K_{13}\,M_{21}\,(\rho_1 + \rho_3) + K_{23}\,M_{11}\,\rho_1, \tag{12.8.18}$$

and

$$D\,w_3 = -K_{12}\,M_{31}\,(\rho_1 + \rho_2) + K_{13}\,M_{21}\,\rho_2 + K_{23}\,M_{11}\,\rho_1. \tag{12.8.19}$$

Any of the last three equations transforms into one of the other two simply by exchanging two indices everywhere in the equation and using the symmetry of the resistance coefficients.

As the number of ions under consideration is increased beyond three or four, a solution by Cramer's rule, although providing insight into the interactions between various ion types, becomes cumbersome, and, to obtain equations for the diffusion velocities in terms of stellar model characteristics, the method of Gaussian elimination described in Section 8.8 of Volume 1 is recommended.

12.9 Gravitational diffusion velocities for helium and iron at the base of the convective envelope of solar models

For quantitative estimates of the gravitational diffusion velocity in a homogeneous medium consisting of two ions and attendant non-degenerate electrons, it is convenient to write eq. (12.7.15) in terms of number abundance parameters Y_s and Y_t:

$$w_s \sim -\left[A_s(1 + Z_t) - A_t(1 + Z_s)\right] \frac{\rho g}{K_{st}} \frac{A_t Y_t}{Y_s(1 + Z_s) + Y_t(1 + Z_t)}\,Y_s\,Y_t. \tag{12.9.1}$$

Adopting K_{st} given by eqs. (12.6.23) and (12.6.20), eq. (12.9.1) becomes

$$w_s(\text{cm s}^{-1}) = -\,2.4301 \times 10^{-14}\,\frac{T_6^{3/2}}{\rho}\,\frac{g}{\log_e\left(1 + 0.187\,69x_{st}^{1.2}\right)}$$

$$\times \frac{A_t Y_t}{\mu_{st}^{1/2}Z_s^2 Z_t^2}\,\frac{A_s(1 + Z_t) - A_t(1 + Z_s)}{Y_s(1 + Z_s) + Y_t(1 + Z_t)}, \tag{12.9.2}$$

where the gravitational acceleration is in cm s^{-2} and ρ is in g cm^{-3}.

As an example, consider hydrogen and helium at the base of the convective envelope of solar model B described in Section 10.2 of Volume 1. There, $T_6 = 2.04$, $\rho = 0.155$ g cm^{-3}, and $g = 4.98 \times 10^4$ cm s^{-2}. Both ions are fully ionized, so $Z_H = 1$ and $Z_{He} = 2$. Setting $Y_{He}/Y_H = 0.1$ and $4Y_{He} + Y_H = 1$, eq. (12.9.2) gives

$$w_{He}(\text{cm s}^{-1}) = -\frac{1.38 \times 10^{-8}}{\log_e\left(1 + 0.187\,96x_{st}^{1,2}\right)}, \tag{12.9.3}$$

where x_{st} is given by eq. (12.6.19). The mean separation between ions relative to the Bohr radius of hydrogen is given by

$$\frac{\lambda_{\text{ions}}}{a_0} = \left(\frac{3}{4\pi}\frac{m_H}{a_0^3}\frac{1}{\rho\sum_s Y_s}\right)^{1/3} = \left(\frac{2.696\,13}{\rho\sum_s Y_s}\right)^{1/3}. \tag{12.9.4}$$

and the Debye radius relative to the Bohr radius is given by

$$\frac{\lambda_D}{a_0} = \left(\frac{kT}{4\pi e^2\left(\sum_s n_s Z_s^2 + n_e\right)}\right)^{1/2}\frac{1}{a_0} = \left(\frac{1}{4\pi}\frac{kT}{e^2/a_0}\frac{m_H}{a_0^3}\frac{1}{\rho}\frac{1}{\sum_s Y_s Z_s(1+Z_s)}\right)^{1/2}$$

$$= 1.687\,02\left(\frac{T_6/\rho}{\sum_s Y_s Z_s(1+Z_s)}\right)^{1/2}. \tag{12.9.5}$$

At the base of the convective envelope in solar model B, $\lambda_{\text{ions}}/a_0 = 2.808$ and $\lambda_D/a_0 = 4.492$. Thus, in eq. (12.6.21), $\lambda = \lambda_D$, and $x_{st}^2 \sim 3369$, so that the logarithmic term in the denominator of eq. (12.9.3) is 3.242 and

$$w_{He} \sim -4.26 \times 10^{-9} \text{ cm s}^{-1} \sim -0.134 \text{ cm yr}^{-1}. \tag{12.9.6}$$

This amounts to a displacement of $\sim 6.2 \times 10^8$ cm or about 9% of the Sun's radius in 4.6×10^9 years, suggesting that, over the course of the Sun's life, gravitational settling leads to a modest increase in the helium–hydrogen abundance ratio below the base of the convective envelope. The estimated diffusion velocity is 12% of 1.123 cm yr^{-1}, the bulk outward velocity of matter at the base of the convective zone.

Another way of describing the effect of diffusion is to define a time scale by $t_{\text{diff}} = H_P/w_s$, where H_P is the pressure scale height. At the base of the convective envelope of solar model B, $H_P \sim 5.81 \times 10^9$ cm, so

$$(t_{\text{diff}})_{He} = \frac{H_P}{w_{He}} \sim 1.36 \times 10^{18} \text{ s} = 4.32 \times 10^{10} \text{ yr}, \tag{12.9.7}$$

which is about nine times larger than the age of the Sun and about four times larger than the nuclear burning time scale of $t_{\text{nuc}} \sim 10^{10}$ yr at the solar center.

As another example, consider the gravitationally-induced diffusion of iron at the base of the Sun's convective envelope. The first task is to determine the number distribution of iron ions. The Saha equation for any two adjacent ions is

$$\frac{n_{Z_s} n_e}{n_{Z_s-1}} = 2\left(\frac{2\pi m_e kT}{h^2}\right)^{3/2}\frac{g_{Z_s}}{g_{Z_s-1}}\exp\left(-\frac{\chi Z_s}{kT}\right), \tag{12.9.8}$$

where the gs are partition functions. Most of the electrons are contributed by fully ionized hydrogen and helium, so the Saha equation for the two ions becomes

$$\frac{n_{Z_s}}{n_{Z_s-1}} = \frac{1}{N_A \rho \, (Y_H + 2Y_{He})} \, 2 \left(\frac{2\pi m_e kT}{h^2}\right)^{3/2} \frac{g_{Z_s}}{g_{Z_s-1}} \exp\left(-\frac{\chi_{Z_s}}{kT}\right)$$

$$\sim \frac{4.04 \left(T_6^{3/2}/\rho\right)}{Y_H + 2Y_{He}} \frac{g_{Z_s}}{g_{Z_s-1}} \exp\left(-\frac{\chi_{Z_s}}{kT}\right) \sim 44.7 \frac{g_{Z_s}}{g_{Z_s-1}} \exp\left(-\frac{\chi_{Z_s}}{kT}\right), \quad (12.9.9)$$

where the final expression comes from setting $Y_H \sim 0.71$, $Y_{He} \sim 0.071$, $T_6 = 2$, and $\rho = 0.3$ g cm^{-3}. Thus, at the base of the solar convective envelope, two iron ions are equally abundant if

$$\chi_{Z_s} = 3.80 \frac{g_{Z_s}}{g_{Z_s-1}} kT = 327.4 \frac{g_{Z_s}}{g_{Z_s-1}} T_6 \text{ eV} = 655 \text{ eV} \frac{g_{Z_s}}{g_{Z_s-1}}. \quad (12.9.10)$$

The outer two electrons in neutral iron are $4s$ electrons, the next inner six are $3d$ electrons, the next six are $3p$ electrons, and the next two are $3s$ electrons. If 15 electrons have been stripped off, the next electron to be stripped off is the remaining $3s$ electron, so a lower limit on the 16th ionization potential is

$$\chi_{16} > \frac{(16)^2}{3^2} \, 13.6 \text{ eV} = 387 \text{ eV}. \quad (12.9.11)$$

When 16 electrons have been stripped off, the next six electrons are $2p$ electrons, so a lower limit on the 17th ionization potential is

$$\chi_{17} > \frac{(17)^2}{2^2} \, 13.6 \text{ eV} = 983 \text{ eV}. \quad (12.9.12)$$

These estimates are lower limits because shielding by inner electrons is incomplete. All of the iron ionization potentials have been measured experimentally and, from the fourth edition of C. W. Allen's book (1999), the two ionization potentials under scrutiny are actually $\chi_{16} = 489$ eV and $\chi_{17} = 1266$ eV. To the extent that the partition functions for ions with $Z_s = 16 \pm 1$ are similar, the measured ionization energies suggest, in the light of eq. (12.9.9), that the dominant ions are characterized by $Z_s = 16$ and $Z_s = 15$.

To explore this possibility further, write eq. (12.9.9) as

$$\frac{n_{Z_s}}{n_{Z_s-1}} = \lambda \frac{g_{Z_s}}{g_{Z_s-1}} \exp\left(-\frac{\chi_{Z_s}}{kT}\right), \quad (12.9.13)$$

and assume that there exists a dominant abundance at $Z_s = Z$. Then number abundances of ions with $Z_s = Z - N$ are related to the dominant number abundance by

$$\frac{n_{Z-N}}{n_Z} = \frac{1}{\lambda^N} \exp\left(\sum_{J=0}^{N-1} \frac{\chi_{Z-J}}{kT}\right) \frac{g_{Z-N}}{g_Z} \quad (12.9.14)$$

and number abundances of ions with $Z_s = Z + N$ are related to the dominant number abundance by

$$\frac{n_{Z+N}}{n_Z} = \lambda^N \exp\left(\sum_{J=1}^{N} -\frac{\chi_{Z+J}}{kT}\right) \frac{g_{Z+N}}{g_Z}. \quad (12.9.15)$$

Using the additional experimentally determined ionization energies $\chi_{12} = 336$ eV, $\chi_{13} = 379$ eV, $\chi_{14} = 411$ eV, $\chi_{15} = 457$ eV, and $\chi_{18} = 1358$ eV, applications of eqs. (12.9.14) and (12.9.15) with $\lambda = 44.7$ and $kT = 172.34$ eV give

$$n_{12} + n_{13} + n_{14} + n_{15} + n_{16} + n_{17} + n_{18}$$

$$= n_{16} \left(0.0059 \, \frac{g_{12}}{g_{16}} + 0.0292 \, \frac{g_{13}}{g_{16}} + 0.121 \, \frac{g_{14}}{g_{16}} + 0.382 \, \frac{g_{15}}{g_{16}} + 1 \right.$$

$$\left. + 0.0296 \, \frac{g_{17}}{g_{16}} + 0.0005 \, \frac{g_{18}}{g_{16}} \right). \tag{12.9.16}$$

For the 17th through 21st ionization potentials, the ejected electron is a $2p$ electron and, to the extent that excitation to $n = 3$ levels can be neglected, the partition functions for the daughter ions are $g_{17} = 6$, $g_{18} = 15$, $g_{19} = 20$, $g_{20} = 15$, and $g_{21} = 6$. For the 16th ionization potential, the daughter ion has six $2p$ electrons and, if excitation to $n = 3$ levels is neglected, a partition function of $g_{16} = 1$. But, excitation to $n = 3$ levels does occur and, in a crude approximation, one may estimate that the partition function is augmented by $\Delta g_{16} \sim 6 \times (2 \times e^{-(1266-552)/172.34} + 6 \times e^{-(1266-515)/172.34} + 10 \times e^{-(1266-437)/172.34}) = 1.22$, giving $g_{16} \sim 2.22$.

For the 15th ionization potential, the ground state of the daughter nucleus is characterized by a valence $3s$ electron and, if excited states are ignored, the associated partition function is $g_{15} = 2$. However, the outer electron can be excited to $3p$ and $3d$ levels with excitation energies which can be of the order of or smaller than kT. If they are small compared with kT, one anticipates a partition function of the order of $g_{15} \sim 2 + 6 + 10 = 18$. At a temperature of $T_6 = 2$ excitation energies are actually comparable with kT and one may estimate $g_{15} \sim 2 + 6 \times e^{-(457-340)/172.34} + 10 \times e^{-(457-340)/172.34} = 11.6$. Obviously, the same considerations imply that the partition functions for other ions with several $n = 3$ electrons are larger than estimated when only the ground state configuration is taken into account.

In summary, while the dominant iron ion at the base of the solar convective zone appears to be characterized by $Z_s = 15$–16, additional ions with, say, $12 \leq Z_s \leq 18$ are certainly present. Fortunately, the diffusion velocity for any given element is not very sensitive to the details of the ion distribution for that element. For example, the diffusion velocity for iron at the base of the solar model convective zone calculated with $Z_s = 16$ and $Z_s = 15$ differ by less than 10%.

The Debye length is the same as before, and with $Z_s = 16$, eq. (12.6.21) gives $x_{st}^2 = 13.16$, so that the logarithmic denominator in eq. (12.9.2) is 0.632. With $Z_s = 16$ and $A_s = 56$, eq. (12.9.2) yields

$$w_{Fe} \sim -2.69 \times 10^{-9} \text{ cm s}^{-1} \sim -0.10 \text{ cm yr}^{-1}, \tag{12.9.17}$$

approximately 60% of the diffusion velocity estimated for helium, as given by eq. (12.9.6). The fact that diffusion acts more effectively in causing helium to diffuse inward is presumably a consequence of the fact that the outward electrical force on the 15–16 times ionized iron is greater than the outward electrical force on alpha particles.

Table 12.9.1 Characteristics at the base of the convective envelope of two solar models

Model	ρ_0	T_6	g	$\Delta R/R_\odot$	$\Delta M/M_\odot$	$h/\Delta R$	w_{He}	w_{Fe}
A	0.09557	1.684	4.723/4	0.2381	0.01167	0.415	6.70/−9	4.23/−9
B	0.15503	2.040	4.977/4	0.2653	0.01990	0.422	4.26/−9	2.69/−9

The characteristics of model B used and derived in these calculations are summarized in row 2 of Table 12.9.1, where g is in units of cm s^{-2} and the ws are in units of cm s^{-1}. The same characteristics for solar model A described in Section 10.1 of Volume 1 are given in row 1 of the table. The quantities ΔR and ΔM are, respectively, the radial width and the mass of the convective envelope.

Making use of the run of temperatures and densities in solar models constructed by Ray Weymann (1957), and using $K_{st} = K_{st}^0$, Lawrence H. Aller and Sydney Chapman (1960) estimate gravitational diffusion velocities for iron at various depths, determining, e.g., that, at depths of 0.3 and 0.2 $R_\odot$ below the surface, the gravitational diffusion velocity varies from 3×10^{-9} cm s^{-1} to 5×10^{-9} cm s^{-1}, not significantly different from the estimates made here. However, they also determine velocity components due to thermal diffusion, which has been neglected here, estimating that the thermal diffusion velocity w_{therm} is approximately 1.4 times larger than w_{grav}, giving a total diffusion velocity $w_{\text{total}} = w_{\text{grav}} + w_{\text{therm}}$, approximately 2.4 times larger than w_{grav}. On the other hand, Thierry Montmerle and George Michaud (1976) find shortcomings in the Aller–Chapman treatment of thermal diffusion and, in their formulation of the diffusion equation as applied to stellar atmosphere models, the thermal diffusion term never exceeds 10% of other terms in the equation.

Aller and Chapman construct a convenient procedure for estimating a time scale for the decrease in the abundance of iron in the convective envelope. Defining a parameter $h(R_{\text{BCE}})$ by

$$M_{\text{CE}} = \int_{R_{\text{BCE}}}^{R\odot} 4\pi r^2 \rho(r)\mathrm{d}r = 4\pi R_{\text{BCE}}^2 \, \rho(R_{\text{BCE}}) \, h(R_{\text{BCE}}), \qquad (12.9.18)$$

where M_{CE} is the mass of the convective envelope and R_{BCE} and $\rho(R_{\text{BCE}})$ are, respectively, the radius and density at the base of the convective envelope, the mass of iron in the convective envelope can be written as

$$M_{\text{Fe}}(R_{\text{BCE}}) = 4\pi R_{\text{BCE}}^2 \, \rho_{\text{Fe}}(R_{\text{BCE}}) \, h(R_{\text{BCE}}). \qquad (12.9.19)$$

The rate at which this mass decreases can be written as

$$\frac{\mathrm{d}M_{\text{Fe}}(R_{\text{BCE}})}{\mathrm{d}t} = 4\pi R_{\text{BCE}}^2 \, \rho_{\text{Fe}}(R_{\text{BCE}}) \, w_{\text{Fe}}(R_{\text{BCE}}), \qquad (12.9.20)$$

where $w_{\text{Fe}}(R_{\text{BCE}})$ is the diffusion velocity of iron at the base of the convective envelope. Thus, the time scale t_0 for the depletion of iron in the convective envelope is given by

$$\frac{1}{t_0} = \frac{1}{M_{\mathrm{Fe}}(R_{\mathrm{BCE}})} \frac{\mathrm{d}M_{\mathrm{Fe}}(R_{\mathrm{BCE}})}{\mathrm{d}t} = \frac{w_{\mathrm{Fe}}(R_{\mathrm{BCE}})}{h(R_{\mathrm{BCE}})}. \tag{12.9.21}$$

In Weymann's models, the parameter $h(R_{\mathrm{BCE}})$ is roughly $1/3$ of the distance between the surface of the Sun and R_{BCE}. For example, when $R_{\mathrm{BCE}} = 0.7\ R_\odot$, $h(R_{\mathrm{BCE}}) = 0.0929$ $R_\odot$ and the Aller–Chapman estimate of the timescale for the depletion of iron is $t_0 \sim 2.78 \times 10^{10}$ yr. Setting

$$\frac{Y_{\mathrm{BCE}}(t)}{Y_{\mathrm{BCE}}(0)} = e^{-t/t_0}, \tag{12.9.22}$$

one has that, over the Sun's evolution up to the present, diffusion has reduced the ratio of iron to hydrogen in the convective envelope by approximately 18 %.

Taking the average of the relevant quantities for models A and B given in Table 12.9.1, one finds $\bar{h} = 7.31 \times 10^9$ cm and $\bar{w}_{\mathrm{grav}} = \bar{w}_{\mathrm{Fe}} = 3.46 \times 10^{-9}$ cm s^{-1}. So, when thermal diffusion is neglected, the models predict $t_0 \sim 6.69 \times 10^{10}$ yr and a reduction by about 7% in the iron to hydrogen ratio in the convective envelope during the evolution of the Sun up to the present. Multiplying $\bar{w}_{\mathrm{grav}}$ by a factor of of 2.4 to estimate the total diffusion velocity in a manner consistent with the Aller–Chapman treatment of diffusion, one recovers the 18% reduction in the surface iron to hydrogen ratio estimated by Aller and Chapman.

12.10   More diffusion velocities below the base of the convective envelope of a solar model

Generalization to an arbitrary set of ions is straightforward. When electron–ion resistance coefficients are neglected, the relationship between diffusion velocities and linear momentum moments can be approximated by

$$\begin{pmatrix} -\sum_t K_{1t} & K_{12} & \cdots & K_{1(N-1)} & K_{1N} \\ K_{21} & -\sum_t K_{2t} & \cdots & K_{2(N-1)} & K_{2N} \\ \vdots & \vdots & \ddots & \vdots & \vdots \\ K_{(N-1)1} & K_{(N-1)2} & \cdots & -\sum_t K_{(N-1)t} & K_{(N-1)N} \\ A_1 Y_1 & A_2 Y_2 & \cdots & A_{N-1} Y_{N-1} & A_N Y_N \end{pmatrix} \begin{pmatrix} w_1 \\ w_2 \\ \vdots \\ w_{N-1} \\ w_N \end{pmatrix} = \begin{pmatrix} M_{11} \\ M_{21} \\ \vdots \\ M_{(N-1)1} \\ 0 \end{pmatrix}, \tag{12.10.1}$$

where the sums of K_{it} over t do not include $t = i$ and, to repeat eq. (12.5.2) once again,

$$M_{i1} = \frac{\mathrm{d}P_i}{\mathrm{d}r} + \frac{n_i Z_i}{n_e} \frac{\mathrm{d}P_e}{\mathrm{d}r} + n_i m_i^{\mathrm{a}} g \tag{12.10.2}$$

is the ith linear momentum moment. This relationship may be written in $N - 1$ additional ways by replacing any of the linear momentum moments in the first $N - 1$ rows of the matrix on the left hand side of the equation by the resistance coefficients for the Nth isotope and replacing the corresponding linear momentum moment in the column vector on the right hand side by the linear momentum moment for the Nth isotope.

In the case of a homogeneous medium, when all particles behave as non-degenerate perfect gases,

$$\frac{1}{P_i}\frac{dP_i}{dr} = \frac{1}{P_e}\frac{dP_e}{dr} = \frac{1}{P}\frac{dP}{dr} = -\frac{g\rho}{P} \tag{12.10.3}$$

and, from eq. (12.10.2) and (12.10.3),

$$M_{i1} = \frac{1}{P}\frac{dP}{dr}\left(P_i + \frac{n_i Z_i}{n_e}P_e\right) + \rho_i g$$

$$= -\frac{g\rho}{P}(n_i kT + n_i Z_i kT) + \frac{n_i m_i^a}{\rho} g\rho$$

$$= -g\rho\frac{n_i(1+Z_i)}{\sum_{\text{all }j} n_j(1+Z_j)} + Y_i A_i g\rho$$

$$= \frac{Y_i}{\sum Y_j(1+Z_j)} g\rho \left[-(1+Z_i)\left(\sum Y_j A_j\right) + A_i\left(\sum Y_j(1+Z_j)\right)\right], \tag{12.10.4}$$

where the term $\sum Y_j A_j = 1$ has been inserted after $(1+Z_s)$. The sums are over all ions. Equation (12.10.4) can also be written as

$$M_{i1} = \rho g\, Y_i\, \frac{A_i\left(\sum_{j\neq i} Y_j(1+Z_j)\right) - (1+Z_i)\left(\sum_{j\neq i} Y_j A_j\right)}{\sum Y_j(1+Z_j)}. \tag{12.10.5}$$

If only two ions, s and t, are present, eq. (12.10.5) becomes

$$M_{s1} = \rho g\, Y_s\, \frac{A_s\, Y_t(1+Z_t) - (1+Z_s)\, Y_t A_t}{Y_s(1+Z_s) + Y_t(1+Z_t)}$$

$$= \rho g\, Y_s\, Y_t\, \frac{A_s\,(1+Z_t) - A_t\,(1+Z_s)}{Y_s(1+Z_s) + Y_t(1+Z_t)} \tag{12.10.6}$$

which is equivalent to eq. (12.5.10).

Consider a situation consisting of the seven elements shown in the first row of Table 12.10.1. The relative number abundances are shown in the second row and the atomic numbers are given in the third row. Although each element other than hydrogen exists in more than one ionization state, there is little loss in accuracy in supposing that all members of each element species are in a single ionization state. The chosen ionization state is given in the fourth row of Table 12.10.1. Somewhat arbitrarily, the density and temperature have been chosen as $\rho = 0.3$ g cm^{-3} and $T_6 = 2$, respectively.

Table 12.10.1 Diffusion velocities at the base of the convective envelope of a solar model

element	hydrogen	helium	lithium	carbon	nitrogen	oxygen	iron
Y	0.7038	0.07038	2.011/ − 9	2.815/−4	7.038/−5	5.329/−4	3.217/−5
A	1	2	7	12	14	16	56
Z	1	2	3	6	6	7	16
w(cm s^{-1})	1.01/−9	−2.44/−9	−2.59/−9	−1.58/−9	−1.95/−9	−1.80/−9	−2.46/−9

The resistance coefficient matrix (using eq. (12.8.17)) and the linear momentum column vector are

$$
\begin{pmatrix}
-715/9 & 655/9 & 3.70/4 & 14.8/9 & 3.73/9 & 35.2/9 & 6.28/9 \\
655/9 & -682/9 & 1.76/4 & 6.75/9 & 1.720/9 & 16.0/9 & 2.64/9 \\
3.70/4 & 1.76/4 & -5.61/4 & 3.82/2 & 9.81/1 & 9.05/2 & 1.46/2 \\
14.8/9 & 6.75/9 & 3.82/2 & -22.0/9 & 3.42/7 & 3.11/8 & 4.85/7 \\
3.73/9 & 1.720/9 & 9.81/1 & 3.42/7 & -5.58/9 & 8.12/7 & 1.29/7 \\
35.2/9 & 16.0/9 & 9.05/2 & 3.11/8 & 8.12/7 & -51.7/9 & 1.18/8 \\
6.28/9 & 2.64/9 & 1.46/2 & 4.85/7 & 1.29/7 & 1.18/8 & -9.09/9
\end{pmatrix}
\begin{pmatrix}
-2428 \\
2275 \\
1.369/\bar{4} \\
32.49 \\
10.24 \\
88.56 \\
21.98
\end{pmatrix},
$$

$$(12.10.7)$$

where $/\bar{4}$ in the third row of the vector is to be interpreted as $\times 10^{-4}$. One of the rows in the matrix and the corresponding row in the vector must be replaced by the mass conservation row:

$$
(0.7038 \quad 0.2815 \quad 1.408/\bar{8} \quad 3.378/\bar{3} \quad 9.853/\bar{4} \quad 8.526/\bar{3} \quad 1.802/\bar{3}), (0.000),
$$

$$(12.10.8)$$

where $/\bar{n} = 10^{-n}$. When normalized, i.e., when in each row i all of the elements a_{ij} in the row and the vector element b_i are divided by the maximum value of $|a_{ij}|$, the matrix and the column vector in eq. (12.10.7) become

$$
\begin{pmatrix}
-1.000 & 0.9160 & 5.176/\bar{8} & 2.077/\bar{2} & 5.220/\bar{3} & 4.923/\bar{2} & 8.781/\bar{3} \\
0.9603 & -1.000 & 2.582/\bar{8} & 9.907/\bar{3} & 2.522/\bar{3} & 2.344/\bar{2} & 3.866/\bar{3} \\
0.6591 & 0.3136 & -1.000 & 6.807/\bar{3} & 1.749/\bar{3} & 1.612/\bar{2} & 2.594/\bar{3} \\
0.6749 & 0.3072 & 1.738/\bar{8} & -1.000 & 1.556/\bar{3} & 1.416/\bar{2} & 2.208/\bar{3} \\
0.6688 & 0.3082 & 1.759/\bar{8} & 6.134/\bar{3} & -1.000 & 1.456/\bar{2} & 2.316/\bar{3} \\
0.6809 & 0.3092 & 1.751/\bar{8} & 6.024/\bar{3} & 1.572/\bar{3} & -1.000 & 2.284/\bar{3} \\
0.6904 & 0.2899 & 1.602/\bar{8} & 5.340/\bar{3} & 1.421/\bar{3} & 1.298/\bar{2} & -1.000
\end{pmatrix}
\begin{pmatrix}
-3.398/\bar{9} \\
3.337/\bar{9} \\
2.440/\bar{9} \\
1.478/\bar{9} \\
1.835/\bar{9} \\
1.714/\bar{9} \\
2.418/\bar{9}
\end{pmatrix}.
$$

$$(12.10.9)$$

The solution of the equations for the w_ss can be accomplished by using the Gaussian elimination or the LU decomposition procedures described in Section 8.8 in Volume 1. It is evident from the normalized matrix in eq. (12.10.9) that the pivoting procedure described in Section 8.8 has essentially already been accomplished: in absolute value, the maximum matrix element in each row is the diagonal element $a(k, k)$ and this automatically becomes the upper left hand matrix element in each step of the Gaussian elimination procedure. The reader may also verify that, if Cramer's rule is employed, the resultant ws depend on which linear momentum equation is replaced by the mass conservation equation and both the signs and the magnitudes of the ws vary wildly with the choice.

The results obtained by using the Gaussian elimination procedure are independent of the selection of equations and are shown in the fifth row of Table 12.10.1. Three features are notable: (1) all of the elements heavier than hydrogen diffuse inward while hydrogen diffuses outward; (2) the inward velocites are very similar; and, (3) the diffusion velocity of iron is nearly the same as estimated in the two ion approximation in the previous subsection.

12.11 Equations for abundance changes due to diffusion and solution algorithms

From the continuity relationships expressed by eqs. (12.3.12) and (12.3.13), one has that, in the center of mass frame of reference moving with velocity u,

$$\left(\frac{d\rho_s}{dt}\right)_u = \frac{\rho_s}{\rho}\frac{d\rho}{dt} - \nabla \cdot (\rho_s \mathbf{w}_s). \tag{12.11.1}$$

Setting $\rho_s = N_A \rho Y_s$, this becomes

$$\frac{dY_s}{dt} = -\frac{1}{\rho} \nabla \cdot (\rho\, Y_s\, \mathbf{w}_s). \tag{12.11.2}$$

The same result follows from eq. (8.9.13) in Volume 1 by setting $f_s = n_s w_s = N_A \rho Y_s w_s$, where f_s is the diffusive flux of ions of type s. In a spherically symmetric situation, eq. (12.11.2) becomes

$$\frac{dY_s}{dt} = -\frac{1}{\rho}\frac{1}{r^2}\frac{d}{dr}\left(r^2 \rho\, Y_s\, w_s\right), \tag{12.11.3}$$

which can also be written as

$$\frac{dY_s}{dt} = -\frac{d}{dM(r)}\left(4\pi r^2 \rho\, Y_s\, w_s\right). \tag{12.11.4}$$

These equations are simply statements that particle numbers are conserved, species by species: the rate of increase in the number abundance of species s in the ith shell is equal to the flux of this species flowing into the shell at its base at M_{i-1} minus the flux outward at its surface at $M_{i-1} + dM_i$.

In many situations, diffusion is such a slow process that the diffusion velocities estimated on the assumption of homogeneity remain reasonable approximations over the course of a calculation and one may use the closed forms for the diffusion velocities in eqs. (12.11.3) and (12.11.4). Since these velocities do not involve spatial gradients of number abundances, the diffusion equations become relatively simple to solve. However, it is clear from the fact that diffusion velocities vary from species to species in an initially homogeneous medium that homogeneity is ephemeral and, in general, it is prudent to solve for diffusion velocities in terms of linear momentum moments which take spatial gradients explicitly into account. If ions are assumed to behave as perfect gases, but no restriction is placed on the equation of state for electrons, eq. (12.5.2) for the sth linear momentum moment may be written as

$$
M_{s1} = \left[kT + \frac{Z_s n_s}{n_e} \left(\frac{\partial P_e}{\partial n_s} \right)_{T, n_t} \right] \frac{dn_s}{dr} + \frac{Z_s n_s}{n_e} \sum_{t \neq s} \left(\frac{\partial P_e}{\partial n_t} \right)_{T, n_u} \frac{dn_t}{dr}
$$

$$
+ \left[n_s m_s^a g + \left(n_s k + \frac{Z_s n_s}{n_e} \left(\frac{\partial P_e}{\partial T} \right)_{n_s, n_t} \right) \frac{dT}{dr} \right]. \tag{12.11.5}
$$

The third term in eq. (12.11.5) may be thought of as a driving force for gravitational settling, whereas the first two terms are both the consequences of gravitational settling and the driving forces for particle-gradient induced diffusion. If electrons also behave as a perfect gas,

$$
M_{s1} = kT \left(1 + \frac{n_s Z_s}{n_e} Z_s \right) \frac{dn_s}{dr} + kT \frac{Z_s n_s}{n_e} \sum_{t \neq s} Z_t \frac{dn_t}{dr} + \left[m_s^a g + (1 + Z_s) k \frac{dT}{dr} \right] n_s. \tag{12.11.6}
$$

A diffusion velocity may be broken into three parts, each associated with one of the main terms in eq. (12.11.5) (or eq. (12.11.6)). Writing a linear momentum moment as

$$
M_{s1} = \alpha_s \frac{dn_s}{dr} + \sum_{t \neq s} \beta_{st} \frac{dn_t}{dr} + \gamma_s n_s, \tag{12.11.7}
$$

one may define velocity coefficients w_s^α by

$$
\begin{pmatrix}
-\sum_t K_{1t} & K_{12} & \cdots & K_{1(N-1)} & K_{1N} \\
K_{21} & -\sum_t K_{2t} & \cdots & K_{2(N-1)} & K_{2N} \\
\vdots & \vdots & \ddots & \vdots & \vdots \\
K_{(N-1)1} & K_{(N-1)2} & \cdots & -\sum_t K_{(N-1)t} & K_{(N-1)N} \\
A_1 Y_1 & A_2 Y_2 & \cdots & A_{N-1} Y_{N-1} & A_N Y_N
\end{pmatrix}
\begin{pmatrix}
w_1^\alpha \\
w_2^\alpha \\
\vdots \\
w_{N-1}^\alpha \\
w_N^\alpha
\end{pmatrix}
=
\begin{pmatrix}
\alpha_1 \\
\alpha_2 \\
\vdots \\
\alpha_{N-1} \\
0
\end{pmatrix},
\tag{12.11.8}
$$

with similar equations for the determination of velocity coefficients w_{st}^β and w_s^γ. After solving for the velocity cefficients, a diffusion velocity is given by

$$
w_s = \alpha_s \frac{dn_s}{dr} + \sum_{t \neq s} \beta_{st} \frac{dn_t}{dr} + \gamma_s n_s. \tag{12.11.9}
$$

In the case of two ions with non-degenerate attendant electrons, one has

$$
\alpha_s = kT \left(1 + \frac{n_s Z_s}{n_e} Z_s \right), \tag{12.11.10}
$$

$$\beta_{st} = kT \frac{n_s Z_s}{n_e} Z_t, \qquad (12.11.11)$$

and

$$\gamma_s = n_s \left(m_s g + (1 + Z_s)k \frac{dT}{dr} \right). \qquad (12.11.12)$$

The matrix equation for w_s^α and w_t^α in this case is

$$\begin{pmatrix} -K_{st} & K_{st} \\ A_s Y_s & A_t Y_t \end{pmatrix} \begin{pmatrix} w_s^\alpha \\ w_t^\alpha \end{pmatrix} = \begin{pmatrix} \alpha_s \\ 0 \end{pmatrix}, \qquad (12.11.13)$$

with the solution

$$w_s^\alpha = -\frac{A_t Y_t}{A_s Y_s} w_t^\alpha = -\frac{A_t Y_t}{A_t Y_t + A_s Y_s} \frac{1}{K_{st}} \alpha_s = -\frac{A_t Y_t}{K_{st}} kT \left(1 + \frac{n_s Z_s}{n_e} Z_s \right),$$
$$(12.11.14)$$

where $A_s Y_s + A_t Y_t = 1$ follows from the conservation of mass and eq. (12.11.10) has been used to replace α_s. Similar equations hold for the other two velocity coefficients. Altogether,

$$w_s = -\frac{A_t Y_t}{K_{st}} kT \left\{ \left(1 + \frac{n_s Z_s}{n_e} Z_s \right) \frac{dn_s}{dr} + \left(\frac{n_s Z_s}{n_e} Z_t \right) \frac{dn_t}{dr} \right.$$
$$\left. + \left(\frac{m_s g}{kT} + (1 + Z_s) \frac{1}{T} \frac{dT}{dr} \right) n_s \right\} \qquad (12.11.15)$$

The term in this equation proportional to the number density is the consequence of gravitational settling. The other two terms are the consequence of the fact that, when left to themselves in the absence of a gravitational field, particles flow from regions of higher to regions of lower particle density.

Noting that

$$\frac{dn_s}{dr} = \frac{d}{dr} (N_A \rho Y_s) = N_A \left(\frac{d\rho}{dr} Y_s + \rho \frac{dY_s}{dr} \right), \qquad (12.11.16)$$

and defining $\tilde{K}_{st}$ by

$$K_{st} = \rho^2 Y_s Y_t \tilde{K}_{st}, \qquad (12.11.17)$$

eq. (12.11.15) may be written as

$$\rho Y_s w_s = -\frac{A_t}{\tilde{K}_{st}} N_A kT$$
$$\times \left(1 + \frac{Y_s Z_s}{Y_e} Z_s \right) \frac{dY_s}{dr} + \left(\frac{Y_s Z_s}{Y_e} Z_t \right) \frac{dY_t}{dr}$$
$$+ \left[\frac{m_s g}{kT} + (1 + Z_s) \left(\frac{1}{\rho} \frac{d\rho}{dr} + \frac{1}{T} \frac{dT}{dr} \right) \right] Y_s. \qquad (12.11.18)$$

Still another form is

$$\rho Y_s w_s = -\frac{A_t}{\tilde{K}_{st}} N_A kT Y_s \left[\left(\frac{1}{Y_s} \frac{dY_s}{dr} + \frac{Z_s}{Y_e} \frac{dY_e}{dr} \right) + \frac{m_s g}{kT} + \frac{1 + Z_s}{(\rho T)} \frac{d(\rho T)}{dr} \right],$$

(12.11.19)

and yet another is

$$\rho Y_s w_s = -\frac{A_t}{\tilde{K}_{st}} N_A kT Y_s \left(\frac{1}{P_s} \frac{dP_s}{dr} + Z_s \frac{1}{P_e} \frac{dP_e}{dr} + \frac{m_s g}{kT} \right).$$

(12.11.20)

Converting eq. (12.11.3) into a difference equation for use in stellar evolution calculations is complicated by the choice made in Chapter 6 of Volume 1 that state and composition variables are defined at the centers of mass shells whereas radius and mass are defined at shell boundaries and by the fact that composition variables to be determined appear in all terms of eq. (12.11.20) for $\rho Y_s w_s$. One approximation is

$$\frac{Y_{s,i} - Y_{s,i}^0}{\delta t} = -\frac{4\pi r_i^2 (\rho Y_s w_s)_{i+1/2} - 4\pi r_{i-1}^2 (\rho Y_s w_s)_{i-1/2}}{dM_i},$$

(12.11.21)

where, on the left hand side of the equation, $Y_{s,i}^0$ is the initial value of Y_s at the center of the ith shell, $Y_{s,i}$ is the value of Y_s in shell i after a time step δt, and quantities with subscripts $i \pm 1/2$ on the right hand side of the equation are estimates at zone boundaries of quantities defined at shell centers. These estimates include both initial composition variables and composition variables to be determined in the course of a calculation. For example, one can approximate

$$(\rho Y_s w_s)_{i+1/2} = -\left(\frac{A_t}{\tilde{K}_{st}} N_A kT \right)^0_{i+1/2}$$

$$\times \left(1 + \frac{Y_s Z_s}{Y_e} Z_s \right)^0_{i+1/2} \frac{Y_{s,i+1} - Y_{s,i}}{(r_{i+1} - r_{i-1})/2}$$

$$+ \left(\frac{Y_s Z_s}{Y_e} Z_t \right)^0_{i+1/2} \frac{Y_{t,i+1} - Y_{t,i}}{(r_{i+1} - r_{i-1})/2}$$

$$+ \left[\frac{m_s^a g_i}{kT_{i+1/2}} + \left(\frac{1 + Z_s}{\rho T} \right)_{i+1/2} \frac{(\rho T)_{i+1} - (\rho T)_i}{(r_{i+1} - r_{i-1})/2} \right]^0 Y_{s,i+1/2},$$

(12.11.22)

where quantities with half integer subscripts are defined by (see, e.g., eq. (8.9.24) in Volume 1)

$$Q_{s,i+1/2} = \frac{Q_{s,i} dM_{i+1} + Q_{s,i+1} dM_i}{dM_i + dM_{i+1}},$$

(12.11.23)

and quantities designated with a null superscript are approximated by their values at the beginning of a time step and remain constant during the calculation for changes due to diffusion. The substitution $i \rightarrow (i-1)$ in eqs. (12.11.22) and (12.11.23) yields the final quantity $(\rho Y_s w_s)_{i-1/2}$ required in eq. (12.11.21).

It is evident that one can construct a series of equations of the form

$$-\mathbf{A}_i \mathbf{y}_{i+1} + \mathbf{B}_i \mathbf{y}_i - \mathbf{C}_i \mathbf{y}_{i-1} = \mathbf{d}_i, \tag{12.11.24}$$

where $\mathbf{A}$, $\mathbf{B}$, and $\mathbf{C}$ are square matrices and $\mathbf{y}$ and $\mathbf{d}$ are vectors, with

$$\mathbf{y}_i = \begin{pmatrix} Y_{s,i} \\ Y_{t,i} \end{pmatrix} \tag{12.11.25}$$

describing the composition variables at the center of shell i and, for example,

$$\mathbf{A}_i = \begin{pmatrix} A_{ss,i} & A_{st,i} \\ A_{ts,i} & A_{tt,i} \end{pmatrix} \tag{12.11.26}$$

describing the coefficients of $Y_{s,i+1}$ and $Y_{t,i+1}$ in eq. (12.11.22).

At the model center (characterized by the index $i = 1$), the diffusion velocity and the radius are zero, so, for the central sphere, there is only one term in the numerator of the fraction on the right hand side of eq. (12.11.21) and, in eq. (12.11.22), only the number abundance parameters for $i = 2$ and $i = 3$ appear. Hence, one can set $\mathbf{C}_2 = 0$ in eq. (12.11.24) to obtain

$$-\mathbf{A}_2 \mathbf{y}_3 + \mathbf{B}_2 \mathbf{y}_2 = \mathbf{d}_2, \tag{12.11.27}$$

or

$$\mathbf{y}_2 = \mathbf{B}_2^{-1} \mathbf{A}_2 \mathbf{y}_3 + \mathbf{B}_2^{-1} \mathbf{d}_2 = \mathbf{E}_2 \mathbf{y}_3 + \mathbf{f}_2, \tag{12.11.28}$$

where $\mathbf{B}_2^{-1}$ is the inverse of $\mathbf{B}_2$ and $\mathbf{E}_2 = \mathbf{B}_2^{-1} \mathbf{A}_2$ and $\mathbf{f}_2 = \mathbf{B}_2^{-1} \mathbf{d}_2$.

If a relationship similar to eq. (12.11.28) holds for other shells, so that

$$\mathbf{y}_i = \mathbf{E}_i \mathbf{y}_{i+1} + \mathbf{f}_i, \tag{12.11.29}$$

where $\mathbf{E}$ and $\mathbf{f}$ are, respectively, a matrix and a vector to be determined, it follows from eq. (12.11.24) that

$$-\mathbf{A}_i \mathbf{y}_{i+1} + (\mathbf{B}_i - \mathbf{E}_{i-1} \mathbf{C}_i) \mathbf{y}_i - \mathbf{C}_i \mathbf{f}_{i-1} = \mathbf{d}_i. \tag{12.11.30}$$

Defining yet another matrix $\mathbf{G}$ by

$$\mathbf{G}_i = \mathbf{B}_i - \mathbf{E}_{i-1} \mathbf{C}_i, \tag{12.11.31}$$

eq. (12.11.30) can be written as

$$\mathbf{y}_i = \mathbf{G}_i^{-1} \mathbf{A}_i \, \mathbf{y}_{i+1} + \mathbf{G}_i^{-1} \left(\mathbf{d}_i + \mathbf{C}_i \mathbf{f}_{i-1} \right), \tag{12.11.32}$$

where $\mathbf{G}_i^{-1}$ is the inverse of $\mathbf{G}_i$. Finally, comparing eq. (12.11.32) with eq. (12.11.29), one has that

$$\mathbf{E}_i = \mathbf{G}_i^{-1} \mathbf{A}_i \tag{12.11.33}$$

and

$$\mathbf{f}_i = \mathbf{G}_i^{-1} \left(\mathbf{d}_i + \mathbf{C}_i \mathbf{f}_{i-1} \right). \tag{12.11.34}$$

In order to find the new abundance parameter in the last interior shell, it is necessary to impose a boundary condition at the outer edge of this last shell, where $i = N$. One possibility is to assume that number abundances in the static envelope are the same as in the last interior shell, so that $\mathbf{y}_{N+1} = \mathbf{y}_N$. From eq. (12.11.24), it follows that

$$(-\mathbf{A}_N + \mathbf{B}_N)\,\mathbf{y}_N - \mathbf{C}_N\mathbf{y}_{N-1} = \mathbf{B}'_N\mathbf{y}_N - \mathbf{C}_N\mathbf{y}_{N-1} = \mathbf{d}_N, \tag{12.11.35}$$

so that, effectively, $\mathbf{A}_N = 0$. Then, from eq. (12.11.33), $\mathbf{E}_N = 0$ and, from eqs. (12.11.32) and (12.11.34), it follows that

$$\mathbf{y}_N = \mathbf{f}_N. \tag{12.11.36}$$

All vectors $\mathbf{y}_i$ for $i = N - 1$ to $i = 2$ follow from eq. (12.11.29).

Generalization to a multicomponent situation involving n ions is straightforward. For a given isotope,

$$Y_{s,i} = Y^0_{s,i} + a_{s,i}Y_{s,i+1} + b_{s,i}Y_{s,i} + c_{s,i}Y_{s,i-1} + \sum_{t \neq s}\left(a_{t,i}Y_{t,i+1} + b_{t,i}Y_{t,i} + c_{t,i}Y_{t,i-1}\right). \tag{12.11.37}$$

Since they make no explicit reference to the number of components, all of the matrix equations derived for a two component system are valid in the general case. The only modification necessary is that matrices have $n \times n$ elements and vectors have n components.

Inverses of matrices appear in the solution of the diffusion equations. For any given matrix $\mathbf{A}$ and its inverse $\mathbf{A}^{-1}$, one has that

$$\mathbf{A}^{-1}\mathbf{A} = \mathbf{A}\,\mathbf{A}^{-1} = \mathbf{I} = \begin{pmatrix} 1 & 0 & \cdots & 0 \\ 0 & 1 & \cdots & 0 \\ \vdots & \vdots & \ddots & 0 \\ 0 & 0 & 0 & 1 \end{pmatrix}, \tag{12.11.38}$$

where $\mathbf{I}$ is the identity matrix, with elements $I_{ij} = \delta_{ij}$ (0 if $i \neq j$ and 1 if $i = j$). Elements in any column k of $\mathbf{A}^{-1}$ can be found by solving

$$\begin{pmatrix} a_{11} & a_{12} & \cdots & a_{1n} \\ a_{21} & a_{22} & \cdots & a_{2n} \\ \vdots & \vdots & \ddots & \vdots \\ a_{21} & a_{22} & \cdots & a_{nn} \end{pmatrix} \begin{pmatrix} a_{1k}^{-1} \\ a_{2k}^{-1} \\ \vdots \\ a_{nk}^{-1} \end{pmatrix} = \begin{pmatrix} \delta_{1k} \\ \delta_{2k} \\ \vdots \\ \delta_{nk} \end{pmatrix}, \tag{12.11.39}$$

which can also be written as

$$\sum_{j=1,n} a_{ij}a_{jk}^{-1} = \delta_{ik}, \ i = 1, n. \tag{12.11.40}$$

These equations can be solved by using the Gaussian elimination or LU decomposition procedures with pivoting, as discussed in Section 8.8 of Volume 1. Since an inverse matrix always appears as a multiplier of another matrix or of a vector, it is possible to avoid finding the inverse explicitly. For example, multiplying both sides of eq. (12.11.33) by $\mathbf{G}$ (and dropping the subscript i), one has the generic equation

$$\mathbf{GE} = \mathbf{A}, \tag{12.11.41}$$

where, in the general case, all three matrices are of dimension $n \times n$. The problem of finding $\mathbf{E}$ becomes one of finding the solution of n sets of linear equations, each set involving all of the elements of $\mathbf{G}$, but only one column in each of the other two matrices. That is,

$$\begin{pmatrix} g_{11} & g_{12} & \cdots & g_{1n} \\ g_{21} & g_{22} & \cdots & g_{2n} \\ \vdots & \vdots & \ddots & \vdots \\ g_{n1} & g_{n2} & \cdots & g_{nn} \end{pmatrix} \begin{pmatrix} e_{1k} \\ e_{2k} \\ \vdots \\ e_{nk} \end{pmatrix} = \begin{pmatrix} a_{1k} \\ a_{2k} \\ \vdots \\ a_{nk} \end{pmatrix}, \; k = 1, n. \tag{12.11.42}$$

After LU-decomposing the matrix $\mathbf{G}$, one next solves the equations

$$\begin{pmatrix} l_{11} & 0 & \cdots & 0 \\ l_{21} & l_{22} & \cdots & 0 \\ \vdots & \vdots & \ddots & \vdots \\ l_{n1} & l_{n2} & \cdots & l_{nn} \end{pmatrix} \begin{pmatrix} y_{1k} \\ y_{2k} \\ \vdots \\ y_{nk} \end{pmatrix} = \begin{pmatrix} a_{1k} \\ a_{2k} \\ \vdots \\ a_{nk} \end{pmatrix}, \; k = 1, n \tag{12.11.43}$$

using forward substitution as prescribed by eq. (8.8.21) in Volume 1:

$$y_{jk} = \frac{1}{l_{jj}} \left(a_{jk} - \sum_{m=1}^{j-1} l_{jm} \, y_{mk} \right), \; j = 1, n. \tag{12.11.44}$$

Finally, one solves the equations

$$\begin{pmatrix} u_{11} & u_{12} & \cdots & u_{1n} \\ 0 & u_{22} & \cdots & u_{2n} \\ \vdots & \vdots & \ddots & \vdots \\ 0 & 0 & \cdots & u_{nn} \end{pmatrix} \begin{pmatrix} e_{1k} \\ e_{2k} \\ \vdots \\ e_{nk} \end{pmatrix} = \begin{pmatrix} y_{1k} \\ y_{2k} \\ \vdots \\ y_{nk} \end{pmatrix}, \; k = 1, n. \tag{12.11.45}$$

using backward substitution as prescribed by eq. (8.8.18) in Volume 1:

$$e_{jk} = \frac{1}{u_{jj}} \left(y_{jk} - \sum_{m=j+1}^{n} u_{jm} \, e_{mk} \right), \; j = n, 1. \tag{12.11.46}$$

Similarly, eq. (12.11.34) may be written generically as

$$\mathbf{G}\,\mathbf{f} = \mathbf{d} + \mathbf{C}\,\mathbf{f}_{-1} = \mathbf{d} + \mathbf{c} \tag{12.11.47}$$

and the solution consists of finding the vector $\mathbf{f}$ in terms of the two vectors on the right hand side of the equation. In the central sphere, one may find $\mathbf{E}_2$ and $\mathbf{f}_2$, respectively, from the equations

$$\mathbf{B}_2 \mathbf{E}_2 = \mathbf{A}_2 \tag{12.11.48}$$

and

$$\mathbf{B}_2\mathbf{f}_2 = \mathbf{d}_2. \tag{12.11.49}$$

Here ends the discussion of solution algorithms (see additionally Iben and MacDonald (1985)).

Bibliography and references

C. W. Allen, *Allen's Astrophysical Quantities*, ed. Arthur N. Cox, (New York: AIP Press, Springer), 1999.

Lawrence H. Aller & Sydney Chapman, *ApJ*, **132**, 461, 1960.

Ludwig Boltzmann, *Sitz. Akad. Wiss.*, **66**, 275, 1872.

J. M. Burgers, *Flow Equations for Composite Gases* (New York: Academic Press), 1969.

Sydney Chapman and T. G. Cowling, *Mathematical Theory of Nonuniform Gases*, (Cambridge: Cambridge University Press), 1939, 1970.

G. Fontaine & G. Michaud, *White Dwarfs and Variable Stars*, eds. Hugh M. Van Horn & Volker Weidemann, Proc. IAU Colloq. 53, (Rochester, N.Y.: University of Rochester), 192, 1979.

Kerson Huang, *Statistical Mechanics* (New York; John Wiley & Sons), 1963.

Icko Iben, Jr. & Jim MacDonald, *ApJ*, **296**, 540, 1985.

Icko Iben, Jr., Masayuki Y. Fujimoto, & Jim MacDonald, *ApJ*, **388**, 521, 1992.

Thierry Montmerle & George Michaud, *ApJS*, **31**, 489, 1976.

David Muchmore, *ApJ*, **278**, 769, 1984.

C. Paquette, C. Pelletier, G. Fontaine, & G. Michaud, *ApJS*, **61**, 197, 1986.

George Gabriel Stokes, *Mathematical and Physical Papers* (Cambridge: Cambridge University Press), 1880.

Ray Weymann, *ApJ*, **126**, 208, 1957.

13 Heat conduction by electrons

Because faster moving particles at higher temperatures transfer energy to more slowly moving particles at lower temperatures, the very existence of a temperature gradient implies a flow of energy in the direction in which the temperature decreases. In the stellar interior, because of their small mass and consequent high velocities, free electrons are the dominant contributors to this mode of thermal energy transfer, which is called thermal or heat conduction.

Thermal conduction does not play a significant role in transporting heat in stars during most of the main sequence phase. However, towards the end of the main sequence phase, as detailed in Chapter 11 of Volume 1 (Section 11.1), low mass stars develop hydrogen-exhausted cores in which electrons become increasingly degenerate, and evolve into red giants with fully electron-degenerate helium cores. Under electron-degenerate conditions, only those electrons with energies within about kT of the Fermi energy ϵ_F participate in transporting heat, but their average cross section for scattering from ions and other electrons is reduced by a factor of the order of $(kT/\epsilon_F)^2$ relative to their average cross section under non-degenerate conditions. Hence, conduction becomes very effective in slowing the rate at which temperatures increase in the electron-degenerate cores and prevents low mass red giants from igniting helium until the degenerate core has grown to almost one-half of a solar mass. As described in Chapters 17 and 18 of this volume, electron conduction plays a similar role in both low and intermediate mass stars after they have exhausted helium at their centers and become asymptotic giant branch stars with electron-degenerate carbon–oxygen or oxygen–neon cores. Such stars never do ignite carbon or oxygen but, as described in Chapter 21, after losing most of their hydrogen-rich envelopes, they evolve into cooling white dwarfs within which conduction is the primary mode of transporting heat from interior to surface layers.

The potential for a thermally induced outward flow of electrons is countered by the presence of an electric field which helps maintain charge neutrality. The primary reason for the existence of the macroscopic electric field is the fact that the gravitational force on an electron is over a thousand times smaller than the gravitational force on an ion; whereas either the pressure-gradient forces which the two types of particles exert are of the same order of magnitude, as is the case when electrons are not degenerate, or the pressure-gradient force exerted by electrons is much larger than the pressure-gradient force exerted by ions, as is the case when electrons are degenerate.

Some of the concepts and issues involved in understanding thermal conduction are explored in Section 13.1 in a very elementary way which ignores the fact that a proper formulation must ensure that the electric field that exists in a star is taken into account and that heat flow does not induce an electric current. The electric field is discussed once again in Section 13.2 from the point of view of satisfying the pressure balance equations for electrons and ions separately. In Section 13.3, the Boltzmann transport equation is used to show how the gravitational and static electric fields, along with spatial gradients in thermodynamic quantities, imply an asymmetry in the distribution function for electrons. On the assumption that there is no net electric current, and with no reliance on the details of the conduction process, it is demonstrated in Section 13.4 that the asymmetry predicts the existence of a macroscopic electric field, the main contribution to which is proportional to the gravitational acceleration and reproduces the results found in Section 13.2. An additional contribution to the electric field is shown to depend on the temperature gradient. In Section 13.5, the asymmetry in the electron distribution function is used to derive expressions for the thermal conductivity in a classical approximation in which scattering of electrons from electrons is neglected, and, in Section 13.6, quantitative estimates in this approximation are presented. Also given in Section 13.6 are analytic fits, for varying degrees of electron degeneracy, to more sophisticated numerical estimates of conductivity which make use of a pair distribution function for ions and which take into account electron–electron scattering.

13.1 The basic physics of thermal diffusion

An elementary kinetic theory approach to understanding thermal conduction begins with the assumption that the flux of thermal energy in the direction in which the temperature gradient dT/dx is the steepest may be approximated by

$$\text{thermal flux} = -n \int_\Omega (\bar{v} \cos\theta) \left(\frac{dE}{dx} \Lambda \cos\theta \right) \frac{d\Omega}{4\pi} \tag{13.1.1}$$

$$= -n\, \bar{v}\, \frac{dE}{dT} \frac{dT}{dx} \Lambda \int_0^{2\pi} \int_0^\pi \cos^2\theta\, \frac{\sin\theta d\theta\, d\phi}{4\pi}, \tag{13.1.2}$$

where n is the number density of conducting particles, $\bar{v}$ and E are, respectively, the average speed and average thermal energy of conducting particles, Λ is a mean free path, θ is the angle made by the velocity vector of a given particle relative to the x axis, and the integration is over all solid angles. Particle motions have been assumed to be isotropic. Performing the integration, one has

$$\text{thermal flux} = -\frac{1}{3}\, n\, \bar{v}\, \frac{dE}{dT} \frac{dT}{dx} \Lambda \tag{13.1.3}$$

$$= -K \frac{dT}{dx}, \tag{13.1.4}$$

where K is called the thermal conductivity.

The result of integrating over all directions θ could also have been achieved by noting that: (1) $\bar{v} \cos\theta = \bar{v}_x$, $\Lambda \cos\theta = \Lambda\, \bar{v}_x/\bar{v}$, $\bar{v}_x^2 = \frac{1}{3}\, \bar{v}^2$; (2) half of the particles are traveling in the $+x$ direction, half in the opposite direction; and (3) the particles moving in the $+x$

direction carry excess energy outward and particles moving in the $-x$ direction carry a deficiency of energy inward equal in absolute magnitude to the excess carried outward.

Defining the heat capacity C by

$$C = n \frac{dE}{dT} \tag{13.1.5}$$

and assuming that the mean free path Λ is related to a scattering cross section σ by

$$\Lambda = \frac{1}{n_s \sigma}, \tag{13.1.6}$$

where n_s is the number density of scattering centers (not necessarily the same as n), one has that

$$K = \frac{1}{3} n \, \bar{v} \, \frac{dE}{dT} \, \Lambda \tag{13.1.7}$$

$$= \frac{1}{3} \bar{v} \, C \, \Lambda \tag{13.1.8}$$

$$= \frac{1}{3} \bar{v} \, \frac{dE}{dT} \, \frac{n}{n_s \sigma}. \tag{13.1.9}$$

For particles of mass m which obey Maxwell–Boltzmann statistics, $E = \frac{3}{2} kT$, $\bar{v} = \sqrt{8kT/\pi m}$, and

$$K = n \sqrt{\frac{2}{\pi} \frac{kT}{m}} \, k \, \Lambda = \sqrt{\frac{2}{\pi} \frac{kT}{m}} \, k \, \frac{n}{n_s \sigma}. \tag{13.1.10}$$

In a fully ionized gas at low density, assuming that the scattering cross sections between electrons and electrons, ions and ions, and electrons and ions are primarily due to Coulomb interactions, it is clear from eq. (13.1.10) that, because of their relatively small mass, electrons are over forty times more effective than ions in transporting heat by conduction. For this reason, in the remainder of this chapter the focus centers on the contribution of free electrons to the thermal conductivity.

In free space, the differential Coulomb cross section for the scattering of an electron of kinetic energy E from an effectively *fixed* target of charge Ze through an angle θ is (see, e.g., Herbert Goldstein, *Classical Mechanics*, 1965)

$$\frac{d\sigma}{d\Omega} = \left(\frac{Ze^2}{4E} \right)^2 \frac{1}{\sin^4 \frac{\theta}{2}} = \left(\frac{Ze^2}{2E} \right)^2 \frac{1}{(1 - \cos\theta)^2}. \tag{13.1.11}$$

Scattering between identical particles (e.g., between electrons) modifies this cross section in a way described later in this section. The impact parameter b and the scattering angle θ are related by

$$b = \frac{Ze^2}{E} \frac{1}{2 \tan(\theta/2)} \rightarrow \frac{Ze^2}{E} \frac{1}{\theta} \left(1 - \frac{1}{12} \theta^2 + \cdots \right) \text{ when } \theta < 1. \tag{13.1.12}$$

In a stellar interior, an electron scatters simultaneously from many ions so the identification of a total scattering cross section between one electron and one ion is problematical. One obvious difficulty is that an integration of eq. (13.1.11) over all angles diverges. The approach adopted by early investigators was to introduce a cutoff angle θ_0, the value of which cannot be unambiguously defined. Multiplying eq. (13.1.11) by the solid angle $d\Omega = d\phi \, \sin \theta \, d\theta$, and integrating, one obtains the total electron–ion scattering cross section

$$\sigma_{ei} = \int_0^{2\pi} d\phi \int_{\theta_0}^{\pi} \sin \theta \, d\theta \, \frac{d\sigma}{d\Omega} = Z^2 \left(\frac{e^2}{2E} \right)^2 \bar{\sigma}, \tag{13.1.13}$$

where

$$\bar{\sigma} = 2\pi \left(\frac{1}{1 - \cos \theta_0} - \frac{1}{2} \right). \tag{13.1.14}$$

For small θ_0, $\cos \theta_0 = 1 - \theta_0^2/2 + \theta_0^4/24 + \cdots$, so

$$\bar{\sigma} = \frac{4\pi}{\theta_0^2} \left(1 - \frac{\theta_0^2}{6} + \cdots \right) \tag{13.1.15}$$

and

$$\sigma_{ei} = \pi \left(\frac{Z_i e^2}{E} \right)^2 \frac{1}{\theta_0^2} \left(1 - \frac{1}{6} \theta_0^2 + \cdots \right). \tag{13.1.16}$$

One possibility for estimating the cutoff angle θ_0 is to assume that the maximum impact parameter is some fraction f of the average distance $2r_{0i}$ between adjacent scattering centers, so that

$$b_{\max} = f \, (2r_{0i}) = f \, 2 \left(\frac{3}{4\pi} \frac{Z_i}{n_e} \right)^{1/3}, \tag{13.1.17}$$

where the second equality defines r_{0i} as the radius of a sphere about an ith ion containing $Z = Z_i$ electrons. Then, from eq. (13.1.12),

$$\theta_{\min} = \frac{Ze^2}{E} \frac{1}{b_{\max}} = \frac{Ze^2}{E} \frac{1}{(2f)r_{0i}}, \tag{13.1.18}$$

and, inserting this in eq. (13.1.16), one has

$$\sigma_{ei} \sim \pi \, b_{\max}^2 = \pi (2f)^2 r_{0i}^2 = \pi (2f)^2 \left(\frac{3}{4\pi} \frac{Z}{n_e} \right)^{2/3} = \pi \, (2f)^2 \left(\frac{3}{4\pi} \frac{\mu_e Z}{\rho} M_{\mathrm{H}} \right)^{2/3}$$

$$= 1.7035 \, (2f)^2 \left(\frac{\mu_e Z}{\rho} \right)^{2/3} \times 10^{-16} \, \mathrm{cm}^2 = 6.0834 \, (2f)^2 \left(\frac{\mu_e Z}{\rho} \right)^{2/3} a_0^2, \tag{13.1.19}$$

where μ_e is the electron molecular weight (number of nucleons per electron), M_{H} is the proton mass, ρ is the density in g cm^{-3}, and a_0 is the radius of the first Bohr orbit in hydrogen. Noting from eq. (13.1.10) that the conductivity is proportional to the inverse of the cross section, this highly oversimplified treatment suggests that the conductivity increases with increasing density and is independent of the temperature. The prediction that the cross section for scattering from an individual ion decreases with increasing density (decreasing ionic separations) makes sense qualitatively in that it recognizes that when an electron passes much more closely to an ion adjacent to the one under consideration, the Coulomb force on the electron from the adjacent ion completely dominates the Coulomb force on the electron from the ion under consideration.

An approach adopted by many early investigators was to invoke the Heisenberg uncertainty principle between the conjugate variables, angular momentum and angle, giving

$$m_e v_e \, b_{max} \, \theta_0 = p_e \, b_{max} \, \theta_0 \gtrsim \hbar, \qquad (13.1.20)$$

from which it may be inferred that

$$\theta_{min} = \frac{\lambdabar_e}{b_{max}} = \frac{\lambdabar_e}{r_{0i}}, \qquad (13.1.21)$$

where $\lambdabar_e = \hbar/m_e v_e$ and b_{max} has been identified with the radius of a sphere containing Z_i electrons. Setting θ_{min} from eq. (13.1.21) in eq. (13.1.16) gives, when electrons are not relativistic,

$$\sigma_{ei} \sim \pi \left(\frac{Ze^2}{E} \right)^2 \left(\frac{r_{0i}}{\lambdabar_e} \right)^2 = 2\pi \left(\frac{e^2}{\hbar c} \right)^2 Z^2 \left(\frac{m_e c^2}{E} \right) r_{0i}^2, \qquad (13.1.22)$$

a result which differs from the one expressed by eq. (13.1.19), not only by the value of the coefficient of r_{0i}^2 but by the fact that the cross section decreases with increasing temperature through the factor E $\left(\text{which is } \frac{3}{2}kT \text{ when electrons are not degenerate}\right)$. The decrease with increasing temperature is qualitatively consistent with the fact that the differential cross section decreases with increasing energy, and this dependence on temperature makes the second approximation perhaps more satisfying than the first. However, the justification for the choice expressed in eq. (13.1.21) is very tenuous and it would be a mistake to assume that eq. (13.1.22) is anything more than a qualitative indication of the dependence of the cross section on conditions in a star.

Still a third approach to estimating the cutoff angle is to introduce a screening radius which incorporates the fact that, as a free electron passes by another charged particle, the Coulomb field of this particle is shielded by the presence of all of the other electrons and ions, as discussed in Section 4.16 of Volume 1. If one chooses a shielded Coulomb potential of the form

$$V(r) = -\frac{Ze^2}{r} \, e^{-r/R}, \qquad (13.1.23)$$

where R is a screening radius, the matrix element of the Hamiltonian for scattering electrons is, in the Born approximation, exactly the same as the matrix element encountered in estimating the free–free radiative transition probabilty in Section 7.5 of Volume 1 (eq. (7.5.27)):

$$H_{12} = \frac{1}{V} \int e^{-i\mathbf{p}_2 \cdot \mathbf{r}/\hbar} \left(-\frac{Ze^2}{r} e^{-r/R} \right) e^{i\mathbf{p}_1 \cdot \mathbf{r}/\hbar} \, d\tau, \tag{13.1.24}$$

where $\mathbf{p}_1$ and $\mathbf{p}_2$ are, respectively, the momenta of the electron before and after scattering. Integration gives

$$H_{12} = -\frac{4\pi Z_0 e^2}{V} \frac{\lambdabar_{12}^2}{1 + (\lambdabar_{12}/R)^2}, \tag{13.1.25}$$

where

$$\lambdabar_{12} = \frac{\hbar}{|\mathbf{p}_2 - \mathbf{p}_1|}. \tag{13.1.26}$$

The differential probability for scattering is

$$d\omega = \frac{2\pi}{\hbar} H_{12}^2 \frac{p^2 dp d\Omega V}{h^3 dE_f},$$

$$= \frac{2\pi}{\hbar} H_{12}^2 \left[\frac{4\pi Z e^2}{V} \frac{\lambdabar_{12}^2}{1 + \left(\dfrac{\lambdabar_{12}}{R} \right)^2} \right]^2 \frac{p^2 dp d\Omega V}{h^3 dE_f}, \tag{13.1.27}$$

where $E = p^2/2m_e$ and $dE_f = p\,dp/m_e$. Note that, since for any initial electron there is only one final electron, the usual factor of 2 associated with the two spin states of an electron does not appear in the phase space factor $p^2 dp d\Omega V/h^3$. The differential cross section is now

$$\frac{d\sigma}{d\Omega} = \frac{d\omega}{d\Omega} \left[\frac{v}{V} \right]^{-1} = \left(\frac{2m_e}{\hbar^2} \lambdabar_{12}^2 \right)^2 \left[\frac{Ze^2}{1 + \left(\dfrac{\lambdabar_{12}}{R} \right)^2} \right]. \tag{13.1.28}$$

Since

$$|\mathbf{p}_2 - \mathbf{p}_1|^2 = p_2^2 - 2p_1 p_2 \cos\theta + p_1^2 = 2p^2(1 - \cos\theta), \tag{13.1.29}$$

one obtains, after some manipulation,

$$\frac{d\sigma}{d\Omega} = \left(\frac{Ze^2}{2E} \right)^2 \left[(1 - \cos\theta) + \left(\frac{\lambdabar}{R\sqrt{2}} \right)^2 \right]^{-2}, \tag{13.1.30}$$

where

$$\lambda = \frac{\hbar}{p}. \tag{13.1.31}$$

Integration of eq. (13.1.30) over azimuthal and polar angles gives a total cross section

$$\sigma_{ei}^{screen} = \pi \left(\frac{Ze^2}{2E}\right)^2 \frac{\left(\frac{2R}{\lambda}\right)^2}{1 + \left(\frac{\lambda}{2R}\right)^2}. \tag{13.1.32}$$

The cross section defined by this equation is a function of energy through both E and λ. However, in keeping with the approach adopted thus far, it is appropriate to identify both E and λ with typical values of these quantities and to interpret the resultant cross section as representative of all scattering interactions. Taking R as the Debye radius defined by eq. (4.17.43) in Volume 1 and $\lambda = \lambda_e$ as defined by eqs. (4.7.16) and (4.7.18), one has

$$\frac{\lambda}{R} \to \frac{\lambda_e}{R_D} = \frac{\hbar}{\sqrt{3m_e kT}} \left(\frac{n_e kT}{P_e}\right)^{1/2} \left(\frac{4\pi e^2 n_e}{kT}\right)^{1/2} \left[\mu_e \sum_j Y_j Z_j^2 + \frac{1}{n_e}\left|\frac{dn_e}{d\alpha}\right|\right]^{1/2}$$

$$= \frac{e\hbar}{km_e}\left(\frac{4\pi}{3}\frac{m_e}{M_H}\right)^{1/2}\left(\frac{n_e kT}{P_e}\right)^{1/2}\left(\frac{\rho}{\mu_e}\right)^{1/2}\frac{1}{T}\left[\mu_e \sum_j Y_j Z_j^2 + \frac{1}{n_e}\left|\frac{dn_e}{d\alpha}\right|\right]^{1/2}$$

$$= 0.192\,326 \left(\frac{n_e kT}{P_e}\right)^{1/2}\left[\mu_e \sum_j Y_j Z_j^2 + \frac{1}{n_e}\left|\frac{dn_e}{d\alpha}\right|\right]^{1/2}\left(\frac{\rho}{\mu_e}\right)^{1/2}\frac{1}{T_6}, \tag{13.1.33}$$

where Y_j is the abundance by number of the jth ion, α is the parameter appearing in the Fermi–Dirac probability distribution function $f = 1/\exp(\alpha + \epsilon/kT)$ and $n_e = n_e(\alpha, kT)$. The second term in square brackets in eq. (13.1.33) is approximately unity when electrons are non-degenerate and becomes smaller as electron degeneracy increases.

For small $\lambda/2R$, the cross section in the screening approximation defined by eq. (13.1.32) with $\lambda \to \lambda_e$ is

$$\sigma_{ei}^{screen} \sim \pi \left(\frac{Ze^2}{E}\right)^2 \left(\frac{\lambda_e}{R}\right)^{-2}. \tag{13.1.34}$$

From eqs. (13.1.22) and (13.1.34), it follows that the integrated cross section estimated in the cutoff approximation with the choice $\theta_{0i} = \lambda/r_{0i}$ and the cross section in the screening approximation are in the ratio

$$\frac{\sigma_{ei}^{cutoff}}{\sigma_{ei}^{screen}} = \left(\frac{r_{0i}}{R_D}\right)^2 = 0.680\,69 \frac{\rho^{1/3}}{T_6}\left(\frac{Z_i^2}{\mu_e}\right)^{1/3}\left[\mu_e \sum_j Y_j Z_j^2 + \frac{1}{n_e}\left|\frac{dn_e}{d\alpha}\right|\right], \tag{13.1.35}$$

where the quantity at the far right comes from eqs. (4.7.47) and (4.7.48) in Volume 1. At the solar center, this ratio is astonishingly of the order of unity.

Equation (13.1.10) may be generalized to describe the conductivity in a fully ionized medium consisting of various types of ions of charge Z_i and number abundance n_i and of non-degenerate free electrons at a number abundance $n_e = \sum_i Z_i \, n_i$. This non-degenerate (ND) conductivity is

$$K_{\text{ND}} = 4 \sqrt{\frac{2}{\pi} \frac{kT}{m_e}} \, k \left(\frac{E}{e^2} \right)^2 \frac{n_e}{\sum_i n_i Z_i^2 \bar{\sigma}_{\text{ei}} + n_e \bar{\sigma}_{\text{ee}}}, \tag{13.1.36}$$

where $\bar{\sigma}_{\text{ei}}$ and $\bar{\sigma}_{\text{ee}}$ refer to scattering of electrons by ions (ei) and by electrons (ee), respectively, and E is the energy of a typical electron.

For comparison with the conductivity estimated in a more sophisticated way in Section 13.5, it is convenient to define a quantity

$$\Phi = \frac{1}{2\pi \, \mu_e} \left[\frac{\sum_i n_i Z_i^2 \bar{\sigma}_{\text{ei}} + n_e \bar{\sigma}_{\text{ee}}}{n_e} \right] \tag{13.1.37}$$

$$= \frac{1}{2\pi} \left[\sum_i Y_i Z_i^2 \bar{\sigma}_{\text{ei}} + \frac{1}{\mu_e} \, \bar{\sigma}_{\text{ee}} \right]. \tag{13.1.38}$$

Setting $E = \frac{3}{2}kT$, and rearranging terms, eq. (13.1.36) becomes

$$K_{\text{ND}} = 9 \sqrt{\frac{2}{\pi}} \left(\frac{ck}{r_e^2} \right) \left(\frac{kT}{m_e c^2} \right)^{5/2} \frac{1}{\mu_e} \frac{1}{2\pi \, \Phi}, \tag{13.1.39}$$

where $r_e = e^2/m_e c^2 = 2.817\,940 \times 10^{-13}$ is the classical radius of an electron. Using $kT/m_e c^2 = T_6/5929.89$ and $ck/r_e^2 = 5.212\,4366 \times 10^{19}$, one has the numerical result

$$K_{\text{ND}} = 1.382\,31 \times 10^{11} \frac{T_6^{5/2}}{\mu_e} \frac{1}{2\pi \, \Phi} \quad \text{(cgs units)}. \tag{13.1.40}$$

Consider next a fully ionized gas in which the electrons are degenerate, but have non-relativistic energies. Equations (4.6.41) and (4.6.43) in Volume 1 give for the average kinetic energy of an electron

$$E = \frac{3}{5} \, \epsilon_F = \frac{3}{5} \, \epsilon_F(0) \left(1 + \frac{5\pi^2}{12} \left(\frac{kT}{\epsilon_F(0)} \right)^2 + \cdots \right), \tag{13.1.41}$$

where

$$\epsilon_F(0) = \frac{h^2}{2m_e} \, p_F^2(0) = \frac{h^2}{2m_e} \left(\frac{3n_e}{8\pi} \right)^{2/3}. \tag{13.1.42}$$

Differentiating with respect to temperature, one obtains

$$\frac{dE}{dT} = \frac{\pi^2}{2} \frac{kT}{\epsilon_F(0)} \, k, \tag{13.1.43}$$

showing that only a fraction of the electrons, those within an energy interval $\sim\pi\,kT/\epsilon_F(0)$ near the top of the Fermi sea, participate in the thermal conduction process. On the other hand, the kinetic energies of these conduction electrons are much larger than ionic kinetic energies, so the electron–electron and electron–ion Coulomb cross sections are smaller by roughly $(kT/\epsilon_F(0))^2$ than in the case of non-degenerate electrons. The increase in the effective mean free path more than offsets the decrease in the number of participating electrons, with the result that the conductivity of electrons increases with the degree of degeneracy.

Since it is only Coulomb scattering of electrons near the top of the Fermi sea that is important, one may set $E = \epsilon_F(0)$ in eq. (13.1.13) and define

$$\sigma'_{ei} = Z_i^2 \left(\frac{e^2}{2\epsilon_F(0)} \right)^2 \bar{\sigma}'_{ei}, \tag{13.1.44}$$

where $\bar{\sigma}_{ei}$ is given by eq. (13.1.14). Similarly, for the mean electron speed of participating electrons, one may choose

$$\bar{v} = \left(\frac{2\epsilon_F(0)}{m_e} \right)^{1/2}. \tag{13.1.45}$$

Using eqs. (13.1.43)–(13.1.45) in eq. (13.1.9), one has that the non-relativistically degenerate (NRD) electron contribution to conductivity is

$$K_{NRD} = \frac{1}{3} \left(\frac{2\epsilon_F(0)}{m_e} \right)^{1/2} \frac{\pi^2}{2} k \frac{kT}{\epsilon_F(0)} \left(\frac{2\epsilon_F(0)}{e^2} \right)^2 \frac{n_e}{\sum_i n_i Z_i^2 \bar{\sigma}'_{ei} + n_e \bar{\sigma}'_{ee}} \tag{13.1.46}$$

or

$$K_{NRD} = \sqrt{2}\, \frac{2\pi^2}{3} \left(\frac{kT}{\epsilon_F(0)} \right) \left(\frac{\epsilon_F(0)}{m_e c^2} \right)^{5/2} \left(\frac{ck}{r_e^2} \right) \frac{1}{\mu_e} \frac{1}{2\pi\,\Phi'}. \tag{13.1.47}$$

The primes on the $\bar{\sigma}$s and on Φ are a recognition of the possibility that the effective Coulomb cross sections are not the same under electron-degenerate conditions as under non-degenerate conditions. With this caveat, Φ' has the same form as in eqs. (13.1.37) and (13.1.38).

Making use of eqs. (4.6.28) and (4.6.29) in Volume 1 which give, respectively,

$$\frac{\epsilon_F(0)}{m_e c^2} = 0.506\,245 \left(\frac{\rho_6}{\mu_e} \right)^{2/3} \quad \text{and} \quad \frac{\epsilon_F(0)}{kT} = 3002.0 \left(\frac{\rho_6}{\mu_e T_6^{3/2}} \right)^{2/3},$$

eq. (13.1.47) becomes

$$K_{NRD} = 2.946\,15 \times 10^{18} \frac{\rho_6\,T_8}{\mu_e^2} \frac{1}{2\pi\,\Phi'} \quad \text{(cgs units)}. \tag{13.1.48}$$

Dividing eq. (13.1.48) by eq. (13.1.36), one has

$$\frac{K_{\mathrm{NRD}}}{K_{\mathrm{ND}}} = \frac{2}{\sqrt{\pi}} \left(\frac{\pi}{3}\right)^3 \left(\frac{\epsilon_{\mathrm{F}}(0)}{kT}\right)^{3/2} \frac{\Phi}{\Phi'}. \tag{13.1.49}$$

Ignoring the differences between Φ and Φ', one has

$$\frac{K_{\mathrm{NRD}}}{K_{\mathrm{ND}}} \sim 213 \, \frac{\rho_6}{\mu_e T_8^{3/2}}, \tag{13.1.50}$$

emphasizing the importance of conduction under electron-degenerate conditions relative to its importance under non-degenerate conditions. The basic reason that conductivity increases with increasing degeneracy is that, because their speed increases as they are compressed into a smaller space, conduction electrons experience less effective Coulomb scattering and their mean free path increases more rapidly than the fraction of electrons in the conduction band decreases.

To compare the effectiveness of energy transport by heat conduction with energy transport by radiation, it is customary to define a conductive opacity κ_{cond} in such a way that the energy flux due to conduction is related to the temperature gradient and to the conductive opacity in the same way that the radiative energy flux is related to the temperature gradient and the radiative opacity. Then, as described in Section 3.4 of Volume 1,

$$\kappa_{\mathrm{cond}} \equiv \frac{4ac}{3} \frac{T^3}{\rho} \frac{1}{K}, \tag{13.1.51}$$

or, numerically,

$$\kappa_{\mathrm{cond}} = 3.024\,21 \times 10^{14} \frac{T_6^3}{\rho} \frac{1}{K} = 3.024\,21 \times 10^{14} \frac{T_8^3}{\rho_6} \frac{1}{K}, \tag{13.1.52}$$

where K is the conductivity as defined by eq. (13.1.4). The sum of the radiative and convective fluxes is

$$F_{\mathrm{cond}} + F_{\mathrm{rad}} = -\frac{4ac}{3} \frac{T^3}{\rho} \left(\frac{1}{\kappa_{\mathrm{cond}}} + \frac{1}{\kappa_{\mathrm{rad}}}\right) \frac{dT}{dx} = -\frac{4ac}{3} \frac{T^3}{\rho} \left(\frac{1}{\kappa_{\mathrm{eff}}}\right) \frac{dT}{dx}, \tag{13.1.53}$$

from which it follows that the effective opacity κ_{eff} is given by

$$\frac{1}{\kappa_{\mathrm{eff}}} = \frac{1}{\kappa_{\mathrm{rad}}} + \frac{1}{\kappa_{\mathrm{cond}}}. \tag{13.1.54}$$

Using eqs. (13.1.40) and (13.1.52), it follows that

$$(\kappa_{\mathrm{cond}})_{\mathrm{ND}} = 2.187\,79 \times 10^3 \frac{T_6^{1/2}}{\rho} \mu_e \, 2\pi \, \Phi, \tag{13.1.55}$$

and using eqs. (13.1.48) and (13.1.52), it follows that

$$(\kappa_{\mathrm{cond}})_{\mathrm{NRD}} = 1.026\,49 \times 10^{-4} \frac{T_8^2}{\rho_6^2} \mu_e^2 \, 2\pi \, \Phi'. \tag{13.1.56}$$

Remembering that the $\bar{\sigma}$s and, therefore, the Φs are large compared with unity, eq. (13.1.55) suggests that electron conduction does not compete effectively with radiation in transporting heat in regions where electrons are not degenerate. Ascertaining whether or not it competes in regions of electron degeneracy requires incorporation of the effective Coulomb cross sections in a more careful manner.

13.2 The macroscopic electrostatic field in an ionized medium in a gravitational field

The treatment of heat conduction in the previous section has ignored the static electric field which has been shown in Chapter 12 to play a role in particle diffusion. The field could in principle induce electrical currents that influence the process of heat transfer. Indeed, as shown in the next section, a more sophisticated treatment of heat conduction using the Boltzmann transfer equation, which requires the inclusion of all macroscopic force fields present, demonstrates that determining the heat conductivity uniquely requires one to stipulate that the electrical current vanishes. The stipulation of zero current leads in turn to a determination of the static electric field strength which is independent of and slightly different from the estimates made in Chapter 12. For this reason it is important to explore even more extensively estimates of the static electric field strength which follow from a consideration of pressure balance forces.

Because of their small mass, electrons experience an inward gravitational force which is at least three orders of magnitude smaller than the gravitational force on nuclei. When the electrons are in bound states, this disparity presents no problems; the bound electrons and the nuclei to which they are bound act as a unit in responding to an external field, whether it be gravitational, electrical, or due to a pressure gradient (collisional forces). The only effect of a macroscopic electric field is to polarize the bound system.

On the other hand, in a fully ionized region, free electrons and ions behave more or less independently. Since charge neutrality must prevail over regions large enough to contain many particles, the sum of the number densities of ions multiplied by their positive charge must equal the number density of free electrons and, therefore, the outward pressure-gradient force contributed by the electrons is at least as large as that contributed by the ions. Thus, absent another force which acts on electrons in the same direction as the gravitational force but with an effectiveness which is three orders of magnitude larger than the gravitational force, the inward force on electrons is considerably out of balance with the pressure-gradient force outward. The standard way out of this difficulty is to postulate the existence of a macroscopic electrostatic field which is directed radially outward and has a strength that is some multiple of $M_H\, g/e$. The consequences of this postulate, touched upon briefly in Section 3.6 in Volume 1 and made use of extensively in Chapter 12, are of such an importance in stellar structure and evolution that they are discussed again here, but from a somewhat different perspective and with an emphasis on several aspects not covered in the previous discussions.

Suppose, first, that there is only one (completely ionized) ionic species of mass m_0, charge $Z_0 e$, and local number density n_0, and that the ions as well as the associated electrons behave as perfect gases. Each electron has a charge $-e$ and mass m_e and the density of electrons is n_e. Because of charge neutrality, $n_e = Z_0 n_0$. In the perfect gas approximation, partial pressures are related to temperatures and number densities by

$$P_0 = n_0 k T \tag{13.2.1}$$

and

$$P_e = n_e k T = Z_0 n_0 k T. \tag{13.2.2}$$

Thus, the ratio of partial pressures is everywhere the same at

$$\frac{P_e}{P_0} = Z_0, \tag{13.2.3}$$

from which it follows that

$$\frac{\mathrm{d}P_e}{\mathrm{d}r} = Z_0 \frac{\mathrm{d}P_0}{\mathrm{d}r}. \tag{13.2.4}$$

Thus, the assumptions of charge neutrality and perfect gas equations of state for ions and electrons separately have the consequences that the pressure-gradient force exerted by electrons exceeds by a factor of Z_0 the pressure-gradient force exerted by ions.

Denoting the electric field strength by E (taken positive in the outward direction), and assuming that, in first approximation, pressure balance holds for the ions and electrons separately, one has that

$$\frac{\mathrm{d}P_0}{\mathrm{d}r} = -n_0 m_0 g + n_0 Z_0 e E = -n_0 (m_0 g - Z_0 e E) \tag{13.2.5}$$

and

$$\frac{\mathrm{d}P_e}{\mathrm{d}r} = -n_e m_e g - n_e e E = -Z_0 n_0 (m_e g + e E). \tag{13.2.6}$$

Making use of eq. (13.2.4), it follows that

$$m_e g + e E = m_0 g - Z_0 e E, \tag{13.2.7}$$

or

$$e E = \frac{m_0 - m_e}{1 + Z_0} g. \tag{13.2.8}$$

Setting eE from eq. (13.2.8) into eqs. (13.2.5) and (13.2.6) gives

$$\frac{\mathrm{d}P_0}{\mathrm{d}r} = -n_0 \frac{m_0 + Z_0 m_e}{Z_0 + 1} g \tag{13.2.9}$$

and

$$\frac{\mathrm{d}P_e}{\mathrm{d}r} = -Z_0 n_0 \frac{m_0 + Z_0 m_e}{Z_0 + 1} g = -n_e \frac{m_0 + Z_0 m_e}{Z_0 + 1} g. \tag{13.2.10}$$

Adding eqs. (13.2.9) and (13.2.10) gives

$$\frac{dP}{dr} = \frac{d}{dr}(P_0 + P_e) = -n_0(m_0 + Z_0 m_e)\, g = -g\rho, \tag{13.2.11}$$

the standard pressure balance equation which is normally written without acknowledging the role of a macroscopic electric field in producing this balance. The inward electrical force on an electron and the outward electrical force on an ion are comparable with the gravitational force on an ion and it is the combination of gravitational and electrical forces that make up the forces which balance the pressure-gradient force.

In the case of pure, completely ionized hydrogen, one has that the inwardly directed electrical force on the electron and the outwardly directed electrical force on the proton are the same and equal to $eE = \frac{1}{2}(m_p - m_e)\, g$, or approximately half the gravitational force exerted on a proton. The net inward force on an electron is the same as the net inward force on a proton and equal to $F_{\text{net inward}} = \frac{1}{2}(m_p + m_e)\, g$. Thus, in agreement with the development in Section 3.6 in Volume 1, the electron and the proton share equally in responding to the gravitational force and contributing to the balance between the pressure-gradient force and the gravitational force.

In the case of pure, completely ionized helium, the strength of the electric field is given by $eE = \frac{1}{3}(M_\alpha - m_e)\, g$, where M_α is the mass of an alpha particle. The net inward force on an alpha particle is $F_\alpha = \frac{1}{3}(M_\alpha + 2\, m_e)\, g$, the same as the net inward force F_e on each free electron. So, under fully ionized conditions, electrons, which at low temperatures would be bound in a neutral atom that responds as a unit to the gravitational force, share equally with alpha particles in the net inward force that is due ultimately to gravity but that, in the star, is composed of gravitational and electrical forces of comparable magnitudes.

The global electric field not only maintains nearly perfect charge neutrality, but also plays a role in determining the conductivity and the rate of element diffusion. Since the electric field is proportional to the gravitational acceleration and since this acceleration is a function of position, the divergence of E is not zero and matter is therefore not precisely charge neutral. Using eq. (13.2.8) and the facts that $g = GM(r)/r^2$ and $dM(r)/dr = 4\pi r^2 \rho = 4\pi r^2 n_0 m_0$, one obtains

$$\nabla \cdot \mathbf{E} = 4\pi\, e\, n_+ = \left(\frac{d}{dr} + \frac{2}{r}\right) E$$

$$= \frac{m_0 - m_e}{1 + Z_0}\, \nabla \cdot \mathbf{g} = \frac{4\pi G\, m_0(m_0 - m_e)}{(1 + Z_0)\, e}\, n_0 \approx \frac{4\pi G\, m_0^2}{(1 + Z_0)\, e}\, n_0, \tag{13.2.12}$$

where n_+ is the number density of hypothetical particles of charge $+e$. Using $m_0 = A_0 M_H$, this result may be rewritten as

$$\frac{n_+}{n_0} = \frac{A_0^2}{1 + Z_0}\, G\left(\frac{M_H}{e}\right)^2 = \frac{A_0^2}{1 + Z_0} \times 8.09 \times 10^{-37}. \tag{13.2.13}$$

For the Sun, this gives a total excess of $\sim 5 \times 10^{20}$ positively charged particles, compared with the 10^{57} or so protons in the Sun! Note that n_+/n_0 is essentially GM_H^2/e^2, the ratio of the strength of the gravitational force between two protons and the strength of the electrostatic force between them.

Consider, next, the situation when an arbitrary mix of ionized species is present. Collisional interactions between ions of different species leads to the phenomenon of diffusion which, in a gravitational field, causes a spatial variation in species abundance ratios. In general, the assumption that pressure balance holds separately for each species is a good approximation only after enough time has elapsed for gravitational settling to have led to an equilibrium spatial distribution which is different for each species. On the other hand, since diffusion coefficients for electrons are over forty times larger than those for ions, eq. (13.2.6) remains a good first approximation under many circumstances. Solving eq. (13.2.6) for the electric field strength gives

$$eE = -\frac{1}{n_e}\frac{dP_e}{dr} - m_e g, \tag{13.2.14}$$

with the second term on the right of the equal sign being three orders of magnitude smaller than the first term in absolute value when electrons are not degenerate, and even smaller in relative magnitude when electrons are degenerate.

If electrons as well as ions behave as perfect gases, one may write

$$\frac{P_e}{P} = \frac{\sum_i Z_i n_i}{\sum_i (1 + Z_i)\, n_i} = \frac{\bar{Z}}{1 + \bar{Z}}, \tag{13.2.15}$$

where

$$\bar{Z} = \frac{\sum_i Z_i n_i}{\sum_i n_i}. \tag{13.2.16}$$

Differentiating eq. (13.2.15) and using $dP/dr = -g\rho$ results in

$$\frac{dP_e}{dr} = \frac{\bar{Z}}{1+\bar{Z}}\frac{dP}{dr} + P\frac{d}{dr}\left(\frac{\bar{Z}}{1+\bar{Z}}\right)$$

$$= -\frac{\bar{Z}}{1+\bar{Z}}\, g\rho + P\frac{1}{(1+\bar{Z})^2}\frac{d\bar{Z}}{dr}. \tag{13.2.17}$$

Setting in eq. (13.2.17)

$$\rho = \sum_i (m_i + Z_i m_e)\, n_i = (\bar{m} + \bar{Z} m_e)\sum_i n_i \tag{13.2.18}$$

and

$$P = kT\sum_i (1 + Z_i)\, n_i = kT\,(1 + \bar{Z})\sum_i n_i \tag{13.2.19}$$

gives

$$\frac{dP_e}{dr} = -\sum_i n_i\left[\frac{\bar{Z}}{1+\bar{Z}}\, g\,(\bar{m} + \bar{Z} m_e) - \frac{kT}{(1+\bar{Z})}\frac{d\bar{Z}}{dr}\right]. \tag{13.2.20}$$

Equations (13.2.14) and (13.2.20) together give

$$eE = \frac{\sum_i n_i}{n_e} \left[\frac{\bar{Z}}{1 + \bar{Z}} g \, (\bar{m} + \bar{Z} m_e) - \frac{kT}{(1 + \bar{Z})} \frac{d\bar{Z}}{dr} \right] - m_e g. \tag{13.2.21}$$

Recognizing that $\sum_i n_i / n_e = 1/\bar{Z}$, eq. (13.2.21) simplifies to

$$eE = \frac{g \, (\bar{m} - m_e)}{1 + \bar{Z}} - \frac{kT}{(1 + \bar{Z})} \frac{1}{\bar{Z}} \frac{d\bar{Z}}{dr}, \tag{13.2.22}$$

which is identical with eq. (13.2.8) when only one species of ion is present (so that $d\bar{Z}/dr = 0$). For a fully ionized mixture of hydrogen and helium, the electrical field strength is

$$eE \sim \left(\frac{4n_\alpha + n_p}{3n_\alpha + 2n_p} \right) M_H g - \frac{kT}{(1 + \bar{Z})} \frac{1}{\bar{Z}} \frac{d\bar{Z}}{dr}, \tag{13.2.23}$$

where

$$\bar{Z} = \frac{n_p + 2n_\alpha}{n_p + n_\alpha}. \tag{13.2.24}$$

Outside of regions of energy production, the temperature-dependent term in eq. (13.2.22) is finite because of diffusion, which produces a negative gradient for $\bar{Z}$ and, hence, a positive contribution to the electric field. In the envelopes of main sequence stars, this term is small compared with the gravity-dependent term because the gravitational potential energy of an ion in the envelope is of the order of the kinetic energy of an ion near the stellar center. As follows from eq. (3.1.8) in Volume 1, the ratio of the second term to the first term can be expressed as

$$\text{ratio} = \frac{T}{T_c} \frac{R}{l_{\text{diff}}}, \tag{13.2.25}$$

where T_c is the central temperature, R is the stellar radius, and $l_{\text{diff}} = \left| \bar{Z}^{-1} \, d\bar{Z}/dr \right|^{-1}$ is a characteristic scale hight for composition changes due to diffusion. Typically, the ratio is much less then 1.

In regions of energy production, the temperature dependent term can become large because of composition gradients induced by nuclear transformations. Since these transformations lead to an increase in $\bar{Z}$ inward, the temperature-dependent term makes a positive contribution to the electric field, just as in the case of diffusion-induced composition gradients.

In another exercise, consider the electric field deduced on the assumption that ions exert a pressure P_0 given by eq. (13.2.1) but that electrons are degenerate to the extent that $kT/\epsilon_F \ll 1$. Then

$$P_e = K(\lambda) \, n_e^\lambda, \tag{13.2.26}$$

where $K(\lambda)$ varies slowly as λ increases from $5/3$ to $4/3$ with increasing degeneracy. Again restricting attention to only one ionic species, it follows that

$$\frac{dP_e}{dP_0} \sim \lambda \, \frac{P_e}{P_0} \, \frac{1}{1 + \dfrac{d \log T}{d \log n_0}}. \tag{13.2.27}$$

Coupling with eqs. (13.2.5) and (13.2.6), one has

$$eE \sim \frac{m_0}{Z_0} g \, \frac{1 - \dfrac{1}{\lambda} \dfrac{m_e}{m_0} \dfrac{P_0}{P_e} Z_0 \left(1 + \dfrac{d \log T}{d \log n_0}\right)}{1 + \dfrac{1}{\lambda} \dfrac{P_0}{P_e} \left(1 + \dfrac{d \log T}{d \log n_0}\right)} \tag{13.2.28}$$

or, better,

$$eE \sim \frac{\dfrac{m_0}{Z_0} g}{1 + \dfrac{1}{\lambda} \dfrac{P_0}{P_e} \left(1 + \dfrac{d \log T}{d \log n_0}\right)}. \tag{13.2.29}$$

In the limit that $P_e/P_0 \gg 1$,

$$eE \to \frac{m_0}{Z_0} g. \tag{13.2.30}$$

Thus, when electrons are highly degenerate, the net inward gravitational force on all charged particles is translated into an inward electrical force on the electrons; the inward gravitational force on the nuclei is effectively canceled by the outward electrical force on them! The balance between gravity and pressure-gradient forces is essentially a balance between the electrical force on electrons and the pressure-gradient force due to electrons acting on electrons.

13.3 Use of the Boltzmann transport equation to find the asymmetry in the electron-distribution function

In Chapter 12, the discussion centered on the Boltzmann transport equation assumes that particles have non-relativistic velocities. This is certainly an adequate approximation for heavy ion motions in many stellar evolution contexts. However, when highly degenerate electrons are involved, it is best to base the discussion on a fully relativistic Boltzmann equation.

Assume that, for electrons, there exists a steady-state distribution function $f(\mathbf{p}, \mathbf{r}, t)$ which depends on electron momentum $\mathbf{p}$, position $\mathbf{r}$, and time t, and that the asymmetry in this function due to the existence of external forces and due to gradients in thermodynamic quantities can be estimated by considering binary collisions between electrons and ions and between electrons and electrons. The differential number density $dn(\mathbf{p}, \mathbf{r}, t)$ of electrons of momentum $\mathbf{p}$ may be written as

$$dn(\mathbf{p}, \mathbf{r}, t) = 2\, \frac{d^3\mathbf{p}}{h^3}\, f(\mathbf{p}, \mathbf{r}, t), \qquad (13.3.1)$$

where $d^3\mathbf{p} = p^2\, dp\, d\Omega$ is a volume element in momentum space and $d\Omega$ is a differential solid angle which defines the direction of the momentum vector.

Assuming steady state, differentiation of the distribution function gives

$$\frac{\Delta f}{\Delta t} = \nabla_\mathbf{p} f \cdot \frac{d\mathbf{p}}{dt} + \nabla_\mathbf{r} f \cdot \frac{d\mathbf{r}}{dt} + \left(\frac{\partial f}{\partial t} \right)_{\text{collisions}} = 0, \qquad (13.3.2)$$

where $\nabla_\mathbf{p} f$ is the gradient of f with respect to momentum, $\nabla_\mathbf{r} f$ is the gradient of f with respect to position, and $(\partial f / \partial t)_{\text{collisions}}$ is the rate at which f changes due to collisions between electrons and other particles. If all external forces and gradients in thermodynamic variables are in the x direction, then

$$\frac{\partial f}{\partial p_x} F_x + \frac{\partial f}{\partial x} v_x + \left(\frac{\partial f}{\partial t} \right)_{\text{collisions}} = 0, \qquad (13.3.3)$$

where F_x is the force on an electron due to the sum of the macroscopic electrical and gravitational forces in the $+x$ direction. Equations (13.3.2) and (13.3.3) are forms of the the Boltzmann transport equation.

A solution of the Boltzmann transport equation begins with the assumption that

$$f(\mathbf{p}, \mathbf{r}, t) = f_0(\alpha, T, p) + f_1(\mathbf{p}, \mathbf{r}, t), \qquad (13.3.4)$$

where f_0 is the Fermi–Dirac distribution function defined in Chapter 4 of Volume 1 and f_1 is an asymmetic term which vanishes in the absence of external forces and of gradients in thermodynamic variables but otherwise embodies the effects of external forces and gradients. Determination of the asymmetric term requires not only an estimate of the third, collision term in eq. (13.3.3) but also makes use of characteristics of the first two terms in this equation.

When electrons are not very degenerate, a convenient formulation for f_0 is

$$f_0 = \frac{1}{\exp\left(\alpha + \dfrac{\epsilon(p)}{kT} \right) + 1}, \qquad (13.3.5)$$

where $\alpha > 0$. When electrons are degenerate, the formulation

$$f_0 = \frac{1}{\exp\left(\dfrac{\epsilon(p) - \epsilon_F}{kT} \right) + 1}, \qquad (13.3.6)$$

where $\epsilon_F > 0$, is more convenient. Obviously, $\alpha = -\epsilon_F / kT < 0$.

The only parameter in f_0 which is a function of p is $\epsilon(p)$. Defining a variable ε by

$$\varepsilon = \frac{\epsilon(p) - \epsilon_F}{kT} = \alpha + \frac{\epsilon(p)}{kT}, \qquad (13.3.7)$$

one has

$$
\frac{\mathrm{d}f_0}{\mathrm{d}\varepsilon} = -\frac{\exp\left(\alpha + \frac{\epsilon(p)}{kT}\right)}{\left[\exp\left(\alpha + \frac{\epsilon(p)}{kT}\right) + 1\right]^2} = -\frac{\exp\left(\frac{\epsilon(p) - \epsilon_F}{kT}\right)}{\left[\exp\left(\frac{\epsilon(p) - \epsilon_F}{kT}\right) + 1\right]^2}. \tag{13.3.8}
$$

The quantity $-\mathrm{d}f_0/\mathrm{d}\varepsilon$ is dimensionless and, under electron-degenerate conditions, acts as an approximation to a delta function, with

$$
\int_0^\infty g\left(\frac{\epsilon}{kT}\right)\left(-\frac{\mathrm{d}f_0}{\mathrm{d}\varepsilon}\right)\mathrm{d}\left(\frac{\epsilon}{kT}\right) = g\left(\frac{\epsilon_F}{kT}\right) \times \left(1 + \text{order of }\left(\frac{kT}{\epsilon_F}\right)^n\right), \tag{13.3.9}
$$

where g is any function of ϵ/kT and $n \geq 1$.

If one supposes that the asymmetric component of f can be neglected in estimating the first of the three terms in eqs. (13.3.2) and (13.3.3), this term becomes

$$
\nabla_{\mathbf{p}}f \to \nabla_{\mathbf{p}}f_0 \to \frac{\partial f_0}{\partial p_x} = \frac{\mathrm{d}f_0}{\mathrm{d}\varepsilon}\frac{\partial}{\partial p_x}\left(\alpha + \frac{\epsilon(p)}{kT}\right) = \frac{\mathrm{d}f_0}{\mathrm{d}\varepsilon}\frac{1}{kT}\frac{\mathrm{d}\epsilon(p)}{\mathrm{d}p_x} = \frac{\mathrm{d}f_0}{\mathrm{d}\varepsilon}\frac{v_x}{kT}. \tag{13.3.10}
$$

Since $n_e = \int 2(\mathrm{d}^3\mathbf{p}/h^3)f_0$, the parameter α in eq. (13.3.5) and the parameter ϵ_F in eq. (13.3.6) are functions of T and the electron density n_e (see, e.g., eqs. (4.5.5), (4.6.25), and (4.9.12) in Volume 1). Their derivatives are related by

$$
\left(\frac{\partial\alpha}{\partial T}\right)_{n_e} = \frac{1}{kT}\left(-\left(\frac{\partial\epsilon_F}{\partial T}\right)_{n_e} + \frac{\epsilon_F}{T}\right) \tag{13.3.11}
$$

and

$$
\left(\frac{\partial\alpha}{\partial n_e}\right)_T = -\frac{1}{kT}\left(\frac{\partial\epsilon_F}{\partial n_e}\right)_T. \tag{13.3.12}
$$

Thus, supposing again that the asymmetric component of f can be neglected in estimating the second of the three terms in eqs. (13.3.2) and (13.3.3), this term becomes

$$
\nabla_{\mathbf{r}}f \to \nabla_r f_0 \to \frac{\partial f_0}{\partial x} = \frac{\mathrm{d}f_0}{\mathrm{d}\varepsilon}\left[\left(\left(\frac{\partial\alpha}{\partial T}\right)_{n_e} - \frac{\epsilon}{kT^2}\right)\frac{\mathrm{d}T}{\mathrm{d}x} + \left(\frac{\partial\alpha}{\partial n_e}\right)_T\frac{\mathrm{d}n_e}{\mathrm{d}x}\right]. \tag{13.3.13}
$$

Noting that

$$
\frac{\mathrm{d}\alpha}{\mathrm{d}x} = \left(\frac{\partial\alpha}{\partial T}\right)_{n_e}\frac{\mathrm{d}T}{\mathrm{d}x} + \left(\frac{\partial\alpha}{\partial n_e}\right)_T\frac{\mathrm{d}n_e}{\mathrm{d}x}, \tag{13.3.14}
$$

one has that

$$
\frac{\partial f_0}{\partial x} = \frac{\mathrm{d}f_0}{\mathrm{d}\varepsilon}\left[\frac{\mathrm{d}\alpha}{\mathrm{d}x} - \frac{\epsilon}{kT}\frac{1}{T}\frac{\mathrm{d}T}{\mathrm{d}x}\right], \tag{13.3.15}
$$

which can also be written as

$$\frac{\partial f_0}{\partial x} = -\frac{d f_0}{d\varepsilon} \frac{1}{kT} \left[\frac{\epsilon - \epsilon_F}{T} \frac{dT}{dx} + \frac{d\epsilon_F}{dx} \right]. \tag{13.3.16}$$

From the experience gained in Section 13.2, one may set

$$F_x = -(e\,E + m_e\,g). \tag{13.3.17}$$

Using this and eqs. (13.3.10), (13.3.15), and (13.3.16) in eq. (13.3.3), one has

$$-\frac{d f_0}{d\varepsilon} \frac{v_x}{kT} \left[eE + m_e g - kT \frac{d\alpha}{dx} + \frac{\epsilon}{T} \frac{dT}{dx} \right] + \left(\frac{\partial f}{\partial t} \right)_{\text{collisions}} = 0 \tag{13.3.18}$$

and

$$-\frac{d f_0}{d\varepsilon} \frac{v_x}{kT} \left[eE + m_e g + \frac{d\epsilon_F}{dx} + \frac{\epsilon - \epsilon_F}{T} \frac{dT}{dx} \right] + \left(\frac{\partial f}{\partial t} \right)_{\text{collisions}} = 0. \tag{13.3.19}$$

The next task is to determine how the term $(\partial f / \partial t)_{\text{collisions}}$ is related to the cross sections for binary collisions. The flux of electrons which have a momentum $\mathbf{p}$ and which attain a new momentum $\mathbf{p}'$ by scattering against ions of type i through an angle θ may be written as

$$\left(\frac{dn}{dt} \right)_{i-} = v\,2\frac{d^3\mathbf{p}}{h^3}\, f(\mathbf{p}, \mathbf{r}, t)\, [1 - f(\mathbf{p}', \mathbf{r}, t)]\, n_i \frac{d\sigma_i}{d\Omega}, \tag{13.3.20}$$

where $d\sigma_i / d\Omega$ can, for example, be estimated by eq. (13.1.11) or eq. (13.1.30). Similarly, the flux of electrons of momentum $\mathbf{p}'$ which achieve momentum $\mathbf{p}$ by scattering against particles of type i through the same angle θ is

$$\left(\frac{dn}{dt} \right)_{i+} = v'\,2\frac{d^3\mathbf{p}'}{h^3}\, f(\mathbf{p}', \mathbf{r}, t)\, [1 - f(\mathbf{p}, \mathbf{r}, t)]\, n_i \frac{d\sigma_i}{d\Omega}. \tag{13.3.21}$$

The factors $[1 - f]$ in eqs. (13.3.20) and (13.3.21) are required by the Pauli exclusion principle.

Balancing gains and losses, the total rate of change in the number of electrons of momentum $\mathbf{p}$ due to binary collisions with particles of type i is the integral over all angles of

$$\left(\frac{dn}{dt} \right)_{i+} - \left(\frac{dn}{dt} \right)_{i-}$$

$$= v\,2\frac{d^3\mathbf{p}'}{h^3}\, \left(f(\mathbf{p}', \mathbf{r}, t)\, [1 - f(\mathbf{p}, \mathbf{r}, t)] - f(\mathbf{p}, \mathbf{r}, t)\, [1 - f(\mathbf{p}', \mathbf{r}, t)] \right)\, n_i \frac{d\sigma_i}{d\Omega}$$

$$= v\,2\frac{d^3\mathbf{p}}{h^3}\, [f(\mathbf{p}', \mathbf{r}, t) - f(\mathbf{p}, \mathbf{r}, t)]\, n_i \frac{d\sigma_i}{d\Omega}. \tag{13.3.22}$$

Since scattering against effectively stationary nuclei results in almost no change in the absolute value of the electron momentum, insofar as scattering by nuclei dominates scattering by electrons, only the asymmetric part of the distribution function expressed by eq. (13.3.4) is of importance in eq. (13.3.22), and one has that

$$
\left(\frac{dn}{dt}\right)_{i+} - \left(\frac{dn}{dt}\right)_{i-} = v' \, 2\frac{d^3\mathbf{p}}{h^3} \, [f_1(\mathbf{p}', \mathbf{r}, t) - f_1(\mathbf{p}, \mathbf{r}, t)] \, n_i \, \frac{d\sigma_i}{d\Omega}. \tag{13.3.23}
$$

Integration over all scattering angles Ω' produces a quantity which is to be identified with the quantity $2\,(d^3\mathbf{p}/h^3)\left(\frac{\partial f}{\partial t}\right)_{\text{collisions}}$. In other words,

$$
\left(\frac{\partial f}{\partial t}\right)_{\text{collisions}} = \int_{\Omega'} v \, [f_1(\mathbf{p}', \mathbf{r}, t) - f_1(\mathbf{p}, \mathbf{r}, t)] \, n_i \, \frac{d\sigma_i}{d\Omega} \, d\Omega'. \tag{13.3.24}
$$

But $(\partial f/\partial t)_{\text{collisions}}$ is also given by eqs. (13.3.18) and (13.3.19) and these equations suggest that $f_1(\mathbf{p}, \mathbf{r}, t)$ is proportional to v_x. Accepting this suggestion, one has that

$$
f_1(\mathbf{p}, \mathbf{r}, t) = A \, v_x, \tag{13.3.25}
$$

where A is a factor to be determined, and

$$
f_1(\mathbf{p}', \mathbf{r}, t) - f_1(\mathbf{p}, \mathbf{r}, t) = A \, (v'_x - v_x) = -A \, v_x \, (1 - \cos\theta), \tag{13.3.26}
$$

where θ is the angle between v_x and v'_x. Equation (13.3.24) becomes

$$
\left(\frac{\partial f}{\partial t}\right)_{\text{collisions}} = -\, v_x \, A \, v \, n_i \int_{\Omega'} (1 - \cos\theta) \, \frac{d\sigma_i}{d\Omega} \, d\Omega', \tag{13.3.27}
$$

and A may now be determined by using eq. (13.3.18) or eq. (13.3.19) in conjunction with eq. (13.3.27), after the indicated integration over angles has been performed.

13.3.1 The cross section integral

Guided by eq. (13.1.11), assume that all cross sections have the form

$$
\frac{d\sigma_i}{d\Omega} = Z_i^2 \left(\frac{e^2}{2\epsilon}\right)^2 \frac{d\bar{\sigma}_i}{d\Omega}, \tag{13.3.28}
$$

where ϵ has been substituted for E, and define a quantity

$$
\Theta_i = \frac{1}{2\pi} \int_{\Omega} (1 - \cos\theta) \, \frac{d\bar{\sigma}_i}{d\Omega} \, d\Omega = \int_{\theta_0}^{\pi} (1 - \cos\theta) \, \frac{d\bar{\sigma}_i}{d\Omega} \, \sin\theta \, d\theta. \tag{13.3.29}
$$

Summing over binary scatterings from all types of particle, eq. (13.3.27) becomes

$$
\left(\frac{\partial f}{\partial t}\right)_{\text{collisions}} = -\, v_x \, A \, v \, n_i \left(\frac{e^2}{2\epsilon}\right)^2 \mu_e \, n_e \, 2\pi \, \Psi, \tag{13.3.30}
$$

where

$$\Psi = \frac{1}{\mu_e} \left[\frac{\sum_i n_i \, Z_i^2 \, \Theta_{ei} + n_e \, \Theta_{ee}}{n_e} \right].$$ (13.3.31)

In eq. (13.3.31), $\Theta_{ei} = \Theta_i$ for electron–ion collisions, and, introducing a contribution from electron–electron scattering, $\Theta_{ee} = \Theta_i$ for electron–electron collisions. Ψ can also be written as

$$\Psi = \sum_i Y_i \, Z_i^2 \, \Theta_{ei} + \frac{1}{\mu_e} \, \Theta_{ee}$$ (13.3.32)

where Y_i is the abundance by number of the ith ion.

All of the details of the scattering processes that contribute to the asymmetric part of the electron distribution function are hidden in the function Ψ. If, for electron–ion scattering, one chooses the differential cross section given by eq. (13.1.11), one obtains

$$\Theta_{ei} = \log_e \left[\frac{2}{1 - \cos\theta_{ei}} \right].$$ (13.3.33)

For a small cutoff angle,

$$\Theta_{ei} = \left[2 \, \log_e \frac{2}{\theta_{ei}} + \frac{\theta_{ei}^2}{12} + \cdots \right].$$ (13.3.34)

Choosing, instead, the differential cross section given by eq. (13.1.30),

$$\Theta_{ei} = \Theta(R,\lambda) = \int_0^\pi \frac{1 - \cos\theta}{\left[(1 - \cos\theta) + \left(\lambda/R\sqrt{2} \right)^2 \right]^2} \, \sin\theta \, d\theta$$

$$= \log_e \left[1 + \left(\frac{2R}{\lambda} \right)^2 \right] - \left[1 + \left(\frac{\lambda}{2R} \right)^2 \right]^{-1}.$$ (13.3.35)

When $(\lambda/2R) < 1$,

$$\Theta(R,\lambda) = \left(2 \, \log_e \frac{2R}{\lambda} - 1 \right) + 2 \left(\frac{\lambda}{2R} \right)^2 - \frac{3}{2} \left(\frac{\lambda}{2R} \right)^4 + \cdots .$$ (13.3.36)

As long as $(\lambda/2R) < 1.144$, $\Theta(R,\lambda) > 0$. Equations (13.3.32) and (13.3.36) demonstrate once again that the cutoff and screening approximations are roughly equivalent to the extent that $\theta_{ei} = \lambda/R \ll 1$. For small electron momenta, λ becomes large, and this illustrates a fundamental difference between the two approaches.

The dependence of the asymmetric term on external forces and on gradients in thermodynamic variables can be found by using eq. (13.3.30) in eq. (13.3.18) to obtain

$$f_1(\mathbf{p}, \mathbf{r}) = -A\,v_x$$

$$= \frac{v_x}{v}\left(\frac{2\epsilon}{e^2}\right)^2 \frac{1}{\mu_e\,n_e}\frac{1}{2\pi}\frac{1}{\Psi}\left(-\frac{\mathrm{d}f_0}{\mathrm{d}\varepsilon}\right)\frac{1}{kT}\left[eE + m_e g - kT\,\frac{\mathrm{d}\alpha}{\mathrm{d}x} + \frac{\epsilon}{T}\frac{\mathrm{d}T}{\mathrm{d}x}\right],$$

$$\tag{13.3.37}$$

where it has been acknowledged that the distribution function does not explicitly depend on time. Combining with eq. (13.3.19), one finds that

$$f_1(\mathbf{p}, \mathbf{r}) = -A\,v_x$$

$$= \frac{v_x}{v}\left(\frac{2\epsilon}{e^2}\right)^2 \frac{1}{\mu_e\,n_e}\frac{1}{2\pi}\frac{1}{\Psi}\left(-\frac{\mathrm{d}f_0}{\mathrm{d}\varepsilon}\right)\frac{1}{kT}\left[eE + m_e g + \frac{\mathrm{d}\epsilon_F}{\mathrm{d}x} + \frac{\epsilon - \epsilon_F}{T}\frac{\mathrm{d}T}{\mathrm{d}x}\right].$$

$$\tag{13.3.38}$$

13.4   Gradients in thermodynamic variables and the electric field

Having obtained an explicit expression for the asymmetrical part of the distribution function for electrons, one may determine the magnitude of the electrical field on the assumption that there is no net flow of electrons. The electrical current associated with electrons is given by

$$j = -e\int 2\frac{\mathrm{d}^3\mathbf{p}}{h^3}\,f(\mathbf{p}, \mathbf{r}, t)\,v_x = -e\int_0^\infty 2\frac{\mathrm{d}\mathbf{p}^3}{h^3}\,f_1(\mathbf{p}, \mathbf{r})\,v_x. \tag{13.4.1}$$

In a first exercise, suppose that electrons are not degenerate or only weakly degenerate, so that the formulation of eq. (13.3.37), with $\mathrm{d}f_0/\mathrm{d}\varepsilon$ given by eq. (13.3.8), is appropriate. Then,

$$j = e\,\frac{1}{n_e}\frac{1}{kT}\left(\frac{2}{e^2}\right)^2$$

$$\times \left\{\int 2\frac{\mathrm{d}^3\mathbf{p}}{h^3}\left(\frac{\mathrm{d}f_0}{\mathrm{d}\varepsilon}\right)\frac{1}{\mu_e\,n_e}\frac{1}{2\pi}\frac{1}{\Psi}\left(\frac{\epsilon^2\,v_x^2}{v}\right)\left[eE + m_e g - kT\,\frac{\mathrm{d}\alpha}{\mathrm{d}x} + \frac{\epsilon}{T}\frac{\mathrm{d}T}{\mathrm{d}x}\right]\right\},$$

$$\tag{13.4.2}$$

Setting $j = 0$, rearranging terms, noting that $v_x^2 = v^2\cos^2\theta$, and integrating over angles, one has that

$$\frac{8\pi}{3h^3}\int_0^\infty p^2\,\mathrm{d}p\left(\frac{\mathrm{d}f_0}{\mathrm{d}\varepsilon}\right)\frac{1}{\mu_e\,n_e}\frac{1}{2\pi}\frac{1}{\Psi}\,(\epsilon^2\,v)\left[eE + m_e g - kT\,\frac{\mathrm{d}\alpha}{\mathrm{d}x}\right]$$

$$= \frac{8\pi}{3h^3}\int_0^\infty p^2\,\mathrm{d}p\left(\frac{\mathrm{d}f_0}{\mathrm{d}\varepsilon}\right)\frac{1}{\mu_e\,n_e}\frac{1}{2\pi}\frac{1}{\Psi}\,(\epsilon^2\,v)\left[-\frac{\epsilon}{T}\frac{\mathrm{d}T}{\mathrm{d}x}\right]. \tag{13.4.3}$$

Given the uncertainty in the appropriate approximation for its value, it is reasonable to assume that Ψ can be replaced by some average value $\langle \Psi \rangle$ and taken out of the integral. Thus,

$$\int_0^\infty p^2 \, dp \left(\frac{df_0}{d\varepsilon} \right) (\epsilon^2 v) \left[eE + m_e g - kT \frac{d\alpha}{dx} \right]$$

$$\sim \int_0^\infty p^2 \, dp \left(\frac{df_0}{d\varepsilon} \right) (\epsilon^2 v) \left[-\frac{\epsilon}{T} \frac{dT}{dx} \right]. \tag{13.4.4}$$

For non-relativistic electrons, $p^2 \, dp \, \epsilon^2 \, v = 2m_e \, \epsilon^3 d\epsilon$, so eq. (13.4.4) can be written as

$$eE + m_e g = kT \frac{d\alpha}{dx} - \frac{J_4}{J_3} \frac{d}{dx} (kT), \tag{13.4.5}$$

where

$$J_m = \int_0^\infty y^m \left(-\frac{\partial f_0}{\partial \varepsilon} \right) dy = \int_0^\infty y^m \left(-\frac{df_0}{dy} \right) dy$$

$$= \int_0^\infty y^m \frac{e^{y+\alpha}}{(e^{y+\alpha} + 1)^2} \, dy$$

$$= \int_0^\infty y^m \frac{e^{-(y+\alpha)}}{(1 + e^{-(y+\alpha)})^2} \, dy, \tag{13.4.6}$$

and $y = \epsilon/kT$. Equation (13.4.6) can be integrated by parts to give

$$J_m = m \int_0^\infty y^{m-1} \, f_0 \, dy$$

$$= m \int_0^\infty y^{m-1} \frac{1}{e^{y+\alpha} + 1} \, dy$$

$$= m \int_0^\infty y^{m-1} \frac{e^{-(y+\alpha)}}{1 + e^{-(y+\alpha)}} \, dy. \tag{13.4.7}$$

When electrons are weakly degenerate ($\alpha > 0$), expansion and integration of eq. (13.4.6) yields

$$J_m = \int_0^\infty y^m \frac{e^{-(y+\alpha)}}{(1 + e^{-(y+\alpha)})^2} dy$$

$$= \sum_{n=1}^\infty (-1)^{n+1} n \, e^{-n\alpha} \int_0^\infty y^m \, e^{-ny} dy$$

$$= \sum_{n=1}^\infty (-1)^{n+1} \frac{1}{n^m} e^{-n\alpha} \int_0^\infty z^m \, e^{-z} \, dz$$

$$= m! \, G_m(\alpha), \tag{13.4.8}$$

where

$$G_m(\alpha) = \sum_{n=1}^{\infty} (-1)^{n+1} \frac{1}{n^m} e^{-n\alpha}. \qquad (13.4.9)$$

Of course, the same result may be obtained by manipulating eq. (13.4.7) in the same way. One can now write

$$\frac{J_4}{J_3} = 4 \frac{G_4(\alpha)}{G_3(\alpha)}, \qquad (13.4.10)$$

Inserting this result into eq. (13.5.5), one has that, to first order ($e^{-\alpha} \ll 1$),

$$eE + m_e g = kT \frac{d\alpha}{dx} - 4 \frac{d(kT)}{dx}. \qquad (13.4.11)$$

From eq. (4.5.5) in Volume 1,

$$n_e = 2 \left(\frac{2\pi m_e kT}{h^2} \right)^{3/2} \sum_{n=1}^{\infty} (-1)^{n+1} \frac{e^{-n\alpha}}{n^{3/2}}$$

$$= 2 \left(\frac{2\pi m_e kT}{h^2} \right)^{3/2} e^{-\alpha} + \cdots . \qquad (13.4.12)$$

Keeping just the first term in this series, and differentiating, one finds

$$\frac{d\alpha}{dx} = -\frac{1}{n_e} \frac{dn_e}{dx} + \frac{3}{2} \frac{1}{T} \frac{dT}{dx}, \qquad (13.4.13)$$

giving, in conjunction with eq. (13.4.11),

$$eE + m_e g = -\frac{kT}{n_e} \frac{dn_e}{dx} - \frac{5}{2} \frac{d(kT)}{dx}. \qquad (13.4.14)$$

Since, to first order, $P_e = n_e kT$, one can write additionally

$$eE + m_e g = -kT \frac{1}{P_e} \frac{dP_e}{dx} - \frac{3}{2} \frac{d(kT)}{dx}. \qquad (13.4.15)$$

To compare with the results in Section 13.2, consider a fully ionized medium with only one type of ion of charge Z_0, mass $m_0 = A_0 M_H$, and number density n_0. One has

$$P_e = \frac{n_e}{n_e + n_0} P = \frac{Z_0}{1 + Z_0} P, \qquad (13.4.16)$$

where P is the total pressure. Equating body forces on the particles to the total pressure gradient, one has

$$\frac{1 + Z_0}{Z_0} \frac{dP_e}{dx} = \frac{dP}{dx} = -g (n_0 m_0 + n_e m_e). \qquad (13.4.17)$$

Note that, because of assumed charge neutrality ($Z_0 n_0 - n_e = 0$), the electric field does not enter into this equation. Using eq. (13.4.17) in eq. (13.4.15), it follows that

$$eE + m_e g = kT \left[\frac{1}{n_e kT} \frac{Z_0}{1 + Z_0} (g \, n_0 m_0) \left(1 + Z_0 \frac{m_e}{m_0} \right) - \frac{3}{2} \frac{1}{T} \frac{dT}{dx} \right]$$

$$= \frac{g \, m_0}{1 + Z_0} \left(1 + Z_0 \frac{m_e}{m_0} \right) - \frac{d}{dx} \left(\frac{3}{2} kT \right),$$
(13.4.18)

or, finally,

$$eE = \frac{g \, m_0}{1 + Z_0} \left(1 - \frac{m_e}{m_0} \right) - \frac{d}{dx} \left(\frac{3}{2} kT \right).$$
(13.4.19)

The first term on the right hand side of eq. (13.4.19) is identical with the electric field estimated in Section 13.2 (eq. (13.2.8)) on the assumption that partial pressure-gradient forces equal partial body forces, and this fact gives support for assuming equality of the partial forces.

The temperature-gradient term in eq. (13.4.19) is presumably due to the effect of electron conduction, a physical process not taken into account in Section 13.2. The sign of this term, predicting as it does an additional electrical force inward on electrons, agrees with intuition. Conduction occurs because faster moving electrons from "upstream" exchange energy with more slowly moving electrons "downstream" and one might anticipate an extra inward electric force ($-e \Delta E < 0$) on the electrons to counter the potential for a thermally induced outward flow of electrons. On the other hand, because of near charge neutrality on a macroscopic scale, the net electrical force on ions and electrons together remains nearly zero and the origin and significance of the temperature-gradient term requires further investigation.

In the interior of a star like the Sun, the contribution of the thermal gradient term to the effective electric field is comparable with that of the gravity-induced term. That is,

$$g M_H \sim \frac{G M_\odot M_H}{R_\odot^2} = 4.59 \times 10^{-20},$$
(13.4.20)

whereas

$$-\frac{3}{2} k \frac{dT}{dx} \sim 1.5 \times 1.38 \times 10^{-16} \frac{1.5 \times 10^7}{R_\odot} = 4.46 \times 10^{-20}.$$
(13.4.21)

It is worth noting that, although the estimate here of the strength of the macroscopic electric field has made no explicit use of the cross section between electrons and ions (other than that the function Ψ, eq. (13.3.31), may be approximated by an energy-independent average), this cross section enters indirectly through the fact that the temperature gradient in a star is directly proportional to the effective opacity (inversely proportional to the effective conductivity). Thus, for a given flux of energy, the larger the cross section between electrons and ions, the larger is the temperature gradient and, hence, the larger is the electric field. In a star like the Sun, of course, conduction is a minor source of opacity, so the portion of the electric field due to a temperature gradient is determined mainly by interactions between matter and photons.

When electrons are non-relativistically degenerate ($\alpha < 0$, $\epsilon_F > 0$), it is convenient, by using eq. (13.3.7) in eq. (13.4.5), to express the force field as

$$eE + m_e g = -\frac{d\epsilon_F}{dx} + \left(\frac{\epsilon_F}{kT} - \frac{J_4}{J_3}\right)\frac{d(kT)}{dx}. \tag{13.4.22}$$

One could next write

$$J_m = \int_0^{y_F} y^m \frac{e^{-(y_F-y)}}{\left(1 + e^{-(y_F-y)}\right)^2}\, dy + \int_{y_F}^{\infty} y^m \frac{e^{-(y-y_F)}}{\left(1 + e^{-(y-y_F)}\right)^2}\, dy, \tag{13.4.23}$$

where

$$y_F = \frac{\epsilon_F}{kT}, \tag{13.4.24}$$

expand the denominators in the integrals, and integrate each term in the expansion, just as was done in arriving at eq. (13.4.7). The sum of the integrals of the first term of the expansion gives the indeterminate result $y_F^m \sum_{n=1}^{\infty} (-1)^{n+1}$. The coefficient of the sum is correct, as is evident from contemplating the nature of the derivative $df_0/d\varepsilon$ given by eq. (13.3.8). When $\epsilon_F/kT \gg 1$, this derivative is effectively a delta function centered on $\epsilon = \epsilon_F$ and the main term in the integral J_m, as described by eq. (13.4.6), is simply the integrand y^m evaluated at $y = y_F$, namely y_F^m.

Given the untidiness of this result, set $\alpha = -y_F$ in the integrals of eq. (13.4.7), and break the integration into three parts:

$$J_m = m \int_0^{y_F} y^{m-1}\, dy - m \sum_{n=1}^{\infty} (-1)^{n+1} \int_0^{y_F} y^{m-1}\, e^{-n(y_F-y)}\, dy$$

$$+ m \sum_{n=1}^{\infty} (-1)^{n+1} \int_{y_F}^{\infty} y^{m-1}\, e^{-n(y-y_F)}\, dy \tag{13.4.25}$$

$$= y_F^m - m \sum_{n=1}^{\infty} (-1)^{n+1} \frac{1}{n^m} \left[e^{-ny_F} \int_0^{ny_F} z^{m-1}\, e^z\, dz - e^{ny_F} \int_{ny_F}^{\infty} z^{m-1}\, e^{-z}\, dz \right]. \tag{13.4.26}$$

Making successive applications of $\int z^m e^{\pm z} dz = \pm z^m e^{\pm z} \mp \int m z^{m-1} e^{\pm z} dz$, integrations yield

$$J_3 = y_F^3 + 6\, (2\, S_2\, y_F + G_3(-\alpha)) \tag{13.4.27}$$

and

$$J_4 = y_F^4 + 24 \left(S_2\, y_F^2 + 2\, S_4 - G_4(-\alpha)\right), \tag{13.4.28}$$

where

$$S_m = \sum_{n=1}^{\infty} (-1)^{n+1} \frac{1}{n^m} \tag{13.4.29}$$

and

$$G_m(-\alpha) = \sum_{n=1}^{\infty} (-1)^{n+1} \frac{1}{n^m} e^{n\alpha} = \sum_{n=1}^{\infty} (-1)^{n+1} \frac{1}{n^m} e^{-ny_F}. \tag{13.4.30}$$

It is worth emphasizing at this point that the argument of the exponents in the definition of the G functions (eqs. (13.4.9) and (13.4.30)) is a negative number.

The procedure beginning with eq. (13.4.23) which was abandoned actually gives correct values for terms in the J_ms beyond the first, and it gives the entire result correctly if one adopts $S_0 = \sum_{n=1}^{\infty} (-1)^{n+1} = 1$.

From eqs. (13.4.27) and (13.4.28), one has

$$\frac{J_4}{J_3} = \frac{\epsilon_F}{kT} \frac{1 + 24 S_2 \left(\frac{kT}{\epsilon_F}\right)^2 + 48 \left(S_4 - \frac{1}{2}G_4(-\alpha)\right) \left(\frac{kT}{\epsilon_F}\right)^4}{1 + 12 \left[S_2 + \frac{1}{2}G_3(-\alpha) \left(\frac{kT}{\epsilon_F}\right)\right] \left(\frac{kT}{\epsilon_F}\right)^2}. \tag{13.4.31}$$

Discarding terms involving $G_3(-\alpha)$, S_4, and $G_4(-\alpha)$ and using $S_2 = \pi^2/12$, one has that

$$\frac{J_4}{J_3} \sim \frac{\epsilon_F}{kT} \left[1 + 12 S_2 \left(\frac{kT}{\epsilon_F}\right)^2\right] = \frac{\epsilon_F}{kT} \left[1 + \pi^2 \left(\frac{kT}{\epsilon_F}\right)^2\right]. \tag{13.4.32}$$

Equation (13.4.22) now becomes

$$eE + m_e g = -\frac{d\epsilon_F}{dx} - \pi^2 \left(\frac{kT}{\epsilon_F}\right) \frac{d(kT)}{dx}, \tag{13.4.33}$$

To make further progress, the explicit relationships between n_e, P_e, T, and ϵ_F derived in Chapter 4 of Volume 1 (eqs. (4.6.25) and (4.6.40)) are needed:

$$n_e = \frac{8\pi}{3} \left(\frac{2m_e\epsilon_F}{h^3}\right)^{3/2} \left[1 + \frac{\pi^2}{8} \left(\frac{kT}{\epsilon_F}\right)^2 + \cdots\right] \tag{13.4.34}$$

and

$$P_e = \frac{2}{5} n_e \epsilon_F \left[1 + \frac{\pi^2}{2} \left(\frac{kT}{\epsilon_F}\right)^2 + \cdots\right]. \tag{13.4.35}$$

Taking derivatives of eqs. (13.4.34) and (13.4.35) and eliminating $(1/n_e)dn_e/dx$ gives

$$\frac{1}{P_e}\frac{dP_e}{dx} = \frac{5}{2} \left[1 - \frac{\pi^2}{2} \left(\frac{kT}{\epsilon_F}\right)^2\right] \frac{1}{\epsilon_F}\frac{d\epsilon_F}{dx} + \frac{5\pi^2}{4} \left(\frac{kT}{\epsilon_F}\right)^2 \frac{1}{T}\frac{dT}{dx}. \tag{13.4.36}$$

Similarly, using

$$P_0 = n_0 kT = \frac{n_e}{Z_0} kT \tag{13.4.37}$$

in conjunction with the derivative of eq. (13.4.34), one finds

$$\frac{1}{P_0}\frac{dP_0}{dx} = \frac{3}{2}\left[1 - \frac{\pi^2}{6}\left(\frac{kT}{\epsilon_F}\right)^2\right]\frac{1}{\epsilon_F}\frac{d\epsilon_F}{dx} + \left[1 + \frac{\pi^2}{4}\left(\frac{kT}{\epsilon_F}\right)^2\right]\frac{1}{T}\frac{dT}{dx}. \quad (13.4.38)$$

Using eqs. (13.4.35) and (13.4.37) in eqs. (13.4.36) and (13.4.38) and adding, one obtains

$$\frac{dP}{dx} = \frac{d(P_e + P_0)}{dx} = \left[n_e + \frac{3}{2}n_0\left(\frac{kT}{\epsilon_F}\right)\right]\frac{d\epsilon_F}{dx} + n_0\left[1 + \frac{\pi^2}{4}\left(\frac{kT}{\epsilon_F}\right)^2\right]\frac{d(kT)}{dx},$$

$$(13.4.39)$$

from which it follows that

$$\frac{d\epsilon_F}{dx} = \frac{1}{Z_0 + \frac{3}{2}\left(\frac{kT}{\epsilon_F}\right)}\frac{1}{n_0}\frac{dP}{dx} - \frac{1 + \frac{\pi^2}{4}\frac{kT}{\epsilon_F}^2}{Z_0 + \frac{3}{2}\left(\frac{kT}{\epsilon_F}\right)}\frac{d(kT)}{dx}. \quad (13.4.40)$$

Inserting this in eq. (13.4.33) and keeping just the leading terms, one arrives at

$$eE = \frac{m_0 g}{Z_0}\left[1 - \frac{3}{2Z_0}\left(\frac{kT}{\epsilon_F}\right)\right] + \frac{1}{Z_0}\frac{d(kT)}{dx}. \quad (13.4.41)$$

The first term on the right hand side of eq. (13.4.41) reproduces the result found in Section 13.2 (see eq. (13.2.30)) that, when electrons are degenerate but the effects of a temperature gradient are not taken into account, the electrical force on the electrons essentially equals the gravitational force on ions, with the consequence that the net external force on the ions effectively vanishes. On the other hand, the sign of the second term on the right hand side, which takes the temperature gradient into account, is exactly opposite to that found when electrons are not degenerate, (see eq. (13.4.19)). The origin of the sign difference is presumably related to the subtleties encountered in estimating the thermal diffusion coefficient. Because of macroscopic charge neutrality, the net static electric force on all electrons and ions is zero, regardless of the sign of the temperature-gradient term, so, fortunately, the pressure balance equation is not affected by the difference in sign.

13.5 Thermal conductivity in the classical approximation

13.5.1 General considerations

The heat flux W is the integral over the asymmetric distribution function, eq. (13.3.37) or eq. (13.3.38), of the kinetic energy times the speed in the x direction (ϵv_x). Thus,

$$W = \int 2\frac{d^3\mathbf{p}}{h^3}f(\mathbf{p}, \mathbf{r}, t)(\epsilon v_x) = \int_0^\infty 2\frac{d\mathbf{p}^3}{h^3}f_1(\mathbf{p}, \mathbf{r})(\epsilon v_x). \quad (13.5.1)$$

Choosing the formulation of eq. (13.3.37), with $d f_0/d\varepsilon$ given by eq. (13.3.8), one has

$$
W = \frac{1}{kT} \left(\frac{2}{e^2}\right)^2
$$

$$
\times \left\{ \int 2\frac{d^3\mathbf{p}}{h^3} \left(-\frac{d f_0}{d\varepsilon}\right) \frac{M_H}{\rho} \frac{1}{2\pi \Psi} \left(\frac{\epsilon^3 v_x^2}{v}\right) \left[e E + m_e g - kT \frac{d\alpha}{dx} + \frac{\epsilon}{T} \frac{dT}{dx}\right]\right\}.
$$
(13.5.2)

Again, replace Ψ defined by eq. (13.3.31) by $\langle\Psi\rangle$ and extract it from under the integral sign. Restricting attention to non-relativistic electrons, so that $p^2 dp\, \epsilon^2 v = 2m_e\epsilon^3 d\epsilon$ and $v^2 = 2m_e\epsilon$, and integrating over angles, one obtains

$$
W = \frac{M_H}{\rho} \frac{1}{2\pi \langle\Psi\rangle} \frac{1}{kT} \left(\frac{2}{e^2}\right)^2
$$

$$
\times \left\{ \frac{8\pi}{3h^3} 2m_e \int_0^\infty \left(-\frac{d f_0}{d\varepsilon}\right) \epsilon^4 d\epsilon \left[e E + m_e g - kT \frac{d\alpha}{dx} + \frac{\epsilon}{T} \frac{dT}{dx}\right]\right\}.
$$
(13.5.3)

Replacing the first three terms in square brackets in eq. (13.5.3) by their equivalent as given by eq. (13.4.5), one has

$$
W = -\frac{M_H}{\rho} \frac{1}{2\pi \langle\Psi\rangle} \frac{1}{kT} \left(\frac{2}{e^2}\right)^2 \frac{8\pi}{3h^3} 2m_e \int_0^\infty \left(-\frac{d f_0}{d\varepsilon}\right) \epsilon^4 d\epsilon\, k \left[\frac{\epsilon}{kT} - \frac{J_4}{J_3}\right]\frac{dT}{dx}.
$$
(13.5.4)

Setting $\epsilon = kT y$, and making use of eq. (13.4.6), one has

$$
W = -K \frac{dT}{dx},
$$
(13.5.5)

where

$$
K = \frac{M_H}{\rho} \frac{1}{2\pi \langle\Psi\rangle} \left(\frac{2}{e^2}\right)^2 \frac{8\pi}{3h^3} 2m_e\, k\, (kT)^4 \left[J_5 - \frac{J_4^2}{J_3}\right]
$$
(13.5.6)

$$
= \frac{32}{3} \left(\frac{ck}{r_e^2}\right) \left[\left(\frac{m_e c}{h}\right)^3 \frac{M_H}{\rho} \frac{1}{\langle\Psi\rangle}\right] \left(\frac{kT}{m_e c^2}\right)^4 J.
$$
(13.5.7)

In writing eq. (13.5.7), yet another function has been defined:

$$
J = J_5 - \frac{J_4^2}{J_3}.
$$
(13.5.8)

Expressing the relationship between conductive opacity and conductivity from eq. (13.1.51) as

$$
\kappa_c = \frac{16\sigma}{3} \left(\frac{m_e c^2}{k}\right)^3 \left(\frac{kT}{m_e c^2}\right)^3 \frac{1}{\rho} \frac{1}{K}
$$
(13.5.9)

and the relationship between the Steffan–Boltzmann constant σ and other fundamental constants from eq. (4.11.28) in Volume 1 as

$$\frac{\sigma}{kc} \left(\frac{ch}{k} \right)^3 = \frac{2\pi^5}{15}, \tag{13.5.10}$$

one has

$$\kappa_c = \frac{\pi^5}{15} r_e^2 \frac{1}{M_H} \langle \Psi \rangle \frac{m_e c^2}{kT} \frac{1}{J} \tag{13.5.11}$$

$$= 5740.29 \frac{1}{T_6} \frac{\langle \Psi \rangle}{J}. \tag{13.5.12}$$

Thus, the task of finding the conductivity and the conductive opacity boils down to calculating J and making the right choice for $\langle \Psi \rangle$.

It is interesting to examine the conductivity in the limiting regimes of very weak degeneracy and of strong degeneracy and to compare with the conductivity estimated in the manner of Section 13.1.

13.5.2   When electrons are not degenerate

Using eq. (13.4.8) in eq. (13.5.8), one has that, when $\alpha > 0$,

$$J = 5! \, G_5(\alpha) - \frac{\left[4! \, G_4^2(\alpha) \right]^2}{3! \, G_3(\alpha)} \tag{13.5.13}$$

$$= 4! \left[5 \, G_5(\alpha) - 4 \frac{G_4^2(\alpha)}{G_3(\alpha)} \right] = 120 \left[G_5(\alpha) - 0.8 \frac{G_4^2(\alpha)}{G_3(\alpha)} \right]. \tag{13.5.14}$$

When electrons are sufficiently non-degenerate that all but the first power of $e^{-\alpha}$ may be neglected, all of the G_ms are the same. Using the relationships defined by eqs. (4.5.6)–(4.5.12) in Volume 1,

$$G_5(\alpha) \sim G_4(\alpha) \sim G_3(\alpha) \sim e^{-\alpha} \sim \frac{1}{2} \left(\frac{h^2}{2\pi m_e kT} \right)^{3/2} n_e \sim 0.123\,730\,\delta, \tag{13.5.15}$$

where the degeneracy parameter is $\delta = \rho / \mu_e T_6^{3/2}$. Thus, to first order in $e^{-\alpha}$ and δ,

$$J = 24 \, e^{-\alpha} = 2.969\,52\,\delta. \tag{13.5.16}$$

Using eqs. (13.5.16) and (13.5.15) in (13.5.7) gives

$$K_{ND} = 16 \left(\frac{2}{\pi} \right)^{3/2} \frac{ck}{r_e^2} \left(\frac{kT}{m_e c^2} \right)^{5/2} \left(\frac{M_H n_e}{\rho} \right) \frac{1}{\langle \Psi \rangle}$$

$$= 16 \left(\frac{2}{\pi} \right)^{3/2} \frac{ck}{r_e^2} \left(\frac{kT}{m_e c^2} \right)^{5/2} \frac{1}{\mu_e} \frac{1}{\langle \Psi \rangle}. \tag{13.5.17}$$

The next task is to choose an appropriate average for the function Ψ defined by eq. (13.3.31). For example, replacing i by e in eqs. (13.3.33) and (13.3.34), replacing θ_{ee} by an average $\langle \theta_{ee} \rangle$, and making the small angle approximation, one can define $\langle \Theta_{ee} \rangle \sim 2 \log_e (2/\langle \theta_{ee} \rangle)$. Similarly, one can define $\langle \Theta(R,\bar{\lambda}) \rangle \sim 2 \log_e 2R/\bar{\lambda}_e$, where $\bar{\lambda}_e$ is defined by eqs. (4.7.16) and (4.7.18) in Volume 1. For $\langle \theta_{ee} \rangle$ one can choose, say, $\langle \theta_{ee} \rangle = \bar{\lambda}_e/r_{0e}$, or even $\langle \theta_{ee} \rangle = \bar{\lambda}_e/R_D$. These definitions give

$$\langle \Psi \rangle \sim 2 \left[\sum_i Y_i \, Z_i^2 \, \log_e \frac{2R}{\bar{\lambda}_e} + \frac{1}{\mu_e} \, \log_e \frac{2}{\langle \theta_{ee} \rangle} \right]. \tag{13.5.18}$$

A corresponding average can be defined for the quantity Φ given by eqs. (13.1.16) and (13.1.17) in Section 13.1. Using eq. (13.1.14), one can set $\langle \bar{\sigma}_{ee} \rangle = 4\pi/\langle \theta_{ee}^2 \rangle$, and using eqs. (13.1.12) and (13.2.17), one can set $\langle \bar{\sigma}_{ei} \rangle = 4\pi \, (R/\bar{\lambda}_e)^2$, so that

$$\langle \Phi \rangle = \frac{1}{2\pi \, \mu_e} \, \frac{\sum_i n_i \, Z_i^2 \, \langle \bar{\sigma}_{ei} \rangle + n_e \, \langle \bar{\sigma}_{ei} \rangle}{n_e}, \tag{13.5.19}$$

$$= 2 \left[\sum_i Y_i \, Z_i^2 \, \left(\frac{R}{\bar{\lambda}_e} \right)^2 + \frac{1}{\mu_e} \, \frac{1}{\langle \theta_{ee}^2 \rangle} \right]. \tag{13.5.20}$$

Comparing eq. (13.5.17) with eq. (13.1.18), and using eqs. (13.5.18) and (13.5.20), one has

$$\frac{K_{simple}}{K_{fancy}} \sim \frac{9\pi}{32} \, \frac{\langle \Psi \rangle}{\langle \Phi \rangle}, \tag{13.5.21}$$

where K_{simple} is the conductivity given by the elementary treatment in Section 13.1 and K_{fancy} is the conductivity given by the somewhat more sophisticated treatment in this section.

Equations (13.5.18)–(13.5.21) show that the conductivity obtained by using the Boltzmann transport equation is larger, by the ratio of the squares of large numbers to the logarithms of these numbers, than is the conductivity estimated in a much more elementary way in Section 13.1 that, among other deficiencies, ignores the asymmetry in the electron distribution function.

13.5.3 When electrons are degenerate but not relativistic

Using the relationship between $\epsilon_F(0)$ and n_e given by eq. (13.1.42) and rearranging, eq. (13.5.7) gives for the conductivity

$$K_{NRD} = 2\sqrt{2} \, \left(\frac{ck}{r_e^2} \right) \frac{1}{\mu_e} \frac{1}{2\pi \langle \Psi \rangle} \left(\frac{m_e c^2}{\epsilon_F(0)} \right)^{3/2} \left(\frac{kT}{m_e c^2} \right)^4 \left(J_5 - \frac{J_4^2}{J_3} \right). \tag{13.5.22}$$

Evaluation of eqs. (13.4.25)–(13.4.26) for $m = 5$ yields

$$J_5 = y_F^5 + 40 \left[S_2 \, y_F^3 + 6 \, S_4 \, y_F + 3 \, G_5(-\alpha) \right]. \tag{13.5.23}$$

Using this and eqs. (13.4.27) for J_3 and (13.4.28) for J_4, one finds

$$J_3 J_5 - J_4^2 = 4\, S_2\, y_F^6 + \left(144\, S_4 - 96\, S_2^2\right) y_F^4 + 576\, S_2\, S_4\, y_F^2 - 2304\, S_4^2$$

$$+ 6\, G_3(-\alpha)\, y_F^5 + 48\, G_4(\alpha)\, y_F^4 + 120\, [G_5(-\alpha) + 2 S_2\, G_3(-\alpha)]\, y_F^3$$

$$+ 1152\, S_2\, G_4(-\alpha)\, y_F^2 + 1440\, [S_2\, G_5(-\alpha) + S_4\, G_3(-\alpha)]\, y_F$$

$$+ 2304\, S_4\, G_4(-\alpha) + 720\, G_3(-\alpha)\, G_5(-\alpha) - 576\, G_4^2(-\alpha). \quad (13.5.24)$$

Using $S_2 = \pi^2/12$ and $S_4 = 7\pi^4/720$, and neglecting the G_ms, one has

$$J_3 J_5 - J_4^2 = \frac{1}{3}\, \pi^2\, y_F^6 + \frac{11}{15}\, \pi^4 y_F^4 + \frac{7}{15}\, \pi^6\, y_F^2 - \left(\frac{7}{15}\right)^2 \pi^8 + \cdots . \quad (13.5.25)$$

Dividing by J_3 from eq. (13.4.27), one gets

$$J = J_5 - \frac{J_4^2}{J_3} = \frac{\pi^2}{3}\, y_F^3\, r_{543}, \quad (13.5.26)$$

where

$$r_{543} = \frac{1 + \dfrac{11}{5}\left(\dfrac{\pi k T}{\epsilon_F}\right)^2 + \dfrac{7}{5}\left(\dfrac{\pi k T}{\epsilon_F}\right)^4 - \dfrac{49}{75}\left(\dfrac{\pi k T}{\epsilon_F}\right)^6}{1 + \left(\dfrac{\pi k T}{\epsilon_F}\right)^2} + \cdots . \quad (13.5.27)$$

Inserting this last result into eq. (13.5.22) and using $y_F = \epsilon_F/kT$ (eq. (13.4.24)) gives

$$K_{\mathrm{NRD}} = \frac{2\sqrt{2}}{3}\, \pi^2 \left(\frac{ck}{r_e^2}\right) \frac{1}{\mu_e}\, \frac{1}{2\pi\langle\Psi\rangle} \left(\frac{\epsilon_F}{\epsilon_F(0)}\right)^{3/2} \left(\frac{kT}{\epsilon_F}\right) \left(\frac{\epsilon_F}{m_e c^2}\right)^{5/2} r_{543},$$

$$(13.5.28)$$

Comparing eq. (13.5.28) with eq. (13.1.47), one has

$$\frac{K_{\mathrm{simple}}}{K_{\mathrm{fancy}}} = \frac{1}{r_{543}} \left(\frac{\epsilon_F(0)}{\epsilon_F}\right)^3 \frac{\langle\Psi\rangle}{\langle\Phi\rangle}, \quad (13.5.29)$$

where $\langle\Psi\rangle$ and $\langle\Phi\rangle$ are given by eqs. (13.5.18) and (13.5.20), respectively. The relationship between $\epsilon_F(0)$ and ϵ_F is given by eq. (4.6.30) in Volume 1. Thus, also in the electron-degenerate regime, the conductivity obtained by using the Boltzmann transport equation is larger, by the ratio of the squares of large numbers to the logarithms of these numbers, than is the conductivity estimated when the asymmetry in the electron distribution function is ignored. It is rather remarkable, however, that, regardless of the degree of degeneracy, the coefficients of the cross-section functions $\langle\Psi\rangle$ and $\langle\Phi\rangle$ are almost identical.

13.6   A quantitative estimate of the conductive opacity

The treatment of conductivity in Section 13.5 has used the Boltzmann transport equation in a manner introduced in the 1930s in estimating the electrical resistance and heat conductivity of metals (see, e.g., M. Abraham & R. Becker, *Theorie Der Electricität, Band II: Electronentheorie*, 1933). It was applied to heat conduction in stellar material by M. E. Marshak (1941; see also 1940), L. Mestel (1950), and T. D. Lee (1950). W. B. Hubbard (1966) suggested replacing the somewhat arbitrarily truncated cross-section integral used by these investigators by an integration involving a pair distribution function based on Debye theory and, following this suggestion, Iben (1975) obtained the quantitative results reported in Table 13.6.1.

As a check on the results of the numerical integrations, one may determine J (eq. (13.5.8)) for $\alpha = 0$ by calculating the $G_m(0)$s from eq. (13.4.9) and inserting them into eq. (13.5.14). The results are $G_5(0) = 0.972\,120$, $G_4(0) = 0.947\,032$, and $G_3(0) = 0.901\,54$, giving $J(0) = 120\left(G_5(0) - 0.8\,G_4^2(0)/G_3(0)\right) = 21.15$, identical with the value in Table (13.6.1) obtained by direct integration.

Expanding the $G(\alpha)$ functions (eq. (13.4.9)) out to terms involving $e^{-3\alpha}$, one finds (eq. (13.5.13))

$$J = 24\,e^{-\alpha}\left(1 - \frac{5}{2^5}\,e^{-\alpha} + \left(\frac{17}{3^5} + \frac{3}{2^6}\right)e^{-2\alpha} + \cdots\right)$$

$$= 24\,e^{-\alpha}\left(1 - \frac{5}{32}\,e^{-\alpha} + \frac{1817}{933\,126}\,e^{-2\alpha} + \cdots\right)$$

$$= 24\,e^{-\alpha}\left(1 - 0.156\,25\,e^{-\alpha} + 0.019\,472\,e^{-2\alpha} + \cdots\right). \tag{13.6.1}$$

Then, using the relationship between δ and $e^{-\alpha}$ given by eq. (4.5.12) in Volume 1, one obtains

$$J = 2.969\,52\,\delta\,(1 + 0.024\,412\,\delta) + \cdots, \tag{13.6.2}$$

which fits the numerical values of J for $\alpha \le 0$ in Table 13.6.1 to better than 0.1%.

At the other end of the degeneracy spectrum, eqs. (13.5.26) may be combined with eqs. (4.6.28) and (4.6.29) in Volume 1 to obtain

$$J = \frac{\pi^2}{3}\left(\frac{\epsilon_F(0)}{kT}\right)^3\left[1 + \frac{19}{20}\,\pi^2\left(\frac{kT}{\epsilon_F(0)}\right)^2 + \frac{1}{12}\,\pi^4\left(\frac{kT}{\epsilon_F(0)}\right)^4\right] \tag{13.6.3}$$

$$= 3.289\,868\left(\frac{\delta}{6.0798}\right)^2\left[1 + 9.376\,124\left(\frac{6.0798}{\delta}\right)^{4/3} + 8.117\,424\left(\frac{6.0798}{\delta}\right)^{8/3}\right]. \tag{13.6.4}$$

		Table 13.6.1 Conductive opacity of a non-relativistic electron gas			
$e^{-\alpha}$	$\dfrac{\rho}{\mu_e T_6^{3/2}}$	$\dfrac{\langle\Psi\rangle}{\kappa_c T_6}$	J	$\left(\dfrac{\bar{\lambda}_e}{2R_D}\right)^2$	$\langle\Theta_{ei}\rangle$	κ_c
0.01	0.08055	4.175/−5	0.2397	2.228/−4	7.410	1775
0.03	0.2400	1.249/−4	0.7170	6.600/−4	6.325	506.4
0.06	0.4751	2.486/−4	1.427	1.296/−3	5.651	227.3
0.10	0.7812	4.118/−4	2.364	2.108/−3	5.166	125.4
0.20	1.513	8.118/−4	4.660	3.981/−3	4.534	55.85
0.30	2.203	1.201/−3	6.894	5.662/−3	4.185	34.85
0.40	2.856	1.580/−3	9.070	7.185/−3	3.950	25.00
0.60	4.063	2.312/−3	13.27	9.849/−3	3.640	15.75
0.80	5.169	3.013/−3	17.30	1.212/−2	3.437	11.40
1.0	6.185	3.685/−3	21.15	1.408/−2	3.291	8.932
1.25	7.354	4.492/−3	25.79	1.621/−2	3.154	7.020
1.5	8.430	5.265/−3	30.22	1.806/−2	3.050	5.793
2.0	10.36	6.726/−3	38.61	2.114/−2	2.898	4.309
2.5	12.05	8.092/−3	41.06	2.363/−2	2.792	3.903
3.0	13.57	9.378/−3	53.83	2.570/−2	2.712	2.892
4.0	16.21	1.176/−2	67.51	2.902/−2	2.597	2.208
5.0	18.46	1.394/−2	80.02	3.159/−2	2.517	1.806
8.0	23.78	1.961/−2	112.6	3.692/−2	2.371	1.209
10.0	26.56	2.289/−2	131.4	3.938/−2	2.311	1.010
20.0	36.21	3.590/−2	206.1	4.659/−2	2.156	0.6005
30.0	42.51	4.580/−2	262.9	5.049/−2	2.083	0.4548
50.0	51.09	6.180/−2	354.7	5.507/−2	2.005	0.3245
100.0	63.77	8.755/−2	502.6	6.074/−2	1.917	0.2189
250.0	82.20	0.1344	771.5	6.743/−2	1.825	0.1358
500.0	97.29	0.1803	1035	7.199/−2	1.768	9.806/−2
1000.0	113.3	0.2364	1357	7.621/−2	1.720	7.276/−2
3000.0	140.4	0.3492	2005	8.234/−2	1.652	4.730/−2
10^4	172.4	0.5110	2933	8.845/−2	1.591	3.114/−2
3×10^4	203.6	0.6986	4010	9.358/−2	1.544	2.210/−2
10^5	239.7	0.9535	5473	9.884/−2	1.498	1.571/−2
3×10^5	274.4	1.236	7095	0.1034	1.461	1.182/−2
10^6	314.2	1.606	9219	0.1081	1.425	8.873/−3
10^7	395.3	2.510	14410	0.1164	1.365	5.438/−3
10^8	482.4	3.708	21280	0.1242	1.313	3.542/−3
10^9	575.2	5.242	30090	0.1315	1.269	2.421/−3
10^{10}	673.3	7.154	41070	0.1383	1.229	1.718/−3

This fits the last value of J in the table to 0.03% but overestimates the value of J at $\delta = 63.77$ by 2.5%. By comparison, eq. (13.6.2) underestimates the value of J at this same δ by 3.7%. At $\delta = 97.29$, eq. (13.6.4) overestimates J by 0.7%. The situation is improved by adding a cubic to eq. (13.6.3) that force fits J at $\delta = 113.3$.

$$J = 2.969\,52\,\delta\,(1 + 0.024\,412\,\delta\,(1 + x.xxx\,\delta)) \tag{13.6.5}$$

Table 13.6.1 contains additional quantities relevant to an estimate of the conductive opacity for pure helium ($\mu_e = 2$, $Z_i = 2$) at a typical helium-burning temperature of $T_6 = 100$. The fifth column gives $(\lambdabar_e / 2R_D)^2$, where $\lambdabar_e$ is the deBroglie wavelength of a typical electron defined by eqs. (4.7.16) and (4.7.18) in Volume I and R_D is the Debye radius defined by eq. (4.13.39) in Volume 1 when electrons are not degenerate and by eqs. (4.17.43)–(4.17.46) in Volume 1 when electrons are partially degenerate. The sixth column gives the cross-sectional factor $\langle \Theta_{ei} \rangle$ for the scattering of a typical electron from a helium nucleus. It has been obtained by setting $\lambda = \lambdabar_e$ in eq. (13.3.35):

$$\langle \Theta_{ei} \rangle = \Theta(R, \lambdabar_e) = \log_e \left[1 + \left(\frac{2R}{\lambdabar_e} \right)^2 \right] - \left[1 + \left(\frac{\lambdabar_e}{2R} \right)^2 \right]^{-1} . \tag{13.6.6}$$

Finally, the seventh column gives a lower limit to the conductive opacity obtained by setting $\langle \Phi \rangle = \langle \Theta_{ei} \rangle$.

13.6.1 Fits to still more sophisticated estimates of the conductive opacity

Treatments of conduction in stars substantially more sophisticated than those described in this monograph are given by W. B. Hubbard (1966), who takes ion–ion correlations into account, and by M. Lampe (1968a, 1968b), who demonstrates the importance and evaluates the contribution of electron–electron collisions. Analytic fits to electron conductivities calculated by Hubbard and Lampe (1969) have been used in constructing the stellar models presented in this book.

The fits (Iben, 1975) make use of the parameters (Iben, 1968)

$$\frac{\langle \Theta \rangle}{\kappa_c T_6}, \tag{13.6.7}$$

$$\delta = \frac{\rho}{\mu_e T_6^{3/2}}, \tag{13.6.8}$$

and

$$10^\alpha = \left(\frac{\langle \lambdabar_e \rangle}{2R_D} \right)^2 = 0.009\,2435\,\delta \left(\frac{n_e kT}{P_e} \right) \frac{1}{T_6^{1/2}} \left[\mu_e \sum_i Y_i Z_i^2 + \frac{1}{n_e} \left| \frac{dn_e}{d\alpha} \right| \right]. \tag{13.6.9}$$

The first parameter is a function of the second parameter:

$$\frac{\langle\Theta\rangle}{\kappa_c T_6} = f_1 = 10^{-3.2862}\,\delta\,(1 + 0.024\,417\,\delta), \ \log\delta < 0.645, \tag{13.6.10}$$

$$\frac{\langle\Theta\rangle}{\kappa_c T_6} = f_2 = 10^{-3.292\,43}\,\delta\,(1 + 0.028\,04\,\delta), \ 0.645 < \log\delta < 2.0, \tag{13.6.11}$$

$$\frac{\langle\Theta\rangle}{\kappa_c T_6} = f_4 = 10^{-4.809\,46}\,\delta^2\left(1 + \frac{9.376}{\eta_0^2}\right), \ \log\delta > 2.5, \tag{13.6.12}$$

and

$$\log\left(\frac{\langle\Theta\rangle}{\kappa_c T_6}\right) = \log f_3 = 2\log f_2(2.5 - \log\delta) + 2\log f_4(\log\delta - 2), \ 2 < \log\delta < 2.5, \tag{13.6.13}$$

where f_2 and f_4 are given by eqs. (13.6.11) and (13.6.12), respectively, and

$$\eta_0 = 10^{-0.522\,55}\,\delta^{2/3}. \tag{13.6.14}$$

In the third parameter,

$$\frac{P_e}{n_e kT} = g_1 = 1 + 0.021\,876\,\delta, \ \log\delta < 1.5, \tag{13.6.15}$$

$$\frac{P_e}{n_e kT} = g_2 = 0.4\,\eta_0\left(1 + \frac{4.1124}{\eta_0^2}\right), \ \log\delta > 2.0, \tag{13.6.16}$$

$$\log\left(\frac{P_e}{n_e kT}\right) = 2\log g_1(2.0 - \log\delta) + 2\log g_2(\log\delta - 1.5), \ 1.5 < \log\delta < 2, \tag{13.6.17}$$

$$\frac{\partial n_e}{\partial\eta} = 1 - 0.01\,\delta\,(2.8966 - 0.034838\,\delta), \ \text{when } \delta < 40, \tag{13.6.18}$$

and

$$\frac{\partial n_e}{\partial\eta} = \frac{1.5}{\eta_0}\left(1 - \frac{0.8225}{\eta_0^2}\right), \ \text{when } \delta > 40. \tag{13.6.19}$$

Fitting to the Hubbard and Lampe (1969) estimates of the conductive opacity for pure hydrogen gives the quantity Θ_X as a function of the parameter α defined by eq. (13.6.9):

$$\log\Theta_X = 1.048 - 0.124\,\alpha, \ \alpha \le -3, \tag{13.6.20}$$

$$\log\Theta_X = 0.13 - \alpha\,(0.745 + 0.105\,\alpha), \ -3 < \alpha \le -1, \tag{13.6.21}$$

and

$$\log\Theta_X = 0.185 - 0.585\,\alpha; \ \alpha > -1. \tag{13.6.22}$$

For pure helium, the fit is described by

$$\log \Theta_Y = 0.937 - 0.111\,\alpha, \quad \alpha \le -3, \tag{13.6.23}$$

$$\log \Theta_Y = 0.24 - \alpha\,(0.55 + 0.0689\,\alpha), \quad -3 < \alpha \le 0, \tag{13.6.24}$$

$$\log \Theta_Y = 0.24 - 0.6\,\alpha, \quad \alpha > 0, \tag{13.6.25}$$

and for pure carbon, the fit gives

$$\log \Theta_C = 1.27 - 0.1\,\alpha, \quad \alpha \le -2.5, \tag{13.6.26}$$

$$\log \Theta_C = 0.727 - \alpha\,(0.511 + 0.0778\,\alpha), \quad -2.5 < \alpha \le 0.5, \tag{13.6.27}$$

$$\log \Theta_C = 0.843 - 0.785\,\alpha, \quad \alpha > 0.5. \tag{13.6.28}$$

In terms of these quantities, the electron conductivity for a mixture of elements is given by

$$\kappa_c = \frac{(X\Theta_X + Y\Theta_Y + Z_C\Theta_C)}{T_6\,f}, \tag{13.6.29}$$

where X is the abundance by mass of hydrogen, Y is the abundance by mass of helium,

$$Z_C = \frac{1}{3}\sum \frac{X_i Z_i^2}{A_i}, \tag{13.6.30}$$

where the sum is over elements other than hydrogen and helium, and

$$f = \frac{\langle\Theta\rangle}{\kappa_c T_6} \tag{13.6.31}$$

is given by eqs. (13.6.10)–(13.6.14).

When electrons are relativistically degenerate, the estimates of V. Canuto (1970) are appropriate. A reasonably good fit is (Iben, 1975)

$$\kappa_c = 6.753 \times 10^{-8}\,\frac{T_6^2}{\epsilon_F^{2.5}(1 + \epsilon_F)}\,\frac{Z_\beta}{G}, \tag{13.6.32}$$

where

$$\log G = \left[(0.873 - 0.298 \log Z_\alpha) + (0.333 - 0.168 \log Z_\alpha)\,M\right]\left(1 - \frac{1}{(1+\gamma)^{0.85}}\right), \tag{13.6.33}$$

$$M = \min(1, 0.5 + \log \epsilon_F), \tag{13.6.34}$$

$$\gamma = 22.76\frac{\rho^{1/3}}{T_6}\,Z_\alpha, \tag{13.6.35}$$

$$Z_\alpha = \sum_i \frac{Z_i^2 X_i}{A_i^{1/3}}, \tag{13.6.36}$$

and

$$Z_\beta = \sum \frac{X_i Z_i^2}{A_i}. \tag{13.6.37}$$

In the models produced for this book, the fit to the Hubbard & Lampe conductive opacity has been used for all densities less than 10^6 g cm^{-3} and the fit to the Canuto conductive opacity has been used for all densities greater than 2×10^6 g cm^{-3}. For 10^6 g cm$^{-3} < \rho < 2 \times 10^6$ g cm^{-3},

$$\log \kappa_c = (1 - f_{\text{int}}) \log \kappa_{\text{Hubbard--Lampe}} + f_{\text{int}} \log \kappa_{\text{Canuto}}, \tag{13.6.38}$$

where

$$f_{\text{int}} = 0.5 \{1 - \cos \left[\pi (\log \rho - 6) / \log 2\right]\}. \tag{13.6.39}$$

Bibliography and references

Max Abraham & Richard Becker, *Theorie Der Electricität, Band II: Electronentheorie* (Leipzig: Teubner), 1933.

Vittorio Canuto, *ApJ*, **159**, 641, 1970.

Herbert Goldstein, *Classical Mechanics*, (Reading, Mass: Addison-Wesley), 1965.

W. B. Hubbard, *ApJ*, **146**, 858, 1966.

W. B. Hubbard & Martin Lampe, *ApJS*, **18**, 297, 1969.

Icko Iben, Jr., *ApJ*, **154**, 557, 1968; **196**, 525, 1975.

Martin Lampe, *Phys. Rev.*, **170**, 306, 1968a; **174**, 276, 1968b.

T. D. Lee, *ApJ*, **111**, 625, 1950.

R. E. Marshak, *ApJ*, **92**, 321, 1940.

R. E. Marshak, *Ann. New York Acad. Sci.*, **41**, 49, 1941.

Leon Mestel, *Proc. Cam. Phil. Soc.*, **46**, 331, 1950.

Beta decay and electron capture in stars at high densities

Beta-decay and electron-capture interactions of the sort that are familiar in terrestrial contexts play interesting roles during several phases of stellar evolution. For example, at the highest densities in the Sun, the rate at which ^{7}Be captures free electrons is much larger than the rate at which the ^{7}Be$(p, \gamma)^8$B reaction occurs and thus prevents the ^{8}B$(e^+ \nu_e)^8$Be reaction, with an attendant loss of $\sim$8.35 MeV in the form of neutrino energy per reaction, from acting as a huge energy drain during the transformation of protons into alpha particles (see Section 6.8 in Volume 1).

In more advanced stages of evolution, in regions where electron degeneracy becomes large enough that electron Fermi energies are in the MeV range, an isotope which is unstable against electron decay in the terrestrial laboratory can become stable in a star and the isotope which is stable under laboratory conditions can become the unstable isotope in the star, capturing from the Fermi sea electrons which have energies larger than the difference between the rest mass energies of the two isotopes in the terrestrial laboratory.

As an example, consider the reaction

$$^{14}\text{C} \rightarrow {}^{14}\text{N} + e^- + \bar{\nu}_e \tag{14.0.1}$$

and its inverse

$$^{14}\text{N} + e^- \rightarrow {}^{14}\text{C} + \nu_e. \tag{14.0.2}$$

In these expressions, ν_e denotes a neutrino, $\bar{\nu}_e$ an antineutrino, and e^- an electron. In the terrestrial laboratory, the half life of the electron-decay reaction is $t_{1/2} = 5690$ yr. The rest mass energy of ^{14}C is 0.667 MeV larger than the rest mass energy of ^{14}N; in the decay process, 0.511 MeV is used up in creating the electron and the remaining 0.156 MeV is shared by the leptons as kinetic energy. In a star, at densities larger than $\sim$10^6 g cm^{-3}, the kinetic energy of an electron at the top of the Fermi sea is larger than 0.156 MeV. Most of the electron levels at energies below the Fermi energy are occupied, so there is very little phase space available for the electron emitted in the ^{14}C $\rightarrow ^{14}$N$ + e^- + \bar{\nu}_e$ reaction. On the other hand, electrons with energies larger than the Fermi energy can be captured by ^{14}N nuclei, and ^{14}C actually becomes more stable than ^{14}N.

In low mass main sequence stars, in matter where hydrogen burning is nearing completion, most of the original CNO nuclei have been converted into ^{14}N. During the subsequent shell hydrogen-burning phase of evolution along the red giant branch, as the ashes of hydrogen burning are deposited into the contracting hydrogen-exhausted core, densities in central regions become large enough that electron captures convert some ^{14}N into ^{14}C which can react with alpha particles to form ^{18}O and a gamma ray, perhaps releasing nuclear energy more rapidly than if only the $^{14}N(\alpha, \gamma)^{17}(e^+\nu_e)^{18}O$ reactions were operative. At one time, it was even thought that the $^{14}C(\alpha, \gamma)^{18}O$ reaction might proceed more rapidly than the triple alpha reaction and trigger the helium-flash episodes that terminate the evolution of low mass stars upward along the first red giant branch (Kaminisi, Arai, & Yoshinaga, 1975). However, as described in Chapter 11 in Volume 1, neutrino losses by the plasma process (see Chapter 15 in this volume) maintain central regions where densities are highest at temperatures significantly below the maximum off center temperatures where the triple alpha process proceeds most rapidly and, as demonstrated by Robert G. Spulak, Jr. (1980), the triple alpha process survives as the terminator of red giant branch evolution.

In the electron-degenerate carbon–oxygen cores of the initially most massive AGB stars (masses in the range $\sim$8.5–10.5 $M_\odot$), with densities in the millions of g cm^{-3} and temperatures in the hundreds of millions of degrees, pairs of isobars can participate in the Urca process (Gamow & Schönberg, 1941). This process occurs when the electron Fermi energy is near the threshold energy for capture on an isobar which is stable against beta decay in the laboratory. In the star, the isobar which is unstable against electron decay in the laboratory deposits electrons into unoccupied energy levels below the Fermi energy and is transformed into the other isobar which can capture electrons from occupied levels above the Fermi energy. The neutrino emitted in the capture and the antineutrino emitted in the decay are lost from the star and the rate of neutrino energy loss is proportional to the square of the threshold energy and to the fourth power of the temperature, as described in Section 14.7.

In electron-degenerate matter that is experiencing carbon burning, the dominant Urca-loss pairs are ^{21}F–^{21}Ne, ^{23}F–^{23}Na, and ^{25}Ne–^{25}Na, the abundances of these Urca-active pairs being increased by one to three orders of magnitude during carbon burning (Iben, 1978). The evolution of AGB models with initial masses in the range 9–11 $M_\odot$ is discussed in a series of papers culminating in a paper by Claudio Ritossa, Iben, and Enrique Garćia Berro (1999). In these papers, it is shown that the beta unstable member of an Urca pair decays in the outer regions of a convective shell and electron capture on the beta-stable member occurs near the base of the shell. The net result is a TPAGB star with an electron-degenerate ONe core. In an earlier investigation (Iben, 1978), it is demonstrated that, as the mass of the electron-degenerate core of a white dwarf accreting matter from a binary companion approaches the Chandrasekhar limit, thermal oscillations involving convective Urca-pair processes set in. Although it is not enunciated explicitly, one may infer that energy losses by the Urca processes cannot prevent a final explosive event that converts the accreting white dwarf into a type Ia supernova.

14.1 The formalism

In this section, an abbreviated version of weak interaction theory is adopted. This version has the virtue of greater transparency than the more rigorous theory sketched in Chapter 15, and it is suitable for use in many stellar astrophysical contexts. The differential probability per unit time $d\omega$ that a nucleus unstable against electron decay will emit an electron and an antineutrino into specific linear momentum states is approximated by

$$d\omega = \frac{2\pi}{\hbar} g_{weak}^2 \left|\langle M_{decay}\rangle\right|^2 d\rho_{lepton}, \tag{14.1.1}$$

where g_{weak} is a coupling constant between nuclei and leptons (see the discussion in Sections 6.1 and 15.1), $d\rho_{lepton}$ is the density of states for the leptons, and

$$\langle M_{decay}\rangle = \int \Psi_{f,nuc}^* \Psi_e^* \Psi_{\bar{\nu}_e}^* \Psi_{i,nuc} \, d\tau_{nuc}, \tag{14.1.2}$$

is a matrix element between the initial and final particle states which are represented by the spatial wave functions Ψ. In eq. (14.1.2), $\Psi_{f,nuc}$ is the final nuclear wave function (^{14}N in eq. (14.0.1)), Ψ_e is the wave function of the electron, $\Psi_{\bar{\nu}_e}$ is the wave function of the antineutrino, $\Psi_{i,nuc}$ is the initial nuclear wave function (^{14}C in eq. (14.0.1)), and $d\tau_{nuc}$ is a nuclear volume element. All lifetime estimates here are proportional to experimental lifetimes and are therefore independent of the size of the coupling constant g_{weak} and of the details of the nuclear wave functions.

In the simplest approximation, the lepton wave functions can be taken as plane waves, with the electron wave function being modified by the Coulomb interaction between the emitted electron and protons in the daughter nucleus. For the antineutrino,

$$\Psi_{\bar{\nu}_e} = \frac{\exp\left(i\mathbf{k}_{\bar{\nu}_e} \cdot \mathbf{r}\right)}{\sqrt{V}}, \tag{14.1.3}$$

where $\mathbf{k}_{\bar{\nu}_e}$ is a wave number and V is a volume of arbitrary size. Within the volume of the daughter nucleus, the electron wave function may be approximated by

$$\Psi_e = \frac{\exp\left(i\mathbf{k}_e \cdot \mathbf{r}\right)}{\sqrt{V}} C(Z, E_e), \tag{14.1.4}$$

where

$$C^2(Z, E_e) = \left|\frac{2\pi\eta}{1 - \exp(-2\pi\eta)}\right|, \tag{14.1.5}$$

$$\eta = Z \frac{e^2}{\hbar v_e} = Z \frac{e^2}{\hbar c} \frac{c}{v_e} \sim \frac{Z}{137} \frac{c}{v_e}, \tag{14.1.6}$$

Z is the atomic number of the daughter nucleus ($Z = 7$ in the prototype, eq.(14.0.1)), and v_e is the speed of the electron.

When the beta transition is of the allowed variety (i.e., not inhibited by spin and parity considerations), the variation of the lepton wave functions over the volume of the overlapping nuclear wave functions may be neglected, and one may write

$$\langle M_{decay}\rangle = \frac{C^2}{V^2} \langle M_{nuc}\rangle, \tag{14.1.7}$$

where

$$\langle M_{nuc}\rangle = \int \Psi_{f,nuc}^* \Psi_{i,nuc} \, d\tau_{nuc}. \tag{14.1.8}$$

The incremental density of final states may be written as

$$d\rho_{\text{lepton}} = V \frac{4\pi}{h^3} p_e^2 dp_e \, V \frac{4\pi}{h^3} p_{\bar{\nu}e}^2 dp_{\bar{\nu}e}/dE_f, \qquad (14.1.9)$$

where

$$E_f = E_{\text{nuc}} = cp_{\bar{\nu}e} + E_e \qquad (14.1.10)$$

is the difference between the rest mass energies of the parent and daughter nuclei, and p_e, E_e, $p_{\bar{\nu}e}$, and $cp_{\bar{\nu}e}$ are, respectively, the momentum and energy of the electron and antineutrino. Since $dp_{\bar{\nu}e}/dE_f = 1/c$,

$$d\rho_{\text{lepton}} = V^2 \left(\frac{4\pi}{h^3}\right)^2 p_e^2 dp_e \, p_{\bar{\nu}}^2 \frac{1}{c} = V^2 \left(\frac{4\pi}{h^3}\right)^2 p_e^2 dp_e \, (E_{\text{nuc}} - E_e)^2 \frac{1}{c^3}. \qquad (14.1.11)$$

Making the conversions

$$\epsilon = \frac{E_e}{m_e c^2} \qquad (14.1.12)$$

and

$$p = \frac{p_e}{m_e c} = \sqrt{\epsilon^2 - 1}, \qquad (14.1.13)$$

and using the fact that $p\,dp = \epsilon\,d\epsilon$, one has

$$d\rho_{\text{lepton}} = V^2 \left(\frac{4\pi}{h^3}\right)^2 m_e^5 c^4 \, (\epsilon_{\text{nuc}} - \epsilon)^2 \left(\epsilon^2 - 1\right)^{1/2} \epsilon \, d\epsilon. \qquad (14.1.14)$$

Putting all of the factors together, and integrating over electron energies, one has finally that the lifetime $t_{\text{decay,lab}}$ for electron decay in the laboratory is

$$\frac{1}{t_{\text{decay, lab}}} = \frac{2\pi}{\hbar} g_{\text{weak}}^2 \, |\langle M_{\text{nuc}}\rangle|^2 \left(\frac{4\pi}{h^3}\right)^2 m_e^5 c^4$$

$$\times \int_1^{\epsilon_{\text{nuc}}} C^2(Z, \epsilon) \, (\epsilon_{\text{nuc}} - \epsilon)^2 \left(\epsilon^2 - 1\right)^{1/2} \epsilon \, d\epsilon. \qquad (14.1.15)$$

In the nuclear physics literature, the integral in eq. (14.1.15) is designated by the letter f and the product $f t_{1/2}$, where $t_{1/2} = \log_e 2 \, t_{\text{decay, lab}}$ is the half life in the laboratory, is an inverse measure of the square of the nuclear matrix element. Here, in order to avoid confusion with the occupation number f which appears in distribution functions for particles in thermodynamic equilibrium, the integral will be designated as F:

$$F_{\text{decay, lab}} = F(Z, \epsilon_{\text{nuc}}) = \int_1^{\epsilon_{\text{nuc}}} C^2(Z, \epsilon) \, (\epsilon_{\text{nuc}} - \epsilon)^2 \left(\epsilon^2 - 1\right)^{1/2} \epsilon \, d\epsilon, \qquad (14.1.16)$$

so that

$$(f \, t_{1/2})_{\text{lab}} = F_{\text{decay, lab}} \times \left[(\log_e 2) \, t_{\text{decay, lab}}\right]. \qquad (14.1.17)$$

In a star, at temperatures and densities such that most of the electrons are free, electron decay is inhibited by the fact that the lowest lying electron states are occupied, reducing the probabilty of a transition which produces an electron of energy ϵ by the factor $[1 - f(\epsilon, \epsilon_F, T)]$, where (see Chapter 4 in Volume 1)

$$f(\epsilon, \epsilon_F, T) = \frac{1}{\exp[(\epsilon - \epsilon_F)/kT] + 1} \tag{14.1.18}$$

is the probability that an electron state is occupied (the electron Fermi energy ϵ_F and kT are in units of $m_e c^2$). The lifetime $t_{decay,star}$ in the star is thus related to the lifetime $t_{decay,lab}$ in the laboratory by

$$\frac{1}{t_{decay,star}} = \frac{1}{t_{decay,lab}} \frac{F_{decay, star}}{F_{decay, lab}} = \frac{\log_e 2}{(ft_{1/2})_{lab}} F_{decay, star}, \tag{14.1.19}$$

where

$$F_{decay, star} = \int_1^{\epsilon_{nuc}} [1 - f(\epsilon, \epsilon_F, T)] \; C^2(Z, \epsilon) \; (\epsilon_{nuc} - \epsilon)^2 \left(\epsilon^2 - 1\right)^{1/2} \epsilon \, d\epsilon. \tag{14.1.20}$$

The inverse process, electron capture on the nucleus which is stable in the laboratory, can occur in a star in regions where ϵ_F is comparable to or larger than the decay energy in the laboratory.

To determine the electron capture rate, one may proceed exactly as before. The matrix element in the capture-rate analogue of eq. (14.1.2) is

$$\langle M_{capture} \rangle = \int \Psi_{f,nuc}^* \; \Psi_{\nu_e}^* \; \Psi_e \; \Psi_{i,nuc} \, d\tau_{nuc}, \tag{14.1.21}$$

where Ψ_{ν_e}, the wave function of the neutrino, has exactly the same form as eq. (14.1.3), the electron wave function Ψ_e has the same form as eq. (14.1.4), and the nuclear wave functions are exactly the same as before except that their roles are reversed. The incremental density of final states for the neutrino is

$$d\rho_{lepton} = V \frac{4\pi}{h^3} \; p_{\nu_e}^2 dp_{\nu_e}/dE_f = V \frac{4\pi}{h^3} \; p_{\nu_e}^2 \frac{1}{c} = V \frac{8\pi}{h^3} \; m_e^2 c \; (\epsilon_{nuc} - \epsilon)^2, \tag{14.1.22}$$

where ϵ is the energy of an electron prior to capture. The number of electrons of energy between ϵ and $\epsilon + d\epsilon$ which can be captured is

$$V dn_e = V \; f(\epsilon, \epsilon_F, T) \frac{8\pi}{h^3} \; p_e^2 \, dp_e = V \; f(\epsilon, \epsilon_F, T) \frac{4\pi}{h^3} \; (m_e c)^3 \; (\epsilon^2 - 1)^{1/2} \epsilon \, d\epsilon, \tag{14.1.23}$$

where $f(\epsilon, \epsilon_F)$ is the electron occupation number given by eq. (14.1.18). Putting all factors together, one has for the capture rate in a star:

$$\frac{1}{t_{capture,star}} = \frac{2\pi}{\hbar} \; g_{weak}^2 \; |\langle M_{nuc} \rangle|^2 \left(\frac{4\pi}{h^3}\right)^2 m_e^5 c^4 \; F_{capture, star}, \tag{14.1.24}$$

where $\langle M_{\text{nuc}} \rangle$ is given by eq. (14.1.8) and

$$F_{\text{capture, star}} = \int_{\epsilon_{\text{nuc}}}^{\infty} f(\epsilon, \epsilon_{\text{F}}, T) \, C^2(Z, \epsilon) \, (\epsilon - \epsilon_{\text{nuc}})^2 \left(\epsilon^2 - 1\right)^{1/2} \epsilon \, d\epsilon. \qquad (14.1.25)$$

Using eqs. (14.1.15)–(14.1.17) and (14.1.23), one has finally that

$$\frac{1}{t_{\text{capture, star}}} = \frac{1}{t_{\text{decay, lab}}} \frac{F_{\text{capture, star}}}{F_{\text{decay, lab}}} = \frac{\log_e 2}{(f t_{1/2})_{\text{lab}}} F_{\text{capture, star}}. \qquad (14.1.26)$$

A comparison of eqs. (14.1.20) and (14.1.25) when $\epsilon_{\text{F}} = \epsilon_{\text{f}}$ reveals the nature of the Urca process. The functions f and $1 - f$ are symmetric about ϵ_{F}; free electrons in the energy interval $\epsilon = \epsilon_{\text{F}}$ to $\epsilon = \epsilon_{\text{F}} + $ a few kT can be captured by isobars characterized by (Z, N) to produce isobars characterized by $(Z - 1, N + 1)$ (plus neutrinos); isobars of type $(Z - 1, N + 1)$ can produce isobars of type (Z, N) (plus antineutrinos) by emitting electrons of energies in the range $\epsilon = \epsilon_{\text{F}} - $ a few kT to ϵ_{F}. Thus, there is a continuous shuttling back and forth between isobars of the two types, and, in each transition, energy is given to neutrinos (of one type or the other) which, in most instances, escape from the star.

Since the energy interval into which decaying isobars can deposit electrons increases with T and since the energy interval from which electrons can be captured increases with T, the rate of neutrino energy loss by the Urca process for any pair of participating isobars increases with T. For a given Urca pair characterised by a threshold energy ϵ_{f}, it is clear that the process is important only over a range of electron Fermi energies such that $|\epsilon_{\text{F}} - \epsilon_{\text{f}}| = $ a few kT. For Fermi energies below this range, the more stable isobar is the one which can capture electrons and, for Fermi energies above this range, the electron-emitting isobar is the more stable one.

A word about the quantity ϵ_{nuc} is in order. The Q values attached to terrestial electron decay reactions do not include the rest mass energy of the electron. Hence, one must make the conversion

$$\epsilon_{\text{nuc}} = \frac{\Delta M_{\text{nuc}} \, c^2}{m_e \, c^2} = \frac{Q(\text{MeV})}{0.511} + 1. \qquad (14.1.27)$$

14.2 Electron capture at high densities

The threshhold energy ϵ_{th} (ϵ_{nuc} in the previous section) for electron capture on a positron-stable nucleus is typically several MeV ($\epsilon_{\text{th}} > 3$) so, to a good approximation, most of the contributions to the integral in eq. (14.1.26) come from electrons for which

$$\frac{v_e}{c} = \frac{p}{\epsilon} = \frac{(\epsilon^2 - 1)^{1/2}}{\epsilon} = \left(1 - \frac{1}{\epsilon^2}\right)^{1/2} = 1 - \frac{1}{2\epsilon^2} - \frac{1}{8\epsilon^4} - \cdots \sim 1. \qquad (14.2.1)$$

This means that the quantity $(p/\epsilon) C^2(Z, \epsilon)$ (see eqs. (14.1.6) and (14.1.7) for $C^2(Z, \epsilon)$) is a very slowly varying function of ϵ which can be reasonably well approximated by an average

$$\left\langle \frac{p}{\epsilon} C^2(Z, \epsilon) \right\rangle, \tag{14.2.2}$$

estimated at one of the limits of the integral and taken outside of the integral. Thus, instead of numerically integrating an integral such as the one in eq. (14.1.26), one may instead evaluate the simpler integral

$$I^{ec} = \int_{\epsilon_{th}}^{\infty} \left(e^{(\epsilon - \epsilon_F)/kT} + 1 \right)^{-1} (\epsilon - \epsilon_{th})^2 \, \epsilon^2 \, d\epsilon, \tag{14.2.3}$$

and multiply it by the quantity given by eq. (14.2.2), which, in most instances, can be reasonably well approximated by its value at an energy near ϵ_{th}.

When $\epsilon_F < \epsilon_{th}$, since $\epsilon > \epsilon_F$ in eq. (14.3.3), one can write

$$\left(e^{(\epsilon - \epsilon_F)/kT} + 1 \right)^{-1} = e^{-(\epsilon - \epsilon_F)/kT} \left(1 - e^{-(\epsilon - \epsilon_F)/kT} + e^{-2(\epsilon - \epsilon_F)/kT} - \cdots \right)$$

$$= \sum_{n=1}^{\infty} (-1)^{n+1} e^{-n(\epsilon - \epsilon_F)/kT}, \tag{14.2.4}$$

and the integral in eq. (14.2.3) can be written as the sum of a series of integrals

$$I^{ec}_{\epsilon_F < \epsilon_{th}} = \sum_{n=1}^{\infty} (-1)^{n+1} I_n, \tag{14.2.5}$$

where

$$I_n = \int_{\epsilon_{th}}^{\infty} e^{-n(\epsilon - \epsilon_F)/kT} (\epsilon - \epsilon_{th})^2 \, \epsilon^2 \, d\epsilon. \tag{14.2.6}$$

Defining

$$x = \frac{n\epsilon}{kT}, \tag{14.2.7}$$

one has

$$I_n = \left(\frac{kT}{n} \right)^5 \int_{x_{th}=n\epsilon_{th}/kT}^{\infty} e^{-(x-x_F)} (x - x_{th})^2 \, x^2 \, dx$$

$$= \left(\frac{kT}{n} \right)^5 e^{x_F} \int_{x_{th}=n\epsilon_{th}/kT}^{\infty} e^{-x} \left(x^4 - 2x_{th}x^3 + x_{th}x^2 \right) \, dx. \tag{14.2.8}$$

After integration, the three terms in the last integral of eq. (14.2.8) become

$$- \left[e^{-x} \left(x^4 + 4x^3 + 12x^2 + 24x + 24 \right) \right]$$

$$+ 2x_{th} \left[e^{-x} \left(x^3 + 3x^2 + 6x + 6 \right) \right]$$

$$- x_{th}^2 \left[e^{-x} \left(x^2 + 2x + 2 \right) \right], \tag{14.2.9}$$

and they are to be evaluated between the limits $x = x_{th} = n\epsilon_{th}/kT$ and $x = \infty$. All terms vanish at the upper limit and, at the lower limit, cancellations leave only

$$e^{-x_{th}} \left(2x_{th}^2 + 12x_{th} + 24 \right). \tag{14.2.10}$$

Thus,

$$I_n = \left(\frac{kT}{n} \right)^5 e^{-(x_{th} - x_F)} \left(2x_{th}^2 + 12x_{th} + 24 \right) \tag{14.2.11}$$

$$= 2 \, \epsilon_{th}^5 \left(\frac{kT}{\epsilon_{th}} \right)^3 e^{-n(\epsilon_{th} - \epsilon_F)/kT} \left(\frac{1}{n^3} + 6 \frac{1}{n^4} \left(\frac{kT}{\epsilon_{th}} \right) + 12 \frac{1}{n^5} \left(\frac{kT}{\epsilon_{th}} \right)^2 \right). \tag{14.2.12}$$

Summing up over n, one obtains

$$I_{\epsilon_F < \epsilon_{th}}^{ec} = 2 \, \epsilon_{th}^5 \left(\frac{kT}{\epsilon_{th}} \right)^3 \left(\Sigma_3 + 6 \, \Sigma_4 \left(\frac{kT}{\epsilon_{th}} \right) + 12 \, \Sigma_5 \left(\frac{kT}{\epsilon_{th}} \right)^2 \right), \tag{14.2.13}$$

where

$$\Sigma_K = \sum_{n=1}^{\infty} (-1)^{n+1} \frac{e^{-n|\epsilon_{th} - \epsilon_F|/kT}}{n^K}. \tag{14.2.14}$$

The absolute value of $(\epsilon_{th} - \epsilon_F)$ is used in the definition of the functions Σ_K because the same functions arise in integrations when $\epsilon_{th} < \epsilon_F$.

When $\epsilon_F = \epsilon_{th}$,

$$I_{\epsilon_F = \epsilon_{th}}^{ec} = 2 \, \epsilon_{th}^5 \left(\frac{kT}{\epsilon_{th}} \right)^3 \left(S_3 + 6 \, S_4 \left(\frac{kT}{\epsilon_{th}} \right) + 12 \, S_5 \left(\frac{kT}{\epsilon_{th}} \right)^2 \right), \tag{14.2.15}$$

where

$$S_K = \sum_{n=1}^{\infty} (-1)^{n+1} \frac{1}{n^K}. \tag{14.2.16}$$

Specifically,

$$S_2 = 0.822\,467, \quad S_3 = 0.901\,543, \quad S_4 = 0.947\,033,$$

$$S_5 = 0.972\,120, \quad \text{and} \quad S_6 = 0.985\,551. \tag{14.2.17}$$

The final approximation to the capture-rate when $\epsilon_F < \epsilon_{th}$ is

$$\frac{1}{t_{\text{capture, star}}} = \frac{\log_e 2}{(ft_{1/2})_{\text{lab}}} F_{\epsilon_F < \epsilon_{th}}^{ec}, \tag{14.2.18}$$

where

$$F_{\epsilon_F < \epsilon_{th}}^{ec} = \left\langle \frac{p}{\epsilon} C^2(Z, \epsilon) \right\rangle I_{\epsilon_F < \epsilon_{th}}^{ec}, \tag{14.2.19}$$

and

$$\left\langle \frac{p}{\epsilon} \, C^2(Z, \epsilon) \right\rangle \sim \frac{p_{\text{th}}}{\epsilon_{\text{th}}} \, C^2(Z, \epsilon_{\text{th}}). \tag{14.2.20}$$

When $\epsilon_{\text{F}} > \epsilon_{\text{th}}$, the integrals in eqs. (14.2.6) and (14.2.3) may be broken into three parts, an integral from $\epsilon = \epsilon_{\text{F}}$ to $\epsilon = \infty$ which makes use of eq. (14.2.4), and two integrals from $\epsilon = \epsilon_{\text{th}}$ to $\epsilon = \epsilon_{\text{F}}$ which make use of the fact that

$$\left(e^{(\epsilon - \epsilon_{\text{F}})/kT} + 1 \right)^{-1} = \left(1 + e^{-(\epsilon_{\text{F}} - \epsilon)/kT} \right)^{-1}$$

$$= \left(1 - e^{-(\epsilon_{\text{F}} - \epsilon)/kT} + e^{-2(\epsilon_{\text{F}} - \epsilon)/kT} - \cdots \right)$$

$$= 1 - \sum_{n=1}^{\infty} (-1)^{n+1} e^{-n(\epsilon_{\text{F}} - \epsilon)/kT}. \tag{14.2.21}$$

Defining

$$I_{\epsilon_{\text{F}} > \epsilon_{\text{th}}}^{\text{ec}} = I_{0>}^{\text{ec}} + I_{1>}^{\text{ec}}, \tag{14.2.22}$$

where

$$I_{0>}^{\text{ec}} = \int_{\epsilon_{\text{th}}}^{\epsilon_{\text{F}}} (\epsilon - \epsilon_{\text{th}})^2 \, \epsilon^2 \, d\epsilon$$

$$= \frac{\epsilon_{\text{F}}^5 - \epsilon_{\text{th}}^5}{5} - 2\epsilon_{\text{th}} \frac{\epsilon_{\text{F}}^4 - \epsilon_{\text{th}}^4}{4} + \epsilon_{\text{th}}^2 \frac{\epsilon_{\text{F}}^3 - \epsilon_{\text{th}}^3}{3}, \tag{14.2.23}$$

and

$$I_{1>}^{\text{ec}} = -\sum_{n=1}^{\infty} (-1)^{n+1} \left(I_n^- - I_n^+ \right), \tag{14.2.24}$$

one has

$$I_n^- = \int_{\epsilon_{\text{th}}}^{\epsilon_{\text{F}}} e^{-n(\epsilon_{\text{F}} - \epsilon)/kT} \, (\epsilon - \epsilon_{\text{th}})^2 \, \epsilon^2 \, d\epsilon, \tag{14.2.25}$$

and

$$I_n^+ = \int_{\epsilon_{\text{F}}}^{\infty} e^{-n(\epsilon - \epsilon_{\text{F}})/kT} \, (\epsilon - \epsilon_{\text{th}})^2 \, \epsilon^2 \, d\epsilon. \tag{14.2.26}$$

An approximation to the component which depends on temperature only through ϵ_{F} which is better than given by eq. (14.2.23) is the integral $\bar{I}_0$ obtained from eq. (14.2.26) without making the approximation $p = \epsilon$:

$$\bar{I}_{0>}^{\text{ec}} = \int_{\epsilon_{\text{th}}}^{\epsilon_{\text{F}}} (\epsilon - \epsilon_{\text{th}})^2 \, p^2 \, dp = \int_{\epsilon_{\text{th}}}^{\epsilon_{\text{F}}} (\epsilon - \epsilon_{\text{th}})^2 \, p \, \epsilon \, d\epsilon. \tag{14.2.27}$$

Making use of the facts that

$$\int p\epsilon^3 d\epsilon = \frac{p^5}{5} + \frac{p^3}{3},$$

(14.2.28)

$$\int p\epsilon^2 d\epsilon = \frac{\epsilon p^3}{4} + \frac{\epsilon p}{8} - \frac{1}{8}\log_e |\epsilon + p|,$$

(14.2.29)

and

$$\int p\epsilon d\epsilon = \frac{p^3}{3},$$

(14.2.30)

one has

$$\bar{I}_{0>}^{ec} = \left(\frac{p_F^5 - p_{th}^5}{5}\right) + \left(\frac{p_F^3 - p_{th}^3}{3}\right)$$

$$- 2\epsilon_{th}\left[\left(\frac{\epsilon_F p_F^3}{4} - \frac{\epsilon_{th} p_{th}^3}{4}\right) + \frac{1}{8}\left(\epsilon_F p_F - \epsilon_{th} p_{th}\right) - \frac{1}{8}\log_e\left(\frac{\epsilon_F + p_F}{\epsilon_{th} + p_{th}}\right)\right]$$

$$+ \epsilon_{th}^2 \frac{p_F^3 - p_{th}^3}{3}.$$

(14.2.31)

After rearrangement, and making use of the fact that $p_F\epsilon_F^3 - p_{th}\epsilon_{th}^3 = p_F^3\epsilon_F - p_{th}^3\epsilon_{th} + p_F\epsilon_F - p_{th}\epsilon_{th}$, eq. (14.2.31) can be written as

$$\bar{I}_{0>}^{ec} = \frac{1}{5}\left(p_F^5 - p_{th}^5\right) + \frac{1}{3}\left(\epsilon_{th}^2 + 1\right)\left(p_F^3 - p_{th}^3\right)$$

$$- \frac{\epsilon_{th}}{2}\left[p_F\epsilon_F\left(\epsilon_F^2 - \frac{1}{2}\right) - p_{th}\epsilon_{th}\left(\epsilon_{th}^2 - \frac{1}{2}\right) - \frac{1}{2}\log_e\left(\frac{\epsilon_F + p_F}{\epsilon_{th} + p_{th}}\right)\right].$$

(14.2.32)

The temperature-dependent quantities defined by eqs. (14.2.25) and (14.2.26) and used in eq. (14.2.24) are best evaluated together since cancellations occur between terms of the same power of ϵ. The result is

$$I_{1>}^{ec} = 4\,\epsilon_{th}^5\left(\frac{kT}{\epsilon_{th}}\right)^2\left(2\frac{\epsilon_F}{\epsilon_{th}} - 1\right)\left(S_2\left(\frac{\epsilon_F}{\epsilon_{th}} - 1\right)\frac{\epsilon_F}{\epsilon_{th}} + 6\,S_4\left(\frac{kT}{\epsilon_{th}}\right)^2\right)$$

$$+ 2\,\epsilon_{th}^5\left(\frac{kT}{\epsilon_{th}}\right)^3\left(\Sigma_3 - 6\,\Sigma_4\left(\frac{kT}{\epsilon_{th}}\right) + 12\,\Sigma_5\left(\frac{kT}{\epsilon_{th}}\right)^2\right),$$

(14.2.33)

where the Σ_K are given by eq. (14.2.14) and the S_K are given by eqs. (14.2.16) and (14.2.17). The terms in eq. (14.2.33) involving even powers of kT are dominated by positive contributions from $\epsilon > \epsilon_F$ which offset negative contributions from $\epsilon < \epsilon_F$. When $\epsilon_F = \epsilon_{th}$, eq. (14.2.33) is identical with eq. (14.2.15). As ϵ_F increases, the terms involving the functions Σ_K fall off exponentially faster than $e^{-(\epsilon_F - \epsilon_{th})/kT}$.

Gathering all terms together, one has that, when $\epsilon_F > \epsilon_{th}$, the electron-capture rate is given by eq. (14.2.18) with $F^{ec}_{\epsilon_F < \epsilon_{th}}$ replaced by

$$\lambda^{ec}_{\epsilon_F > \epsilon_{th}} = \langle C^2(Z, \epsilon) \rangle_0 \, \bar{I}^{ec}_{0>} + \left\langle \frac{p}{\epsilon} C^2(Z, \epsilon) \right\rangle_1 I^{ec}_{1>}, \tag{14.2.34}$$

where

$$\langle C^2(Z, \epsilon) \rangle_0 \sim C^2(Z, \epsilon_F) \tag{14.2.35}$$

and

$$\left\langle \frac{p}{\epsilon} C^2(Z, \epsilon) \right\rangle_1 \sim \frac{p_F}{\epsilon_F} C^2(Z, \epsilon_F). \tag{14.2.36}$$

It is of interest to examine in detail the rates for the $^{14}C \to {}^{14}N + e^- + \bar{\nu}_e$ reaction, which is actually quite unusual in that, at threshhold, p differs not inconsequentially from ϵ. With $\epsilon_{th} = 1.3053$, one has $p_{th} = \sqrt{\epsilon_{th}^2 - 1} = 0.8389$ and $p_{th}/\epsilon_{th} = 0.6427$. Expanding $p = \epsilon\sqrt{1 - 1/\epsilon^2}$ in powers of $1/\epsilon^2$, eq. (14.2.1) gives $p_{th} = 1.3053 \, (1 - 0.29347 - 0.04306 - 0.012637 - 0.004635 - \cdots)$. Adding successively one term at a time, we have $p_{th} \sim 1.3053, 0.9222, 0.8660, 0.8495, 0.8435$, etc. From eqs. (14.2.7), (14.2.6), and (14.2.1), one has that $2\pi \eta_{th} = 0.4995$ and $C^2(Z, \epsilon_{th}) = 1.2704$. Equation (14.2.20) gives $\left\langle \frac{p}{\epsilon} C^2(Z, \epsilon) \right\rangle \sim 0.8165$, and eqs. (14.2.15)–(14.2.19) then give

$$F^{ec}_{\epsilon_F = \epsilon_{th}} \sim 1.2027 \times 10^{-5} \, T_8^3 \, \left(1 + 0.0814 \, T_8 + 0.00216 \, T_8^2\right). \tag{14.2.37}$$

Using $\log_{10}\left(f t_{1/2}\right)_{lab} = 9.03$, eq. (14.2.18) gives finally, when $\epsilon_F = \epsilon_{th}$,

$$t^{\epsilon_F = \epsilon_{th}}_{capture,\, star} = \frac{4.073 \times 10^6 \, yr}{T_8^3} \, \left(1 + 0.0814 \, T_8 + 0.00216 \, T_8^2\right)^{-1}, \tag{14.2.38}$$

where T_8 is the temperature in units of 10^8 K. At $T_8 = 1$, $t_{capture,\, star}$ is about $540 \times t_{decay,\, lab}$.

14.3 Electron decay at high densities

To examine the electron-decay rate in a star when free electrons are degenerate and $\epsilon_F > \epsilon_{th}$, use eq. (14.2.21) in eq. (14.1.20) to write

$$F^{ed}_{\epsilon_F > \epsilon_{th}} \sim + \left\langle \frac{p}{\epsilon} C^2(Z, \epsilon) \right\rangle_{1>} J^{ed}_{1>}, \tag{14.3.1}$$

where

$$J^{ed}_{1>} = \sum_{n=1}^{\infty} (-1)^{n+1} J_n, \tag{14.3.2}$$

$$J_n = \int_1^{\epsilon_{th}} e^{-n(\epsilon_F - \epsilon)/kT} \, (\epsilon_{th} - \epsilon)^2 \, \epsilon^2 \, d\epsilon, \tag{14.3.3}$$

and the coefficient $\left\{(p_{th}/\epsilon_{th})C^2(Z, \epsilon_{th})\right\}_{1>}$ of $J_{1>}^{ed}$ in eq. (14.3.1) is given approximately by eq. (14.2.20). Making the transformation given by eq. (14.2.7),

$$J_n = \left(\frac{kT}{n}\right)^5 \int_{x_1}^{x_{th}=n\epsilon_{th}/kT} e^{-(x_F-x)} (x_{th}-x)^2 x^2 \, dx$$

$$= \left(\frac{kT}{n}\right)^5 e^{-x_F} \int_{x_1=n/kT}^{x_{th}=n\epsilon_{th}/kT} e^{+x} \left(x^4 - 2x_{th}x^3 + x_{th}x^2\right) dx. \tag{14.3.4}$$

After integration, the three terms in the last integral, obtained by changing the sign of x in eq. (14.2.9) and then reversing the sign of all terms, are

$$e^{+x} \left(x^4 - 4x^3 + 12x^2 - 24x + 24\right)$$

$$- 2x_{th} \, e^{+x} \left(x^3 - 3x^2 + 6x - 6\right)$$

$$+ x_{th}^2 \, e^{+x} \left(x^2 - 2x + 2\right), \tag{14.3.5}$$

and they are to be evaluated between the limits $x = x_1 = n/kT$ and $x = x_{th} = n\epsilon_{th}/kT$. The upper limit produces

$$2 \, \epsilon_{th}^2 \left(\frac{n}{kT}\right)^2 e^{n\epsilon_{th}/kT} \left(1 - \frac{6}{n}\frac{kT}{\epsilon_{th}} + \frac{12}{n^2}\left(\frac{kT}{\epsilon_{th}}\right)^2\right) \tag{14.3.6}$$

and the leading term at the lower limit produces

$$e^{n/kT} \epsilon_{th}^2 \left(\frac{n}{kT}\right)^4 \left(1 - \frac{1}{\epsilon_{th}}\right)^2. \tag{14.3.7}$$

Altogether, the result of performing the integration in eq. (14.3.4) can be written as

$$J_n = 2 \, e^{n\epsilon_{th}/kT} \epsilon_{th}^2 \left(\frac{n}{kT}\right)^2 \left(1 - \frac{6}{n}\frac{kT}{\epsilon_{th}} + \frac{12}{n^2}\right)(1 - O(r_n)), \tag{14.3.8}$$

where $O(r_n)$ means of the order of r_n and

$$r_n = \frac{1}{2} \left(1 - \frac{1}{\epsilon_{th}}\right)^2 \left(\frac{n}{kT}\right)^2 e^{-n(\epsilon_{th}-1)/kT}$$

$$= \frac{1}{2} \frac{1}{\epsilon_{th}^2} \left(\frac{5.930 \, (\epsilon_{th} - 1)}{T_9} n\right)^2 \exp\left(-\frac{5.930 \, (\epsilon_{th} - 1)}{T_9} n\right). \tag{14.3.9}$$

In this last expression ϵ_{th} is still in in units of $m_e c^2$, but T_9 is the temperature in units of 10^9 K. In all decays of interest in stellar interiors, r_n is effectively zero, so in eq. (14.3.3) and in subsequent integrations with the same lower limit, the lower limit of integration can be set equal to $-\infty$. Using eqs. (14.3.2), (14.3.3), and (14.3.5), one can write

$$J^{ed}_{\epsilon_F > \epsilon_{th}} = J^{ed}_{1>} = 2\,\epsilon^5_{th} \left(\frac{kT}{\epsilon_{th}}\right)^3 \left(\Sigma_3 - 6\,\Sigma_4 \left(\frac{kT}{\epsilon_{th}}\right) + 12\,\Sigma_5 \left(\frac{kT}{\epsilon_{th}}\right)^2\right), \quad (14.3.10)$$

where the Σ_K are defined by eq. (14.2.14).

The electron decay rate for $\epsilon_F > \epsilon_{th}$ is given by

$$\frac{1}{t_{decay,\,star}} = \frac{\log_e 2}{(ft_{1/2})_{lab}}\, F^{ed}_{\epsilon_F > \epsilon_{th}}, \quad (14.3.11)$$

where $F^{ed}_{\epsilon_F > \epsilon_{th}}$ is defined by eqs. (14.3.1), (14.3.20), and (14.3.10).

Note that $J^{ed}_{\epsilon_F > \epsilon_{th}}$ given by eq. (14.3.10) differs from $I^{ec}_{\epsilon_F < \epsilon_{th}}$ given by eq. (14.2.13) only by the sign of the coefficient of the second term proportional to the function Σ_4. Thus, to first order, the electron-decay probability for the beta-unstable isobar of an Urca pair is the same function of $|\epsilon_F - \epsilon_{th}|$ as is the probability for electron capture by the beta-stable isobar of the pair. In the paradigm reaction $^{14}C \to {}^{14}N + e^- + \bar{\nu}_e$, when $\epsilon_F = \epsilon_{th}$,

$$F^{ed}_{\epsilon_F = \epsilon_{th}} \sim 1.2027 \times 10^{-5}\, T_8^3 \left(1 - 0.0814\, T_8 + 0.002\,16\, T_8^2\right), \quad (14.3.12)$$

and

$$t_{decay,\,star} = \frac{4.073 \times 10^6\,\text{yr}}{T_8^3} \left(1 - 0.0814\, T_8 + 0.002\,16\, T_8^2\right)^{-1}, \quad (14.3.13)$$

which, apart from the sign of the middle term in parentheses, is almost identical with the capture time scale (eq. (14.2.38)). At $T_8 = 1$, $t_{decay,\,star}$ is ~ 460 times larger than the lifetime for decay in the terrestrial laboratory, $t_{decay,\,lab} = 8209$ yr.

When $\epsilon_F < \epsilon_{th}$, with the help of eqs. (14.2.4) and (14.2.21), eq. (14.2.20) yields

$$F^{ed}_{\epsilon_F < \epsilon_{th}} = J^{ed}_{0<} + \left\langle \frac{p}{\epsilon}\, C^2(Z, \epsilon) \right\rangle_1 J^{ed}_{1<}, \quad (14.3.14)$$

where

$$J^{ed}_{0<} = \int_{\epsilon_F}^{\epsilon_{th}} C^2(Z, \epsilon)\, (\epsilon_{th} - \epsilon)^2 \left(\epsilon^2 - 1\right)^{1/2} \epsilon\, d\epsilon \quad (14.3.15)$$

and

$$J^{ed}_{1<} = \sum_{n=1}^{\infty} (-1)^{n+1} \left(J_n^- - J_n^+\right), \quad (14.3.16)$$

with

$$J_n^- = \int_{-\infty}^{\epsilon_F} e^{-n(\epsilon_F - \epsilon)/kT}\, (\epsilon_{th} - \epsilon)^2\, \epsilon^2\, d\epsilon, \quad (14.3.17)$$

and

$$
J_n^+ = \int_{\epsilon_F}^{\epsilon_{th}} e^{-n(\epsilon - \epsilon_F)/kT} \, (\epsilon_{th} - \epsilon)^2 \, \epsilon^2 \, d\epsilon. \tag{14.3.18}
$$

The coefficient of $J_{1<}^{ed}$ in eq. (14.3.14) is given approximately by eq. (14.2.36). Making the transformation given by eq. (14.2.7), and following the argument culminating in eq. (14.3.9), the lower limit on the integral for J_n^- has been set equal to $-\infty$. After integrating and summing, one has

$$
J_{1<}^{ed} = -4 \, \epsilon_{th}^5 \left(\frac{kT}{\epsilon_{th}}\right)^2 \left(2\frac{\epsilon_F}{\epsilon_{th}} - 1\right) \left(\left(\frac{\epsilon_F}{\epsilon_{th}} - 1\right) \frac{\epsilon_F}{\epsilon_{th}} S_2 + 6 \, S_4 \left(\frac{kT}{\epsilon_{th}}\right)^2\right)
$$

$$
+ 2 \, \epsilon_{th}^5 \left(\frac{kT}{\epsilon_{th}}\right)^3 \left(\Sigma_3 + 6 \, \Sigma_4 \left(\frac{kT}{\epsilon_{th}}\right) + 12 \, \Sigma_5 \left(\frac{kT}{\epsilon_{th}}\right)^2\right). \tag{14.3.19}
$$

When $\epsilon_F = \epsilon_{th}$, eq. (14.3.19) reduces to eq. (14.3.10), as it should.

With one exception, the terms for $J_{1<}^{ed}$ in eq. (14.3.19) are of opposite sign to those for $I_{1>}^{ec}$ in eq. (14.2.33). The one exception is the term proportional to Σ_4. When making the comparison between the two equations, however, keep in mind that eq. (14.2.33) is valid only for $\epsilon_F > \epsilon_{th}$, whereas eq. (14.3.19) is valid only for $\epsilon_F < \epsilon_{th}$.

When the electron-degeneracy is high, J_0 can be approximated by

$$
J_{0<}^{ed} \sim \left\langle C^2(Z, \epsilon)\right\rangle_0 \bar{J}_{0<}^{ed}, \tag{14.3.20}
$$

where, with

$$
\bar{\epsilon} = \frac{1}{2} \, (\epsilon_F + \epsilon_{th}), \tag{14.3.21}
$$

one has

$$
\left\langle C^2(Z, \epsilon)\right\rangle_0 \sim C^2(Z, \bar{\epsilon}), \tag{14.3.22}
$$

and $\bar{J}_{0<}^{ed}$ is obtained from $\bar{I}_{0>}^{ec}$ in eq. (14.2.32) simply by exchanging the integration limits:

$$
\bar{J}_{0<}^{ed} = \frac{1}{5} \left(p_{th}^5 - p_F^5\right) + \frac{1}{3} \left(\epsilon_F^2 + 1\right) \left(p_{th}^3 - p_F^3\right)
$$

$$
- \frac{\epsilon_{th}}{2} \left[p_{th}\epsilon_{th} \left(\epsilon_{th}^2 - \frac{1}{2}\right) - p_F\epsilon_F \left(\epsilon_F^2 - \frac{1}{2}\right) - \frac{1}{2} \log_e \left(\frac{\epsilon_{th} + p_{th}}{\epsilon_F + p_F}\right)\right]. \tag{14.3.23}
$$

When the electron-degeneracy is not large, $J_{0<}^{ed}$ approaches the laboratory estimate, $F(Z, \epsilon_f)$ given by eq. (14.1.16), which makes no approximation with regard to the Coulomb factor $C^2(Z, \epsilon)$, and it is advisable to construct a smooth match between the laboratory value and the stellar estimate over some range in a degeneracy parameter.

14.4 Positron decay and general considerations concerning electron capture on a positron emitter

Electron capture on a positron-stable nucleus requires that the electron energy exceeds a threshold energy which, in most instances, is several MeV in magnitude. In contrast, positron-unstable nuclei can capture electrons of any energy, with the added feature that there is a large phase space available for the emitted neutrino, enhancing the probability of capture and the rate of energy loss to neutrinos.

As prototypes, choose the reaction

$$^{13}\text{N} \rightarrow {}^{13}\text{C} + e^+ + \nu_e, \tag{14.4.1}$$

which plays a prominent role in the CN cycle, and its inverse

$$^{13}\text{N} + e^- \rightarrow {}^{13}\text{C} + \nu_e. \tag{14.4.2}$$

In eq. (14.4.1), e^+ denotes a positron. In most circumstances, there is no restriction on the phase space available for positrons, so the positron-decay rate is the same in a star as in the laboratory. Proceeding as before, one obtains

$$\frac{1}{t_{e^+,\,\text{star}}} = \frac{1}{t_{e^+,\,\text{lab}}} = \frac{2\pi}{\hbar}\, g_{\text{weak}}^2 \, |\langle M_{\text{nuc}}\rangle|^2 \left(\frac{4\pi}{h^3}\right)^2 m_e^5 c^4 \, \frac{1}{F_{e^+,\,\text{lab}}}, \tag{14.4.3}$$

where

$$F_{e^+,\,\text{lab}} = \int_1^{\epsilon_{\text{nuc}}} C^2(-Z,\epsilon)\,(\epsilon_{\text{nuc}} - \epsilon)^2 \left(\epsilon^2 - 1\right)^{1/2} \epsilon \, d\epsilon. \tag{14.4.4}$$

In these equations, ϵ is the energy of the positron, ϵ_{nuc} is the difference between the rest mass energies of the two nuclei involved in the transformation, and $\epsilon_{\text{nuc}} - \epsilon$ is the energy of the neutrino. All energies are in units of $m_e c^2$. The factor $C^2(-Z,\epsilon)$ (see eqs. (14.1.5) and (14.1.6)) takes into account the Coulomb repulsion between the ejected positron and the protons in the daughter nucleus of charge Z. Apart from the sign of Z, eq. (14.4.4) is identical with eq. (14.1.16).

In the nuclear physics literature, the Q value quoted for a positron-decay reaction is the kinetic energy of the emitted leptons plus the rest mass energy of *two* electrons. The basic reason for this convention is that, in the laboratory, the emitted positron annihilates with an electron to form two or three γ rays which are ultimately converted into heat. Thus,

$$Q = \Delta M_{\text{nuc}}\, c^2 + m_e\, c^2 = \left(E_{e^+} + E_{\nu_e} + m_e\, c^2\right) + m_e\, c^2 = E_{\text{kinetic}} + 2\, m_e\, c^2, \tag{14.4.5}$$

where $\Delta M_{\text{nuc}} c^2$ is the difference in the rest mass energies of the related nuclei. In eq. (14.4.4), where the rest mass energy of the positron has been included in the definitions of ϵ and ϵ_{nuc},

$$\epsilon_{\text{nuc}} = \frac{\Delta M_{\text{nuc}}\, c^2}{m_e\, c^2} = \frac{Q(\text{MeV})}{0.511} - 1, \tag{14.4.6}$$

and

$$\epsilon_{\nu_e}^{max} = \epsilon_{nuc} - 1. \qquad (14.4.7)$$

In the case of the prototype, $^{13}N \rightarrow {}^{13}C + e^+ + \nu_e$, $Q = 2.221$ MeV and $\epsilon_{nuc} = 3.346$. The total kinetic energy of the emitted leptons is $\epsilon_{kin} = \epsilon_{nuc} - 1 = 2.346$ and the energy released in γ rays when the emitted positron annihilates with an electron is $\epsilon_{annihilation} = 1.022/0.551 = 2$.

In the laboratory, electron–positron annihilation is sometimes preceeded by the formation of positronium. The singlet state decays into two photons and the triplet state decays into three photons. In a star, at locations where positron-unstable nuclei are formed, the density is typically too large for positronium to exist. The radius of the electron cloud in the ground state of positronium is of the order of 10^{-8} cm, whereas the inter-electron separation is of the order of $(g\ cm^{-3}/\rho)^{1/3}\ 10^{-8}$ cm. Thus, when the density is larger than $1\ g\ cm^{-3}$, positronium cannot form and an emitted positron annihilates with an electron directly into two or three gamma rays.

Through a series of Compton scatterings, the positron loses kinetic energy and comes into thermal equilibrium with heavy nuclei at the ambient temperature. However, the electrons are still distributed according to the demands of the exclusion principle. Due to electrostatic attraction, the wave function describing a thermalized positron and a given electron in the Fermi sea peaks at zero separation, with the amplitude of the peak increasing with decreasing relative velocity. Thus, the probability for the annihilation of an electron–positron pair increases with decreasing electron kinetic energy; electrons near the bottom of the Fermi sea are preferentially destroyed. The Fermi sea readjusts to a new equilibrium as the remaining electrons cascade to again fill all of the lowest lying levels, with the net energy release being the Fermi energy, ϵ_F. This is the same as the additional energy which would be released if an electron at the top of the Fermi sea were annihilated and no cascading to fill lower levels occurred. In fact, the net additional energy released is independent of which electron in the Fermi sea is annihilated. The contribution to heat exchange brought about by a change dn_e in the number abundance of electrons is thus given by $\delta Q_{out} = - \epsilon_F dn_e$. This is a microscopic justification for the inclusion (see Sections 8.1 and 8.2 in Volume 1) of $\bar{\mu}_e = \epsilon_F$ in the thermodynamic equation $dQ_{out} = -n_e T d\bar{s}_e - \bar{\mu}_e dn_e$, where $\bar{s}_e$ is the specific entropy of an electron and $\bar{\mu}_e$ is a creation-destruction potential (see the derivation of eq. (8.2.32) in Volume 1).

When the Coulomb interaction is neglected, $C^2 = 1$ and the integrands in eqs. (14.4.4) and (14.1.16) are simple functions of ϵ and p. Integration gives

$$F(0, \epsilon_{nuc}) = \int_1^{\epsilon_{nuc}} (\epsilon_{nuc} - \epsilon)^2 (\epsilon^2 - 1)^{1/2}\, \epsilon\, d\epsilon$$

$$= p_{nuc} \left(\frac{1}{30} p_{nuc}^4 - \frac{1}{12} p_{nuc}^2 - \frac{1}{4} \right) + \frac{\epsilon_{nuc}}{4} \log_e(\epsilon_{nuc} + p_{nuc})$$

$$= \frac{1}{60} \left(\epsilon_{nuc}^2 - 1 \right)^{1/2} \left(2\epsilon_{nuc}^4 - 9\epsilon_{nuc}^2 - 8 \right) + \frac{\epsilon_{nuc}}{4} \log_e \left[\epsilon_{nuc} + \left(\epsilon_{nuc}^2 - 1 \right)^{1/2} \right],$$

$$(14.4.8)$$

agreeing with the result obtained in Section 6.2 in Volume 1 in a calculation of the rate of the pp reaction (eqs. (6.2.6)–(6.2.8)).

For the $^{13}\text{N} \to {}^{13}\text{C} + e^+ + \nu_e$ reaction, $F(0, 3.346) = 9.124$, while the exact integral is $F_{\text{lab}} = F(-Z, \epsilon_{\text{nuc}}) = F(-6, 3.346) = 7.712$. Thus, in this particular instance, neglecting the Coulomb interaction between the positron and the product nucleus overestimates F_{lab} by about 18 percent. The half life of ^{13}N is $t_{1/2} \sim 598$ s, so

$$\log_{10}(f \, t_{1/2}) = \log_{10}(F_{\text{lab}} \, t_{1/2}) = 3.664. \tag{14.4.9}$$

Since there is no threshold energy which must be surmounted, electrons of all energies can be captured by a positron emitter. The formalism is identical with that developed in the case of electron captures on isotopes stable against electron decay, except that the energy of the emitted neutrino is at least $m_e c^2$ larger than the difference in rest mass between the participating nuclei and, hence, is at least $2 \, m_e c^2$ larger than the energy of the neutrino emitted in the positron decay. That is,

$$E_{\nu_e} = \Delta M_{\text{nuc}} \, c^2 + E_{e^-} + m_e \, c^2 \geq \Delta M_{\text{nuc}} \, c^2 + m_e \, c^2 \tag{14.4.10}$$

or, equivalently,

$$\epsilon_{\nu_e} = \epsilon_{\text{nuc}} + \epsilon_e \geq \epsilon_{\text{nuc}} + 1. \tag{14.4.11}$$

By inspection, the lifetime $t_{\text{capture, star}}$ for electron capture is related to the lifetime $t_{e^+, \text{lab}}$ for positron decay by

$$\frac{t_{e^+, \text{lab}}}{t_{\text{capture, star}}} = \frac{F_{\text{capture, star}}}{F_{e^+, \text{lab}}}, \tag{14.4.12}$$

or

$$\frac{1}{t_{\text{capture, star}}} = F_{\text{capture, star}} \frac{\log_e 2}{(f t_{1/2})_{\text{lab}}}, \tag{14.4.13}$$

where

$$F_{\text{capture, star}} = \int_1^\infty f(\epsilon, \epsilon_F, T) \, C^2(Z, \epsilon) \, (\epsilon_{\text{nuc}} + \epsilon)^2 \, (\epsilon^2 - 1)^{1/2} \, \epsilon \, d\epsilon. \tag{14.4.14}$$

In eq. (14.4.14), ϵ is the energy of a free electron and Z is the charge of the positron emitter, and not that of its daughter.

14.5 Electron capture on a positron emitter when electrons are not degenerate

It is of interest to estimate the electron-capture rate on positron-unstable nuclei when free electrons are not degenerate. Then the Fermi energy is negative and one may use the formalism developed in Section 4.5 in Volume 1. Equation (14.4.14) becomes

$$F_{\text{capture, star}} = \int_0^\infty \sum_{n=1}^\infty (-1)^{n+1} e^{-n(\alpha + \epsilon'/kT)} \, C^2(Z, \epsilon) \, (\epsilon_{\text{nuc}} + \epsilon)^2 \, p^2 \, dp, \tag{14.5.1}$$

where ϵ' is the kinetic energy of an electron, ϵ and ϵ_f include the rest mass energy of the electron and are in units of $m_e c^2$, and p is in units of $m_e c$. Assuming that the electrons are not relativistic, one has

$$\frac{\epsilon'}{kT} = \frac{m_e c^2}{2kT} p^2. \qquad (14.5.2)$$

and (see eqs. (14.1.5) and (14.1.6))

$$C^2(Z, \epsilon) \sim 2\pi\eta \sim 2\pi Z \frac{e^2}{\hbar c} \frac{1}{p}. \qquad (14.5.3)$$

With these approximations, one has

$$F_{\text{capture, star}} = \int_0^\infty \sum_{n=1}^\infty (-1)^{n+1} e^{-n\alpha} \, e^{-n(m_e c^2/2kT)p^2} \, 2\pi \frac{Z e^2}{\hbar c} \frac{1}{p} (\epsilon_{\text{nuc}} + \epsilon)^2 \, p^2 \, dp \qquad (14.5.4)$$

$$= 2\pi \frac{Z e^2}{\hbar c} \frac{2kT}{m_e c^2} \sum_{n=1}^\infty (-1)^{n+1} \frac{e^{-n\alpha}}{n} \int_0^\infty (\epsilon_{\text{nuc}} + \epsilon)^2 \, e^{-x^2} x \, dx, \qquad (14.5.5)$$

where

$$x = n \left(\frac{m_e c^2}{2kT} \right)^{1/2} p. \qquad (14.5.6)$$

Rewriting eq. (4.5.5) as

$$\frac{1}{2} \left(\frac{h}{m_e c} \right)^3 n_e = \left(2\pi \frac{kT}{m_e c^2} \right)^{3/2} \sum_{n=1}^\infty (-1)^{n+1} \frac{e^{-n\alpha}}{n^{3/2}}, \qquad (14.5.7)$$

eq. (14.5.5) becomes

$$F_{\text{capture, star}} = \frac{1}{2} n_e \left(\frac{h}{m_e c} \right)^3 \frac{Z e^2}{\hbar c} \left(\frac{m_e c^2}{2\pi kT} \right)^{1/2}$$

$$\times \sum_{n=1}^\infty (-1)^{n+1} \frac{e^{-n\alpha}}{n} \int_0^\infty (\epsilon_{\text{nuc}} + \epsilon)^2 \, e^{-x^2} \, dx^2 \div \sum_{n=1}^\infty (-1)^{n+1} \frac{e^{-n\alpha}}{n^{3/2}}. \qquad (14.5.8)$$

Making the approximation

$$(\epsilon_{\text{nuc}} + \epsilon)^2 = (\epsilon_{\text{nuc}} + 1)^2 \left(1 + \frac{p^2}{\epsilon_{\text{nuc}} + 1} + \cdots \right), \qquad (14.5.9)$$

using the relationship between p^2 and x^2 given by eq. (14.5.6), and integrating over dx^2, eq. (14.5.8) becomes

$$F_{\text{capture, star}} = \frac{1}{2} \left[n_e \left(\frac{h}{m_e c} \right)^3 \right] \left[\left(\frac{m_e c^2}{2\pi kT} \right)^{1/2} \frac{Z e^2}{\hbar c} \right] \left[(\epsilon_{\text{nuc}} + 1)^2 \right] \gamma(\alpha), \qquad (14.5.10)$$

where

$$\gamma(\alpha) = \sum_{n=1}^{\infty} (-1)^{n+1} e^{-n\alpha} \left[\frac{1}{n} + \frac{1}{n^3} \frac{kT}{m_e c^2} \frac{2}{\epsilon_{nuc} + 1} \right]$$

$$\div \sum_{n=1}^{\infty} (-1)^{n+1} \frac{e^{-n\alpha}}{n^{3/2}}. \tag{14.5.11}$$

The first term in square brackets in eq. (14.5.10) gives the number of electrons in a cube one Compton wavelength on a side. The second term in square brackets gives the effect of the Coulomb interaction between the captured electron and the protons in the parent nucleus. The final term in square brackets is a consequence of the phase space available to the emitted neutrino. One may also write eq. (14.5.11) as

$$\gamma(\alpha) = \left[\Sigma_1(\alpha) + \frac{kT}{m_e c^2} \frac{2}{\epsilon_{nuc} + \epsilon} \Sigma_3(\alpha) \right] \div \Sigma_{3/2}, \tag{14.5.12}$$

where the Σs are given by eq. (14.2.14) with $|\epsilon_{th} - \epsilon_F|$ replaced by α.

Finally, using eq. (14.4.13),

$$\frac{1}{t_{capture, star}} = F_{capture, star} \frac{\log_e 2}{(ft_{1/2})_{lab}}.$$

$$= \frac{1}{2} n_e \left(\frac{h}{m_e c} \right)^3 \left(\frac{m_e c^2}{2\pi kT} \right)^{1/2} \frac{Ze^2}{\hbar c} (\epsilon_{nuc} + 1)^2 \gamma(\alpha) \frac{\log_e 2}{(ft_{1/2})_{lab}}. \tag{14.5.13}$$

Inserting numerical values for the constants, eq. (14.5.13) becomes

$$\frac{1}{t_{capture, star}} = 6.399 \times 10^{-7} \gamma(\alpha) \left(\frac{\rho}{\mu_e T_7^{1/2}} \right) Z (\epsilon_{nuc} + 1)^2 \frac{\log_e 2}{(ft_{1/2})_{lab}}. \tag{14.5.14}$$

When α is very large, $\gamma(\alpha) \to 1$. In the limit $\alpha \to 0$,

$$\gamma(\alpha) \to \left[\log_e 2 + \frac{\pi^2}{12} \frac{2kT}{m_e c^2} \frac{1}{\epsilon_{nuc} + 1} \right] \div 0.7649 = 0.906 + 0.00363 \frac{T_7}{\epsilon_{nuc} + 1}, \tag{14.5.15}$$

and, using the fact that $\rho/\mu_e T_6^{3/2} \to 6.182$ as $\alpha \to 0$ (see Section 4.5 in Volume 1),

$$\frac{t_{e^+, lab}}{t_{capture, star}} \to 1.133 \times 10^{-3} T_7 Z (\epsilon_{nuc} + 1)^2 \frac{1}{g_{e^+, lab}}. \tag{14.5.16}$$

In the case of electron capture on ^{13}N, with $Z = 7$, $\epsilon_{nuc} + 1 = 4.346$, and $F_{e^+, lab} = 7.712$, one has

$$\frac{t_{e^+, lab}}{t_{capture, star}} = 1.097 \times 10^{-6} \gamma(\alpha) \left(\frac{\rho}{\mu_e T_7^{1/2}} \right), \tag{14.5.17}$$

and, when $\alpha \rightarrow 0$,

$$\frac{t_{e^+, \text{lab}}}{t_{\text{capture, star}}} \rightarrow 1.939 \times 10^{-2} T_7. \tag{14.5.18}$$

As discussed in Section 6.8 in Volume 1, in hydrogen-burning regions in the Sun, the probability that ^{7}Be (which is stable against both positron and electron emission) captures a K-shell electron is 20% or so of the probability of the capture of a free electron. It is of interest to examine the probabilty of K-capture for positron-unstable isotopes of importance in CN-burning regions, where typically one to three out of ten CNO isotopes have at least one K-shell electron (Iben, 1969).

Over the nuclear volume, the wave function of a K-shell electron may be approximated by

$$\Psi_e = \frac{1}{\sqrt{\pi a_K^3}}, \tag{14.5.19}$$

where

$$a_K = \frac{\hbar^2}{m_e e^2} \frac{1}{Z}. \tag{14.5.20}$$

Following the standard procedure, one has

$$\frac{1}{t_{\text{K-capture}}} = \frac{2\pi}{\hbar} M_{\text{nuc}}^2 \left(\frac{4\pi}{h^3}\right)^2 \frac{1}{c^3} E_\nu^2 \frac{1}{\pi a_K^3}$$

$$= \frac{2\pi}{\hbar} M_{\text{nuc}}^2 \left(\frac{4\pi}{h^3}\right)^2 m_e^2 c (\epsilon_{\text{nuc}} + 1)^2 \frac{1}{\pi a_K^3}, \tag{14.5.21}$$

so that

$$\frac{t_{e^+, \text{lab}}}{t_{\text{K-capture}}} = \frac{1}{4\pi} \left(\frac{h}{m_e c}\right)^3 \frac{1}{a_K^3} \frac{(\epsilon_{\text{nuc}} + 1)^2}{f_{e^+ \text{lab}}},$$

$$= \left(\frac{e^2}{\hbar c} Z\right)^3 \frac{(\epsilon_{\text{nuc}} + 1)^2}{f_{e^+ \text{lab}}}, \tag{14.5.22}$$

In the case of the reaction $^{13}\text{N} + e^- \rightarrow ^{13}\text{C} + \nu_e$,

$$\frac{t_{e^+, \text{lab}}}{t_{\text{K-capture}}} = \left(\frac{7}{137}\right)^3 \frac{(4.346)^2}{7.712} = 3.267 \times 10^{-4}. \tag{14.5.23}$$

The probability that a nucleus has at least one K-shell electron can be written as (see Chapter 4 in Volume 1)

$$P_K \sim \frac{y}{1 + y}, \tag{14.5.24}$$

where

$$y = n_e \, 2 \left(\frac{h^2}{2\pi m_e kT} \right)^{3/2} \exp\left(\frac{e^2}{2a_K kT} \right) = 0.015\,78 \left(\frac{\rho}{\mu_e T_7^{3/2}} \right) \exp\left(\frac{0.015\,77\,Z}{T_7} \right).$$

(14.5.25)

Given that, under typical hydrogen-burning conditions in intermediate mass stars, P_K is of the order of 0.1–0.3, comparison of eq. (14.5.25) with eq. (14.5.14), shows that, when electrons are not degenerate, K-capture is typically more frequent than the capture of a free electron. Although neither electron-capture process is competitive with positron emission when electrons are not degenerate, it has been worthwhile to quantify this fact.

14.6 Electron capture on a positron emitter when electrons are degenerate

When electrons become degenerate, the phase space available for the emitted neutrino increases faster than the square of the electron Fermi energy. To investigate reaction rates in this regime, one may employ the techniques developed in Sections 14.1 and 14.3 to find

$$F_{\text{capture, star}} \sim C^2(Z, \epsilon_F) \, K_0^{ec} + \frac{p_F}{\epsilon_F} \, C^2(Z, \epsilon_F) \, K_1^{ec},$$

(14.6.1)

where

$$K_0^{ec} = \int_1^{\epsilon_F} (\epsilon_{\text{nuc}} + \epsilon)^2 \, (\epsilon^2 - 1)^{1/2} \, \epsilon \, d\epsilon,$$

(14.6.2)

$$K_1^{ec} = -K_{1<} + K_{1>},$$

(14.6.3)

$$K_{1<} = \sum_{n=1}^{\infty} (-1)^{n+1} \int_1^{\epsilon_F} e^{-n(\epsilon_F - \epsilon)/kT} \, (\epsilon_{\text{nuc}} + \epsilon)^2 \, \epsilon^2 \, d\epsilon,$$

(14.6.4)

and

$$K_{1>} = \sum_{n=1}^{\infty} (-1)^{n+1} \int_{\epsilon_F}^{\infty} e^{-n(\epsilon - \epsilon_F)/kT} \, (\epsilon_{\text{nuc}} + \epsilon)^2 \, \epsilon^2 \, d\epsilon.$$

(14.6.5)

Integrations give

$$K_0^{ec} = \frac{1}{5} \, p_F^5 + \frac{1}{3} \left[1 + \epsilon_{\text{nuc}}^2 + \frac{3}{2} \epsilon_{\text{nuc}} \epsilon_F \right] p_F^3$$

$$+ \frac{\epsilon_{\text{nuc}}}{4} \left[p_F \, \epsilon_F - \log_e(\epsilon_F + p_F) \right]$$

(14.6.6)

Table 14.6.1 Characteristics of the $^{13}\text{N} + e^- \rightarrow {}^{13}\text{C} + \nu_e$ reaction as a function of ϵ_F

ϵ_F	p_F	K_0^{ec}	v_F/c	$2\pi\eta$	F	F_{star}
1.01	0.142	0.0184	0.140	2.293	2.551	0.0066
1.05	0.320	0.209	0.305	1.053	1.617	0.103
1.1	0.458	0.623	0.417	0.771	1.434	0.372
1.2	0.663	1.943	0.553	0.581	1.318	1.417
1.3	0.831	3.926	0.639	0.502	1.272	3.191
1.4	0.980	6.402	0.698	0.460	1.248	5.575
1.45	1.050	9.287	0.724	0.443	1.238	8.324
1.5	1.118	10.13	0.745	0.431	1.231	9.295
3.346	3.193	388.3	0.954	0.336	1.176	435.6

and

$$K_1^{ec} = 4\,\epsilon_{nuc}^2\,(2\epsilon_F + \epsilon_{nuc})\left(\frac{kT}{\epsilon_{nuc}}\right)^2\left(\epsilon_F\,(\epsilon_F + \epsilon_{nuc})\,S_2 + 6\left(\frac{kT}{\epsilon_{nuc}}\right)^2 S_4\right),$$

$$+ (kT)\,\Sigma_1\,(\epsilon_{nuc} + 1)^2 - (kT)^2\,\Sigma_2(\epsilon_{nuc} + 1)(\epsilon_{nuc} + 2)$$

$$+ (kT)^3\,\Sigma_3\,2\left(\epsilon_{nuc}^2 + 6\,\epsilon_{nuc} + 6\right)$$

$$- (kT)^4\,\Sigma_4\,12\,(\epsilon_{nuc} + 2) + (kT)^5\,\Sigma_5\,24. \tag{14.6.7}$$

The S_K in eq. (14.6.7) are given by eqs. (14.2.16) and (14.2.17), and

$$\Sigma_K = \sum_{n=1}^{\infty}(-1)^{n+1}\frac{e^{-n(\epsilon_F-1)/kT}}{n^K}, \tag{14.6.8}$$

which is the same as eq. (14.2.14) when one sets $\epsilon_{th} = 1$. The terms in eq. (14.6.7) containing Σs come from the lower limit of integration of eq. (14.6.4). They have been included because they permit comparison with the results of Section 14.5 in the weakly degenerate regime.

In Table 14.6.1, $F_{star} = F_{capture,star} \sim K_0\,F\,v_{\epsilon_F}/c$ is shown as a function of ϵ_F for the reaction $^{13}\text{N} + e^- \rightarrow {}^{13}\text{C} + \nu_e$. It is evident that electron capture on ^{13}N becomes more important than positron decay ($g_{e^+,lab} = 7.712$) when the Fermi energy increases beyond $\epsilon_F \sim 1.44$, corresponding to an electron kinetic energy of ~ 0.24 MeV, or an electron "density" of about $\rho_e = \rho/\mu_e \sim 10^6$ g cm^{-3}. The last row in Table 14.6.1 shows that, when $\epsilon_F = \epsilon_{nuc}$, the probabilty of electron capture is over fifty times larger than the probability of positron decay.

14.7 Urca neutrino energy-loss rates

The neutrino energy-loss rate for a particular beta process is obtained simply by inserting the energy of the emitted neutrino into the appropriate integrand given in previous sections. As an example, consider an Urca pair of isobars at densities such that the electron Fermi energy is close to the threshhold energy for capture on the beta-unstable member of the pair.

In the case of electron capture on a positron-stable nucleus when $\epsilon_F < \epsilon_{th}$, one inserts $\epsilon - \epsilon_{th}$ in the integrand of eq. (14.2.6) and calculates

$$L^{ec}_{\epsilon_F < \epsilon_{th}} = \sum_{n=1}^{\infty} (-1)^{n+1} L_n, \tag{14.7.1}$$

where

$$L_n = \int_{\epsilon_{th}}^{\infty} e^{-n(\epsilon - \epsilon_F)/kT} \; (\epsilon - \epsilon_{th})^3 \; \epsilon^2 \, d\epsilon. \tag{14.7.2}$$

After integration, eq. (14.7.2) becomes

$$L_n = \epsilon_{th}^6 \left(\frac{kT}{\epsilon_{th}}\right)^4 e^{-n(\epsilon_{th} - \epsilon_F)/kT} \left(\frac{6}{n^4} + \frac{48}{n^5}\frac{kT}{\epsilon_{th}} + \frac{120}{n^6}\left(\frac{kT}{\epsilon_{th}}\right)^2\right), \tag{14.7.3}$$

so that

$$L^{ec}_{\epsilon_F < \epsilon_{th}} = 6\,\epsilon_{th}^6 \left(\frac{kT}{\epsilon_{th}}\right)^4 \left(\Sigma_4 + 8\,\Sigma_5 \frac{kT}{\epsilon_{th}} + 20\,\Sigma_6 \left(\frac{kT}{\epsilon_{th}}\right)^2\right), \tag{14.7.4}$$

where the Σ_K are given by eq. (14.2.14).

The average energy of the neutrino emitted in the capture process can now be written as

$$\bar{E}_{\nu_e} = m_e c^2 \frac{L^{ec}_{\epsilon_F < \epsilon_{th}}}{I^{ec}_{\epsilon_F < \epsilon_{th}}}, \tag{14.7.5}$$

where $I^{ec}_{\epsilon_F < \epsilon_{th}}$ is given by eq. (14.2.13). Thus,

$$\bar{E}^{\epsilon_F < \epsilon_{th}}_{\nu_e} = 3kT \frac{\left(\Sigma_4 + 8\,\Sigma_5 \dfrac{kT}{\epsilon_{th}} + 20\,\Sigma_6 \left(\dfrac{kT}{\epsilon_{th}}\right)^2\right)}{\left(\Sigma_3 + 6\,\Sigma_4 \dfrac{kT}{\epsilon_{th}} + 12\,\Sigma_5 \left(\dfrac{kT}{\epsilon_{th}}\right)^2\right)}. \tag{14.7.6}$$

In a similar fashion, when $\epsilon_F > \epsilon_{th}$, the average energy of the antineutrino emitted in an electron decay is

$$\bar{E}_{\bar{\nu}_e} = m_e c^2 \frac{L^{ed}_{\epsilon_F > \epsilon_{th}}}{J^{ed}_{\epsilon_F > \epsilon_{th}}}, \tag{14.7.7}$$

where

$$L_{\epsilon_F > \epsilon_{th}}^{ed} = 6\,\epsilon_{th}^6 \left(\frac{kT}{\epsilon_{th}}\right)^4 \left(\Sigma_4 - 8\,\Sigma_5\,\frac{kT}{\epsilon_{th}} + 20\,\Sigma_6 \left(\frac{kT}{\epsilon_{th}}\right)^2\right) \tag{14.7.8}$$

and $J_{\epsilon_F > \epsilon_{th}}^{ed}$ is given by eq. (14.3.10). Thus,

$$\bar{E}_{\bar{\nu}_e}^{\epsilon_F > \epsilon_{th}} = 3kT\,\frac{\left(\Sigma_4 - 8\,\Sigma_5\,\dfrac{kT}{\epsilon_{th}} + 20\,\Sigma_6 \left(\dfrac{kT}{\epsilon_{th}}\right)^2\right)}{\left(\Sigma_3 - 6\,\Sigma_4\,\dfrac{kT}{\epsilon_{th}} + 12\,\Sigma_5 \left(\dfrac{kT}{\epsilon_{th}}\right)^2\right)}. \tag{14.7.9}$$

It is interesting that the average energy of an emitted neutrino or antineutrino ($\sim 3.15\,kT$ when $\epsilon_F = \epsilon_{th}$) is not too different from the energy of a typical photon (eq. (4.11.21) gives $\langle h\nu \rangle = 2.70\,kT$).

Combining eqs. (14.2.18), (14.2.19), (14.2.20), (14.2.13) and (14.7.6), one has that the rate of neutrino energy loss per electron-capturing isobar is

$$\epsilon_{\nu_e}\,^{\epsilon_F < \epsilon_{th}} = \frac{\bar{E}_{\nu_e}^{\epsilon_F < \epsilon_{th}}}{t_{capture,\,star}} = \frac{\log_e 2}{(ft_{1/2})_{lab}}\,2\,\epsilon_{th}^2 \left(\frac{kT}{m_e c^2}\right)^3 3\,kT$$

$$\times\,\frac{p_{th}}{\epsilon_{th}}\,C^2(Z, \epsilon_{th}) \left(\Sigma_4 + 8\,\Sigma_5\,\frac{kT}{\epsilon_{th}} + 20\,\Sigma_6 \left(\frac{kT}{\epsilon_{th}}\right)^2\right). \tag{14.7.10}$$

Similarly, eqs. (14.3.11), (14.2.10), (14.2.20), (14.3.1) and (14.7.9) may be combined to give the rate of antineutrino energy loss per electron-decaying isobar:

$$\epsilon_{\bar{\nu}_e}\,^{\epsilon_F > \epsilon_{th}} = \frac{\bar{E}_{\bar{\nu}_e}^{\epsilon_F > \epsilon_{th}}}{t_{decay,\,star}} = \frac{\log_e 2}{(ft_{1/2})_{lab}}\,2\,\epsilon_{th}^2 \left(\frac{kT}{m_e c^2}\right)^3 3\,kT$$

$$\times\,\frac{p_{th}}{\epsilon_{th}}\,C^2(Z, \epsilon_{th}) \left(\Sigma_4 - 8\,\Sigma_5\,\frac{kT}{\epsilon_{th}} + 20\,\Sigma_6 \left(\frac{kT}{\epsilon_{th}}\right)^2\right). \tag{14.7.11}$$

At this juncture, one can calculate exactly the neutrino energy-loss rate for an Urca pair when $\epsilon_F = \epsilon_{th}$. Then the Σ_K reduce to the S_K given by eq. (14.2.16) and the abundances of the participating nuclei are approximately the same. Thus, the neutrino energy-loss rate per pair is given approximately by the sum of eqs. (14.7.9) and (14.7.10). On multiplying this sum by Avogadro's number N_A and the abundance by the number y_{pair} of the Urca pair, and neglecting all but the terms involving S_4, one obtains the first order approximation to the neutrino energy-loss rate per gram per second:

$$\epsilon_{Urca}^{max,0} = N_A\,y_{pair}\,\frac{\log_e 2}{(ft_{1/2})_{lab}}\,2\,\epsilon_{th}^2 \left(\frac{kT}{m_e c^2}\right)^3 3\,kT\,\frac{p_{th}}{\epsilon_{th}}\,C^2(Z, \epsilon_{th})\,S_4$$

$$= 1.570 \times 10^6\,y_{pair}\,\frac{10^5}{(ft_{1/2})_{lab}}\,\epsilon_{th}\,p_{th}\,C^2(Z, \epsilon_{th})\,T_8^4\,\text{erg g}^{-1}\,\text{s}^{-1}. \tag{14.7.12}$$

Table 14.7.1 Characteristics of several Urca pairs when $\epsilon_F = \epsilon_{th}$ and $T_8 = 1$

Urca pair	ϵ_{ex}	Q	$\log(ft_{1/2})$	$\frac{v_{th}}{c}$	$2\pi\eta$	$C^2(Z, \epsilon_F)$	$\epsilon_{Urca}^{max,0}$
$^{21}F-^{21}Ne$	0.00	5.69	5.21	0.989	0.464	1.25	4.37/7
	0.35	5.34	4.74	0.988	0.464	1.23	1.17/8
$^{23}Ne-^{23}Na$	0.00	4.38	5.27	0.983	0.513	1.28	2.50/7
	0.43	3.94	5.39	0.980	0.515	1.28	1.60/7
$^{25}Ne-^{25}Na$	0.00	7.40	∞				
	0.09	7.31	4.44	0.993	0.508	1.28	4.07/8
	1.07	6.33	4.76	0.991	0.509	1.28	1.51/8
$^{25}Na-^{25}Mg$	0.00	3.84	5.25	0.978	0.563	1.31	2.16/7
	0.97	2.87	5.05	0.966	0.570	1.31	2.17/7
	1.61	2.23	5.03	0.951	0.579	1.32	1.56/7

In the case of the $^{14}C-^{14}N$ Urca pair,

$$\epsilon_{Urca}^{max,0} \sim 204 \, y_{14} \, T_8^4 \text{ erg g}^{-1} \text{ s}^{-1}. \tag{14.7.13}$$

At temperatures and abundances typical at the start of helium burning, this amounts, at most, to a fraction of an erg per gram per second. More potent Urca pairs appear during carbon burning. Selected characteristics of several pairs are listed in Table 14.7.1. In this table, ϵ_{ex} is the excitation energy of the isobar which electron decays (in MeV), Q is in MeV, $\log(ft_{1/2}) = \log_{10}(ft_{1/2})$, $\epsilon_{Urca}^{max,0}$ is in units of erg g^{-1} s-1, and it has been assumed that $T_8 = 1$ and $y_{pair} = 1$. In the carbon–oxygen core of a massive AGB star about to ignite carbon burning at the center ($\rho \sim 10^9$ g cm^{-3} and $T_8 \sim 2$), the Urca pairs $^{23}Ne-^{23}Na$ and $^{25}Na-^{25}Mg$ produce peak neutrino energy-loss rates of ~ 5000 erg g^{-1} s^{-1}, compared with a current–current induced neutrino energy-loss rate of ~ 300 erg g^{-1} s^{-1} (Iben, 1978).

When $\epsilon_F = \epsilon_{th}$, the number abundance y_{ec} of the isobar which can capture electrons is slightly smaller than the number abundance y_{ed} of the other isobar. This is because electrons of all energies larger then ϵ_{th} can be captured, whereas the maximum energy of an emitted electron is limited to $\leq \epsilon_{th}$. The rate of electron capture is proportional to $y_{ec}(c + d)$ and the rate of electron decay is proportional to $y_{ed}(c - d)$, where, from eqs. (14.2.15) and (14.3.10),

$$c = S_3 + 12 \, S_5 \left(\frac{kT}{\epsilon_{th}}\right)^2 \tag{14.7.14}$$

and

$$d = 6 \, S_4 \left(\frac{kT}{\epsilon_{th}}\right). \tag{14.7.15}$$

This means that, when the reaction rates are in equilibrium,

$$y_{ec} = \frac{1}{2} \left(1 - \frac{d}{c} \right) y_{pair} \tag{14.7.16}$$

and

$$y_{ed} = \frac{1}{2} \left(1 + \frac{d}{c} \right) y_{pair} \tag{14.7.17}$$

The rate of energy loss due to neutrinos can be written as

$$\epsilon_{Urca} = e \, A \, (\, y_{ec} \, (a + b) + y_{ed} \, (a - b))$$

$$= e \, A \, y_{pair} \left(a - b \frac{d}{c} \right), \tag{14.7.18}$$

where, from eqs. (14.7.10) and (14.7.11),

$$a = S_4 + 20 \, S_6 \left(\frac{kT}{\epsilon_{th}} \right)^2, \tag{14.7.19}$$

$$b = 8 \, S_5 \left(\frac{kT}{\epsilon_{th}} \right), \tag{14.7.20}$$

and e is the coefficient of the expressions in large parentheses in eqs. (14.7.10) and (14.7.11). After some manipulation, one finds

$$\epsilon_{Urca}^{max} = \epsilon_{Urca}^{max,0} \frac{1 - 18.0048 \left(\frac{kT}{\epsilon_{th}} \right)^2 + 269.229 \left(\frac{kT}{\epsilon_{th}} \right)^4}{1 + 12.9394 \left(\frac{kT}{\epsilon_{th}} \right)^2}, \tag{14.7.21}$$

where $\epsilon_{Urca}^{max,0}$ is the first order approximation given by eq. (14.7.12).

14.8 Additional neutrino energy-loss rates for beta-decay reactions involving positron-stable isotopes

In this section, the neutrino energy-loss rate associated with electron capture on a positron-stable isotope when $\epsilon_F > \epsilon_{th}$ and the neutrino energy-loss rate associated with the electron decay of the same isotope when $\epsilon_F < \epsilon_{th}$ are examined. In the first instance, the relevant integral which depends on the temperature only through ϵ_F is obtained by inserting $(\epsilon - \epsilon_{th})$ into the integrand of eq. (14.2.27):

$$\bar{L}_{0>}^{ec} = \int_{\epsilon_{th}}^{\epsilon_F} (\epsilon - \epsilon_{th})^3 \, p^2 \, dp = \int_{\epsilon_{th}}^{\epsilon_F} (\epsilon - \epsilon_{th})^3 \, p \, \epsilon \, d\epsilon. \tag{14.8.1}$$

Integration gives

$$\bar{L}_{0>}^{ec} = \left[\frac{\epsilon^3 p^3}{6} + \frac{\epsilon p^3}{8} + \frac{\epsilon p}{16} - \frac{1}{16} \log_e(p + \epsilon) \right]_{\epsilon_{th}}^{\epsilon_F} - 3\epsilon_{th} \left[\frac{p^5}{5} + \frac{p^3}{3} \right]_{\epsilon_{th}}^{\epsilon_F}$$

$$+ 3\epsilon_{th}^2 \left[\frac{\epsilon p^3}{4} + \frac{\epsilon p}{8} - \frac{1}{8} \log_e(\epsilon + p) \right]_{\epsilon_{th}}^{\epsilon_F} - \epsilon_{th}^3 \left[\frac{p^3}{3} \right]_{\epsilon_{th}}^{\epsilon_F} \quad (14.8.2)$$

$$= \frac{1}{6} \left(\epsilon_F^3 \, p_F^3 - \epsilon_{th}^3 \, p_{th}^3 \right) + \frac{1}{8} \left(\epsilon_F \, p_F^3 - \epsilon_{th} \, p_{th}^3 \right) + \frac{1}{16} \left(\epsilon_F \, p_F - \epsilon_{th} \, p_{th} \right)$$

$$+ \frac{3}{4} \epsilon_{th}^2 \left(p_F^3 \, \epsilon_F - p_{th}^3 \, \epsilon_{th} \right) + \frac{3}{8} \epsilon_{th}^2 (p_F \, \epsilon_F - p_{th} \, \epsilon_{th})$$

$$- \frac{3}{5} \epsilon_{th} \left(p_F^5 - p_{th}^5 \right) - \epsilon_{th} \left(1 + \frac{\epsilon_{th}^2}{3} \right) \left(p_F^3 - p_{th}^3 \right)$$

$$- \frac{3}{8} \left(\epsilon_{th}^2 + \frac{1}{6} \right) \log_e \left(\frac{p_F + \epsilon_F}{p_{th} + \epsilon_{th}} \right). \quad (14.8.3)$$

Guided by eqs. (14.2.24)–(14.2.26), the relevant quantities which depend explicitly on the temperature are

$$L_{1>}^{ec} = -\sum_{n=1}^{\infty} (-1)^{n+1} \left(L_n^- - L_n^+ \right), \quad (14.8.4)$$

where

$$L_n^- = \int_{\epsilon_{th}}^{\epsilon_F} e^{-n(\epsilon_F - \epsilon)/kT} \, (\epsilon - \epsilon_{th})^3 \, \epsilon^2 \, d\epsilon \quad (14.8.5)$$

and

$$L_n^+ = \int_{\epsilon_F}^{\infty} e^{-n(\epsilon - \epsilon_F)/kT} \, (\epsilon - \epsilon_{th})^3 \, \epsilon^2 \, d\epsilon. \quad (14.8.6)$$

After integration, one has

$$L_{1>}^{ec} = -6 \, (kT)^4 \, \epsilon_{th}^2 \left[\Sigma_4 - 8 \, \Sigma_5 \left(\frac{kT}{\epsilon_{th}} \right) + 20 \, \Sigma_6 \left(\frac{kT}{\epsilon_{th}} \right)^2 \right]$$

$$+ 2 \, (kT)^2 \, \epsilon_{th}^4 \left(\frac{\epsilon_F}{\epsilon_{th}} \right) S_2 \left[\left(5 \left(\frac{\epsilon_F}{\epsilon_{th}} \right)^3 - 12 \left(\frac{\epsilon_F}{\epsilon_{th}} \right)^2 + 9 \left(\frac{\epsilon_F}{\epsilon_{th}} \right) - 2 \right) \right]$$

$$+ 6 \, (kT)^2 \, \epsilon_{th}^4 \, S_4 \left[20 \left(\frac{\epsilon_F}{\epsilon_{th}} \right)^2 - 24 \left(\frac{\epsilon_F}{\epsilon_{th}} \right) + 6 \right] \left(\frac{kT}{\epsilon_{th}} \right)^2$$

$$+ 6 \, (kT)^2 \, \epsilon_{th}^4 \, S_6 \left[40 \left(\frac{kT}{\epsilon_{th}} \right)^4 \right], \quad (14.8.7)$$

where the first term, which involves the Σ_K functions (eq. (14.2.14)), comes from the integration and summing of the L_n^- in eq. (14.8.5), whereas the terms involving the functions S_K have contributions from both L_n^+ and L_n^-.

The terms in eq. (14.8.7) can be rearranged to read

$$L_{1>}^{ec} = (kT)^2 \; \epsilon_{th}^4$$

$$\times \left\{ 2 \left(\frac{\epsilon_F}{\epsilon_{th}} \right) S_2 \left[5 \left(\frac{\epsilon_F}{\epsilon_{th}} \right)^3 - 12 \left(\frac{\epsilon_F}{\epsilon_{th}} \right)^2 + 9 \left(\frac{\epsilon_F}{\epsilon_{th}} \right) - 2 \right] \right.$$

$$+ 6 \left[S_4 \left(20 \left(\frac{\epsilon_F}{\epsilon_{th}} \right)^2 - 24 \left(\frac{\epsilon_F}{\epsilon_{th}} \right) + 6 \right) - \Sigma_4 \right] \left(\frac{kT}{\epsilon_{th}} \right)^2$$

$$\left. + 6 \left[8 \, \Sigma_5 + 20 \, (2 \, S_6 - \Sigma_6) \left(\frac{kT}{\epsilon_{th}} \right) \right] \left(\frac{kT}{\epsilon_{th}} \right)^3 \right\}, \tag{14.8.8}$$

which shows transparently that, as $\epsilon_F \to \epsilon_{th}$, $L_{1>}^{ec} \to L_{\epsilon_F < \epsilon_{th}}^{ec}$ (eq. (14.7.4)).

The net neutrino energy-loss rate for electron capture on a positron-stable isotope when $\epsilon_F > \epsilon_{th}$ is

$$\epsilon_{\nu_e} = L_{\epsilon_F > \epsilon_{th}}^{ec} \frac{\log_e 2}{(f t_{1/2})_{\text{lab}}}, \tag{14.8.9}$$

where

$$L_{\epsilon_F > \epsilon_{th}}^{ec} = \left\langle C^2(Z, \epsilon) \right\rangle_0 L_{0>}^{ec} + \left\langle \frac{p}{\epsilon} C^2(Z, \epsilon) \right\rangle_1 L_{1>}^{ec}, \tag{14.8.10}$$

and $\left\langle C^2(Z, \epsilon) \right\rangle_0$ is approximated by eq. (14.2.35) and $\left\langle (p/\epsilon) \, C^2(Z, \epsilon) \right\rangle_1$ is approximated by eq. (14.2.36).

One may obtain the relevant integrals for antineutrino losses due to electron-decay reactions when $\epsilon_F < \epsilon_{th}$ by inserting the quantity $(\epsilon_{th} - \epsilon)$ into the integrands of eqs. (14.3.15)–(14.3.18). The integral which depends on the temperature only through ϵ_F is

$$\bar{L}_{0<}^{ed} = \int_{\epsilon_F}^{\epsilon_{th}} (\epsilon_{th} - \epsilon)^3 \, p^2 \, dp = \int_{\epsilon_F}^{\epsilon_{th}} (\epsilon_{th} - \epsilon)^3 \, p \, \epsilon \, d\epsilon. \tag{14.8.11}$$

The sign of the integrand in this equation is opposite to that in eq. (14.8.1) but the limits are reversed. Thus, the integrals are identical. However, for eq. (14.8.11), $\epsilon_F < \epsilon_{th}$, so one rearranges eq. (14.8.3) to read

$$\bar{L}_{0<}^{ed} = +\frac{3}{5} \epsilon_{th} \left(p_{th}^5 - p_F^5 \right) + \epsilon_{th} \left(1 + \frac{\epsilon_{th}^2}{3} \right) \left(p_{th}^3 - p_F^3 \right)$$

$$+ \frac{3}{8} \left(\epsilon_{th}^2 + \frac{1}{6} \right) \log_e \left(\frac{p_{th} + \epsilon_{th}}{p_F + \epsilon_F} \right)$$

$$- \frac{1}{6} \left(\epsilon_{th}^3 \, p_{th}^3 - \epsilon_F^3 \, p_F^3 \right) - \frac{1}{8} \left(\epsilon_{th} \, p_{th}^3 - \epsilon_F \, p_F^3 \right) - \frac{1}{16} \left(\epsilon_{th} \, p_{th} - \epsilon_F \, p_F \right)$$

$$- \frac{3}{4} \epsilon_{th}^2 \left(p_{th}^3 \, \epsilon_{th} - p_F^3 \, \epsilon_F \right) - \frac{3}{8} \epsilon_{th}^2 \left(p_{th} \, \epsilon_{th} - p_F \, \epsilon_F \right) . \tag{14.8.12}$$

Guided by eqs. (14.3.14) and (14.3.16)–(14.3.26), the relevant integrals which depend explicitly on the temperature are

$$L_{1<}^{ed} = \sum_{n=1}^{\infty} (-1)^{n+1} \left(L_n^- - L_n^+ \right), \tag{14.8.13}$$

where

$$L_n^- = \int_{-\infty}^{\epsilon_F} e^{-n(\epsilon_F - \epsilon)/kT} \, (\epsilon_{th} - \epsilon)^3 \, \epsilon^2 \, d\epsilon \tag{14.8.14}$$

and

$$L_n^+ = \int_{\epsilon_{th}}^{\epsilon_F} e^{-n(\epsilon - \epsilon_F)/kT} \, (\epsilon_{th} - \epsilon)^3 \, \epsilon^2 \, d\epsilon . \tag{14.8.15}$$

After integration and rearrangement, one finds

$$L_{1>}^{ed} = (kT)^2 \, \epsilon_{th}^4$$

$$\times \left\{ 2 \left(\frac{\epsilon_F}{\epsilon_{th}} \right) S_2 \left[5 \left(\frac{\epsilon_F}{\epsilon_{th}} \right)^3 - 12 \left(\frac{\epsilon_F}{\epsilon_{th}} \right)^2 + 9 \left(\frac{\epsilon_F}{\epsilon_{th}} \right) - 2 \right] \right.$$

$$+ 6 \left[S_4 \left(20 \left(\frac{\epsilon_F}{\epsilon_{th}} \right)^2 - 24 \left(\frac{\epsilon_F}{\epsilon_{th}} \right) + 6 \right) - \Sigma_4 \right] \left(\frac{kT}{\epsilon_{th}} \right)^2$$

$$\left. \times 6 \left[-8 \, \Sigma_5 + 20 \left(2 \, S_6 - \Sigma_6 \right) \left(\frac{kT}{\epsilon_{th}} \right) \right] \left(\frac{kT}{\epsilon_{th}} \right)^3 \right\} , \tag{14.8.16}$$

which is identical with eq. (14.8.8) except that the sign of the term proportional to Σ_5 is reversed.

The net neutrino energy-loss rate for electron decay when $\epsilon_F < \epsilon_{th}$ is

$$\epsilon_{\nu_e} = L_{\epsilon_F < \epsilon_{th}}^{ec} \frac{\log_e 2}{(ft_{1/2})_{lab}} , \tag{14.8.17}$$

where

$$L_{\epsilon_F < \epsilon_{th}}^{ed} \sim C^2(Z, \epsilon_F) \, L_{0<}^{ed} + \frac{p_F}{\epsilon_F} C^2(Z, \epsilon_F) \, L_{1<}^{ed} . \tag{14.8.18}$$

Here, $\langle C^2(Z, \epsilon) \rangle_0$ is approximated by eq. (14.2.35) and $\langle (p/\epsilon) \, C^2(Z, \epsilon) \rangle_1$ is approximated by eq. (14.2.36).

14.9 Neutrino energy-loss rates for electron capture on a positron emitter

Neutrino energy-loss rates associated with electron capture on a positron emitter are obtained by inserting the factor $\epsilon_{\text{nuc}} + \epsilon$ under the integral sign in the relevant integrals for electron capture rates. When electrons are degenerate, define

$$L_{\nu_e} \sim C^2(Z, \epsilon_F) \frac{p_F}{\epsilon_F} (L_0 + L_1), \tag{14.9.1}$$

where

$$L_0 = \int_1^{\epsilon_F} (\epsilon + \epsilon_{\text{nuc}})^3 \, p \, \epsilon \, d\epsilon \tag{14.9.2}$$

$$= \frac{\epsilon_F^3 p_F^3}{6} + \frac{\epsilon_F p_F^3}{8} + \frac{1}{16}[\epsilon_F p_F - \log_e(\epsilon_F + p_F)] + 3\epsilon_{\text{nuc}} \left[\frac{p_F^5}{5} + \frac{p_F^3}{3} \right]$$

$$+ 3\epsilon_{\text{nuc}}^2 \left[\frac{\epsilon_F p_F^3}{4} + \frac{\epsilon_F p_F}{8} - \frac{1}{8}\log_e(\epsilon_F + p_F) \right] + \epsilon_{\text{nuc}}^3 \frac{p_F^3}{3} \tag{14.9.3}$$

and

$$L_1 = -L_{1<} + L_{1>}, \tag{14.9.4}$$

with

$$L_{1<} = \sum_{n=1}^{\infty} (-1)^{n+1} \int_1^{\epsilon_F} e^{-n(\epsilon_F - \epsilon)/kT} (\epsilon_{\text{nuc}} + \epsilon)^3 \, p \, \epsilon \, d\epsilon \tag{14.9.5}$$

and

$$L_{1>} = \sum_{n=1}^{\infty} (-1)^{n+1} \int_{\epsilon_F}^{\infty} e^{-n(\epsilon - \epsilon_F)/kT} (\epsilon_{\text{nuc}} + \epsilon)^3 \, p \, \epsilon \, d\epsilon. \tag{14.9.6}$$

After integrating and collecting terms, one has

$$L_1 = (kT)^2 \, S_2 \left[10\epsilon_F^4 + 24\epsilon_{\text{nuc}}\epsilon_F^3 + 18\epsilon_{\text{nuc}}^2\epsilon_F^2 + 4\epsilon_{\text{nuc}}^3\epsilon_F \right]$$

$$+ (kT)^4 \, S_4 \left[120\epsilon_F^2 + 144\epsilon_{\text{nuc}}\epsilon_F + 36\epsilon_{\text{nuc}}^2 \right] + (kT)^6 \, S_6 \, 240$$

$$+ (kT) \, \Sigma_1 \, (\epsilon_{\text{nuc}} + 1)^3 - (kT)^2 \, \Sigma_2 \, (\epsilon_{\text{nuc}} + 1)[3 + 2(\epsilon_{\text{nuc}} + 1)]$$

$$+ (kT)^3 \, \Sigma_3 \, (\epsilon_{\text{nuc}} + 1)[6 + 12(\epsilon_{\text{nuc}} + 1) + (\epsilon_{\text{nuc}} + 1)^2]$$

$$- (kT)^4 \, \Sigma_4 \, [6 + 36(\epsilon_{\text{nuc}} + 1) + (\epsilon_{\text{nuc}} + 1)^2]$$

$$+ (kT)^5 \, \Sigma_5 \, [48 + 72(\epsilon_{\text{nuc}} + 1)] - (kT)^6 \, \Sigma_6 \, 120, \tag{14.9.7}$$

where the S_Ks are given by eqs. (14.2.16) and (14.2.17) and the Σ_Ks are given by eq. (14.6.8). The neutrino loss rate is given by

$$\epsilon_{\nu_e} = L_{\nu_e} \frac{\log_e 2}{(ft_{1/2})_{\text{lab}}}, \tag{14.9.8}$$

When electrons are not degenerate and not relativistic, one may use eqs. (14.5.8)–(14.5.14) to find

$$L_{\nu_e} = \frac{1}{2} n_e \left(\frac{h}{m_e c}\right)^3 \left(\frac{m_e c^2}{2\pi kT}\right)^{1/2} \frac{Ze^2}{\hbar c} (\epsilon_{\text{nuc}} + 1)^3 \lambda(\alpha) \frac{1}{F_{e^+,\,\text{lab}}\, t_{e^+,\,\text{lab}}} \tag{14.9.9}$$

$$= 6.399 \times 10^{-7} \lambda(\alpha) \left(\frac{\rho}{\mu_e T_7^{1/2}}\right) Z (\epsilon_{\text{nuc}} + 1)^3 \frac{\log_e 2}{(ft_{1/2})_{\text{lab}}}, \tag{14.9.10}$$

$$\lambda(\alpha) = \left[\Sigma_1(\alpha) + \frac{kT}{m_e c^2} \frac{3}{\epsilon_{\text{nuc}} + 1} \Sigma_3(\alpha)\right] \div \Sigma_{3/2}(\alpha), \tag{14.9.11}$$

where the Σs are given by eq. (14.3.14) with $|\epsilon_{\text{th}} - \epsilon_F|$ replaced by α. The neutrino energy-loss rate is given by eq. (14.9.8).

14.10 Higher order beta transitions and experimental properties of beta-decay reactions

When the nuclear matrix element given by eq. (14.1.8) is finite, a beta transition is said to be allowed. When, because of parity or spin considerations, the nuclear matrix element vanishes, the beta transition is called forbidden and, before evaluating the integral in eq. (14.1.2), it is necessary to expand the exponentials in the lepton wave functions given by eqs. (14.1.3) and (14.1.4), using

$$\exp(i\Delta\mathbf{k} \cdot \mathbf{r}) = 1 + (i\Delta\mathbf{k} \cdot \mathbf{r}) + (i\Delta\mathbf{k} \cdot \mathbf{r})^2 + \cdots, \tag{14.10.1}$$

where $\Delta\mathbf{k} = \mathbf{k}_{\bar{\nu}_e} - \mathbf{k}_e$ is the difference in the lepton $\mathbf{k}$ vectors in the exponents of eqs. (14.1.3) and (14.1.4). If the integral

$$\langle M_{\text{nuc}} \rangle = \int \Psi_{f,\text{nuc}}^* (\Delta\mathbf{k} \cdot \mathbf{r}) \Psi_{i,\text{nuc}} \, d\tau_{\text{nuc}} \tag{14.10.2}$$

is finite, the transition is called first forbidden. If it is zero and the integral involving $(\Delta\mathbf{k}\cdot\mathbf{r})^2$ is finite, the transition is called second forbidden. Details of the calculation of beta decay and capture rates under stellar conditions for both allowed and forbidden transitions are described by Iben (1978).

Fundamental experimental properties of beta decay reactions are summarized and analyzed by F. Ajzenberg-Selove & T. Lauritsen (1959) for atomic numbers A = 5–20, by T. Lauritsen & F. Ajzenberg-Selove (1966) for atomic numbers A = 5–10, and by F. Ajzenberg-Selove & T. Lauritsen (1968) for atomic numbers A = 11–12, as well as by F. Ajzenberg-Selove (1970), (1971), (1972), and (1975), and by P.M. Endt and C. van der Leun, (1973). The summary and analysis is continued by F. Ajzenberg-Selove (1986a) for atomic numbers A = 13–15, (1986b) for A = 16–17, (1987) for A = 18–20, (1988) for A = 5–10, and (1989) for A = 11–12.

Bibliography and references

Fay Ajzenberg-Selove & T. Lauritsen, *Nucl. Phys.*, **11**, 1, 1959; **114**, 1, 1968.

Fay Ajzenberg-Selove, *Nucl. Phys.*, **152**, 1, 1970; **166**, 1, 1971; **190**, 1, 1972; **248**, 1, 1975; **449**, 1, 1986a; **460**, 1, 1986b; **475**, 1, 1987; **490**, 1, 1988; **506**, 1, 1989.

Hans Albrecht Bethe, & Philip Morrison, *Elementary Nuclear Theory*: second edition, Dover, 2006.

P. M. Endt & C. van der Leun, *Nucl. Phys.*, **214**, 1, 1973.

Enrico Fermi, Zeitzschrift für Physik, **88**, 161, 1934.

George Gamow & Mario Schönberg, *Phys. Rev.*, **59**, 539, 1941.

Icko Iben, Jr., *ApJ*, **158**, 1033, 1969; **219**, 213, 1978.

K. Kaminisi, K. Arai, & K. Yoshinaga, *Prog. Theoret. Phys.*, **53**, 1855, 1975.

T. Lauritzen & F. Ajzenberg-Selove, *Nucl. Phys.*, **78**, 1, 1966.

Claudio Ritossa, Icko Iben, Jr, & Enrique García Berro, *ApJ*, **515**, 381, 1999.

Robert G. Spulak, Jr., *ApJ*, **235**, 565, 1980.

15 Current–current weak interactions and the production of neutrino–antineutrino pairs

By the end of the third decade of the twentieth century, it had become clear that, in nuclear beta-decay events, beta particles are emitted in a continuous energy spectrum rather than with a unique energy equal to the change in energy of the emitting nucleus (although the maximum energy of the beta particle is equal to the change in nuclear energy), that the change in the electrical charge of the nucleus is exactly equal in absolute value but of opposite sign to the charge of the emitted beta particle, and that beta decay events often involve a unit change in the spin of the nucleus. In 1930, Wolfgang Pauli began communicating to other physicists the idea that, in order to account for these facts, another, previously unknown, "penetrating" particle must also be emitted in beta decay events with the properties: mass much smaller than the electron mass, no electrical charge, and an intrinsic spin equal to that of the electron. According to M. Mladjenović (1998), for over three years Pauli considered the idea too speculative to publish, and his first formal account appeared in the proceedings of an international conference on physics held in London in 1934. Enrico Fermi dubbed the hypothetical particle the "neutrino" (little neutral one) and formulated a mathematical theory of beta interactions involving neutrinos (Fermi, 1934) which has guided experimental and theoretical work on the weak interaction up to the present time.

The extensive body of experimental information about beta interactions which involve the transformation, within a nucleus, of a neutron into a proton or a proton into a neutron has been utilized in Chapter 6 in Volume 1 and in Chapter 14 in this volume to calculate, under conditions found in stellar interiors, the rates of various processes (involving nuclei, electrons, antineutrinos, positrons, and neutrinos) of the sort that take place in laboratory experiments and are well described by Fermi's theory of weak interactions. Under conditions found in main sequence, core helium-burning stars, and core carbon-burning stars, many relevant beta-decay rates and neutrino energy-loss rates are directly tied to experiment.

Over the years, the theory of beta interactions has been (1) refined to conform with the discovery that electrons and neutrinos are created with a definite helicity (i.e., spin and linear momentum parallel or antiparallel), (2) incorporated into a more general theory of weak interactions which includes muons and muon neutrinos, and (3) incorporated into a still more general theory which unites electromagnetic and weak interactions. The first advance clarifies the mathematical character of the beta interaction as consisting of a vector component and an axial vector component (the so-called V-A formulation). The second advance is an extention of the theory based on a charged-current interaction Hamiltonion which predicts, in addition to nuclear beta decay reactions, reactions involving only leptons (Richard P. Feynman and Murray Gell-Mann, 1958).

The third extension is a union of quantum electrodynamics and weak interaction theory which introduces a neutral current associated with the existence of an intermediate vector boson and which, as related by the recipients of the Nobel prize awarded for this extension, was the consequence of concerted efforts by many physicists (S. Weinberg 1967, 1980; A. Salam 1968, 1980; S. L. Glashow, 1980).

The second and third extensions lead to predictions of processes, such as the conversion of electron–positron pairs into neutrinos and antineutrinos, which have not been encountered directly in laboratory experiments but which play important roles in stars during several phases of stellar evolution. Two of these processes, which are not envisioned in the Fermi theory, but are predicted by the theoretical extentions of this theory, involve the conversion of an electron–positron pair into a neutrino–antineutrino pair. At very high temperatures, the electron–positron pairs are real. They are formed predominantly by an electromagnetic interaction between photons and charged particles, and an equilibrium distribution of pairs (see Section 4.10 in Volume 1) is maintained by a balance between electromagnetic formation and electromagnetic annihilation. Although the predicted rate at which electron–positron pairs annihilate by emitting neutrinos and antineutrinos is quite small compared with the rate at which pairs annihilate by emitting gamma rays, the neutrinos in most instances escape directly from the star without interacting with matter in the star and the consequent cooling affects the temperature distribution in the star. At temperatures of the order of 5×10^8 K or larger and at densities such that electrons are not very degenerate, the conversion of real electron–positron pairs into neutrino–antineutrino pairs is the dominant neutrino-loss process. As described in Chapter 20, this process controls the rate of core contraction in massive stars during the core carbon-burning phase through more advanced stages of core burning, continuing all the way to the formation of iron-peak elements.

The formation of real electron–positron pairs is inhibited when electrons are degenerate and another process, involving plasmons which transform into virtual e^+e^- pairs that decay into $\nu_e \bar{\nu}_e$ pairs, becomes the dominant source of neutrino energy losses. Under appropriate conditions, electrons can oscillate coherently in response to the electric field of a traveling electromagnetic wave to form what is called a transverse plasmon, which is essentially a photon with an effective mass given by $\hbar \omega_p / c^2$, where ω_p is the plasma frequency, a frequency below which an electromagnetic wave cannot propagate. Another type of plasmon consists of electrons and an electric field which oscillate with respect to one another in such a way that there is no associated magnetic field. Such oscillations are called longitudinal plasmons and they too have an effective mass given by $\hbar \omega_p / c^2$.

Since both types of plasmon have an effective mass, their decay into $\nu_e \bar{\nu}_e$ pairs is energetically favorable. The rate at which the plasma neutrino process occurs is proportional to a high power of the plasma frequency which is in turn proportional to the square root of the electron density. In regions where it is an important energy-loss process, its rate also increases with increasing temperature. It is the primary neutrino-loss mechanism in the electron-degenerate helium cores of low mass red giants, as described in Chapter 11 of Volume 1, in the electron-degenerate carbon–oxygen cores of AGB stars, as is elaborated in Chapters 17 and 18 of this volume, and in the electron-degenerate oxygen–neon cores of TPAGB stars which descend from stars of initial mass in the range $\sim$8.5–10.5 $M_\odot$ (e.g. C. Ritossa, I. Iben, Jr., & E. García-Berro, 1999).

The photoneutrino process, whereby an electron or positron is transformed into an intermediate state by absorbing a photon and then decays into a neutrino–antineutrino pair is a modestly important neutrino-loss mechanism at temperatures extending below where the electron–positron pair annihilation process is dominant. A final process, whereby an electron (or positron) is scattered by a heavy charged particle into an intermediate state which then decays into a neutrino–antineutrino pair, is called the neutrino bremsstrahlung process. It has been found to be of only secondary importance in stellar evolution.

The primary objective of this chapter is to examine aspects of weak interaction theory that are relevant for understanding the neutrino–antineutrino pair production processes. In Section 15.1, the Feynman–Gell-Mann theory of weak interactions is introduced and arguments are given for two coupling constants being required when nucleons are involved in the interaction. In Section 15.2, it is shown that only one coupling constant is required to explain the decay of a muon into an electron, an electron antineutrino, and a muon neutrino. In Section 15.3, an approximation to the cross section for the conversion of an electron and positron into a neutrino and antineutrino is obtained and the rate of energy loss by this process is estimated at temperatures and densities such that electrons are not degenerate. In Section 15.4, the Dirac equation and its solutions are discussed in preparation for obtaining, in Section 15.5, the exact cross section for the electron–positron annihilation process using the V-A theory. Along the way, a form for the weak interaction Hamiltonian is constructed (eq. (15.5.9)) which is used in Section 10.6 in Volume 1 to understand how electron neutrinos formed near the center of the Sun can be transformed into muon neutrinos on their passage out of the Sun.

In Section 15.6, a brief overview is given of the history and nature of weak interaction induced neutrino–antineutrino production processes. In Section 15.7, plasma oscillations are examined from the point of view of classical physics. In Section 15.8, the concept of plasmons as quanta is introduced and an outline is given of the rates at which energy is lost due to the decay of transverse and longitudinal plasmons into neutrino–antineutrino pairs.

15.1 The charged-current interaction Hamiltonian and the necessity for two coupling constants in nuclear beta decay

In its simplest form, the current in the Feynman–Gell-Mann theory may be written schematically as

$$J = (\bar{n}p) + (\bar{e}\nu_e) + (\bar{\mu}\nu_\mu), \tag{15.1.1}$$

where the terms in parentheses may, for conceptualization purposes, be thought of as creation and destruction operators and/or as wave functions of pairs of particles. The quantities $\bar{n}, \bar{e}$, and $\bar{\mu}$ are associated, respectively, with neutrons, electrons, and muons, whereas p, ν_e, and ν_μ are associated, respectively, with protons, electron neutrinos, and muon neutrinos. The interaction Hamiltonian density may be written as

$$H_{\text{int}} = g_{\text{weak}} \, J^\dagger J, \tag{15.1.2}$$

where g_{weak} is an interaction strength and $J^\dagger$ is the Hermitian conjugate of J. As an example, one may think of the cross product

$$(\bar{n}p)^\dagger (\bar{e}\nu_e) = (\bar{p}n)\,(\bar{e}\nu_e) \tag{15.1.3}$$

as an operator which is responsible for the reaction

$$n \to p + e^- + \bar{\nu}_e, \tag{15.1.4}$$

with the interpretation of $\bar{p}$, $\bar{e}$, and ν_e in eq. (15.1.3) as creation operators for, respectively, protons, electrons, and antineutrinos, and of n as a neutron destruction operator. The associated transition matrix element is

$$\int H_{\text{int}} \, d\tau = g_{\text{weak}} \, M_{\text{decay}}, \tag{15.1.5}$$

where

$$M_{\text{decay}} = \int \Psi_p^* \, \Psi_n \, \Psi_e^* \, \Psi_{\bar{\nu}_e}^* \, d\tau_{\text{nuc}}. \tag{15.1.6}$$

Here, the Ψs are spatial wave functions; the spin dependence of the interaction is taken into account in a statistical factor introduced in calculating the transition probability. Choosing, as in Section 6.1 in Volume 1 and in Section 14.1, a plane wave for the neutrino and a Coulomb-distorted plane wave for the positron, one has

$$M_{\text{decay}} = \frac{C}{V} \, M_{\text{nuc}}, \tag{15.1.7}$$

where C is the Coulomb function given by eqs. (14.1.5) and (14.1.6) and

$$M_{\text{nuc}} = \int \Psi_p^* \, \Psi_n \, d\tau_{\text{nuc}}. \tag{15.1.8}$$

The decay probability now becomes

$$\omega_{\text{decay}} = \frac{2\pi}{\hbar} \, g_{\text{weak}}^2 \left[\int |M_{\text{nuc}}|^2 \, \Sigma \, C^2 \left(\frac{4\pi p_e^2}{h^3} \right) \left(\frac{4\pi p_{\bar{\nu}_e}^2}{c \, h^3} \right) dp_e \right], \tag{15.1.9}$$

where the statistical factor Σ is unity when there is no change in spin and 3 if there is a unit change in spin. The last three terms in square brackets give the density of final lepton states for electron momentum in the range p_e to $p_e + dp_e$.

Assuming that the nuclear matrix element does not depend on p_e, one has finally that

$$\frac{1}{t_{\text{decay}}} = \omega_{\text{decay}} = \frac{1}{2\pi^3} \left(\frac{m_e c^2}{\hbar} \right) g_{\text{weak}}^2 \, |M_{\text{nuc}}|^2 \, \Sigma \left[\left(\frac{m_e c}{\hbar} \right)^3 \frac{1}{m_e c^2} \right]^2 F(Z, \epsilon_f), \tag{15.1.10}$$

where ϵ_f is the maximum energy of the electron, including rest mass energy, in units of $m_e c^2$, and $F(Z, \epsilon_f)$ is defined by eq. (14.1.16). Equation (15.1.10) can also be written as

$$\frac{1}{t_{\text{decay}}} = \frac{1}{2\pi^3} \left(\frac{m_e c^2}{\hbar} \right) G_{\text{weak}}^2 \ |M_{\text{nuc}}|^2 \ \Sigma \ F(Z, \epsilon_f), \tag{15.1.11}$$

where the dimensionless constant G_{weak} is related to g_{weak} by

$$g_{\text{weak}} = G_{\text{weak}} \left[m_e c^2 \left(\frac{\hbar}{m_e c} \right)^3 \right] = 4.714\,44 \times 10^{-38} \ \text{erg cm}^3 \ G_{\text{weak}}. \tag{15.1.12}$$

Another dimensionless constant often encountered is

$$\bar{G}_{\text{weak}} = \left(\frac{M_p}{m_e} \right)^2 G_{\text{weak}} = 3.371\,45 \times 10^6 \ G_{\text{weak}}, \tag{15.1.13}$$

where M_p/m_e is the ratio of the proton mass to the electron mass.

Solving eq. (15.1.11) for g_{weak}^2 gives

$$g_{\text{weak}}^2 = 2\pi^3 \left(\frac{\hbar/m_e c^2}{t_{\text{decay}}} \right) \frac{1}{\left(|M_{\text{nuc}}|^2 \ \Sigma \ F(Z, \epsilon_f) \right)} \left[m_e c^2 \left(\frac{\hbar}{m_e c} \right)^3 \right]^2 \tag{15.1.14}$$

$$= \left(1.3324 \times 10^{-47} \ \text{erg cm}^3 \right)^2 \frac{1 \ \text{second}}{t_{\text{decay}}} \frac{1}{\left(|M_{\text{nuc}}|^2 \ \Sigma \ F(Z, \epsilon_f) \right)}. \tag{15.1.15}$$

The neutron has a half life of $t_{1/2} = 10.24$ min, so $t_{\text{decay}} = 886.6$ s. The neutron–proton mass difference is $M_n - M_p = 2.3056 \times 10^{-27}$ g $= 1.293\,33$ MeV, so $\epsilon_f = 2.531$ and $\eta_f = (\epsilon_f^2 - 1)^{1/2} = 2.325$. The average energy of the electron in the decay is large enough that C^2 is only a few percent larger than 1, so, instead of using eq. (14.1.16), one can use eq. (6.2.7) in Volume 1 or eq. (14.4.8) to estimate $F(Z = 1, 2.531) \sim F(2.531) = 1.636$, giving $\log (F \ t_{1/2}) \sim 3.00$. Assuming perfect overlap between the neutron and proton wave functions, $M_{\text{nuc}} = 1$, and eq. (15.1.15) yields

$$g_{\text{weak}} \sim \frac{3.50 \times 10^{-49} \ \text{erg cm}^3}{(\Sigma)^{1/2}}. \tag{15.1.16}$$

If one supposes that the neutron and the proton must have the same spin orientations, $\Sigma = 1$. However, since the spins of the emitted leptons can couple to unity, there is no reason to suppose that, as it changes from a neutron into a proton, the nucleon cannot flip its spin, leading to the choice $\Sigma = 3$. Going one step further, note that, since both the spin flip and the no spin flip transitions can occur, one should actually choose $\Sigma = 1 + 3 = 4$, giving

$$g_{\text{weak}} \sim 1.75 \times 10^{-49} \ \text{erg cm}^3. \tag{15.1.17}$$

It is of interest to explore a bit further the consequences of assuming the existence of a universal weak coupling constant in an interaction Hamiltonian that does not depend on the nuclear environment. In Table 15.1.1 are given the experimentally determined

Table 15.1.1 Lower limits on an assumed universal weak coupling constant g_{weak}

Reaction	ϵ_f	$t_{1/2}(s)$	Σ	F_0	F_Z	$f t_{1/2}$	$g_{weak} >$
$n \rightarrow p + e^- + \bar{\nu}_e$	2.531	614.4	4	1.64	1.69	1038	1.72
^{3}H(1/2+) $\rightarrow$ ^{3}He(1/2+) + e^- + $\bar{\nu}_e$	1.0364	3.888/8	4	2.0/-6	2.9/-6	1128	1.65
^{6}He(0+) $\rightarrow$ ^{6}Li(1+) + e^- + $\bar{\nu}_e$	7.863	0.8067	6	926	995	802.7	1.60
^{11}C(3/2-) $\rightarrow$ ^{11}B(3/2-) + e^+ + ν_e	2.879	1223	1	3.70	3.15	3853	1.79
^{13}N(1/2+) $\rightarrow$ ^{13}C(1/2+) + e^+ + ν_e	3.346	597.9	1	9.12	7.62	4556	1.37
^{14}O(0+) $\rightarrow$ ^{14}N*(0+) + e^+ + ν_e	4.540	70.61	2	50.92	42.00	2966	1.44
^{15}O(1/2-) $\rightarrow$ ^{15}N*(1/2-) + e^+ + ν_e	4.389	122.2	1	42.3	34.8	4253	1.70
^{17}F(5/2+) $\rightarrow$ ^{17}O(5/2+) + e^+ + ν_e	4.403	64.49	4	43.0	34.5	2225	1.18
^{18}F(1+) $\rightarrow$ ^{18}O(0+) + e^+ + ν_e	2.239	6586	3	0.713	0.530	3491	1.08
^{19}Ne(1/2+) $\rightarrow$ ^{19}F(1/2+) + e^+ + ν_e	5.336	17.22	2	122	95.7	1648	1.36
^{21}Na(3/2+) $\rightarrow$ ^{21}Ne(3/2+) + e^+ + ν_e	5.95	22.49	1	217	167	3756	1.81
^{23}Mg(3/2+) $\rightarrow$ ^{23}Na(3/2+) + e^+ + ν_e	6.95	12.7	2	488	368	4674	1.13

characteristics of twelve "superallowed" transitions, so called because they are characterized by $ft_{1/2}$ values which are large enough that one may expect substantial spatial overlap between the wave functions of the nuclei participating in the transition. The first three transitions involve the conversion of a neutron into a proton and the last nine involve the conversion of a proton into a neutron. References to the literature on the chosen reactions may be found in the compilations cited in Section 14.10. In the table, entries in the columns labeled F_0 and F_Z give, respectively, the values of the phase space integral $F(Z, \epsilon_f)$ for the choices $Z = 0$ (no Coulomb effects) and $Z = $ the charge of the daughter nucleus. The lower limits on g_{weak} given in the last column in units of 10^{-49} erg cm^3 follow from eq. (15.1.10) on the assumptions that (1) $M_{\text{nuc}} = 1$, (2) the statistical factor Σ is the one given in the fourth column of the table, and (3) Z in $F(Z, \epsilon_f)$ is the charge of the daughter nucleus.

In an attempt to achieve as much consistency as possible, choices for Σ in several cases have been made which are difficult to defend. In spite of this, given that many nuclear structure characteristics enter into the actual matrix element ($\overline{M_{\text{nuc}}} < 1$), it is remarkable that the estimates are as close to one another as they are. Another factor which may account for some of the differences is the possibility that (as it is in fact) the Hamiltonian is more complicated than the one adopted here.

From the very beginning of beta-decay theory, it was recognized that five different terms, each with a different coupling constant, could be present in a relativistically invariant interaction Hamiltonian. As the experimental data base grew and the theory developed, it became evident that, of the five relativistically invariant terms (called scalar, pseudoscalar, vector, axial vector, and tensor), the vector and axial vector terms dominate. For the vector interaction, the selection rules for allowed transitions are zero parity change and zero spin change. For the axial vector interaction, the selection rules for allowed transitions are zero parity change and unit spin change.

It is appropriate to next explore the consequences of assuming two coupling constants g_0 and g_1, pertaining, respectively, to vector transitions and to axial vector transitions. The expressions for decay rates are modified in the following way. When there is no spin change and no parity change,

$$g_{\text{weak}}^2 \Sigma \rightarrow g_0^2 \times (1 \text{ or } 2). \qquad (15.1.18)$$

When there is a unit spin change and no parity change,

$$g_{\text{weak}}^2 \Sigma \rightarrow 3g_1^2 \times (1 \text{ or } 2). \qquad (15.1.19)$$

When both possibilities can occur,

$$g_{\text{weak}}^2 \Sigma \rightarrow (g_0^2 + 3g_1^2) \times (1 \text{ or } 2). \qquad (15.1.20)$$

In eqs. (15.1.18)–(15.1.20), the factor of 2 enters when it is reasonable to assume that either of two nucleons in the nucleus can undergo the same beta transition.

Two historically popular reactions for estimating the coupling constants are the decays $^{14}\text{O}(0+) \rightarrow {}^{14}\text{N}^*(0+) + e^+ + \nu_e$, where $^{14}\text{N}^*$ denotes the first excited state of ^{14}N, and

Table 15.1.2 Zeroth order lower limits on the vector weak coupling constant g_0 from 0^+–0^+ superallowed nuclear beta decays

Reaction	ϵ_f	$t_{1/2}(s)$	F_0	F_Z	$f\,t_{1/2}$	g_0	F'_Z	$f'\,t_{1/2}$
$^{14}O(e^+\nu_e)^{14}N^*$	4.539	70.61	50.92	42.00	2966	1.44	42.69	3037
$^{26m}Al(e^+\nu_e)^{26}Mg$	7.282	6.345	622.6	457.2	2901	1.46	477.8	3034
$^{34}Cl(e^+\nu_e)^{32}Ar$	9.748	1.526	2786	1760	2682	1.51	1997	3050
$^{38m}K(e^+\nu_e)^{38}Ca$	10.827	0.9240	4755	2852	2635	1.53	3296	3048
$^{42}Sc(e^+\nu_e)^{42}Ti$	11.574	0.6799	6673	3793	2576	1.55	4468	3041
$^{46}V(e^+\nu_e)^{42}Cr$	12.798	0.4224	11113	6001	2535	1.56	7198	3043
$^{50}Mn(e^+\nu_e)^{50}Fe$	13.936	0.2831	17106	8750	2479	1.58	10726	3039
$^{54}Co(e^+\nu_e)^{54}Ni$	15.130	0.1932	25863	12542	2423	1.59	15741	3045

$n \to p + e^- + \bar{\nu}_e$. Using eq. (15.1.18) with $M_{nuc} = 1$ and $\Sigma = 2$, the transition from ^{14}O to $^{14}N^*$ suggests that

$$g_0 \sim 1.44 \times 10^{-49} \text{ erg cm}^3. \tag{15.1.21}$$

Using eqs. (15.1.20) and (15.1.21) with $M_{nuc} = 1$ and $\Sigma = 4$, the neutron decay reaction then suggests

$$g_1 \sim 1.80 \times 10^{-49} \text{ erg cm}^3 = 1.25\, g_0. \tag{15.1.22}$$

The lower limit on g_1 suggested by the reaction $^6He(0+) \to {}^6Li(1+) + e^- + \bar{\nu}_e$ is $g_1 \geq 1.60 \times 10^{-49}$ erg cm^3. One interpretation of the two different estimates for g_1 is that the matrix element for the ^6He decay is $\sim 12\%$ smaller than unity. But, of course, the matrix element for the ^{14}O decay could also be less than unity, implying that $g_0 > 1.44 \times 10^{-49}$ erg cm^3 and that, if indeed the neutron and proton wave functions overlap perfectly, $g_1 < 1.805 \times 10^{-49}$ erg cm^3.

This uncertainty has been to a large extent removed by J. C. Hardy *et al.* (1990) who take into account several higher order effects, including the fact that the nuclear charge is distributed over a finite volume, and determine that the effective $f\,t_{1/2}$ values for eight superallowed $0^+ \to 0^+$ transitions are essentially identical. The decay properties of the selected transitions are given in Table 15.1.2.

The entries in the first seven columns have already been defined. The values of g_0 in the seventh column follow from the $f\,t_{1/2}$ value in the sixth column when $\Sigma = 2$ and $M_{nuc} = 1$. The fact that these g_0 values increase monotonically with increasing atomic number suggests that some effect associated with charge and/or mass has not been properly taken into account. In fact, when the finite size of the nuclear charge distribution and the charge distribution of orbital electrons are taken into account in obtaining solutions of the Dirac equation for the positrons, Hardy *et al.* (1990) obtain the phase-space functions F'_Z given in the eighth column of Table 15.1.2. The ratio F'_Z/F_Z increases monotonically with Z, in agreement with the intuitive expectation that spreading the nuclear charge over a larger volume increases the amplitude of the positron over the nuclear volume. It is interesting that

$$\frac{F'_Z}{F_Z} = 1.05 + 0.011\,(Z - 10) \tag{15.1.23}$$

and

$$\frac{F'_Z}{F_Z} = 1.07 + 0.0056\,(A - 22) \tag{15.1.24}$$

provide good fits to seven of the eight decays, the aluminum decay being the exception.

The values of $ft_{1/2}$ which follow from the choice of the phase space factor F'_Z are given in the ninth column of Table 15.1.2. These values agree with one another to within $\pm 0.2\%$ at $ft_{1/2} = 3042$ s and the related value of the vector coupling constant is $g_0 = 1.42 \times 10^{-49}$ erg cm^3. When additional higher order corrections having to do with nuclear structure are made, Hardy et $al.$ (1990) find $ft_{1/2} = 3073.5 \pm 0.12\%$ s, corresponding to $g_0 = 1.415 \times 10^{-49}$ erg cm^3. The value of the axial vector coupling constant derived by using this value of g_0 in connection with neutron decay properties is $g_1 = 1.812 \times 10^{-49}$ erg cm$^3 = 1.28g_0$.

The full blown theory of beta interactions predicts, as functions of the coupling constants, correlations between electron–neutrino directions, electron direction and spin, and neutrino direction and spin. Correlations found in experiments on neutron decay can therefore be used in the context of the full blown theory to determine both coupling constants directly (D. Dubbers, W. Mampe, & J. Döhner, 1990). In one experiment, B. G. Erozolimskii & Yu. A. Mostovoi (1991) obtain $g_0 = 1.425 \times 10^{-49}$ erg cm^3 and $g_1 = -1.793 \times 10^{-49}$ erg cm$^3 = -1.26g_0$. The minus sign is a consequence of the fact that, as first predicted by T. D. Lee and C. N. Yang (1956) and experimentally supported by C. S. Wu et $al.$ (1957) and directly demonstrated by H. Frauenfelder et $al.$ (1957), parity conservation is completely violated and leptons are emitted with opposite helicities (see also T. D. Lee & C. S. Wu, 1965). Further discussion and references may be found in Hikasa et $al.$ (1992).

In summary, if one accepts as established that, in nuclear beta decay transitions, there are essentially only two universal coupling constants g_0 and g_1 that operate, then

$$g_0 \sim (1.42 - 1.43) \times 10^{-49} \text{ erg cm}^3 \tag{15.1.25}$$

and

$$g_1 \sim -(1.79 - 1.81) \times 10^{-49} \text{ erg cm}^3 \sim -(1.25 - 1.27)\,g_0. \tag{15.1.26}$$

If one equates g_{weak} with g_0, eq. (15.1.12) gives

$$G_{\text{weak}} = (3.01 - 3.03) \times 10^{-12}, \tag{15.1.27}$$

and eq. (15.1.13) gives

$$\bar{G}_{\text{weak}} = (1.01 - 1.02) \times 10^{-5}. \tag{15.1.28}$$

15.2 The charged-current interaction and muon decay

Identification of a positively charged penetrating particle in cosmic rays of mass of the order of 240 times the mass of an electron (eventually established as 207 times the mass of an electron), and now called an antimuon, was made by Seth H. Neddermeyer and Carl D. Anderson (1938). The particle decayed into a positron and what has subsequently been determined to be two particles, one of which is an electron neutrino and the other of which has been classified as a muon antineutrino. Muons, electrons, and neutrinos are members of the lepton family of elementary particles, with properties described by, e.g., Donald H. Perkins in *Introduction to High Energy Physics* (1987).

The cross terms in the Feynman–Gell-Mann Hamiltonian which involve only leptons associated with electrons and muons are

$$(\bar{e}v_e)^\dagger(\bar{\mu}v_\mu) + (\bar{\mu}v_\mu)^\dagger(\bar{e}v_e) = (\bar{v}_e e)(\bar{\mu}v_\mu) + (\bar{v}_\mu \mu)(\bar{e}v_e) \tag{15.2.1}$$

and they may be thought of as operators which are responsible for the reactions

$$\mu^- \rightarrow e^- + \bar{v}_e + v_\mu \tag{15.2.2}$$

and

$$\mu^+ \rightarrow e^+ + v_e + \bar{v}_\mu. \tag{15.2.3}$$

In the heuristic approach, the associated matrix elements for both decays are the same:

$$M_{\text{decay}} = \int \Psi_{v_\mu}^* \, \Psi_\mu \, \Psi_e^* \, \Psi_{\bar{v}_e}^* \, d\tau = \left(\frac{1}{\sqrt{V}}\right)^4 V = \frac{1}{V}. \tag{15.2.4}$$

Because three particles are involved, construction of the appropriate phase space factor is not completely trivial. In the barycentric system, the sum of the linear momenta of the three particles is zero, so

$$\mathbf{p}_1 + \mathbf{p}_2 + \mathbf{p}_3 = 0, \tag{15.2.5}$$

where $\mathbf{p}_1$ and $\mathbf{p}_2$ are the momenta of the neutrinos and $\mathbf{p}_3$ is the momentum of the electron. Conservation of energy requires that

$$E_f = m_\mu c^2 = c \left[p_1 + p_2 + \sqrt{(m_e c)^2 + p^2} \right], \tag{15.2.6}$$

where $E_f = m_\mu c^2$ is the rest mass energy of the muon, p_1 and p_2 are the absolute values of the linear momenta of the neutrinos, and $p = p_3$ is the absolute value of the linear momentum of the electron.

The three vectors $\mathbf{p}_1$, $\mathbf{p}_2$, and $\mathbf{p}$ lie in a plane. For a given value of p, consider an ellipse such that the quantities p_1 and p_2 are the distances from the foci to a point on the ellipse. The two foci are separated by the distance $p = 2ae$, where a is the semimajor axis of the ellipse and e is the eccentricity. From eq. (15.2.6),

$$2a = p_1 + p_2 = \frac{E_f}{c} - \sqrt{p^2 + (m_e c)^2}. \tag{15.2.7}$$

Since the mass of the muon and the mass of the electron are related by $m_\mu \sim 207\, m_e$, for most values of p, $\sqrt{(p)^2 + (m_e c)^2} \sim p$, so, in good approximation,

$$2a \sim \frac{E_f}{c} - p, \tag{15.2.8}$$

and

$$e = \frac{p}{2a} \sim \frac{cp}{E_f - cp}. \tag{15.2.9}$$

The semiminor axis b of the ellipse is related to the semimajor axis a by

$$b^2 = a^2(1 - e^2) = a^2 \left[1 - \frac{1}{4} \left(\frac{p}{a} \right)^2 \right] = a^2 - \frac{1}{4}\, p^2$$

$$\sim \frac{1}{4c^2} \left[(E_f - cp)^2 - (cp)^2 \right]. \tag{15.2.10}$$

Rotating the ellipse about its major axis defines a spheroid of volume in momentum space:

$$P_{1,2} = \frac{4\pi}{3} b^2 a \sim \frac{\pi}{6c^3} \left[(E_f - cp)^3 - (E_f - cp)\,(cp)^2 \right]$$

$$= \frac{\pi}{6c^3} \left[E_f^3 - 3\, E_f^2\,(cp) + 2\, E_f\,(cp)^2 \right]. \tag{15.2.11}$$

The derivative of the volume $P_{1,2}$ with respect to E_f is

$$\frac{\partial P_{1,2}}{\partial E_f} = \frac{\partial}{\partial E_f} \left[\frac{4\pi}{3} \left(a^2 - \frac{p^2}{4} \right) a \right] = \frac{4\pi}{3} \left(3a^2 - \frac{p^2}{4} \right) \frac{\partial a}{\partial E_f}$$

$$= \frac{2\pi}{c} \left(a^2 - \frac{p^2}{12} \right)$$

$$= \frac{\pi}{2c^2} \left[E_f^2 + (m_e c^2)^2 - 2\, E_f \sqrt{(cp)^2 + (m_e c^2)^2} + \frac{2}{3}(cp)^2 \right]$$

$$\sim \frac{\pi}{2c^2} \left[E_f^2 - 2\, E_f\,(cp) + \frac{2}{3}(cp)^2 \right]. \tag{15.2.12}$$

As a first approximation to the differential density in phase space for the three leptons, one can now write

$$d\rho_f = \frac{V\, 4\pi p^2\, dp}{h^3} \frac{V\, \partial P_{1,2}/\partial E_f}{h^3}$$

$$\sim 2\pi^2\, V^2 \left(\frac{1}{ch^2} \right)^3 p^2\, dp \left[E_f^2 - 2\, E_f\,(cp) + \frac{2}{3}\,(cp)^2 \right]. \tag{15.2.13}$$

The differential decay probability is

$$d\omega_\mu = \frac{2\pi}{\hbar}\, g_\mu^2\, M_{\text{decay}}^2\, d\rho_f\, \Sigma, \tag{15.2.14}$$

where Σ is a statistical factor related to spin changes and, as a precaution, g_{weak} has been temporarily replaced by g_μ. Performing the integration over electron momentum ($cp = 0 \to (E_f/2)[1 - (m_e c^2/E_f)^2]^{1/2}$, or $cp = 0 \to \sim E_f/2$) one obtains

$$\frac{1}{\tau_\mu} = \omega_\mu = \frac{7\pi^3}{120} \frac{(m_e c^2)^5}{\hbar \, (ch)^6} \, g_\mu^2 \, \Sigma \, (\epsilon_f - 1)^3 \left[\epsilon_f^2 + \frac{11}{7} \epsilon_f + \frac{2}{7} \right], \tag{15.2.15}$$

where $\epsilon_f = m_\mu/m_e \sim 207$. Setting $(\epsilon_f - 1)^3 \sim \epsilon_f^3$ and neglecting the second two terms in square brackets, one has

$$\frac{1}{\tau_\mu} \sim \left(\frac{0.7 \, \Sigma}{4} \right) \frac{1}{192\pi^3} \left(\frac{g_\mu}{(c\hbar)^3} \right)^2 \frac{E_f^5}{\hbar} \tag{15.2.16}$$

$$= \left(\frac{0.7 \, \Sigma}{4} \right) \frac{1}{192\pi^3} \frac{m_e c^2}{\hbar} G_\mu^2 \, \epsilon_f^5, \tag{15.2.17}$$

where G_μ and g_μ are related in the same way as G_{weak} and g_{weak} in eq. (15.1.12). In his Silliman lectures, Fermi (1951) used eq. (15.2.16) with the choice $\Sigma = 1$ to estimate $g_\mu = 3.3 \times 10^{-49}$ erg cm^3.

Equation (15.2.17) can be manipulated further to yield eq. (15.1.15) with the definition

$$F(Z, \epsilon_f) \, \Sigma = \frac{1}{96} \left(\frac{0.7 \, \Sigma}{4} \right) \left(\frac{m_\mu}{m_e} \right)^5 = \frac{7 \, \Sigma}{3840} \left(\frac{m_\mu}{m_e} \right)^5. \tag{15.2.18}$$

Inserting numbers,

$$F(Z, \epsilon_f) \sim \frac{7}{3840} (207)^5 = 6.93 \times 10^8. \tag{15.2.19}$$

With $\tau_\mu = 2.20 \times 10^{-6}$ s,

$$F(Z, \epsilon_f) \, \tau_\mu = 1524 \text{ s} \tag{15.2.20}$$

or

$$f t_{1/2} \sim 1056 \text{ s}. \tag{15.2.21}$$

Using eq. (15.2.20) in eq. (15.1.15) yields

$$g_\mu \sim \frac{3.41 \times 10^{-49}}{\Sigma^{1/2}} \text{ erg cm}^3. \tag{15.2.22}$$

Once again, both zero and unit spin changes are permitted, so it is appropriate to adopt the algorithm given by eq. (15.1.20):

$$\Sigma \, g_\mu^2 = g_0^2 \left(1 + 3 \left(\frac{g_1}{g_0} \right)^2 \right). \tag{15.2.23}$$

If the axial vector and vector coupling constants for muon decay are related in the same way as in nuclear beta decay ($g_1 \sim 1.26\, g_0$), $\Sigma = 5.763$ and

$$g_\mu = g_0 \sim 1.42 \times 10^{-49} \text{ erg cm}^3 = g_{\text{weak}}. \tag{15.2.24}$$

It would thus appear that the weak interaction coupling constants for muon decay are identical with those for nuclear beta decay.

However, the heuristic estimate of the volume in momentum space available to the two neutrinos for a given electron momentum is only approximately correct. One might just as well have estimated this volume by rotating the ellipse defined by the three momentum vectors of the emitted leptons about the minor axis of the ellipse to obtain

$$
\begin{aligned}
P'_{1,2} = \frac{4\pi}{3} a^2 b &\sim \frac{4\pi}{3} \left(\frac{E_f - cp}{2c} \right)^2 \left(\frac{E_f^{1/2} (E_f - cp)^{1/2}}{2c} \right) \\
&= \frac{\pi}{6c^3} E_f^{1/2} (E_f - cp)^{5/2},
\end{aligned} \tag{15.2.25}
$$

which yields the derivative

$$
\begin{aligned}
\frac{\partial P'_{1,2}}{\partial E_f} &= \frac{\pi}{12c^3} \frac{1}{E_f^{1/2}} \left[(E_f - cp)^{5/2} + 5E_f (E_f - cp)^{3/2} \right] \\
&= \frac{5\pi}{12c^3} \frac{(E_f - cp)^{3/2}}{E_f^{1/2}} \left[E_f + \frac{1}{5} (E_f - cp) \right]
\end{aligned} \tag{15.2.26}
$$

To evaluate the transition probability with this differential phase space factor, one requires the definite integrals

$$\int_0^{E_f/2} (E_f - x)^{3/2} x^2 \, dx = \frac{2}{5} \left(\frac{E_f}{2} \right)^{9/2} \left(1 + \frac{4}{7} \left(1 + \frac{2}{9} \right) \right) \tag{15.2.27}$$

and

$$\int_0^{E_f/2} (E_f - x)^{5/2} x^2 \, dx = \frac{2}{7} \left(\frac{E_f}{2} \right)^{11/2} \left(1 + \frac{4}{9} \left(1 + \frac{2}{11} \right) \right). \tag{15.2.28}$$

Then, for the density of final states one finds

$$
\begin{aligned}
\rho'_f &= V^2 \frac{\pi^2}{24} \frac{E_f^5}{(ch)^6} \frac{1}{\sqrt{2}} \left[\left(1 + \frac{44}{63} \right) + \frac{1}{14} \left(1 + \frac{52}{99} \right) \right] \\
&\sim V^2 \frac{\pi^2}{24} \frac{E_f^5}{(ch)^6} \frac{1.70 + 0.11}{\sqrt{2}}
\end{aligned} \tag{15.2.29}
$$

and for the decay probability one has

$$\frac{1}{\tau'_\mu} \sim \left(\frac{1.28\, \Sigma}{4} \right) \frac{1}{192\pi^3} \left(\frac{g_\mu}{(c\hbar)^3} \right)^2 \frac{E_f^5}{\hbar}. \tag{15.2.30}$$

The average of the two estimates expressed by eqs. (15.2.30) and (15.2.16) is, to within 1%,

$$\frac{1}{\tau_\mu} = \left(\frac{\Sigma}{4}\right) \frac{1}{192\pi^3} \left(\frac{g_\mu}{(c\hbar)^3}\right)^2 \frac{E_f^5}{\hbar}. \tag{15.2.31}$$

Repeating the development which leads to eq. (15.2.22), but making the replacement

$$\left(\frac{0.7\ \Sigma}{4}\right) \rightarrow \frac{\Sigma}{4}, \tag{15.2.32}$$

one obtains

$$g_\mu \sim \frac{2.85 \times 10^{-49}}{\Sigma^{1/2}}\ \text{erg cm}^3. \tag{15.2.33}$$

From this it follows that, if g_μ for muon decay is the same as g_0 for nuclear beta decay, then $\Sigma \sim 4$, or

$$g_1 = g_0 = g_\mu = g_{\text{weak}} \sim 1.42 \times 10^{-49}\ \text{erg cm}^3. \tag{15.2.34}$$

Thus, it appears that, in contrast with nuclear beta decay, the axial vector and vector coupling constants for muon decay have the same absolute value.

This result has been achieved in a roundabout way with a very simplistic, but instructive, version of weak interaction theory. However, eq. (15.2.30) with $\Sigma = 4$ is, in fact, the exact answer for the muon decay rate predicted by the full blown V-A theory when the axial vector and vector coupling strengths are taken to be equal in absolute magnitude and of opposite sign (see, e.g., J. D. Bjorken & S. D. Drell, *Relativistic Quantum Mechanics*, 1964).

The fact that the vector coupling constant for a pure leptonic decay appears to be the same as that for nuclear beta decay lends a "universality" status to the vector portion of the weak interaction. The fact that, for muon decay, the axial vector and vector coupling constants are the same in absolute magnitude enhances the universality status of the complete theory. In this interpretation, the fact that the effective axial vector coupling constant for nuclear beta decay differs from that for lepton decay is to be attributed to the influence of the quark–pion makeup of nucleons.

15.3   Annihilation of electron–positron pairs into neutrino–antineutrino pairs and the associated energy-loss rate when electrons are not degenerate

Statistical equilibrium requires that positrons exist at a well defined equilibrium abundance in the stellar interior (see Section 4.10 in Volume 1). Positrons are created as partners with newly created electrons, primarily by way of an electromagnetic interaction between high energy photons and existing charged particles. Just as the reaction $\gamma + Ze \rightarrow e^+ + e^-$

creates electron–positron pairs, so the reaction $e^+ + e^- \rightarrow \gamma + \gamma$ is the primary means by which pair equilibrium is maintained.

However, from eqs. (15.1.1) and (15.1.2), one of the square products in the charged-current Hamiltonian density is

$$g_{\text{weak}} \, (\bar{e} \nu_e)^\dagger (\bar{e} \nu_e) = g_{\text{weak}} \, (\bar{\nu}_e e) \, (\bar{e} \nu_e). \tag{15.3.1}$$

Adopting the principles of particle number conservation and of lepton number conservation, the expression in eq. (15.3.1) may be thought of as an operator which is responsible for the reaction

$$e^+ + e^- \rightarrow \nu_e + \bar{\nu}_e, \tag{15.3.2}$$

leading to the interpretation of e and $\bar{e}$ as destruction operators for electrons and positrons, respectively, and ν_e and $\bar{\nu}_e$ as creation operators for antineutrinos and neutrinos, respectively.

One can guess that the matrix element associated with e^+–e^- annihilation into a neutrino–antineutrino pair looks something like

$$M_{\nu_e \bar{\nu}_e \leftarrow e^+ e^-} = \int \Psi_{\nu_e}^* \, \Psi_{\bar{\nu}_e}^* \, \Psi_{e^+} \, \Psi_{e^-} \, d\tau_e = \frac{1}{\sqrt{V}} \frac{1}{\sqrt{V}} \left(\frac{C}{\sqrt{V}\sqrt{V}} \right) V = \frac{C}{V}, \tag{15.3.3}$$

where C, which takes into account the Coulomb attraction between the electron and the positron, is given by eqs. (13.1.5) and (13.1.6) with $Z = 1$. That is, $C^2 = 2\pi\eta/(1 - e^{-2\pi\eta})$, where $\eta = e^2/\hbar v = (e^2/\hbar c)(c/v)$. For non-relativistic velocities $C^2 \rightarrow 2\pi\eta$ and for relativistic velocities $C^2 \rightarrow 1$.

In the barycentric (zero linear momentum) system, the neutrino and antineutrino are constrained to move exactly opposite to each other, and each neutrino has the same unique energy,

$$cp_{\nu_e} = \frac{1}{2} E_f, \tag{15.3.4}$$

where E_f is the total energy liberated in the destruction of the electron–positron pair. The phase space available for the neutrinos is twice that available for each neutrino and the density of final states is then

$$\rho_f = \frac{d}{dE_f} \left(2 \, V \, \frac{4\pi \, p_{\nu_e}^2 \, dp_{\nu_e}}{h^3} \right) = \frac{d}{d(2cp_{\nu_e})} \left(V \, \frac{8\pi \, p_{\nu_e}^2 \, d(cp_{\nu_e})}{h^3 c} \right)$$

$$= V \, \frac{4\pi \, p_{\nu_e}^2}{h^3 c} = V \, \pi \, \frac{E_f^2}{(hc)^3}. \tag{15.3.5}$$

The probability of a transition is

$$\omega_{\nu_e \bar{\nu}_e \leftarrow e^+ e^-} = \frac{2\pi}{\hbar} \, g_{\text{weak}}^2 \, |M_{\nu_e \bar{\nu}_e \leftarrow e^+ e^-}|^2 \, V \, \pi \, \frac{E_f^2}{(hc)^3} \, \Sigma, \tag{15.3.6}$$

where Σ is the statistical weight for the transition, and the related cross section is

$$\sigma_{\nu_e \bar{\nu}_e \leftarrow e^+ e^-} = \frac{\omega_{\nu_e \bar{\nu}_e \leftarrow e^+ e^-}}{(v/V)} = \frac{2\pi}{\hbar} g^2_{\text{weak}} |M_{\nu_e \bar{\nu}_e \leftarrow e^+ e^-}|^2 V \pi \frac{E_f^2}{(hc)^3} \Sigma \frac{V}{v}. \quad (15.3.7)$$

From eqs. (15.3.3), (15.3.4), and (15.3.7), one has

$$\sigma_{\nu_e \bar{\nu}_e \leftarrow e^+ e^-} = \frac{1}{4\pi} g^2_{\text{weak}} C^2 \left(\frac{E_f}{m_e c^2}\right)^2 \left(\frac{m_e}{\hbar^2}\right)^2 \Sigma \frac{c}{v} \quad (15.3.8)$$

or, using the relationship between g_{weak} and G_{weak} defined by eq. (15.1.12),

$$\sigma_{\nu_e \bar{\nu}_e \leftarrow e^+ e^-} = \frac{1}{4\pi} G^2_{\text{weak}} \left(\frac{\hbar}{m_e c}\right)^2 C^2 \epsilon_f^2 \Sigma \frac{c}{v}. \quad (15.3.9)$$

In these equations,

$$\epsilon_f = \frac{E_f}{m_e c^2}. \quad (15.3.10)$$

Since electrons and positrons are not identical particles, the Pauli exclusion principle does not place any restriction on the allignment of electron and positron spins. Hence, one quarter of the time e^+–e^- pairs interact in the singlet spin state and three quarters of the time they interact in the triplet spin state. However, in the barycentric system, the neutrino and antineutrino move in opposite directions. Since neutrinos are created left handed and antineutrinos are created right handed, the spin of the final lepton state is unity. If one assumes that, when the electron and positron interact in the spin zero state, $\Sigma = 3$, and, when they interact in the triplet state, $\Sigma = 1$, then in all equations for the cross section one has that

$$\Sigma = \frac{1}{4} 3 + \frac{3}{4} 1 = \frac{3}{2}, \quad (15.3.11)$$

giving

$$\sigma_{\nu_e \bar{\nu}_e \leftarrow e^+ e^-} = \frac{3}{8\pi} G^2_{\text{weak}} \left(\frac{\hbar}{m_e c}\right)^2 C^2 \epsilon_f^2 \frac{c}{v}. \quad (15.3.12)$$

Choosing $g_{\text{weak}} \sim 1.43 \times 10^{-49}$ erg cm^3, or $G_{\text{weak}} \sim 3.03 \times 10^{-12}$, the cross section becomes

$$\sigma_{\nu_e \bar{\nu}_e \leftarrow e^+ e^-} \sim 1.634 \times 10^{-45} \text{ cm}^2 C^2 \epsilon_f^2 \frac{c}{v} = 1.634 \times 10^{-21} \text{ barn } C^2 \epsilon_f^2 \frac{c}{v}. \quad (15.3.13)$$

The methodology that has been adopted to estimate the cross section given by eq. (15.3.12) is entirely heuristic. Solving the Dirac equation to obtain relativistically correct wave functions for the leptons and adopting the relativistically invariant V-A Hamiltonian for the weak interaction (see Sections 15.4 and 15.5), H. Y. Chiu & R. C. Stabler (1961) and M. J. Levine (1963) find

$$\sigma_{\nu_e \bar{\nu}_e \leftarrow e^+ e^-} = \sigma_{\text{weak}} (\epsilon_f^2 - 1) \frac{c}{v}, \quad (15.3.14)$$

where

$$\sigma_{\text{weak}} = \frac{1}{3\pi} \, G^2_{\text{weak}} \left(\frac{\hbar}{m_e c} \right)^2 = 1.45 \times 10^{-45} \, \text{cm}^2. \tag{15.3.15}$$

The steps involved in arriving at this version of the cross section are presented in Section 15.5.

The ratio of the cross section $\sigma_{\text{heuristic}}$ given by eq. (15.3.12) and the cross section σ_{fancy} given by eqs. (15.3.14) and (15.3.15) is

$$\frac{\sigma_{\text{heuristic}}}{\sigma_{\text{fancy}}} = \frac{9}{8} C^2 \frac{\epsilon_f^2}{\epsilon_f^2 - 1}. \tag{15.3.16}$$

At non-relativistic velocities, the energy in the center of momentum system is nearly the same as in a stationary frame of reference, so $\epsilon_f \sim 2$ and the two cross sections are related by $\sigma_{\text{heuristic}} \sim 1.5 \, C^2 \, \sigma_{\text{fancy}}$. At relativistic velocities, $(\epsilon_f^2 - 1) \to \epsilon_f^2$ and $C^2 \to 1$, so $\sigma_{\text{heuristic}} \sim (9/8) \, \sigma_{\text{fancy}} = 1.125 \, \sigma_{\text{fancy}}$.

To estimate the total rate at which neutrinos and antineutrinos are produced in the stellar context, one may make use of (1) eq. (4.10.7) in Volume 1 which defines the product of electron–positron number densities as the square of a quantity I which depends only on the temperature and (2) eqs. (4.10.13) and (4.10.31) in Volume 1 which, respectively, provide estimates of I in the low temperature limit and in the high temperature limit. In the low temperature limit, the rate of neutrino pair production in the heuristic approximation is

$$r_{\nu_e \bar{\nu}_e} = n_- n_+ \langle \sigma \, v \rangle$$

$$= \frac{3}{4\pi} \left(\frac{2\pi m_e \, kT}{h^2} \right)^3 e^{-2m_e c^2 / kT} \, G^2_{\text{weak}} \left(\frac{\hbar}{m_e c} \right)^2 c \left\langle \epsilon_f^2 \, \frac{2\pi \eta}{1 - e^{-2\pi \eta}} \right\rangle, \tag{15.3.17}$$

where angle brackets denote an average over the relative velocity distribution and the Coulomb factor C^2 has been entered as an explicit function of the quantity $2\pi \eta$.

A determination of $\langle \epsilon_f^2 \, C^2 \rangle$ in the general case requires a numerical integration. However, at non-relativistic velocities, $\epsilon_f \sim 2$, and a reasonably good approximation is achieved by taking the Coulomb factor outside of the integral and replacing it by

$$\langle C^2 \rangle \sim \frac{\langle 2\pi \eta \rangle}{1 - \exp\left(-\langle 2\pi \eta \rangle \right)}, \tag{15.3.18}$$

where $\langle 2\pi \eta \rangle$ is an average over the distribution of electrons relative to positrons in the non-relativistic limit. In the center of mass system, where the reduced mass is $\mu_e = m_e/2$, one has

$$\left\langle \frac{c}{v} \right\rangle = \left(\frac{2}{\pi} \frac{\mu_e c^2}{kT} \right)^{1/2} = \left(\frac{m_e c^2}{\pi kT} \right)^{1/2} = \frac{1.374}{T_9^{1/2}}, \tag{15.3.19}$$

so that

$$\langle 2\pi \eta \rangle = 2\pi \, \frac{e^2}{\hbar c} \left(\frac{m_e c^2}{\pi kT} \right)^{1/2} = \frac{0.0630}{T_9^{1/2}}. \tag{15.3.20}$$

Hence,

$$\langle C^2 \rangle \sim \frac{0.063/\sqrt{T_9}}{1 - \exp\left(-0.063/\sqrt{T_9}\right)}. \tag{15.3.21}$$

For temperatures such that $(\langle 2\pi\eta \rangle) \gg 1$, or $T_6 \ll 4$,

$$\langle C^2 \rangle \sim \langle 2\pi\eta \rangle = 0.063/\sqrt{T_9}. \tag{15.3.22}$$

For temperatures such that $(\langle 2\pi\eta \rangle) < 1$, or $T_6 > 4$,

$$\langle C^2 \rangle \sim \left(1 - \frac{1}{2}\langle 2\pi\eta \rangle + \frac{1}{6}(\langle 2\pi\eta \rangle)^2 + \cdots\right)^{-1}$$

$$= \left(1 - \frac{0.0315}{\sqrt{T_9}} + \frac{0.000\,6615}{T_9} + \cdots\right)^{-1}. \tag{15.3.23}$$

Adopting now the cross section given by eqs. (15.3.14) and (15.3.15), the equivalent of eq. (15.3.17) becomes

$$r_{\nu_e \bar{\nu}_e} = n_- n_+ \langle \sigma \, v \rangle \sim 4 \left(\frac{2\pi m_e \, kT}{h^2}\right)^3 e^{-2m_e c^2/kT} \sigma_{\text{weak}} \, c \, \langle \epsilon_f^2 - 1 \rangle \langle C^2 \rangle, \tag{15.3.24}$$

where the factor $\langle C^2 \rangle$ has been introduced to take into account the Coulomb interaction between the charged leptons. Inserting values of the fundamental constants, one has, as a final approximation to the neutrino-production rate in the non-relativistic regime,

$$r_{\nu_e \bar{\nu}_e} = n_- n_+ \langle \sigma \, v \rangle \sim 1.016 \times 10^{24} \, T_9^3 \, e^{-11.86/T_9} \langle \epsilon_f^2 - 1 \rangle \langle C^2 \rangle \, \text{cm}^{-3} \, \text{s}^{-1}. \tag{15.3.25}$$

The rate at which energy is lost from a star due to neutrino–antineutrino pair production at low temperatures is $2m_e c^2 (= 1.637 \times 10^{-6}$ erg) times the rate of neutrino-pair production, so, in general,

$$\epsilon_{\nu_e \bar{\nu}_e} \sim 1.663 \times 10^{18} \, T_9^3 \, e^{-11.86/T_9} \left(\frac{\epsilon_f}{2}(\epsilon_f^2 - 1)\right) \langle C^2 \rangle \, \text{erg cm}^{-3} \, \text{s}^{-1}. \tag{15.3.26}$$

To the extent that $\epsilon_f = 2$ is an adequate approximation,

$$r_{\nu_e \bar{\nu}_e} \sim 3.05 \times 10^{24} \, T_9^3 \, e^{-11.86/T_9} \langle C^2 \rangle \, \text{cm}^{-3} \, \text{s}^{-1} \tag{15.3.27}$$

and

$$\epsilon_{\nu_e \bar{\nu}_e} \sim 4.99 \times 10^{18} \, T_9^3 \, e^{-11.86/T_9} \langle C^2 \rangle \, \text{erg cm}^{-3} \, \text{s}^{-1}. \tag{15.3.28}$$

At a temperatures of $T_9 \sim 0.3, 0.5, 0.8$, and 1.0, eq. (15.3.23) gives $\langle C^2 \rangle = 1.06, 1.045,$ 1.036, and 1.032, respectively, and eq. (15.3.28) gives $\epsilon_{\nu_e \bar{\nu}_e} = 9.67, 3.12 \times 10^7, 9.65 \times 10^{11},$ and 3.64×10^{13} erg cm^{-3} s^{-1}, respectively. This numerical exercise demonstrates that, at temperatures such that the rate at which neutrinos produced by $e^+ e^-$ annihilation carry off energy is important, the Coulomb interaction between charged leptons affects the loss rate by only a few percent.

At temperatures large enough that the approximation $\epsilon_f \sim 2$ is no longer valid, it is necessary to take into account the fact that number densities and cross sections in the center of momentum (CM) frame of reference which has been used to obtain an estimate of the cross section are not the same as in the stationary or laboratory (L) frame of reference in which it is convenient to define number densities. Let E_+ and $\mathbf{p}_+$ be, respectively, the energy and momentum of the positron and E_- and $\mathbf{p}_-$ be the energy and momentum of the electron in the L frame, and let E_T and $\mathbf{P}_T(=0)$ be the total energy and momentum in the CM frame. The velocity of the CM frame of reference with respect to the L frame is related to particle momenta and energies in the L frame by

$$\frac{\mathbf{v}_c}{c} = \frac{\mathbf{p}_+ + \mathbf{p}_-}{E_+ + E_-}, \tag{15.3.29}$$

where the $\mathbf{p}$s are in units such that the velocity of light $c = 1$.

In the laboratory frame of reference, differential particle densities of positrons and electrons are given by (see Section 4.10 in Volume 1)

$$\mathrm{d}n_{\pm} = 2 \, \frac{p_{\pm}^2 \, \mathrm{d}p_{\pm} \, \mathrm{d}\Omega_{\pm}}{h^3} \, f_{\pm} = 2 \, \frac{p_{\pm}^2 \, \mathrm{d}p_{\pm} \, \mathrm{d}\Omega_{\pm}}{h^3} \, \frac{1}{\mathrm{e}^{\mp\alpha + E_{\pm}/kT} + 1}, \tag{15.3.30}$$

where α is a chemical potential and $\mathrm{d}\Omega$ is a solid angle in the direction of the momentum vector. The quantity $(\mathrm{d}n_+, \mathrm{d}n_+\mathbf{v}_+)$ is a relativistic four vector, so one may perform the Lorentz transformation

$$(\mathrm{d}n_+)_{\mathrm{CM}} = \frac{1}{\sqrt{1 - v_c^2/c^2}} \left((\mathrm{d}n_+)_{\mathrm{L}} - \frac{\mathbf{v}_c}{c^2} \cdot (\mathrm{d}n_+)_{\mathrm{L}} \, \mathbf{v}_+ \right)$$

$$= (\mathrm{d}n_+)_{\mathrm{L}} \, \frac{1}{\sqrt{1 - v_c^2/c^2}} \left(1 - \frac{\mathbf{v}_c}{c^2} \cdot \mathbf{v}_+ \right), \tag{15.3.31}$$

and similarly for $\mathrm{d}n_-$. Since the quantity $(\mathrm{d}n_+\mathrm{d}n_-\sigma v)$ is an invariant, it is also true that

$$(\sigma v)_{\mathrm{L}} = (\sigma v)_{\mathrm{CM}} \, \frac{1}{1 - v_c^2/c^2} \left(1 - \frac{\mathbf{v}_c}{c^2} \cdot \mathbf{v}_+ \right) \left(1 - \frac{\mathbf{v}_c}{c^2} \cdot \mathbf{v}_- \right). \tag{15.3.32}$$

Next, note that

$$1 - \frac{v_c^2}{c^2} = 1 - \frac{|\mathbf{p}_+ + \mathbf{p}_-|^2}{(E_+ + E_-)^2}$$

$$= \frac{E_+^2 + E_-^2 + 2E_+E_- - (p_+^2 + p_-^2 + 2p_+p_- \cos\theta_{\pm})}{(E_+ + E_-)^2}$$

$$= \frac{2\,m^2 + 2E_+E_- - 2p_+p_- \cos\theta_{\pm}}{(E_+ + E_-)^2}, \tag{15.3.33}$$

where m is the mass of the electron in units where $c = 1$ and $\theta_\pm$ is the angle formed by the two momentum vectors. The scalar product of any two four vectors is an invariant. In particular,

$$E_T^2 - \mathbf{P}_T \cdot \mathbf{P}_T = (E_+ + E_-)^2 - (\mathbf{p}_+ + \mathbf{p}_-) \cdot (\mathbf{p}_+ + \mathbf{p}_-)$$

$$= E_+^2 + 2E_+ E_- + E_-^2 - (p_+^2 + p_-^2 + 2\mathbf{p}_+ \cdot \mathbf{p}_-) \tag{15.3.34}$$

or, since $\mathbf{P}_T = 0$,

$$E_T^2 = 2(m^2 + E_+ E_- - p_+ p_- \cos \theta_m). \tag{15.3.35}$$

Equations (15.3.33) and (15.3.35) give

$$1 - \frac{v_c^2}{c^2} = \frac{E_T^2}{(E_+ + E_-)^2}. \tag{15.3.36}$$

Further,

$$1 - \frac{\mathbf{v}_c}{c} \cdot \frac{\mathbf{v}_\pm}{c} = 1 - \frac{\mathbf{p}_+ + \mathbf{p}_-}{E_+ + E_-} \cdot \frac{\mathbf{p}_\pm}{E_\pm} = \frac{m^2 + E_+ E_- - p_+ p_- \cos \theta_\pm}{(E_+ + E_-) E_\pm} = \frac{E_T^2/2}{(E_+ + E_-) E_\pm}. \tag{15.3.37}$$

Thus, using eqs. (15.3.36) and (15.3.37) in eq. (15.3.32), it follows that

$$(\sigma v)_L = (\sigma v)_{CM} \frac{1}{4} \frac{E_T^2}{E_+ E_-}. \tag{15.3.38}$$

One can now write

$$r_{\nu_e \bar{\nu}_e} = \int \int (\sigma v)_{CM} \frac{1}{4} \frac{E_T^2}{E_+ E_-} \, dn_+ \, dn_-$$

$$= \sigma_{\text{weak}} \, c \int \int \left(\frac{E_T^2 - m^2}{m^2} \right) \frac{1}{4} \frac{E_T^2}{E_+ E_-} \, dn_+ \, dn_-$$

$$= \sigma_{\text{weak}} \, c \int \int (\epsilon_T^2 - 1) \frac{\epsilon_T^2}{\epsilon_+ \epsilon_-} \frac{1}{4} \left(\frac{2}{h^3} \right)^2 f_+ \, p_+^2 \, dp_+ \, d\Omega_+ \, f_- \, p_-^2 \, dp_- \, d\Omega_-, \tag{15.3.39}$$

where $\epsilon_\pm = (E_\pm / m_e c^2)$, σ_{weak} is given by eq. (15.3.15), and f_+ and f_- are given by eq. (15.3.30). Define an integral I_Ω over angles by

$$I_\Omega = \int \int (\epsilon_T^2 - 1) \, \epsilon_T^2 \, d\Omega_+ \, d\Omega_-. \tag{15.3.40}$$

Choosing the z axis along one of the momentum vectors and integrating over the solid angle associated with this vector, one has that

$$I_\Omega = 4\pi \int (\epsilon_T^2 - 1)\, \epsilon_T^2\, d\Omega_\pm$$

$$= 4\pi \int_0^{2\pi} \int_{-1}^1 (1 + 2\epsilon_+\epsilon_- - 2p_+p_-\cos\theta_\pm)$$

$$\times (2 + 2\epsilon_+\epsilon_- - 2p_+p_-\cos\theta_\pm)\, d\cos\theta_\pm\, d\phi_\pm. \qquad (15.3.41)$$

Since $\int_{-1}^1 \cos^N\theta\, d\cos\theta = 0$ when N is an odd integer, it follows that

$$I_\Omega = 4\pi \times 2 \times 2\pi \int_{-1}^1 [(1 + 2\epsilon_+\epsilon_-)(1 + \epsilon_+\epsilon_-) + 2p_+^2 p_-^2 \cos^2\theta_\pm]\, d\cos\theta_\pm$$

$$= 32\pi^2 \left[(1 + 2\epsilon_+\epsilon_-)(1 + \epsilon_+\epsilon_-) + \frac{2}{3}\, p_+^2 p_-^2 \right]$$

$$= \frac{(16\pi)^2}{3} \left[\epsilon_+^2 \epsilon_-^2 + \frac{1}{8}(9\epsilon_+\epsilon_- - 2\epsilon_+^2 - 2\epsilon_-^2) + \frac{5}{8} \right]. \qquad (15.3.42)$$

Combining eqs. (15.3.39), (15.3.40), and (15.3.42), one has

$$r_{\nu_e\bar{\nu}_e} = \sigma_{\text{weak}}\, c\, \frac{(16\pi)^2}{3} \left(\frac{kT}{hc}\right)^6 \int_0^\infty \int_0^\infty \frac{1}{\epsilon_+\epsilon_-}$$

$$\times \left[\epsilon_+^2 \epsilon_-^2 + \frac{1}{8}(9\epsilon_+\epsilon_- - 2\epsilon_+^2 - 2\epsilon_-^2) + \frac{5}{8} \right] f_+\, x_+^2\, dx_+ f_-\, x_-^2\, dx_-, \qquad (15.3.43)$$

where

$$x_\pm = \frac{cp_\pm}{kT}. \qquad (15.3.44)$$

Similarly,

$$\epsilon_{\nu_e\bar{\nu}_e} = \sigma_{\text{weak}}\, c\, \frac{(16\pi)^2}{3} \left(\frac{kT}{hc}\right)^6 m_e c^2 \int_0^\infty \int_0^\infty \frac{\epsilon_+ + \epsilon_-}{\epsilon_+\epsilon_-}$$

$$\times \left[\epsilon_+^2 \epsilon_-^2 + \frac{1}{8}(9\epsilon_+\epsilon_- - 2\epsilon_+^2 - 2\epsilon_-^2) + \frac{5}{8} \right] f_+\, x_+^2\, dx_+ f_-\, x_-^2\, dx_-, \qquad (15.3.45)$$

Obviously, exact solutions for $r_{\nu_e\bar{\nu}_e}$ and $\epsilon_{\nu_e\bar{\nu}_e}$ as functions of the chemical potential α and the temperature require numerical calculations. Here, attention is limited to an examination of just the leading term in each equation on the assumptions that (1) the electrons are non-degenerate, (2) energies are highly relativistic, and (3) the chemical potential is small enough that it is sufficient to keep just the first term in the Taylor expansions of $f_\pm$. Setting

$$f_\pm \sim e^{\pm\alpha - m_e c^2 \epsilon_\pm / kT} \qquad (15.3.46)$$

and

$$\epsilon = \sqrt{1 + \left(\frac{kT}{m_e c^2}\right)^2 x^2} \sim \frac{kT}{m_e c^2} x, \tag{15.3.47}$$

one obtains

$$r_{\nu_e \bar{\nu}_e} \sim \sigma_{\text{weak}} \, c \, \frac{(16\pi)^2}{3} \left(\frac{kT}{hc}\right)^6 \int_0^\infty \int_0^\infty \frac{1}{\epsilon_+ \epsilon_-} \, \epsilon_+^2 \epsilon_-^2 \, e^{-m_e c^2 (\epsilon_+ + \epsilon_-)/kT} \, x_+^2 \, dx_+ \, x_-^2 \, dx_-$$

$$\sim \sigma_{\text{weak}} \, c \, \frac{(16\pi)^2}{3} \left(\frac{kT}{hc}\right)^6 \left(\frac{kT}{m_e c^2}\right)^2 \int_0^\infty \int_0^\infty e^{-(x_+ + x_-)} \, x_+^3 \, dx_+ \, x_-^3 \, dx_-$$

$$= \sigma_{\text{weak}} \, c \, \frac{(16\pi)^2}{3} \left(\frac{kT}{hc}\right)^6 \left(\frac{kT}{m_e c^2}\right)^2 (3!)^2. \tag{15.3.48}$$

Similarly,

$$\epsilon_{\nu_e \bar{\nu}_e} \sim \sigma_{\text{weak}} \, c \, \frac{(16\pi)^2}{3} \left(\frac{kT}{hc}\right)^6 \left(\frac{kT}{m_e c^2}\right)^3 m_e c^2$$

$$\times \int_0^\infty \int_0^\infty e^{-(x_+ + x_-)} \, (x_+ + x_-) \, x_+^3 \, dx_+ \, x_-^3 \, dx_-.$$

$$= \sigma_{\text{weak}} \, c \, \frac{(16\pi)^2}{3} \left(\frac{kT}{hc}\right)^6 \left(\frac{kT}{m_e c^2}\right)^3 m_e c^2 \, 2 \, (4!)(3!). \tag{15.3.49}$$

Comparing these rates, one has

$$\epsilon_{\nu_e \bar{\nu}_e} = 8 \, kT \, r_{\nu_e \bar{\nu}_e}. \tag{15.3.50}$$

Equations (4.10.7) and (4.10.37) in Volume 1 give

$$n_- n_+ \sim (16\pi)^2 \left(\frac{kT}{hc}\right)^6 = 2.848 \times 10^{56} \, T_9^6 \, \text{cm}^{-6}$$

$$= (1.688 \times 10^{28} \, T_9^3 \, \text{cm}^{-3})^2. \tag{15.3.51}$$

Using this result in eq. (15.3.48), one has

$$r_{\nu_e \bar{\nu}_e} = (\sigma_{\text{weak}} \, c) \, \frac{4}{3} \left(\frac{3 \, kT}{m_e c^2}\right)^2 n_+ n_- = 4.237 \times 10^{21} \, T_9^8 \, \text{cm}^{-3} \, \text{s}^{-1}, \tag{15.3.52}$$

where σ_{weak} from eq. (15.3.15) has been used to obtain the numerical value on the rightmost side of the equation. Finally,

$$\epsilon_{\nu_e \bar{\nu}_e} \sim 2 \left(\frac{4 \, kT}{m_e c^2}\right) m_e c^2 \, r_{\nu_e \bar{\nu}_e} = 4.68 \times 10^{15} \, T_9^9 \, \text{erg} \, \text{cm}^{-3} \, \text{s}^{-1}. \tag{15.3.53}$$

In summary, in the general case, evaluation of $r_{\nu_e \bar{\nu}_e}$ and $\epsilon_{\nu_e \bar{\nu}_e}$ involves very tedious integrations described by eqs. (15.3.43)–(15.3.45). In the extreme relativistic limit, estimates of the rates are given by eqs. (15.3.48) and (15.3.49). Estimates in the non-relativistic limit are given by eqs. (15.3.27) and (15.3.28). To obtain estimates of rates when charged leptons are moderately relativistic, begin with eqs. (15.3.25) and (15.3.26), which require estimates of the averages over the electron–positron distributions of $(\epsilon_f^2 - 1)$ and $\epsilon_f (\epsilon_f^2 - 1)$, and adopt the approximation that

$$n_- n_+ = \left(\frac{2\pi m_e \, kT}{h^2} \right)^3 e^{-2m_e c^2 / kT} \, S_1^2, \tag{15.3.54}$$

where S_1 is a first order relativistic correction given by eq. (4.10.64) in Volume 1 as $S_1 \sim 1 + 0.316\,189\, T_9 (1 + 0.073\,777\, T_9)$. Altogether,

$$r_{\nu_e \bar{\nu}_e} = n_- n_+ \langle \sigma \, v \rangle \sim 4 \left(\frac{2\pi m_e \, kT}{h^2} \right)^3 e^{-2m_e c^2 / kT} \, S_1^2 \, \sigma_{\text{weak}} \, c \, \langle \epsilon_f^2 - 1 \rangle, \tag{15.3.55}$$

$$\epsilon_{\nu_e \bar{\nu}_e} = n_- n_+ \langle \sigma \, v \rangle \sim 4 \left(\frac{2\pi m_e \, kT}{h^2} \right)^3 e^{-2m_e c^2 / kT} \, S_1^2 \, \sigma_{\text{weak}} \, c \, \langle \epsilon_f (\epsilon_f^2 - 1) \rangle, \tag{15.3.56}$$

and the problem boils down to making estimates of $\langle \epsilon_f^2 - 1 \rangle$ and $\langle \epsilon_f (\epsilon_f^2 - 1) \rangle$ without going through the tedium of numerical integrations in the charged lepton barycentric system. The obvious and conservative first choices are

$$\langle \epsilon_f \rangle \sim 2 + \frac{3}{2} \frac{kT}{m_e c^2}, \tag{15.3.57}$$

$$\langle \epsilon_f^2 \rangle \sim 4 + 6 \frac{kT}{m_e c^2} + \frac{9}{4} \left(\frac{kT}{m_e c^2} \right)^2, \tag{15.3.58}$$

and

$$\langle \epsilon_f (\epsilon_f^2 - 1) \rangle \sim 6 \left(1 + \frac{3}{4} \frac{kT}{m_e c^2} \right) \left(1 + 2 \frac{kT}{m_e c^2} + \frac{3}{4} \left(\frac{kT}{m_e c^2} \right)^2 \right)$$

$$\sim 6 \left(1 + \frac{11}{4} \frac{kT}{m_e c^2} \right) = 6 \, (1 + 0.464 \, T_9). \tag{15.3.59}$$

So, for example, a first order relativistic approximation to the neutrino energy-loss rate due to $e^+ e^-$ pair annihilation is

$$\epsilon_{\nu_e \bar{\nu}_e} \sim 4.99 \times 10^{18} \, T_9^3 \, e^{-11.86 / T_9} \, (1 + 1.10 \, T_9) \, \text{erg cm}^{-3} \, \text{s}^{-1}. \tag{15.3.60}$$

15.4 The Dirac equation, plane-wave solutions, helicity eigenfunctions, and gamma matrices

Before embarking on a rigorous derivation of the cross section for neutrino pair formation using the formal V-A theory, it is useful to review some of the properties of the Dirac wave equation for electrons, some of the solutions of this equation, and some of the properties of these solutions. The equation is (Dirac, 1926, 1928)

$$-\frac{\hbar}{i}\frac{\partial \Psi}{\partial t}\left(\vec{\alpha}\cdot\left[\frac{\hbar}{i}\nabla-\frac{e}{c}\mathbf{A}\right]+e\phi+\beta\,m\right)\Psi, \tag{15.4.1}$$

where ϕ and $\mathbf{A}$ are, respectively, the scalar and vector potential of the electromagnetic field, $m = m_e$, β is the 4×4 matrix

$$\beta = \begin{pmatrix} 1 & 0 & 0 & 0 \\ 0 & 1 & 0 & 0 \\ 0 & 0 & -1 & 0 \\ 0 & 0 & 0 & -1 \end{pmatrix}, \tag{15.4.2}$$

and

$$\vec{\alpha} = \begin{pmatrix} 0 & \vec{\sigma} \\ \vec{\sigma} & 0 \end{pmatrix} \tag{15.4.3}$$

is a 4×4 matrix made up of elements which are the 2×2 Pauli spin matrices

$$\sigma_x = \begin{pmatrix} 0 & 1 \\ 1 & 0 \end{pmatrix},\ \sigma_y = \begin{pmatrix} 0 & -i \\ i & 0 \end{pmatrix},\ \text{and}\ \sigma_z = \begin{pmatrix} 1 & 0 \\ 0 & -1 \end{pmatrix}, \tag{15.4.4}$$

which satisfy the anticommutation relations

$$\sigma_i\sigma_j + \sigma_j\sigma_i = \begin{cases} 0 & \text{if } i \neq j, \\ 2 & \text{if } i = j, \end{cases} \tag{15.4.5}$$

where both i and $j = 1, 2, 3 = x, y, z$. The convention $c = 1$ has been adopted.

Solutions of eq. (15.4.1) are column matrices:

$$\Psi_{\text{col}} = \begin{pmatrix} \Psi_1 \\ \Psi_2 \\ \Psi_3 \\ \Psi_4 \end{pmatrix}. \tag{15.4.6}$$

When $\mathbf{A} = \phi = 0$, plane wave solutions may be written as

$$\Psi_{\text{col}} = \begin{pmatrix} u_1 \\ u_2 \\ u_3 \\ u_4 \end{pmatrix} \exp\frac{i}{\hbar}(\mathbf{p}\cdot\mathbf{r} - E\,t) = \begin{pmatrix} U \\ V \end{pmatrix} \exp\frac{i}{\hbar}(\mathbf{p}\cdot\mathbf{r} - E\,t), \tag{15.4.7}$$

where

$$U = \begin{pmatrix} u_1 \\ u_2 \end{pmatrix}, \tag{15.4.8}$$

and

$$V = \begin{pmatrix} u_3 \\ u_4 \end{pmatrix}. \tag{15.4.9}$$

Inserting eq. (15.4.7) in eq. (15.4.1), one obtains

$$E \begin{pmatrix} U \\ V \end{pmatrix} = \mathbf{p} \cdot \vec{\sigma} \begin{pmatrix} V \\ U \end{pmatrix} + m \begin{pmatrix} U \\ -V \end{pmatrix}, \tag{15.4.10}$$

from which it follows that, when $E = +\sqrt{p^2 + m^2} > 0$,

$$V = +\frac{\mathbf{p} \cdot \vec{\sigma}}{E + m} U \tag{15.4.11}$$

and

$$U = +\frac{\mathbf{p} \cdot \vec{\sigma}}{E - m} V. \tag{15.4.12}$$

Putting these two equations together and using the properties of the Pauli matrices (eqs. (15.4.4)) gives

$$V = \frac{\mathbf{p} \cdot \vec{\sigma}}{E + m} \frac{\mathbf{p} \cdot \vec{\sigma}}{E - m} V = \frac{p^2}{E^2 - m^2} V, \tag{15.4.13}$$

which produces the familiar relativistic relationship between mass, momentum, and energy:

$$E^2 = p^2 + m^2. \tag{15.4.14}$$

Using the properties of the Pauli matrices in eq. (15.4.11), one has

$$V = \frac{1}{E + m} \begin{pmatrix} p_- u_2 + p_z u_1 \\ p_+ u_1 - p_z u_2 \end{pmatrix} U, \tag{15.4.15}$$

where $p_\pm = p_x \pm i p_y$. It is conventional to choose as basis vectors the two component spinors

$$U_\uparrow = \begin{pmatrix} 1 \\ 0 \end{pmatrix} \tag{15.4.16}$$

and

$$U_\downarrow = \begin{pmatrix} 0 \\ 1 \end{pmatrix}, \tag{15.4.17}$$

which are eigenfunctions of σ_z corresponding, respectively, to "spin up" and "spin down" in the z direction. Altogether, the general plane wave solution for an electron may be written as

$$\Psi_{e^-} = N \left[A \begin{pmatrix} 1 \\ 0 \\ \frac{p_z}{E+m} \\ \frac{p_+}{E+m} \end{pmatrix} + B \begin{pmatrix} 0 \\ 1 \\ \frac{p_-}{E+m} \\ -\frac{p_z}{E+m} \end{pmatrix} \right] \exp \frac{i}{\hbar} (\mathbf{p} \cdot \mathbf{r} - E t), \tag{15.4.18}$$

where A and B are arbitrary numbers and N is a normalization constant. Normalization to one electron in a box of volume V is achieved by first forming the row vector equivalent of the column vector solution (eq. (15.4.6)),

$$\Psi_{\text{row}} = (\Psi_1 \ \Psi_2 \ \Psi_3 \ \Psi_4), \qquad (15.4.19)$$

next constructing the quantity $\Psi_{\text{row}}^* \Psi_{\text{col}}$, and finally setting $\int \Psi_{\text{row}}^* \Psi_{\text{col}} \, dV = 1$. When $|A|^2 + |B|^2 = 1$,

$$N = \frac{1}{\sqrt{V}} \left(\frac{E + m}{2E} \right)^{1/2}. \qquad (15.4.20)$$

In spherical polar coordinates, $p_x = p \ \sin\theta \ \cos\phi$, $p_y = p \ \sin\theta \ \sin\phi$, and $p_z = p \ \cos\theta$, where $p = |\mathbf{p}| \geq 0$. Using these relationships and setting

$$r = \frac{p}{|E| + m} \leq 1, \qquad (15.4.21)$$

eq. (15.4.18) may be written in the more compact form

$$\Psi_{e^-} = N \left[A \begin{pmatrix} 1 \\ 0 \\ r \cos\theta \\ r \sin\theta \ e^{i\phi} \end{pmatrix} + B \begin{pmatrix} 0 \\ 1 \\ r \sin\theta \ e^{-i\phi} \\ -r \cos\theta \end{pmatrix} \right] \exp \frac{i}{\hbar} (\mathbf{p} \cdot \mathbf{r} - E \, t), \qquad (15.4.22)$$

Negative energy solutions of eq. (15.4.1) are known to be appropriate descriptions of positrons. When $E = -\sqrt{p^2 + m^2} < 0$,

$$U = -\frac{\mathbf{p} \cdot \vec{\sigma}}{|E| + m} V, \qquad (15.4.23)$$

or

$$U = -\frac{1}{|E| + m} \begin{pmatrix} p_- u_2 + p_z u_1 \\ p_+ u_1 - p_z u_2 \end{pmatrix} V. \qquad (15.4.24)$$

As basis vectors, choose the two component spinors

$$V_\uparrow = \begin{pmatrix} 1 \\ 0 \end{pmatrix} \qquad (15.4.25)$$

and

$$V_\downarrow = \begin{pmatrix} 0 \\ 1 \end{pmatrix}, \qquad (15.4.26)$$

and write the general plane wave solution for negative energy as

$$\Psi_{e^+} = N \left[C \begin{pmatrix} \frac{-p_z}{|E|+m} \\ \frac{-p_+}{|E|+m} \\ 1 \\ 0 \end{pmatrix} + D \begin{pmatrix} \frac{-p_-}{|E|+m} \\ \frac{p_z}{|E|+m} \\ 0 \\ 1 \end{pmatrix} \right] \exp \frac{i}{\hbar} (\mathbf{p} \cdot \mathbf{r} - E \, t), \qquad (15.4.27)$$

where the normalization constant N is given by eq. (15.4.20) and C and D are constrained by $|C|^2 + |D|^2 = 1$. In spherical polar coordinates, the wave function becomes

$$\Psi_{e^+} = N \left[C \begin{pmatrix} -r\cos\theta \\ -r\sin\theta\, e^{i\phi} \\ 1 \\ 0 \end{pmatrix} + D \begin{pmatrix} -r\sin\theta\, e^{-i\phi} \\ r\cos\theta \\ 0 \\ 1 \end{pmatrix} \right] \exp\frac{i}{\hbar}(\mathbf{p}\cdot\mathbf{r} - E\,t),$$

$$(15.4.28)$$

where r is given by eq. (15.4.21).

If one supposes that the neutrino and antineutrino wave functions are also solutions of the Dirac equation with $\mathbf{A} = \phi = 0$ and $m = 0$, the plane wave solution for the antineutrino wave function may be taken as the linear combination of the two negative energy solutions:

$$\Psi_{\bar{\nu}_e} = \frac{1}{\sqrt{2V}} \left[C \begin{pmatrix} -p_z \\ -p_+ \\ 1 \\ 0 \end{pmatrix} + D \begin{pmatrix} -p_- \\ p_z \\ 0 \\ 1 \end{pmatrix} \right] \exp\frac{i}{\hbar}(\mathbf{p}\cdot\mathbf{r} - E\,t), \qquad (15.4.29a)$$

$$= \frac{1}{\sqrt{2V}} \left[C \begin{pmatrix} -\cos\theta \\ -\sin\theta\, e^{i\phi} \\ 1 \\ 0 \end{pmatrix} + D \begin{pmatrix} -\sin\theta\, e^{-i\phi} \\ \cos\theta \\ 0 \\ 1 \end{pmatrix} \right] \exp\frac{i}{\hbar}(\mathbf{p}\cdot\mathbf{r} - E\,t),$$

$$(15.4.29b)$$

where $p_\pm = \mathbf{p}\cdot(\hat{x}\pm i\hat{y})/|\mathbf{p}|$, $p_z = \mathbf{p}\cdot\hat{z}/|\mathbf{p}|$, and $\hat{x}$, $\hat{y}$, and $\hat{z}$ are unit vectors in the x, y, z directions. In other words, p_z and $p_\pm$ are also components of a unit vector in the x, y, z directions. Similarly, the general plane wave solution for the neutrino can be written as the sum of two independent positive energy wave functions:

$$\Psi_{\nu_e} = \frac{1}{\sqrt{2V}} \left[A \begin{pmatrix} 1 \\ 0 \\ p_z \\ p_+ \end{pmatrix} + B \begin{pmatrix} 0 \\ 1 \\ p_- \\ -p_z \end{pmatrix} \right] \exp\frac{i}{\hbar}(\mathbf{p}\cdot\mathbf{r} - E\,t) \qquad (15.4.30a)$$

$$= \frac{1}{\sqrt{2V}} \left[A \begin{pmatrix} 1 \\ 0 \\ \cos\theta \\ \sin\theta\, e^{i\phi} \end{pmatrix} + B \begin{pmatrix} 0 \\ 1 \\ \sin\theta\, e^{-i\phi} \\ -\cos\theta \end{pmatrix} \right] \exp\frac{i}{\hbar}(\mathbf{p}\cdot\mathbf{r} - E\,t). \quad (15.4.30b)$$

Knowing that leptons are created with a definite helicity provides motivation for constructing eigenfunctions of the helicity operator,

$$S = \begin{pmatrix} \vec{\sigma}\cdot\mathbf{p} & 0 \\ 0 & \vec{\sigma}\cdot\mathbf{p} \end{pmatrix}, \qquad (15.4.31)$$

which commutes with the Hamiltonian and is thus a conserved quantity. That is, one wants to find solutions for which the spin is quantized along the axis defined by the direction of motion. One may begin by finding solutions of the equation

$$(\vec{\sigma} \cdot \mathbf{p}) \, U_\lambda = \lambda \, p \, U_\lambda, \tag{15.4.32}$$

where U_λ is a two component spinor, p is the absolute value of the momentum, and λ is an eigenvalue. Each 2×2 component of the helicity operator may be written as

$$\vec{\sigma} \cdot \mathbf{p} = \big[(\sin\theta \cos\phi) \, \sigma_x + (\sin\theta \sin\phi) \, \sigma_y + \cos\theta \, \sigma_z \big] \, p$$

$$= \begin{pmatrix} \cos\theta & \sin\theta \, e^{-i\phi} \\ \sin\theta \, e^{i\phi} & -\cos\theta \end{pmatrix} p, \tag{15.4.33}$$

so that eq. (15.4.32) becomes

$$\begin{pmatrix} \cos\theta & \sin\theta \, e^{-i\phi} \\ \sin\theta \, e^{i\phi} & -\cos\theta \end{pmatrix} \begin{pmatrix} u_1 \\ u_2 \end{pmatrix} = \lambda \begin{pmatrix} u_1 \\ u_2 \end{pmatrix}. \tag{15.4.34}$$

From this equation it follows that

$$u_1 \, (\lambda - \cos\theta) = u_2 \, \sin\theta \, e^{-i\phi} \tag{15.4.35}$$

and that

$$u_1 \, \sin\theta \, e^{i\phi} = u_2 \, (\cos\theta + \lambda) = \left(\frac{\lambda - \cos\theta}{\sin\theta \, e^{-i\phi}} u_1 \right) (\cos\theta + \lambda). \tag{15.4.36}$$

Thus, $\sin^2\theta = -\cos^2\theta + \lambda^2$, or

$$\lambda = \pm 1. \tag{15.4.37}$$

When $\lambda = 1$,

$$\frac{u_2}{u_1} = \frac{1 - \cos\theta}{\sin\theta} e^{i\phi} = \frac{2 \, \sin^2(\theta/2)}{2 \, \sin(\theta/2) \, \cos(\theta/2)} e^{i\phi} = \frac{\sin(\theta/2)}{\cos(\theta/2)} e^{i\phi}. \tag{15.4.38}$$

Normalizing the spinor by

$$|u_1|^2 + |u_2|^2 = |u_1|^2 \left(\frac{\sin^2(\theta/2)}{\cos^2(\theta/2)} + 1 \right) = \frac{|u_1|^2}{\cos^2(\theta/2)} = 1, \tag{15.4.39}$$

and combining with eq. (15.4.38), one finds

$$U_{\lambda=+1} = \begin{pmatrix} u_1 \\ u_2 \end{pmatrix}_{\lambda=+1} = \begin{pmatrix} \cos(\theta/2) \\ \sin(\theta/2) \, e^{i\phi} \end{pmatrix} e^{i\delta_+}, \tag{15.4.40}$$

where δ_+ is an arbitrary phase factor. Similar arithmetic gives, when $\lambda = -1$,

$$U_{\lambda=-1} = \begin{pmatrix} u_1 \\ u_2 \end{pmatrix}_{\lambda=-1} = \begin{pmatrix} -\sin(\theta/2)\,e^{-i\phi} \\ \cos(\theta/2) \end{pmatrix} e^{i\delta_-}, \qquad (15.4.41)$$

where δ_- is another arbitrary phase factor. Note that the two eigenstates are orthogonal in the sense that $\tilde{U}^*_{\lambda=+1}\,U_{\lambda=-1} = 0$.

One may further ask for solutions of the Dirac equation which are also eigenfunctions of the 16 element helicity operator, eq. (15.4.31). The Dirac equation gives

$$E \begin{pmatrix} U \\ V \end{pmatrix} = \lambda\,p \begin{pmatrix} V \\ U \end{pmatrix} + m \begin{pmatrix} U \\ -V \end{pmatrix}. \qquad (15.4.42)$$

When $E > 0$,

$$V = \lambda\,\frac{p}{E+m}\,U = \lambda\,r\,U, \qquad (15.4.43)$$

and, when $E < 0$,

$$U = -\lambda\,\frac{p}{|E|+m}\,V = -\lambda\,r\,V, \qquad (15.4.44)$$

where r is defined by eq. (15.4.21).

If one assumes that a lepton created in a weak interaction has negative helicity and that an antilepton created in a weak interaction has positive helicity, then, for the antilepton,

$$\Psi_{\text{antilepton}\uparrow} = N \begin{pmatrix} -r\cos(\theta/2) \\ -r\sin(\theta/2)\,e^{i\phi} \\ \cos(\theta/2) \\ \sin(\theta/2)\,e^{i\phi} \end{pmatrix} \exp\frac{i}{\hbar}(\mathbf{p}\cdot\mathbf{r} - E\,t), \qquad (15.4.45)$$

and, for the lepton,

$$\Psi_{\text{lepton}\downarrow} = N \begin{pmatrix} -\sin(\theta/2)\,e^{-i\phi} \\ \cos(\theta/2) \\ r\sin(\theta/2)\,e^{-i\phi} \\ -r\cos(\theta/2) \end{pmatrix} \exp\frac{i}{\hbar}(\mathbf{p}\cdot\mathbf{r} - E\,t). \qquad (15.4.46)$$

In preparation for a calculation of the cross section for the conversion of an electron–positron pair into a neutrino–antineutrino pair, it is useful to cast the Dirac equation into a form which is more transparently invariant to special relativistic transformations. The elements of the column vector produced by the operation $\alpha^*\,\Psi^*_{\text{col}}$ are identical with the elements of the row vector produced by the operation $\Psi^*_{\text{row}}\,\alpha$. Also, the elements of $\beta\,\Psi^*_{\text{col}}$ are identical with the elements of $\Psi^*_{\text{row}}\,\beta$. Thus, writing the Dirac equation as

$$i\hbar\frac{\partial\Psi_{\text{col}}}{\partial t} = -i\hbar\nabla\cdot(\vec{\alpha}\,\Psi_{\text{col}}) + \left(-\frac{e}{c}\,\mathbf{A}\right)\cdot(\vec{\alpha}\,\Psi_{\text{col}}) + e\phi\,\Psi_{\text{col}} + m(\beta\,\Psi_{\text{col}}), \qquad (15.4.47)$$

it is also true that

$$-i\hbar\frac{\partial \Psi^*_{\text{row}}}{\partial t} = i\hbar\nabla \cdot (\Psi^*_{\text{row}}\vec{\alpha}) + \left(-\frac{e}{c}\mathbf{A}\right) \cdot (\Psi^*_{\text{row}}\vec{\alpha}) + e\phi\,\Psi^*_{\text{row}} + m(\Psi^*_{\text{row}}\beta). \quad (15.4.48)$$

Multiplying eq. (15.4.47) on the left by Ψ^*_{row}, multiplying eq. (15.4.48) on the right by Ψ^*_{col}, and taking the difference of the resulting two equations gives

$$\frac{\partial}{\partial t}(\Psi^*_{\text{row}}\Psi_{\text{col}}) + \nabla \cdot (\Psi^*_{\text{row}}\vec{\alpha}\Psi_{\text{col}}) = \frac{\partial\rho}{\partial t} + \nabla\cdot\mathbf{j} = 0, \quad (15.4.49)$$

which is the equation for the conservation of the probability density if

$$\rho = (\Psi^*_{\text{row}}\Psi_{\text{col}}) \quad (15.4.50)$$

is interpreted as this density and

$$\mathbf{j} = (\Psi^*_{\text{row}}\vec{\alpha}\Psi_{\text{col}}) \quad (15.4.51)$$

is interpreted as the current associated with this density.

Noting that β^2 is the unit matrix and defining the "adjoint" of the wave function Ψ_{col} as

$$\bar{\Psi} = \Psi^*_{\text{row}}\beta, \quad (15.4.52)$$

one has that

$$\frac{\partial}{\partial t}(\Psi^*_{\text{row}}\beta\beta\Psi_{\text{col}}) + \nabla \cdot (\Psi^*_{\text{row}}\beta\beta\vec{\alpha}\Psi_{\text{col}}) = \frac{\partial}{\partial t}(\bar{\Psi}\beta\Psi) + \nabla\cdot(\bar{\Psi}\beta\vec{\alpha}\Psi)$$

$$= \frac{\partial}{\partial t}(\bar{\Psi}\gamma_4\Psi) + \nabla\cdot(\bar{\Psi}\vec{\gamma}\Psi) = 0, \quad (15.4.53)$$

where

$$\gamma_4 = \beta \quad (15.4.54)$$

and

$$\vec{\gamma} = \beta\vec{\alpha} = \begin{pmatrix} 0 & \vec{\sigma} \\ -\vec{\sigma} & 0 \end{pmatrix}. \quad (15.4.55)$$

The four γs defined by γ_4 and $\vec{\gamma}$ satisfy the anticommutation relations

$$\gamma_\mu\gamma_\nu + \gamma_\nu\gamma_\mu = \begin{cases} 0 & \text{if } \mu \neq \nu, \\ 2 & \text{if } \mu = \nu = 4 = t, \\ -2 & \text{if } \mu = \nu = i = 1, 2, 3 = x, y, z, \end{cases} \quad (15.4.56)$$

Comparing eqs. (15.4.49) and (15.4.53), one has that

$$\rho = \bar{\Psi}\beta\Psi = \bar{\Psi}\gamma_4\Psi \quad (15.4.57)$$

and

$$\mathbf{j} = \bar{\Psi}\beta\vec{\alpha}\Psi = \bar{\Psi}\vec{\gamma}\Psi \qquad (15.4.58)$$

constitute a four vector. Here and hereinafter $\bar{\Psi}$ and Ψ are, respectively, row and column vectors.

The continuity equation for the probability density may be written as

$$\frac{\partial}{\partial x_4}(\bar{\Psi}\gamma_4\Psi) + \frac{\partial}{\partial x_i}(\bar{\Psi}\gamma_i\Psi) = 0, \qquad (15.4.59)$$

where $x_4 = t$ and the second term on the right hand side of the equation is to be understood as the sum over the space-like indices $i = 1, 2, 3 = x, y, z$. Defining the scalar product of two four vectors as

$$a_\mu b_\mu = a_4 b_4 - a_j b_j, \qquad (15.4.60)$$

where the term on the left hand side of the equation is the sum over the indices $\mu = 4, 1, 2, 3 = t, x, y, z$ and the second term on the right hand side is the sum over the indices $j = 1, 2, 3 = x, y, z$, eq. (15.4.57) may be thought of as the scalar product of two four vectors, the gradient four vector

$$\frac{\partial}{\partial x_\mu} = \left(\frac{\partial}{\partial x_4}, -\frac{\partial}{\partial x_1}, -\frac{\partial}{\partial x_2}, -\frac{\partial}{\partial x_3}\right), \qquad (15.4.61)$$

and the γ four vector

$$\left((\bar{\Psi}\gamma_4\Psi),\ (\bar{\Psi}\gamma_x\Psi),\ (\bar{\Psi}\gamma_y\Psi),\ (\bar{\Psi}\gamma_z\Psi)\right). \qquad (15.4.62)$$

The continuity equation becomes

$$\frac{\partial}{\partial x_4}(\bar{\Psi}\gamma_4\Psi) - \left(-\frac{\partial}{\partial x_i}\right)(\bar{\Psi}\gamma_i\Psi)) = \frac{\partial}{\partial x_\mu}(\bar{\Psi}\gamma_\mu\Psi) = 0, \qquad (15.4.63)$$

and it is evident that the quantities $(\bar{\Psi}\gamma_\mu\Psi)$ are elements of a relativistic four vector.

Next, rearrange the Dirac equation to read

$$\left(i\hbar\frac{\partial}{\partial t} - e\phi\right)\Psi_{\text{col}} - \vec{\alpha}\cdot\left(-i\hbar\nabla - \frac{e}{c}\mathbf{A}\right)\Psi_{\text{col}} = \beta\, m\,\Psi_{\text{col}}. \qquad (15.4.64)$$

Multiplying both sides of this equation by β produces

$$\beta\left(i\hbar\frac{\partial}{\partial t} - e\phi\right)\Psi_{\text{col}} - \beta\vec{\alpha}\cdot\left(-i\hbar\nabla - \frac{e}{c}\mathbf{A}\right)\Psi_{\text{col}} = m\,\Psi_{\text{col}}$$

$$= \gamma_4\left(i\hbar\frac{\partial}{\partial x_4} - e\phi\right)\Psi_{\text{col}} - \vec{\gamma}\cdot\left(-i\hbar\nabla - \frac{e}{c}\mathbf{A}\right)\Psi_{\text{col}} = m\,\Psi_{\text{col}}, \qquad (15.4.65)$$

which, in turn, can be written as

$$\gamma_\mu\left(i\hbar\frac{\partial}{\partial x_\mu} - \frac{e}{c}A_\mu\right)\Psi_{\text{col}} = m\,\Psi_{\text{col}}, \qquad (15.4.66)$$

where $(e/c)A_4 = e\phi$. Since m is a constant, this equation shows that the scalar product of the two vector operators on the left hand side of the equation is effectively a constant and that the Dirac equation is therefore relativistically invariant.

An additional 12 independent 4×4 matrices can be formed from the four vector-like γs, each having different transformation properties. A particularly important one is the pseudoscalar

$$\gamma_5 = \gamma_x \gamma_y \gamma_z \gamma_4 = \mathrm{i} \begin{pmatrix} 0 & 1 \\ 1 & 0 \end{pmatrix}, \tag{15.4.67}$$

which anticommutes with the γ_μs ($\mu = 1, 4$) and satisfies $\gamma_5^2 = -1$. The four quantities $\gamma_5 \gamma_\mu$ form what is called an axial vector and the six quantities $\gamma_\mu \gamma_\nu$, for $\mu \neq \nu$, form what is called a tensor, which can be written in various forms including the antisymmetric form

$$T_{\mu\nu} = \frac{1}{2} (\gamma_\mu \gamma_\nu - \gamma_\nu \gamma_\mu). \tag{15.4.68}$$

Properties of these 16 matrices, including the fact that they are independent and form a complete set, are discussed at length by, e.g., J. D. Bjorken & S. D. Drell (*Relativistic Quantum Mechanics*, 1964). It should be noted that there is a considerable variation in the literature in the explicit definitions of the matrices, both with respect to sign and the presence or absence of $\mathrm{i} = \sqrt{-1}$ as a factor. For example, γ_5 defined by Bjorken and Drell is $-\mathrm{i}$ times γ_5 defined by Feynman. Fortunately, for any two complete sets of matrices γ and γ' that satisfy the anticommutation relations expressed in eq. (15.4.56), there is a similarity transformation S such that $\gamma' = S^{-1}\gamma S$.

15.5 Derivation of the cross section for electron–positron pair annihilation in the V-A theory

Using standard V-A theory, in this section a simple derivation is provided of the cross section for the process by which a real electron–positron pair annihilates to produce a neutrino–antineutrino pair.

The charged current density in the V-A Hamiltonian can be written as

$$J_\mu = \bar{\Psi}_p \gamma_\mu a_1 \Psi_n + \bar{\Psi}_{\nu_e} \gamma_\mu a_0 \Psi_e + \bar{\Psi}_{\nu_\mu} \gamma_\mu a_0 \Psi_\mu, \tag{15.5.1}$$

where

$$a_1 = \frac{1}{\sqrt{2}} (1 + \lambda \mathrm{i} \gamma_5) \tag{15.5.2}$$

and

$$a_0 = \frac{1}{\sqrt{2}} (1 + \mathrm{i} \gamma_5). \tag{15.5.3}$$

In eq. (15.5.2), λ is the absolute value of the ratio of the axial vector coupling constant to the vector coupling constant in nuclear beta decay. The minus sign in the designation V-A comes from the fact that, in a common representation, γ_5' is related to the γ_5 used here by $\gamma_5' = -i\gamma_5$.

In a practical calculation, J_μ and $J_\mu^\dagger$ are just numbers. They do not have transposes, only complex conjugates. It is easy to show that $(\bar{\Psi}_a \gamma_\mu \Psi_b)^* = (\bar{\Psi}_b \gamma_\mu \Psi_a)$ where Ψ_a and Ψ_b are two arbitrary column vectors and $\bar{\Psi}_a$ and $\bar{\Psi}_b$ are their adjoints. Furthermore, $(\bar{\Psi}_a \gamma_\mu \gamma_\nu \Psi_b)^* = (\bar{\Psi}_b \gamma_\nu \gamma_\mu \Psi_a)$ and $(\bar{\Psi}_a i \Psi_b)^* = -(\bar{\Psi}_b i \Psi_a)$. So, for example, with

$$J_{\mu,2} = \bar{\Psi}_{\nu_e} \gamma_\mu (1 + i\gamma_5) \Psi_e, \tag{15.5.4}$$

one also has

$$J_{\mu,2}^\dagger = J_{\mu,2}^* = \bar{\Psi}_e (1 - i\gamma_5) \gamma_\mu \Psi_{\nu_e} = \bar{\Psi}_e \gamma_\mu (1 + i\gamma_5) \Psi_{\nu_e}. \tag{15.5.5}$$

Hence, the Hermitian conjugate of eq. (15.5.1) is

$$J_\mu^\dagger = \bar{\Psi}_n \gamma_\mu a_1 \Psi_p + \bar{\Psi}_e \gamma_\mu a_0 \Psi_{\nu_e} + \bar{\Psi}_\mu \gamma_\mu a_0 \Psi_{\nu_\mu}. \tag{15.5.6}$$

The coefficient $1/\sqrt{2}$ in the definition of a_1 and a_0 comes from the fact that, in the product $J_\mu^\dagger J_\mu = J_\mu J_\mu^\dagger$, cross terms come in pairs which have the identical effect so that, e.g., for nuclear beta decay, $H_{\text{int}}^{\text{nuc}} = 2 (\bar{\Psi}_n \gamma_\mu a_1 \Psi_p)(\bar{\Psi}_{\nu_e} \gamma_\mu a_0 \Psi_e)$.

The Hamiltonian for the process whereby an electron–positron pair is converted into a neutrino–antineutrino pair is

$$H_{\text{int}} = g_{\text{weak}} (\bar{\Psi}_e \gamma_\mu a_0 \Psi_{\nu_e}) (\bar{\Psi}_{\nu_e} \gamma_\mu a_0 \Psi_e). \tag{15.5.7}$$

Direct calculation shows that, for any four wave functions Ψ_a, Ψ_b, Ψ_c, and Ψ_d,

$$(\bar{\Psi}_a \gamma_\mu a_0 \Psi_b)(\bar{\Psi}_c \gamma_\mu a_0 \Psi_d) = (\bar{\Psi}_c \gamma_\mu a_0 \Psi_b)(\bar{\Psi}_a \gamma_\mu a_0 \Psi_d), \tag{15.5.8}$$

which is known as a Fierz transformation. Thus,

$$H_{\text{int}} = g_{\text{weak}} (\bar{\Psi}_{\nu_e} \gamma_\mu a_0 \Psi_{\nu_e}) (\bar{\Psi}_e \gamma_\mu a_0 \Psi_e) \tag{15.5.9}$$

is equivalent to eq. (15.5.7). By pairing created leptons and pairing destroyed leptons, this last formulation greatly simplifies the calculation of the cross section. It has been used in Section 10.6 in Volume 1 to determine the effective mass of an electron neutrino created by an interaction with free electrons and to establish, as a function of an adopted mass difference between electron and muon neutrinos, the scale length over which electron and muon neutrinos can be converted into one another in the presence of free electrons in the solar interior.

In the center of momentum frame of reference, choose the momenta of both positron and electron to be along the z axis, and choose, for the electron, the wave functions

$$\Psi_{e\uparrow} = N \begin{pmatrix} 1 \\ 0 \\ r_z \\ 0 \end{pmatrix} \exp \frac{i}{\hbar} (\mathbf{p} \cdot \mathbf{r} - E t) \tag{15.5.10}$$

and

$$
\Psi_{e\downarrow} = N \begin{pmatrix} 0 \\ 1 \\ 0 \\ -r_z \end{pmatrix} \exp \frac{i}{\hbar}(\mathbf{p} \cdot \mathbf{r} - E\,t). \tag{15.5.11}
$$

For the positron, choose the wave functions

$$
\Psi_{p\uparrow} = N \begin{pmatrix} -\bar{r}_z \\ 0 \\ 1 \\ 0 \end{pmatrix} \exp \frac{i}{\hbar}(\bar{\mathbf{p}} \cdot \mathbf{r} - \bar{E}\,t) \tag{15.5.12}
$$

and

$$
\Psi_{p\downarrow} = N \begin{pmatrix} 0 \\ \bar{r}_z \\ 0 \\ 1 \end{pmatrix} \exp \frac{i}{\hbar}(\bar{\mathbf{p}} \cdot \mathbf{r} - \bar{E}\,t). \tag{15.5.13}
$$

In these wave functions,

$$
N = \frac{1}{\sqrt{V}} \left(\frac{|E| + m}{2|E|} \right)^{1/2}, \tag{15.5.14}
$$

$$
r_z = \frac{p_z}{|E| + m} = \frac{p}{|E| + 1}, \tag{15.5.15}
$$

and, because $\bar{p}_z = -p_z = -|\bar{p}| = -p$, where p is the absolute value of the electron momentum,

$$
\bar{r}_z = \frac{\bar{p}_z}{|E| + m} = -r_z = -\frac{p}{|E| + m}. \tag{15.5.16}
$$

From eqs. (15.4.46) and (15.4.45), one obtains for the neutrino and antineutrino, respectively,

$$
\Psi_{\nu\downarrow} = \sqrt{\frac{1}{2V}} \begin{pmatrix} -\sin(\theta/2)\,e^{-i\phi} \\ \cos(\theta/2) \\ \sin(\theta/2)\,e^{-i\phi} \\ -\cos(\theta/2) \end{pmatrix} \exp \frac{i}{\hbar}(\mathbf{p} \cdot \mathbf{r} - E\,t) \tag{15.5.17}
$$

and

$$
\Psi_{\bar{\nu}\uparrow} = \sqrt{\frac{1}{2V}} \begin{pmatrix} -\cos(\bar{\theta}/2) \\ -\sin(\bar{\theta}/2)\,e^{i\bar{\phi}} \\ \cos(\bar{\theta}/2) \\ \sin(\bar{\theta}/2)e^{i\bar{\phi}} \end{pmatrix} \exp \frac{i}{\hbar}(\bar{\mathbf{p}} \cdot \mathbf{r} - \bar{E}\,t). \tag{15.5.18}
$$

Setting

$$H_{\text{int}} = g_{\text{weak}} \frac{1}{2} \sum_{\mu=1}^{4} P_{e\mu} \, P_{n\mu}, \tag{15.5.19}$$

where the subscript e refers to the electron and positron and the subscript n refers to the neutrino and antineutrino, one may write

$$P_{e\mu} = \bar{\Psi}_e \gamma_\mu (1 + i\gamma_5) \Psi_e \tag{15.5.20}$$

and

$$P_{n\mu} = \bar{\Psi}_\nu \gamma_\mu (1 + i\gamma_5) \Psi_\nu. \tag{15.5.21}$$

For convenience, construct the matrices

$$\gamma_4 \, (1 + i\gamma_5) = \begin{pmatrix} 1 & 0 & 0 & 0 \\ 0 & 1 & 0 & 0 \\ 0 & 0 & -1 & 0 \\ 0 & 0 & 0 & -1 \end{pmatrix} \begin{pmatrix} 1 & 0 & -1 & 0 \\ 0 & 1 & 0 & -1 \\ -1 & 0 & 1 & 0 \\ 0 & -1 & 0 & 1 \end{pmatrix} = \begin{pmatrix} 1 & 0 & -1 & 0 \\ 0 & 1 & 0 & -1 \\ 1 & 0 & -1 & 0 \\ 0 & 1 & 0 & -1 \end{pmatrix},$$

$$\gamma_x \, (1 + i\gamma_5) = \begin{pmatrix} 0 & 0 & 0 & 1 \\ 0 & 0 & 1 & 0 \\ 0 & -1 & 0 & 0 \\ -1 & 0 & 0 & 0 \end{pmatrix} \begin{pmatrix} 1 & 0 & -1 & 0 \\ 0 & 1 & 0 & -1 \\ -1 & 0 & 1 & 0 \\ 0 & -1 & 0 & 1 \end{pmatrix} = \begin{pmatrix} 0 & -1 & 0 & 1 \\ -1 & 0 & 1 & 0 \\ 0 & -1 & 0 & 1 \\ -1 & 0 & 1 & 0 \end{pmatrix},$$

$$\gamma_y \, (1 + i\gamma_5) = \begin{pmatrix} 0 & 0 & 0 & -i \\ 0 & 0 & i & 0 \\ 0 & i & 0 & 0 \\ -i & 0 & 0 & 0 \end{pmatrix} \begin{pmatrix} 1 & 0 & -1 & 0 \\ 0 & 1 & 0 & -1 \\ -1 & 0 & 1 & 0 \\ 0 & -1 & 0 & 1 \end{pmatrix} = \begin{pmatrix} 0 & i & 0 & -i \\ -i & 0 & i & 0 \\ 0 & i & 0 & -i \\ -i & 0 & i & 0 \end{pmatrix},$$

and

$$\gamma_z \, (1 + i\gamma_5) = \begin{pmatrix} 0 & 0 & 1 & 0 \\ 0 & 0 & 0 & -1 \\ -1 & 0 & 0 & 0 \\ 0 & 1 & 0 & 0 \end{pmatrix} \begin{pmatrix} 1 & 0 & -1 & 0 \\ 0 & 1 & 0 & -1 \\ -1 & 0 & 1 & 0 \\ 0 & -1 & 0 & 1 \end{pmatrix} = \begin{pmatrix} -1 & 0 & 1 & 0 \\ 0 & 1 & 0 & -1 \\ -1 & 0 & 1 & 0 \\ 0 & 1 & 0 & -1 \end{pmatrix}. \tag{15.5.22}$$

The next objective is to determine matrix elements for four different initial conditions corresponding to positron and electron spins in the z direction (1) both up, (2) both down, (3) up and down, and (4) down and up.

When both electron and positron spins are up, eqs. (15.5.10) and (15.5.12) are relevant and

$$P_{e\mu} = N^2 \begin{pmatrix} -\bar{r}_z & 0 & -1 & 0 \end{pmatrix} \gamma_\mu (1 + i\gamma_5) \begin{pmatrix} 1 \\ 0 \\ r_z \\ 0 \end{pmatrix}. \tag{15.5.23}$$

When $\mu = 4$,

$$P_{e4} = -N^2 (1 + \bar{r}_z)(1 - r_z). \tag{15.5.24}$$

In the center of momentum system, $\bar{r} = -r$, hence

$$(1 + \bar{r}_z)(1 - r_z) = (1 - r_z)^2 = \left(1 - \frac{p}{|E| + m}\right)^2 = \frac{2|E|}{|E| + m}\left(1 - \frac{v}{c}\right). \tag{15.5.25}$$

Using this result and N from eq. (15.5.14), one has

$$P_{e4} = -\frac{1}{V}\left(1 - \frac{v}{c}\right). \tag{15.5.26}$$

Similar calculations give

$$P_{ez} = -P_{e4} \tag{15.5.27}$$

and

$$P_{ex} = P_{ey} = 0. \tag{15.5.28}$$

Since the neutrino and antineutrino momenta are oppositely directed, $\bar{\phi} = \phi + \pi$, so $e^{i\bar{\phi}} = -e^{i\phi}$. Also, $\bar{\theta}/2 = \pi/2 - \theta/2$, so $\cos(\bar{\theta}/2) = \sin(\theta/2)$ and $\cos(\bar{\theta}/2) = \sin(\theta/2)$. Therefore,

$$\bar{\Psi}_{\bar{\nu}\uparrow} = \frac{1}{\sqrt{2V}}\begin{pmatrix} -\sin(\theta/2) & \cos(\theta/2)\,e^{-i\phi} & -\sin(\theta/2) & \cos(\theta/2)\,e^{-i\phi} \end{pmatrix} \tag{15.5.29}$$

and

$$P_{n\mu} = (\bar{\Psi}_\nu \gamma_\mu a_0 \Psi_\nu) = \frac{1}{2V}\begin{pmatrix} -\sin(\theta/2) & \cos(\theta/2)\,e^{-i\phi} & -\sin(\theta/2) & \cos(\theta/2)\,e^{-i\phi} \end{pmatrix}$$

$$\times \gamma_\mu (1 + i\gamma_5)\begin{pmatrix} -\sin(\theta/2)\,e^{-i\phi} \\ \cos(\theta/2) \\ \sin(\theta/2)\,e^{-i\phi} \\ -\cos(\theta/2) \end{pmatrix}. \tag{15.5.30}$$

From this equation and the first identity in eq. (15.5.22),

$$P_{n4} = \frac{2}{V} e^{-i\phi},$$ (15.5.31)

so, combining with eq. (15.5.26), one has

$$P_{e4} \, P_{n4} = -\frac{2}{V^2} \left(1 - \frac{v}{c}\right) e^{-i\phi}.$$ (15.5.32)

Similarly,

$$P_{nz} = \frac{2}{V} (\cos^2(\theta/2) - \sin^2(\theta/2)) e^{-i\phi} = \frac{2}{V} \cos\theta \, e^{-i\phi}$$ (15.5.33)

and

$$P_{ez} \, P_{nz} = \frac{2}{V^2} \left(1 - \frac{v}{c}\right) \cos\theta \, e^{-i\phi}.$$ (15.5.34)

Performing the sum over μ and integrating over volume, one obtains the relevant matrix element

$$M_{\uparrow\uparrow} = \int H_{\text{int}} \, dV = g_{\text{weak}} \frac{1}{2} \sum_{\mu=1}^{4} P_{e\mu} \, P_{n\mu} \, V = g_{\text{weak}} \frac{1}{2} (P_{e4} \, P_{n4} - P_{ez} \, P_{nz}) \, V$$

$$= -g_{\text{weak}} \frac{1}{V} \left(1 - \frac{v}{c}\right)(1 + \cos\theta) \, e^{-i\phi}.$$ (15.5.35)

When both electron and positron spins are down, eqs. (15.5.11) and (15.5.13) are relevant. One finds

$$P_{e4} = N^2 \begin{pmatrix} 0 & \bar{r}_z & 0 & -1 \end{pmatrix} \gamma_\mu (1 + i\gamma_5) \begin{pmatrix} 0 \\ 1 \\ 0 \\ -r_z \end{pmatrix}.$$

$$= -N^2 (1 - \bar{r}_z)(1 + r_z) = -N^2 (1 + r_z)^2 = -\frac{1}{V}\left(1 + \frac{v}{c}\right),$$ (15.5.36)

$$P_{ez} = +P_{e4},$$ (15.5.37)

and

$$P_{ex} = P_{ey} = 0.$$ (15.5.38)

Altogether,

$$M_{\downarrow\downarrow} = g_{\text{weak}} \frac{1}{2} (P_{e4} \, P_{n4} - P_{ez} \, P_{nz}) \, V$$

$$= -g_{\text{weak}} \frac{1}{V}\left(1 + \frac{v}{c}\right)(1 - \cos\theta) \, e^{-i\phi}.$$ (15.5.39)

When the positron has spin up and the electron has spin down,

$$P_{e\mu} = N^2 \begin{pmatrix} -\bar{r}_z & 0 & -1 & 0 \end{pmatrix} \gamma_\mu (1 + i\gamma_5) \begin{pmatrix} 0 \\ 1 \\ 0 \\ -r_z \end{pmatrix}. \tag{15.5.40}$$

One finds that $P_{e4} = P_{ez} = 0$ and that

$$P_{ex} = N^2 (1 + \bar{r}_z)(1 + r_z) = N^2 (1 - r_z^2). \tag{15.5.41}$$

Now,

$$1 - r_z^2 = 1 - \frac{p_z^2}{(E + m)^2} = \frac{2m}{E + m}, \tag{15.5.42}$$

so

$$P_{ex} = \frac{1}{V} \frac{E + m}{2E} \frac{2m}{E + m} = \frac{1}{V} \sqrt{\left(\frac{m^2}{E^2}\right)} = \frac{1}{V} \sqrt{\left(1 - \frac{v^2}{c^2}\right)}. \tag{15.5.43}$$

Similarly,

$$P_{ey} = -i\, P_{ex}. \tag{15.5.44}$$

From eq. (15.5.30) and (15.5.22), one has further that

$$P_{nx} = \frac{2}{V} \sin(\theta/2)\, \cos(\theta/2)\, (1 + e^{-2i\phi}) = \frac{2}{V} \sin\theta\, \cos\phi\, e^{-i\phi} \tag{15.5.45}$$

and that

$$P_{ny} = -\frac{2}{V} \sin(\theta/2)\, \cos(\theta/2)\, (1 - e^{-2i\phi}) = -\frac{2}{V} \sin\theta\, i\, \sin\phi\, e^{-i\phi}. \tag{15.5.46}$$

Altogether, then,

$$M_{\uparrow\downarrow} = g_{\text{weak}} \frac{1}{2} \sum_{\mu=1}^{4} P_{e\mu}\, P_{n\mu}\, V = -g_{\text{weak}} \frac{1}{2} (P_{ex}\, P_{nx} + P_{ey}\, P_{ny})\, V$$

$$= -g_{\text{weak}} \frac{1}{V} \sqrt{1 - \left(\frac{v}{c}\right)^2}\, \sin\theta\, e^{-i\phi}\, (\cos\phi - \sin\phi). \tag{15.5.47}$$

When positron spin is down and electron spin is up, the result is the same as when spins are reversed, so

$$M_{\downarrow\uparrow} = M_{\uparrow\downarrow}. \tag{15.5.48}$$

Averaging over the four possibilities, one has

$$\langle |M_{ij}|^2 \rangle = \frac{1}{4} \sum_{ij} |M_{ij}|^2 = \frac{1}{4} \left(|M_{\uparrow\uparrow}|^2 + |M_{\downarrow\downarrow}|^2 + |M_{\uparrow\downarrow}|^2 + |M_{\downarrow\uparrow}|^2 \right)$$

$$= \frac{1}{4V^2} g_{\text{weak}}^2 \left[\left(1 - \frac{v}{c}\right)^2 (1 + \cos\theta)^2 + \left(1 + \frac{v}{c}\right)^2 (1 - \cos\theta)^2 \right.$$

$$\left. + 2\left(1 - \frac{v^2}{c^2}\right) \sin^2\theta (1 - 2\cos\phi \sin\phi) \right]$$

$$= \frac{1}{V^2} g_{\text{weak}}^2 \left[\left(1 + \frac{v^2}{c^2}\cos^2\theta\right) - \left(1 - \frac{v^2}{c^2}\right) \sin^2\theta \cos\phi \sin\phi \right]. \quad (15.5.49)$$

On integrating eq. (15.5.49) over a solid angle $d\Omega$, the second term vanishes and one is left with

$$\int \langle |M_{ij}|^2 \rangle \, d\Omega = \frac{1}{V^2} g_{\text{weak}}^2 \, 2\pi \left(2 + \frac{2}{3}\frac{v^2}{c^2}\right)$$

$$= \frac{g_{\text{weak}}^2}{V^2} \frac{4\pi}{3} \left(3 + \frac{v^2}{c^2}\right). \quad (15.5.50)$$

The cross section may be written as

$$\sigma = \frac{2\pi}{\hbar} \frac{1}{4\pi} \int \langle |M_{ij}|^2 \rangle \, d\Omega \, \rho_f \, \frac{V}{v}, \quad (15.5.51)$$

where ρ_f is the density of final states given by eq. (15.3.5). After a little arithmetic, one finds

$$\sigma = g_{\text{weak}}^2 \frac{1}{3\pi} \left(\frac{m_e}{\hbar}\right)^2 \left(\frac{E}{m}\right)^2 \left(3 + \frac{v^2}{c^2}\right) \frac{c}{v}.$$

$$= G_{\text{weak}}^2 \frac{1}{3\pi} \left(\frac{\hbar}{m_e c}\right)^2 \left(\frac{E}{m}\right)^2 \left(3 + \frac{v^2}{c^2}\right) \frac{c}{v}. \quad (15.5.52)$$

Noting that

$$\frac{v^2}{c^2} = 1 - \frac{m^2}{E^2}, \quad (15.5.53)$$

so that

$$3 + \frac{v^2}{c^2} = 4 - \frac{m^2}{E^2} = \left(\frac{m}{E}\right)^2 \left[\left(\frac{2E}{m}\right)^2 - 1\right], \quad (15.5.54)$$

one obtains, finally,

$$\sigma = G_{\text{weak}}^2 \frac{1}{3\pi} \left(\frac{\hbar}{m_e c}\right)^2 \left[\left(\frac{2E}{m}\right)^2 - 1\right] \frac{c}{v}, \quad (15.5.55)$$

which is identical with the result quoted by Chiu and Morrison (1960), as expressed by eqs. (15.3.14) and (15.3.15). This result has been used in Section 15.3 to estimate the neutrino–antineutrino energy-loss rate for the electron–positron pair annihilation process in three approximations: eq. (15.3.28) when electrons are non-relativistic, eq. (15.3.53) when electrons are highly relativistic, and eq. (15.3.60) when electrons are modestly relativistic.

15.6   A brief overview of the history and the nature of weak-interaction induced neutrino–antineutrino production processes

Bruno Pontecorvo (1959, 1960) was the first to suggest in the literature that, in the framework of the current–current formulation of the weak interaction Hamiltonian, neutrino–antineutrino pairs could be formed in stars and that the energy carried directly out of the star by the neutrinos could play an important role during certain phases of stellar evolution.

It soon became evident that, at high temperatures and at densities such that electrons are not degenerate, the primary mechanism for neutrino energy loss which does not involve nucleons is the conversion of real electron–positron pairs into neutrino–antineutrino pairs. The fact that the electron–positron pair annihilation process plays such a crucial role in the evolution of massive stars has been a primary motivation for the effort expended in Section 15.5 in calculating the cross section for the process. Hong-Yee Chiu and Philip Morrison (1960) were the among first to discuss the implications of the process and Chiu (1961a, 1961b) made the first calculation of the cross section for the process and of the energy-loss rate associated with the process.

As temperatures are decreased, the energy-loss rate by the electron–positron annihilation process drops off because of a decrease in the equilibrium number abundance of positrons relative to the abundance of electrons. This is because, the smaller the temperature, the less thermal energy there is to contribute to the rest mass energy of new electrons and positrons. At any given temperature, as the density is increased and electrons become more degenerate, the larger is the energy required to produce the kinetic energy of an electron at the top of the electron Fermi sea in addition to the energy required for the rest masses of the new electron and its positron companion; thus, with increasing density, the number density of positrons decreases relative to the number density of electrons, and the neutrino energy-loss rate associated with the electron–positron pair annihilation process decreases accordingly.

In contrast with the electron–positron pair annihilation process in which the initial state passes directly into a final state, the additional processes which lead to the formation of a neutrino–antineutrino pair require passage through an intermediate state involving the operation of the electromagnetic field, making the calculation of rates somewhat more complicated. The additional processes are called the photoneutrino, the bremsstrahlung, and the plasma processes.

In the photoneutrino process, which competes with the electron–positron pair production process at intermediate temperatures and dominates at lower temperatures, a photon interacts with an electron (or positron) through the electromagnetic field to produce an electron (or positron) in an intermediate state from which it is then transformed through the weak interaction into a neutrino–antineutrino pair and a final electron (or positron). Alternatively, the electron (or positron) interacts through the weak interaction to produce a neutrino–antineutrino pair and is then transformed into an intermediate electron (or positron) which is transformed by a final electromagnetic interaction into the final electron (or positron). The overall rate of this process is proportional to the sum of the electron and positron densities rather than to the product of the two densities, as in the case of the real electron–positron annihilation process, and, because the electron density does not explicitly decrease with decreasing temperature, the effectiveness of the photoneutrino process relative to the effectiveness of the real electron–positron pair-annihilation process increases with decreasing temperature.

The earliest calculations of the rate of the photoneutrino $\nu_e \bar{\nu}_e$ production process and of the energy-loss rate due to the process were performed by H.-Y. Chiu & R. C. Stabler (1961), M. J. Levine (1963), V. I. Ritus (1962), and V. Petrosian, G. Beaudet, & E. E. Salpeter (1967).

In the bremsstrahlung process, an electron (or positron) is scattered into an intermediate state by an electromagnetic interaction with a positively charged ion and is then transformed into a final electron (or positron) and a neutrino–antineutrino pair through the mediation of the weak interaction. Alternatively the electron (or positron) can be first transformed through the weak interaction into an intermediate state in conjunction with a neutrino–antineutrino pair and then tranformed by an electromagnetic interaction with a positive ion into a final electron (or positron). The energy-loss rate due to the bremsstrahlung process was first calculated by G. C. Festa & M. A. Ruderman (1969). In the context of stellar evolution, this process is not of major importance.

The plasma process relies on the fact that an imposed oscillating electric field in a medium containing free electrons forces the electrons to oscillate out of phase with the field. The electrons move in synchronism and the electric field associated with the coherently moving electrons combines with the imposed electric field to produce what is called a plasmon. The composite oscillation can persist only if the frequency is equal to or larger than the plasma frequency ω_p introduced in Section 7.12 in Volume 1. There are two principle modes of oscillation, depending on the orientation of the propagation vector $\mathbf{k}$ relative to the orientation of the electric field vector $\mathbf{E}$. When $\mathbf{k}$ and $\mathbf{E}$ are orthogonal, the composite oscillation is a travelling wave which may be thought of as having an effective mass equal to $\hbar\omega_p/c^2$; there is no restriction on the wave number k and the oscillation frequency ω can take on any value $\geq \omega_p$. When $\mathbf{k}$ and $\mathbf{E}$ are parallel, there is no associated magnetic field, and the composite oscillation may be thought of as an electrostatic oscillation of effective mass also equal to $\hbar\omega_p/c^2$, but with a wave number restricted to be less than or equal to ω_p/c and with a frequency restricted to be very close to ω_p. In both cases, the fact that an effective mass is associated with the coupled oscillation of the electric field and free electrons makes the plasma emission of neutrino–antineutrino pairs energetically more favorable than the photoneutrino process.

The decays of transverse and longitudinal plasmons into neutrino–antineutrino pairs are by far the dominant neutrino energy-loss mechanisms in the hydrogen-exhausted cores of low mass red giants and in the helium-exhausted cores of asymptotic giant branch stars. The first calculations of the energy-loss rates by these processes were made by J. B. Adams, M. A. Ruderman, & C. H. Woo (1963), C. L. Inman & M. A. Ruderman (1964), V. N. Tsytovich (1963, 1964), and M. H. Zaidi (1965). A description of some of the complexities involved in the calculations is given in Chapter 6 of Hong-Yee Chiu's book *Stellar Physics* (1968).

15.7 On the character of classical plasma oscillations

In this section, the symbol E is used both for the electric field strength and for the electron Fermi energy and the symbol k is used both for the propagation vector of a plane wave and for Boltzmann's constant. The reader should have no difficulty in recognizing from the context which meaning is being used. For example, Boltzmann's constant always appears in conjunction with temperature T, as in kT.

A transverse electromagnetic wave can propagate without attenuation through a medium containing free electrons at a number density of n_e only if its frequency ω is larger than the plasma frequency which, as demonstrated in Section 7.12 in Volume 1, is given by

$$\omega_p = \sqrt{\frac{4\pi e^2 n_e}{m_e}} \tag{15.7.1}$$

when the electrons have non-relativistic velocities. The magnitude of the propagation vector **k** of the electromagnetic wave modified by interaction with free electrons is, in first approximation, related to the oscillation frequency by the dispersion relationship

$$\omega^2 = \omega_p^2 + (ck)^2, \tag{15.7.2}$$

where k and wavelength λ are related by $k = 2\pi/\lambda$.

The possibility of plasma oscillations was discovered in the early part of the twentieth century in laboratory investigations by Irving Langmuir (1925, 1926). Pioneering theoretical interpretations of these oscillations were discussed by Lewi Tonkas and Langmuir (1929) and an insightful summary of the physics involved is given by J. D. Jackson in Section 10.9 of his book *Classical Electrodynamics* (1962). Repeating the steps outlined by Jackson is useful for understanding the physics of plasma waves and for understanding how the dispersion relationships for these waves come about.

At the high frequencies at which plasma oscillations occur, positive ions are essentially motionless when compared with the motions of free electrons, and the electric current density may be approximated by

$$\mathbf{j} = -e \int_{\text{electrons}} \mathbf{v}\, dn_e(\mathbf{v}_0) = -e \int (\mathbf{v}_0 + \mathbf{u})\, dn_e(\mathbf{v}_0) = -e\, n_e\, \mathbf{u}, \tag{15.7.3}$$

where n_e is the electron number density, $\mathbf{v}_0$ is the thermal velocity and $\mathbf{u}$ is the bulk velocity of $dn_e(\mathbf{v}_0)$ electrons, and

$$\frac{1}{n_e} \int \mathbf{v}_0 \, dn_e(\mathbf{v}_0) = 0. \tag{15.7.4}$$

That is, the average over thermal velocities vanishes, and only forcing by high frequency variations in the electromagnetic field contributes to the motion of electrons over and above thermal motions. A caveat on the validity of these equations is that the integrals must be over volumes small compared with the dimensions of variations in electromagnetic field quantities.

Assume that all variable quantities in Maxwell's equations, as described by eqs. (7.12.36) in Volume 1, vary according to

$$\mathbf{Q} = \mathbf{Q}_0 \, e^{i(\omega t - \mathbf{k} \cdot \mathbf{r})}, \tag{15.7.5}$$

where $\mathbf{k}$ and $\mathbf{Q}_0$ point in arbitrary directions. Inserting eq. (15.7.3) into the first relationship in eq. (7.12.36),

$$\nabla \times \mathbf{H} = \frac{4\pi}{c} \, \mathbf{j} + \frac{1}{c} \, \frac{\partial \mathbf{E}}{\partial t},$$

adopting eq. (15.7.5) for all variables, and taking derivatives produces

$$-i \, \mathbf{k} \times \mathbf{H} = \frac{i\omega}{c} \, \mathbf{E} - 4\pi \, e n_e \, \frac{\mathbf{u}}{c} \tag{15.7.6}$$

which may be rearranged to read

$$\mathbf{u} = i \, \frac{c}{4\pi \, e n_e} \left(\mathbf{k} \times \mathbf{H} + \frac{\omega}{c} \, \mathbf{E} \right). \tag{15.7.7}$$

Inserting dependences of the form given by eq. (15.7.5) into the second equation of eq. (7.12.36), $\nabla \times \mathbf{E} = -(1/c) \, \partial \mathbf{H}/\partial t$, and taking derivatives produces

$$\mathbf{k} \times \mathbf{E} = \frac{\omega}{c} \, \mathbf{H}, \tag{15.7.8}$$

Determining $\mathbf{k} \times \mathbf{H}$ in terms of vector products of $\mathbf{k}$ and $\mathbf{E}$ from eq. (15.7.8), making use of the identity

$$\mathbf{k} \times (\mathbf{k} \times \mathbf{E}) = (\mathbf{k} \cdot \mathbf{E}) \, \mathbf{k} - k^2 \mathbf{E}, \tag{15.7.9}$$

and inserting the result into eq. (15.7.7) produces

$$\mathbf{u} = i \, \frac{c}{4\pi \, e n_e} \left\{ \frac{c}{\omega} \left((\mathbf{k} \cdot \mathbf{E}) \, \mathbf{k} - k^2 \mathbf{E} \right) + \frac{\omega}{c} \, \mathbf{E} \right\}. \tag{15.7.10}$$

Finally, using eq. (15.7.1), one may write

$$\mathbf{u} = i \, \frac{e}{\omega m_e} \, \frac{1}{\omega_p^2} \, \{ c^2 ((\mathbf{k} \cdot \mathbf{E}) \, \mathbf{k} - k^2 \mathbf{E}) + \omega^2 \, \mathbf{E} \}$$

$$= i \, \frac{e}{\omega m_e} \, \frac{1}{\omega_p^2} \, \{ c^2 (\mathbf{k} \cdot \mathbf{E}) \, \mathbf{k} + (\omega^2 - c^2 k^2) \mathbf{E} \}. \tag{15.7.11}$$

The body force on free electrons exerted directly by the electromagnetic field may be approximated by

$$\left(n_e\, m_e\, \frac{d\mathbf{u}}{dt}\right)_{\text{EM}} = -e \int \left(\mathbf{E} + \frac{\mathbf{v}}{c} \times \mathbf{H}\right) dn_e(\mathbf{v_0}) = e\, n_e \left(\mathbf{E} + \frac{\mathbf{u}}{c} \times \mathbf{H}\right), \quad (15.7.12)$$

where eq. (15.7.4) has been used in arriving at the expression to the right of the rightmost equal sign. There is an additional pressure body force due to the small scale spatial variations in the electron density brought about by interaction with the electromagnetic field. This force may be written as

$$\left(m_e\, n_e\, \frac{d\mathbf{u}}{dt}\right)_{\text{pressure}} = -\nabla P, \quad (15.7.13)$$

where

$$\nabla P = \frac{\partial P}{\partial n_0}\, \nabla n \quad (15.7.14)$$

and

$$n = n_e - n_0. \quad (15.7.15)$$

Here, P is the component of the total pressure due to motions in the direction of the electric field, n_0 is the average electron density, n is the excess in the electron density due to forcing by the electric field, and it has been assumed that the distance scale over which n varies is small compared with the distance scale over which n_0 varies, so that variations in n_0 may be neglected.

Combining the two body forces gives

$$\frac{d\mathbf{u}}{dt} = \frac{\partial \mathbf{u}}{\partial t} + \mathbf{u} \cdot (\nabla \mathbf{u}) = -\frac{e}{m_e} \left(\mathbf{E} + \frac{\mathbf{u}}{c} \times \mathbf{H}\right) - \frac{1}{m_e n_e} \frac{\partial P}{\partial n_0} \nabla n. \quad (15.7.16)$$

Making use of eq. (15.7.5), eq. (15.7.16) becomes

$$i\omega \mathbf{u} - i(\mathbf{u} \cdot \mathbf{k})\mathbf{u} = -\frac{e}{m_e} \left(\mathbf{E} + \frac{\mathbf{u}}{c} \times \mathbf{H}\right) + i \frac{\mathbf{k}}{m_e n_e} \frac{\partial P}{\partial n_0} n. \quad (15.7.17)$$

From the third equation in eq. (7.12.36), $\nabla \cdot \mathbf{E} = 4\pi \rho_e$, and eq. (15.7.5), one has that

$$\nabla \cdot \mathbf{E} = -4\pi e n = -i\mathbf{k} \cdot \mathbf{E}, \quad (15.7.18)$$

which may be used to replace n in eq. (15.7.17) to give

$$i\omega \mathbf{u} - i(\mathbf{u} \cdot \mathbf{k})\mathbf{u} = -\frac{e}{m_e} \left(\mathbf{E} + \frac{\mathbf{u}}{c} \times \mathbf{H}\right) + \frac{\mathbf{k}}{m_e n_e} \left(\frac{\partial P}{\partial n_0}\right) \frac{\mathbf{k} \cdot \mathbf{E}}{4\pi e}. \quad (15.7.19)$$

Rearranging and then using eq. (15.7.1) to eliminate n_e, one may write

$$\mathbf{u}\left(1 - \frac{\mathbf{u}}{c} \cdot \frac{c\mathbf{k}}{\omega}\right) = i\frac{e}{\omega m_e}\left(\mathbf{E} + \frac{\mathbf{u}}{c} \times \mathbf{H}\right) + i\frac{\mathbf{k}}{\omega m_e n_e}\left(\frac{\partial P}{\partial n_0}\right)\frac{\mathbf{k} \cdot \mathbf{E}}{4\pi e}$$

$$= i\frac{e}{\omega m_e}\left(\mathbf{E} + \frac{\mathbf{u}}{c} \times \mathbf{H}\right) + i\frac{e}{\omega m_e}\frac{1}{\omega_p^2 m_e}\left(\frac{\partial P}{\partial n_0}\right)(\mathbf{k} \cdot \mathbf{E})\,\mathbf{k}. \quad (15.7.20)$$

Using eq. (15.7.8) and an appropriate decomposition of a triple cross product into simple vectors, one has that

$$\frac{\mathbf{u}}{c} \times \mathbf{H} = \frac{1}{\omega}\mathbf{u} \times (\mathbf{k} \times \mathbf{E}) = \frac{1}{\omega}\left((\mathbf{u} \cdot \mathbf{E})\mathbf{k} - (\mathbf{u} \cdot \mathbf{k})\mathbf{E}\right) = \left(\frac{\mathbf{u}}{c} \cdot \mathbf{E}\right)\frac{c\mathbf{k}}{\omega} - \left(\mathbf{u} \cdot \frac{c\mathbf{k}}{\omega}\right)\mathbf{E}.$$

$$(15.7.21)$$

Replacing $(\mathbf{u}/c) \times \mathbf{H}$ in eq. (15.7.20) by the last expression in eq. (15.7.21) and comparing the result with eq. (15.7.11) produces

$$\frac{1}{\omega_p^2}\{c^2(\mathbf{k} \cdot \mathbf{E})\,\mathbf{k} + (\omega^2 - c^2 k^2)\mathbf{E}\}\left(1 - \frac{\mathbf{u}}{c} \cdot \frac{c\mathbf{k}}{\omega}\right)$$

$$= \mathbf{E}\left(1 - \frac{\mathbf{u}}{c} \cdot \frac{c\mathbf{k}}{\omega}\right) + \left(\frac{\mathbf{u}}{c} \cdot \mathbf{E}\right)\frac{c\mathbf{k}}{\omega} + \frac{1}{m_e}\frac{1}{\omega_p^2}\left(\frac{\partial P}{\partial n_0}\right)(\mathbf{k} \cdot \mathbf{E})\,\mathbf{k}, \quad (15.7.22)$$

which is the master equation from which the dispersion relationships for both transverse and longitudinal oscillations can be deduced.

When $\mathbf{k} \cdot \mathbf{E} = 0$, the composite electromagnetic–electron oscillation is transverse, and eq. (15.7.22) gives

$$(\omega^2 - c^2 k^2 - \omega_p^2)\mathbf{E}\left(1 - \frac{\mathbf{u}}{c} \cdot \frac{c\mathbf{k}}{\omega}\right) - \omega_p^2\left(\frac{\mathbf{u}}{c} \cdot \mathbf{E}\right)\frac{c\mathbf{k}}{\omega} = 0. \quad (15.7.23)$$

Combining eqs. (15.7.10) and (15.7.1) gives

$$\mathbf{u} = i\frac{e}{\omega m_e}\mathbf{E}, \quad (15.7.24)$$

showing that the induced current given by eq. (15.7.4) is entirely in the direction of the electric field and is exactly 180° out of phase with this field. Further, since $\mathbf{u} \propto \mathbf{E}$, the product $\mathbf{k} \cdot \mathbf{u}$ in eq. (15.7.23) vanishes. Taking the dot product of eq. (15.7.23) with $\mathbf{E}$ and using $\mathbf{k} \cdot \mathbf{E} = 0$ once more, eq. (15.7.2) is reproduced.

From eq. (15.7.18), it follows that $n = 0$, which means that there is no variation in the electron density due to the transverse oscillation. Noting that $\langle|\mathbf{v}_0 + \mathbf{u}|^2\rangle = \langle v_0^2\rangle + \langle u^2\rangle$, eqs. (15.7.24), (15.7.2), and (15.7.1) can be used to relate the kinetic energy density associated with the action of the electric field on the electrons to the energy density of the electric field. Thus,

$$\frac{1}{2}m_e|\mathbf{u}|^2 n_e = \frac{1}{2}m_e u^2 n_e = \frac{1}{2}m_e\,n_e\frac{e^2}{\omega^2 m_e^2}E^2 = \frac{E^2}{8\pi}\left(\frac{\omega_p}{\omega}\right)^2 = \frac{E^2}{8\pi}\left(1 + \left(\frac{ck}{\omega_p}\right)^2\right)^{-1}.$$

$$(15.7.25)$$

For small wave numbers such that frequencies are near the plasma frequency, the kinetic energy density of electrons oscillating out of phase with the electric field is nearly the same as the energy density in the electric field. For large wave numbers such that $\omega \gg \omega_p$, the energy density associated with high frequency electron oscillations becomes a negligible fraction of the energy density in the electric field. In both limits, the amplitude of the variations in the magnetic field is the same as the amplitude of the variations in the electric field, so the energy densities of the electric field and of the magnetic field are identical at all frequencies. This may be demonstrated by combining eqs. (15.7.6), (15.7.24), (15.7.2), and (15.7.1) to obtain

$$|\mathbf{k} \times \mathbf{H}|^2 = k^2 \, H^2 = \left| \frac{\omega}{c} \mathbf{E} - 4\pi e n_e \frac{e}{c\omega m_e} \mathbf{E} \right|^2 = \frac{\omega^2}{c^2} \left(1 - \frac{\omega_p^2}{\omega^2} \right) E^2 = k^2 \, E^2.$$

$$(15.7.26)$$

Thus, for $ck \ll \omega_p$, energy is shared almost equally between the electric field, the magnetic field, and high frequency motions of free electrons. For $ck \gg \omega_p$, energy is shared almost equally between the electric field and the magnetic field, and the kinetic energy associated with the forced high frequency motions of free electrons is negligible.

When $\mathbf{k} \times \mathbf{E} = 0$, the composite wave formed by the interaction between the electro-magnetic field and the free electrons is longitudinal. Equation (15.7.8) shows that $\mathbf{H} = 0$, and, setting $\mathbf{k} \cdot \mathbf{E} = kE$ and $\mathbf{u} \cdot \mathbf{k} = uk$ in eq. (15.7.22) gives

$$\omega^2 = \left(1 - \frac{u}{c} \frac{ck}{\omega} \right)^{-1} \left(\omega_p^2 + \frac{1}{m_e} \left(\frac{\partial P}{\partial n_0} \right) k^2 \right). \tag{15.7.27}$$

The fraction of the pressure due to motion in one dimension is given by

$$P = m_e n_0 \langle v_x^2 \rangle,$$

where v_x is the total velocity of an electron in the direction of the oscillations forced by the electric field and angle brackets about v_x^2 denote an average over all electrons. The kinetic energy associated with motion in one dimension in a box of volume V is

$$E_{\mathrm{kin}} = \frac{1}{2} m_e n_0 \langle v_x^2 \rangle V,$$

where $n_0 V = \mathrm{constant}$. For an adiabatic change,

$$dQ = PdV + dE_{\mathrm{kin}} = m_e n_0 \langle v_x^2 \rangle dV + \frac{1}{2} m_e n_0 V d\langle v_x^2 \rangle = 0,$$

from which it follows that

$$\frac{d\langle v_x^2 \rangle}{\langle v_x^2 \rangle} = -2 \frac{dV}{V} = 2 \frac{dn_0}{n_0}.$$

Thus,

$$dP = d(m_e n_0 \langle v_x^2 \rangle) = m_e (dn_0 \langle v_x^2 \rangle + n_0 \, d\langle v_x^2 \rangle)$$

$$= m_e dn_0 (\langle v_x^2 \rangle + 2\langle v_x^2 \rangle) = m_e dn_0 (3\langle v_x^2 \rangle),$$

so that

$$\left(\frac{\partial P}{\partial n_0}\right) = 3\, m_e \langle v_x^2 \rangle \approx m_e \langle v_0^2 \rangle,$$

where $\langle v_0^2 \rangle$ is the average of the three dimensional velocity squared.

In first approximation, then, the dispersion relationship for longitudinal plasmons is

$$\omega^2 = \left(1 - \frac{u}{c}\frac{ck}{\omega}\right)^{-1}\left(\omega_p^2 + \left\langle \frac{v_0^2}{c^2}\right\rangle (ck)^2\right). \tag{15.7.28}$$

Since ck/ω is always smaller than unity and, since one can expect $u \ll c$,

$$\omega^2 \approx \omega_p^2 + \left\langle \frac{v_0^2}{c^2}\right\rangle (ck)^2. \tag{15.7.29}$$

When electron velocities are not relativistic, and $ck < \omega_p$, longitudinal oscillations are effectively all at the plasma frequency.

From eq. (15.7.7), it follows that, for longitudinal oscillations,

$$\mathbf{u} = i\,\frac{\omega}{4\pi\, e n_e}\,\mathbf{E}, \tag{15.7.30}$$

which says that the amplitude of the coherent electron velocity variations is proportional to the amplitude of the electric field oscillations and is 180° out of phase with these oscillations, just as in the case of transverse oscillations. Using eqs. (15.7.30), (15.7.1), and (15.7.2), one finds that

$$\frac{1}{2}\, m_e\, u^2\, n_e = \frac{E^2}{8\pi}\left(1 + \left\langle \frac{v_0^2}{c^2}\right\rangle \frac{(ck)^2}{\omega_p^2}\right), \tag{15.7.31}$$

which says that, for $ck < \omega_p$, the kinetic energy density associated with high frequency longitudinal oscillations of free electrons forced by the electric field is essentially equal to the energy density of the electric field.

Setting $dx/dt = \mathbf{u}$ and and using eq. (15.7.5), eq. (15.7.30) can be solved for $\mathbf{x}$, the amplitude of the spatial variations forced by the electric field, with the result that

$$\mathbf{x} = \frac{\mathbf{E}}{4\pi\, e n_e}. \tag{15.7.32}$$

Equation (15.7.18) gives

$$n = i\,\frac{\mathbf{k}\cdot\mathbf{E}}{4\pi\, e} = i\,\frac{kE}{4\pi\, e}, \tag{15.7.33}$$

and, combining with eq. (15.7.32), one obtains

$$n = i\, n_e\, k\, x = i\, n_e\, 2\pi\,\frac{x}{\lambda}. \tag{15.7.34}$$

Since, by hypothesis, $n/n_e \ll 1$, eq. (15.7.34) shows that the maximum electron displacement in a longitudinal oscillation is much smaller than the wavelength of the oscillation.

Further insight into the nature of the two oscillation modes follows from an examination of phase and group velocities. For the transverse mode,

$$v_{\text{phase}} = \frac{\omega}{k} = c \sqrt{(\omega_p/ck)^2 + 1} > c, \qquad (15.7.35)$$

and

$$v_{\text{group}} = \frac{\partial \omega}{\partial k} = c^2 \frac{k}{\omega} = \frac{c}{\sqrt{(\omega_p/ck)^2 + 1}} < c, \qquad (15.7.36)$$

showing that $v_{\text{group}} v_{\text{phase}} = c^2$, as expected for a traveling electromagnetic wave.

Setting

$$v_{\text{rms}} = \sqrt{\langle v_0^2 \rangle} = \sqrt{3kT/m_e} , \qquad (15.7.37)$$

for the longitudinal mode one has

$$v_{\text{phase}} = \frac{\omega}{k} = \sqrt{(\omega_p/k)^2 + v_{\text{rms}}^2} = \sqrt{c^2 (\omega_p/ck)^2 + v_{\text{rms}}^2} > v_{\text{rms}} \qquad (15.7.38)$$

and

$$v_{\text{group}} = \frac{\partial \omega}{\partial k} = v_{\text{rms}}^2 \frac{k}{\omega} = \frac{v_{\text{rms}}^2}{\sqrt{(\omega_p/k)^2 + v_{\text{rms}}^2}} < v_{\text{rms}}, \qquad (15.7.39)$$

showing that the product $v_{\text{group}} v_{\text{phase}}$ ($= v_{\text{rms}}^2$) has nothing to do with a traveling electromagnetic wave and is rather a function only of the thermal properties of the electrons. As long as $ck < \omega_p$, $v_{\text{phase}} > c$ and, therefore, when electron motions are non-relativistic, $v_{\text{phase}} \gg v_{\text{rms}}$.

On the other hand, for $v_{\text{rms}} k = \omega_p$, $v_{\text{phase}} = \sqrt{2} \, v_{\text{rms}} \sim v_{\text{rms}}$. This means that electrons move substantially in random directions during the passage of several wave crests, implying that electrons can be trapped between crests, with energy being transferred from the electric field to the thermal motion of the electrons. This transfer is made plausible by eq. (15.7.31), which shows that, under the conditions assumed, twice as much energy resides in the kinetic energy of electron oscillatory motion as in the electric field, in contrast with the near equality of the two energies when $ck = \omega_p$.

The upshot of these considerations is that a very strong upper limit on the wave number for persistent longitudinal plasma oscillations is given by

$$k_{\text{max}}^2 = \left(\frac{2\pi}{\lambda_{\text{min}}} \right)^2 \ll \frac{\omega_p^2}{v_{\text{rms}}^2} = \frac{4\pi e^2 n_e}{m_e} \frac{m_e}{3kT} = \frac{1}{3} \frac{4\pi e^2 n_e}{kT}. \qquad (15.7.40)$$

It is interesting to compare the corresponding lower limit on wavelength with the Debye shielding radius discussed in Section 4.17 in Volume 1 and given by eq. (4.17.12) as

$$\left(\frac{1}{R_{\text{Debye}}} \right)^2 = \frac{4\pi e^2 \left(n_e + \sum_{\text{ions}} n_{0j} \, Z_j^2 \right)}{kT} = \frac{4\pi e^2 n_e}{kT} \left(1 + \frac{1}{n_e} \sum_{\text{ions}} n_{0j} \, Z_j^2 \right), \qquad (15.7.41)$$

where $n_e = \sum n_{0j}\, Z_j$. For pure, completely ionized, helium, the term in parentheses on the far right of eq. (15.7.41) is 3, and $\lambda_{\min} \gg 6\pi\, R_{\text{Debye}}$. For pure, completely ionized, oxygen, $\lambda_{\min} \gg 54\pi\, R_{\text{Debye}}$.

In estimating the rate of neutrino–antineutrino production by quantized longitudinal plasmons, it has become conventional to suppose that the maximum wave number and the corresponding minimum wavelength satisfy the equations

$$k_{\max}^2 = \left(\frac{2\pi}{\lambda_{\min}}\right)^2 = \frac{\omega_{\text{p}}^2}{c^2}. \tag{15.7.42}$$

Although, in the case of non-relativistic electrons, these choices formally satisfy the limits defined by eq. (15.7.40), they are not a fundamental consequence of energy and momentum conservation, as has on occasion been asserted in the literature.

The derivations in this section have assumed that each electron can move freely in response to the force $-e\mathbf{E}$ without any restrictions as to available phase space. When electrons are degenerate, one might expect the plasma frequency to be smaller than in the non-degenerate case, given the oft cited mantra that only electrons in a range $\Delta E \sim 2\,kT$ near the top of the Fermi sea are free to move in response to an external force. Subscribing to this mantra, one might expect n_e in eq. (15.7.1) to be multiplied by a factor of the order of $kT/(p_F^2/2m_e)$, which is a measure of the fraction of electrons which are mobile and contribute, for example, to electron electrical and thermal conductivity.

On the other hand, it is almost self evident that the response of a degenerate electron to the gravitational field is affected by its being part of an electron-degenerate sea only to the extent that its effective mass is larger than its rest mass by the factor $1/\sqrt{1 - v^2/c^2}$. The consensus opinion in the literature appears to be that this is also the case when the imposed force is due to a local electromagnetic field. That is, it is tacitly assumed that, when all electrons are subjected to the same electromagnetic field, movement of any electron from one energy state into an adjacent higher energy state is made possible by the movement of the erstwhile occupier of the adjacent state into a still higher energy state. However, since the effective mass of an electron is given by its energy divided by c^2, every electron responds differently to a given force, depending on its velocity, so a proper explanation must be somewhat more complex than normally assumed.

Nevertheless, it is accepted in the literature as a fact that, as long as most electrons do not have excessively relativistic velocities, the plasma frequency is given by

$$\omega_{\text{p}}^2 = 4\pi e^2 \int dn_e\, \frac{c^2}{E}\left(1 - \frac{1}{3}\frac{v^2}{c^2}\right), \tag{15.7.43}$$

where $E = \sqrt{(m_e c^2)^2 + (cp)^2}$ is the total energy of an electron, not to be confused with $|\mathbf{E}|$, where $\mathbf{E}$ is the electric field strength. In an effort to understand this relationship, suppose that a degenerate electron with velocity $\mathbf{v}$ is accelerated by an imposed electromagnetic field according to the equation

$$\frac{d}{dt}\left(m_e \frac{\mathbf{v}}{\sqrt{1 - v^2/c^2}}\right) = -e\left(\mathbf{E} + \frac{\mathbf{v}}{c} \times \mathbf{H}\right). \tag{15.7.44}$$

Suppose that the fields $\mathbf{E}$ and $\mathbf{H}$ are orthogonal, that the $\mathbf{E}$ field is in the $\mathbf{y}$ direction, and that the only acceleration is in the $\mathbf{y}$ direction. Carrying out the differentiation with respect to t on the left hand side of eq. (15.7.33) and neglecting the cross term involving the magnetic field on the right hand side gives

$$\frac{m_e}{\sqrt{1-v^2/c^2}}\left(\frac{1+v_y^2/c^2}{1-v^2/c^2}\right)\frac{dv_y}{dt} = -eE_y. \tag{15.7.45}$$

On average, $v_x^2 = v_y^2 = v_z^2 = v^2/3$ and $dv_y/dt = d^2y/dt^2$, so

$$\frac{m_e}{\sqrt{1-v^2/c^2}}\left(1+\frac{(1/3)\,v^2/c^2}{1-v^2/c^2}\right)\frac{dv_y}{dt} = \frac{m_e}{\sqrt{1-v^2/c^2}}\left(\frac{1-(2/3)v^2/c^2}{1-v^2/c^2}\right)\frac{d^2y}{dt^2} = -eE_y. \tag{15.7.46}$$

To first order in v^2/c^2, eq. (15.7.46) can be written as

$$\frac{m_e}{\sqrt{1-v^2/c^2}}\left(1-\frac{1}{3}\frac{v^2}{c^2}\right)^{-1}\frac{d^2y}{dt^2} = -eE_y, \tag{15.7.47}$$

but it could just as well have been written with $\left(1-\frac{1}{3}\frac{v^2}{c^2}\right)^{-1}$ replaced by $\left(1+\frac{1}{3}\frac{v^2}{c^2}\right)$.
    With the replacement

$$m_e \rightarrow \frac{E}{c^2}\left(1-\frac{1}{3}\frac{v^2}{c^2}\right)^{-1}, \tag{15.7.48}$$

eq. (7.12.42) for a single electron becomes

$$\frac{dy}{dt} = -\frac{eE_y}{i\omega\,E/c^2}\left(1-\frac{1}{3}\frac{v^2}{c^2}\right), \tag{15.7.49}$$

and integration over the electron number distribution converts eq. (7.12.43) for the current into

$$j_y = \frac{e^2 E_y}{i\omega}\int\frac{dn_e}{E/c^2}\left(1-\frac{1}{3}\frac{v^2}{c^2}\right). \tag{15.7.50}$$

Since going from eq. (15.7.45) to (15.7.46) is very much a sleight of hand, one could defer the trick of setting $v_x^2 = v_y^2 = v_z^2 = v^2/3$ until performing the average in eq. (15.7.50). Repeating the steps outlined by eqs. (7.12.42)–(7.12.48) in Volume 1 leads to eq. (15.7.43).
    It is interesting to explore the consequence of performing the integrations in eq. (15.7.43) in the limiting case of zero temperature. Writing the equation as

$$\omega_p^2 = 4\pi e^2\int_0^\infty\frac{8\pi}{h^3}\frac{p^2\,dp}{e^{(E-E_F)/kT}+1}\frac{c^2}{E}\left(1-\frac{1}{3}\frac{v^2}{c^2}\right), \tag{15.7.51}$$

setting $kT = 0$ and $c = 1$, one has

$$
\begin{aligned}
\omega_p^2 &= 4\pi e^2 \frac{8\pi}{h^3} \int_0^{p_F} \frac{p^2}{(p^2 + m_e^2)^{1/2}} \left(1 - \frac{1}{3} \frac{p^2}{p^2 + m_e^2} \right) dp \\
&= 4\pi e^2 \frac{8\pi}{h^3} \left\{ \frac{1}{2} p_F E_F - \frac{m_e^2}{2} \log_e \frac{p_F + E_F}{m_e} \right\} \\
&\quad - 4\pi e^2 \frac{8\pi}{h^3} \frac{1}{3} \left\{ \frac{1}{2} p_F E_F + \frac{m_e^2 p_F}{E_F} - \frac{3 m_e^2}{2} \log_e \frac{p_F + E_F}{m_e} \right\} \\
&= 4\pi e^2 \frac{8\pi}{h^3} \frac{1}{3} p_F \left(E_F - \frac{m_e^2}{E_F} \right) = 4\pi e^2 \frac{8\pi}{h^3} \frac{1}{3} \frac{p_F^3}{E_F} = \frac{4\pi e^2}{E_F/c^2} n_e,
\end{aligned}
\tag{15.7.52}
$$

where the temporarily adopted convention of setting $c = 1$ has been dropped in the expression to the right of the very last equal sign. Since, by assumption, electrons are not terribly relativistic, $E_F \sim m_e c^2$ and this last expression is essentially the same as the non-degenerate electron result. The more important lesson is that, even when $(v/c)^2$ is small, the first order relativistic correction in the form given by the standard integral expression for ω_p^2 is essential for obtaining the simple result expressed by the final expression in eq. (15.7.52) when electrons are degenerate.

Dispersion relationships which take into account first order relativistic corrections have been constructed by, among others, G. Beaudet, V. Petrosian, & E. E. Salpeter (1967). These have been extended by Eric Braaten & Daniel Segel (1993) into the domain where electrons are fully relativistic. As an example, a first approximation to the dispersion relationship for longitudinal waves under electron-degenerate conditions is

$$
\omega^2 = \omega_p^2 + \frac{3}{5} \left\langle \frac{v_F^2}{c^2} \right\rangle (ck)^2,
\tag{15.7.53}
$$

where v_F is the Fermi velocity.

It is somewhat ironic to note that, in deriving the expression for the plasma frequency under electron-degenerate conditions, classical physics has been used to treat the quantum mechanical interaction between photons and electrons as if it were an interaction between a classical electromagnetic field and a classical electron, whereas quantum physics which treats electrons as waves has been used to describe the distribution of electrons with respect to energy.

15.8 Quantized plasma oscillations and the neutrino–antineutrino energy-loss rate due to plasmon decay

In order to establish a relationship between classical plasma oscillations and the production of neutrino–antineutrino pairs, it is necessary to quantize the plasma oscillations. This was first done in a series of papers by J. M. Jauch & K. M. Watson (Jauch & Watson, 1948a; 1948b and Watson & Jauch, 1949).

Whereas a quantized longitudinal oscillation is sufficiently different in character from a photon in free space to be called a plasmon, it is sometimes argued (e.g., Braaten & Segel, 1993) that the proper designation of a quantized transverse oscillation should be a photon with mass rather than a transverse plasmon. In most of the literature, however, both modes of oscillation are called plasmons, and that is the convention adopted here.

Assuming that transverse plasmons with an effective mass are bosons with a statistical weight of $g_t = 2$, the number density of such plasmons is

$$n_t = \frac{8\pi}{h^3} \int_0^\infty \frac{p^2\,dp}{e^{\hbar\omega/kT} - 1} = \frac{1}{\pi^2} \int_0^\infty \frac{k^2\,dk}{e^{\hbar\omega/kT} - 1}, \tag{15.8.1}$$

just as in the case of photons in free space, and their energy density is

$$U_t = \frac{8\pi}{h^3} \int_0^\infty \hbar\omega \frac{p^2\,dp}{e^{\hbar\omega/kT} - 1} = \frac{1}{\pi^2} \int_0^\infty \hbar\omega \frac{k^2\,dk}{e^{\hbar\omega/kT} - 1} > n_t\,\hbar\omega_p. \tag{15.8.2}$$

Defining a variable x by

$$k = \frac{\omega_p}{c}\,x = \left(\frac{\hbar\omega_p}{m_e c^2}\right)\left(\frac{m_e c}{\hbar}\right) x, \tag{15.8.3}$$

and using eq. (15.7.2) as the relationship beteen ω and k, one has

$$\int_0^\infty \frac{k^2\,dk}{e^{\hbar\omega/kT} - 1} = \left(\frac{\hbar\omega_p}{m_e c^2}\right)^3 \left(\frac{m_e c}{\hbar}\right)^3 \int_0^\infty \frac{x^2\,dx}{e^{(\hbar\omega_p/kT)\sqrt{1+x^2}} - 1}, \tag{15.8.4}$$

When $\hbar\omega_p/kT \gg 1$, the maximum in the integrand in eq. (15.8.4) occurs at $x \sim \sqrt{2}$ $(kT/\hbar\omega_p) \ll 1$, so, in good approximation,

$$\begin{aligned}
\int_0^\infty \frac{x^2\,dx}{e^{(\hbar\omega_p/kT)\sqrt{1+x^2}} - 1} &= \int_0^\infty e^{-(\hbar\omega_p/kT)(1+x^2/2)}\,x^2\,dx + \cdots \\
&= e^{-\hbar\omega_p/kT} \int_0^\infty e^{-(\hbar\omega_p/2kT)\,x^2}\,x^2\,dx + \cdots \\
&= e^{-\hbar\omega_p/kT} \left(\frac{2kT}{\hbar\omega_p}\right)^{3/2} \int_0^\infty e^{-y^2}\,y^2\,dy + \cdots \\
&= e^{-\hbar\omega_p/kT} \left(\frac{2kT}{\hbar\omega_p}\right)^{3/2} \frac{\sqrt{\pi}}{4} + \cdots \\
&= e^{-\hbar\omega_p/kT} \left(\frac{kT}{\hbar\omega_p}\right)^{3/2} \sqrt{\frac{\pi}{2}} + \cdots.
\end{aligned} \tag{15.8.5}$$

Using this result in eq. (15.8.1), one has, for $\frac{\hbar\omega_p}{kT} \gg 1$,

$$n_t \approx \frac{1}{\pi^2} \left(\frac{\omega_p}{c}\right)^3 e^{-\hbar\omega_p/kT} \left(\frac{kT}{\hbar\omega_p}\right)^{3/2} \sqrt{\frac{\pi}{2}}. \tag{15.8.6}$$

From eq. (4.11.7) in Volume 1, the number density of photons is

$$n_{\text{ph}} = \frac{1}{\pi^2} \left(\frac{kT}{\hbar c} \right)^3 \sum_{n=1}^{\infty} \frac{2}{n^2} = \frac{1}{\pi^2} \left(\frac{kT}{\hbar c} \right)^3 2.403\,95, \qquad (15.8.7)$$

giving the ratio of transverse plasmons to photons as

$$\frac{n_{\text{t}}}{n_{\text{ph}}} = \frac{\sqrt{\pi}}{4.8079} \left(\frac{kT}{\hbar \omega_{\text{p}}} \right)^{3/2} e^{-\hbar \omega_{\text{p}}/kT}$$

$$= 0.368\,72 \left(\frac{kT}{\hbar \omega_{\text{p}}} \right)^{3/2} e^{-\hbar \omega_{\text{p}}/kT} \ll 1 \text{ when } \frac{kT}{\hbar \omega_{\text{p}}} \ll 1. \qquad (15.8.8)$$

From eqs. (15.8.2) and (15.8.3), the energy density of transverse plasmons can be written as

$$U_{\text{t}} = \frac{1}{\pi^2} \left(\frac{\omega_{\text{p}}}{c} \right)^3 \hbar \omega_{\text{p}} \int_0^{\infty} \frac{\sqrt{1+x^2}\, x^2 \, dx}{e^{(\hbar \omega_{\text{p}}/kT)\sqrt{1+x^2}} - 1}, \qquad (15.8.9)$$

and, for $\hbar \omega_p/kT \gg 1$,

$$U_{\text{t}} = \frac{1}{\pi^2} \left(\frac{\omega_{\text{p}}}{c} \right)^3 \hbar \omega_{\text{p}} \, e^{-\hbar \omega_{\text{p}}/kT} \left(\frac{kT}{\hbar \omega_{\text{p}}} \right)^{3/2} \sqrt{\frac{\pi}{2}} + \cdots. \qquad (15.8.10)$$

Comparing with eq. (15.8.6),

$$U_{\text{t}} \approx \hbar \omega_{\text{p}} n_{\text{t}} \text{ when } \frac{kT}{\hbar \omega_{\text{p}}} \ll 1. \qquad (15.8.11)$$

From eq. (4.11.21) in Volume 1, $U_{\text{ph}} = 2.701\,36\, kT$, so

$$\frac{U_{\text{t}}}{U_{\text{ph}}} = \frac{n_{\text{t}}}{n_{\text{ph}}} \frac{\hbar \omega_{\text{p}}}{2.701\,36\, kT}$$

and using eq. (15.8.8),

$$\frac{U_{\text{t}}}{U_{\text{ph}}} = 0.136\,50 \left(\frac{kT}{\hbar \omega_{\text{p}}} \right)^{1/2} e^{-\hbar \omega_{\text{p}}/kT} \ll 1 \text{ when } \frac{kT}{\hbar \omega_{\text{p}}} \ll 1. \qquad (15.8.12)$$

In the limit $\hbar \omega_{\text{p}} \ll kT$, transverse plasmons behave almost identically to photons, with $n_{\text{t}} \approx n_{\text{ph}}$ and $U_{\text{t}} \approx U_{\text{ph}}$. Since the two quanta are essentially indistinguishable, the potential for double counting exists and it makes sense to simply deny the existence of transverse plasmons.

When the free electrons have non-relativistic velocities, the dispersion relation for longitudinal plasmons is given by eq. (15.7.29) when electrons are not degenerate, and by eq. (15.7.53) when they are degenerate. Adopting the conventional choice for the maximum wave number given by eq. (15.7.42), the wavelength of a longitudinal plasmon varies by only a factor of $\sqrt{2}$ over the permissible range in wave number, and all longitudinal plasmons oscillate at essentially the plasma frequency ω_p. Thus, the number density of longitudinal plasmons (statistical weight $g_\mathrm{l} = 1$) is estimated as

$$n_\mathrm{l} \approx \frac{1}{2\pi^2} \int_0^{\omega_\mathrm{p}/c} \frac{k^2 \, dk}{\mathrm{e}^{\hbar\omega_\mathrm{p}/kT} - 1} = \frac{1}{6\pi^2} \frac{1}{\mathrm{e}^{\hbar\omega_\mathrm{p}/kT} - 1} \left(\frac{\omega_\mathrm{p}}{c}\right)^3 \qquad (15.8.13)$$

and the energy density associated with the plasmons is estimated as

$$U_\mathrm{l} \approx \hbar\omega_\mathrm{p} \, n_\mathrm{l}, \qquad (15.8.14)$$

independent of the magnitude of $kT/\hbar\omega_\mathrm{p}$.

Assuming that

$$n_\mathrm{l} \, \hbar\omega_\mathrm{p} = n_\mathrm{e} \frac{1}{2} m_\mathrm{e} \, u^2 + \frac{E^2}{8\pi}, \qquad (15.8.15)$$

and using eq. (5.7.31), one has that

$$n_\mathrm{l} \, \hbar\omega_\mathrm{p} \approx n_\mathrm{e} \, m_\mathrm{e} \, u^2. \qquad (15.8.16)$$

Using eqs. (15.7.1) and (15.8.13), this equation becomes

$$m_\mathrm{e} \, u^2 = \frac{2}{3\pi} \frac{e^2}{\hbar c} \frac{\hbar\omega_\mathrm{p}}{m_\mathrm{e} c^2} \frac{\hbar\omega_\mathrm{p}}{\mathrm{e}^{\hbar\omega_\mathrm{p}/kT} - 1}. \qquad (15.8.17)$$

Then, using eq. (15.7.37),

$$\begin{aligned}
\left(\frac{u}{v_\mathrm{rms}}\right)^2 &= \frac{2}{3\pi} \frac{e^2}{\hbar c} \frac{\hbar\omega_\mathrm{p}}{m_\mathrm{e} c^2} \frac{\hbar\omega_\mathrm{p}}{m_\mathrm{e} v_\mathrm{rms}^2} \frac{1}{\mathrm{e}^{\hbar\omega_\mathrm{p}/kT} - 1} \\
&= \frac{2}{9\pi} \frac{e^2}{\hbar c} \frac{\hbar\omega_\mathrm{p}}{m_\mathrm{e} c^2} \frac{\hbar\omega_\mathrm{p}}{kT} \frac{1}{\mathrm{e}^{\hbar\omega_\mathrm{p}/kT} - 1} < \frac{2}{9\pi} \frac{e^2}{\hbar c} \frac{\hbar\omega_\mathrm{p}}{m_\mathrm{e} c^2} = 0.000\,516 \, \frac{\hbar\omega_\mathrm{p}}{m_\mathrm{e} c^2}.
\end{aligned} \qquad (15.8.18)$$

The equal sign in eq. (15.8.18) obtains when $kT/\hbar\omega_\mathrm{p} \ll 1$, and the inequality increases with increasing $\hbar\omega_\mathrm{p}/kT$. Thus, the assumption that velocity increments due to forcing by the electric field are small relative to mean square thermal velocities is satisfied.

It is instructive to estimate the number N_e of electrons that make up a typical longitudinal plasmon. Setting

$$\hbar\omega_\mathrm{p} = m_\mathrm{e} \, u^2 N_\mathrm{e} = N_\mathrm{e} \frac{2}{3\pi} \frac{e^2}{\hbar c} \frac{\hbar\omega_\mathrm{p}}{m_\mathrm{e} c^2} \frac{\hbar\omega_\mathrm{p}}{\mathrm{e}^{\hbar\omega_\mathrm{p}/kT} - 1}, \qquad (15.8.19)$$

one finds

$$\begin{aligned}
N_\mathrm{e} &= \frac{3\pi}{2} \frac{\hbar c}{e^2} \frac{m_\mathrm{e} c^2}{\hbar\omega_\mathrm{p}} (\mathrm{e}^{\hbar\omega_\mathrm{p}/kT} - 1) \\
&= 645.60 \frac{m_\mathrm{e} c^2}{\hbar\omega_\mathrm{p}} (\mathrm{e}^{\hbar\omega_\mathrm{p}/kT} - 1) = 645.60 \frac{m_\mathrm{e} c^2}{kT} \frac{kT}{\hbar\omega_\mathrm{p}} (\mathrm{e}^{\hbar\omega_\mathrm{p}/kT} - 1).
\end{aligned} \qquad (15.8.20)$$

Similar exercises can be performed for transverse plasmons. From eqs. (15.7.26) and (15.7.25), one has that the total energy density associated with a transverse plasmon is

$$\frac{1}{2}m_e u^2 n_e + \frac{E^2}{8\pi} + \frac{H^2}{8\pi} = \frac{1}{2}m_e u^2 n_e + \frac{E^2}{4\pi} = \frac{1}{2}m_e u^2 n_e \left(1 + 2\frac{\omega^2}{\omega_p^2}\right). \tag{15.8.21}$$

Defining

$$\langle \omega \rangle = \frac{1}{n_t}\int \omega \, dn_t \text{ and } \langle \omega^2 \rangle = \frac{1}{n_t}\int \omega^2 \, dn_t, \tag{15.8.22}$$

one can write

$$\hbar\langle\omega\rangle \sim \frac{1}{2}m_e u^2 n_e \left(\frac{1 + 2\langle\omega^2\rangle}{\omega_p^2}\right). \tag{15.8.23}$$

When $\hbar\omega_p \gg kT$, $\langle\omega\rangle \sim \omega_p$ and $\langle\omega^2\rangle \sim \omega_p^2$, so

$$\hbar\omega_p \approx \frac{3}{2}m_e u^2 n_e. \tag{15.8.24}$$

After some maneuvering, one finds

$$\begin{aligned}
\left(\frac{u}{v_{\rm rms}}\right)^2 &= \frac{2}{9\pi^2}\sqrt{\frac{\pi}{2}}\frac{e^2}{\hbar c}\frac{\hbar\omega_p}{m_e c^2}\left(\frac{\hbar\omega_p}{kT}\right)^{1/2} e^{-\hbar\omega_p/kT} \\
&= 0.000\,2060\,\frac{e^2}{\hbar c}\frac{\hbar\omega_p}{m_e c^2}\left(\frac{\hbar\omega_p}{kT}\right)^{1/2} e^{-\hbar\omega_p/kT}
\end{aligned} \tag{15.8.25}$$

and

$$N_e = \frac{3}{\pi}\sqrt{\frac{2}{\pi}}\frac{\hbar c}{e^2}\frac{m_e c^2}{\hbar\omega_p}\left(\frac{\hbar\omega_p}{kT}\right)^{3/2} e^{\hbar\omega_p/kT} = 104.4\,\frac{m_e c^2}{kT}\left(\frac{\hbar\omega_p}{kT}\right)^{1/2} e^{\hbar\omega_p/kT}. \tag{15.8.26}$$

Having examined some of the relationships between classical plasma oscillations and quantum mechanical plasmons, it is appropriate to discuss the energy-loss rates due to the decay of plasmons into $\nu_e\bar{\nu}_e$ pairs. Authors of the first calculations of these rates are enumerated at the end of Section 15.6. The probability per unit of time that a transverse plasmon decays into a neutrino and an antineutrino can be inferred from these authors to be approximately

$$\begin{aligned}
\frac{1}{\tau_t} &= \frac{1}{\hbar}\frac{1}{3}\left(\frac{g}{4\pi e}\right)^2 \frac{c}{\omega}\left(\left(\frac{\omega}{c}\right)^2 - k^2\right)\left(\frac{\omega_p}{c}\right)^4 \\
&= \frac{1}{\hbar}\frac{1}{3}\left(\frac{g}{4\pi e}\right)^2\left(\frac{\omega_p}{c}\right)^6 \frac{c}{\omega},
\end{aligned} \tag{15.8.27}$$

where the explicit insertion of the velocity of light and Planck's constant have been dictated by the choice of $g = g_{\rm weak}$, where $g_{\rm weak}$ is the weak interaction coupling constant in units defined in Section 15.2.

The overall rate at which energy is lost to neutrino–antineutrino pairs by transverse plasmons is obtained by inserting the transition probability given by eq. (15.8.27) into the integrand of the integral in eq. (15.8.2), giving

$$
\begin{aligned}
\epsilon_{\nu_e \bar{\nu}_e, t} &= \frac{1}{\pi^2} \int_0^\infty \frac{1}{\tau_t} \frac{\hbar\omega}{e^{\hbar\omega_p/kT} - 1} k^2 \, dk \\
&= \frac{1}{\hbar} \frac{1}{3} \left(\frac{g}{4\pi e}\right)^2 \left(\frac{\omega_p}{c}\right)^6 \frac{1}{\pi^2} \int_0^\infty \frac{\hbar c}{e^{\hbar\omega/kT} - 1} k^2 \, dk \\
&= \frac{1}{6\pi} \frac{1}{(2\pi)^3} \frac{1}{\hbar} \frac{g^2}{e^2/\hbar c} \left(\frac{\omega_p}{c}\right)^6 \int_0^\infty \frac{k^2 \, dk}{e^{\hbar\omega/kT} - 1}.
\end{aligned}
\tag{15.8.28}
$$

Using eqs. (15.8.3) and (15.8.4), eq. (15.8.28) becomes

$$
\epsilon_{\nu_e \bar{\nu}_e, t} = \frac{1}{6\pi} \frac{1}{(2\pi)^3} \frac{1}{\hbar} \frac{g^2}{e^2/\hbar c} \left(\frac{m_e c}{\hbar}\right)^9 \left(\frac{\hbar\omega_p}{m_e c^2}\right)^9 \int_0^\infty \frac{x^2 \, dx}{e^{(\hbar\omega_p/kT)\sqrt{1+x^2}} - 1}.
\tag{15.8.29}
$$

Adopting the value $g_{weak} = 1.42 \times 10^{-49}$ erg cm^3 given by eq. (15.2.33), the constant coefficient of the integral in eq. (15.8.29) is

$$
\begin{aligned}
\frac{1}{6\pi} \frac{1}{(2\pi)^3} \frac{1}{\hbar} \frac{g^2}{e^2/\hbar c} \left(\frac{m_e c}{\hbar}\right)^9 &= 1.4556 \times 10^{119} \text{ erg}^{-1} \text{ s}^{-1} \text{ cm}^{-9} g_{weak}^2 \\
&= 2.9351 \times 10^{21} \text{ erg cm}^{-3} \text{ s}^{-1}.
\end{aligned}
\tag{15.8.30}
$$

When $\hbar\omega_p \gg kT$, the integrand in eq. (15.8.28) is given by eq. (15.8.5) and, combining with eq. (15.8.30), one obtains

$$
\epsilon_{\nu_e \bar{\nu}_e, t} \sim 3.6786 \times 10^{21} \left(\frac{\hbar\omega_p}{m_e c^2}\right)^{7.5} \left(\frac{kT}{m_e c^2}\right)^{3/2} e^{-\hbar\omega_p/kT} \text{ erg cm}^{-3} \text{ s}^{-1}.
\tag{15.8.31}
$$

When $(\hbar\omega_p/kT) \ll 1$, the maximum in the integrand in eq. (15.8.29) occurs at $x \sim 2$ $(kT/\hbar\omega_p) \gg 1$, so

$$
\int_0^\infty \frac{x^2 \, dx}{e^{(\hbar\omega_p/kT)\sqrt{1+x^2}} - 1} \sim \int_0^\infty \frac{x^2 \, dx}{e^{(\hbar\omega_p/kT) x} - 1}
\tag{15.8.32}
$$

$$
\begin{aligned}
&= \int_0^\infty x^2 \, dx \frac{e^{-(\hbar\omega_p/kT)x}}{1 - e^{-(\hbar\omega_p/kT)x}} = \int_0^\infty x^2 \, dx \sum_{n=1}^\infty e^{-n(\hbar\omega_p/kT)x} \\
&= \left(\frac{kT}{\hbar\omega_p}\right)^3 \int_0^\infty y^2 \, dy \sum_{n=1}^\infty e^{-ny} = \left(\frac{kT}{\hbar\omega_p}\right)^3 \sum_{n=1}^\infty \frac{1}{n^3} \int_0^\infty e^{-z} z^2 \, dz \\
&= \left(\frac{kT}{\hbar\omega_p}\right)^3 1.202\,057 \times 2.
\end{aligned}
\tag{15.8.33}
$$

Using this result in eq. (15.8.29) and making use of eq. (15.8.30), one obtains finally

$$\epsilon_{\nu_e\bar{\nu}_e,t} \sim 7.0563 \times 10^{21} \left(\frac{\hbar\omega_p}{m_e c^2}\right)^6 \left(\frac{kT}{m_e c^2}\right)^3 \text{erg cm}^{-3} \text{ s}^{-1}. \qquad (15.8.34)$$

The probability that a longitudinal plasmon decays into a neutrino–antineutrino pair may be deduced from the early literature presented in Section 15.6 to be approximately

$$\frac{1}{\tau_l} = \frac{1}{\hbar}\frac{1}{3}\left(\frac{g}{4\pi e}\right)^2 \left(\left(\frac{\omega_p}{c}\right)^2 - k^2\right)^2 \frac{\omega_p}{c}. \qquad (15.8.35)$$

Inserting $\hbar\omega/\tau_l$ into the integrand of the integral in eq. (15.8.13), and taking into account that $\omega \approx \omega_p$ over the entire range of integration, the rate at which energy is lost to neutrinos by longitudinal plasmon decay is

$$\begin{aligned}
\epsilon_{\nu_e\bar{\nu}_e,l} &= \frac{1}{2\pi^2}\int_{k=0}^{k=\omega_p/c}\frac{1}{\tau_l}\frac{\hbar\omega}{e^{\hbar\omega/kT}-1}k^2\,dk \\
&\approx \frac{1}{2\pi^2}\frac{\hbar\omega_p}{e^{\hbar\omega_p/kT}-1}\int_{k=0}^{k=\omega_p/c}\frac{1}{\tau_l}k^2\,dk.
\end{aligned} \qquad (15.8.36)$$

Using the decay rate given by eq. (15.8.35), one has

$$\begin{aligned}
\epsilon_{\nu_e\bar{\nu}_e,l} &\approx \frac{1}{\hbar}\frac{1}{3}\left(\frac{g}{4\pi e}\right)^2\frac{\hbar\omega_p}{e^{\hbar\omega_p/kT}-1}\frac{1}{2\pi^2}\int_{k=0}^{k=\omega_p/c}\left(\left(\frac{\omega_p}{c}\right)^2 - k^2\right)^2\frac{\omega_p}{c}k^2\,dk \\
&= \frac{1}{\hbar}\frac{1}{3}\left(\frac{g}{4\pi e}\right)^2\frac{\hbar\omega_p}{e^{\hbar\omega_p/kT}-1}\frac{1}{2\pi^2}(\omega_p/c)\int_0^{\omega_p/c}((\omega_p/c)^2 - k^2)^2\,k^2\,dk \\
&= \frac{1}{\hbar}\frac{1}{3}\left(\frac{g}{4\pi e}\right)^2\frac{\hbar\omega_p}{e^{\hbar\omega_p/kT}-1}\frac{1}{2\pi^2}(\omega_p/c)^8\left(\frac{1}{3} - \frac{2}{5} + \frac{1}{7}\right) \\
&= \frac{1}{\hbar}\frac{1}{3}\frac{1}{(4\pi)^2}\frac{g^2}{e^2/\hbar c}\frac{1}{e^{\hbar\omega_p/kT}-1}\frac{1}{2\pi^2}\left(\frac{\omega_p}{c}\right)^9\frac{8}{105} \\
&= \frac{1}{12\pi}\frac{1}{(2\pi)^3}\frac{8}{105}\frac{1}{\hbar}\frac{g^2}{e^2/\hbar c}\frac{1}{e^{\hbar\omega_p/kT}-1}\left(\frac{m_e c}{\hbar}\right)^9\left(\frac{\hbar\omega_p}{m_e c^2}\right)^9.
\end{aligned} \qquad (15.8.37)$$

Making use of eq. (15.8.30), one has finally that

$$\epsilon_{\nu_e\bar{\nu}_e,l} = 1.118 \times 10^{20}\frac{1}{e^{\hbar\omega_p/kT}-1}\left(\frac{\hbar\omega_p}{m_e c^2}\right)^9 \text{erg cm}^{-3} \text{ s}^{-1}. \qquad (15.8.38)$$

It is evident that there are many issues involved in obtaining neutrino–antineutrino energy-loss rates adequate for stellar interior calculations. The dispersion relationships quoted and used in this section are only first approximations, the energy-loss rates derived for two limiting conditions are inadequate between the two limits, and the quoted results do not take into account the contribution of neutral currents, as predicted by the unified Weinberg–Salam theory. The inadequacy of the limiting estimates of energy-loss rates can be overcome by numerical calculations and the use either of tables of results or of analytic fits to the results. Analytic fits to energy-loss rates for all relevant neutrino–antineutrino production processes based on the current–current interaction of Feynman and Gell-Mann are given by G. Beaudet, V. Petrosian, and E. E. Salpeter (1967). They have been used extensively in stellar evolution calculations.

Neutral current contributions predicted by the Weinberg–Salam formulation have been taken into account in a number of studies which have also introduced refinements in the estimation of dispersion relationships and of decay rates. Work extending over a decade by Naoki Itoh and Yasuharu Kohyama and their coworkers has been summarized by N. Itoh, H. Hayashi, A. Nishikawa, & Y. Kohyama (1996), who also provide analytic fits to their results and those of others. Insightful contributions have been made by E. Braaten & D. Segel (1993), by M. Haft, G. Raffelt, & A. Weiss (1994), and by D. B. Melrose, J. I. Weise, & J. McOrist (2006), among others.

Bibliography and references

J. B. Adams, M. A. Ruderman, & C. H. Woo, *Phys. Rev.*, **129**, 1383, 1963.

G. Beaudet, Vahe Petrosian, & Edwin E. Salpeter, *ApJ*, **150**, 979, 1967.

J. D. Bjorken & S. D. Drell, *Relativistic Quantum Mechanics* (San Francisco: McGraw-Hill), p. 257–264, 1964.

Eric Braaten & Daniel Segel, *Phys. Rev. D*, **48**, No. 4, 1478, 1993.

Hong-Yee Chiu, *Ann. Phys. (NY)*, **15**, 1961a; **16**, 321, 1961b.

Hong-Yee Chiu, *Stellar Physics* (Waltham, Mass: Blaisdell), 1968.

Hong-Yee Chiu and Philip Morrison, *Phys. Rev. Lett.*, **5**, 573, 1960.

Hong-Yee Chiu & R. C. Stabler, *Phys. Rev.*, **122**, 1317, 1961.

P. A. M. Dirac, *Proc. Roy. Soc.*, **A112**, 661, 1926; **A117**, 610, 1928.

D. Dubbers, W. Mampe, & J. Döhner, *Europhys. Lett.*, **11**, 195, 1990.

B. G. Erozolimskii & Yu. A. Mostovoi, *Sov. Journ. Nucl. Phys.*, **53**, 260, 1991.

Enrico Fermi, *Zeitschrift für Physik*, **88**, 161, 1934.

Enrico Fermi, *Elementary Particles*, (New Haven: Yale University Press), 1951.

G. C. Festa & Malvin A. Ruderman, *Phys. Rev.*, **180**, 1227, 1969.

Richard P. Feynman & Murray Gell-Mann, *Phys. Rev.*, **109**, 193, 1958.

Hans Frauenfelder, R. Bobone, E. von Goeler, *et al.*, *Phys. Rev*, **106**, 386, 1957.

Sheldon Lee Glashow, *Rev. Med. Phys.*, **52**, 539, 1980.

M. Haft, G. Raffelt, & A. Weiss, *ApJ*, **425**, 222, 1994.

J. C. Hardy, I. S. Towner, V. T. Koslowsky, E. Hagberg & K. Schmeing, *Nucl. Phys. A*, **509**, 429, 1990.

K. Hikasa, K. Hagiwara, S. Kawabata, *et al.*, *Phys Rev. D*, **45**, 7–10, 1992.

C. L. Inman & Malvin A. Ruderman, *ApJ*, **140**, 1025, 1964.

N. Itoh, H. Hayashi, A. Nishikawa, & Y. Kohyama, *ApJS*, **102**, 411, 1996.

J. David Jackson, *Classical Electrodynamics* (New York: John Wiley & Sons), 1962.

J. M. Jauch & K. M. Watson, *Phys. Rev.*, **74**, 950, 1948a; **74**, 1485, 1948b; **75**, 1249, 1949.

Irving Langmuir, *Phys. Rev.*, **26**, 585, 1925.

Irving Langmuir, *Proc. Nat. Acad. Sci.*, **14**, 627, 1926.

T. D. Lee & C. N. Yang, *Phys. Rev.*, **104**, 254, 1956.

T. D. Lee & C. S. Wu, *Ann. Rev. Nucl. Sci.*, **15**, 381, 1965.

M. J. Levine, PhD. thesis, Cal Tech, 1963.

D. B. Melrose, J. I. Weise, & J. McOrist, *J. Phys. A., Math. Gen.*, **39**, 8727, 2006.

M. Mladjenović, *The Defining Years of Nuclear Physics 1932-1960s* (Bristol and Philadelphia: Inst. of Phys.), 1998.

Seth H. Neddermeyer & Carl D. Anderson, *Phys. Rev. Lett.*, **54**, L88, 1938.

Donald H. Perkins, *Introduction to High Energy Physics*, (Reading, Mass: Addison Wesley), 1987.

Vahe Petrosian, G. Beaudet, & Edwin E. Salpeter, *Phys. Rev.*, **154**, 1445, 1967.

Bruno Pontecorvo, *Zh. Eksp. Teor. Fiz.*, **36**, 615, 1959; *Soviet Phys., JETP*, **9**, 1148, 1960.

V. I. Ritus, *Soviet Phys., JETP*, **14**, 915, 1962.

Abdus Salam, in *Elementary Particle Physics*, ed. N. Svartholm (Stockholm: Almquist and Wiksell), 367, 1968.

Abdus Salam, *Rev. Mod. Phys.*, **52**, 525, 1980.

Lewi Tonkas & Irwin Langmuir, *Phys. Rev.*, **33**, 195, 1929.

V. N. Tsytovich, *Zh. Expr. Teor. Fiz.*, **45**, 1183, 1963; *Sov. Phys., JETP*, **18**, 816, 1964.

K. M. Watson & J. M. Jauch, *Phys. Rev.*, **75**, 8, 1249, 1949.

Steven Weinberg, *Phys. Rev. Lett.*, **19**, 1264, 1967

Steven Weinberg, *Rev. Mod. Phys.*, **52**, 515, 1980.

C. S. Wu, E. Ambler, R. W. Hayward, D. D. Hoppes, & R. P. Hudson, *Phys. Rev.*, **107**, 641, 1957.

M. H. Zaidi, *Nuovo Cimento*, **40**, 502, 1965.

Helium-burning nuclear reactions and energy-generation rates

Because of the large Coulomb barrier between an alpha particle and another, heavier nucleus, temperatures at which helium-burning reactions occur in stars at interesting rates are considerably larger than temperatures at which hydrogen-burning reactions occur. Under appropriate conditions, a substantial fraction of interacting alpha particle, heavy nucleus pairs have relative kinetic energies such that the compound nucleus formed during a collision has an excitation energy close to the energy of a discrete level in the compound nucleus. For a reaction in which the compound nucleus decays with the emission of a gamma ray or a nucleon to form a stable nucleus, the cross section can become very large and the reaction is said to be a resonant one. In Section 16.1, the formation and decay of a compound nucleus is examined and the Breit–Wigner form for a single level resonant cross section is derived heuristically. In Section 16.2, the set of processes whereby three ^{4}He nuclei combine to form ^{12}C through resonant states of intermediate compound nuclei is examined. It is conventional to call the set of processes the triple-alpha process.

After the exhaustion of hydrogen over the inner 13% or so of its mass, a low mass star develops an electron-degenerate helium core and evolves upward in the HR diagram along a red giant branch burning hydrogen quiescently in a shell. The core grows in mass until, when it reaches a mass of about 0.5 $M_\odot$, it has become sufficiently hot for the triple alpha process to terminate further core growth. Evolution up to this point is described in Chapter 11 of Volume 1 (Section 11.1). The subsequent evolution is described in Chapter 17 in this volume. A series of thermonuclear helium-burning runaways is initiated, with electron degeneracy being lifted in layers successively closer to the center. Once enough heat has been added to remove electron degeneracy throughout the core, the triple alpha process, followed by (α, γ) captures on ^{12}C, continues quiescently in a convective core. After the exhaustion of helium in central regions, the model evolves into an AGB star with an electron-degenerate carbon–oxygen core.

As described in Chapter 11 of Volume 1, in intermediate mass (Section 11.2) and massive stars (Section 11.3), the activation of the triple-alpha process in central regions of a non-electron-degenerate helium core marks the onset of a quiescent core helium-burning phase which almost immediately follows the core hydrogen-burning main sequence phase. During this quiescent burning phase, as the abundance of ^{12}C nuclei increases, substantial quantities of ^{16}O are formed in consequence of ^{12}C$(\alpha, \gamma)^{16}$O reactions. In massive stars, as described in Chapter 20, after the exhaustion of helium in central regions, helium burning continues steadily in a shell outside of a core in which carbon burning prevents the establishment of electron degeneracy. In contrast, as described in Chapter 18, after helium is exhausted in central regions, an intermediate mass star evolves into an   AGB configuration with an electron-degenerate core composed of carbon and oxygen (if initial mass is in the range ~ 2.25–$8.5 \ M_\odot$) or of oxygen and neon (if initial mass is in the range ~ 8.5–$10.5 \ M_\odot$).

In both low and intermediate mass AGB stars, nuclear burning alternates between quiescent hydrogen burning in a shell and helium burning in a shell. Helium burning remains dormant until temperatures in the helium ashes deposited by the hydrogen-burning shell become large enough to initiate a helium shell flash. Responsible for the flash phenomenon is the enormous temperature sensitivity of the triple-alpha reaction (rate $\propto T^{40}$). Following a flash, some of the matter which during the flash has been partially processed into carbon by the operation of triple alpha reactions (with only a modest admixture of oxygen produced by (α, γ) reactions on carbon) is dredged up into the convective envelope. As described in Chapter 21, matter in the convective envelope is blown by a superwind into the interstellar medium and, consequently, low and intermediate mass stars are potent sources of carbon in the Universe, if not the most important sources. After a thermonuclear shell flash subsides, quiescent helium burning in a shell continues, and adds to the hot white dwarf interior ashes consisting primarily of oxygen and carbon, with oxygen being more abundant than carbon. Thus, while their AGB progenitors have contributed primarily to the enrichment of carbon in the Universe, most white dwarfs in the Universe are composed of oxygen and carbon, with a small admixture composed of oxygen and neon.

In massive stars, after a core and shell carbon burning phase described in Chapter 20, the core becomes dynamically unstable and collapses into a neutron star or black hole, but thick layers of carbon and oxygen have been expelled from the star in a wind prior to a supernova explosion and by a shock wave generated by a supernova explosion which accompanies core collapse. Thus, massive stars are a major source of oxygen (and some carbon) in the Universe.

At low enough temperatures, the number densities of the relevant compound nuclei that are in near statistical equilibrium at the resonant energies may be small compared with the number densities of compound nuclei at excitation energies considerably less than the resonant excitation energies. In this circumstance, reactions are said to proceed through the tails of the resonances. In Section 16.3, what is involved in calculating reaction rates at low temperatures is explored, with a focus on non-resonant triple alpha processes that are important in white dwarfs and neutron stars accreting hydrogen-rich or helium-rich matter from a companion and in low mass first-generation stars that, during hydrogen-burning phases, rely on proton capture, α capture, and β decay reactions to produce ^{8}Be which can capture α particles to produce the ^{12}C which leads to the activation of the CN cycle as a major contributor to the hydrogen-burning rate.

The buildup of nuclei by successive (α, γ) reactions is discussed in Section 16.4. As already noted, as the abundance of ^{12}C increases during the core helium-burning phase, alpha-particle captures by ^{12}C produce ^{16}O nuclei to the extent that the final abundance of oxygen in the core is greater than the abundance of carbon, with the abundances of other alpha-particle nuclei being modest in comparison. In matter which has experienced complete hydrogen burning, most of the initial CNO elements have been converted into ^{14}N. In regions where the triple alpha process and the reaction ^{12}C$(\alpha, \gamma)^{16}$ are active, most of the ^{14}N nuclei initially in the regions are converted into ^{22}Ne by the ^{14}N(α, γ)

^{18}F$(\beta^+, \nu_e)^{18}$O$(\alpha, \gamma)^{22}$Ne reactions. Only a modest amount of helium (an abundance by number equal to twice the initial abundance of CNO nuclei) is consumed by these processes, while most helium nuclei are converted into ^{12}C and ^{16}O.

The reactions ^{13}C$(\alpha, n)^{16}$O and ^{22}Ne$(\alpha, n)^{25}$Mg, which are the primary sources of neutrons for the production of s-process elements in stars, are examined in Section 16.5. The ^{22}Ne$(\alpha, n)^{25}$Mg reaction requires fairly high temperatures. It occurs in the convective cores of massive stars during the main core helium-burning phase and capture by seed nuclei of neutrons released by the reaction is responsible for the production of a number of s-process elements of modest atomic number, as described in Chapter 20. At the higher temperatures occurring in the helium-burning convective shells of intermediate mass TPAGB stars during helium shell flashes, capture of neutrons released by the ^{22}Ne$(\alpha, n)^{25}$Mg reaction contributes to the production of heavier s-process elements, as described in Chapter 19.

The ^{13}C$(\alpha, n)^{16}$O reaction occurs in light and intermediate mass stars during the TPAGB phase and is possibly primarily responsible for the production of heavy s-process elements in the solar system distribution. The necessary ^{13}C is produced during the interpulse phase by the ^{12}C$(p, \gamma)^{13}$N$(e^+ \nu_e)^{13}$C reactions in two regions: just outside of the carbon-rich layer formed during a helium shell flash, where fresh carbon mixes with hydrogen and hydrogen burning produces a ^{13}C abundance peak, and behind the advancing hydrogen-burning shell during the interpulse phase. The ^{13}C in the abundance peak and over a large fraction of the region between the C-rich interior and the hydrogen-burning shell burns alpha particles and produces s-process elements before the next helium shell flash occurs, but a substantial quantity of unburned ^{13}C is ingested and burned in the convective shell formed during the next shell flash. These events are also described in Chapter 19.

In hydrogen-burning regions of first generation metal-free stars, the ^{8}Be which plays a role in creating ^{12}C is produced not only by binary interactions between alpha particles but also as the end result of one channel of hydrogen-burning reactions. The reactions $p(p, e^+\nu_e)d(p, \gamma)^3He(\alpha, \gamma)^7$Be are sometimes followed by the ^{7}Be$(e^-, \nu_e)^7$Li reaction and by the ^{7}Li$(p, \gamma)^8$Be reaction which proceeds through the fifth excited state of the ^{8}Be compound nucleus. Some of the time, the excited ^{8}Be nucleus decays by gamma emission to the ground state of ^{8}Be. Conditions in low mass metal-poor stars under which this avenue of ^{8}Be production can compete with the classical triple-alpha mechanism are not explored in this book.

16.1 Some basic physics of resonant reactions

In principle, the reaction between two nuclei which results in the formation of a third nucleus plus a gamma ray, a nucleon, or an alpha particle is an extremely complicated many-body problem. However, the concept of an intermediate structure called a compound nucleus (Niels Bohr, 1936), when coupled with observed characteristics of reactions in the laboratory, allows for a considerably simplified description of the results of such reactions, particularly when the excitation energy of the compound nucleus is close to the energy of a discrete level. A very complex problem is replaced by a simple description in terms of parameters which, though difficult if not impossible to calculate from first principles, can be determined experimentally. The cornerstone of the concept of a compound nucleus is the observation that lifetimes of the postulated intermediate structure for decay into the various observed exit channels can be measured and are found to be orders of magnitude longer than the typical crossing time for a nucleon in a nucleus. This circumstance suggests that the compound nucleus forgets how it was formed and adopts an equilibrium configuration specified only by its excitation, as measured by the difference between the sum of the rest mass energies plus the relative kinetic energy of the initial particles and the rest mass energy of the compound nucleus in its ground state.

To initiate a heuristic mathematical description, assume that a compound nucleus has been magically prepared in a resonant state at energy E_c and that this state can decay into a remnant nucleus r plus a particle p of relative kinetic energy E_p and reduced mass $\mu_{p,r}$ or into a lower state of the compound nucleus plus a gamma ray of energy E_γ. Borrowing from the results of time-dependent perturbation theory as developed in Section 7.2 in Volume 1, the amplitude a_c of the compound state c decays according to

$$a_c = \exp\left(-\frac{\Gamma}{2\hbar}t\right), \tag{16.1.1}$$

where

$$\Gamma = \Gamma_p + \Gamma_\gamma, \tag{16.1.2}$$

Γ_p is the partial energy width for the decay into particle p and remnant nucleus r, and Γ_γ is the partial energy width for decay into a lower state of the compound nucleus and a gamma ray. The Γs are related to corresponding lifetimes τ, to the matrix elements of an interaction Hamiltonian, and to the kinematic properties of the emitted particle and gamma ray by

$$\frac{\Gamma_p}{\hbar} = \frac{1}{\tau_p} = \frac{2\pi}{\hbar}|H_{pc}|^2\frac{4\pi\,p_p^2\,dp_p}{h^3\,dE_p} = \frac{2\pi}{\hbar}|H_{pc}|^2\frac{4\pi\,p_p\,\mu_{p,r}}{h^3} \tag{16.1.3}$$

and

$$\frac{\Gamma_\gamma}{\hbar} = \frac{1}{\tau_\gamma} = \frac{2\pi}{\hbar}|H_{\gamma c}|^2\frac{4\pi\,p_\gamma^2\,dp_\gamma}{h^3\,dE_\gamma} = \frac{2\pi}{\hbar}|H_{\gamma c}|^2\frac{4\pi\,E_\gamma^2}{c^3 h^3}. \tag{16.1.4}$$

Here, H_{pc} is the matrix element of the interaction Hamiltonian taken between the prepared state c of the compound nucleus and the final state consisting of the nucleus r and the emitted particle p and $H_{\gamma c}$ is the matrix element of the interaction Hamiltonian taken between the compound nucleus state c and the final state consisting of the compound nucleus in a lower energy state and the emitted gamma ray.

The next question to address is what happens when the compound nucleus is formed at the resonance energy in consequence of a reaction between a projectile particle of the same type p that can be emitted by the compound nucleus and a target nucleus t which is identical to the nucleus r which is left when the compound nucleus ejects the particle p. Temporarily neglecting the decay of the compound nucleus and borrowing again from

Section 7.2 in Volume 1, the amplitude of the compound state wave function increases initially according to

$$\frac{da_c}{dt} = -\frac{i}{\hbar} H_{c0} \, e^{\frac{i}{\hbar}(E_c - E_0)t}, \tag{16.1.5}$$

where E_0 and E_c are, respectively, the total energies of the initial system and of the compound nucleus and H_{c0} is the matrix element of the interaction Hamiltonian taken between the initial state (0) of the system and the state of the compound nucleus (c). The energy E_0 is equal to the sum of the rest mass energies of the projectile particle p and of the target nucleus t plus the relative kinetic energy E of the projectile and target ($E_0 = E + M_p c^2 + M_t c^2$) and E_c is equal to the rest mass energy of the compound nucleus in its resonant state $\left(E_c = M_c^* c^2\right)$. Thus,

$$E_c - E_0 = \left[M_c^* c^2 - (M_p c^2 + M_t c^2)\right] - E = E_{res} - E, \tag{16.1.6}$$

where E_{res} is the relative kinetic energy of the projectile and target which, when added to the rest mass energies of the interacting pair, equals the rest mass energy of the compound nucleus in state c.

Differentiating eq. (16.1.1) and adding the result to eq. (16.1.5) gives an approximation to the rate at which the amplitude a_c changes in consequence of both the buildup and the decay of the compound nucleus (*Nuclear Physics by Enrico Fermi*, notes compiled by J. Orear, A. H. Rosenfeld, & R. A. Schluter, 1949):

$$\frac{da_c}{dt} = -\frac{\Gamma}{2\hbar} a_c - \frac{i}{\hbar} H_{c0} \, e^{\frac{i}{\hbar}(E_c - E_0)t}. \tag{16.1.7}$$

Multiplying both sides of this equation by $\exp\left(\Gamma t / 2\hbar\right)$, and rearranging,

$$e^{\frac{\Gamma}{2\hbar}t} \left(\frac{da_c}{dt} + \frac{\Gamma}{2\hbar} a_c\right) = \frac{d}{dt} \left(e^{\frac{\Gamma}{2\hbar}t} a_c\right) = -\frac{i}{\hbar} H_{c0} \, e^{\frac{i}{\hbar}(E_c - E_0)t} \, e^{\frac{\Gamma}{2\hbar}t}. \tag{16.1.8}$$

Setting

$$\lambda = \frac{i}{\hbar}(E_c - E_0) + \frac{\Gamma}{2\hbar}, \tag{16.1.9}$$

eq. (16.1.8) has the solution

$$e^{\frac{\Gamma}{2\hbar}t} a_c = -\frac{i}{\hbar} H_{c0} \frac{e^{\lambda t}}{\lambda} + C, \tag{16.1.10}$$

where the constant C is determined by setting $a_c = 0$ at $t = 0$, giving $C = (i/\hbar)\,(H_{c0}/\lambda)$ and

$$a_c = \frac{i}{\hbar} \frac{H_{c0}}{\lambda} \left(e^{\frac{i}{\hbar}(E_c - E_0)t} - e^{-\frac{\Gamma}{2\hbar}t}\right) = \frac{i}{\hbar} \frac{H_{c0}}{\lambda} \, e^{\frac{i}{\hbar}(E_c - E_0)t} \left(1 - e^{-\lambda t}\right). \tag{16.1.11}$$

The second term in parentheses decays rapidly with time and the absolute square of the amplitude of the compound nucleus approaches the asymptotic value

$$|a_{\rm c}|^2 \to \frac{|H_{\rm c0}|^2}{|\lambda|^2} = \frac{|H_{\rm c0}|^2}{(E_{\rm c} - E_0)^2 + (\Gamma/2)^2}. \tag{16.1.12}$$

The fact that the final amplitude is finite is a consequence of having assumed that the amplitude of the initial configuration – target plus projectile – remains constant at unity. In short, a steady state solution has been constructed which corresponds to the laboratory situation in which a collection of target nuclei is exposed to a steady, monochromatic beam of particles. The cross sections for the reactions are related to the amplitude by

$$\sigma_{\rm p,p}\, v = |a_{\rm c}|^2\, \frac{\Gamma_{\rm p}}{\hbar} \tag{16.1.13}$$

and

$$\sigma_{\rm p,\gamma}\, v = |a_{\rm c}|^2\, \frac{\Gamma_\gamma}{\hbar}, \tag{16.1.14}$$

where v is the relative velocity corresponding to the total energy E_0. From eq. (16.1.3), one has

$$|H_{\rm pc}|^2 = \pi\, \frac{\hbar^3}{\mu_{\rm p,t}^2}\, \frac{\Gamma_{\rm p}}{v_{\rm p}}, \tag{16.1.15}$$

where $v_{\rm p} = p_{\rm p}/\mu_{\rm p,t}$ is the relative velocity corresponding to the total energy $E_{\rm c}$. Combining eqs. (16.1.12)–(16.1.15), one can write

$$\sigma_{\rm p,p} = \frac{|H_{\rm c0}|^2}{|H_{\rm pc}|^2}\, \pi\, \lambdabar_{\rm p}\, \lambdabar\, \frac{\Gamma_{\rm p}^2}{(E_{\rm res} - E)^2 + (\Gamma/2)^2} \tag{16.1.16}$$

and

$$\sigma_{\rm p,\gamma} = \frac{|H_{\rm c0}|^2}{|H_{\rm pc}|^2}\, \pi\, \lambdabar_{\rm p}\, \lambdabar\, \frac{\Gamma_{\rm p}\,\Gamma_\gamma}{(E_{\rm res} - E)^2 + (\Gamma/2)^2}, \tag{16.1.17}$$

where

$$\lambdabar_{\rm p} = \frac{\hbar}{\mu_{\rm p,t}\, v_{\rm p}} \tag{16.1.18}$$

is the deBroglie wavelength of the particle p emitted by the compound nucleus in the resonant state c and

$$\lambdabar = \frac{\hbar}{\mu_{\rm p,t}\, v} \tag{16.1.19}$$

is the deBroglie wavelength of the projectile p interacting with the target nucleus t to produce the compound nucleus in the resonant state c. In eq. (16.1.17), $E_{\rm c} - E_0$ has been replaced by its equivalent $E_{\rm res} - E$ as given by eq. (16.1.6).

When $E \sim E_{\rm res}$, $|H_{\rm c0}|^2 \sim |H_{\rm pc}|^2$ and $\lambdabar_{\rm p} \sim \lambdabar$. Thus, near resonance,

$$\sigma_{1,2} = \pi\, \lambdabar^2\, \omega_{1,j}\, \frac{\Gamma_1\,\Gamma_2}{(E_{\rm res} - E)^2 + (\Gamma/2)^2}, \tag{16.1.20}$$

where the subscript 1 denotes the projectile particle, j denotes the target nucleus, 2 denotes the particle or gamma ray ejected by the compound nucleus, and level width $\Gamma = \Gamma_1 + \Gamma_2$. One last refinement has been introduced: a statistical factor $\omega_{1,j} = \frac{g_{1,j}}{g_1\, g_j}$, where $g_i = 2J_i + 1$ and the J_is are the spins of the projectile, the target, and the compound nucleus, respectively. Equation (16.1.20) is the Breit–Wigner (1936) expression for the cross section when the energy of the interacting pair lies near a resonance which is separated from other possible resonances by an energy large compared with the level width Γ.

It is a straightforward task to estimate the contribution of a resonance to the rate of the associated reaction in the stellar interior. From the development in Chapter 6 in Volume 1, the rate at which a reaction between a particle of type 1 and a nucleus of type 2 takes place may be written as

$$R_{1,j} = n_1\, n_j\, \langle \sigma_{1,2}\, v \rangle, \qquad (16.1.21)$$

where

$$\langle \sigma_{1,2}\, v \rangle = \frac{1}{n} \int \sigma_{1,2}\, v\, \mathrm{d}n = \int_0^\infty \sigma_{1,2}\, v\, \left(\frac{1}{2\pi\mu\, kT} \right)^{3/2} 4\pi p^2 \mathrm{d}p\, \mathrm{e}^{-E/kT}, \qquad (16.1.22)$$

v is the relative velocity, and $\mu = M_i M_j / (M_i + M_j)$ is the reduced mass of the particle–nucleus pair. Using $\lambda = \hbar/p$, $v = p/\mu$, and $E = p^2/2\mu$, eqs. (16.1.20)–(16.1.22) give

$$\langle \sigma_{1,2}\, v \rangle = \omega_{1,j} \left(\frac{\hbar^2}{2\pi\mu\, kT} \right)^{3/2} 4\pi^2 \frac{\Gamma_1 \Gamma_2}{\hbar} \int_0^\infty \frac{\mathrm{e}^{-E/kT}}{(E - E_{\mathrm{res}})^2 + (\Gamma/2)^2} \mathrm{d}E \qquad (16.1.23)$$

$$= \omega_{1,j} \left(\frac{h^2}{2\pi\mu\, kT} \right)^{3/2} \frac{1}{2\pi} \frac{\Gamma_1 \Gamma_2}{\hbar} \mathrm{e}^{-E_{\mathrm{res}}/kT} \int_{-E_{\mathrm{res}}}^\infty \frac{\mathrm{e}^{-x/kT}}{x^2 + (\Gamma/2)^2} \mathrm{d}x, \qquad (16.1.24)$$

where $x = E - E_{\mathrm{res}}$. Replacing the lower limit of the integral by $-\infty$, using the fact that

$$\int_{-\infty}^\infty \frac{1}{x^2 + (\Gamma/2)^2}\, \mathrm{d}x = \frac{2\pi}{\Gamma}, \qquad (16.1.25)$$

and treating the integrand in eq. (16.1.25) as a delta function (which peaks at $x = 0$, $E = E_{\mathrm{res}}$), one has

$$\langle \sigma_{1,2}\, v \rangle = \left(\frac{h^2}{2\pi\mu\, kT} \right)^{3/2} \omega_{1,j} \frac{\Gamma_1}{\Gamma} \frac{\Gamma_2}{\hbar} \mathrm{e}^{-E_{\mathrm{res}}/kT}. \qquad (16.1.26)$$

Numerically,

$$N_{\mathrm{A}}\, \langle \sigma_{1,2}\, v \rangle = \frac{4.8127 \times 10^6}{A_{1j}^{3/2}} \left[\omega_{1,j} \frac{\Gamma_1}{\Gamma} \Gamma_2(\mathrm{eV}) \right] \frac{1}{T_8^{3/2}} \exp\left(-0.116\,05\, \frac{E_{\mathrm{res}}(\mathrm{keV})}{T_8} \right),$$

$$(16.1.27)$$

where N_A is Avogadro's number, $A_{1j} = A_1 A_j / (A_1 + A_j)$, and A_1 and A_j are atomic mass numbers.

In rare circumstances, it is possible that, before decaying into the remnant-particle pair which formed it or decaying to the ground state by emission of a gamma ray (or gamma cascade), the compound nucleus interacts with a third particle to form a heavier compound nucleus. The rarity of this circumstance is evident from the smallness of the lifetime $\tau_{mode}(i)$ of an unstable compound nucleus i against decay by a particular mode:

$$\tau_{mode}(i) = \frac{\hbar}{\Gamma_{mode}(i)} = \frac{6.58 \times 10^{-16} \text{ s}}{\Gamma_{mode}(i) \text{ (eV)}}, \qquad (16.1.28)$$

where $\Gamma_{mode}(i)$ is the width associated with the decay mode.

On the other hand, a relatively frequent situation in the stellar environment arises when the temperature is small enough that the resonant contribution to the reaction rate is smaller than the contribution made by reactions at kinetic energies considerably smaller than required for resonance and sometimes even smaller than those for which cross section data can be obtained directly in the laboratory. In such cases it is necessary to resort to an extrapolation based in part on the properties of the nearest resonance. To show that an extrapolation is potentially feasible, it is instructive to explore the case in which the projectile is a neutron. In this case, the prediction of an extrapolation can be tested by comparison with experimentally obtained cross sections. Since there is no Coulomb barrier involved, one may guess that $|H_{c0}|^2 / |H_{pc}|^2 = 1$, and, from eq. (16.1.17),

$$\sigma_{n,\gamma} = \pi \lambdabar_n \lambdabar \, \omega_{n,j} \frac{\Gamma_n \Gamma_\gamma}{(E_{res} - E)^2 + (\Gamma/2)^2} = \frac{v_{res}}{v} \pi \lambdabar_n^2 \, \omega_{n,j} \frac{\Gamma_n \Gamma_\gamma}{(E_{res} - E)^2 + (\Gamma/2)^2}, \qquad (16.1.29)$$

where v_{res} is the relative velocity at resonance, $\lambdabar_n$ is the deBroglie wavelength at resonance, and Γ_n and Γ_γ are, respectively, the neutron and gamma widths evaluated at the resonance. It is well known that, at low energies encountered in terrestrial laboratories, neutron cross sections often vary as $1/v$, a fact with which eq. (16.1.29) agrees.

If the projectile is a charged particle, one may guess that, at zero separation, the square of the amplitude of the wave function describing the initial state in the Hamiltonian H_{c0} and the square of the amplitude of the wave function describing the final state in the Hamiltonian H_{pc} are those of Coulomb-distorted plane waves, so that, therefore,

$$\frac{|H_{c0}|^2}{|H_{pc}|^2} = \frac{2\pi \eta(E) \, e^{-2\pi \eta(E)}}{2\pi \eta(E_{res}) \, e^{-2\pi \eta(E_{res})}} = \frac{v_{res}}{v} \frac{e^{-2\pi \eta(E)}}{e^{-2\pi \eta(E_{res})}}, \qquad (16.1.30)$$

where $\eta(E) = Z_p Z_j e^2 / \hbar v$. Then, from eq. (16.1.17),

$$\sigma_{p,\gamma}(E) = \pi \lambdabar^2 \, \omega_{p,j} \frac{\Gamma_p \Gamma_\gamma}{(E_{res} - E)^2 + (\Gamma/2)^2} \frac{e^{-2\pi \eta(E)}}{e^{-2\pi \eta(E_{res})}}, \qquad (16.1.31)$$

where $\lambdabar$ is the deBroglie wavelength at the energy E and Γ_p and Γ_γ are, respectively, the proton and gamma widths evaluated at the resonance. The convention used here departs from the traditional convention of including the Coulomb transmission factor in the definition of charged particle width at energies far from resonance.

Another way of writing eq. (16.1.31) is

$$\sigma_{p,\gamma}(E) = \pi \,\hat{\lambda}^2 \, e^{-2\pi\eta(E)} \left[\frac{\sigma_{p,\gamma}(E_{res}) \, e^{2\pi\eta(E_{res})}}{\pi \,\hat{\lambda}^2_{res}} \right] \frac{(\Gamma/2)^2}{(E_{res} - E)^2 + (\Gamma/2)^2}, \quad (16.1.32)$$

allowing a determination of the cross section factor $\mathcal{S}_0(E)$ defined by eq. (6.7.5) in Volume 1 in terms of experimental quantities characterizing the resonance:

$$\mathcal{S}_0(E) = \left[\frac{\sigma_{p,\gamma}(E_{res}) \, e^{2\pi\eta(E_{res})}}{\pi \,\hat{\lambda}^2_{res}} \right] \frac{(\Gamma/2)^2}{(E_{res} - E)^2 + (\Gamma/2)^2}. \quad (16.1.33)$$

Then S(E), the cross section factor needed in the reaction rate equation can be found from $\mathcal{S}_0(E)$ given by eq. (6.7.11) in Volume 1.

16.2 The triple-alpha reactions in the classical approximation

Early in the development of nuclear astrophysics, it became apparent that, because there are no stable isotopes of atomic mass $A = 5$ and $A = 8$, it is not possible in stars to build stable elements of large atomic mass simply by a succession of proton capture and beta decay reactions. Hans Bethe (1939) noted that interactions involving three alpha particles might lead to the synthesis of ^{12}C, and, in the 1950s, this idea was developed into a quantitative model in which, in consequence of a collision between two alpha particles, an unstable ^{8}Be nucleus is formed and lives long enough that, under appropriate circumstances, it interacts with another alpha particle to form an excited ^{12}C compound nucleus which can emit a gamma ray to become a bound nucleus (E. J. Öpik, 1951; E. E. Salpeter, 1952a, 1952b). At the time, the level structure of ^{12}C was poorly known and it was assumed that the ^{8}Be$(\alpha, \gamma)^{12}$C reaction is non-resonant.

The luminosity of low mass stars defining the tip of the red giant branch in galactic globular clusters is almost the same in all clusters and this suggests that helium burning is initiated at temperatures and densities prevailing near the center of stars at the tip of the giant branch. Models of low mass red giants constructed by Fred Hoyle and Martin Schwarzschild (1955) suggested central densities of the order of several times 10^5 g cm^{-3} and central temperatures in the neighborhood of 0.8×10^8 K. Under such conditions, it was clear that the nuclear energy-generation rate predicted on the assumption that the ^{8}Be$(\alpha, \gamma)^{12}$C reaction is not resonant was small compared with the rate of gravothermal energy generation and prompted Hoyle (1954) to suggest that there must be an excited energy level in ^{12}C with an energy near the rest mass energy of a system consisting of a ^{8}Be nucleus and an alpha particle, both in the ground state. At Kellogg Radiation Laboratory at Cal Tech, the predicted level in ^{12}C was found to be the second excited state (C. W. Cook, W. A. Fowler, C. C. Lauritsen, & T. Lauritsen, 1957). Using the experimental results, concrete reaction and energy-generation rates were estimated for a variety of density and temperature regimes by E. E. Salpeter (1957) and by A. G. W. Cameron (1959).

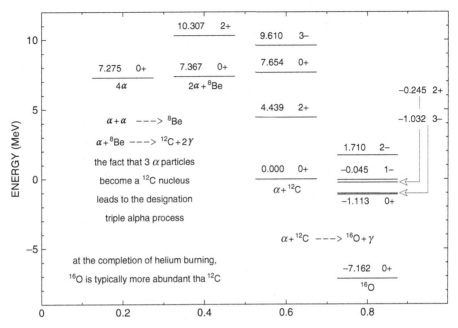

Fig. 16.2.1 Energy levels relevant for determining reaction and energy generation rates for the triple alpha reactions and the $^{12}C(\alpha,\gamma)^{16}O$ reaction

The energy levels in the three nuclei involved in the triple-alpha reactions are shown in Fig. 16.2.1 where the zero point in energy is taken as the sum of the rest mass energies of a ^{12}C nucleus in the ground state and of an alpha particle in the ground state. It is apparent that the rest mass energy of two alpha particles in the ground state is 92 keV less than the rest mass energy of a ^{8}Be nucleus in the ground state and the sum of the rest mass energies of a ^{8}Be nucleus and an alpha particle, both in the ground state, is 287 keV less than the energy of the ^{12}C nucleus in the second excited state.

Experiments involving the reaction $^{7}Li(p,\gamma)^{8}Be^{*****} \rightarrow ^{8}Be \rightarrow 2\alpha$ establish that (1) $^{8}Be^{*****}$, the fifth excited state of ^{8}Be, is 440 keV more energetic than the rest mass energies of a proton and a ^{7}Li nucleus in its ground state; (2) the emission by $^{8}Be^{*****}$ of a gamma ray of energy 17.5 MeV leads to the ground state of ^{8}Be; (3) the rest mass energy of the ^{8}Be nucleus in the ground state is larger than the sum of the rest mass energies of two alpha particles by

$$E_{r8} = 92 \text{ keV}; \qquad (16.2.1)$$

and (4) the lifetime of ^{8}Be for decay into two alpha particles is $t_{2\alpha}(^{8}Be) = 1.18 \times 10^{-16}$ s, corresponding to an energy width of

$$\Gamma_{2\alpha} = 5.6 \text{ eV}. \qquad (16.2.2)$$

Given that the characteristic crossing time of a nucleon in a ^{8}Be nucleus is $\sim 10^{-22}$ s, a typical nucleon in this nucleus experiences $\sim 10^{6}$ crossings before decay occurs, thus ensuring that the concept of a compound nucleus is applicable.

Adopting the Breit–Wigner expression given by eq. (16.1.17), the cross section for the reaction between two alpha particles to form a ^{8}Be compound nucleus which then decays back into two alpha particles is

$$\sigma(\alpha,\alpha) = \pi \, \lambdabar_{r8}^2 \, \omega_{1,2} \left[\frac{\Gamma_1 \Gamma_2}{(E - E_{r8})^2 + (\Gamma_{2\alpha}/2)^2} + \frac{\Gamma_1 \Gamma_1}{(E - E_{r8})^2 + (\Gamma_{2\alpha}/2)^2} \right], \quad (16.2.3)$$

where Γ in eq. (16.1.17) has been replaced by $\Gamma_{2\alpha}$ and E_{res} has been replaced by E_{r8}. Because the two exit channels are indistinguishable and are the same as the entrance channel, $\Gamma_1 = \Gamma_2 = \Gamma_{2\alpha}$, and, because both the alpha particle and the ^{8}Be nucleus in the ground state have spin zero, $\omega_{1,2} = \omega_{\alpha,\alpha} = 1$. Altogether,

$$\sigma_{\alpha,\alpha} = 2 \, \pi \, \lambdabar_{r8}^2 \, \frac{\Gamma_{2\alpha}^2}{(E - E_{r8})^2 + (\Gamma_{2\alpha}/2)^2}. \quad (16.2.4)$$

Another way of describing the appearance of the factor of 2 in this equation is that the consequence of elastic scattering, wherein the projectile is ejected, is indistinguishable from the consequence of a reaction in which the target nucleus is ejected.

In analogy with the treatment of the pp reaction (see eqs. (6.6.5)–(6.6.7) in Volume 1), the rate at which ^{8}Be is produced and decays may be written as

$$R_{\alpha\alpha} = \frac{1}{2} \, n_\alpha^2 \, \langle \sigma_{\alpha\alpha} \, v \rangle, \quad (16.2.5)$$

where n_α is the abundance by number per cm^3 of alpha particles and $R_{\alpha\alpha}$ is in units of number per cm^3 per second. From eqs. (16.2.4) and (16.1.26), one has

$$\langle \sigma_{\alpha\alpha} \, v \rangle = 2 \left(\frac{h^2}{2\pi \, \mu_{\alpha,\alpha} kT} \right)^{3/2} \frac{\Gamma_{2\alpha}}{\hbar} \, e^{-E_{r8}/kT}, \quad (16.2.6)$$

and insertion into eq. (16.2.5) gives

$$R_{\alpha\alpha} = n_\alpha^2 \left(\frac{h^2}{2\pi \, \mu_{\alpha,\alpha} kT} \right)^{3/2} \frac{\Gamma_{2\alpha}}{\hbar} \, e^{-E_{r8}/kT}. \quad (16.2.7)$$

Using eq. (16.1.28), the lifetime of ^{8}Be against α decay is

$$\tau_{2\alpha}(^8\text{Be}) = \frac{6.58 \times 10^{-16} \text{ s}}{5.6} = 1.18 \times 10^{-16} \text{ s}. \quad (16.2.8)$$

As will become evident, the probability that ^{8}Be decays back into two alpha particles is much, much larger than the probability with which it interacts with another alpha particle to form the second excited state of ^{12}C. Hence, the rate at which the abundance of ^{8}Be nuclei changes with time may be written as

$$\frac{dn_8}{dt} = R_{\alpha\alpha} - n_8 \, \frac{\Gamma_{2\alpha}}{\hbar}, \quad (16.2.9)$$

where n_8 is the abundance (number per cm^3) of the ^{8}Be compound nucleus in the ground state. Using eq. (16.2.7) for $R_{\alpha\alpha}$, and assuming that $dn_8/dt = 0$, it follows that

$$n_8 = n_\alpha^2 \left(\frac{h^2}{2\pi \mu_{\alpha,\alpha} kT} \right)^{3/2} e^{-E_{r8}/kT} \tag{16.2.10}$$

is the steady state, or equilibrium, abundance of ^{8}Be in the ground state.

Using the definition $y_i = n_i / \rho \, N_A$, where n_i is the abundance of the ith nucleus in units of number per cm^3, ρ is the density in g cm^{-3}, and N_A is Avogadro's number, this becomes

$$y_8 = y_\alpha^2 \, \rho \, N_A \left(\frac{h^2}{2\pi \mu_{\alpha,\alpha} kT} \right)^{3/2} e^{-E_{r8}/kT}. \tag{16.2.11}$$

Numerically,

$$y_8 = \left(y_\alpha^2 \rho \right) 1.13 \times 10^{-9} \frac{1}{T_8^{3/2}} e^{-10.68/T_8} = \left(y_\alpha^2 \rho \right) \frac{1}{T_8^{3/2}} \exp\left(-20.60 - \frac{10.68}{T_8} \right). \tag{16.2.12}$$

Assuming that ^{8}Be nuclei and alpha particles are in thermodynamic equilibrium, one can use the techniques developed in Sections 4.12–4.15 in Volume 1 to find a relationship between the abundances of alpha particles and of ^{8}Be nuclei in the ground state. Thus,

$$n_\alpha = e^{-\alpha} g_\alpha \left(\frac{2\pi M_\alpha kT}{h^2} \right)^{3/2} \tag{16.2.13}$$

and

$$n_8 = e^{-2\alpha} g_8 \left(\frac{2\pi M_8 kT}{h^2} \right)^{3/2} e^{-E_{r8}/kT}, \tag{16.2.14}$$

where M_8 is the mass of the ^{8}Be nucleus, the gs are statistical weights, and α in the exponential terms is the Lagrangian multiplier of the condition that the sum of all alpha particles, whether free or in ^{8}Be nuclei, is constant. From eqs. (16.2.13) and (16.2.14) it follows that

$$n_8 = n_\alpha^2 \frac{g_8}{g_\alpha^2} \left(\frac{h^2}{2\pi M_\alpha kT} \right)^{3/2} \left(\frac{M_8}{M_\alpha} \right)^{3/2} e^{-E_{r8}/kT}. \tag{16.2.15}$$

Using the facts that all relevant particles have spin zero and that $M_8/M_\alpha^2 = 1/\mu_{\alpha,\alpha}$, it is evident that eqs. (16.2.15) and (16.2.10) are identical. Thus, at temperatures large enough for the resonant contribution to the alpha–alpha reaction rate to dominate, the ^{8}Be compound nucleus exists primarily in the ground state and is in thermodynamic equilibrium.

The next step is to explore the reaction between an alpha particle and the ^{8}Be compound nucleus. From experiment, the rest mass energy of ^{12}C in its second excited state is greater than the sum of the rest mass energies of three alpha particles by

$$E_{r12} = 7.654 \text{ MeV} - 7.275 \text{ MeV} = 379 \text{ keV} \tag{16.2.16}$$

and therefore greater than the sum of the rest mass energies of an alpha particle and a ^{8}Be nucleus in its ground state by

$$E_r = E_{r12} - E_{r8} = (379 - 92) \text{ keV} = 287 \text{ keV}. \tag{16.2.17}$$

The second excited state (spin-parity designation 0^+) decays primarily into an alpha particle and a ^{8}Be nucleus. It can also decay to the first excited state (spin-parity designation 2^+) by emitting a gamma ray of energy 3.126 MeV or to the ground state (spin-parity designation 0^+) by emitting an electron–positron pair. The level widths for these decay processes are, respectively,

$$\Gamma_\alpha \sim 8.5 \text{ eV}, \tag{16.2.18}$$

$$\Gamma_\gamma = (3.7 \pm 0.5) \times 10^{-3} \text{ eV}, \tag{16.2.19}$$

and

$$\Gamma_{e^+e^-} \sim 6 \times 10^{-5} \text{ eV}. \tag{16.2.20}$$

Assuming that the reaction between a ^{8}Be nucleus and an alpha particle leads to the formation of a compound ^{12}C nucleus at an excitation energy near the energy of the second excited state of this nucleus, and noting that the spins of all of the particles involved are zero, the cross section for the formation of the compound nucleus followed by decay into a bound state of ^{12}C can be written as (eq. (16.1.20))

$$\sigma_{\alpha,\gamma+e^+e^-} = \pi \lambda^2 \frac{\Gamma_\alpha \Gamma_{\text{rad}}}{(E - E_r)^2 + (\Gamma/2)^2}, \tag{16.2.21}$$

where

$$\Gamma_{\text{rad}} = \Gamma_\gamma + \Gamma_{e^+e^-} \tag{16.2.22}$$

and

$$\Gamma = \Gamma_\alpha + \Gamma_{\text{rad}}. \tag{16.2.23}$$

The total cross section, scattering plus capture, is

$$\sigma_{\alpha,\alpha+\gamma+e^+e^-} = \pi \lambda^2 \frac{\Gamma_\alpha \Gamma}{(E - E_r)^2 + (\Gamma/2)^2}, \tag{16.2.24}$$

and the total rate at which reactions occur is

$$R_{\alpha,\alpha+\gamma+e^+e^-} = n_\alpha \, n_8 \, \langle \sigma_{\alpha,\alpha+\gamma+e^+e^-} \, v \rangle, \tag{16.2.25}$$

where, using eq. (16.1.26) with $\omega_{1,j} = 1$,

$$\langle \sigma_{\alpha,\alpha+\gamma+e^+e^-} \, v \rangle = \left(\frac{h^2}{2\pi \mu_{\alpha,8} kT} \right)^{3/2} e^{-E_r/kT} \frac{\Gamma_\alpha}{\Gamma} \frac{\Gamma}{\hbar}. \tag{16.2.26}$$

The time rate of change of the abundance of the ^{12}C** compound nucleus is given by

$$\frac{dn_{12}^{**}}{dt} = R_{\alpha,\alpha+\gamma+e^+e^-} - n_{12}^{**} \frac{\Gamma}{\hbar}, \tag{16.2.27}$$

so, in steady state,

$$n_{12}^{**} = n_\alpha \, n_8 \left(\frac{h^2}{2\pi \mu_{\alpha,8} kT} \right)^{3/2} e^{-E_r/kT} \frac{\Gamma_\alpha}{\Gamma}. \tag{16.2.28}$$

Using eq. (16.2.10) to replace n_8 in this last equation, one has

$$n_{12}^{**} = n_\alpha^3 \left(\frac{h^2}{2\pi \mu_{\alpha,\alpha} kT} \right)^{3/2} e^{-E_{r8}/kT} \left(\frac{h^2}{2\pi \mu_{\alpha,8} kT} \right)^{3/2} e^{-E_r/kT} \frac{\Gamma_\alpha}{\Gamma}. \tag{16.2.29}$$

Making use of eq. (16.2.17), one obtains the final result that the steady state abundance of $^{12}C^{**}$ is

$$\left(n_{12}^{**} \right)_{ss} = n_\alpha^3 \left(\frac{h^2}{2\pi M_\alpha kT} \right)^3 \left(\frac{M_{12}}{M_\alpha} \right)^{3/2} e^{-E_{r12}/kT} \frac{\Gamma_\alpha}{\Gamma}. \tag{16.2.30}$$

As in the case of ^8Be, one can determine the abundance by number of $^{12}C^{**}$ in thermodynamic equilibrium with respect to alpha particles. By induction from eqs. (16.2.13) and (16.2.14), one has

$$\left(n_{12}^{**} \right)_{eq} = e^{-3\alpha} g_{12}^{**} \left(\frac{2\pi M_{12} kT}{h^2} \right)^{3/2} e^{-E_{r12}/kT}, \tag{16.2.31}$$

Combining eqs. (16.2.31) and (16.2.13),

$$\left(n_{12}^{**} \right)_{eq} = n_\alpha^3 \frac{g_{12}^{**}}{g_\alpha^3} \left(\frac{h^2}{2\pi M_\alpha kT} \right)^3 \left(\frac{M_{12}}{M_\alpha} \right)^{3/2} e^{-E_{r12}/kT}$$

$$= n_\alpha^3 \left(\frac{h^2}{2\pi M_\alpha kT} \right)^3 \left(\frac{M_{12}}{M_\alpha} \right)^{3/2} e^{-E_{r12}/kT}, \tag{16.2.32}$$

where, to obtain the final expression, the fact has been used that all of the gs are unity.

From eqs. (16.2.18)–(16.2.23), one has that $\Gamma_\alpha / \Gamma = 1 - \Gamma_{rad}/\Gamma \sim 1 - 0.0004$, and comparing eq. (16.2.32) with eq. (16.2.30), it is evident that the steady state abundance is smaller than the thermodynamic equilibrium abundance by less than 0.05%. The two estimated abundances are not exactly identical because, instead of always decaying back into an alpha particle and a ^8Be nucleus, a $^{12}C^{**}$ compound nucleus is now and then converted into a stable carbon nucleus (either $^{12}C^*$ or ^{12}C). Thus, in steady state, there are fewer $^{12}C^{**}$ compound nuclei than in thermodynamic equilibrium.

The rate at which $^{12}C^{**}$ nuclei decay to bound states is given by

$$R_{\alpha,\gamma+e^+e^-} = n_{12}^{**} \frac{\Gamma_{rad}}{\hbar} = n_\alpha^3 \left(\frac{h^2}{2\pi M_\alpha kT} \right)^3 \left(\frac{M_{12}}{M_\alpha} \right)^{3/2} e^{-E_{r12}/kT} \frac{\Gamma_\alpha}{\Gamma} \frac{\Gamma_{rad}}{\hbar}. \tag{16.2.33}$$

The most remarkable aspect of this result is that essentially the only nuclear properties appearing are those of the ^{12}C compound nucleus; no details of the manner in which the ^{8}Be nucleus is made are present other than that the rate of formation of ^{8}Be is proportional to n_α^2. The result (apart from the factor of Γ_α/Γ) can be obtained in just three steps: (1) find the energy (relative to the rest mass energy of three alpha particles), the radiative lifetime, and the spin of ^{12}C**; (2) determine the abundance of ^{12}C** in thermodynamic equilibrium with respect to alpha particles; and (3) divide the abundance of ^{12}C** by the radiative lifetime ($\hbar$ over radiative width). Nevertheless, the fact that the result depends upon the third power of n_α serves as a powerful reminder that the ^{8}Be compound nucleus plays a pivotal role in the formation of ^{12}C.

It is interesting to examine quantitatively the abundance predicted by eq. (16.2.32) in the coordinate system defined by $y_i = N_A n_i/\rho$. Setting $y_{12}^{**} = N_A n_{12}^{**}/\rho$ and $y_\alpha = N_A n_\alpha/\rho$, one has

$$y_{12}^{**} = y_\alpha^3\, \rho^2\, N_A^2 \left(\frac{h^2}{2\pi M_\alpha kT}\right)^3 \left(\frac{M_{12}}{M_\alpha}\right)^{3/2} e^{-E_{r12}/kT}. \tag{16.2.34}$$

Numerically,

$$y_{12}^{**} = \left(y_\alpha^3\, \rho^2\right) 8.33 \times 10^{-19}\, \frac{1}{T_8^3}\, e^{-43.98/T_8} = \left(y_\alpha^3\, \rho^2\right) \frac{1}{T_8^3}\, \exp\left(-41.63 - \frac{43.98}{T_8}\right). \tag{16.2.35}$$

Combining with eq. (16.2.12), one has that the abundances of the ^{12}C** compound nucleus and the ^{8}Be compound nucleus are related by

$$y_{12}^{**} = y_8\, (y_\alpha\, \rho)\, 7.37 \times 10^{-10}\, \frac{1}{T_8^{3/2}}\, e^{-33.30/T_8}$$

$$= y_8\, (y_\alpha\, \rho)\, \frac{1}{T_8^{3/2}}\, \exp\left(-21.03 - \frac{33.30}{T_8}\right), \tag{16.2.36}$$

demonstrating that the abundance of ^{12}C** nuclei is tiny compared to the abundance of ^{8}Be nuclei.

In complete thermodynamic equilibrium, the populations of the second and first excited states of ^{12}C would be related by

$$\left(\frac{y_{12}^{**}}{y_{12}^{*}}\right)_{\text{eq.}} = \frac{1}{5}\, e^{-373/T_8}. \tag{16.2.37}$$

However, in the envisioned stellar environment with a finite number of alpha particles present, the ratio of the two states is determined by the decay lifetimes of the states to be

$$\left(\frac{y_{12}^{**}}{y_{12}^{*}}\right)_{3\alpha} = \frac{\Gamma_\gamma^{*}}{\Gamma_\gamma^{**}} \sim \frac{10.8 \times 10^{-3}\ \text{eV}}{3.7 \times 10^{-3}\ \text{eV}} = 2.9. \tag{16.2.38}$$

If carbon nuclei in the second excited state were in thermodynamic equilibrium with respect to carbon nuclei in the ground state,

$$\frac{\left(y_{12}^{**}\right)_{\rm eq}}{y_{12}} \sim e^{-(7.654 \ {\rm MeV}/kT)} = e^{-888.2/T_8}, \tag{16.2.39}$$

where $\left(y_{12}^{**}\right)_{\rm eq}$ is the abundance of ^{12}C nuclei in the second excited state due to gamma capture and collisional excitation and y_{12} is the abundance of ^{12}C nuclei in the ground state. Comparing with eq. (16.2.35), one has that

$$\frac{\left(y_{12}^{**}\right)_{\rm eq}}{\left(y_{12}^{**}\right)_{3\alpha}} = \frac{y_{12}}{y_\alpha^3} \frac{T_8^3}{\rho^2} \exp\left(-41.63 - \frac{844.2}{T_8}\right), \tag{16.2.40}$$

where $\left(y_{12}^{**}\right)_{3\alpha}$ is the abundance of ^{12}C nuclei in the second excited state due to the triple-alpha processes. Under typical helium-burning conditions ($\rho \sim 10^4$ g cm^{-3}, $T_8 \sim 2$), the ratio is exceedingly small as long as y_{12}/y_α^3 remains finite, demonstrating that the dominant mechanism for populating the ^{12}C** state is the triple-alpha process rather than excitation by collisions or gamma capture. On the other hand, the ratio can exceed unity when $T_8 > 20$. This means that, at very high temperatures, the gamma-capture and collisional excitation rates must be explicitly taken into account. Thus, a proper treatment at very high temperatures requires a detailed consideration of all reaction channels.

Numerical values of the lifetimes for the various processes are instructive. Using eqs. (16.1.28) and (16.2.19), the lifetime of ^{12}C** against γ decay to the first excited state plus electron–positron pair decay to the ground state is

$$\tau_{\rm rad}(^{12}{\rm C}^{**}) = \frac{6.58 \times 10^{-16} \ {\rm s}}{3.7 \times 10^{-3}} = 1.78 \times 10^{-13} \ {\rm s}. \tag{16.2.41}$$

Using eqs. (16.1.28) and (16.2.18), the lifetime of ^{12}C** against decay into ^{8}Be and an alpha particle is

$$\tau_\alpha(^{12}{\rm C}^{**}) = \frac{6.58 \times 10^{-16} \ {\rm s}}{8.5} = 7.74 \times 10^{-17} \ {\rm s}. \tag{16.2.42}$$

Defining $\tau_{\alpha+^8{\rm Be}\to^{12}{\rm C}^{**}}$ as the lifetime of a ^{8}Be nucleus against capture of an alpha particle to form ^{12}C**, one has that, in good approximation,

$$\frac{n_8}{\tau_{\alpha+^8{\rm Be}\to^{12}{\rm C}^{**}}} = \frac{n_{12}^{**}}{\tau_\alpha(^{12}{\rm C}^{**})}. \tag{16.2.43}$$

Using eqs. (16.2.38) and (16.2.42), one finds

$$\tau_{\alpha+^8{\rm Be}\to^{12}{\rm C}^{**}} = \frac{y_8}{y_{12}^{**}} \tau_\alpha(^{12}{\rm C}^{**}) = \frac{9.98 \times 10^{-6}}{y_\alpha \ \rho} \left[T_8^{3/2} \exp\left(\frac{33.3}{T_8}\right)\right] \ {\rm s}. \tag{16.2.44}$$

Comparing eqs. (16.2.8) and (16.2.44), it is clear that the probability that ^{8}Be decays back into two alpha particles is very much larger than the probability that it interacts with another alpha particle to form the second excited state of ^{12}C, thus explicitly justifying the use of eq. (16.2.9) to find the steady state abundance of ^{8}Be.

The lifetime of the first excited state of ^{12}C against gamma decay to the ground state ($\Gamma_\gamma \sim 10.8 \times 10^{-3}$ eV) is comparable to the lifetime of the second excited state against decay to the first excited state ($\Gamma_\gamma \sim 3.7 \times 10^{-3}$ eV). Hence, decay from the second to the first excited state is followed very quickly by decay to the ground state and the rate at which the ground state is populated is given by eq. (16.2.33). In terms of the fractional number abundances y, eq. (16.2.33) can be written as

$$\frac{dy_{12}}{dt} = \frac{y_{12}^{**}}{\tau_{\mathrm{rad}}(^{12}\mathrm{C}^{**})} = y_{12}^{**} \frac{\Gamma_{\mathrm{rad}}(^{12}\mathrm{C}^{**})}{\hbar}. \tag{16.2.45}$$

Using eqs. (16.2.35) and (16.2.41), one has

$$\frac{dy_{12}}{dt} = y_\alpha^3 \frac{\rho^2}{T_8^3} 4.68 \times 10^{-6} e^{-44.0/T_8} = y_\alpha^3 \frac{\rho^2}{T_8^3} \exp\left(-12.27 - \frac{44.0}{T_8}\right) \tag{16.2.46}$$

and

$$N_A \frac{dy_{12}}{dt} = y_\alpha^3 \frac{\rho^2}{T_8^3} \exp\left(42.6 - \frac{44.0}{T_8}\right). \tag{16.2.47}$$

The rate of energy generation by the triple-alpha process in the classical approximation is

$$\epsilon_{3\alpha} = \Delta E_{3\alpha} N_A y_{12}^{**} \frac{\Gamma_\gamma}{\hbar} = \Delta E_{3\alpha} N_A \frac{dy_{12}}{dt}, \tag{16.2.48}$$

where

$$\Delta E_{3\alpha} = 7.275 \text{ MeV} = 11.66 \times 10^{-6} \text{ erg} \tag{16.2.49}$$

is the energy liberated by the conversion of three alpha particles into ^{12}C in the ground state. Using eq. (16.2.47),

$$\epsilon_{3\alpha} = y_\alpha^3 \frac{\rho^2}{T_8^3} \exp\left(31.12 - \frac{44.0}{T_8}\right) \text{ erg g}^{-1} \text{ s}^{-1}. \tag{16.2.50}$$

At $T_8 = 1$ and $\rho = 10^4$ g^{-1} cm^{-3}, $\epsilon_{3\alpha} = y_\alpha^3$ 255 erg g^{-1} s$^{-1} = (4y_\alpha)^3$ 4 erg g^{-1} s^{-1}. Since

$$\left(\frac{d\log \epsilon_{3\alpha}}{d\log T}\right)_{\rho=\mathrm{const}} = -3 + \frac{44}{T_8} \tag{16.2.51}$$

it is evident that $\epsilon_{3\alpha}$ varies by about the fortieth power of temperature at $T_8 \sim 1$.

16.3 Triple-alpha reactions at low temperatures

When the temperature is low enough, the triple-alpha reactions proceed primarily through the low energy wings of the resonances and, therefore, a fundamental prerequisite for a simple calculation of an abundance in thermodynamic equilibrium – occupation of a well defined energy level – is not met.

At relative kinetic energies near the resonance energy, the Coulomb transmission factor $e^{-2\pi\eta(E)}$ may be thought of as being hidden in the level width for the entrance channel. When defined in this way, at relative kinetic energies small compared with the resonance energy, the level width of the entrance channel is reduced by the ratio of penetration factors $e^{-2\pi\eta(E)}/e^{-2\pi\eta(E_r)}$ and, since, in general $\Gamma \ll E_r - E$, the cross section will, to first order, behave as $e^{-2\pi\eta(E)}/(E - E_r)^2$, which becomes $e^{-2\pi\eta(E)}/E_r^2$ when $E \ll E_r$. Thus, one may use the formalism of Section 6.7 in Volume 1 to approximate the rate at which a reaction between nuclei of type i and type j proceeds in the low energy tail of a resonance (see eq. (6.7.15)):

$$\frac{dy_i}{dt} = N_A \langle \sigma \, v \rangle_{ij} = \frac{4.35 \times 10^5}{1 + \delta_{ij}} \, \rho \, y_i \, y_j \, S_{ij}(E_0)(\text{keV barn}) \, \frac{\tau_{ij}^2 \, e^{-\tau_{ij}}}{Z_i Z_j A_{ij}} \, s^{-1}, \quad (16.3.1)$$

where y_i is the fractional abundance by number of the ith nucleus, τ_{ij} is defined by eq. (6.7.16) and the center of mass cross section factor $S_{ij}(E)$ should be evaluated (in units of keV barn) at the energy where the integrand leading to the stated rate peaks, namely, at (see eq. (6.7.18))

$$E_0 = 26.28 \left(Z_i^2 Z_j^2 A_{ij} \, T_8^2 \right)^{1/3} \text{keV.} \quad (16.3.2)$$

The width ΔE_0 of the peak in the integrand (the Gamow peak), is related to E_0 by

$$\frac{\Delta E_0}{E_0} = \frac{4}{\sqrt{3}} \left(\frac{kT}{E_0} \right)^{1/2} = 1.322 \left(\frac{T_8}{Z_i^2 Z_j^2 A_{ij}} \right)^{1/6}. \quad (16.3.3)$$

If $S_{ij}(E)$ varies appreciably over an energy of the order of ΔE_0, $S_{ij}(E_0)$ in eq. (16.3.1)) should be replaced by an average $\langle S_{ij}(E) \rangle$ determined by a numerical integration.

Defining a reduced cross section factor S by

$$\sigma = \pi \lambdabar^2 \, S \, e^{-2\pi\eta}, \quad (16.3.4)$$

eq. (16.3.1) can also be written as

$$\frac{dy_i}{dt} = \frac{2.84 \times 10^8}{1 + \delta_{ij}} \, \rho \, y_i \, y_j \, S_{ij}(E_0) \, \frac{1}{Z_i Z_j A_{ij}^2} \, \tau_{ij}^2 \, e^{-\tau_{ij}} \, s^{-1}, \quad (16.3.5)$$

where ρ is in units of g cm^{-3}.

In the non-resonant approximation, the creation rate of the ^{8}Be compound nucleus by the collision of two alpha particles is then

$$\left(\frac{dy_8}{dt} \right)_+^{\text{non-res}} = \frac{1.78 \times 10^7}{2} \, \rho \, y_\alpha^2 \, S_{\alpha\alpha}(E_0) \, \tau_{\alpha\alpha}^2 e^{-\tau_{\alpha\alpha}} \, s^{-1}, \quad (16.3.6)$$

where

$$\tau_{\alpha\alpha} = \frac{29.1}{T_8^{1/3}}, \quad (16.3.7)$$

$$E_0 = 83.45 \, T_8^{2/3} \text{ keV,} \quad (16.3.8)$$

and

$$\frac{\Delta E_0}{E_0} = 0.742 \, T_8^{1/6}. \tag{16.3.9}$$

To estimate $\mathcal{S}_{\alpha\alpha}(E_0)$, suppose that

$$\sigma_{\alpha\alpha}(E) = \pi \, \lambdabar^2 \, \mathcal{S}_{\alpha\alpha} e^{-2\pi\eta(E)} = \pi \, \lambdabar^2 \, 2 \, \frac{\Gamma_\alpha^{\mathrm{entrance}}(E) \, \Gamma_\alpha^{\mathrm{exit}}(E)}{(E - E_{r8})^2 + \left(\Gamma_\alpha^{\mathrm{exit}}(E)/2\right)^2}, \tag{16.3.10}$$

where

$$\Gamma_\alpha^{\mathrm{entrance}}(E) = \frac{e^{-2\pi\eta(E)}}{e^{-2\pi\eta(E_{r8})}} \, \Gamma_\alpha^{\mathrm{exit}}(E_{r8}) \tag{16.3.11}$$

and

$$\Gamma_\alpha^{\mathrm{exit}}(E) = \Gamma_\alpha^{\mathrm{exit}}(E_{r8}) = \Gamma_{2\alpha}. \tag{16.3.12}$$

By making the choice expressed by eq. (16.3.12), it has been assumed that the probability for the decay of the compound ^{8}Be nucleus near threshold is independent of excitation. With these approximations, one has

$$\mathcal{S}_{\alpha\alpha}(E_0) = \frac{2 \, \Gamma_{2\alpha}^2 \, e^{2\pi\eta(E_{r8})}}{(E_0 - E_{r8})^2 + (\Gamma_{2\alpha}/2)^2}. \tag{16.3.13}$$

In general,

$$\eta(E) = \frac{Z_i Z_j \, e^2}{\hbar v} = Z_i Z_j \, \frac{e^2}{\hbar c} \, \frac{c}{v} = Z_i Z_j \, \frac{e^2}{\hbar c} \left(\frac{A_{i,j} M_{\mathrm{H}} c^2}{2E}\right)^{1/2}$$

$$= \frac{Z_i Z_j A_{i,j}^{1/2}}{137.036} \left(\frac{938.78 \times 500}{E\,(\mathrm{keV})}\right)^{1/2} = 0.2236 \, Z_i Z_j A_{i,j}^{1/2} \left(\frac{500\mathrm{keV}}{E}\right)^{1/2}. \tag{16.3.14}$$

For the α-α system, $v_{r8}/c = 0.009\,90$, $\eta(E_{r8}) = 2.949$, and $e^{2\pi\eta(E_{r8})} = 1.1116 \times 10^8$. Using eqs. (16.3.6)–(16.3.8) and (16.3.13) and inserting numbers for E_{r8} and $\Gamma_{2\alpha}$, one obtains

$$\left(\frac{d y_8}{dt}\right)_+^{\mathrm{non\text{-}res}} = 1.51 \times 10^{18} \, \frac{(5.6)^2}{10^6 \left(83.5 \, T_8^{2/3} - 92\right)^2 + (2.8)^2}$$

$$\times \, (\rho \, y_\alpha^2) \, \frac{\exp\left(-29.1/T_8^{1/3}\right)}{T_8^{2/3}} \, \mathrm{s}^{-1}. \tag{16.3.15}$$

In the resonant approximation, the rate at which ^{8}Be is created may be obtained by dividing n_8 from eq. (16.2.12) by $\tau_{2\alpha}$ from eq. (16.28), yielding

$$\left(\frac{d y_8}{dt}\right)_+^{\mathrm{res}} = \frac{n_8}{\tau_{2\alpha}} = \rho \, y_\alpha^2 \, \frac{0.97 \times 10^7}{T_8^{3/2}} \, \exp\left(\frac{-10.7}{T_8}\right) \, \mathrm{s}^{-1}, \tag{16.3.16}$$

where, again, ρ is in units of g cm^{-3}.

The resonant and non-resonant approximations give the same result when

$$1.56 \times 10^{11} \frac{(5.6)^2}{10^6 \left(83.5 \, T_8^{2/3} - 92\right)^2 + (2.8)^2} = \frac{1}{T_8^{5/6}} \exp\left(\frac{29.1 \, T_8^{2/3} - 10.7}{T_8}\right).$$

(16.3.17)

The solution of this equation is $T_8 \sim 0.28$, very close to the result of a similar calculation by K. Nomoto, F.-K. Thielmann, & S. Miyaji (1985) who estimate equality occurs at $T_8 \sim 0.27$. The ratio of the term on the right hand side to the term on the left hand side of eq. (16.3.17) increases rapidly enough with increasing temperature that, for all intents and purposes, one may adopt

$$\left(\frac{dy_8}{dt}\right)_+ = \left(\frac{dy_8}{dt}\right)_+^{\text{non-res}} + \left(\frac{dy_8}{dt}\right)_+^{\text{res}}$$

(16.3.18)

as a reasonable first approximation. Setting $dy_8/dt = (\Gamma_{2\alpha}/\hbar) \, y_8$ and solving for y_8 gives

$$y_8 = \left(\frac{5.39 \times 10^{-3}}{\left(83.5 \, T_8^{2/3} - 92\right)^2 + 7.4 \times 10^{-6}} \frac{e^{-29.1/T_8^{1/3}}}{T_8^{2/3}} + 1.13 \times 10^{-9} \frac{e^{-10.7/T_8}}{T_8^{3/2}}\right) \rho \, y_\alpha^2.$$

(16.3.19)

Similarly, one may assume that, at low enough temperatures, the rate at which ^{12}C is created by the collision of an alpha particles with a ^{8}Be nucleus can be written as

$$\left(\frac{dy_{12}}{dt}\right)_+^{\text{non-res}} = 1.78 \times 10^7 \, \rho \, y_\alpha \, y_8 \, S_{\alpha^8\text{Be}}(E_0) \, \tau_{\alpha^8\text{Be}}^2 e^{-\tau_{\alpha^8\text{Be}}} \, \text{s}^{-1},$$

(16.3.20)

where

$$\tau_{\alpha^8\text{Be}} = \frac{50.77}{T_8^{1/3}},$$

(16.3.21)

$$E_0 = 145.8 \, T_8^{2/3} \, \text{keV},$$

(16.3.22)

and

$$\frac{\Delta E_0}{E_0} = 0.561 \, T_8^{1/6}.$$

(16.3.23)

To estimate $S_{\alpha^8\text{Be}}(E_0)$, assume that

$$\sigma_{\alpha^8\text{Be}}(E) = \pi \, \lambdabar^2 \, S_{\alpha^8\text{Be}}(E) e^{-2\pi\eta(E)} = \pi \, \lambdabar^2 \frac{\Gamma_\alpha^{\text{entrance}}(E) \, \Gamma_\gamma^{\text{exit}}(E)}{(E - E_r)^2 + \left(\Gamma_\alpha^{\text{exit}}(E)/2\right)^2},$$

(16.3.24)

where

$$\Gamma_\alpha^{\text{entrance}}(E) = \frac{e^{-2\pi\eta(E)}}{e^{-2\pi\eta(E_r)}} \Gamma_\alpha^{\text{entrance}}(E_r) = e^{-2\pi\eta(E)} \, e^{2\pi\eta(E_r)} \, \Gamma_\alpha,$$

(16.3.25)

with $\Gamma_\alpha = 8.5$ eV (eq. 6.3.18)). Further, set

$$\Gamma_\gamma^{\text{exit}}(E) = \Gamma_\gamma(E_r) \, g = 3.7 \times 10^{-3} \, \text{eV} \, g,$$

(16.3.26)

where

$$g = \left(\frac{(E_{12}^{**} - E_{12}^*) - (E_r - E)}{E_{12}^{**} - E_{12}^*}\right)^{2M+1} = \left(\frac{2.938 + E(\text{MeV})}{3.216}\right)^5. \qquad (16.3.27)$$

In the intermediate form of eq. (16.3.27), M is the multipolarity of the gamma transition, ^{12}C* has the angular momentum and parity designation 2^+, and ^{12}C** has the designation 0^+, so $M = 2$, giving the final form of the equation.

For the α-^{8}Be system, $v_r/c = 0.015\,15$, $\eta(E_r) = 3.853$, and $e^{2\pi \eta(E_r)} = 3.27 \times 10^{10}$. Altogether,

$$S_{\alpha^8\text{Be}}(E_0) = 3.27 \times 10^{10}\, g \, \frac{8.5 \times (3.7 \times 10^{-3})}{10^6 \left(146\, T_8^{2/3} - 287\right)^2 + (4.25)^2}. \qquad (16.3.28)$$

Inserting this in eq. (16.3.20) and using eqs. (16.3.21) and (16.3.22), one has

$$\left(\frac{dy_{12}}{dt}\right)_+^{\text{non-res}} = \frac{4.72 \times 10^{13}\, g}{(146\, T_8^{2/3} - 287)^2 + (4.25)^2}\, (\rho\, y_\alpha\, y_8)\, \frac{\exp\left(-50.765/T_8^{1/3}\right)}{T_8^{2/3}}\, \text{s}^{-1}. \qquad (16.3.29)$$

In the resonant approximation, the equivalent expression is

$$\left(\frac{dy_{12}}{dt}\right)_+^{\text{res}} = N_A\, (\rho\, y_\alpha\, y_8)\, \langle\sigma_{\alpha^8\text{Be}} v\rangle$$

$$= N_A\, (\rho\, y_\alpha\, y_8)\, \left(\frac{h^2}{2\pi\, \mu_{\alpha,8} kT}\right)^{3/2} e^{-E_r/kT}\, \frac{\Gamma_{\text{rad}}}{\hbar}, \qquad (16.3.30)$$

where eq. (16.3.26) has been used for $\langle\sigma_{\alpha^8\text{Be}} v\rangle$ and it has been assumed that $\Gamma_\alpha/\Gamma = 1$. Numerically,

$$\left(\frac{dy_{12}}{dt}\right)_+^{\text{res}} = (\rho\, y_\alpha\, y_8)\, 4.0865 \times 10^3\, \frac{\exp\left(-33.3/T_8\right)}{T_8^{3/2}}. \qquad (16.3.31)$$

The resonant and non-resonant approximations give the same result when

$$\frac{1.15 \times 10^{10}\, g}{(146\, T_8^{2/3} - 287)^2 + 10^{-6}(4.25)^2} = \frac{1}{T_8^{5/6}}\, \exp\left(\frac{50.77\, T_8^{2/3} - 33.3}{T_8}\right). \qquad (16.3.32)$$

The solution of this equation is $T_8 \sim 0.78$, compared with the value of $T_8 \sim 0.74$ estimated by Nomoto *et al.* (1985). Just as in the case of the α-α reaction, the ratio of the term on the right hand side to the term on the left hand side of eq. (16.3.32) increases rapidly enough with increasing temperature that one may adopt for the creation rate of y_{12} an equation analogous to eq. (16.3.18):

$$\frac{dy_{12}}{dt} = \left(\frac{4.72 \times 10^{13}\ g}{(146\ T_8^{2/3} - 287)^2 + 1.81 \times 10^{-5}} \frac{e^{-50.77/T_8^{1/3}}}{T_8^{2/3}} \right.$$

$$\left. +4.09 \times 10^3\ \frac{e^{-33.3/T_8}}{T_8^{3/2}} \right)\ \rho\ y_\alpha\ y_8. \tag{16.3.33}$$

The story is completed by using eq. (16.3.19) to replace y_8 in eq. (16.3.33).

Using a formalism similar to that described by Cameron (1959), Nomoto *et al.* (1985) and Angulo *et al.* (1999) have estimated the triple-alpha rate over the temperature range $T_6 = 1$ to $T_{10} = 1$ by calculating all relevant integrals numerically and have produced fitting formulae to fit the numerical results. It is somewhat worrisome that, for temperatures above $T_8 \sim 1$, when the resonant rate expressed by eq. (16.2.42) should be applicable, the leading term in the Angulo *et al.* (1999) analytic fit is

$$\frac{dy_{12}}{dt} = y_\alpha^3\ \frac{\rho^2}{T_8^3}\ \exp\left(-11.60 - \frac{43.9}{T_8} \right)\ s^{-1}, \tag{16.3.34}$$

which is approximately twice the rate given by eq. (16.2.46). This difference may be a consequence of their introduction of a resonant contribution at very high temperatures from a theoretically expected 2^+ excited state whose existence has as yet not been experimentally confirmed (S. Austin, 2004). From the point of view of aesthetics, this is a serious discrepancy. From the point of view of stellar structure, it is a minor discrepancy; at $T_8 = 1$–3, a change in temperature of only $3(1 \pm 0.5)\%$ produces the same result as a change of a factor of 2 in the multiplicative rate constant. More troublesome is the extrapolation required for estimating the cross sections at low temperatures where one or both of the relevant compound nuclei are not formed predominantly at the resonance energies. The derived rates could be wrong by one or more orders of magnitude.

For discussion of the triple-alpha rate in the pycnonuclear regime (high densities and strong screening), see Cameron (1959) and I. Fushiki & D. Q. Lamb (1987).

16.4   The formation of ^{16}O by alpha capture on ^{12}C and the conversion of ^{14}N into ^{22}Ne

As helium burning continues and the abundance of ^{12}C increases, (α, γ) reactions on ^{12}C produce ^{16}O. As the abundance of ^{16}O increases, (α, γ) reactions on ^{16}O produce ^{20}Ne, and so on, but these heavier alpha-particle nuclei are not produced at very interesting abundances. Once helium has been exhausted, by far the most abundant nuclei in the final mix are ^{12}C and ^{16}O, and, when it is expelled from stars, matter in which nucleosynthesis has terminated after complete helium burning contains far fewer alpha-particle nuclei such as ^{20}Ne and ^{24}Mg than necessary to account for the abundances of these nuclei in the

Universe. In this section, reaction rate estimates for the production of alpha-particle nuclei are quoted only for the ^{12}C$(\alpha, \gamma)^{16}$O and ^{16}O$(\alpha, \gamma)^{20}$Ne reactions.

For the reaction ^{12}C$(\alpha, \gamma)^{16}$O, G. R. Caughlan and W. A. Fowler (1988) estimate that, at temperatures $T_8 \leq 3$,

$$N_A \langle \sigma v \rangle = \frac{1}{T_8^2} \left(1 + \frac{0.227}{T_8^{2/3}}\right)^{-2} \exp\left(23.07 - \frac{69.20}{T_8^{1/3}} - \left(\frac{T_8}{34.96}\right)^2\right)$$

$$+ \frac{1}{T_8^2} \left(1 + \frac{1.232}{T_8^{2/3}}\right)^{-2} \exp\left(23.59 - \frac{69.20}{T_8^{1/3}}\right)$$

$$+ \frac{1}{T_8^{3/2}} \exp\left(10.59 - \frac{275.0}{T_8}\right) + T_8^5 \exp\left(7.265 - \frac{155.4}{T_8}\right), \quad (16.4.1)$$

whereas Angulo $et\ al.$ (1999) estimate that

$$N_A \langle \sigma v \rangle = \frac{A}{T_9^2} \exp\left(22.62 - \frac{69.21}{T_8^{1/3}} - \left(\frac{T_8}{46}\right)^2\right) + \frac{A}{T_8^{3/2}} \exp\left(10.61 - \frac{289.3}{T_8}\right)$$

$$+ \frac{B}{T_8^2} \exp\left(22.61 - \frac{69.21}{T_8^{1/3}} - \left(\frac{T_8}{13}\right)^2\right) + B\,T_8^2 \exp\left(-1.68 - \frac{269}{T_8}\right), \quad (16.4.2)$$

where

$$A = 1 + 0.254\,T_8 + 0.0104\,T_8^2 - 0.000226\,T_8^3 \quad (16.4.3)$$

and

$$B = 1 + 0.923\,T_8 - 0.137\,T_8^2 + 0.0074\,T_8^3. \quad (16.4.4)$$

The energy liberated in the reaction is $Q = 7.162$ MeV, so the energy generation rate is given by

$$\epsilon\left(^{12}\text{C}(\alpha, \gamma)^{16}\text{O}\right) = 1.1475 \times 10^{-5}\,N_A \langle \sigma v \rangle \text{ erg g}^{-1}\,\text{s}^{-1}, \quad (16.4.5)$$

where $N_A \langle \sigma v \rangle$ is given by eq. (16.4.1) or by eqs. (16.4.2)–(16.4.4).

For the ^{16}O$(\alpha, \gamma)^{20}$Ne reaction, Caughlan & Fowler (1988) give

$$N_A \langle \sigma v \rangle = \frac{1}{T_8^{2/3}} \exp\left(24.50 - \frac{85.65}{T_8^{1/3}} - \left(\frac{T_8}{15.86}\right)^2\right)$$

$$+ \frac{1}{T_8^{3/2}} \left[\exp\left(7.58 - \frac{103.0}{T_8}\right) + \exp\left(9.74 - \frac{122.3}{T_8}\right)\right]$$

$$+ T_8 \exp\left(4.87 - \frac{200.9}{T_8}\right) \quad (16.4.6)$$

whereas Angulo *et al.* (1999) give

$$N_A \langle \sigma\, v \rangle = \frac{1}{T_8^{2/3}} \exp\left(25.55 - \frac{84.75}{T_8^{1/3}} - \left(\frac{T_8}{16}\right)^2\right)$$

$$+ \frac{1}{T_8^{3/2}} \left[\exp\left(7.39 - \frac{103.2}{T_8}\right) + \exp\left(9.88 - \frac{122.0}{T_8}\right)\right]$$

$$+ T_8^{2.966} \exp\left(-7.79 - \frac{119.0}{T_8}\right). \tag{16.4.7}$$

The energy liberated in the reaction is $Q = 4.730$ MeV, so the energy generation rate is

$$\epsilon \left(^{12}O(\alpha, \gamma)^{20}Ne\right) = 7.578 \times 10^{-6}\, N_A \langle \sigma\, v \rangle \text{ erg g}^{-1}\text{ s}^{-1}, \tag{16.4.8}$$

where $N_A \langle \sigma\, v \rangle$ is given by eq. (16.4.6) or by eq. (16.4.7).

In matter in which all of the initial hydrogen has been converted into helium, most of the carbon and oxygen isotopes which were present at the start of hydrogen burning have been converted into ^{14}N. The helium- and nitrogen-rich matter is ultimately compressed and heated and, when temperatures become of the order of 10^8 K, the reactions

$$^{14}N(\alpha, \gamma)^{18}F(e^+ \nu_e)^{18}O(\alpha, \gamma)^{22}Ne \tag{16.4.9}$$

convert nitrogen into the most massive stable isotope of neon. As discussed in Section 16.5, ^{22}Ne ultimately captures alpha particles to be converted into ^{25}Mg and a neutron. The neutron is easily captured by iron-group elements as well as by light elements. Thus, the CNO nuclei present in a star at birth contribute to the formation of many neutron-rich elements in the Universe.

For the $^{14}N(\alpha, \gamma)^{18}F$ reaction, the four lowest lying resonances occur at excitation energies of 237 (4^+), 435 (1^+), 535 (2^+), and 883 (4^+) keV. Caughlan and Fowler (1988) estimate that, at temperatures $T_8 \le 3$,

$$N_A \langle \sigma_{4,14}\, v \rangle = \frac{1}{T_8^{2/3}} \exp\left(24.31 - \frac{77.63}{T_8^{1/3}} - \left(\frac{T_8}{8.81}\right)^2\right)$$

$$+ \frac{1}{T_8^{3/2}} \left[\exp\left(-18.71 - \frac{27.98}{T_8}\right) + \exp\left(4.16 - \frac{50.54}{T_8}\right)\right]. \tag{16.4.10}$$

Angulo *et al.* (1999) construct a fit to the data which, for the same temperature range, is approximately

$$N_A \langle \sigma_{4,14}\, v \rangle = \frac{1}{T_8^{2/3}} \exp\left(28.93 - \frac{77.64}{T_8^{1/3}} - \left(\frac{T_8}{0.7}\right)^2\right)$$

$$+ \frac{1}{T_8^{3/2}} \left[\exp\left(-18.96 - \frac{27.50}{T_8}\right) + \exp\left(4.42 - \frac{50.45}{T_8}\right)\right]. \tag{16.4.11}$$

The energy emitted in the $^{14}N(\alpha, \gamma)^{18}F$ reaction is 4.415 MeV. The $^{18}F(e^+\nu_e)^{18}O$ reaction is characterized by $\tau_{1/2} = 110$ minutes, $Q = 1.655$ MeV, and $\log(ft) = 3.554$. The beta decay lifetime of ^{18}F is short enough that the overall contribution of both reactions to the energy-generation rate can be approximated by multiplying the energy released by both reactions, 6.070 MeV, by the rate of the $^{14}N(\alpha, \gamma)^{18}F$ reaction to give

$$\epsilon_{4,14} = 9.726 \times 10^{-6} \, N_A \langle \sigma_{4,14} \, v \rangle \text{ erg g}^{-1} \text{ s}^{-1}, \tag{16.4.12}$$

where $N_A \langle \sigma_{4,14} \, v \rangle$ is given by eq. (16.4.10) or eq. (16.4.11).

For the $^{18}O(\alpha, \gamma)^{22}Ne$ reaction, the three lowest lying resonances occur at excitation energies of 178 (2^+), 385 (0^+), and 542 (1^-) keV, and $Q = 9.668$ MeV. Caughlan and Fowler (1988) give

$$N_A \langle \sigma_{4,18} \, v \rangle = \frac{1}{T_8^{2/3}} \exp\left(29.76 - \frac{86.3}{T_8^{1/3}} - \left(\frac{T_8}{3.43}\right)^2\right)$$
$$+ \frac{1}{T_8^{3/2}} \left[\exp\left(5.474 - \frac{62.3}{T_8}\right) + \exp\left(7.003 - \frac{73.0}{T_8}\right)\right]$$
$$+ \frac{1}{T_8^{3/2}} \left[\exp\left(9.89 - \frac{169.9}{T_8}\right) T_8 + \exp\left(-24.18 - \frac{19.94}{T_8}\right)\right], \tag{16.4.13}$$

whereas Angulo *et al.* (1999) give

$$N_A \langle \sigma_{4,18} \, v \rangle = \frac{1}{T_8^{2/3}} \left[\exp\left(-25.81 - \frac{20.69}{T_8}\right) + \exp\left(-0.707 - \frac{44.62}{T_8}\right)\right]$$
$$+ \frac{1}{T_9^{3/2}} \left[\exp\left(5.766 - \frac{63.91}{T_8}\right) + \exp\left(7.240 - \frac{73.89}{T_8}\right)\right]$$
$$+ \frac{1}{T_8^{1/2}} \exp\left(13.90 - \frac{221.03}{T_8}\right). \tag{16.4.14}$$

The energy-generation rate associated with the reaction is

$$\epsilon_{4,18} = 1.5490 \times 10^{-5} \, N_A \langle \sigma_{4,18} \, v \rangle \text{ erg g}^{-1} \text{ s}^{-1}, \tag{16.4.15}$$

where $N_A \langle \sigma_{4,18} \, v \rangle$ is given by eq. (16.4.13) or eq. (16.4.14).

16.5 Neutron production by (α, n) reactions on ^{13}C and ^{22}Ne

A. G. W. Cameron was the first to recognize that the $^{13}C(\alpha, n)^{16}O$ reaction (Cameron, 1960) and the $^{22}Ne(\alpha, n)^{25}Mg$ reaction (Cameron, 1955) might be important stellar neutron sources. Subsequent research has shown that, in fact, they are the two most important sources of neutrons for s-process nucleosynthesis in stars. A summary of the experimental information available for these reactions as of 1999 is given by Angulo *et al.* (1999).

The ^{13}C$(\alpha, n)^{16}$O reaction is exothermic by 2.216 MeV. The cross section has been measured from $E \sim 275$ keV (center of mass system) to $E > 4$ MeV, and, over this range, the center of mass cross section factor fluctuates by over four orders of magnitude. At $E \sim 810$ keV, the cross section factor attains a maximum of $S \sim 4 \times 10^7$ MeV barn and, at $E \sim 3$ MeV, it descends to a minimum of $S \sim 10^3$ MeV barn. For energies below the resonance at 810 keV, the cross section factor decreases monotonically, reaching a minimum of $S \sim 0.8 \times 10^6$ keV barn at $E \sim 375$ keV. Over the interval $E \sim 300$–500 keV, the center of mass cross section factor is nearly constant at $S = (0.9 \pm 0.1) \times 10^6$ MeV barn. At the lowest energies so far achievable in the laboratory, the cross section factor increases with decreasing energy, and it is suspected that a sub-threshhold resonance at -3 keV is responsible for this increase.

From the point of view of stellar evolution, the important range of energies is evident from the relationship between the energy at the Gamow peak and the temperature. Inserting appropriate values for nuclear charge and atomic number in eq. (16.3.2), one has that the Gamow peak occurs at

$$E_0 = 200 \, T_8^{2/3} \text{ keV}. \tag{16.5.1}$$

Assuming that $S \geq 0.8 \times 10^9$ keV barn, a lower limit to the rate at which the relative abundance of neutrons would increase if neutrons did not beta decay and neutron-capture reactions did not occur is given by

$$\frac{\mathrm{d}y_n}{\mathrm{d}t} = (\rho \, y_\alpha y_{13}) \, N_A \langle \sigma \, v \rangle \tag{16.5.2}$$

where y_α and y_{13} are, respectively, the fractional number abundances of alpha particles and ^{13}C nuclei and

$$N_A \langle \sigma \, v \rangle = \frac{1}{T_8^{2/3}} \exp\left(38.36 - \frac{69.63}{T_8^{1/3}}\right). \tag{16.5.3}$$

Fitting a quadratic to the experimental points given by Drotleff *et al.* (1993),

$$S(E) = 0.8 \times 10^9 \left[1 + \left(\frac{E - 375 \text{ keV}}{217.5 \text{ keV}}\right)^2\right] \text{ keV barn.} \tag{16.5.4}$$

Thus,

$$\langle S(E) \rangle \sim 0.8 \times 10^9 \left[1 + \left(\frac{200 \, T_8^{2/3} - 375}{217.5}\right)^2\right] \text{ keV barn.} \tag{16.5.5}$$

Caughlan and Fowler (1988) approximate the reaction rate at stellar temperatures by

$$
N_A \langle \sigma v \rangle = \frac{1}{T_8^{2/3}} \exp\left(37.99 - \frac{69.65}{T_8^{1/3}} - \left(\frac{T_8}{12.84} \right)^2 \right)
$$

$$
+ \frac{1}{T_8^{3/2}} \left[\exp\left(16.31 - \frac{93.73}{T_8} \right) + \exp\left(17.61 - \frac{118.7}{T_8} \right) \right]
$$

$$
+ \frac{1}{T_8^{3/2}} \left[\exp\left(24.87 - \frac{204.1}{T_8} \right) + \exp\left(25.25 - \frac{292.8}{T_8} \right) \right] \tag{16.5.6}
$$

whereas Drotleff *et al.* (1993) estimate

$$
N_A \langle \sigma v \rangle = \frac{1 + 0.225\, T_8^{1/3} - 1.712\, T_8^{2/3} + 1.073\, T_8}{T_8^2} \exp\left(41.06 - \frac{71.30}{T_8^{1/3}} \right)
$$

$$
+ \frac{1}{T_8^{3/2}} \left[\exp\left(10.38 - \frac{62.59}{T_8} \right) + \exp\left(16.21 - \frac{84.30}{T_8} \right) \right]. \tag{16.5.7}
$$

The $1/T^2$ factor in the first term of eq. (16.5.7) is the consequence of an assumed sub-threshold resonance at $-3\,\text{KeV}$ and the second two terms reflect the occurrence of resonances at $600\,\text{keV}$ and $810\,\text{keV}$. Angulo *et al.* (1999) provide yet another estimate:

$$
N_A \langle \sigma v \rangle = \frac{1 + 4.68\, T_8 - 2.92\, T_8^2 + 0.738\, T_8^3}{T_8^2} \exp\left(38.17 - \frac{69.66}{T_8^{1/3}} \right)
$$

$$
+ T_8^{0.45} \exp\left(15.915 - \frac{130.3}{T_8} \right). \tag{16.5.8}
$$

Fortunately, the leading term in all four analytic approximations is nearly the same and it is this leading term that is the one which dominates in stellar models at the onset of ^{13}C burning.

The $^{22}\text{Ne}(\alpha, n)^{25}\text{Mg}$ reaction is endothermic, with a threshold energy of $478\,\text{keV}$. Only upper limits exist for the cross section for center of mass energies $E < 690\,\text{keV}$, but there is a strong resonance centered at $E = 703 \pm 3\,\text{keV}$, where the cross section reaches $\sim 5.5 \times 10^{-8}$ barn, dropping to $\sim 6 \times 10^{-10}$ barn at distances of $\sim \pm 15\,\text{keV}$ from the peak in the cross section (Drotleff *et al.*, 1993).

There are a total of 31 resonances in the energy range 700–1900 keV and properties for the first few are given in Table 16.5.1. The energy in the center of mass system is given in the first column of the table and an estimate of the energy width of the resonance is given in the second column. The quantity ω_γ in the third column of the table is called the "strength" of the resonance and it is defined by the equivalent of the square bracketed term in eq. (16.1.27), except that it is in units of $10^{-3}\,\text{eV}$ instead of eV. That is

$$
\omega_\gamma = \omega_{1,2} \frac{\Gamma_1 \Gamma_2}{\Gamma}. \tag{16.5.9}
$$

Table 16.5.1 Properties of resonances in the ^{22}Ne$(\alpha,n)^{25}$Mg reaction		
energy (keV)	Γ (keV)	ω_γ (10^{-3} eV)
703 ± 3	< 3	0.18 ± 0.03
835 ± 10	30 ± 15	0.16 ± 0.1
910 ± 10	9 ± 5	1.9 ± 0.5
1017 ± 4	19 ± 3	10.6 ± 1.5
1142 ± 10	65 ± 20	55 ± 20
$1171 + \pm 5$	25 ± 10	23 ± 15
1213 ± 3	< 3	1105 ± 120

It is interesting to examine how the strength and width of a resonance can be derived from experimental results. An integration of the cross section over a resonance gives

$$\int_{\text{res}} \sigma(E)\, dE = \int \pi \lambdabar^2 \frac{\omega_{1,2}\, \Gamma_1 \Gamma_2}{(E - E_{\text{res}})^2 + (\Gamma/2)^2} dE$$

$$\sim \pi\, \lambdabar_{\text{res}}^2 \left(\omega_{1,2}\Gamma_1\Gamma_2\right) \frac{2\pi}{\Gamma} = 2\pi^2 \lambdabar_{\text{res}}^2 \left(\omega_{1,2} \frac{\Gamma_1\Gamma_2}{\Gamma}\right) = 2\pi^2 \lambdabar_{\text{res}}^2\, \omega_\gamma. \tag{16.5.10}$$

The integral at the far left of eq. (16.5.10) can be performed numerically using experimental cross section data and, using

$$\lambdabar_{\text{res}}^2 = \frac{1}{A_{1,2}} \frac{M_{\text{H}}c^2}{2E_{\text{res}}} \left(\frac{\hbar}{M_{\text{H}}c}\right)^2 = \frac{0.415 \text{ barn}}{A_{1,2}} \left(\frac{500 \text{ keV}}{E_{\text{res}}}\right) \tag{16.5.11}$$

on the far right of eq. (16.5.10), the equation can be solved for ω_γ.

At the resonance energy, the cross section is related to the strength and width by

$$\sigma(E_{\text{res}}) = \pi\lambdabar_{\text{res}}^2 \left(\omega_{1,2}\Gamma_1\Gamma_2\right) \frac{4}{\Gamma^2} = \pi\lambdabar_{\text{res}}^2\, \omega_\gamma \frac{4}{\Gamma}. \tag{16.5.12}$$

Thus, combining eqs. (16.5.10) and (16.5.12), one finds

$$\Gamma = \frac{2}{\pi} \frac{\int_{\text{res}} \sigma(E)\, dE}{\sigma(E_{\text{res}})} = \frac{4\pi\, \lambdabar_{\text{res}}^2\, \omega_\gamma}{\sigma(E_{\text{res}})}. \tag{16.5.13}$$

In the case of the resonance at 703 keV, $\lambdabar_{\text{res}}^2 = 0.0872$ barn and, using $\omega_\gamma = 0.18 \times 10^{-3}$ eV from Table 16.5.1, one obtains

$$\Gamma = \frac{4\pi \times 0.0872 \text{ barn} \times 0.18 \times 10^{-3} \text{ eV}}{5.5 \times 10^{-8} \text{ barn}} = 3.6 \text{ keV} \tag{16.5.14}$$

compared with $\Gamma < 3$ keV given in the table.

Using eq. (16.1.27), the direct contribution from the first resonance to the reaction rate is

$$N_A \langle \sigma \, v \rangle_1 = \frac{1}{T_8^{3/2}} \exp\left(4.93 - \frac{81.58}{T_8}\right). \tag{16.5.15}$$

At $T_8 = 3$, this expression gives $N_A \langle \sigma \, v \rangle = \exp(-23.91)$.

Since there is no experimental data below the resonance and since there can be no contribution from the resonance wing below the threshold at 478 keV, the estimation of contributions from the 478–700 keV interval is extremely problematic. The formal Gamow peak occurs at

$$E_0 = 291 \, T_8^{2/3} \text{ keV}, \tag{16.5.16}$$

and this does not exceed the threshold energy unless $T_8 > 2.1$ and does not reach the energy at which data exist until $T_8 \sim 3.65$. One can form a vague idea of the contribution to the reaction rate from energies in the wings of the first resonance by assuming that the center of mass cross section factor $S(E)$ in the low energy tail is related to the center of mass cross section factor $S(E_{\text{res}})$ at resonance peak by

$$S(E) = S(E_{\text{res}}) \frac{(\Gamma/2)^2}{(E - E_{\text{res}})^2 + (\Gamma/2)^2}. \tag{16.5.17}$$

At the peak, $2\pi\eta = 51.69 \, (500/703)^{1/2} = 43.59$, so the center of mass cross section factor is

$$S(E_{\text{res}}) = E_{\text{res}} \, \sigma(E_{\text{res}}) \, e^{2\pi\eta(E_{\text{res}})}$$

$$= 703 \text{ keV} \times 5.5 \times 10^{-8} \text{ barn} \times e^{43.59} = 3.3 \times 10^{14} \text{ keV barn}, \tag{16.5.18}$$

which is huge. Eventually, one arrives at the estimate

$$N_A \langle \sigma \, v \rangle_{\text{wing}} \sim \frac{3.24}{\left(1 - 0.414 \, T_8^{2/3}\right)^2} \frac{1}{T_8^{2/3}} \exp\left(39.3 - \frac{101.2}{T_8^{1/3}}\right), \tag{16.5.19}$$

which, at $T_8 = 3$, yields $N_A \langle \sigma \, v \rangle_{\text{wing}} \exp(-27.4)$.

The Cauglan and Fowler (1988) estimate, evaluated at temperatures in the neighborhood of $T_8 = 3$, is

$$N_A \langle \sigma \, v \rangle = \frac{1}{T_8^{2/3}} \exp\left(46.71 - \frac{101}{T_8^{1/3}} - \left(\frac{2.00}{T_8}\right)^{4.82}\right) + \exp\left(-8.846 - \frac{55.77}{T_8}\right). \tag{16.5.20}$$

At $T_8 = 3$, this gives $N_A \langle \sigma \ v \rangle = \exp(-24.19) + \exp(-27.44)$. Angulo *et al.* (1999) estimate, for $T_9 < 2$,

$$N_A \langle \sigma \ v \rangle = \exp\left(2.001 - \frac{77.9}{T_8}\right) + T_8^{0.83} \exp\left(-10.86 - \frac{55.2}{T_8}\right)$$

$$+ T_8^{2.78} \exp\left(2.748 - \frac{117}{T_8}\right) + T_8^{0.892} \exp\left(13.912 - \frac{244}{T_8}\right). \quad (16.5.21)$$

At $T_8 = 3$, this gives $N_A \langle \sigma \ v \rangle = \exp(-23.97) + \exp(-28.35) + \exp(-33.20) + \exp(-94.27)$, almost the same as the Caughlan and Fowler estimate and almost the same as the estimate given by eq. (16.5.15).

16.6 On the contribution of the $^7\text{Li}(p,\gamma)^8\text{Be}$ reaction to the production of carbon in metal-free stars

The information in this section is not made use of in subsequent discussions, but is included for the sake of completeness. In first generation stars, because of the availability of protons and ^3He, the ^8Be compound nucleus that interacts with alpha particles to make the ^{12}C compound nucleus is produced by two channels other than the collision of two alpha particles. The reactions $p(p, e^+ \nu_e)d(p, \gamma)^3\text{He}(\alpha, \gamma)^7\text{Be}$ are followed by the reactions $^7\text{Be}(p, \gamma)^8\text{B}(e^+ \nu_e)^8\text{Be}$ and the reactions $^7\text{Be}(e^-, \nu_e)^7\text{Li} \ (p, \gamma)^8\text{Be}^{*****}$. $^8\text{Be}^{*****}$ can beta decay to the first excited state of ^8Be at an excitation energy of ~ 2.9 MeV and this state decays directly into two alpha particles, but, some of the time, this state decays by gamma emission to the ground state of ^8Be, thus contributing to the ^8Be that can interact with an alpha particle to form the ^{12}C compound nucleus.

Bibliography and references

C. Angulo, M. Arnould, M. Rayet, *et al.*, *Nucl. Phys. A*, **656**, 3, 1999.

Sam M. Austin, arXiv:astro-ph/0411783v1 30Nov, 2004.

Hans Bethe, *Phys. Rev.*, **55**, 434, 1939.

Niels Bohr, *Nature*, **137**, 344, 1936.

Gregory Breit & Eugene Wigner, *Phys. Rev.*, **49**, 519, 1936.

Alastair G. W. Cameron, *ApJ*, **121**, 144, 1955; **130**, 916, 1959.

Alastair G. W. Cameron, *AJ*, **65**, 485, 1960.

Georgeanne R. Caughlan & William A. Fowler, *Atom. Data Nucl. Data Tables*, 1988.

C. W. Cook, William A. Fowler, Charles C. Lauritsen, & Thomas Lauritsen, *Phys. Rev.*, **107**, 508, 1957.

H. W. Drotleff, A. Denker, H. Knee, *et al.*, *ApJ*, **414**, 735, 1993.

I. Fushiki & D. Q. Lamb, *ApJ*, **317**, 368, 1987.

Fred Hoyle, *ApJS*, **1**, 121, 1954.

Fred Hoyle and Martin Schwarzschild, *ApJ*, **138**, 239, 1955.

Ken'ichi Nomoto, F.-K. Thielmann, & S. Miyaji, *A&A*, **149**, 239, 1985.

Ernst Julius Öpik, *Proc. Roy. Irish Acad.*, **A54**, 49, 1951.

J. Orear, A. H. Rosenfeld, & R. A. Schluter, *Nuclear Physics by Enrico Fermi*, (Chicago: University of Chicago Press), 1949.

Edwin E. Salpeter, *Phys. Rev.*, **88**, 547, 1952a.

Edwin E. Salpeter, *ApJ*, **115**, 326, 1952b.

Edwin E. Salpeter, *Phys. Rev.*, **107**, 516, 1957.

PART V

EVOLUTION DURING HELIUM-BURNING PHASES

17 Evolution of a low mass model burning helium and hydrogen

The evolution of a 1 $M_\odot$ population I model star of initial composition $(Z, Y) = (0.015,$ 0.275), begun in Volume 1 (Section 11.1) and carried there to the ignition of helium on the red giant branch, is continued in this chapter through four distinct helium- and hydrogen-burning phases. In Section 17.1, evolution is followed from the off-center ignition of helium at the tip of the red giant branch through a series of helium shell flashes which lift electron degeneracy in shells successively closer to the center of the hydrogen-exhausted core. Once helium burning reaches the center, shell flashes of this sort no longer occur. As described in Section 17.2, the model metamorphoses into a horizontal branch star, converting helium quiescently into carbon and oxygen at the base of a convective core which grows in mass, while hydrogen burning continues to convert hydrogen quiescently into helium in a shell outside of the core.

Once helium is exhausted at the center, the model continues to burn helium quiescently in a shell. The helium exhausted core contracts until electrons in the core become degenerate, converting the core into a hot white dwarf composed of carbon and oxygen. The envelope of the model expands to giant dimensions, the strength of the hydrogen-burning shell at the base of the envelope declines significantly, and the surface luminosity is provided primarily by a helium-burning shell which increases in strength as the model climbs upward in the HR diagram. The structure of the model becomes very similar to that of a first RGB model except that, instead of having an electron-degenerate core composed of helium and having a hydrogen-burning shell as the primary source of surface luminosity, it has an electron-degenerate core composed of carbon and oxygen and a helium-burning shell as the primary source of surface luminosity. To distinguish it from models in the following TPAGB phase, during which helium and hydrogen burning alternately as primary sources of surface luminosity, the model is said to be an early AGB (EAGB) star. Evolution through the EAGB phase is followed in Section 17.3. At the end of this phase, the helium-burning luminosity passes through a relative minimum and then accelerates to a relative maximum as the model experiences a first thermal pulse.

In Section 17.4 the systematics of the thermally pulsing asymptotic giant branch (TPAGB) phase are explored. During this phase, shell hydrogen burning, gravothermal energy generation, and shell helium burning alternate semiperiodically as the primary source of surface luminosity. With each successive thermal pulse episode, the strength of the helium flash which initiates the episode increases, with ramifications for mixing outward into surface layers products of nucleosynthesis in the helium-burning region.

In Section 17.5, a summary is given of how nuclear burning, convective mixing, and gravothermal activity influence the major composition changes that occur in a low mass TPAGB model, and an outline is presented of how, once thermal pulses become violent enough, a TPAGB star ultimately contributes to the enrichment of the interstellar medium in carbon and s-process elements.

17.1 Helium shell flashes during evolution from the red giant branch to the horizontal branch

The initial model is the most evolved model discussed in Section 11.1 in Volume 1. Several structure variables in this model are shown in Fig. 17.1.1. Also shown in the figure are the abundances of CNO isotopes with equal numbers of protons and neutrons. The discontinuities in the composition variables occur at convective zone boundaries which are marked by thin vertical lines. The labels BCS and ECS denote, respectively, the base and outer edge of the convective shell sustained by fluxes from helium burning and the label BCE denotes the base of the convective envelope for which the large fluxes generated by the hydrogen-burning shell and large envelope opacities are primarily responsible.

The luminosity and temperature profiles through the hydrogen-burning shell centered at $M \sim 0.4689 \, M_\odot$ are too narrow to be resolved in the figure, but this region has been treated exhaustively in Section 11.1. Suffice it to note that the luminosity increases outward precipitously, the temperature decreases outward precipitously, and composition variables are effectively discontinuous along a common, essentially vertical, thick solid line just to

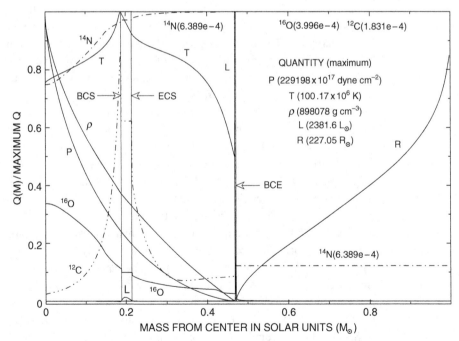

Fig. 17.1.1 Structure variables and CNO abundances in a 1 $M_\odot$ model beginning shell helium burning as a luminous red giant ($Z = 0.015, Y = 0.275, \Delta t = -818.36 \, \text{yr}$)

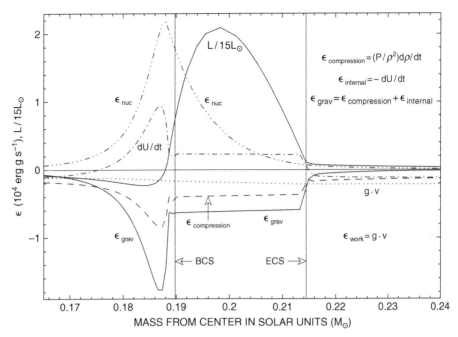

Fig. 17.1.2 Gravothermal and nuclear energy-generation rates in a 1 $M_\odot$ model beginning shell helium burning as a luminous red giant ($Z = 0.015$, $Y = 0.275$, $\Delta t = -818.36$ yr)

the left of the BCE. Although helium burns at a rate fast enough to engender a convective shell, essentially all of the energy produced in the shell is gravothermally absorbed in the shell, as demonstrated by the fact that the tiny portion of the luminosity profile within the convective shell vanishes at the outer boundary of the shell and nearly vanishes at the base of the shell. Over most of the region between the BCS and the center, temperature decreases inward. The luminosity in this region is therefore negative, with energy being carried inward by photon diffusion and electron conduction to compensate for the energy carried directly out of the model by neutrinos generated by the plasma-neutrino process discussed in Section 15.8.

The CNO abundance profiles in the inner third of the initial model are a consequence of the operation of helium-burning reactions. The ^{12}C abundance reflects the operation of the triple-alpha process described in Section 16.1, the ^{16}O abundance reflects the operation of the ^{12}C$(\alpha, \gamma)^{16}$O reaction described in Section 16.4 and the ^{14}N abundance reflects the operation of the ^{14}N$(\alpha, \gamma)^{18}$O$(\alpha, \gamma)^{22}$Ne reactions, also described in Section 16.4.

Gravothermal and nuclear energy-generation rates centered on the location of the convective shell are shown in Fig. 17.1.2. Comparison between the profile of ϵ_{nuc} in Fig. 17.1.2 and the temperature profile in Fig. 17.1.1 demonstrates that ϵ_{nuc} increases rapidly with temperature. From the profiles of $\epsilon_{compression} = (P/\rho^2)\,(d\rho/dt)$ and $\epsilon_{internal} = -dU/dt$ in Fig. 17.1.2, it follows that everywhere in the region where ϵ_{nuc} is large, the density is decreasing and the internal energy U is increasing. Since density is decreasing, the contribution of electron degeneracy to U is decreasing. Hence, the increase with time of U in the region where ϵ_{nuc} is large means that temperatures there are also increasing

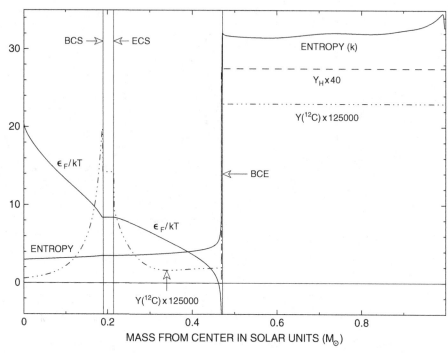

Fig. 17.1.3 Entropy, ϵ_F/kT, and ^{12}C and ^{1}H abundances in a 1 $M_\odot$ model beginning shell helium burning as a luminous red giant ($Z = 0.015$, $Y = 0.275$, $\Delta t = -818.36$ yr)

and that, therefore, the helium-burning luminosity L_{He}, which is the integral over mass of ϵ_{nuc}, is increasing with time. The time increment $\Delta t = -818.36$ yr in the caption of Figs. 17.1.1–17.1.2 gives the time prior to when the peak helium-burning luminosity is achieved.

Since energy flows both inward and outward from the position where $L = 0$, it is appropriate to call this position the luminosity watershed. It coincides with the position of maximum temperature in Fig. 17.1.1 and also with the maximum in the nuclear energy-generation rate ϵ_{nuc} in Fig. 17.1.2. Almost all of the nuclear energy produced to the right of the luminosity watershed is absorbed in the convective shell where matter is expanding outward and heating. Nuclear energy produced to the left of the luminosity watershed goes into heating and expanding matter between the watershed and the center over a mass range similar to the mass range defined by the convective shell.

Two quantities which are useful in describing the evolution of the model during the helium-burning phase are the entropy S per nucleon and the degree of electron degeneracy as measured by ϵ_F/kT. Profiles of these quantities in the initial model are shown in Fig. 17.1.3. In the hydrogen-exhausted core, entropy and the degeneracy indicator are inversely related. Over all but a few hundredths of a solar mass below the hydrogen-burning shell, electrons are significantly degenerate $\left(1\overset{\sim}{<}\epsilon_F/kT\overset{\sim}{<}20\right)$. In the convective shell, both entropy and the electron-degeneracy indicator are spatially constant. Over most of the convective envelope, departures from constancy in the entropy are due to the fact that the degree of ionization of all isotopic species decreases outward; in the very outer portion of

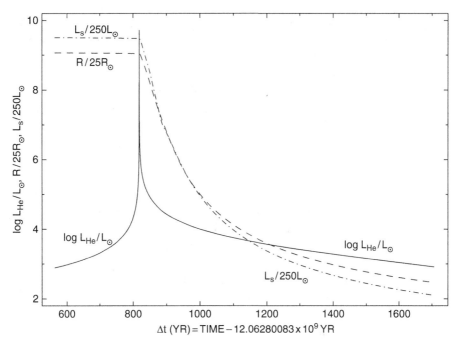

Fig. 17.1.4 Helium-burning and surface luminosities and radius of a 1 $M_\odot$ model during the helium shell flash ending luminous red giant branch evolution ($Z = 0.015, Y = 0.275$)

the envelope, some of the responsibility for the variation in the entropy is with a superadiabatic temperature gradient.

The time variation of L_{He} is shown in Fig. 17.1.4, where time is recorded in years from the initial model. At peak energy-production rate, helium burning releases energy at a rate comparable with the luminosity of a small galaxy. The time scale for the variation of L_{He} near maximum is very short compared with the overall time scale for the helium-burning episode. Whereas the time scale for the overall build-up and decay of the episode is measured in hundreds of years, the time scale for the build-up and decay near maximum L_{He} is measured in days. For example, the time for L_{He} to increase to $L_{He}^{max} = 5.21 \times 10^9\ L_\odot$ from $L_{He} = (1/10)\ L_{He}^{max}$ is 3.61 days and the time to decrease from L_{He}^{max} to $(1/10)\ L_{He}^{max}$ is 7.64 days. Luminosity doubling from $L_{max}/2$ to L_{He}^{max} occurs in 1.19 days and luminosity halving from the maximum takes 1.46 days. Near peak luminosity, time steps required to follow evolution properly are measured in minutes. The phrase helium-burning flash is commonly used to describe the entire helium-burning episode, even though the phrase is strictly appropriate only for evolution near peak helium-burning luminosity.

Also shown in Fig. 17.1.4 are the time variations of surface luminosity L_s and surface radius R. Over all but the very last portion of the 818.36 years separating the initial model from the model at flash peak, the surface radius decreases monotonically by about 1 $R_\odot$, from ~227.1 $R_\odot$ in the initial model to ~226.1 $R_\odot$ at flash peak. Not resolvable in the figure is the fact that, over a period lasting from several days prior to flash peak to about a month after flash peak, matter in the model envelope is accelerated outward and surface radius increases by ~0.05 $R_\odot$ before again shrinking monotonically, but on a much longer time scale than during the approach to flash peak. Over a period of 900 years, the radius decreases by a factor of $\sim$3.6 to $\sim$62.5 $R_\odot$. Over the same period, surface luminosity decreases by a factor of $\sim$4.8 to $L_s \sim 500\ L_\odot$. The decrease in L_s is due to the fact that, at and beyond the helium-hydrogen interface, the outward acceleration of matter produces expansion and cooling to the extent that shell hydrogen burning is effectively extinguished. The radius shrinks because the only energy source available for supplying the demand for an energy flux at the surface is gravitational potential energy released by a contracting envelope. Thus, the first helium shell flash forces the model to evolve in the HR diagram downward along the red giant branch to a position far below the most luminous position achieved on the branch.

The distributions in the HR diagram defined by stars in globular clusters in our Galaxy offer strong support for this theoretical result. That is, in all clusters there is a well defined red giant branch and the luminosity of the most luminous star on the branch does not vary much from one cluster to the next. This most luminous star defines what is called the tip of the red giant branch. The existence of a tip and the fact that the luminosity at the tip is basically the same in all globular clusters is very convincing evidence that a helium shell flash is responsible for the termination of upward evolution along the red giant branch.

Several state and composition variables in the model approximately 12.694 hours after flash peak, when the helium-burning luminosity has declined to $L_{He} = 3.05 \times 10^9\ L_\odot$, are shown in Fig. 17.1.5. As in the initial model, the discontinuities in composition variables occur at the boundaries of convective zones. The abundance of ^{12}C in the convective shell of the more evolved model is an order of magnitude larger than the abundance of ^{12}C in the convective shell in the initial model, indicating that the triple-alpha reactions ^{4}He(^{4}He, γ)^{8}Be(^{4}He, γ)^{12}C have been active. On the other hand, the ^{14}N and ^{16}O abundance distributions in the more evolved model do not differ significantly from those in the initial model, indicating that the ^{12}C(α, γ)^{16}O and ^{14}N(α, γ)^{18}O reactions operate at much slower rates.

As in the initial model, there is a region of negative luminosity which extends inward from just below the BCS, but it is not resolvable on the scale in Fig. 17.1.5. Also, as in the initial model, essentially all of the energy produced by helium burning to the right of the watershed in luminosity, where $L = 0$, is absorbed in the convective shell. The difference in maximum energy-generation rates by helium burning in the two models is enormous, being $\epsilon_{He}^{max} \sim 2.18 \times 10^4$ erg g^{-1} s^{-1} in the initial model and $\epsilon_{He}^{max} \sim 6.65 \times 10^{11}$ erg g^{-1} s^{-1} in the model near flash peak.

Comparison of the temperature profile in Fig. 17.1.5 with the temperature profile in Fig. 17.1.1 shows that the maximimum temperature in the more evolved model is approximately double the maximum temperature in the initial model, whereas in the region interior to the location of the temperature maximum, the mean temperature, at $T_{mean} \sim 53 \times 10^6$ K, is significantly smaller than $T_{mean} \sim 85 \times 10^6$ K in the initial model. In the region below the base of the convective shell, the temperature profile is flatter in the more evolved model than in the initial model and the mean density is much smaller, with the density at the center of the more evolved model being only about half the density at the center of the initial model. In the model near flash peak, both density and temperature exhibit essentially

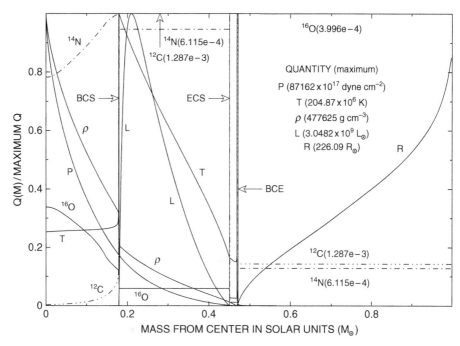

Fig. 17.1.5 Structure variables and CNO abundances in a 1 $M_\odot$ model 12.694 HR after the peak of the first helium shell flash ($Z = 0.015, Y = 0.275$)

discontinuous jumps across the base of the convective shell, with temperature increasing from $\sim 57 \times 10^6$ K to a maximimum of $T_{max} = 204.9 \times 10^6$ K and density decreasing from $\sim 153\,000$ g cm^{-3} to $\sim 95\,500$ g cm^{-3}. Thus, in the region directly affected by the helium-burning flash, the release of nuclear energy has led to heating and expansion, and in the region interior to the flashing zone, the reduction in the pressure exerted by overlying layers has led to expansion and cooling.

Gravothermal energy generation rates in the model shortly after flash peak are shown in Fig. 17.1.6. The rate $\mathbf{g} \cdot \mathbf{v}$ at which gravity does work is negative over the entire region from the center to just below the BCE, demonstrating that, everywhere in this region, matter is flowing outward. Over this same region, the rate $(P/\rho^2)\,(d\rho/dt)$ at which compressional work is being done is also negative, demonstrating that, everywhere in the region, matter is expanding. In the region below the base of the convective shell, $(P/\rho^2)\,(d\rho/dt)$ is indistinguishable from dU/dt, demonstrating that expansion at any point in the electron-degenerate core is for all intents and purposes adiabatic. Actually, since the temperature gradient in this region is finite and negative (see Fig. 17.1.5), the net gravothermal energy-production rate in the region does not vanish precisely; that is, $\epsilon_{grav} = (P/\rho^2)\,(d\rho/dt) - (dU/dt) \gtrsim 0$. Nevertheless, the departure from local adiabaticity during expansion and cooling of the electron-degenerate core is so slight that the entropy at every mass point in the core remains essentially constant. Since ϵ_F/kT is a single valued function of the entropy, this means that the degree of electron degeneracy at any mass point in the core does not change with time.

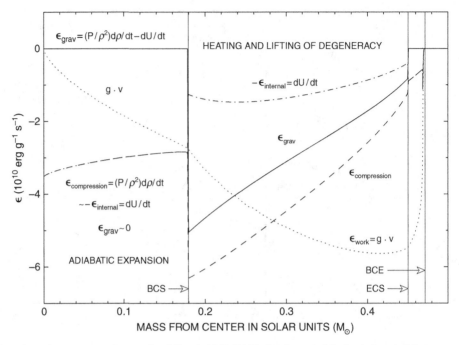

Fig. 17.1.6 Gravothermal energy-generation rates in a 1 $M_\odot$ model 12.694 HR after the peak of the first helium shell flash ($Z = 0.015$, $Y = 0.275$)

In order to understand the significance of gravothermal energy-generation rates in the convective shell, across the base of which helium is burning furiously, it is important to distinguish between the rate of change of temperature and the rate of change of the total internal energy per unit mass. During the build-up to the maximum helium-burning luminosity, and for a short time thereafter, temperatures in the helium-burning region increase monotonically while, thanks to the decrease in densities as well as to the increase in temperatures, the contribution of degenerate electrons to the internal energy decreases. Thus, the fact that dU/dt in the convective shell is negative means not that temperatures are decreasing but, rather, that the energy of degenerate electrons and hence the degree of electron-degeneracy is decreasing. Temperatures in the convective shell attain maximum values approximately a day after the helium-burning luminosity reaches its maximum.

In Fig. 17.1.7, the profile of the electron-degeneracy indicator ϵ_F/kT corroborates the lessons learned from the analysis of the gravothermal energy-generation rates described in Fig. 17.1.6: interior to the flashing zone, electrons remain quite degenerate while, in the convective shell, electron degeneracy is in the process of being lifted. In the electron-degenerate core, the pressure exerted by electrons at any point is $P_e \sim (2/5)\, n_e\, \epsilon_F$, where n_e is the electron number density, and the pressure exerted by alpha particles is $P_\alpha \sim n_e kT/2$. Thus, $P_e/P_\alpha \sim (4/5)\, (\epsilon_F/kT)$. From Fig. 17.1.7, one deduces that, in the region interior to the flashing zone, the contribution of electrons to pressure is from 8 to 16 times larger than the contribution of alpha particles. In contrast, in the flashing zone, not only

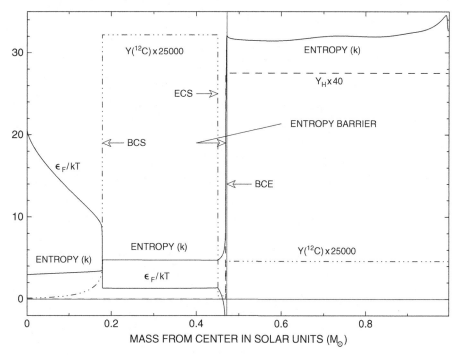

Fig. 17.1.7 Entropy, ϵ_F/kT, and ^{12}C and ^{1}H abundances in a 1 M$_\odot$ model 12.694 HR after the peak of the first helium shell flash ($Z = 0.015, Y = 0.275$)

are the contributions of the two particle species comparable but, because electrons are not strongly degenerate ($\epsilon_F/kT \sim 1.2$), the total pressure is approximately proportional to the temperature. As evolution progresses, further expansion leads to cooling and, as the rate of nuclear energy production decreases, thermal energy contributes to the supply of energy necessary to do the work of expansion.

The only essential difference between the entropy profiles in Figs. 17.1.7 and 17.1.3 is in the convective shell where, in consequence of the decrease in the degree of degeneracy, the entropy has increased. An important characteristic of the entropy distributions is the jump in entropy which coincides with the steepest portion of the hydrogen-abundance profile established during the ascent of the red giant branch. Given the fact that, in first approximation, the entropy in a fully convective zone is spatially constant, the jump may be viewed as a barrier which inhibits mixing between matter in the convective shell, which is maintained by the flux of energy due to helium burning, and hydrogen-rich matter just below the base of the convective envelope. Despite the intense burning that has taken place in the neighborhood of the BCS, the amount by which the entropy per nucleon in the convective shell has increased is only $\sim 2k$, small compared with the $\sim 27k$ entropy jump between the convective shell and the convective envelope.

On the other hand, the entropy per nucleon at a point where the hydrogen abundance is $Y_H \sim 0.1$ is only $\sim 3k$ larger than the entropy in the convective shell. Thus, the potential for contact between the ECS and the base of the hydrogen-rich profile warrants additional

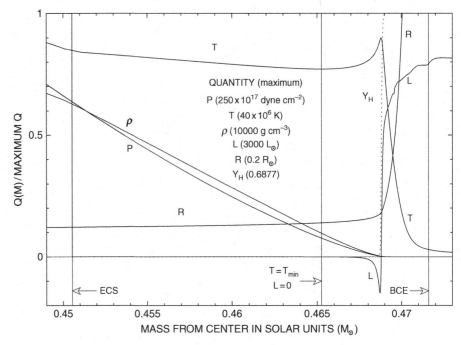

Fig. 17.1.8 Structure variables between the ECS and the BCE in a 1 $M_\odot$ model 12.694 HR after the peak of the first helium shell flash ($Z = 0.015, Y = 0.275$)

scrutiny. Structure variables in the region between the ECS and the BCE are shown in Fig. 17.1.8. Regarding the question of mixing, the most relevant characteristic is the existence of a region between the ECS and the hydrogen profile in which the temperature gradient and therefore the luminosity are negative. Such a region is stable against convection and, if a region of these characteristics persists throughout the helium-flash phase, contact between the ECS and hydrogen-rich material can be excluded.

Given the huge gravothermal energy-generation rates generated in the process of preventing the nuclear energy released by helium-burning reactions from reaching into hydrogen-rich regions, it is of interest to examine the possibility that hydrodynamical motions may be generated in the model. Velocity profiles centered on the region of maximum nuclear and gravothermal energy-production rates for eight models constructed in evolutionary runs of 60 models each are shown in Fig. 17.1.9 for a time period leading up to and following flash peak. In the left hand portion of the figure, beside each curve for the 60th model in a given run, the age of the model relative to the time of maximum L_{He} is shown. Models are labeled in the order of increasing age from $m = 8$ through $m = 15$. In the right hand portion of the figure, the time between successive 60th models is given in seconds and, as a measure of the average acceleration between models, the quantity $(v_m - v_{m-1})/(t_m - t_{m-1})$ is recorded, where m designates the 60th model in the mth run and the estimate is made at a mass $M \sim 0.4705\ M_\odot$ where the maximum velocities along the velocity profiles are located.

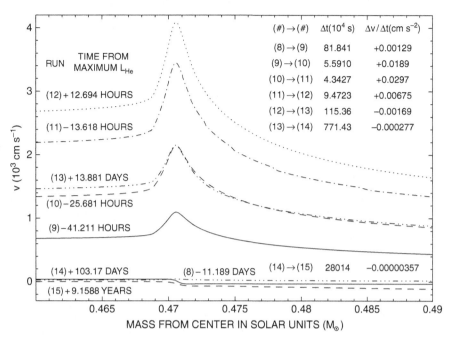

$(\#) \to (\#)$	$\Delta t(10^4\,s)$	$\Delta v/\Delta t(cm\,s^{-2})$
$(8) \to (9)$	81.841	+0.00129
$(9) \to (10)$	5.5910	+0.0189
$(10) \to (11)$	4.3427	+0.0297
$(11) \to (12)$	9.4723	+0.00675
$(12) \to (13)$	115.36	−0.00169
$(13) \to (14)$	771.43	−0.000277
$(14) \to (15)$	28014	−0.00000357

Fig. 17.1.9 Velocity and acceleration history of a 1 $M_\odot$ model near the temporal and spatial peak of the first helium shell flash ($Z = 0.015, Y = 0.275$)

Since the acceleration in the spherically symmetric approximation is rigorously

$$\frac{dv}{dt} = g - \frac{1}{\rho}\frac{dP}{dr}, \qquad (17.1.1)$$

and since the quasistatic approximation assumes the right hand side of this equation to be zero, the relevant comparison between the acceleration found in the quasistatic approximation, e.g., estimated by $\Delta v/\Delta t$ shown in Fig. 17.1.9, and the rigorously correct acceleration is a comparison between $\Delta v/\Delta t$ and g (or $(1/\rho)\,dP/dr$). Fig. 17.1.10 gives the logarithm of g as a function of mass over the same mass range adopted in Fig. 17.1.9. It is evident that, at the positions of largest velocity, the ratio of the quasistatically estimated accelerations to g are, at maximum, of the order of $(\dot{v}/g)_{max} \sim 5 \times 10^{-7}$, where $\dot{v} = \Delta v/\Delta t$. Over the entire mass range shown in Fig. 17.1.9, the separations between chronologically adjacent velocity profiles at a given mass are smaller than the separations at maximum velocity by factors ranging from about 1.5 (at the smallest masses) to less than 3 (at the largest masses). Thus, accelerations estimated in the quasistatic approximation vary only modestly over the entire mass range shown. However, over the same mass range, the gravitational acceleration varies by a factor of $\sim 3 \times 10^5$. Thus, for example, at $M = 0.46\,M_\odot$, $(\dot{v}/g)_{max} \sim 1.5 \times 10^{-9}$, whereas, at $M = 0.49\,M_\odot$, $(\dot{v}/g)_{max} \sim 10^{-4}$. These comparisons suggest that hydrodynamical effects are not very large in the region where nuclear and grovothermal energy-generation rates are the largest and that the results of the quasistatic

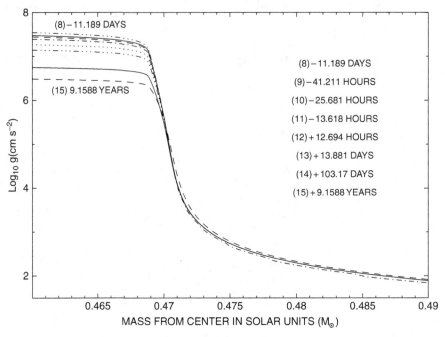

Fig. 17.1.10 Gravitational acceleration history of a 1 $M_\odot$ model near the temporal and spatial peak of the first helium shell flash ($Z = 0.015, Y = 0.275$)

calculations provide a reasonably accurate picture of the development of the helium shell flash in the region of most active helium burning.

On the other hand, the accelerations engendered by the injection of energy into the nuclear active region extend outward to the surface of the model at values similar to those in the region already explored, while the acceleration due to gravity continues to fall off fairly steeply. Velocity profiles and the logarithms of the gravitational acceleration over most of the interior are shown in Figs. 17.1.11 and 17.1.12, respectively, for six of the models described in the previous two figures as well as for three more evolved models, labeled 15, 16, and 17. It is most remarkable that, for models 9–14, the velocities and the estimated quasistatic accelerations near the outer edges of the models are almost identical to the velocities and estimated quasistatic accelerations in the highly nuclearly and gravothermally active region. One infers that the pressure changes generated by the rapid release of nuclear energy near the BCS are communicated outward through the rest of the model star as fast as possible (i.e., at the speed of sound).

At the outer edges of models 8–15, the gravitational acceleration drops to approximately 0.75 cm s^{-2}, compared with a maximum quasistatically estimated acceleration of $\sim$0.03 cm s^{-2} for models within a day's time of flash peak. Thus, $(\dot{v}/g)_{max} \sim 0.04$, suggesting that hydrodynamical effects play a modest role in the outer portions of the models near flash peak. Assuming that, once $(\dot{v}/g)_{max}$ drops below $\sim 10^{-4}$, hydrodynamical effects are unimportant, the phase of mild hydrodynamical activity in envelope regions lasts for of the order of a month.

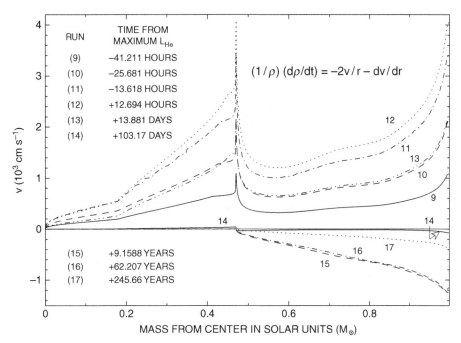

Fig. 17.1.11   Velocity history of a 1 M$_\odot$ model from near the peak into the decay phase of the first helium shell flash ($Z = 0.015$, $Y = 0.275$)

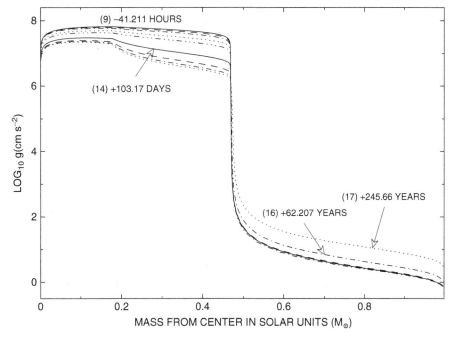

Fig. 17.1.12   Gravitational acceleration history of a 1 M$_\odot$ model from near the peak into the decay phase of the first helium shell flash ($Z = 0.015$, $Y = 0.275$)

The velocity curves in Fig. 17.1.11 carry information regarding expansion and contraction. Writing the continuity equation as

$$\frac{1}{\rho}\frac{d\rho}{dt} = -\frac{2v}{r} - \frac{dv}{dr}, \tag{17.1.2}$$

it is evident by inspection that matter between the centers of all models and the maxima in the velocity curves at $M \sim 0.4705\ M_\odot$ is expanding and that all matter in models 9–13 beyond the minima in the velocity curves at $M \sim 0.575\ M_\odot$ and the surfaces of the models is also expanding. These inferences follow from the fact that, in the stated ranges, both v and dv/dr are positive. In the region in each of models 9–13 where dv/dr is negative, matter is also expanding because, in this region, r is small and $2v/r > -dv/dr$.

Further inspection of the velocity profiles in Fig. 17.1.11 shows that matter in the electron-degenerate core of models 14–17 is essentially static, neither expanding nor contracting, that matter in the outer portion of model 14 is contracting, and that all matter between $M \sim 0.4705$ and the surface in models 14–17 is contracting. In Fig. 17.1.12, the increase in the gravitational acceleration outside of the helium-burning regions in models 16 and 17 corroborates that the hydrogen-rich envelopes of these models are contracting. Expansion and cooling of matter at the interface between the helium-rich interior and the base of the hydrogen-rich envelope has led to the effective elimination of hydrogen burning as a contributor to the surface luminosity, and the liberation of gravitational potential energy by the contracting envelope is responsible for the surface luminosity.

As the model evolves towards flash peak, the energy liberated by helium burning continues to cause temperatures in the burning region to increase. Increasing temperatures not only accelerate the burning rate, they decrease the degree of electron degeneracy and matter in the burning region expands. At some point, the rate of nuclear energy release and the rate of absorption by expanding matter equilibrate and, as densities in the region continue to decrease, temperatures in the convective shell also begin to decrease and the rate of energy production by helium burning drops off with time.

Although the helium-burning luminosity decreases, the continued release of nuclear energy results in a continued increase in the entropy in the convective shell and, for roughly a century following flash peak, the ECS creeps outward in mass, with the entropy at the ECS rising slowly upward along the entropy barrier. Over the years, there has been considerable speculation about whether or not the ECS and the hydrogen profile make contact during the decay phase of the helium shell flash, with overshooting of convection into regions formally stable against convection being invoked as a promising process for establishing contact. In the present study, the outward creep of the ECS terminates before contact with the hydrogen profile is achieved and reasons can be adduced for why contact is not made. The closest approach between the ECS and the hydrogen profile is reached at 103.65 yr following flash peak and amounts to a separation of $M_{10} - M_{\rm ECS} = 0.002\,89\ M_\odot$, where M_{10} is the mass near the base of the hydrogen profile and $M_{90} - M_{10} \sim 0.0008\ M_\odot$ is a measure of the width of the profile.

Conditions in a model in which the ECS is in the process of receding from the helium-hydrogen interface are shown in Figs. 17.1.13–17.1.19. The model is 245.66 yr beyond flash peak but is qualitatively similar to the model 62.207 yr beyond flash peak. The

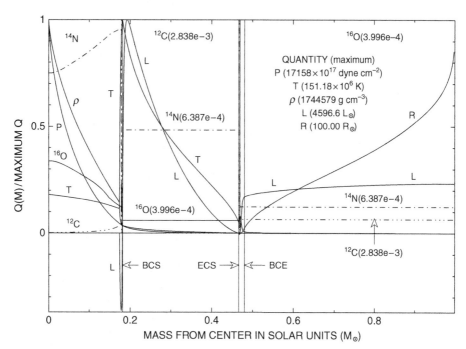

Fig. 17.1.13 Structure variables and CNO abundances in a 1 $M_\odot$ model when the ECS and the H-rich profile nearly touch ($Z = 0.015, Y = 0.275, \Delta t = 245.66$ yr)

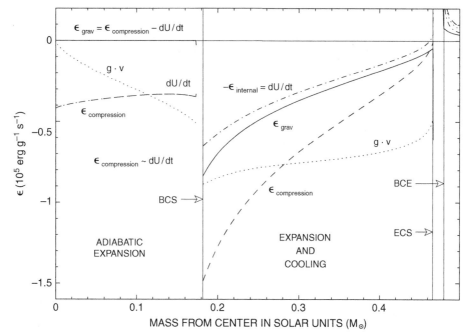

Fig. 17.1.14 Gravothermal energy-generation rates in a 1 $M_\odot$ model 245.66 yr after the peak of the first helium shell flash ($Z = 0.015, Y = 0.275$)

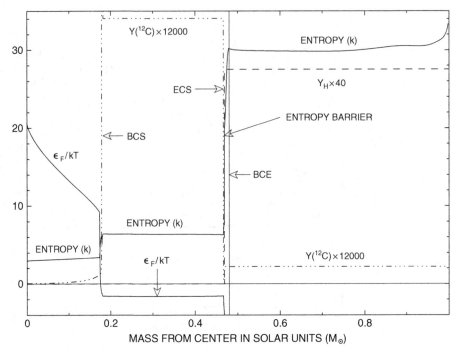

Fig. 17.1.15 Entropy, ϵ_F/kT, and ^{12}C and ^1H abundances in a 1 $M_\odot$ model when the ECS and the H-rich profile nearly touch ($Z = 0.015$, $Y = 0.275$, $\Delta t = 245.66$ yr)

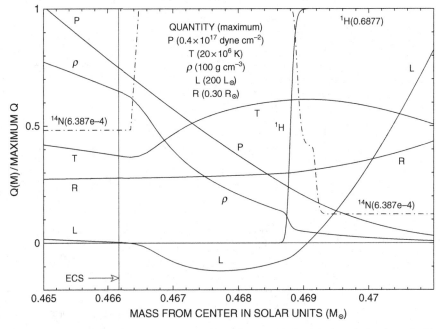

Fig. 17.1.16 Conditions in a region containing the ECS and the hydrogen profile in a 1 $M_\odot$ model 245.66 yr after the peak of the first helium shell flash ($Z = 0.015$, $Y = 0.275$)

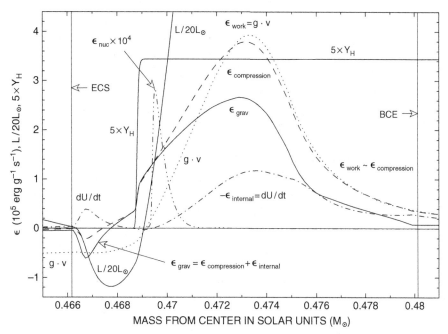

Fig. 17.1.17 Gravothermal energy-generation rates between the ECS and the BCE in a 1 $M_\odot$ model 245.66 yr after the peak of the first helium shell flash ($Z = 0.015$, $Y = 0.275$)

acceleration of matter in the outer portion of the model may be deduced from the curves labeled 17 and 16 in Fig. 17.1.11 and the time difference $\Delta t = 183.45$ yr between models 17 and 16. The acceleration due to gravity in model 17 is described by the curve labeled by the same number in Fig. 17.1.12. Figures 17.1.13–17.1.15 display for the model $\sim$246 yr beyond flash peak the same characteristics that are shown in Figs. 17.1.5–17.1.7 for the model 12.694 hr beyond flash peak. Figures 17.1.16–17.1.17 provide details near the BCS in the more evolved model and Figs. 17.1.18–17.1.19 provide details between the ECS and the BCE in the more evolved model.

In Fig. 17.1.13, which describes structure variables and CNO abundances throughout the model, many curves crowd together in the neighborhoods of both the BCS and the ECS. Although it is difficult to distinguish one curve from another with the unaided eye, disentanglement of curves is possible with the help of a magnifying glass. An example of the perspective provided by crowded curves which are nevertheless resolvable with a suitable viewing aid is that the ECS is at a mass which is ever so slightly smaller than its location at maximumum mass defined by the vertical portion of the profile of the ^{12}C abundance. From Fig. 17.1.15 one can determine that the ^{12}C and ^{1}H abundance profiles in the neighborhood of the ECS are actually separated, but by an intriguingly small amount.

Comparing variables in Figs. 17.1.13 and 17.1.5, it is evident that, over the roughly two and one half centuries which have elapsed between the two models, the density at the center has decreased by almost a factor of 3 and the temperature has decreased by a factor of about 2. And yet, as anticipated by the discussion of Fig. 17.1.6 and as demonstrated by a comparison between the profiles of ϵ_F/kT in Figs. 17.1.15 and 17.1.7, the degree of electron degeneracy in the electron-degenerate core remains essentially unchanged as a function of mass. In the electron-degenerate core, which extends from the center to just below the BCS where the degree of electron degeneracy plummets downward and the entropy jumps upward, the temperature gradient has switched signs, reflecting the fact that energy loss due to escaping plasma neutrinos is no longer an important factor in the core.

Both models have in common that none of the energy produced by helium burning reaches the helium–hydrogen interface. A major difference between the two models is the source of surface luminosity. As shown in Fig. 17.1.8, in the model near flash peak, hydrogen burning is the primary source, producing a surface luminosity $L_s \sim 2460 \, L_\odot$. Applying a magnifying glass to the region between the ECS and the BCE in Fig. 17.1.13, it is evident that the steepest portion of the luminosity profile in this region is well beyond the steepest portion of the hydrogen-abundance profile. So, the increase in luminosity outward must be due to gravothermal energy released above the helium–hydrogen interface. The release of gravothermal energy is, of course, also responsible for the increase in luminosity through the convective envelope. Thus, the surface luminosity of the 246 year old model, $L_s \sim 1060 \, L_\odot$, instead of being due to the release of nuclear energy, either by hydrogen burning across the helium–hydrogen interface or by helium burning in the deep interior, is due to the release of gravitational potential energy by the contracting hydrogen-rich envelope.

At the BCS, the temperature in the model 246 yr past flash peak is smaller by a factor of ~ 1.35 than in the model near flash peak and the density is smaller by a factor of about 2. In the convective shell, triple-alpha burning has approximately doubled the abundance of ^{12}C (compare Figs. 17.1.5 and 17.1.13), the entropy has increased by about 40% (compare Figs. 17.1.7 and 17.1.15), and, as evidenced by the fact that ϵ_F / kT between the curves labeled BCS and ECS in Fig. 17.1.15 is negative by a substantial amount, electrons in the convective shell have become thoroughly non-degenerate.

In the model 246 yr past flash peak, in a region of mass $\sim 0.01 \, M_\odot$ below the BCS and in the region between the ECS and the BCE, variations in gravothermal energy-generation rates are sufficiently complicated that an initially more instructive comparison with the gravothermal energy-generation rates in the model 12.694 yr after flash peak is achieved by omitting the variations in the identified regions in the more evolved model. The omissions account for the gaps in the curves for rates in Fig. 17.1.14. Comparison with the gravothermal energy-generation rates in Fig. 17.1.6 shows that, although the rates differ in magnitude by factors of the order of $\sim 10^5$, the variations with mass of the rates in the two models are qualitatively similar in most of the electron-degenerate core and in the convective shell. In both models, matter in the electron-degenerate core is locally expanding and cooling adiabatically and matter in the convective shell is expanding. However, instead of heating, as in the model near flash peak, matter in the convective shell of the more evolved model is cooling. That is, since electrons in the convective shell of the more evolved model are not degenerate, the internal energy U is closely proportional to the temperature, and the fact that $dU/dt < 0$ means that temperatures are decreasing.

Applying a magnifying glass to examine structure variables in the region between the ECS and the helium-hydrogen interface in the model 246 yr beyond flash peak reveals in Fig. 17.1.16 that, independent of the entropy-based argument against mixing between matter in the convective shell and hydrogen-rich matter, mixing is prevented by the existence of a region of negative luminosity which extends from the ECS to beyond the helium–hydrogen interface. The maximum outward extent of the ECS during prior evolution is marked in Fig. 17.1.16 by the discontinuity in the ^{14}N abundance profile at $M \sim 0.4664$ $M_\odot$. The fact that the ECS in the model 246 yr beyond flash peak is at the smaller mass of $M \sim 0.4662$ $M_\odot$ shows that the ECS is in the process of receding in mass. The coup de grâce for the mixing hypothesis is the fact that luminosity nearly vanishes at the ECS and, from slightly beyond the ECS to slightly beyond the variable portion of the hydrogen-abundance profile centered at $M \sim 0.4687$ $M_\odot$, the temperature gradient is of opposite sign to the pressure gradient and the luminosity is negative. Since the flux of energy is nearly zero at the ECS, convective velocities there are formally very small and, since the radiative temperature gradient and the adiabatic temperature gradient are of opposite sign in the region of negative luminosity, convection currents cannot arise there.

In all models participating in the most nuclear active part of the helium-flash phase, a region of negative luminosity separates the ECS from the helium–hydrogen interface, preventing contact between the ECS and the interface. In contrast with the situation during the ascent of the red giant branch prior to the onset of helium burning, when gravothermal energy liberated by the contracting core maintains a positive luminosity at the base of the helium–hydrogen interface, the expansion induced by helium burning modifies conditions below the interface in such a way that some of the energy produced by whatever means in the neighborhood of the interface flows inward.

As illustrated in Fig. 17.1.8, while CN-cycle burning is still important locally relative to the rate of gravothermal energy generation, the region in which $L < 0$ extends inward from slightly above the base of the helium–hydrogen interface to considerably below this interface. The region within which the flow of energy increases inward, i.e., where $L < 0$ and $dL/dM > 0$, more or less coincides with the base of the hydrogen profile. Thus, the source of energy flowing inward is nuclear energy released at the base of the hydrogen-burning shell.

In the model 246 yr beyond flash peak, the source of inflowing energy is entirely gravothermal, as may be seen by inspection of the curves in Fig. 17.1.17. The region within which $L < 0$ and $dL/dM > 0$ extends approximately from where $\mathbf{g} \cdot \mathbf{v} = 0$ at $M \sim 0.469$ $M_\odot$ to where ϵ_{grav} vanishes at $M \sim 0.46775$ $M_\odot$. This is the region within which contributions to the inward flow of energy occur. The steepest portion of the hydrogen-abundance profile lies within the region and the fact that the profile of the nuclear energy-generation rate lies outside of the region demonstrates both that burning by CN-cycle reactions is defunct and that nuclear energy does not contribute to the inward flow of energy, aside from the fact that, at maximum, ϵ_{nuc} is four orders of magnitude smaller than typical values of ϵ_{grav}. Thus, the source of energy flowing inward is compressional energy released between the stationary point where $\mathbf{g} \cdot \mathbf{v} = 0$ and the point where ϵ_{grav} first vanishes. The absorption that occurs in the remainder of the negative luminosity region which extends approximately to the ECS is due primarily to heating and secondarily to the work of expansion.

It is ironic that, in spite of the nuclear burning pyrotechnics near the BCS, the surface luminosity in the model 246 yr beyond flash peak is essentially entirely due to the liberation of compressional energy by matter beyond the helium–hydrogen interface. Roughly 75%

of the surface luminosity is due to compressional energy liberated in the region extending to the BCE from the near discontinuity in the hydrogen profile centered at $M \sim 0.4687\ M_\odot$, the other 25% being due to compressional energy liberated in the convective envelope. Interestingly, in contrast with the usual situation, over most of this region, the rate $\mathbf{g} \cdot \mathbf{v}$ at which gravity does work is locally very similar to the rate at which compressional work is being done. Over the outer three quarters of the region, the two rates are locally almost identical.

To complete the story of the model 246 yr after flash peak, application of a magnifying glass to the region centered on the BCS yields the distributions of structure variables shown in Fig. 17.1.18 and the distributions of gravothermal and nuclear energy-production rates shown in Fig. 17.1.19. The first lesson from these figures has to do with aesthetics: some portions of several of the distributions consist of straight line segments of different slopes, a situation which could have been avoided with more careful zoning.

The second lesson has to do with physics. Everywhere in the region shown, matter is moving outward ($\mathbf{g} \cdot \mathbf{v} < 0$) and expanding $((P/\rho^2)\,(d\rho/dt) < 0)$; in most of the region where the rate of nuclear energy generation is large, matter is also cooling, and this accounts for the continued decline with time in the helium-burning luminosity. The luminosity watershed (the outer border of the region within which the luminosity and the temperature gradient are negative) is at $M \sim 0.1812\ M_\odot$. The most negative luminosity occurs at $M \sim 0.1783\ M_\odot$. In the region between these two masses, the net rate of energy generation is positive ($dL/dM > 0$) and is responsible for the fact that the energy flux increases

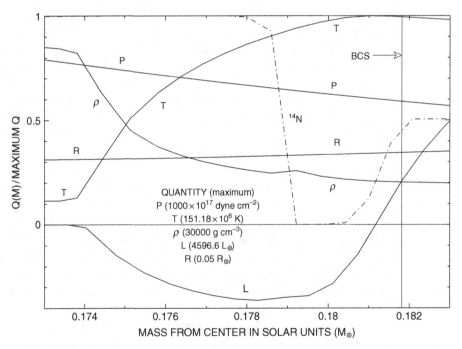

Fig. 17.1.18 Conditions below the BCS in a 1 $M_\odot$ model when the ECS and the hydrogen-rich profile nearly touch ($Z = 0.015$, $Y = 0.275$, $\Delta t = 245.66$ yr)

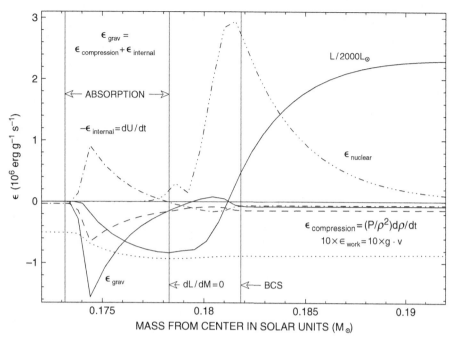

Fig. 17.1.19 Gravothermal and nuclear energy-generation rates near the BCS in a 1 $M_\odot$ model 245.66 yr after the peak of the first helium shell flash ($Z = 0.015$, $Y = 0.275$)

inward. By far the major contributor to the inward flow of energy is the energy released by nuclear burning.

In the region where the luminosity gradient is negative, between $M \sim 0.1732\ M_\odot$ and $M \sim 0.1783\ M_\odot$ (indicated by the two vertical lines in Fig. 17.1.19 labeled ABSORPTION), matter is heating and expanding and the fact that the slope of the luminosity profile is negative indicates that energy is being absorbed from the inward flow of energy to do the heating and the work of expansion. Thus, even during the declining portion of the helium shell flash phase, some of the nuclear energy liberated near the BCS contributes to heating and expanding matter in the region below the BCS, consolidating the region as one in which electrons are not degenerate.

As the helium-burning luminosity declines still further, the structure of and gravothermal activity in the model continue to entertain. Examples of the complexities in a model 25 110 years after flash peak are shown in Fig. 17.1.20, where the gyrations in the luminosity profile are the most interesting feature, and in Fig. 17.1.21, where manifold variations in all indicators of gravothermal activity are evident. The surface luminosity of the model is $L_s = 74.728\ L_\odot$ and the radius of the model is $R = 15.643\ R_\odot$, placing the model far below the red giant tip, roughly midway along the giant branch in the HR diagram shown in Fig. 11.1.51 in Volume 1. Nuclear burning plays an insignificant role in the evolution at this point. The helium-burning luminosity, at $L_{He} \sim 0.0825\ L_\odot$, is near its absolute minimum and the hydrogen-burning luminosity is only $L_H \sim 3.005\ L_\odot$.

There are two major peaks in the profile for ϵ_{grav} in Fig. 17.1.21, one positive and one negative. Calling the area under the positive peak L_g^+ and the absolute value of the

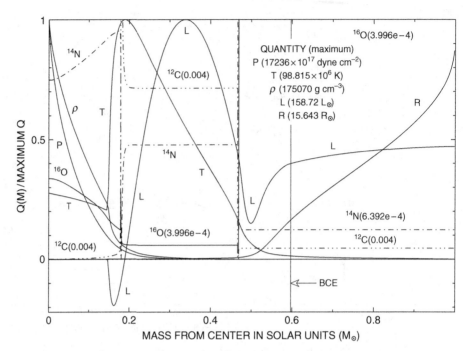

Fig. 17.1.20 Structure variables and CNO abundances in a 1 $M_\odot$ model near minimum L_{He} after the first helium shell flash ($Z = 0.015$, $Y = 0.275$, $\Delta t = 25110$ yr)

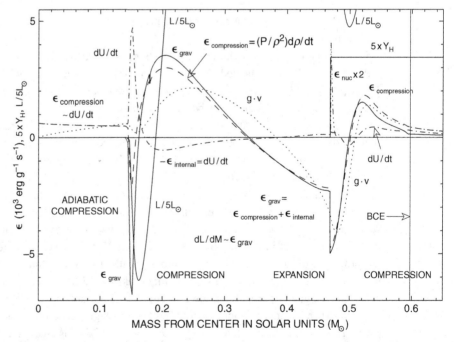

Fig. 17.1.21 Gravothermal energy-generation rates in a 1 $M_\odot$ model near minimum L_{He} after the first helium shell flash ($Z = 0.015$, $Y = 0.275$, $\Delta t = 25110$ yr)

area under the negative peak L_g^-, it is evident that $L_g^+ \sim L_g^-$. Since $\epsilon_{grav} \sim \epsilon_{compression}$ in the region encompassed by the two peaks, one can interpret the near identity of L_g^+ and L_g^- to mean that excess gravitational potential energy stored in the neighborhood of the helium-burning shell during the flash is being transferred to a region encompassing the hydrogen-burning shell. The existence of a region just outside of the hydrogen-burning shell where $\epsilon_{compression}$ and ϵ_{grav} are positive accounts for the increase in luminosity outward beyond the relative minimum in L at $M \sim 0.5 \, M_\odot$, and may be interpreted as a response by the model to the decrease in the strength of the hydrogen-burning shell in the form of contraction of envelope material to release gravitational Potential energy.

With the help of the profiles for $\mathbf{g} \cdot \mathbf{v}$ and $(P/\rho^2) \, (d\rho/dt)$ in Fig. 17.1.21, the model can be broken loosely into three parts according to the direction in which matter is moving and according to whether density is decreasing or increasing. The inner approximately one third of the model is moving inward and contracting and the outer half of the model is doing the same. Most of the matter between the inward moving and contracting regions is moving outward and expanding, exhibiting the behavior prevalent during the height of the helium flash.

In contrast with the situation during the peak of the flash, the source of energy which drives expansion outward against gravity is, instead of nuclear energy, the compressional energy liberated by matter between the luminosity watershed at $M = 0.189 \, M_\odot$ and the base of the expanding region at $M \sim 0.34 \, M_\odot$. The luminosity profile in Fig. 17.1.20 shows that only about 15% of the released compressional energy survives absorption in the expanding region to contribute to the surface luminosity, to the tune of about 30% of the surface luminosity. The other 70% of the surface luminosity comes from the compressional energy liberated near the base of the contracting region beyond the mass $M \sim 0.51 \, M_\odot$.

In the region extending from the center to $M \sim 0.145 \, M_\odot$, dU/dt and $(P/\rho^2) \, d\rho/dt$ are essentially the same, identifying this region as the electron-degenerate core in which local changes occur adiabatically. However, the two gravothermal rates are positive, demonstrating that, instead of expanding as do the electron-degenerate cores in models still experiencing the helium shell flash, the core in this model is contracting.

Interestingly, in the region of negative luminosity, the place where $dL/dM = 0$ occurs, at $M \sim 0.162 \, M_\odot$, coincides with the position where $\epsilon_{grav} = 0$. By far the dominant contributor to the increase in the inward energy flow between the luminosity watershed at $M \sim 0.189 \, M_\odot$ and $M \sim 0.162 \, M_\odot$ is compressional energy. In the region of negative luminosity between $M = 0.162 \, M_\odot$ and the innermost boundary of the region at $M \sim 0.145 \, M_\odot$, the major mode of absorption is heating, with roughly 20% of the energy absorbed being used to do the work of expansion.

Distributions of entropy and the degree of electron degeneracy in the model 25 110 yr after flash peak are shown in Fig. 17.1.22. Comparing with these same distributions in Fig. 17.1.15 for the model 246 yr past flash peak, it is apparent that the mass of the electron-degenerate core has decreased from $\sim 0.174 \, M_\odot$ to $\sim 0.145 \, M_\odot$, that the entropy distribution in the core and the mean entropy in the region between the core and the helium–hydrogen interface have remained the same, while, in consequence of envelope contraction, the entropy in the convective envelope has decreased.

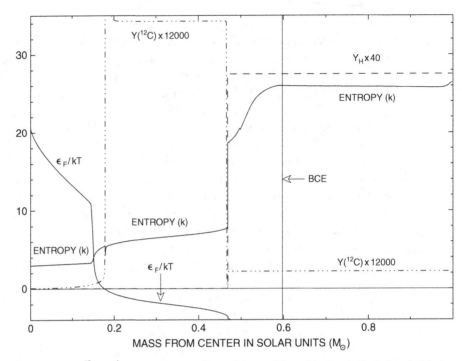

Fig. 17.1.22 Entropy, ϵ_F/kT, and ^{12}C and ^{1}H abundances in a 1 $M_\odot$ model near minimum L_{He} after the first helium shell flash ($Z = 0.015$, $Y = 0.275$, $\Delta t = 25110$ yr)

With regard to progress toward reaching the quiescent core helium-burning horizontal branch phase, the main result of helium burning thus far has been a decrease in the mass of the electron-degenerate core from an initial mass of $M_{core} \sim 0.47\ M_\odot$ to $M_{core} \sim 0.145\ M_\odot$. As evolution progresses still further, additional, relatively mild helium-burning flashes ignite at positions successively closer in mass to the model center. During each flash, matter in the region directly affected by the flash heats and expands to the extent that electron degeneracy is lifted. In the region interior to the flashing zone, matter initially expands and cools adiabatically and then contracts and heats adiabatically, with the degree of electron degeneracy, as measured by the ratio ϵ_F/kT, remaining almost constant. Ultimately, helium burning reaches the center and the model embarks on an extended phase of quiescent helium burning in a convective core above which hydrogen burning continues in a shell.

The variations with time of various global and local characteristics during most of flashing phase of evolution are shown in Figs. 17.1.23–17.1.27. From Fig. 17.1.23 it is evident that, with each successive flash, the peak helium-burning luminosity decreases, the duration of the largest L_{He} portion of the flash increases, and the time between successive flashes increases. In Fig. 17.1.24 are shown the time variations of the global hydrogen-burning luminosity L_H and the model surface velocity L_s. Comparing with Fig. 17.1.23, it is evident that the absolute minimum in L_H occurs after the absolute minimum in L_{He} and that, after climbing to a relative maximum, L_H experiences an abrupt decline as L_{He}

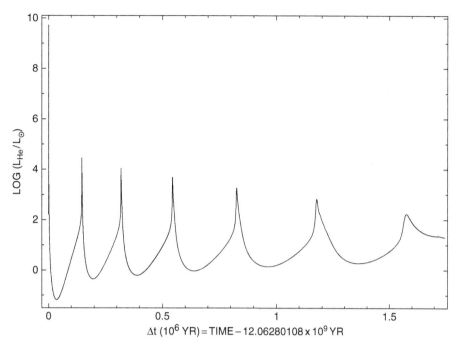

Fig. 17.1.23   Helium-burning luminosity during helium shell flashes in a 1 $M_\odot$ model evolving from the red giant tip to the horizontal branch ($Z = 0.015, Y = 0.275$)

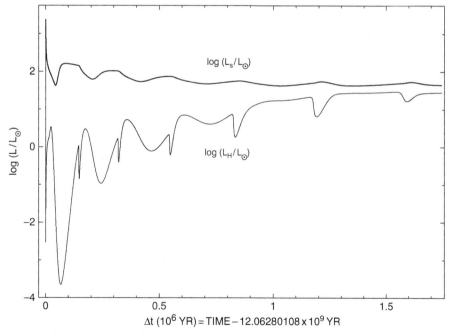

Fig. 17.1.24   H-burning and surface luminosities during helium shell flashes in a 1 $M_\odot$ model evolving from the red giant tip to the horizontal branch ($Z = 0.015, Y = 0.275$)

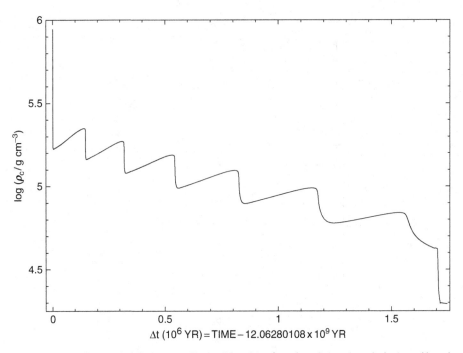

Fig. 17.1.25 Central density during helium shell flashes in a 1 $M_\odot$ model evolving from the red giant tip to the horizontal branch ($Z = 0.015$, $Y = 0.275$)

shoots to a relative maximum, but then climbs to and recedes gradually from a second and slightly larger relative maximum. During the period that the six secondary helium-flash episodes occur, the mean hydrogen-burning luminosity increases until, after the last flash, L_H is roughly 60% of the surface luminosity and L_{He} accounts for the other 40%.

The density and temperature at model center are shown as functions of time in Figs. 17.1.25 and 17.1.26, respectively. Over the entire time shown, the central density drops by a factor of $\sim$21 from $\sim$890 000 g cm^{-3} to $\sim$42 000 g cm^{-3} and central temperature decreases by a factor of $\sim$7 from $\sim$77 $\times$ 10^6 K to $\sim$11 $\times$ 10^6 K. In the approximation that electrons are non-relativistically degenerate, the ratio $\rho^{2/3}/T$ is proportional to the quantity ϵ_F/kT. As recorded in Fig. 17.1.26, remarkably, but quite consistent with the analysis of individual models already conducted, this ratio remains almost constant at the center of the model.

In Fig. 17.1.27 are shown the outlines of the convective shells produced by helium burning during the helium-flashing episodes. The gaps along the curves outlining the last five convective shells occur because convection can begin and/or disappear over several zones simultaneously. Connecting the boundaries of the (almost) closed curves closest to the center defines a mass border above which electrons are either not degenerate or are in the process of becoming non-degenerate during the flashing episodes and between which and the model center the degree of electron degeneracy increases rapidly toward the center.

In the following, attention is limited to an examination of structure variables, thermodynamic conditions, and gravothermal activity in six models: (1) near maximum L_{He} during

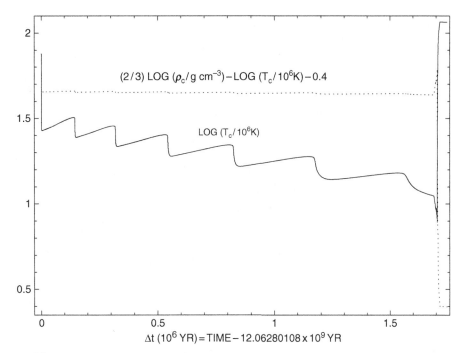

Fig. 17.1.26 T and $\rho^{2/3}/T$ at the center during helium shell flashes in a 1 M$_\odot$ model evolving from the red giant tip to the horizontal branch (Z = 0.015, Y = 0.275)

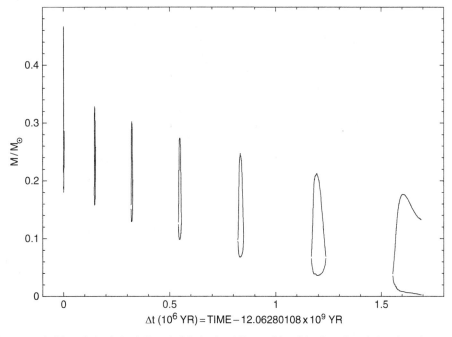

Fig. 17.1.27 Convective shell boundaries during helium shell flashes in a 1 M$_\odot$ model evolving from the red giant tip to the horizontal branch (Z = 0.015, Y = 0.275)

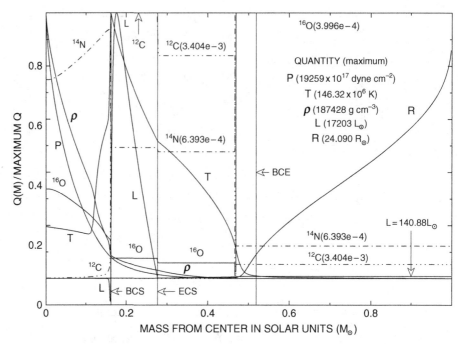

Fig. 17.1.28 Structure variables and CNO abundances in a 1 $M_\odot$ model 72.214 yr after the peak of the second helium shell flash ($Z = 0.015$, $Y = 0.275$, $\Delta t = 145970$ yr)

the second flash, (2) near minimum L_{He} after the second flash, (3) near maximum L_{He} during the sixth flash, (4) between minimum L_{He} after the sixth flash and maximum L_{He} during the seventh flash, (5) after the seventh flash, and (6) during the final stage of shell helium burning as convection extends to the center.

Conditions in the model 72.214 years after the peak of the second helium shell flash are shown in Figs. 17.1.28–17.1.30. The helium-burning luminosity, at $L_{He} \sim 1.72 \times 10^4$ $L_\odot$, is only 5% smaller than at the second flash peak. The fact that, at and near the second flash peak, L_{He} is almost 2×10^5 times smaller than during the first flash peak can be understood as a consequence of the fact that, during evolution between flashes, the degree of electron degeneracy has been reduced in a considerable region below the temperature maximum by absorption from the inward energy flow driven by the release of compressional energy. The decrease in the degree of electron degeneracy is evident from a comparison between the curve for ϵ_F/kT in Fig. 17.1.30 and the same curve in Fig. 17.1.7. At the height of the first flash, matter in the convective shell is still somewhat electron degenerate, and, everywhere below the BCS, it is highly electron degenerate. At the height of the second flash, matter within the convective shell between $M \sim 0.16$ $M_\odot$ and $M \sim 0.28$ $M_\odot$ is emphatically non-electron degenerate and the degree of electron degeneracy does not reach values characteristic of the electron-degenerate core, within which the distribution of $\frac{\epsilon_F}{kT}$ with mass is invariant, until $M \sim 0.11$ $M_\odot$. In contrast with behavior during the first flash, most of the energy liberated by nuclear burning during the second flash goes immediately into increasing the thermal energy of both nuclei and electrons rather than into lifting

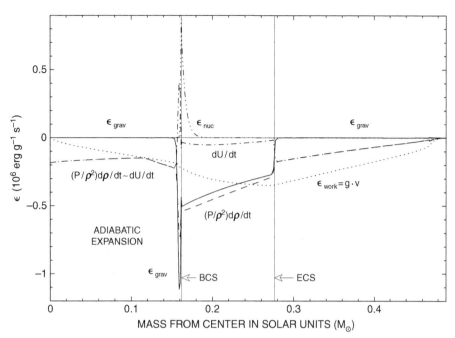

Fig. 17.1.29   Gravothermal energy-generation rates in a 1 $M_\odot$ model 72.214 yr after the peak of the second helium shell flash ($Z = 0.015, Y = 0.275, \Delta t = 145970$ yr)

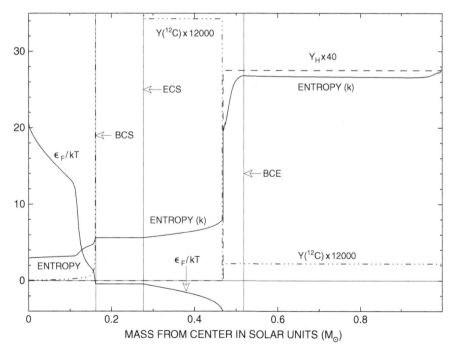

Fig. 17.1.30   Entropy, ϵ_F/kT, and ^{12}C and ^{1}H abundances in a 1 $M_\odot$ model 72.214 yr after the peak of the second helium shell flash ($Z = 0.015, Y = 0.275, \Delta t = 145970$ yr)

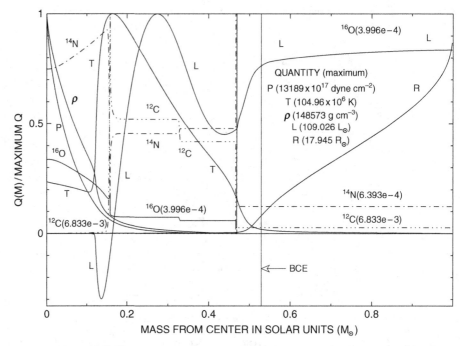

Fig. 17.1.31   Structure variables and CNO abundances in a 1 $M_\odot$ model 20702 yr after the peak of the second helium shell flash ($Z = 0.015$, $Y = 0.275$, $\Delta t = 167180$ yr)

degeneracy. Hence, the pressure gradients and the concomitant cooling that are needed to contain the thermonuclear runaway are called into play immediately, resulting in a much less violent flash.

Conditions in the model 20 702 yr after the peak in L_{He} reached during the second helium shell flash are shown in Figs. 17.1.31–17.1.33. The helium-burning luminosity, at $L_{He} = 1.2244\ L_\odot$, is approaching the second relative minimum of $L_{He}^{min} \sim 0.41\ L_\odot$. The distributions in the various variables in the model are, not surprisingly, both qualitatively and in many respects quantitatively very similar to those shown in Figs. 17.1.20–17.1.22 for the model near the relative minimum of L_{He} after the peak in L_{He} reached during the first flash.

In both cases, nuclear burning plays an exceedingly minor role and the luminosity profile exhibits two well defined relative minima and two well defined relative maxima which are explicable in terms of gravothermal activity. In particular, the division of the model into three segments according to whether expansion or contraction is occurring is common to both models and the magnitudes of gravothermal energy-generation rates are quite comparable.

Comparison between the profiles of ϵ_{grav} in Figs. 17.1.32 and 17.1.21 demonstrates that, just as in the first model, gravothermal energy stored in the neighborhood of the helium-burning shell during the preceding helium shell flash is being transferred outward

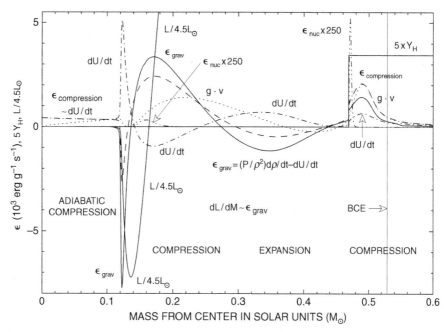

Fig. 17.1.32 Gravothermal energy-generation rates in a 1 $M_\odot$ model 20702 yr after the peak of the second helium shell flash ($Z = 0.015$, $Y = 0.275$, $\Delta t = 167180$ yr)

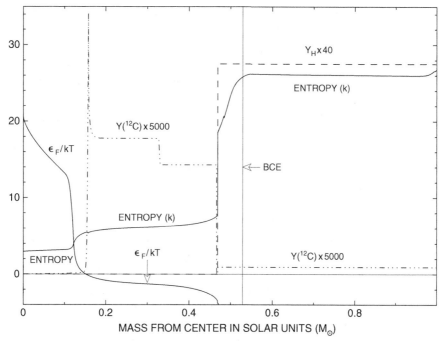

Fig. 17.1.33 Entropy, ϵ_F/kT, and ^{12}C and ^{1}H abundances in a 1 $M_\odot$ model 20702 yr after the peak of the second helium shell flash ($Z = 0.015$, $Y = 0.275$, $\Delta t = 167180$ yr)

and stored in a region which extends from midway between the very weak helium- and hydrogen-burning shells to a point near the hydrogen-burning shell, with the difference that this region extends beyond the helium–hydrogen interface in the first case but does not quite reach this interface in the second case. Another difference is that the ratio of heat energy to compressional energy in the gravothermal energy taking part in the transfer is larger in the second case than in the first. Finally, in both cases, compression and heating occur in a region extending from just above the region in which transfer is taking place and the gravitational potential energy that is released goes into local heating and into a contribution to the surface luminosity.

From the point of view of progress toward the horizontal branch phase of quiescent core helium burning and quiescent shell hydrogen burning, the most significant difference between the two models is that the mass of the electron-degenerate core has formally decreased from $\sim 0.145\ M_\odot$ in the earlier model to $\sim 0.11\ M_\odot$ in the more evolved model.

Because subsequent flashing episodes and their aftermath follow many of the same patterns exhibited by the first and second flashes and their aftermath, it might be thought that nothing could be gained by exploring model characteristics during additional flash episodes. However, the circumstance that each flash is less violent than its predecessor, with a concomitant decrease in the amplitudes of spatial variations in structure and gravothermal characteristics through any given model, means that further insights about the nature of evolution during flash episodes may be uncovered by exploring details during a milder flash.

In Figs. 17.1.34–17.1.37 are shown characteristics of the model 448 years after the peak of the sixth flash. The helium-burning luminosity, at $L_{He} \sim 653\ L_\odot$, is only ~ 9 $L_\odot$ smaller than at the sixth flash peak and the hydrogen-burning luminosity is $L_H \sim 12.0$ $L_\odot$. From the luminosity profile in Fig. 17.1.34, it is evident that the surface luminosity, at $L_s \sim 50.1\ L_\odot$, is made up of four components: $L_1 \sim 10\ L_\odot$ leaking out of the convective shell, $L_2 \sim 20\ L_\odot$ contributed by the expanding and cooling region between the ECS and the helium–hydrogen interface, $L_3 \sim 12\ L_\odot$ contributed by the hydrogen-burning shell, and $L_4 \sim 10\ L_\odot$ contributed by gravothermal action in and slightly beyond the hydrogen-burning shell. There is yet a fifth gravothermal component which extends further outward toward the surface and contributes to a modest decrease in luminosity of about $\sim 2\ L_\odot$.

The nature of these contributions is made clearer in Fig. 17.1.35 where gravothermal energy-generation characteristics are shown for a region extending from somewhat below the ECS to somewhat beyond the hydrogen-burning shell. It is evident that, over most of the region between $M = 0.065\ M_\odot$ and just above the ECS, at $M \sim 0.175\ M_\odot$, the rate at which work is done to expand matter $-(P/\rho^2)\,(d\rho/dt)$ is considerably larger than the rate $-dU/dt$ at which cooling is contributing to this work. Thus, the luminosity in this region decreases outward because absorption of energy from the ambient flow is required to do the work of expansion. Over the region between the relative minimum in the luminosity profile at $M \sim 0.18\ M_\odot$ and the base of the hydrogen-burning shell at $M \sim 0.47\ M_\odot$, the rate at which the internal energy decreases almost exactly matches the rate at which work is done to expand matter, but not quite. That is, $-dU/dt \gtrsim (P/\rho^2)$, and the slight, almost mass

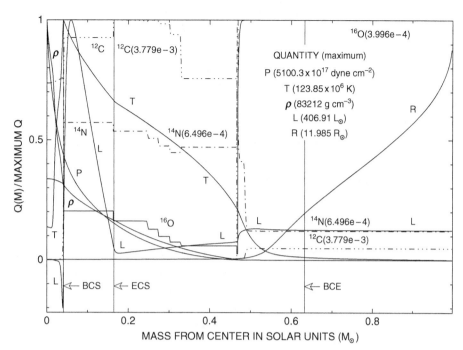

Fig. 17.1.34 Structure variables and CNO abundances in a 1 M$_\odot$ model 448.04 yr after the peak of the sixth helium shell flash ($Z = 0.015$, $Y = 0.275$, $\Delta t = 1.1719 \times 10^6$ yr)

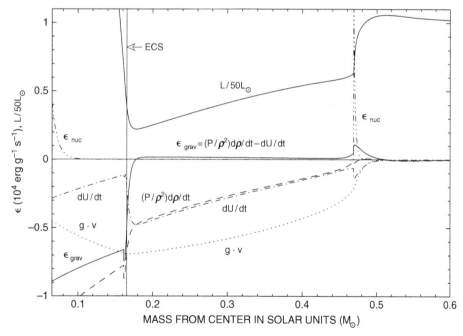

Fig. 17.1.35 Gravothermal and nuclear energy-generation rates in a 1 M$_\odot$ model 448.04 yr after the peak of the sixth helium shell flash ($Z = 0.015$, $Y = 0.275$, $\Delta t = 1.1719 \times 10^6$ yr)

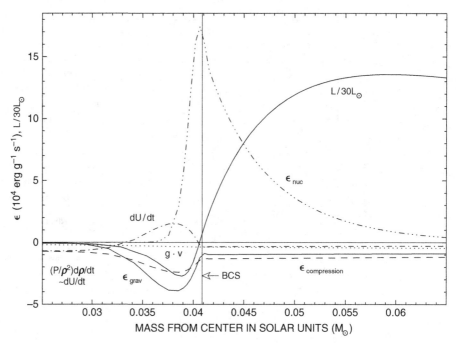

Fig. 17.1.36 Gravothermal and nuclear energy-generation rates in a 1 $M_\odot$ model 448.04 yr after the peak of the sixth helium shell flash ($Z = 0.015$, $Y = 0.275$, $\Delta t = 1.1719 \times 10^6$ yr)

independent, difference between the two rates accounts for the increase in the luminosity outward at nearly constant slope.

The next region of interest extends from just below the hydrogen-burning shell at $M \sim 0.46\ M_\odot$ to $M \sim 0.51\ M_\odot$. The hydrogen-burning shell accounts for most of the steep rise in luminosity between $M \sim 0.47\ M_\odot$ and $M \sim 0.48\ M_\odot$. In this same region, the rate of expansion decreases to nominal values, but the rate of cooling reaches a maximum in the hydrogen-burning shell and produces approximately half of the luminosity increase over the mass range $M \sim 0.46\ M_\odot$ to $M \sim 0.51\ M_\odot$. The final region in which gravothermal activity has a modest effect on the surface luminosity extends from $\sim 0.51\ M_\odot$ to the surface, with most of the activity occurring below the BCE which is located at $M \sim 0.6324$ $M_\odot$. In this region, energy is absorbed from the ambient flow at the total rate of $\sim 2\ L_\odot$ to contribute the energy required for expansion not supplied by cooling.

The distribution of nuclear and gravothermal energy-generation rates in the neighborhood of the BCS that are shown in Fig. 17.1.36 for the model shortly after the sixth flash peak may be compared with these same characteristics shown in Fig. 17.1.19 for the model 246 yr after the first flash peak. The most important lessons from the comparison are that: (1) qualitatively, the gravothermal responses to nuclear burning activity near the BCS are the same in the two models, and (2) the inner edge of the negative portion of the luminosity profile has decreased from $\sim 0.173\ M_\odot$ in the earlier model to $\sim 0.03\ M_\odot$ in the more evolved model, demonstrating that the region within which electron degeneracy

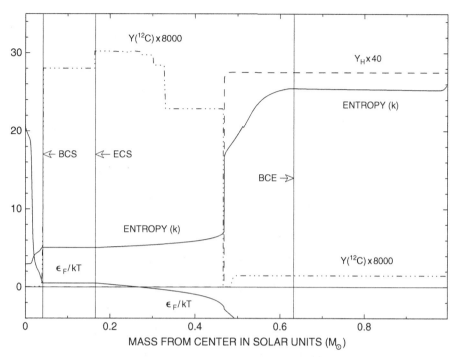

Fig. 17.1.37 Entropy, ϵ_F/kT, and ^{12}C and ^1H abundances in a 1 $M_\odot$ model 448.04 yr after the peak of the sixth helium shell flash ($Z = 0.015, Y = 0.275, \Delta t = 1.1719 \times 10^6$ yr)

is being reduced by absorption of energy from the helium-burning shell has nearly reached the center.

Entropy, the degree of electron degeneracy, and the ^{12}C and ^1H abundance profiles for the model shortly after the sixth pulse peak are shown as functions of mass in Fig. 17.1.37. Comparison with the same distributions in Fig. 17.1.30 for the model shortly after the second flash peak show that the mass of the electron-degenerate core, defined as the mass at which the electron-degeneracy parameter ϵ_F/kT declines precipitously with mass, has decreased from ∼0.122 $M_\odot$ in the earlier model to ∼0.015 $M_\odot$ in the more evolved model.

The next four figures, Figs. 17.1.38–17.1.41, describe conditions in a model which has passed the relative minimum in L_{He} following the sixth helium shell flash and is approximately half way in time between this relative minimum and the peak of the seventh flash. It is instructive to compare with characteristics of the model near minimum L_{He} following the first flash peak (Figs. 17.1.20–17.1.22) and with those of the model approaching minimum L_{He} following the second flash peak (Figs. 17.1.31–17.1.33). A major difference between the three models is that, in the most evolved model, energy produced by hydrogen burning contributes significantly to the surface luminosity and energy produced by helium burning is the dominant contributor to the heating and expansion of matter just outside of the electron-degenerate core. In the other two models, energy generation by neither hydrogen burning nor helium burning plays a role in the energy budget and the energy for heating

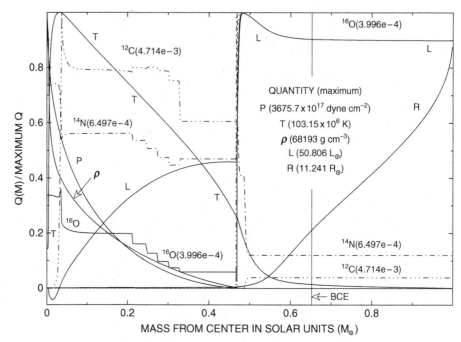

Fig. 17.1.38   Structure variables and CNO abundances in a 1 $M_\odot$ model between the 6th minimum and 7th maximum in L_{He} ($Z = 0.015$, $Y = 0.275$, $\Delta t = 1.48748 \times 10^6$ yr)

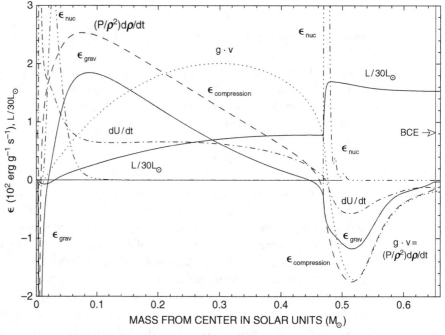

Fig. 17.1.39   Gravothermal and nuclear energy-generation rates in a 1 $M_\odot$ model between the 6th minimum and 7th maximum in L_{He} ($Z = 0.015$, $Y = 0.275$, $\Delta t = 1.48748 \times 10^6$ yr)

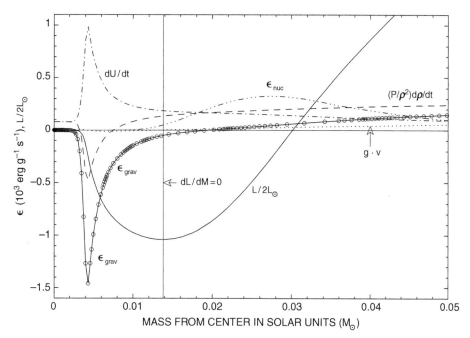

Fig. 17.1.40 Central gravothermal and nuclear energy-generation rates in a 1 $M_\odot$ model between the 6th minimum and 7th maximum in L_{He} ($Z = 0.015$, $Y = 0.275$, $\Delta t = 1.48748 \times 10^6$ yr)

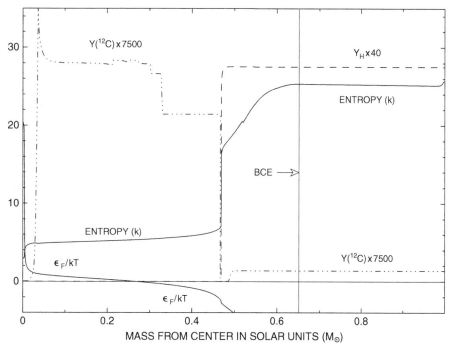

Fig. 17.1.41 Entropy, ϵ_F/kT, and ^{12}C and ^1H abundances in a 1 $M_\odot$ model between the 6th minimum and 7th maximum in L_{He} ($Z = 0.015$, $Y = 0.275$, $\Delta t = 1.4875 \times 10^6$ yr)

and expanding matter just outside of the electron-degenerate core comes primarily from compression between the luminosity watershed and the luminosity minimum.

From the luminosity profile in Fig. 17.1.38, it is clear that approximately half of the energy flux emanating from the outer edge of the hydrogen-burning shell in the most evolved model is due to the release of gravitational potential energy by matter below the shell, the other half being due to hydrogen burning in the shell. Over a region extending from approximately the middle of the hydrogen-burning shell to the BCE, absorption from the ambient flow helps cooling provide the energy to expand matter outward. The luminosity profiles in the other two models, shown in Figs. 17.1.20 and 17.1.31, respectively, are quite similar to one another, are entirely a consequence of gravothermal activity, and differ in many respects from the luminosity profile in the most evolved of the three models. The differences can be best understood by examining the profiles of the gravothermal energy-generation rates for the most evolved model shown in Figs. 17.1 .39 and 17.1.40 and profiles of these quantities for the least evolved model in Fig. 17.1.21 and for the model of intermediate age in Fig. 17.1.32.

A major difference occurs in the region between the luminosity watershed (which coincides roughly with the location of composition discontinuities associated with the maximum inward extent of the BCS) and the helium–hydrogen interface. In the most evolved model, some of the energy of compression released by contracting matter in this region goes into heating, but most of it contributes to the outward flux of energy and the luminosity increases monotonically outward throughout the entire region. In the two less evolved models, matter in the outer portion of the region between the luminosity watershed and the hydrogen–helium interface is expanding and the compressional energy released in the inner portion of the region is countered by absorption from the ambient flow of energy in the outer region. Hence, only a fraction ($\sim$15% in the least evolved model and $\sim$45% in the model of intermediate age) of the energy produced by compression in the inner portion of the region contributes to the surface luminosity. In the least evolved model, the absorbed energy is converted into the work of expansion; in the model of intermediate age, the absorbed energy is converted into thermal energy.

Another pronounced difference in luminosity profiles occurs in the region extending from near the helium–hydrogen interface to the BCE and from there to the surface. In the two less evolved models, an outward increase in the luminosity is due to the release of compressional energy in a contracting envelope but, in the most evolved model, the luminosity decreases outward for a small distance in mass due to absorption of energy required for expansion.

A penultimate difference and a final similarity worth noting occurs in the region of negative luminosity that extends from the luminosity watershed where $L = 0$ and $dL/dM > 0$ to near the outer edge of the electron-degenerate core where $L = 0$ and $dL/dM < 0$. Gravothermal characteristics in this region in the most evolved model are highlighted in Fig. 17.1.40, where the spacing of open circles along the curve for ϵ_{grav}, which give the gravothermal energy-generation rate at zone centers, provides information about resolution in both mass and time; the spacing in the horizontal direction describes the prevailing zoning in mass, and the spacing in the vertical direction is a measure of the rapidity with which changes are occurring in time.

In all three models, the absorption which occurs between the position of minimum luminosity where $dL/dM = 0$ and the outer edge of the electron-degenerate core is due to heating and expansion. In all three models, the source of the energy which drives this heating and expansion is the region between the luminosity watershed and the luminosity minimum. In the less evolved models, the primary source is compressional energy. In the most evolved model, most of the compressional energy released goes into heating matter locally and the primary source contributing to the increase in the inward flux of energy is energy supplied by helium burning.

The distributions of entropy and the degree of electron-degeneracy shown in Fig. 17.1.41 for the model between the sixth and seventh helium shell flash peaks demonstrate that the electron-degenerate core has incredible survival instincts. The mass of the region within which ϵ_F/kT increases from ~ 1 to ~ 20 is only about three times the mass of Jupiter. On the other hand, as revealed by a comparison of the electron-degeneracy indicators in Figs. 17.1.41 and 17.1.33, between the outer edge of the electron-degenerate core and the helium–hydrogen interface, the average value of ϵ_F/kT has increased from ~ -1 to ~ 0, suggesting that electron degeneracy will be considerably larger in the helium-burning core of the final zero age horizontal branch model than in the hydrogen-burning core of its main sequence precursor.

The high density of zones in the electron-degenerate core in Fig. 17.1.40 has been chosen in anticipation of the inward movement of the steep profile in density manifested just outside of the core. Despite the attempt to determine zoning in advance in such a way that difference equations are reasonably good approximations to differential equations, variations in ϵ_{grav} as well as in density and temperature from zone to zone in the region just outside of the electron-degenerate core continue to be uncomfortably large. Since discontinuities are a fact of life, the discomfort must either be tolerated or dealt with by, for example, rezoning after every calculation and repeating the calculation, perhaps several times.

Conditions in a model which has experienced the seventh helium shell flash and is burning both helium and hydrogen quiescently, but has not yet managed to remove its electron-degenerate core, are described in Figs. 17.1.42–17.1.47. The first three of the figures describe conditions through most of the model and the second three focus on conditions between the center and the BCS. The model is 1.1520×10^5 yr past the seventh flash peak, where $L_{He}^{max}(7) = 169.24$, and the helium-burning luminosity has declined to $L_{He} = 23.58 \ L_\odot$. The hydrogen-burning luminosity is $L_H = 27.37 \ L_\odot$, so that $L_{nuc} = 50.95 \ L_\odot$. Since the surface luminosity is $L_s = 45.65 \ L_\odot$, the gravothermal luminosity is $L_{grav} = 5.30 \ L_\odot$.

Evidence for the occurrence of seven flashes is imprinted in the seven-tiered abundance profiles for ^{16}O and ^{14}N in Fig. 17.1.42 over the region between the ECS and the hydrogen-burning shell. The abundance-profile staircases demonstrate that the maximum outward extent of the ECS during each flash decreases with each flash, consistent with the decline of maximum L_{He} with each flash.

From the luminosity profile in Fig. 17.1.42, it is evident that the model is still in the process of recovering from the last flash, with gravothermal absorption occurring in the inner approximately two-thirds of the mass between the outer edge of the active

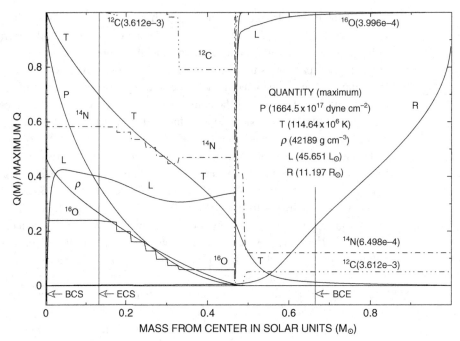

Fig. 17.1.42   Structure variables and CNO abundances in a 1 $M_\odot$ model 1.1520×10^5 yr after the 7th maximum in L_{He} ($Z = 0.015$, $Y = 0.275$, $\Delta t = 1.69077 \times 10^6$ yr)

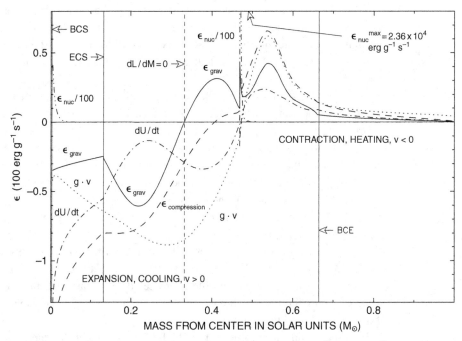

Fig. 17.1.43   Gravothermal and nuclear energy-generation rates in a 1 $M_\odot$ model 1.1520×10^5 yr after the 7th maximum in L_{He} ($Z = 0.015$, $Y = 0.275$, $\Delta t = 1.69077 \times 10^6$ yr)

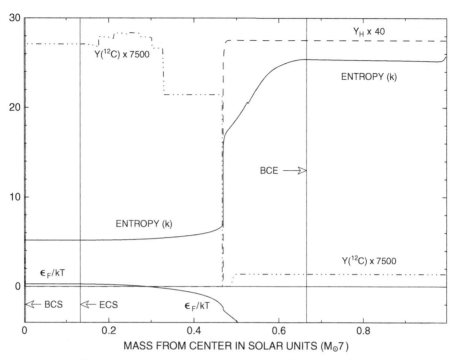

Fig. 17.1.44 Entropy, ϵ_F/kT, and ^{12}C and ^{1}H abundances in a 1 $M_\odot$ model 1.152×10^5 yr after the 7th maximum in L_{He} ($Z = 0.015, Y = 0.275, \Delta t = 1.69077 \times 10^6$ yr)

helium-burning region and the hydrogen-burning shell and release of gravothermal energy occurring in the outer third of this region as well as in the region between the hydrogen-burning shell and the surface.

The nature of the gravothermal activity through most of the model is shown in Fig. 17.1.43. Matter in the entire region from the BCS to the hydrogen-burning shell is moving outward and cooling. Except for a small region extending inward by ~0.075 $M_\odot$ from the hydrogen-burning shell, it is also expanding. The position where $(dU/dt) = (P/\rho^2)(d\rho/dt)$ and therefore $dL/dM = \epsilon_{\text{gravothermal}} = 0$ is flagged by the dashed vertical line at $M = 0.332\,46\ M_\odot$. To the left of this position, cooling provides approximately half of the energy required for expansion, with absorption from the ambient flow providing the other half. Between this position and the base of the hydrogen-burning shell, most of the cooling energy contributes to the ambient flow of energy outward. In the region to the right of the hydrogen-burning shell, that portion of the compressional energy released which is not used up in heating matter locally contributes to the ambient outward flow of energy.

Distributions of entropy, ϵ_F/kT, and the ^{1}H and ^{12}C abundance profiles are shown in Fig. 17.1.44. Comparison with these same profiles in Fig. 17.1.41 for the model approximately 10^5 yr before the seventh flash peak shows that, in the region beyond the ECS in the more evolved model, most quantities have stabilized. The major differences appear interior to this region, the most pronounced being a reduction in the degeneracy

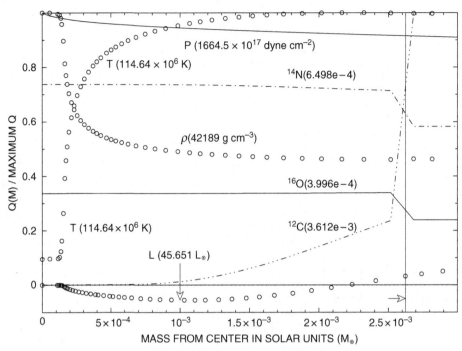

Fig. 17.1.45   Structure variables and CNO abundances near the center of a 1 $M_\odot$ model 1.1520×10^5 yr after the 7th maximum in L_{He} ($Z = 0.015$, $Y = 0.275$, $\Delta t = 1.69077 \times 10^6$ yr)

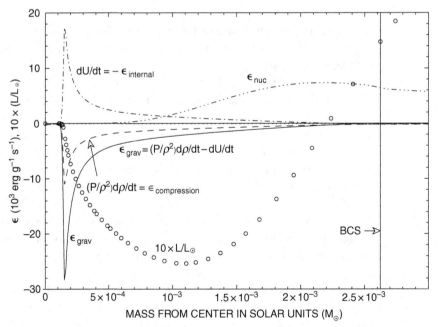

Fig. 17.1.46   Central gravothermal and nuclear energy-generation rates in a 1 $M_\odot$ model 1.1520×10^5 yr after the 7th maximum in L_{He} ($Z = 0.015$, $Y = 0.275$, $\Delta t = 1.69077 \times 10^6$ yr)

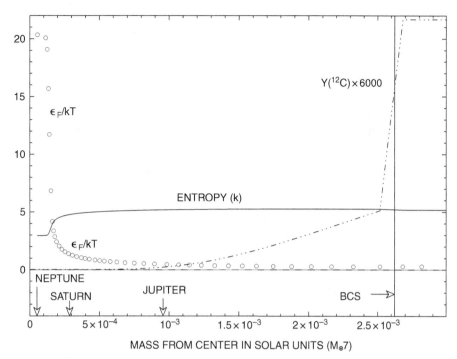

Fig. 17.1.47 Entropy, ϵ_F/kT, and the ^{12}C abundance near the center of a 1 $M_\odot$ model 1.1520×10^5 yr after the 7th maximum in L_{He} ($Z = 0.015$, $Y = 0.275$, $\Delta t = 1.69077 \times 10^6$ yr)

indicator in the convective shell of the more evolved model to $\epsilon_F/kT \sim 1/3$ and a further shrinking of the mass of the electron-degenerate core.

How small the electron-degenerate core has become is dramatized in Figs. 17.1.45–17.1.47, where conditions between the BCS and the center of the model are shown. In Fig. 17.1.45 temperatures and densities at shell centers are indicated by open circles; in both Figs. 17.1.45 and 17.1.46 luminosities at shell boundaries are indicated by open circles, and in Fig. 17.1.47 values of ϵ_F/kT at shell centers are indicated by open circles. From the open circle distributions, it is evident that the electron-degenerate core is essentially coincident with the central spherical shell outside of which all structure variables vary almost discontinuously with respect to mass. Comparison in Fig. 17.1.47 of the mass of the central sphere with the masses of gaseous planets about the Sun demonstrates that, whereas an isolated object of planetary mass cannot become a white dwarf, the imposition of a large enough external pressure can induce an object of such a mass to exhibit white dwarf characteristics.

In the model chosen for display, the central white dwarf and the central spherical shell essentially coincide in mass. This is an accident, as may be verified by forcibly refining the rezoning in central regions and repeating the calculation. Continuing calculations with the ever finer zoning near the center required to resolve core structure shows that the central electron-degenerate core persists to astonishingly small masses. For example, as shown in

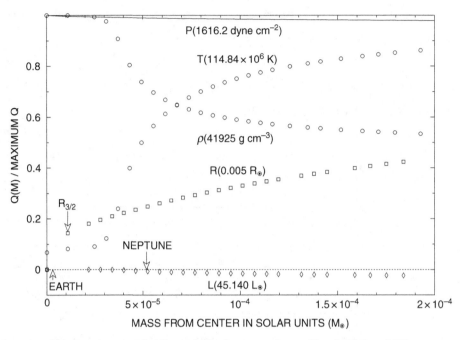

Fig. 17.1.48 Structure variables near the center of a 1 M$_\odot$ model forming a convective core (Z = 0.015, Y = 0.275, Δt = 1.70457 $\times$ 10^6 yr)

Fig. 17.1.48, after another ~14 000 years of evolution, the mass of the electron-degenerate core is approximately ten times the mass of the Earth.

Ultimately, the maximum in the temperature profile and the maximum in the helium-burning energy-generation rate make their way to the center and the model settles into a stable configuration in which the flux of energy produced by quiescent helium burning supports a convective core and quiescent hydrogen-burning in a shell supports a deep convective envelope. Structure variables in the model near the beginning of this phase are shown in Fig. 17.1.49, gravothermal energy-generation rates are shown in Fig. 17.1.50 and the entropy and ϵ_F/kT distributions are shown in Fig. 17.1.51. Comparing the distributions of structure variables in Fig. 17.1.49 with those in Fig. 17.1.42 for the model with a small electron-degenerate core, only one difference between the two models stands out. In the model without an electron-degenerate core, the density increases inward at modest slope to ~2 × 10^4 g cm^{-3} at the center while, in the model with an electron-degenerate core, over a very small distance near the center the density increases by over a factor of two, to ~4 × 10^4 g cm^{-3} at the center. The fact that the temperature at the center has increased by over a factor of ten between the two models is not resolved in the figures.

The model described in Figs. 17.1.49–17.1.51 is close to what has traditionally been called a zero age horizontal branch (ZAHB) model: a model in which (1) hydrogen burns quiescently in a shell, (2) helium burns quiescently in a hydrogen-exhausted core in which (3) the helium abundance has been depleted enough to account for the energy required to

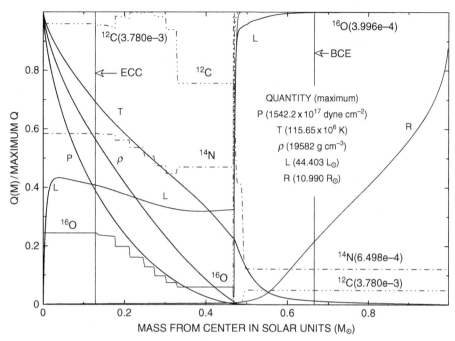

Fig. 17.1.49 Structure variables and CNO abundances in a 1 $M_\odot$ model approaching the zero age horizontal branch
($Z = 0.015, Y = 0.275, \Delta t = 1.74837 \times 10^6$ yr)

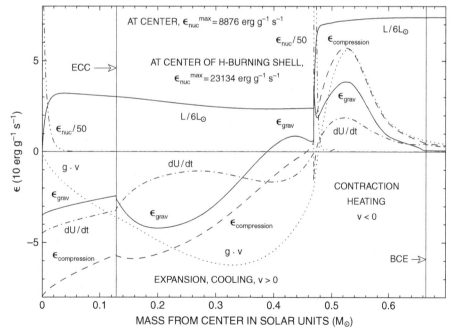

Fig. 17.1.50 Gravothermal and nuclear energy-generation rates in a 1 $M_\odot$ model approaching the zero age horizontal branch
($Z = 0.015, Y = 0.275, \Delta t = 1.74837 \times 10^6$ yr)

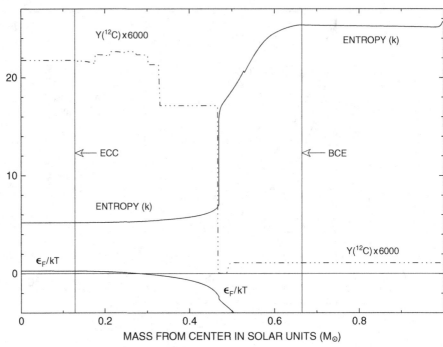

Fig. 17.1.51 Entropy, ϵ_F/kT, and the ^{12}C abundance in a 1 $M_\odot$ model approaching the zero age horizontal branch ($Z = 0.015, Y = 0.275, \Delta t = 1.74837 \times 10^6$ yr)

lift the electron-degeneracy prevailing at the red giant tip, and (4) the rate of gravothermal energy production is negligible compared with the rate of nuclear energy production. It is evident from the luminosity profile in Fig. 17.1.49 that conditions (1) and (2) are fulfilled but that condition (4) is not: helium and hydrogen burning make comparable contributions to the surface luminosity, but approximately 25% of the energy produced by helium burning near the center is used up in doing work to expand matter between the center and the hydrogen-burning shell and approximately 5% of the surface luminosity is due to energy liberated by contraction in the hydrogen-rich envelope. Gravothermal and nuclear energy-generation rates are compared in Fig. 17.1.50.

Whether or not condition (3) is met can be tested by finding the total amount of ^{12}C and ^{16}O in the hydrogen-exhausted core, determining how much energy has been required to produce these isotopes from helium, and comparing with the binding energy of the model at the red giant tip. From Fig. 17.1.49, one has that the average number abundances of ^{12}C and ^{16}O in the core are, respectively, $\langle Y_{12} \rangle \sim 3.3 \times 10^{-3}$ and $\langle Y_{16} \rangle \sim 6.0 \times 10^{-5}$. The number of nucleons in the core, of mass $M_{\text{core}} \sim 0.47 \ M_\odot$, is $\sim 0.56 \times 10^{57}$. An energy of 7.275 MeV is released in making every ^{12}C nucleus via the triple-alpha reactions and an energy of 7.161 MeV is released in making every ^{16}O nucleus via the ^{12}C$(\alpha, \gamma)^{16}$O reaction. Hence, the total amount of energy released in producing the final abundances in the core is

$$\Delta E \sim 0.57 \times 10^{57} \times \left(3.3 \times 10^{-3} \times 7.275 \text{ MeV} + 6.0 \times 10^{-5} \times (7.275 + 7.161) \text{ MeV}\right)$$

$$= 1.42 \times 10^{55} \text{ MeV} = 2.27 \times 10^{49} \text{ erg}. \tag{17.1.3}$$

From Fig. 11.1.50 in Volume 1, the binding energy of the precursor at the red giant tip is $\sim 1.94 \times 10^{49}$ erg. The binding energy in the model approaching the horizontal branch is $\sim 0.66 \times 10^{49}$ erg, which means that, in the process of evolving from the red giant tip to the horizontal branch, the binding energy of the model has been reduced by $\sim 1.28 \times 10^{49}$ erg. Thus, almost twice as much helium has been burned as is required to lift core electron degeneracy. The extra energy has been radiated from the star and it may be said that the process of lifting degeneracy by the conversion of nuclear energy into gravitational potential energy occurs with an efficiency slightly larger than 50%.

17.2 Horizontal branch and early asymptotic giant branch evolution

In less than 2×10^5 yr of further evolution, all four criteria for a ZAHB model are met. Conditions in the ZAHB model are described in Figs. 17.2.1–17.2.4. Examination of the gravothermal energy-generation rates and of the luminosity profile in Fig. 17.2.2 shows that the integral $\int \epsilon_{\text{grav}} \, dM$ over the mass interval $M = 0 \to 0.2 \; M_\odot$ is almost exactly

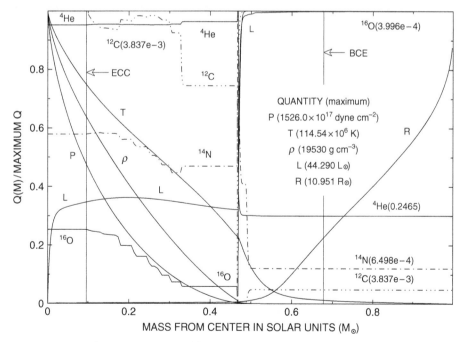

Fig. 17.2.1 Structure variables and ^{4}He and CNO abundances in a 1 $M_\odot$ model on the zero age horizontal branch ($Z = 0.015, Y = 0.275, \Delta t = 1.92335 \times 10^6$ yr)

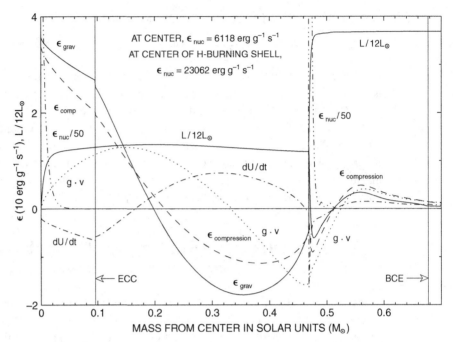

Fig. 17.2.2 Gravothermal and nuclear energy-generation rates in a 1 $M_\odot$ model on the zero age horizontal branch ($Z = 0.015$, $Y = 0.275$, $\Delta t = 1.92335 \times 10^6$ yr)

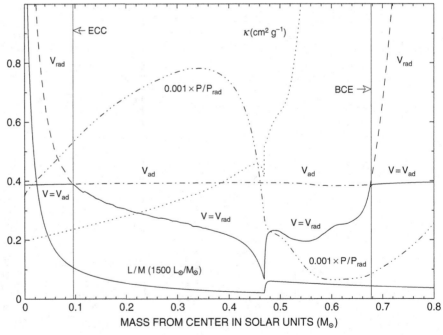

Fig. 17.2.3 V_{rad}, V_{ad}, V and ingredients of V_{rad} in a 1 $M_\odot$ model on the zero age horizontal branch ($Z = 0.015$, $Y = 0.275$, $\Delta t = 1.92335 \times 10^6$ yr)

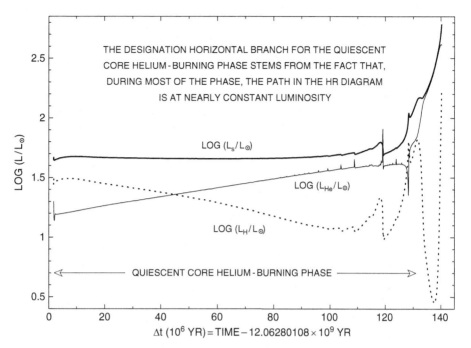

Fig. 17.2.4 Helium- and hydrogen-burning luminosities in a 1 $M_\odot$ model evolving on the horizontal branch and early AGB ($Z = 0.015, Y = 0.275$)

balanced by the integral $\int \epsilon_{\rm grav}\, dM$ over the mass interval $M = 0.2 \to 0.47\ M_\odot$. Similarly, the integral $\int \epsilon_{\rm grav}\, dM$ over the mass interval $M = 0.47 \to 0.505\ M_\odot$ offsets approximately half of the integral $\int \epsilon_{\rm grav}\, dM$ over the mass interval $M = 0.505 \to 0.47$ $M_\odot$. The net gravothermal energy-generation rate is $L_{\rm grav} = \int_0^{M_\odot} \epsilon_{\rm grav}\, dM \sim 0.51\ L_\odot$, only $\sim$1% of the total nuclear energy-generation rate of $L_{\rm nuc} \sim 43.77\ L_\odot$. The hydrogen-burning luminosity $L_{\rm H} = 29.90\ L_\odot$ is approximately twice the helium-burning luminosity $L_{\rm He} = 13.87\ L_\odot$.

From the ^{4}He abundance profile in Fig. 17.2.1, it is evident that approximately 4% of the initial helium in the core of the precursor red giant branch model has been burned in the process of lifting core electron degeneracy and contributing to radiation emitted during evolution from the red giant tip to the horizontal branch. Since the mass at the center of the hydrogen-burning shell increases by only $\sim$0.0018 $M_\odot$ during the transition from the red giant tip to the horizontal branch, only a small fraction of energy radiated during the transition is a consequence of energy produced by hydrogen burning. This means that, since most of the energy produced by helium burning during helium-flashing episodes is converted locally into gravothermal energy, most of the light emitted during the transition is a consequence of the release of stored gravothermal energy.

Logarithmic temperature-pressure gradients and the ingredients of the radiative gradient $V_{\rm rad}$ in the ZAHB model are shown in Fig. 17.2.3. The ingredient most important for understanding mixing by convection during the subsequent horizontal branch phase is the opacity. It is evident that, at the center of the ZAHB model, electron scattering is the primary contributor to the opacity. The increase in opacity outward from the center over the inner 0.45 $M_\odot$ of the ZAHB model is due to free–free absorption. This may be understood by comparing the opacity profile in Fig. 17.2.3 with the profile of $P/P_{\rm rad}$ in the same figure and with the temperature profile in Fig. 17.2.1. As described in Section 7.6 of Volume 1 (see eq. 7.6.43), the free–free contribution to the opacity is proportional to $\rho/T^{3.5}$, which is in turn proportional to $(P/P_{\rm rad})\,(1/T^{1/2})$. Over the inner 0.35 $M_\odot$ of the ZAHB model, $P/P_{\rm rad}$ increases outward and, over the next $\sim$0.1 $M_\odot$, the decrease in $P/P_{\rm rad}$ is offset by the continued decrease in temperature. The abrupt increase in opacity which coincides with the increase in luminosity through the hydrogen-burning shell is due to an increase in the electron-scattering contribution to the opacity associated with an increase in the electron number density per gram through the shell.

Variations with time in the surface luminosity and in the hydrogen- and helium-burning contributions to the surface luminosity during the core helium-burning and shell hydrogen-burning phase are shown in Fig. 17.2.4, along with variations occurring during the first part of a subsequent quiescent shell helium-burning and shell hydrogen-burning phase. The phase of quiescent core helium-burning and shell hydrogen-burning lasts for approximately 1.25×10^8 yr and, during most of this time, the surface luminosity remains almost constant, despite large variations in the relative contributions of hydrogen and helium burning to the luminosity.

The near constancy of the luminosity means that the evolutionary track in the HR diagram is nearly horizontal, and this helps understand why this phase is called the horizontal branch phase. The primary motivation for the designation horizontal branch is the fact that, as the adopted model metallicity is decreased, the evolutionary track during the nearly constant luminosity, quiescent core helium-burning phase spreads out in the surface temperature coordinate and that, in many metal poor globular clusters, the observational counterpart to this phase is in fact a horizontal branch of bolometric luminosity $L \sim 40$–50 $L_\odot$ that extends from the red giant branch almost to the main sequence.

In the metal rich model being examined in this chapter, variations in surface temperature during most of the quiescent core helium-burning phase are quite small and, in the HR diagram, the model evolves during most of the HB phase within a region of minuscule dimensions compared with the regions traversed during prior and subsequent evolutionary phases. This contrast is highlighted in Fig. 17.2.5 where the circle in the lower left hand corner of the figure encompasses the region where the HB model spends $\sim$95% of its $\sim$1.25 $\times$ 10^8 yr HB lifetime. The observational counterpart of the model in a metal-rich cluster or in the field is appropriately called a clump star.

By comparison, over a period of $\sim$6 $\times$ 10^7 yr, the precursor red giant model evolves along the rightmost track in Fig. 17.2.5 by over a factor of 60 in luminosity and by $\sim$60% in surface temperature as the mass of its hydrogen-exhausted core grows from $M_{\rm H} \sim 0.25$ $M_\odot$ to $M_{\rm H} \sim 0.47$ $M_\odot$ at the red giant tip (see Figs. 11.1.1 and 11.1.51). Open circles along the rightmost (first) red giant branch track identify the mass $M_{\rm H}$ at the center of the hydrogen-burning shell at intervals of 0.05 $M_\odot$. In less than 2 $\times$ 10^6 yr, the model evolves from the red giant tip to the HB (clump) branch along a track (not shown) which remains very close to the first red giant branch.

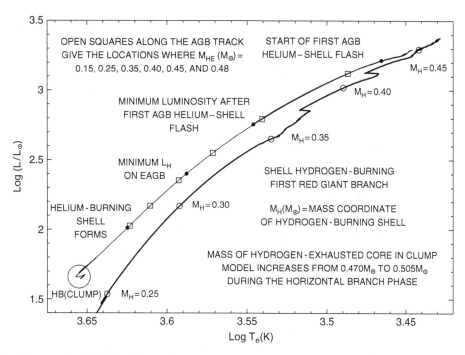

Fig. 17.2.5 Evolutionary track of a 1 $M_\odot$ model as a red giant, horizontal branch (clump) star, and asymptotic giant branch star ($Z = 0.015, Y = 0.275$)

As helium becomes exhausted at the center of the HB (clump) model, helium burning continues in an off-center shell, the center of which evolves outward in mass. The model follows a path in the HR diagram (the leftmost track in Fig. 17.2.5) which, as the mass at the center of the helium-burning shell increases, becomes increasingly closer to the first red giant branch. This convergence of tracks explains the designation asymptotic giant branch (AGB) for the shell helium-burning track. Open squares along the track identify the mass M_{He} at the center of the helium-burning shell at intervals described in the upper left hand corner of the figure.

The first portion of the AGB track, during which quiescent helium burning provides most of the surface luminosity, is called the early AGB (EAGB) phase. The solid disks along the EAGB track in Fig. 17.2.5 mark where the interior occurrences described by the text to the left of the track are taking place. Since the energy release per gram of fuel processed by helium burning is between 10.3 and 6.9 times less than is released by hydrogen burning (the exact ratio is smaller, the larger the ratio of oxygen to carbon in the final product), the rate at which the helium shell mass increases along the EAGB at any given luminosity is over seven times larger than the rate at which the hydrogen shell mass increases along the first red giant branch. However, the relationships between luminosity and shell mass along the two branches differ in such a way that the total times to traverse the same luminosity interval differ by less than a factor of seven.

The total time which elapses from when the helium-burning shell first forms (at $M_{He} \sim 0.1~M_\odot$) and when the first helium shell flash occurs (at $M_{He} \sim 0.48~M_\odot$) is

$\sim 1.3 \times 10^7$ yr, compared with the $\sim 5 \times 10^7$ yr which elapses during evolution over roughly the same luminosity interval along the first red giant branch as the mass of the hydrogen-burning shell increases from $M_H \sim 0.26\ M_\odot$ to $M_H \sim 0.47\ M_\odot$. Thus, the amount of mass processed by helium burning along the EAGB is approximately 1.8 times the amount of mass processed by hydrogen burning along the first red giant branch over the same luminosity interval, accounting for the factor of approximately $7/1.8 \sim 4$ difference in lifetimes. In a metal rich cluster, at luminosities between those of clump stars and the first red giant branch tip, approximately one out of every five stars is an EAGB star and four out of five are red giant branch stars. The time for evolution along the first red giant branch up to the luminosity level of clump stars is $\sim 5 \times 10^8$ yr.

The relative theoretical lifetimes for models in the different evolutionary phases translate into predictions regarding relative numbers of stars in corresponding branches in globular clusters. For example, because the lifetime in the HB phase in metal poor globular clusters is comparable with the lifetime along the first red giant branch and the EAGB combined at luminosities larger than the luminosity of the HB, one would expect the number of red giants in a cluster at luminosities as large as or larger than the luminosity level of HB stars to be comparable with the number of stars on the HB. Since the precise theoretical ratio is a function of the adopted initial helium to hydrogen ratio, the observed ratio of numbers can be used to estimate the initial helium to hydrogen mass ratio in metal poor globular clusters (Iben, 1968). The fact that the best estimates of the initial abundance by mass of helium (A. Buzzoni, F. Fusi-Pecci, R. Buononno, & C. E. Corsi, 1983) are close ($X_{He} = 0.23 \pm 10\%$) to the abundance by mass produced by theoretical models of Big Bang nucleosynthesis as well as to the abundance by mass of helium in interstellar matter of the lowest metallicity is consistent with the fact that globular cluster stars of the lowest metallicity are among the oldest stars in the Universe.

Time dependences of various interior characteristics in the model during the horizontal branch phase are shown in Figs. 17.2.6–17.2.8. Figure 17.2.6 shows that the mass of the convective core initially increases monotonically, but then proceeds to oscillate about a mean value of ~ 0.16–$0.20\ M_\odot$, with brief excursions of increasing amplitude. Connecting the relative maxima in the convective core mass which appear during the oscillatory phase produces a relatively smooth curve which increases monotonically over most of the oscillatory phase and appears as an extention of the initial almost monotonic curve. This interesting behavior is a consequence of the composition dependence of the opacity and can be understood by examining the structure of several adjacent models in detail. It turns out that, during the oscillatory phase, convection occurs in a shell outside of the growing convective core and, when the edge of the convective core reaches the base of the convective shell, the two convective regions merge and the consequent readjustment in the abundance distributions alters the opacity distribution in such a way that the single convective zone again splits into two zones.

Number abundances in the convective core of several pertinent isotopes are shown as functions of time in Fig. 17.2.7. The number abundances of ^{12}C and of ^{16}O are multiplied, respectively, by factors of 3 and 4, so that the curves for these isotopes and for ^{4}He actually reflect mass abundances, with the sum $Y(^4\text{He}) + 3 \times Y(^{12}\text{C}) + 4 \times Y(^{16}\text{O})$ remaining constant. It is interesting that, with one exception, the fluctuations in the abundance of

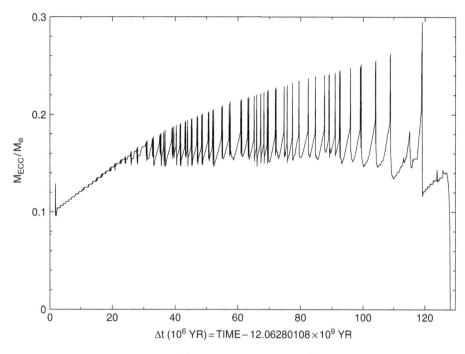

Fig. 17.2.6
Mass of the convective core in a 1 $M_\odot$ horizontal branch model ($Z = 0.015$, $Y = 0.275$)

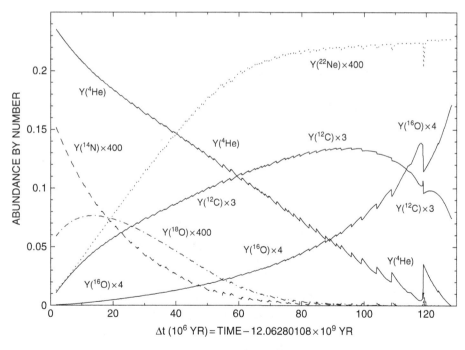

Fig. 17.2.7
Abundances at the center of a 1 $M_\odot$ horizontal branch model ($Z = 0.015$, $Y = 0.275$)

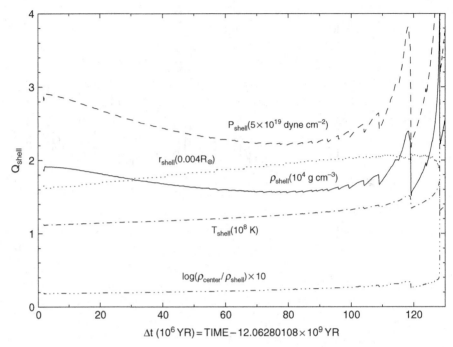

Fig. 17.2.8
Characteristics at the middle of the helium-burning region in a 1 $M_\odot$ model during the horizontal branch phase ($Z = 0.015$, $Y = 0.275$)

any given isotope are modest compared with the much larger fluctuations in the mass of the convective core. The one exception occurs toward the end of the core helium-burning phase, when the largest outward excursion of the edge of the convective core leads to an increase by almost an order of magnitude in the abundance of helium in the core. Just prior to this last ingestion of fuel, the abundances by mass of carbon and oxygen reach parity; after the ingestion, the abundance by mass of ^{16}O in the inner ~ 0.1 $M_\odot$ of the model becomes more than double the abundance by mass of ^{12}C.

Recognizing that the sum $Y(^{14}N) + Y(^{18}O) + Y(^{22}Ne)$ remains constant, the number abundances of the isotopes ^{14}N, ^{18}O, and ^{22}Ne in Fig. 17.2.7 have all been multiplied by the same factor of 400. It is evident that, at the center of a low mass star at the end of core helium burning, the CNO isotopes with which the star is born have been converted into the isotope ^{22}Ne.

Properties of matter at the center of the helium-burning region are shown as functions of time in Fig. 17.2.8. Comparing the curve for $\log(\rho_{center}/\rho_{shell})$ in Fig. 17.2.8 with the curve for $Y(^4He)$ in Fig. 17.2.7, it is evident that the center of the helium-burning region remains very close to the center of the model as long as helium is present at the center of the model ($\rho_{center}/\rho_{shell}$ increases slowly from ~ 1.6 to ~ 2.5), but moves out abruptly to much lower densities once helium vanishes at the center (at $\Delta t \sim 1.28 \times 10^8$ yr). This contrasts with the situation at the end of the main sequence phase when, once the prevailing nuclear fuel is exhausted at the center, the center of the nuclear burning region

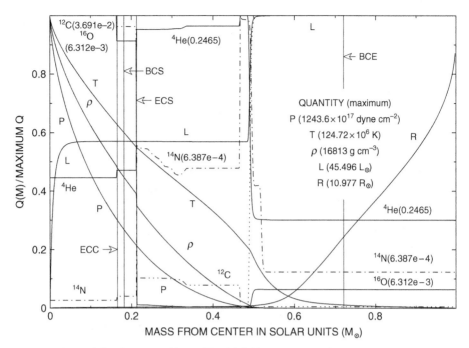

Structure variables and abundances in a 1 $M_\odot$ model nearly halfway through the horizontal branch phase ($Z = 0.015, Y = 0.275, \Delta t = 59.572 \times 10^6$ yr)

moves outward gradually. The difference is due to the fact that, in the helium-burning case, the fuel vanishes suddenly over a large region about the center whereas, in the hydrogen-burning case, the fuel increases gradually outward from the center.

The oscillatory behavior of the mass of the convective core can be understood by a consideration of the characteristics of two models near the middle of the horizontal branch phase. Characteristics of the model at $\Delta t = 0.596 \times 10^8$ yr are shown in Figs. 17.2.9–17.2.11. Structure and composition variables are shown in Fig. 17.2.9, gravothermal energy-generation rate characteristics are shown in Fig. 17.2.10, and temperature–pressure gradients and the ingredients of the radiative gradient V_{rad} are shown in Fig. 17.2.11. Of particular interest are the ^{4}He abundance distribution in Fig. 17.2.9, the variations in V_{rad} and its ingredients κ, L/M, and P/P_{rad} in Fig. 17.2.11, and the locations of convective-radiative zone boundaries shown in all three figures. Note that V_{rad} decreases steadily through the convective core primarily due to the decrease in L/M and becomes smaller than the adiabatic gradient V_{ad} at the edge of the core at $M_{CC} \sim 0.165\ M_\odot$. V_{rad} reaches a minimum just outside the convective core and begins to increase due to increases in both κ and P/P_{rad}. V_{rad} again exceeds V_{ad} at $M \sim 0.182\ M_\odot$ and continues to be larger than V_{ad} over a region of constant ^{4}He abundance until a discontinuous jump in the ^{4}He abundance is encountered at $M \sim 0.213\ M_\odot$. The associated abrupt drop in the opacity can be understood from the fact that the contribution to the opacity by free–free absorption is proportional to (see eq. (7.6.43) in Section 7.6 in Volume 1) $S = \sum_i (X_i/A_i)\, Z_i^2$, where X_i, A_i, and Z_i are, respectively, abundances by mass, atomic number, and atomic charge,

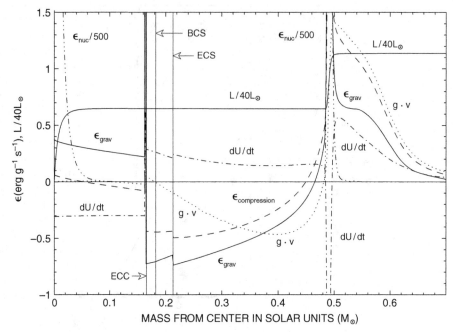

Fig. 17.2.10 Gravothermal and nuclear energy-generation rates in a 1 $M_\odot$ model nearly halfway through the horizontal branch phase ($Z = 0.015$, $Y = 0.275$, $\Delta t = 59.572 \times 10^6$ yr)

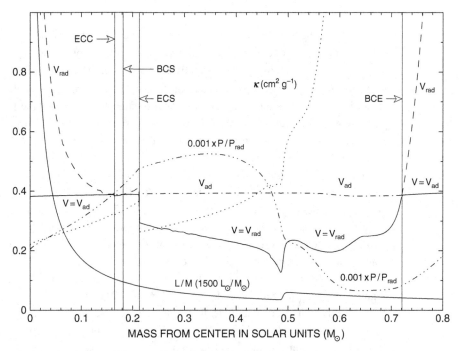

Fig. 17.2.11 V_{rad}, V_{ad}, V and ingredients of V_{rad} in a 1 $M_\odot$ model nearly halfway through the horizontal branch phase ($Z = 0.015$, $Y = 0.275$, $\Delta t = 59.572 \times 10^6$ yr)

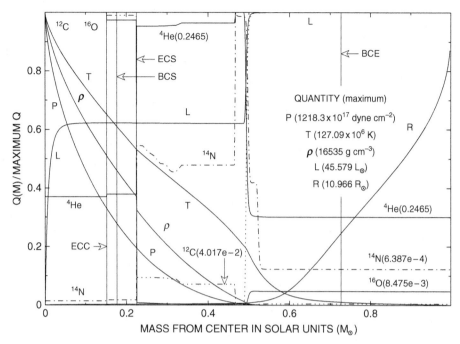

Fig. 17.2.12 Structure variables and abundances in a 1 $M_\odot$ model slightly over halfway through the horizontal branch phase
($Z = 0.015, Y = 0.275, \Delta t = 70.2107 \times 10^6$ yr)

and the sum is over all isotopes. For a mixture of completely ionized ^{4}He and ^{12}C at equal
abundances by mass, $S = 2$, whereas $S = 1$ for pure completely ionized ^{4}He. Thus, at the
outer edge of the convective core across which the abundance by mass of ^{4}He approximately doubles, from ~0.48 to ~0.95, and the ^{12}C abundance by mass drops from ~0.42
to ~0.044, the contribution of free–free absorption to the opacity decreases by almost a
factor of 2 and V_{rad} drops considerably below V_{ad}. Convection is not again encountered
until far beyond the hydrogen-burning shell, after P/P_{rad} reaches a minimum and begins
to increase outward.

The same story is told by the second model at $\Delta t = 0.702 \times 10^8$ yr, with characteristics
described in Figs. 17.2.12–17.2.14, analogues of Figs. 17.2.9–17.2.11 which describe the
first model. Small but significant differences are discernable. For example, the mass of the
convective core has decreased from 0.165 $M_\odot$ to 0.152 $M_\odot$ and the mass of the convective shell has increased from 0.031 $M_\odot$ to 0.048 $M_\odot$, extending both further inward and
further outward in mass. Comparing the ^{4}He abundance profile in Fig. 17.2.12 with the
corresponding profile in Fig. 17.2.9, it is evident that, during the evolution between the
two models, the ^{4}He abundance in the convective core has decreased by about 16% while
the ^{4}He abundance in the region between the outer edge of the convective core and the
outer edge of the convective shell has decreased by ~19%. From the profiles of nuclear
energy-generation rates in Figs. 17.2.10 and 17.2.13 it is clear that helium-burning is confined to the inner part of the convective core, which means that the decrease in the ^{4}He
abundance between the edges of the two convective regions has been the consequence of

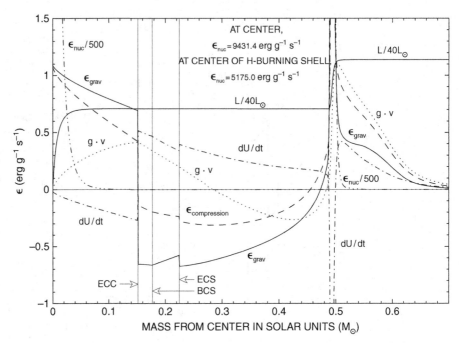

Fig. 17.2.13   Gravothermal and nuclear energy-generation rates in a 1 $M_\odot$ model slightly over halfway through the horizontal branch phase ($Z = 0.015$, $Y = 0.275$, $\Delta t = 70.211 \times 10^6$ yr)

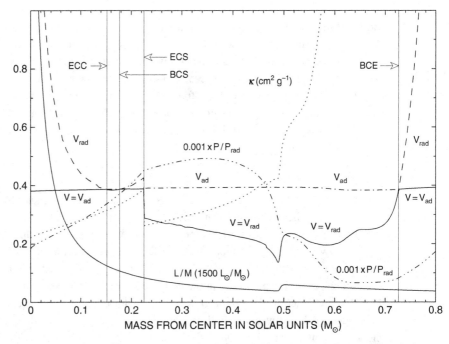

Fig. 17.2.14 V_{rad}, V_{ad}, V and ingredients of V_{rad} in a 1 $M_\odot$ model slightly over halfway through the horizontal branch phase ($Z = 0.015$, $Y = 0.275$, $\Delta t = 70.211 \times 10^6$ yr)

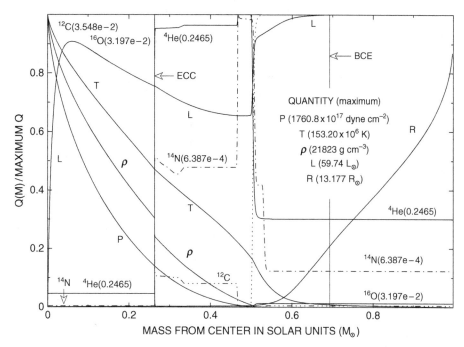

Fig. 17.2.15 Structure variables and abundances in a 1 $M_\odot$ model experiencing a final spurt of convective core growth ($Z = 0.015$, $Y = 0.275$, $\Delta t = 118.978 \times 10^6$ yr)

mixing all the way from the center to the outer edge of the convective shell at least once during the evolution between the two models. That is, at one or more times, the convective core and the convective shell have merged into one large convective core and then broken up again into two convective regions. Inspection of Fig. 17.2.6 reveals that, in fact, mergers and detachments have occurred approximately 20 times during the course of evolution between the two models.

It is evident from Fig. 17.2.6 that, after each detachment of convective regions, the mass of the convective core drops to roughly the same value (in the neighborhood of $M_{CC} \sim 0.15$ $M_\odot$) and then increases steadily with time until reaching the base of the convective shell, at which time another merger of convective zones occurs. The growth of the convective core between mergers is encouraged by taking convective overshoot into account in an elementary way: it is assumed that, after each time step, mixing extends one zone further outward than given by the formal requirement that $V_{rad} > V_{ad}$ in every zone in the mixed region. It is evident from Fig. 17.2.7 that, toward the end of the horizontal branch phase, the factor by which the central helium abundance increases temporarily in consequence of the short lived merger between convective zones increases. In the final episode of central helium enhancement, the enhancement factor is about 8.

Characteristics of the model shortly after the beginning of the final fuel-ingestion episode are shown in Figs. 17.2.15–17.2.17. The model age (measured from the peak of the helium shell flash on the first red giant branch) is $\Delta t \sim 119 \times 10^6$ yr. The edge of the convective core has reached the immediately previous relative maximum of ~ 0.263 $M_\odot$,

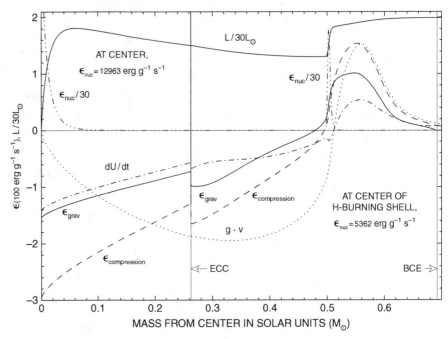

Fig. 17.2.16 Gravothermal and nuclear energy-generation rates in a 1 $M_\odot$ model during a final spurt of convective core growth ($Z = 0.015$, $Y = 0.275$, $\Delta t = 118.978 \times 10^6$ yr)

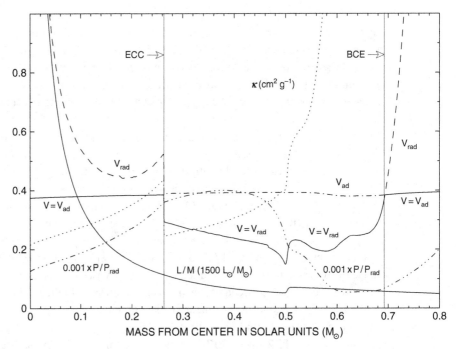

Fig. 17.2.17 V_{rad}, V_{ad}, V and ingredients of V_{rad} in a 1 $M_\odot$ model during a final spurt of convective core growth ($Z = 0.015$, $Y = 0.275$, $\Delta t = 118.978 \times 10^6$ yr)

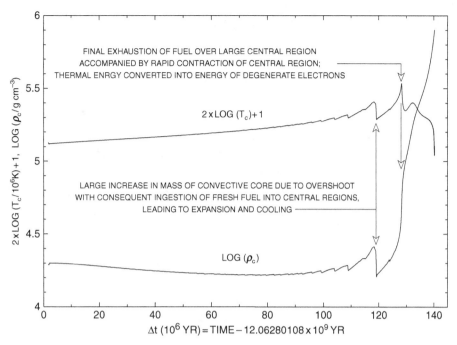

Fig. 17.2.18 Central temperature and density in a 1 M$_\odot$ model evolving on the horizontal branch and on the early AGB
($Z = 0.015, Y = 0.275$)

and the number abundance of helium at the center has increased from a minimum of
$Y(^4\text{He}) \sim 0.004$ to $Y(^4\text{He}) \sim 0.0113$. The increase in the abundance of helium in the core
leads to an increase in the helium-burning luminosity which has the effect of increasing
V_{rad} throughout the convective core. Particularly enlightening is a comparison of the profiles for L/M and V_{rad} in Fig. 17.2.17 with these same profiles in Fig. 17.2.14 for a model
midway in the HB phase. Although the central helium abundance in the more evolved
model is much smaller than in the less evolved model, the central temperature is much
larger, with the result that both L/M and V_{rad} are larger everywhere in the convective
core of the more evolved model than in the less evolved model. Instead of falling below
V_{ad} at the edge of the convective core as in the less evolved model, V_{rad} in the convective core of the more evolved model reaches a minimum considerably larger than V_{ad}
and then continues to increase outward, not falling below V_{ad} until the discontinuities in
abundances are reached at the edge of the convective core. The result is that, in the more
evolved model, convective overshoot plays an important role in forcing the growth of the
convective core.

The sudden increase in the rate of nuclear energy release due to the rapid ingestion of
fresh fuel by the convective core leads to expansion and cooling over the entire region
between the center of the model and the hydrogen-burning shell, as is demonstrated by
the profiles of gravothermal energy-generation rates in Fig. 17.2.16. The expansion and
cooling at the center of the model are displayed in Fig. 17.2.18 where the time dependences of the central density and the central temperature during the HB and EAGB phases

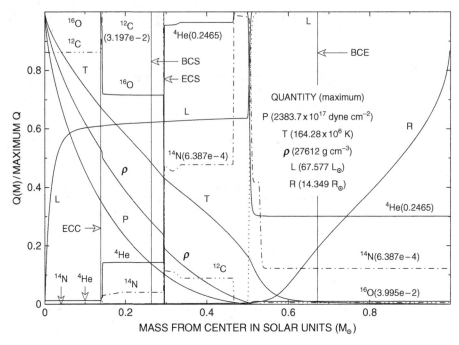

Structure variables and abundances in a 1 $M_\odot$ model near the end of the core helium-burning phase
($Z = 0.015$, $Y = 0.275$, $\Delta t = 126.889 \times 10^6$ yr)

are shown. Very shortly into the last ingestion episode, the rate of decrease in nuclear
reaction rates associated with expansion and cooling exceeds the rate of increase in the
abundance of freshly ingested fuel, causing the mass of the convective core to decrease
as rapidly as it has grown. At maximum, the mass of the convective core is $M_{CC} \sim$
0.296 $M_\odot$ and the number abundance of helium at the center is $Y(^4\mathrm{He}) \sim 0.036$. At
the end of the ingestion episode, the mass of the convective core has stabilized at
$M_{CC} \sim 0.12\ M_\odot$.

As evolution progresses over the next $\sim 10^7$ yr, the central helium abundance decreases
monotonically (see Fig. 17.2.7) and, over most of this time, the mass of the convective
core increases steadily, from $\sim 0.12\ M_\odot$ to $\sim 0.14\ M_\odot$ (see Fig. 17.2.6), before rapidly
decreasing to zero. Conditions in the model at $\Delta t = 126.889 \times 10^6$ yr, just as the mass
of the convective core is beginning its final decline to zero are shown in Figs. 17.2.19–
17.2.21. The number abundance of $^4\mathrm{He}$ in the convective core, of mass $M_{CC} \sim 0.14\ M_\odot$,
is $Y(^4\mathrm{He}) = 3.074 \times 10^{-3}$, the nuclear energy-generation rate at the center is $\epsilon_{nuc} = 6975$
erg g s^{-1}, the midpoint of the helium-burning region is at $M = 0.009\,50\ M_\odot$, and the total
helium-burning luminosity is $L_{He} = 40.295\ L_\odot$. The center of the hydrogen-burning shell
is at $M_H = 0.5047$, where $\epsilon_{nuc} = 9292$ erg g^{-1} s^{-1}, and $L_H = 25.15\ L_\odot$.

As shown in Fig. 17.2.19, the number abundance of $^{16}\mathrm{O}$ exceeds that of $^{12}\mathrm{C}$ in the
convective core and is comparable to that of $^{12}\mathrm{C}$ over the remainder of the region occupied by the convective core at its maximum outward extent (note that the maximum num-
ber abundances of $^{16}\mathrm{O}$ and $^{12}\mathrm{C}$ are, respectively, $Y(^{16}\mathrm{O}) = 3.995 \times 10^{-2}$ and $Y(^{12}\mathrm{C}) =$

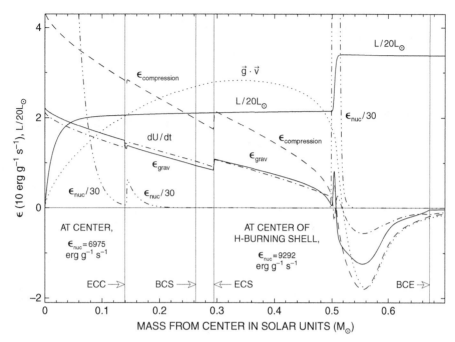

Fig. 17.2.20 Gravothermal and nuclear energy-generation rates in a 1 M$_\odot$ model near the end of the horizontal branch phase ($Z = 0.015, Y = 0.275, \Delta t = 126.889 \times 10^6$ yr)

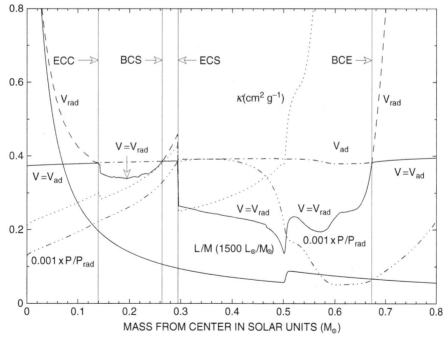

Fig. 17.2.21 V_{rad}, V_{ad}, V and ingredients of V_{rad} in a 1 M$_\odot$ model near the end of the core helium-burning phase ($Z = 0.015, Y = 0.275, \Delta t = 126.889 \times 10^6$ yr)

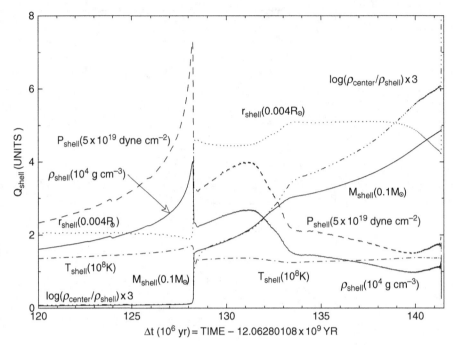

Fig. 17.2.22 Characteristics at the middle of the helium-burning shell in a 1 $M_\odot$ model near the end of the HB phase and during the EAGB phase ($Z = 0.015$, $Y = 0.275$)

3.197×10^{-2}). The profile of the nuclear energy-generation rate in Fig. 17.2.20 shows that a helium-burning shell is beginning to form just beyond the outer edge of the convective core in the current model. Comparing the profile of $V_{\rm rad}$ with the location of the outer edge of the convective core in Fig. 17.2.20, it is evident that the mass of the convective core is becoming smaller. The same inference follows from the fact that the innermost discontinuity in the abundances of each of the isotopes shown in Fig. 17.2.19 occurs just outside of the outer edge of the convective core.

The convective core persists for another 1.38×10^6 yr of evolution. As the convective core shrinks in mass, the secondary peak in the profile of $\epsilon_{\rm nuc}$ located at $M \sim 0.14\ M_\odot$ in Fig. 17.2.20 increases in magnitude while the peak at the center decreases in magnitude until the center of the helium burning region shifts abruptly to $M \sim 0.14\ M_\odot$ as convection ceases at the center. When this occurs, the number abundance of helium at the center is $Y(^4{\rm He}) \sim 1.6 \times 10^{-5}$.

The time dependences of characteristics at the center of the helium-burning region are shown in Fig. 17.2.22 and the time dependences of the helium- and hydrogen-burning luminosities are shown in Fig. 17.2.23. The abrupt changes found in all profiles at time $\Delta t \sim 128.27 \times 10^6$ yr identify where the off center helium-burning shell is created. These changes mark the formal end of the horizontal branch phase and the beginning of the AGB phase. The position in the HR diagram when this occurs is identified by the lowest luminosity solid disk along the leftmost track in Fig. 17.2.5. At this point, the mass at

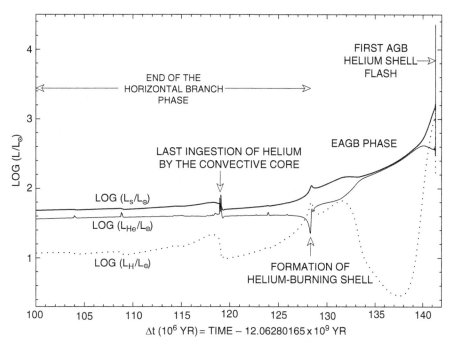

Fig. 17.2.23 Helium- and hydrogen-burning luminosities and surface luminosity in a 1 $M_\odot$ model transitioning from the horizontal branch to the AGB ($Z = 0.015, Y = 0.275$)

the center of the hydrogen-burning shell is $M_H = 0.5051$ $M_\odot$, compared with a mass of $M_H = 0.4705$ $M_\odot$ at the beginning of the horizontal branch phase.

The time required to evolve in the HR diagram along the EAGB (early asymptotic giant branch) prior to the occurrence of the first AGB helium shell flash can be determined by identifying the times at which values of M_{shell} in Fig. 17.2.22 correspond to values of M_{He} associated with the open boxes along the AGB track in Fig. 17.2.5. For example, approximately 3.7×10^6 yr are required to evolve between the two lowest luminosity boxes and approximately 1.3×10^7 yr are required to evolve from the lowest luminosity box to the highest luminosity box just prior to the onset of the first helium shell flash on the AGB.

Conditions in a model when L_H is near its absolute minimum during the EAGB phase are shown in Figs. 17.2.24–17.2.28. At an age of $\Delta t = 137.98 \times 10^6$ yr and at a location in the HR diagram given by $\log L_s(L_\odot) = 2.453$ and $\log T_e(K) = 3.582$, the model lies slightly above the second solid disk and approximately midway between the third and fourth open boxes along the EAGB track in Fig. 17.2.5. Structure and composition variables are shown in Fig. 17.2.24 for the entire model and in Fig. 17.2.25 for a region in the model encompassing just the helium-burning shell and the helium–hydrogen interface. Scaling of variables for both figures is given in Fig. 17.2.24. Both figures show that, as the helium-burning shell moves outward in mass through the region occupied by the convective core at its maximum outward extent ($M_{CC}^{max} = 0.296$ $M_\odot$), the ^{16}O abundance by number left behind is roughly double the ^{12}C abundance by number. Thus, in the entire

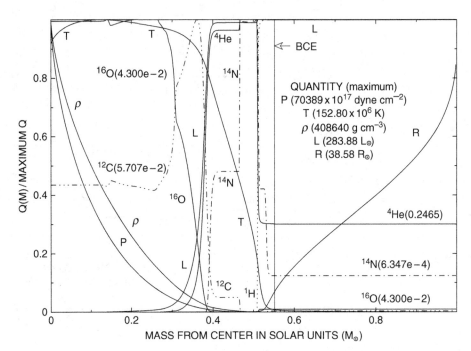

Fig. 17.2.24   Structure variables and abundances in a 1 $M_\odot$ model near minimum L_H during the early AGB phase
($Z = 0.015$, $Y = 0.275$, $\Delta t = 137.98 \times 10^6$ yr)

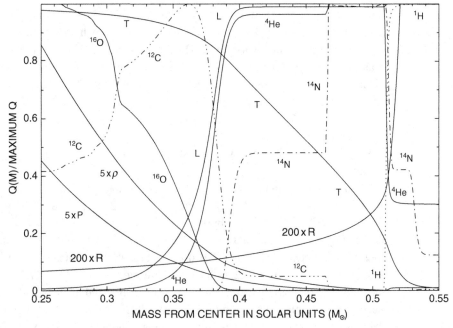

Fig. 17.2.25   Structure variables and abundances in a 1 $M_\odot$ model near minimum L_H during the early AGB phase
($Z = 0.015$, $Y = 0.275$, $\Delta t = 137.98 \times 10^6$ yr)

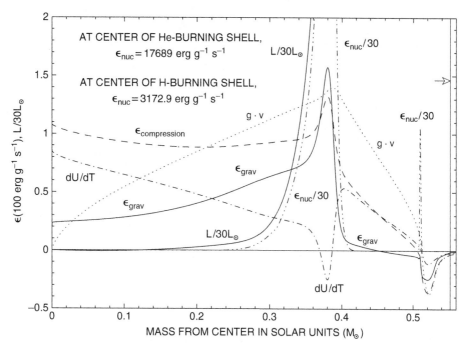

Fig. 17.2.26 Gravothermal and nuclear energy-generation rates in a 1 $M_\odot$ model near minimum L_H during the early AGB phase ($Z = 0.015$, $Y = 0.275$, $\Delta t = 137.98 \times 10^6$ yr)

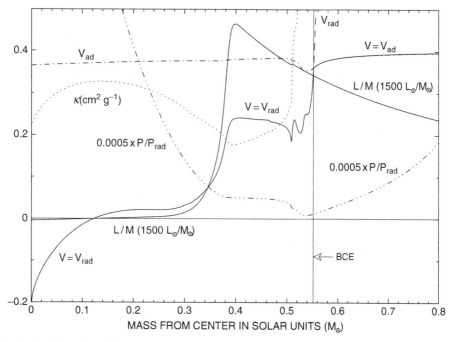

Fig. 17.2.27 V_{rad}, V_{ad}, V and ingredients of V_{rad} in a 1 $M_\odot$ model near minimum L_H during the early AGB phase ($Z = 0.015$, $Y = 0.275$, $\Delta t = 137.98 \times 10^6$ yr)

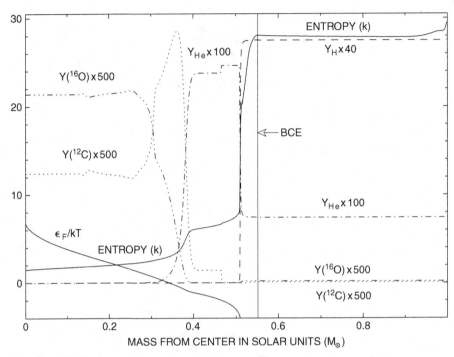

Fig. 17.2.28 Entropy, ϵ_F/kT, ^{12}C, ^{16}O, ^{4}He, and ^{1}H abundances in a 1 $M_\odot$ model near minimum L_H during the early AGB phase ($Z = 0.015, Y = 0.275, \Delta t = 137.98 \times 10^6$ yr)

helium-exhausted, carbon–oxygen core, of mass $M_{CO} \sim 0.3\ M_\odot$, ^{16}O is almost twice as abundant as ^{12}C. On the other hand, as the center of the helium-burning shell continues to move outward in mass, the ratio of ^{16}O to ^{12}C in the helium-depleted matter left behind soon drops below unity and continues to decline steadily.

The main insights from Fig. 17.2.24 which are not evident in Fig. 17.2.25 are that the densities and temperatures in the CO core are such that electrons in the core are degenerate and that neutrino losses are responsible for a decrease in temperature inward from a maximum of $T_6 \sim 153$ at $M \sim 0.12\ M_\odot$. Figure (17.2.25) highlights the fact that, in contrast with the hydrogen-burning shell along the first red giant branch, the helium-burning shell along the EAGB is very broad. In the model under consideration, the masses below which 10%, 50%, and 90% of the helium-burning luminosity are generated are, respectively, $M_{He,10} = 0.3400\ M_\odot$, $M_{He,50} = 0.3751\ M_\odot$, and $M_{He,90} = 0.3886\ M_\odot$, giving a total width of $\Delta M_{He} \equiv M_{He,90} - M_{He,10} = 0.0486\ M_\odot$, two orders of magnitude larger than the width of the hydrogen-burning shell along the first red giant branch at comparable surface luminosity. From Fig. 11.1.32 in Volume 1, the first red giant branch model at $t_9 = 12.05$ has a luminosity given by $\log L(L_\odot) = 2.42$ (compared with $\log L(L_\odot) = 2.45$ for the model described by Fig. 17.2.25) and, from Fig. 11.1.27 in Volume 1, the width of the hydrogen-burning shell is $\Delta M_H \sim 0.00035\ M_\odot$.

In Fig. 17.2.26, gravothermal energy-generation rates are compared with the nuclear energy-generation rate. It is evident from the curves for $\mathbf{g} \cdot \mathbf{v}$, $\epsilon_{\text{compression}}$, and dU/dt, that all matter below the very weak hydrogen-burning shell is moving inward, releasing compressional energy, and, with the exception of a relatively tiny region centered at the helium-burning shell, increasing in internal energy. This means that, as evolution continues, both the helium-burning luminosity and the hydrogen-burning luminosites increase, consistent with the curves for L_{He} and L_H in Fig. 17.2.23 beyond the minimum in L_H at $\Delta t \sim 137.5 \times 10^6$ yr.

Logarithmic temperature–pressure derivatives and ingredients of V_{rad} are shown in Fig. 17.2.27 and entropy, ϵ_F/kT, and number abundances of four composition variables are shown in Fig. 17.2.28. The fact that the curve for V_{rad} passes through zero at $M = 0.12$ $M_\odot$ is useful in identifying the position below which energy flows inward to supply neutrino losses. The curve for ϵ_F/kT in Fig. 17.2.28 along with the helium-abundance profile shows that, throughout the helium-exhausted CO core of mass $M_{CO} \sim 0.3$ $M_\odot$, electrons are degenerate. Finally, the entropy profile exhibits the barrier coincident with the hydrogen-helium interface that was introduced in Section 17.1 to help understand why convection during a helium shell flash does not extend into hydrogen-rich material.

As shown in Fig. 17.2.23, after the absolute minimum in L_H, L_H increases more rapidly with time than does L_{He}, and L_H exceeds L_{He} at $\Delta t \sim 140.65 \times 10^6$ yr. Since the fuel content of one gram of helium is $\sim$6.7–10 times smaller than the fuel content of one gram of hydrogen (the exact factor being a function of the final ratio of ^{16}O to ^{12}C), the center of the helium-burning shell approaches the center of the hydrogen-burning shell ever more closely in mass. When $L_H \sim 3\,L_{He}$, the rate of helium burning accelerates exponentially and the model experiences a first helium-burning shell flash on the AGB. The changes with time of the global helium- and hydrogen-burning luminosities and of the surface luminosity as this first flash develops and decays are described in the next section.

Conditions in a model approaching the beginning of the flash, when $\Delta t = 141.3544 \times 10^6$ yr, are shown in Figs. 17.2.29–17.2.33. As seen by comparing structure variables in Figs. 17.2.30 and 17.2.25, in the course of evolution between the model near minimum L_H and the model approaching flash conditions, the centers of the two nuclear burning shells have increased, respectively, from $M_{He} = 0.375\,09$ $M_\odot$ and $M_H = 0.509\,71$ $M_\odot$ in the model near minimum L_H to $M_{He} = 0.486\,78$ $M_\odot$ and $M_H = 0.521\,04$ $M_\odot$ in the model approaching flash conditions. Thus, during evolution between the two models, the helium-burning shell has processed almost exactly ten times as much mass as the hydrogen-burning shell, and the separation between the centers of the nuclear burning shells has decreased from 0.134 62 $M_\odot$ to 0.034 26 $M_\odot$. The width in mass of the helium-burning shell has decreased from $\Delta M_{He} = 4.86 \times 10^{-2}$ $M_\odot$ to $\Delta M_{He} = 1.84 \times 10^{-2}$ $M_\odot$ and the width of the hydrogen-burning shell in the more evolved model is $\Delta M_H = 3.33 \times 10^{-4}$ $M_\odot$.

Comparison between structure variables in Figs. 17.2.29 and 17.2.24 shows that the central temperature has decreased by $\sim$30% and the central density has increased by a factor of $\sim$3, and comparison between thermodynamic variables in Figs. 17.2.32 and 17.2.27 shows that, correspondingly, the degeneracy indicator ϵ_F/kT at the center has increased by a factor of about 2.6.

That the global rate of helium burning in the model approaching flash conditions is increasing is corroborated in Fig. 17.2.31 by the fact that temperatures are increasing $(dU/dt > 0)$ in and above a broad region through the helium-burning region outlined by

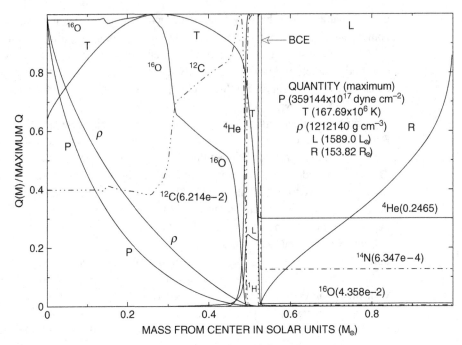

Fig. 17.2.29 Structure variables and abundances in a 1 M$_\odot$ model beginning the first helium shell flash on the AGB ($Z = 0.015$, $Y = 0.275$, $\Delta t = 141.3544 \times 10^6$ yr)

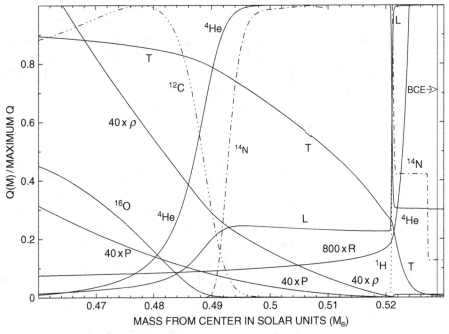

Fig. 17.2.30 Structure variables and abundances in a 1 M$_\odot$ model beginning the first helium shell flash on the AGB ($Z = 0.015$, $Y = 0.275$, $\Delta t = 141.3544 \times 10^6$ yr)

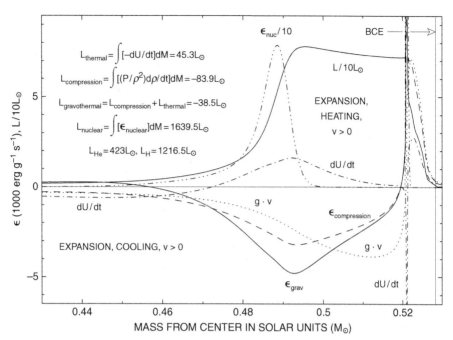

Gravothermal and nuclear energy-generation rates in a 1 $M_\odot$ model beginning the first helium shell flash on the AGB ($Z = 0.015, Y = 0.275, \Delta t = 141.3544 \times 10^6$ yr)

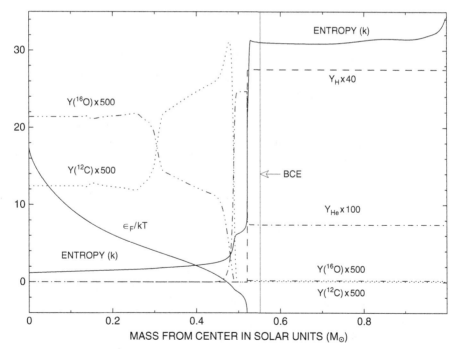

Entropy, ϵ_F/kT, ^{12}C, ^{16}O, ^{4}He, and ^{1}H abundances in a 1 $M_\odot$ model beginning the first helium shell flash on the AGB ($Z = 0.015, Y = 0.275, \Delta t = 141.3544 \times 10^6$ yr)

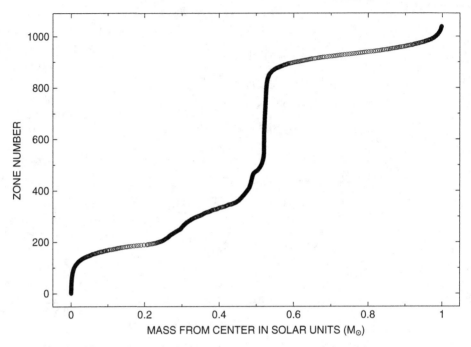

Fig. 17.2.33   Zoning in a 1 $M_\odot$ model beginning the first helium shell flash on the AGB ($Z = 0.015$, $Y = 0.275$, $\Delta t = 141.3544 \times 10^6$ yr)

the nuclear energy-generation profile centered at $M \sim 0.488\ M_\odot$. That excess energy is being injected into the model by helium burning is evident from the fact that, over the same region where temperatures are increasing, matter is expanding ($\epsilon_{\text{compression}} < 0$) outward ($\mathbf{g} \cdot \mathbf{v} < 0$). Actually, matter is expanding and moving outward all the way to the center and all matter below $M \sim 0.47\ M_\odot$, where the abundance by mass of helium is $X_{\text{He}} \sim 0.05$, is cooling and it is cooling to the extent that the global rate of change in the thermal energy content of the entire model is negative, in spite of the fact that temperatures are increasing over most of the helium-burning region.

That the flash will be much less extreme than the first flash which terminates evolution upward along the first red giant branch is evident from the fact demonstrated by the positioning of the profiles for ϵ_{F}/kT and $Y(^4\text{He})$ in Fig. 17.2.32 that electrons in the helium-burning region are not degenerate.

Several regions of negative luminosity arise in the model during the first and subsequent helium shell flashes. Since it is not advisable to employ the magnitude of luminosity variations in determining zoning in the neighborhood of places where luminosity changes signs, an attempt has been made to anticipate where the sign changes will occur and to introduce fine zoning in the associated neighborhoods ahead of time. Zoning in the model described in Figs. 17.2.29–17.2.32 is shown in Fig. 17.2.33, where zone number is plotted against mass.

17.3 The first helium shell flash on the asymptotic giant branch

In Fig. 17.3.1, the global nuclear burning energy-generation rates L_H and L_{He}, the global gravothermal energy-generation rate L_g, and the surface luminosity L_s and surface radius R_s are described as functions of time over a time period extending from considerably before the beginning of the first AGB helium shell flash to considerably past the first flash, when hydrogen burning has again become the dominant nuclear energy source as well as the prime contributor to the surface luminosity. These same quantities are described in Fig. 17.3.2 for a time period centered on the first flash, from shortly before helium burning becomes the dominant nuclear energy source to shortly before hydrogen burning again becomes the dominant nuclear energy source. In both figures, the gravothermal luminosity is described by the function $\log\left(1 + |L_g|\right)$ multiplied by the sign of L_g and it is understood that the global gravothermal luminosity L_g is in solar units. The function has the property that, for small L_g, it is approximately equal to $L_g/2.3$ and, for large $|L_g|$ it is approximately $\pm \log\left(|L_g|\right)$. Thus, the function always has the same sign as L_g and, for large $|L_g|$, it offers a meaningful comparison with $\log L_H$, $\log L_{He}$, and $\log L_s$.

From the fact that the AGB track in Fig. 17.2.5 has no structure, one may deduce that the model moves upward and downward along a single path in the HR diagram, with surface radius and surface luminosity increasing and decreasing in lockstep. The tight correlation

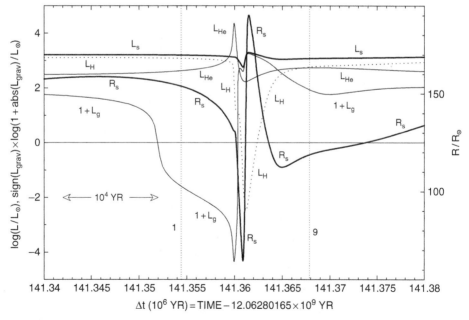

Fig. 17.3.1   Global luminosities and radius of a 1 $M_\odot$ model during the first thermal pulse on the asymptotic giant branch ($Z = 0.015, Y = 0.275$)

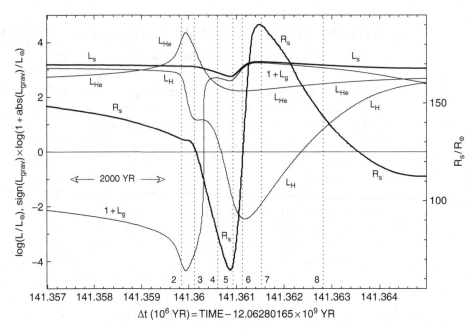

Fig. 17.3.2 Global luminosities and radius of a 1 $M_\odot$ model during the first thermal pulse on the asymptotic giant branch
($Z = 0.015, Y = 0.275$)

is exhibited explicitly in Fig. 17.3.3. A straight line fit to the AGB track in Fig. 17.2.5
yields

$$\frac{\mathrm{d}\log L_s}{\mathrm{d}\log T_e} \sim -6, \tag{17.3.1}$$

or

$$L_s \overset{\sim}{\propto} \frac{1}{T_e^6}. \tag{17.3.1a}$$

Coupled with the relationship $T_e^4 \propto L_s/R_s^2$, one has that

$$L_s \overset{\sim}{\propto} R_s^{6/5}, \tag{17.3.2}$$

accounting for the almost linear relationship betweee L_s and R_s exhibited in Fig. 17.3.3.

Luminosity profiles centered on the nuclear active region in one model before (model 1),
two models during (models 2 and 3), and one model after (model 4) the peak helium-
burning phase are shown in Fig. 17.3.4. A number beside each profile corresponds to the
same number beside a dotted vertical line in Fig. 17.3.1 or Fig. 17.3.2 which identifies the
age of the model and allows one to correlate the shape of the luminosity profile in the model
with the global rates of nuclear and gravothermal energy generation in the model and with
the directions in which the surface radius and luminosity of the model are changing.

Local properties of model 1 have been described in Figs. 17.2.29–17.2.33. The luminosity profile for this model in Fig. 17.3.4 serves as a reference standard for the other three
profiles. The gentle rise in the profile for model 1 over a mass interval $\Delta M \sim 0.02\ M_\odot$

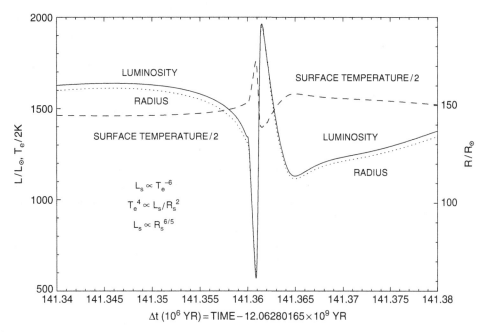

Fig. 17.3.3 Luminosity, radius, and surface temperature of a 1 $M_\odot$ model before, during and after the first thermal pulse on the asymptotic giant branch ($Z = 0.015$, $Y = 0.275$)

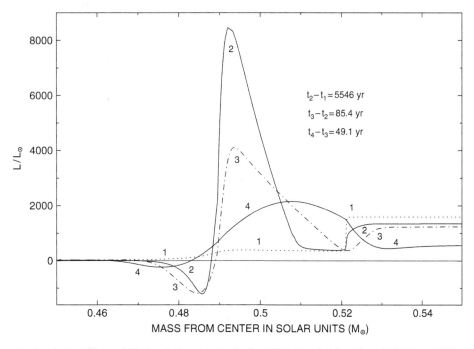

Fig. 17.3.4 Luminosity in a 1 $M_\odot$ model before, during, and after the first AGB helium shell flash ($Z = 0.015$, $Y = 0.275$)

centered at $M \sim 0.48\ M_\odot$ identifies the base of the helium-rich shell and the steep rise in the profile at $M \sim 0.52\ M_\odot$ identifies the location of the helium-hydrogen interface. The increase in luminosity through the helium-burning shell and the increase in luminosity through the hydrogen-burning shell are in essentially the same ratio ($\sim 1/3$) as is given by the ratio of the intersections of the curves L_{He} and L_H with the vertical dotted line labeled 1 in Fig. 17.3.1. The luminosity profile exhibits no detectable evidence of gravothermal activity, consistent with the fact deduced from Fig. 17.3.1 that $L_{grav} \sim -50\ L_\odot$, compared with a surface luminosity of $L_s \sim 1600\ L_\odot$. However, the fact that the gravothermal energy-generation rate is negative and is in absolute magnitude approximately 10% of $L_{He} \sim 500\ L_\odot$ suggests that a fraction of the nuclear energy liberated by helium burning is being converted into stored heat or gravitational potential energy or both. The fact that L_{He} is increasing is evidence that heat is being stored locally in the region where helium is burning and the fact that L_H is decreasing (not obvious from the figures but evident from a detailed inspection of the models) is evidence that matter in the vicinity of the hydrogen-burning shell is expanding or cooling or both. Inspection of the curves for dU/dt, $\mathbf{g} \cdot \mathbf{v}$, and $\epsilon_{compression}$ in the neighborhood of the helium-hydrogen interface in Fig. 17.2.31 shows that matter at the base of the hydrogen-burning region is cooling, but still moving inward and contracting.

In approximately 550 yr following model 1, the helium-burning luminosity reaches a relative maximum given by $\log L_{He}^{max}(L_\odot) = 4.3564$, very similar to the maximum achieved in the second flash on the first red giant branch (see Fig. 17.1.23). In both cases, the flux of energy produced in the flash causes the formation of a convective shell, the outer edge of which does not reach the hydrogen–helium interface. In the case of the first AGB flash, the outer edge reaches a maximum mass of $M_{ECS}^{max} = 0.515\,78\ M_\odot$, $0.005\,38\ M_\odot$ shy of the base of the hydrogen-burning shell. The convective shell lasts for approximately 450 yr, disappearing when the helium-burning luminosity has dropped to about $L_{He} \sim 4000\ L_\odot$.

Luminosity profiles in two models which flank the model of maximum L_{He} are given in Fig. 17.3.4 by the curves labeled 2 and 3, at times identified in Fig. 17.3.2 by the vertical dotted lines labeled by the same numbers. From luminosity profiles 1 and 2 in the neighborhood of $M \sim 0.52\ M_\odot$, it is evident that the hydrogen-burning luminosity has been approximately halved during evolution between the two models and this is consistent with the values given by the intersections of the curve for L_H with the vertical lines labeled 1 and 2 in Figs. 17.3.1 and 17.3.2, respectively. The increase in luminosity outward beyond the location of the hydrogen-burning shell along curve 2 indicates that the reduction in the flux of energy flowing into the model envelope due to the reduction in the rate of hydrogen burning is being partially compensated for by the liberation of gravothermal energy from matter which is contracting inward. From the intersection of the curve for $\log L_H$ with the dotted curve labeled 3 in Fig. 17.3.2 it is evident that, in model 3, the hydrogen-burning shell is well on its way to extinction as an effective energy source and the luminosity profile 3 in Fig. 17.3.3 in the region beyond $M = 0.52\ M_\odot$ demonstrates that gravothermal energy liberated by contracting matter makes up for most of the decrease in the contribution of hydrogen burning to the luminosity flowing toward the surface.

From Fig. 17.3.2, it is evident by inspection that, for both models 2 and 3, $L_{grav} \sim -L_{He} \sim 10^4\ L_\odot$, demonstrating that almost all of the energy liberated by helium burning is converted locally into gravothermal energy. Since, in model 2, L_{He} is increasing with time, implying that temperatures in the helium-burning zone are increasing, one may infer that some of the nuclear energy released goes into stored heat. However, an examination of model details shows that most of the nuclear energy released goes into doing work against gravity. Since, in model 3, L_{He} is decreasing with time, one may infer that temperatures in the helium-burning region in this model are decreasing and that, therefore, not only is all of the nuclear energy released converted into gravitational potential energy, but some heat energy is also converted into gravitational potential energy. From Fig. 17.3.3, one may infer that most of the conversion occurs between the most active helium-burning region and the hydrogen–helium interface which was once the home of the major source of nuclear energy release in the model. The fact that the luminosity profiles for models 2 and 3 are negative over most of the region defined by the helium-burning shell in model 1 demonstrates that nuclear energy released near the base of the helium-rich region in models 2 and 3 flows inward. An examination of model details reveals that, again, most of the nuclear energy released in the region of negative luminosity goes into expanding matter and doing work against gravity.

The phase during which most of the energy generated by helium burning is converted into gravothermal energy stored in the model lasts of the order of 1000 yr. As L_{He} decreases below L_s, the global rate of gravothermal energy production shoots upward from a large negative value to a large positive value and, as L_{He} continues to decline, the global rate of gravothermal energy production exceeds the global rate of nuclear energy production. In the HR diagram of Fig. 17.2.5, models approach the solid disk along the AGB red giant branch designated as the position of minimum luminosity after the first helium shell flash.

Because the local rate of gravothermal energy production can be negative as well as positive, the global rate of gravothermal energy production can be a considerable underestimate of the extent of gravothermal activity. That is, writing

$$L_{grav} = L_g^+ - L_g^-, \tag{17.3.3}$$

where $L_g^+ > 0$ and $L_g^- > 0$, situations arise in which

$$L_g^+ \sim L_g^- \gg |L_{grav}|. \tag{17.3.4}$$

This is the case during an extended period following the first helium shell flash. For example, in model 4, which is approximately 770 yr older than the model at flash peak, the rate of nuclear energy generation is everywhere locally small compared with the absolute value of the rate of gravothermal energy generation, as shown in Fig. 17.3.5. The integral under the main positive peak in the distribution for ϵ_{grav} is given by

$$L_g^+ = L_{max} - L_{\text{ first min}} - (L_{He} - \delta) = 2122.2 \, L_\odot, \tag{17.3.5}$$

where $L_{max} = 2154.5 \, L_\odot$ is the maximum luminosity at $M = 0.5076 \, M_\odot$, $L_{\text{ first min}} = -228.9 \, L_\odot$ is the (negative) minimum luminosity at $M = 0.4759 \, M_\odot$, $L_{He} = 271.2 \, L_\odot$ is the helium-burning luminosity, and $\delta \sim 10 \, L_\odot$ is that portion of the integral under the first peak in the ϵ_{nuc} (helium-burning) profile centered at $M = 0.4896 \, M_\odot$ which lies to

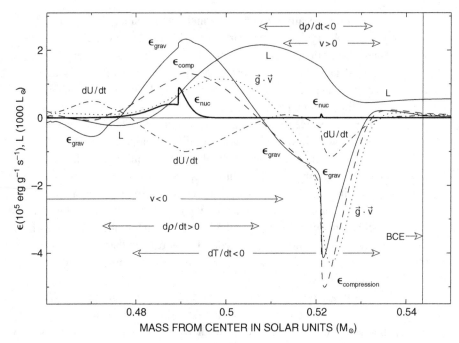

Fig. 17.3.5   Gravothermal and nuclear energy-generation rates in a 1 $M_\odot$ model relaxing after the first AGB helium shell flash ($Z = 0.015$, $Y = 0.275$, $\Delta t = 141.3606 \times 10^6$ yr)

the left of the minimum luminosity. Similarly, the integral under the main negative peak in the distribution of ϵ_{grav} is given by

$$L_g^- = L_{\mathrm{max}} - L_{\text{ second min}} + L_{\mathrm{H}} = 1712.1\ L_\odot, \qquad (17.3.6)$$

where $L_{\text{ second min}} = 444.5\ L_\odot$ is the minimum (positive) luminosity at $M = 0.5315\ M_\odot$ and $L_{\mathrm{H}} = 2.06\ L_\odot$ is the hydrogen-burning luminosity. Thus, L_g^+ and L_g^- are, respectively, 7.8 and 6.3 times larger than L_{nuc}, whereas $L_g^+ - L_g^- = 410.1$ is only 1.5 times larger than L_{nuc}.

The final contribution to the gravothermal energy-generation rate in model 4 occurs in the contracting envelope and amounts to

$$L_g^{\mathrm{env}} = L_s - L_{\text{ second min}} = 313.8\ L_\odot, \qquad (17.3.7)$$

where $L_s = 758.3\ L_\odot$ is the surface luminosity. Altogether, gravothermal energy contributes at the net rate

$$L_{\mathrm{grav}} = L_{\text{ first min}} + L_g^+ - L_g^- + L_g^{\mathrm{env}} = 495.0\ L_\odot, \qquad (17.3.8)$$

which agrees with the value of L_{grav} given by the intersection of the curve $\log(1 + L_g)$ and the dotted line labeled 4 in Fig. 17.3.2.

The directions in which thermodynamic variables and the radius at Lagrangian mass points are changing with time in model 4 are detailed explicitly in Fig. 17.3.5. The fact that temperature is decreasing over essentially the entire region from the base of the helium-burning shell to approximately half-way between the hydrogen-burning shell and the base of the convective envelope accounts for the fact that, as shown in Fig. 17.3.2, both L_{He} and L_H are decreasing with time.

Not surprisingly, matter over the entire region under the positive peak in the curve for ϵ_{grav} in Fig. 17.3.5 is both moving inward and increasing in density, whereas, matter over the entire region under the negative peak in the curve for ϵ_{grav} is both moving outward and decreasing in density. What is basically occurring is that some of the gravothermal energy that has been stored in a region centered on the location of the helium-burning shell during the peak of the helium flash is being transferred outward and stored as gravothermal energy in a region centered on the location of the remnant hydrogen-burning shell.

It is truly remarkable how closely the profiles of gravothermal energy-generation rates in Fig. 17.3.5 resemble those in Fig. 17.1.21 for a model near minimum L_{He} after the helium shell flash that occurs in the electron-degenerate core at the tip of the first red giant branch. The resemblance occurs despite the fact that the range over which energy-generation rates vary is of the order of 100 times larger and the range in mass over which the resemblance occurs is of the order of 10 times smaller in the model which is on the AGB than in the model which has just descended from the red giant tip.

It would appear that, in both cases, the nuclear energy that is liberated during a helium shell flash is converted into surface luminosity in three stages. First, during the flash, the nuclear energy released is converted into gravothermal energy stored in the neighborhood of the helium–burning shell. Second, after the flash has died down, this stored energy is transferred into a region whose base is roughly midway in mass between the location of the once active helium–burning shell and the location of the helium–hydrogen interface and whose outer edge moves gradually outward in mass to slightly beyond the helium–hydrogen interface. In this second storage region, expansion and cooling lead to a decline in the strength of the hydrogen-burning shell which is compensated for by contraction in the region between the outer edge of the second storage region and the surface, with a consequent release of gravitational potential energy which helps maintain an adequate flow of energy to the surface to satisfy the surface boundary condition. Finally, in the third stage, energy is transferred from the second storage region into the envelope and the rate of release of gravitational potential energy in the envelope declines.

Structure and composition variables centered on the region of most intense gravothermal activity in model 4 are shown in Fig. 17.3.6. Maximum values of variables are given in Fig. 17.3.7: for structure variables, maxima are listed in the legend in the upper right hand portion of the figure; for isotope abundances, maxima are listed in parentheses beside isotope symbols. The location of the maximum inward extent of the base of the convective shell during the flash coincides with the steepest portion of the ^{4}He abundance profile at $M \sim 0.4895 \, M_\odot$ and the location of the maximum outward extent of the outer edge of the convective shell coincides with the steepest portion of the ^{14}N abundance profile at $M \sim 0.5158 \, M_\odot$. The most interesting feature of the abundance distributions is that, after the helium-burning flash has ended, there is a region of mass $\sim 0.0263 \, M_\odot$ (corresponding roughly to the maximum mass of the convective shell during the flash) in which by far the dominant product of helium burning is ^{12}C at an abundance $Y(^{12}C) \cong 0.004$. That

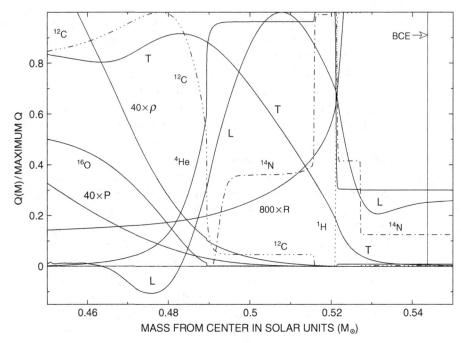

Fig. 17.3.6 Structure variables and abundances in a 1 $M_\odot$ model relaxing after the first AGB helium shell flash
($Z = 0.015$, $Y = 0.275$, $\Delta t = 141.3606 \times 10^6$ yr)

is, following the flash, the processed matter that remains just below the helium–hydrogen interface is the product of incomplete helium burning. This result has profound implications for the production of carbon in the Universe, as is elaborated in later sections.

To complete the story of model 4, structure and composition variables in the entire model are shown in Fig. 17.3.7. The increase in the luminosity between the minimum in luminosity at $M \sim 0.531$ $M_\odot$ and the surface corroborates the estimate made in eq. (17.3.7) of the rate at which gravothermal energy is being released in the envelope.

The steep decrease in surface radius which is evident in Fig. 17.3.2 as evolution proceeds from shortly before model 3 to shortly before model 5 is a consequence of the need for the model to compensate for the decline in the contribution of hydrogen burning to the flux of energy flowing into the hydrogen-rich envelope. The excess gravothermal energy that is initially stored in the region where helium burns and is now being stored in a region that is centered on the helium–hydrogen interface and extends to approximately 0.01 $M_\odot$ above this interface takes time to diffuse into the envelope, and the model has to rely at first on gravothermal energy released in the envelope to make up the deficit created by the decline in strength of the hydrogen-burning shell. Thus, the envelope contracts and converts gravitational potential energy locally into a contribution to the outgoing convective energy flow. The decline in surface luminosity is a consequence of the fact that, in the photosphere, the opacity (which is dominated by the contribution of the H^- ion) increases with increasing surface temperature, with the consequence that surface luminosity and surface radius are proportional to one another (see the discussion in Section 9.2). The phase of most rapid envelope contraction lasts for approximately 800 yr, ending shortly before model 5.

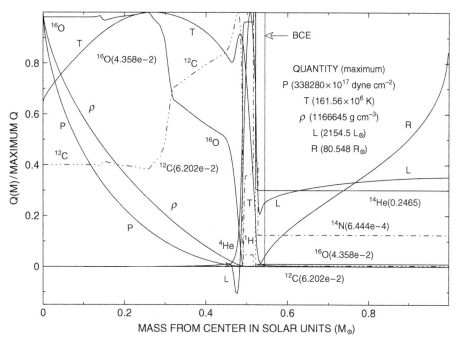

Fig. 17.3.7 Structure variables and abundances in a 1 $M_\odot$ model relaxing after the first AGB helium shell flash ($Z = 0.015, Y = 0.275, \Delta t = 141.3606 \times 10^6$ yr)

Eventually, some of the energy once stored entirely below the base of the convective envelope diffuses into the envelope and the luminosity at the base of and throughout the envelope increases. The coupling between surface radius and luminosity forces the envelope to expand. Expansion against gravity requires work, so energy is abstracted from the outward flow of energy to do this work. Instead of increasing outward, as during the phase of envelope contraction, the luminosity now decreases outward. This change in behavior is illustrated in Figs. 17.3.8 and 17.3.9 by the luminosity profiles for models 5 and 6, as compared with the luminosity profile for model 4. Note that, in the inner part of the model envelope, the luminosity profile for model 5 lies above the luminosity profile for model 4, whereas, near the surface, the situation is reversed, consistent with the fact that the surface radius of model 5 is smaller than that of model 4.

The location of the base of the convective envelope is indicated by the open circles along the luminosity profiles in Fig. 17.3.8. The location of the hydrogen-helium interface is indicated by the discontinuity at $M \sim 0.52\ M_\odot$ in the luminosity for model 9, which can be viewed as marking the end of the thermally pulsing episode when quiescent hydrogen and helium burning have replaced gravothermal energy generation as the major source of surface luminosity. In model 9, the gravothermal energy-generation rate is only $\sim$8 % of the surface luminosity and hydrogen burning and helium burning contribute almost equally to the surface luminosity.

It is evident that, for models 5 through 7, in the region beyond $M \sim 0.53\ M_\odot$ where the rate of change with mass of the luminosity has moderated, the larger the luminosity, the closer in mass is the base of the convective envelope to the hydrogen–helium interface.

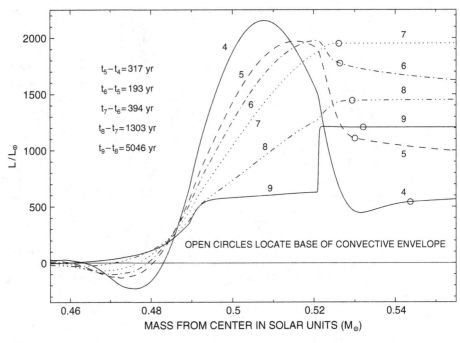

Fig. 17.3.8   Luminosity inside a 1 M$_\odot$ model powering down after the first AGB helium shell flash (Z = 0.015, Y = 0.275)

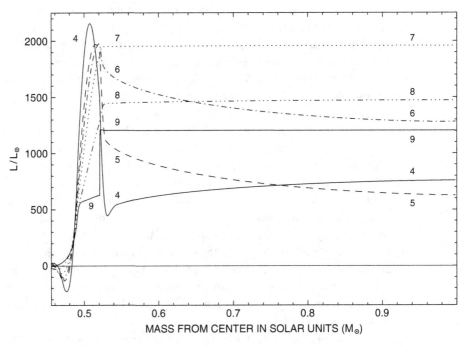

Fig. 17.3.9   Luminosity inside a 1 M$_\odot$ model powering down after the first AGB helium shell flash (Z = 0.015, Y = 0.275)

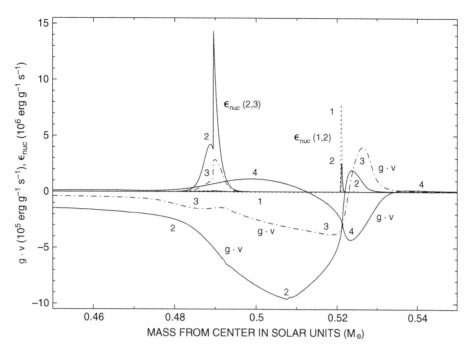

Fig. 17.3.10 Nuclear and gravitational energy-generation rates in a 1 $M_\odot$ model before, during, and after the first AGB helium shell flash ($Z = 0.015, Y = 0.275$)

This is a consequence of the facts that the adiabatic temperature gradient V_{ad} is nearly constant and the radiative temperature gradient V_{rad} is proportional to L/M. In subsequent episodes of increasingly stronger helium shell flashes, the base of the convective envelope can extend inward beyond the helium–hydrogen interface, leading to the dredge-up into the envelope of matter which has been partially processed by helium-burning reactions.

The phase of radial expansion and luminosity increase lasts for approximately 600 yr. As illustrated by the luminosity profiles for models 5 through 8 in Fig. 17.3.8, the drop in luminosity between the maximum luminosity and the luminosity at the base of the convective envelope decreases with time until, at maximum surface radius (just prior to model 7), the convective envelope becomes a passive conduit for the gravothermal energy liberated below its base. This situation persists for the roughly 1300 yr between models 7 and 8 and for several hundred additional years during which gravothermal energy remains the dominant contributor to the surface luminosity. Over this time, the rate of release of gravothermal energy gradually decreases and both the surface luminosity and the surface radius decrease apace.

Fig. 17.3.10 shows as a function of the mass coordinate the rate at which nuclear energy is being liberated by helium burning during the first helium shell flash (curves labeled by numbers 2 and 3 as well as by $\epsilon_{nuc}(2, 3)$) and the rate at which nuclear energy is being liberated by the hydrogen-burning shell (curves labeled by numbers 1 and 2 as well as by $\epsilon_{nuc}(1, 2)$). The four curves which have portions which are negative and are labeled by model numbers as well as by $\mathbf{g} \cdot \mathbf{v}$ show the rate at which work is being done against gravity (when $\mathbf{g} \cdot \mathbf{v} < 0$) or by gravity (when $\mathbf{g} \cdot \mathbf{v} > 0$). The scale for the nuclear energy-generation rates is ten times larger than the scale for the rates at which work is done by or against gravity, with the consequence that the rate of energy generation by helium burning is not identifiable for models 1 and 4 and the rate of energy generation by hydrogen burning is not identifiable for models 3 and 4.

The information in Fig. 17.3.10 is best understood in conjunction with the information in Fig. 17.3.4. It follows from the curves for $\mathbf{g} \cdot \mathbf{v}$ for models 2 and 3 in Fig. 17.3.10 (along with the knowledge that these curves remain negative from just above the H-burning shell all the way to the center), that the flux of energy generated by the release of nuclear energy during the flashing episode causes matter to move outward over the entire region from the center to a mass point approximately 0.001–0.002 $M_\odot$ beyond the helium–hydrogen interface. It is convenient to define a mass $M_{int} \sim M_{YX} + 0.01 \ M_\odot \sim 0.53 \ M_\odot$, where M_{YX} ($\sim 0.521 \ M_\odot$) is the location of the helium–hydrogen interface. For both models, $\int \mathbf{g} \cdot \mathbf{v} \ dM$ is of the order of $\int \epsilon_{nuc} \ dM$, where the first integral extends from the center to M_{int} and the second integral extends over the helium-burning region. Since the first integral is also a fair measure of the work of expansion over the region below M_{int}, and since, as follows from the luminosity profiles for models 2 and 3 in Fig. 17.3.4, essentially all of the energy flowing out from the helium-burning region is absorbed in the region which extends out to M_{int}, one may infer that matter over the entire model below M_{int} is expanding and that most of the nuclear energy released goes into the work of expansion against gravity and is stored as gravitational potential energy for future use.

The fact that the peak in the hydrogen-burning energy-generation rate shown in Fig. 17.3.10 drops by about a factor of 2 between models 1 and 2 and is undetectable for models 3 and 4 demonstrates that cooling accompanies expansion and the fact that $\mathbf{g} \cdot \mathbf{v}$ for models 2 and 3 is positive over a region of the order of $\sim 0.01 \ M_\odot$ beyond the helium–hydrogen interface demonstrates that matter over this region is moving inward; the increase in luminosity over this region that is shown in Fig. 17.3.4 demonstrates that matter in the region is also contracting and releasing gravitational potential energy.

The distributions of $\mathbf{g} \cdot \mathbf{v}$ with respect to mass in Fig. 17.3.11 for models 4–8 are to be viewed in conjunction with the luminosity profiles in Fig. 17.3.8 for these same models. It is evident from the two figures that, as evolution progresses, less and less of the excess gravothermal energy stored in the neighborhood of the helium-burning shell during the helium-flashing phase is used up in doing work against gravity in the second storage region which extends to M_{int} and more and more of the energy in both storage regions leaks outward into the model envelope and makes its way to the surface; that is, as evolution progresses, the region over which $\mathbf{g} \cdot \mathbf{v}$ is positive increases in mass and the integral $\int \mathbf{g} \cdot \mathbf{v} \ dM$ over the inner region where $\mathbf{g} \cdot \mathbf{v}$ is positive gradually dominates the integral over the outer region where $\mathbf{g} \cdot \mathbf{v}$ is is negative until the second integral vanishes.

Some aspects of the energetics of the nine models selected for explicit study are described in Table 17.3.1. In this table,

$$L_{work} = \int \mathbf{g} \cdot \mathbf{v} \ dM \qquad (17.3.9)$$

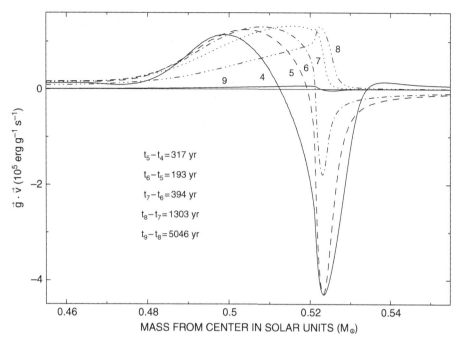

Fig. 17.3.11   Rate at which gravity does work in a 1 $M_\odot$ model after the first AGB helium shell flash ($Z = 0.015, Y = 0.275$)

is the rate at which gravity does work,

$$L_{\text{comp}} = \int \frac{P}{\rho^2} \frac{d\rho}{dt} \, dM \tag{17.3.10}$$

is the net rate at which compression delivers energy flow,

$$L_{\text{therm}} = -\int \frac{dU}{dt} \, dM \tag{17.3.11}$$

is the net rate at which changes in internal energy occur,

$$L_{\text{grav}} = L_{\text{comp}} + L_{\text{therm}} \tag{17.3.12}$$

is the net rate at which gravothermal energy changes occur,

$$L_{\text{neut}} = \int \epsilon_{\text{neutrino}} \, dM \tag{17.3.13}$$

is the rate at which neutrino losses are occurring, and

$$L_{\text{nuc}} = \int \epsilon_{\text{nucear}} \, dM \tag{17.3.14}$$

is the rate at which energy is emitted by nuclear reactions. All integrals in these equations extend over the entire model.

It is gratifying that, in all models, the numerical estimate of the rate at which work is done by gravity agrees to within essentially three significant figures with the numerical estimate of the rate at which compressional work is being done. The two estimates differ

Table 17.3.1 Global luminosities of models 1–9

model	δt (yr)	L_{work}	L_{comp}	L_{therm}	L_{grav}	L_{nuc}	L_{neut}	$L_{0.53}$	L_s	L_g^{env}	L_g^{int}
1	−5513.7	−84.0	−83.9	45.3	−38.6	1639.5	−12.2	1588	1589	1	−39.5
2	−67.9	−31110	−31050	16430	−14620	15983	−11.2	1341	1349	8	−14630
3	202.7	−8728	−8714	4620	−4094	5393	−10.0	1206	1290	84	−4177
4	693.8	1063	1078	583	495	273	−10.0	458	758	298	197
5	1010.7	1019	1033	−572	461	176	−10.1	1107	627	−480	941
6	1203.6	2355	2370	−1245	1124	169	−10.2	1735	1283	−452	1576
7	1597.8	3803	3806	−2025	1781	187	−10.4	1951	1958	7	1774
8	2901.1	2515	2518	−1353	1166	316	−10.9	1444	1471	27	1139
9	7947.3	185.7	185.8	−101.1	84.7	1131	−11.4	1207	1204	−3	87.6

because the integrands and mass intervals for the two integrals are defined differently. Whereas ϵ_{grav} is defined at zone centers, $\mathbf{g} \cdot \mathbf{v}$ is defined at zone boundaries; whereas a mass interval in the integral for L_{grav} is the difference between masses at adjacent zone boundaries, a mass interval in the integral for L_{work} is half of the mass interval between the two zones which flank the zone where $\mathbf{g} \cdot \mathbf{v}$ is defined.

As is evident from Fig. 17.3.8, in all of the models, the luminosity varies quite slowly with mass in the neighborhood of the mass $M = M_{int} \sim 0.53\ M_{\odot}$, which lies beyond where nuclear energy in the models is generated. Thus, for accounting purposes, the mass M_{int} can be regarded as the interface between the interior and the hydrogen-rich envelope and the luminosity $L_{0.53}$ at this mass can be defined as the luminosity at the base of the envelope, so that

$$L_g^{env} = L_s - L_{0.53}, \tag{17.3.15}$$

where L_s is the surface luminosity, is a measure of the rate at which gravothermal energy is either produced or absorbed in the model envelope.

The net rate L_g^{int} at which gravothermal energy is generated below M^{int} may be written as

$$L_g^{int} = L_{0.53} - L_{nuc} - L_{neut}, \tag{17.3.16}$$

or as

$$L_g^{int} = L_{grav} - L_g^{env}. \tag{17.3.17}$$

That the two formulations are equivalent follows from the fact that equating the two formulations results in $L_s = L_{nuc} + L_{neut} + L_{grav}$, as required by energy conservation. Energy conservation also stipulates that

$$L_g^{core} = L_{grav} - L_g^{int} - L_g^{env}, \tag{17.3.18}$$

where the core is defined as the point where $L = 0$ at the left hand side of the negative luminosity profile which extends inward through the base of the helium-burning region into the helium-free interior. From relevant entries in Table 17.3.1, it follows that

$$L_g^{core} \equiv 0, \tag{17.3.19}$$

in agreement with an earlier observation that the core responds adiabatically to all nuclear burning and gravothermal activity above the core.

The basic lessons from Table 17.3.1 are threefold: (1) during quiescent nuclear burning phases, as exemplified by models 1 and 9, the global rate of gravothermal energy generation is small compared with the global rate of nuclear energy generation, (2) during the helium-flashing phase, as exemplified by models 2 and 3, the bulk of the nuclear energy released is stored as gravothermal energy below the hydrogen-rich envelope in the neighborhood of the helium-burning shell, and (3) after the flash, as exemplified by models 4–8, gravothermal energy is released from the interior into the envelope. The time scale for process (3) is several times larger than that of process (2), indicating that the time scale for the

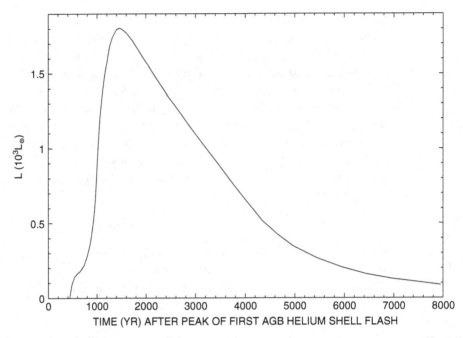

TIME (YR) AFTER PEAK OF FIRST AGB HELIUM SHELL FLASH

Fig. 17.3.12 Rate at which energy of gravothermal origin flows outward at $M = 0.53 \, M_\odot$ in a $1 \, M_\odot$ model after peak of first AGB helium shell flash ($Z = 0.015$, $Y = 0.275$)

outward diffusion of excess gravothermal energy is several times larger than the time scale for the injection of nuclear energy during the helium shell flash.

The time for energy liberated by helium burning to diffuse from above the helium-burning region where it is first stored as gravothermal energy to the base of the envelope is described quantitatively in Fig. 17.3.12, where the luminosity $L_{0.53}$ is given as a function of time following the peak of the first AGB helium shell flash (as defined by $L_{He} = L_{He}^{max}$). Approximately 1400 yr are required for the peak rate at which nuclear energy is liberated to manifest itself at the base of the envelope in the form of an ambient flux of energy of gravothermal origin. During most of this time (particularly during the roughly 900 yr corresponding to the interval between the vertical dotted lines in Fig 17.3.2 labeled 3 and 5), the primary source of surface luminosity is gravothermal energy liberated by the shrinking envelope. For the roughly 3000 or so years thereafter that the rate of energy generated by hydrogen burning remains small compared with the surface luminosity (with, say, $L_H < 100 \, L_\odot$), the dominant source of surface luminosity is the release of gravothermal energy stored in the interior during the helium flash.

Energy generated by helium burning becomes and remains the primary direct source of surface luminosity for a remarkably short time. In Fig. 17.3.1, the three luminosities L_{He}, L_H, and L_g become comparable at approximately the same time (at $\Delta t \sim 141.365 \times 10^6$ yr or $\sim$5000 yr after peak L_{He}) and L_{He} remains the largest of the three luminosities for only $\sim$2000 yr.

17.4 Systematics of thermal pulses along the asymptotic giant branch

The evolution of global luminosities during the first thermal pulse episode and for an extended period beyond this episode is shown in Fig. 17.4.1, where time zero has been reset to coincide with the time of the first relative maximum in L_{He}. It is evident that helium-burning runaways are a recurrent phenomenon. That a runaway is the consequence of the bottling up of released nuclear energy is demonstrated by the fact that, at maximum L_{He}, $|L_g|$ and L_{He} are almost identical, indicating that essentially all of the nuclear energy released is converted into gravothermal energy locally. That a runaway is the manifestation of an instability is evidenced by the fact that L_g drops precipitously through zero before it becomes apparent from the slope of the curve for L_{He} that a runaway is occurring.

The time between the onset of the first and second runaways is ~78 000 yr. After the first thermal pulse has run its course, the time which elapses while L_H remains larger than L_{He} is ~68 000 yr. The second helium-burning runaway is followed by a third, as evidenced by the two spikes in the curve for L_{He} at $\Delta t \sim 74\,000$ yr and $\Delta t \sim 79\,000$ yr. The two spikes coincide in time with the two negative spikes in the curves for $1 + L_g$. As during the first thermal pulse, the coincidences in time and the near coincidences in absolute magnitude

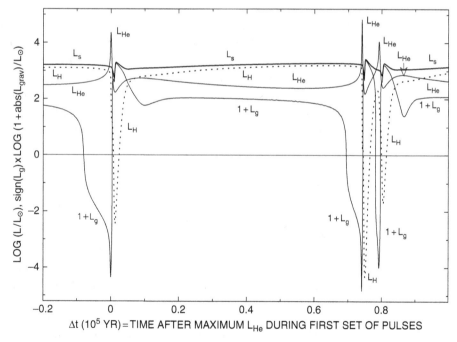

Fig. 17.4.1 Global luminosities of a 1 $M_\odot$ TPAGB model from before the first to beyond the second thermal pulse triggered by helium burning ($Z = 0.015$, $Y = 0.275$)

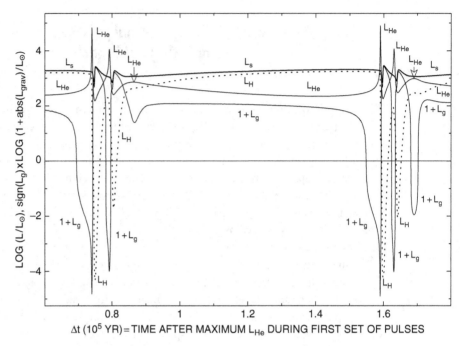

Fig. 17.4.2   Global luminosities of a 1 $M_\odot$ TPAGB model from before the second to beyond the third set of thermal pulses triggered by helium burning ($Z = 0.015, Y = 0.275$)

of the nuclear energy and gravothermal energy spikes demonstrate that the nuclear energy released during the second and third runaways is converted locally into stored gravothermal energy. The behavior of the curves for L_{He} and $\log(1 + |L_g|)$ following the third runaway suggest that yet a fourth thermal runaway is narrowly averted.

As evolution progresses, the structure of thermal pulse episodes increases in complexity. As illustrated in Fig. 17.4.2, the second thermal pulse episode is followed by a third thermal pulse episode which consists of a major pulse followed by a secondary pulse and then by what may be called a tertiary pulse, given that the third relative maximum in L_{He} is preceeded by a third steep decline in $\text{sign}(L_g) \log(1 + |L_g|)$. Thus, it is appropriate to describe the thermal pulse episodes as sets of thermal pulses separated by periods of quiescent hydrogen and helium burning. At least during early TPAGB evolution, the duration of a pulse episode is of the order of a quarter of the interval between pulse episodes.

The evolution of global luminosities during the fourth set of thermal pulses is shown in Fig. 17.4.3 and the evolution of these luminosities during the fifth set is shown in Fig. 17.4.4. One can argue that three instabilites leading to helium-burning runaways occur during the fourth thermal pulse episode and that four occur during the fifth episode.

Luminosity profiles in the model at various times during and between the first two of the fifth set of thermal pulses are shown in Fig. 17.4.5. Global luminosities of the model at the times listed in Fig. 17.4.5 are given by the intersections between the vertical dotted curves and the luminosity curves in Fig. 17.4.6. Exactly 120 time steps have been taken between two consecutive numbered models. The model radius is also shown in Fig. 17.4.6. Some

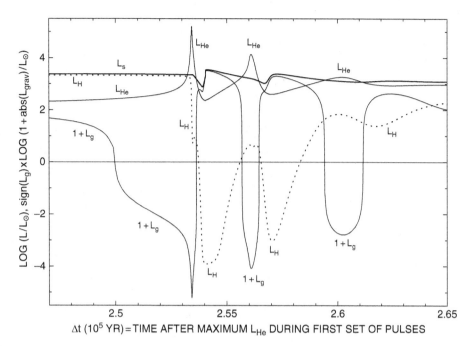

Fig. 17.4.3 Global luminosities of a 1 M$_\odot$ TPAGB model before, during, and after the fourth set of thermal pulses triggered by helium burning ($Z = 0.015$, $Y = 0.275$)

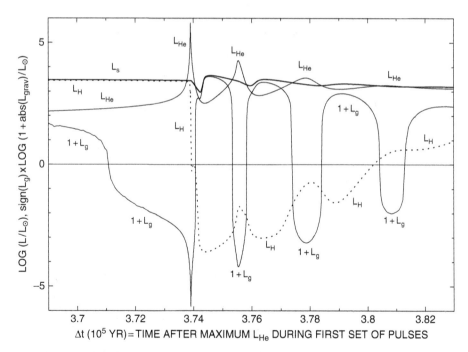

Fig. 17.4.4 Global luminosities of a 1 M$_\odot$ TPAGB model before, during, and after the fifth set of thermal pulses triggered by helium burning ($Z = 0.015$, $Y = 0.275$)

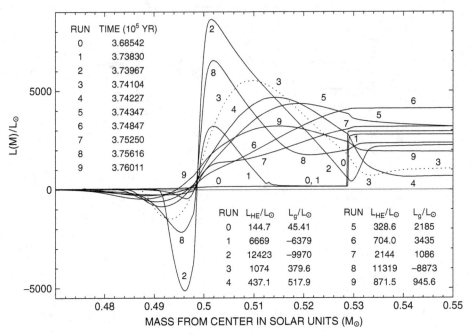

Luminosity profiles in a 1 M$_\odot$ TPAGB model during and between the first two of the fifth set of thermal pulses
($Z = 0.015$, $Y = 0.275$)

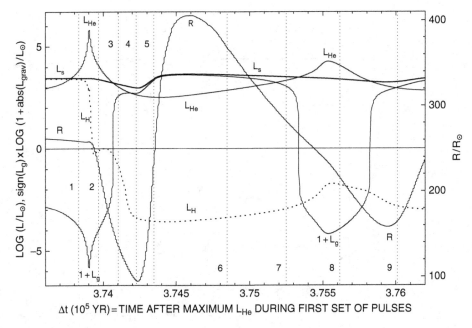

Global luminosities of a 1 M$_\odot$ TPAGB model during and between the first two of the fifth set of thermal pulses
triggered by helium burning ($Z = 0.015$, $Y = 0.275$)

model	δt(yr)	L_s	L_{He}	L_{grav}	L_{min1}	L_{max}	L_{min2}	L_{env}
				Table 17.4.1 Some luminosity characteristics of models 0–9				
0	368542	2947.5	2915.9	45.41		2947.9	L_s	$\sim$0
1	373830	2795.6	6699.3	−6378.8	−465.2	3202.3	149.6	309
2	373967	2447.7	12423	−9969.7	−5128.2	8630.7	408.6	2039
3	374104	1444.4	1074.4	379.6	−1496.2	5511.4	700.4	744
4	374227	945.06	437.1	517.9	−863.2	4661.1	640.0	305
5	374347	2503.1	328.6	2184.5	−611.0	4347.7	2448.0	−2300
6	374847	4128.8	704.0	3435.3	−304.6	4139.7	L_s	$\sim$0
7	375250	3219.3	2143.8	1085.8	−263.6	3224.0	L_s	$\sim$0
8	375616	2437.3	11319	−8873.3	−2159.0	6534.9	1744.1	693
9	376001	1807.2	871.5	945.6	−477.7	3215.4	L_s	$\sim$0

luminosity characteristics of the models are listed in Table 17.4.1 where table entries are given in solar units. The quantity L_{min1} is the minimum luminosity along the luminosity profile just below $M \sim 0.499\, M_\odot$, L_{max} is the maximum luminosity above this mass point, and L_{min2} is the minimum luminosity between the maximum luminosity and the surface.

Since in models 0 and 1 the surface luminosity L_s and the hydrogen-burning luminosity L_H are essentially identical and since in models 2–9, L_H is insignificant, L_H is not shown explicitly in the table. In models 0–3, $L_{neutrino} \sim 14\, L_\odot$, and, in the other models, $L_{neutrino} \sim 10.3 \pm 0.3\, L_\odot$.

Analysis in Section 17.3 of gravothermal and nuclear burning characteristics of models during the first TPAGB thermal pulse episode relied heavily on distributions of local energy generation rates as a function of mass. In this section, analysis of these characteristics makes use only of luminosity profiles which are the integrals of the sum of the local rates.

The global helium-burning luminosity for model 1 is approximately 100 times smaller than at the peak of the flash. The luminosity profile for model 1 in Fig. 17.4.5 demonstrates that all of the energy generated by helium burning in the model is absorbed within a region of width $\sim$0.03 $M_\odot$ encompassing the luminosity watershed at $M \sim 0.498\, M_\odot$. The positive portion of the profile remains above the profile for model 0 over the mass range $\sim$0.498 $M_\odot$ to $\sim$0.515 $M_\odot$, or slightly more than half way from the watershed in luminosity to the location of the hydrogen-burning shell at $M \sim 0.529\, M_\odot$. The hydrogen-burning shell has moved slightly beyond its location $\sim$5300 yr earlier in model 0 and its strength has diminished somewhat, indicating that expansion and cooling have begun to extend into the hydrogen-burning region. The modest increase in luminosity which extends outward beyond the hydrogen-burning shell shows that the base of the model envelope is contracting and heating in response to the decrease in luminosity flux from the interior. Since $L_s = 2795.6\, L_\odot$, $L_H = 2486.7\, L_\odot$, and the luminosity entering the base of the hydrogen-burning shell is $\sim$150 $L_\odot$, the rate at which matter extending from the hydrogen-burning shell to the surface contributes gravothermal energy to the surface luminosity is $L_{grav}^{env} \sim L_s - L_H - 150\, L_\odot = 309\, L_\odot$.

The global peak helium-burning rate occurs approximately midway between models 1 and 2, and L_{He} in model 2 is $\sim$1.86 times larger than in model 1. However, expansion and cooling has extended all the way from the luminosity watershed to beyond the hydrogen–helium interface and the hydrogen-burning shell has become effectively extinguished. Absorption from outwardly flowing energy extends from the peak in the luminosity profile at $\sim$0.506 $M_\odot$ out to $\sim$0.5295 $M_\odot$, with $\sim$409 $L_\odot$ remaining to contribute to the surface luminosity. This remaining flux is a direct contribution from helium burning (reduced by the neutrino energy-loss rate from near the model center). The major contribution to the surface luminosity comes from the release of gravothermal energy between $\sim$0.5295 $M_\odot$ and $\sim$0.534 $M_\odot$ in the amount of $\sim$1710 $L_\odot$. An additional contribution to the surface luminosity of $\sim$330 $L_\odot$ comes from the release of gravothermal energy between $\sim$0.534 $M_\odot$ and the surface. All in all, gravothermal energy contributes $\sim$2039 $L_\odot$ and helium-burning energy (minus neutrino loss) directly contributes $\sim$409 $L_\odot$ to the surface luminosity.

In models 3–5, the gravothermal energy stored initially in the helium-burning region at the height of the helium-burning flash is steadily diffusing outward beyond the flashing region, as evidenced by the steady motion outward in mass of the maximum in the luminosity profile. To the left of the maximum, gravothermal energy is contributing to the ambient energy flow and to the right of the maximum, energy from the ambient flow is being converted into stored gravothermal energy to be released later.

Model 3 is approximately 2000 yr older than the model at peak helium-burning luminosity and the helium-burning luminosity, at $L_{He} = 1074\ L_\odot$, is small compared with the maximum luminosity $L_{max} \sim 5500\ L_\odot$ in the model, so most of both the positive and negative contributions to the outward energy flux are due to gravothermal action. The bump in the luminosity profile at $M \sim 0.513\ M_\odot$ is a numerical artifact associated with rezoning about the outer edge of a convective shell which extends from $M \sim 0.499\ M_\odot$ to $M \sim 0.513\ M_\odot$.

As seen from Fig. 17.4.6, model 4 is near minimum surface luminosity and radius and, as in the case of model 3, most of both the positive and negative contributions to the outward energy flux are due to gravothermal action. This time, however, the transfer of gravothermal energy extends beyond the hydrogen–helium interface, with over half of the gravothermal energy being absorbed beyond the interface.

In model 5, the transfer of gravothermal energy beyond the hydrogen–helium interface has been nearly completed. In model 6, the surface luminosity is due predominantly to the release of gravothermal energy from below the hydrogen–helium interface, with helium burning making a minor contribution and essentially no absorption occurring in the envelope of the model.

In model 7, gravothermal energy still contributes substantially to the energy flux outward beyond the region of helium burning and again essentially no absorption occurs in the envelope. However, the rate of energy production by helium burning has approximately tripled over the rate in model 6, and much of the nuclear energy released is being stored as gravothermal energy in and near the helium-burning region, with the consequence that the global rate of gravothermal energy production L_{grav} for model 7 is substantially smaller than the actual contribution of gravothermal energy release to the surface luminosity. An estimate of this latter contribution is given by taking the difference between the surface luminosity and the luminosity in the model at $\sim 0.508\ M_\odot$ and amounts to $L^+_{grav} \sim 1700$ $L_\odot$. Combining this estimate with the global rate $L_{grav} \sim 1100\ L_\odot$ shows that energy is being absorbed in and near the helium-burning region at the rate $L^-_{grav} \sim 600\ L_\odot$ and that, therefore, nuclear energy release contributes about $1500\ L_\odot$ to the surface luminosity, or slightly less than the contribution of gravothermal energy release. It is interesting to note that one might come to exactly the opposite conclusion from an untutored inspection of Fig. 17.4.6 where $L_{He} > L_g$.

The precipitous drop in L_g from large positive values to large negative values which occurs in Fig. 17.4.6 shortly after model 7 indicates that a second helium-burning thermal runaway is beginning. The peak burning rate occurs shortly before model 8, achieving a value, $L^{max}_{He} = 1.827 \times 10^4\ L_\odot$, which is about 35 times smaller than the maximum rate of $L^{max}_{He} = 6.455 \times 10^5\ L_\odot$ achieved between models 1 and 2 during the first thermal pulse. The luminosity profile for model 8 in Fig. 17.4.5 shows that gravothermal energy is still being released (in the region beyond the minimum in luminosity at $\sim 0.52\ M_\odot$) after the fashion occurring in models 6 and 7 and that, therefore, gravothermal energy stored after the first thermal pulse is still in the process of diffusing outward.

Model 9 occurs at a position relative to the second thermal pulse peak which is very similar to the position of model 4 relative to the first thermal pulse peak. In both cases $L_g \sim L_{He}$ and the corresponding luminosity profiles in Fig. 17.4.5 resemble one another in that the maximum luminosity along each profile occurs at nearly the same mass and that an abrupt steepening in slope along each profile occurs at the hydrogen–helium interface.

Analysis of this sort for any given set of thermal pulses could continue almost indefinitely, but there are other aspects of evolution to explore. Figures 17.4.7–17.4.10 show the variations with time of global luminosities during the sixth through ninth sets of thermal pulse episodes. On comparison with the analogous distributions in Figs. 17.4.3 and 17.4.4 for thermal pulse episodes four and five, respectively, these additional distributions show primarily that, while the general character of the distributions does not change, variations in details from one set of pulses to the next are not completely predictable. For example, whereas both the fifth and sixth pulse sets show the occurrence of four unstable thermal runaways, the fourth and the seventh through ninth sets show only three.

Variations with time of various global characteristics are fairly systematic, as is demonstrated in Figs. 17.4.11–17.4.14 which show, respectively, variations with time of L_{He}, L_{grav}, L_H, and L_s. For example, as shown in Fig. 17.4.11, the maximum helium-burning luminosity achieved during the major pulse in a set of pulses increases monotonically with pulse set number and the mean helium-burning luminosity during interpulse evolution decreases monotonically. The increase with time in the amplitude of the major pulse is related to the increase in mass of the helium-exhausted core: the larger the core mass, the smaller is the core radius and the greater is the energy required to expand matter during the major flash episode in a set of pulses. Comparing Figs. 17.4.12 and 17.4.11, it is apparent that the increase with time in the strength of a major helium flash is reflected by an increase in the maximum gravothermal absorption rate.

From Fig. 17.4.13 it is apparent that the mean hydrogen-burning luminosity between sets of pulses increases with increasing pulse set number and this is consistent with the behavior

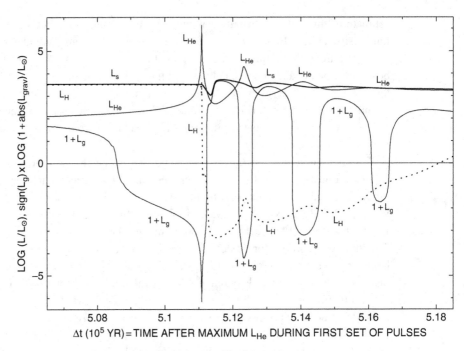

Fig. 17.4.7 Global luminosities of a 1 M$_\odot$ TPAGB model before, during, and after the sixth set of thermal pulses triggered by helium burning (Z = 0.015, Y = 0.275)

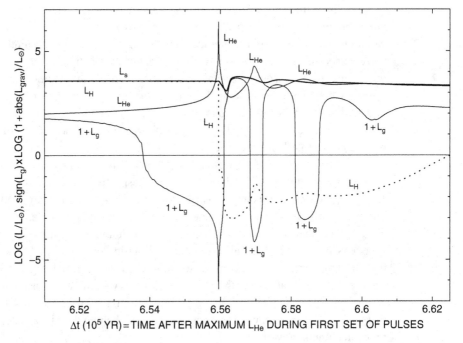

Fig. 17.4.8 Global luminosities of a 1 M$_\odot$ TPAGB model before, during, and after the seventh set of thermal pulses triggered by helium burning (Z = 0.015, Y = 0.275)

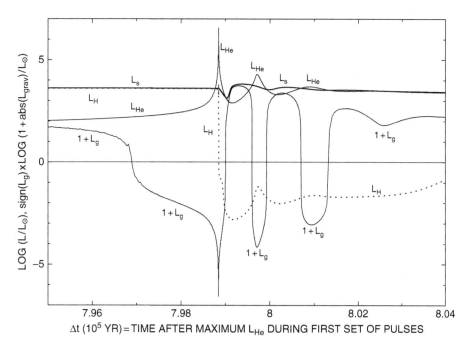

Fig. 17.4.9 Global luminosities of a 1 $M_\odot$ TPAGB model before, during, and after the eighth set of thermal pulses triggered by helium burning ($Z = 0.015$, $Y = 0.275$)

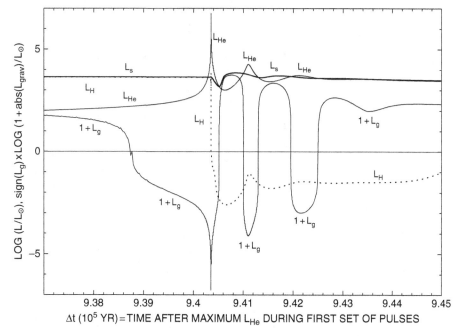

Fig. 17.4.10 Global luminosities of a 1 $M_\odot$ TPAGB model before, during, and after the ninth set of thermal pulses triggered by helium burning ($Z = 0.015$, $Y = 0.275$)

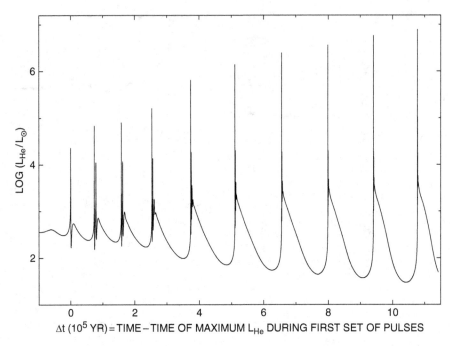

Fig. 17.4.11   Helium-burning luminosity of a 1 $M_\odot$ TPAGB model experiencing multiple thermal pulses triggered by helium burning ($Z = 0.015, Y = 0.275$)

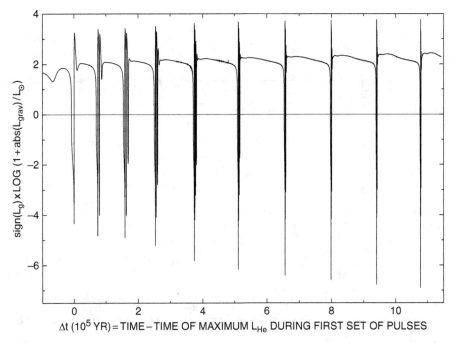

Fig. 17.4.12   Gravothermal luminosity of a 1 $M_\odot$ TPAGB model experiencing multiple thermal pulses triggered by helium burning ($Z = 0.015, Y = 0.275$)

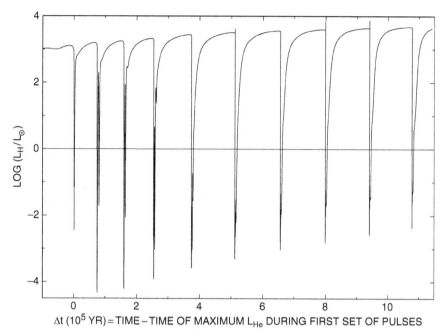

Fig. 17.4.13 Hydrogen-burning luminosity of a 1 $M_\odot$ TPAGB model experiencing multiple thermal pulses triggered by helium burning ($Z = 0.015, Y = 0.275$)

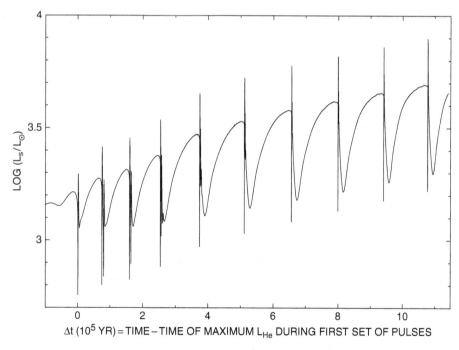

Fig. 17.4.14 Surface luminosity of a 1 $M_\odot$ TPAGB model experiencing multiple thermal pulses triggered by helium burning ($Z = 0.015, Y = 0.275$)

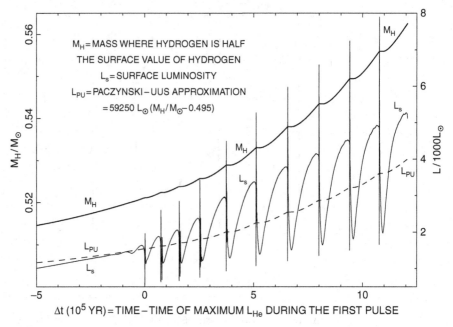

Fig. 17.4.15 Luminosity and mass at half maximum X_H in a 1 $M_\odot$ TPAGB model experiencing multiple thermal pulses triggered by helium burning ($Z = 0.015, Y = 0.275$)

of models along the first red giant branch which become brighter as the electron-degenerate helium core increases in mass. It is interesting that the minimum in the hydrogen-burning luminosity which is achieved during a helium shell flash increases with each successive pulse set. An explanation of this fact is not straighforward. Finally, from Fig. 17.4.14 it is evident that the mean surface luminosity during interpulse evolution increases after each pulse episode, consistent with the increase in the mean hydrogen-burning luminosity, but the total amplitude of the change in surface luminosity during a pulse episode remains fairly constant.

In Fig. 17.4.15, the surface luminosity is compared with the mass at the hydrogen–helium interface, defined as the mass M_H where the hydrogen abundance is half the abundance of hydrogen at the surface. The dashed curve, defined by

$$L_{PU} = 59250 \, L_\odot \left(\frac{M_H}{M_\odot} - 0.495 \right), \tag{17.4.1}$$

is an approximation prepared by Bohdan Paczyñski (1970) to fit the relationship between core mass and luminosity of models produced by him and by U. Uus (1970) in which thermal pulses have been suppressed by assuming that the rate at which hydrogen processes matter is at all times exactly equal to the rate at which helium processes matter and by assuming that the local rate of gravothermal energy production is given by $\epsilon_{grav} = \dot{M}_H \, T \, \partial S / \partial M$, where S is the entropy. This approximation is a remarkably good fit to the average surface luminosity defined by the thermally pulsing models. At the same time, it is evident that the core mass of a typical real low mass TPAGB star cannot be

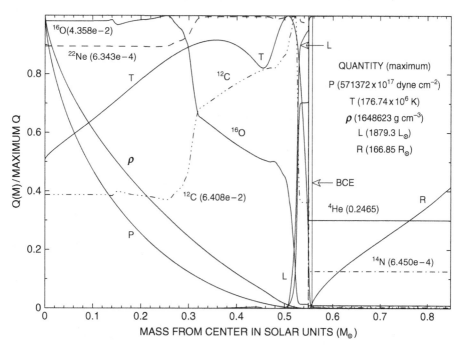

Fig. 17.4.16 Structure variables and abundances in a 1 $M_\odot$ model near minimum L_s after the ninth set of thermal pulses ($Z = 0.015, Y = 0.275, \Delta t = 9.54561 \times 10^5$ yr)

estimated on the basis of its luminosity to better than 0.01–0.02 $M_\odot$ during interpulse phases or to better than 0.02–0.04 during pulsing phases.

Conditions in a model as its surface luminosity approaches a relative minimum after the ninth thermal pulse episode has subsided are shown in Figs. 17.4.16 and 17.4.17, and conditions in a model in which L_H is near a relative maximum prior to the tenth thermal pulse episode are shown in Figs. 17.4.18 and 17.4.19. The two models are separated in time by 87 000 yr, which is approximately two thirds of the time between two helium shell-flash episodes. The first of each pair of figures shows conditions over most of the model and describes the normalization of structure and composition variables. The second figure of each pair shows conditions over the region within which hydrogen and helium burning take place. Comparing the helium profiles in Figs. 17.4.17 and 17.4.19, it is evident that the amount of helium produced by hydrogen burning during the interpulse phase is not dramatically larger than the amount of helium destroyed by helium burning. That is, although the hydrogen-burning luminosity is large compared with the helium-burning luminosity during most of the interpulse period, helium burning processes matter at an average rate comparable with the average rate at which hydrogen burning processes matter.

Thus, both helium and hydrogen burning during the interpulse period contribute to the establishment of conditions leading to the next thermal pulse episode: the mass of both the helium-exhausted core and the mass of the helium-rich region below the hydrogen–helium interface increase with time. During evolution between the two models described in Figs. 17.4.16–17.4.19, the mass of the helium-exhausted core increases by slightly less

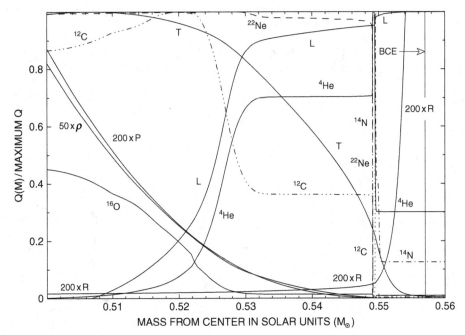

Fig. 17.4.17 Structure variables and abundances in a 1 $M_\odot$ model near minimum L_s after the ninth set of thermal pulses ($Z = 0.015$, $Y = 0.275$, $\Delta t = 9.54561 \times 10^5$ yr)

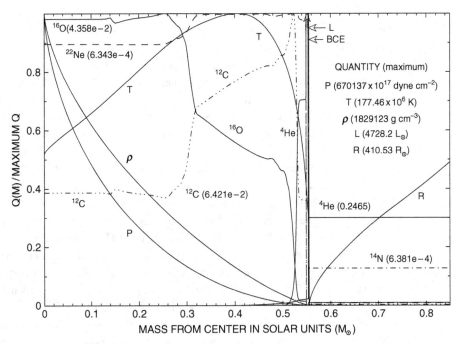

Fig. 17.4.18 Structure variables and abundances in a 1 $M_\odot$ model near maximum L_H before the tenth set of thermal pulses ($Z = 0.015$, $Y = 0.275$, $\Delta t = 10.41518 \times 10^5$ yr)

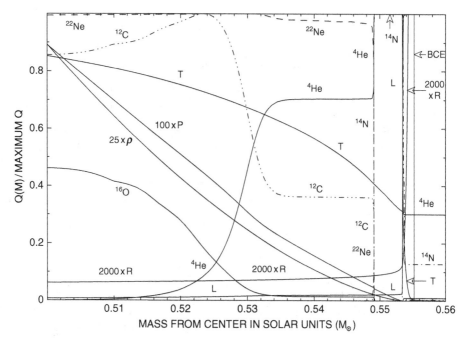

Fig. 17.4.19   Structure variables and abundances in a 1 $M_\odot$ model near maximum L_H before the tenth set of thermal pulses ($Z = 0.015$, $Y = 0.275$, $\Delta t = 10.41518 \times 10^5$ yr)

than 0.003 $M_\odot$ and the mass of the hydrogen-exhausted core increases by slightly more than 0.004 $M_\odot$, which means that slightly more than 0.001 $M_\odot$ of helium has been burned during the helium-flash episode that preceeds the first of the two models. It is instructive to note that approximately three times as much helium is burned quiescently during the interpulse phase as is burned during helium shell flashes.

The fact that the total mass of hydrogen converted into helium over an entire pulse cycle is nearly equal to the total mass of helium converted into carbon and oxygen can be used to demonstrate that the average hydrogen-burning luminosity is substantially larger than the average helium-burning luminosity. At the temperatures prevailing in the hydrogen-burning shells of TPAGB models, CN-cycle reactions are far more frequent than pp-chain reactions, so

$$L_{\mathrm{nuc}} \sim L_{\mathrm{CN}} = 24.97 \text{ MeV} \left(\frac{dN_4}{dt}\right)_{\mathrm{CN}}, \tag{17.4.2}$$

where N_4 is the total abundance of alpha particles in the model. The rate at which alpha particles are destroyed can be described by

$$L_{\mathrm{He}} = -24.97 \text{ MeV} \left[\frac{1}{10.3}\left(\frac{dN_4}{dt}\right)_{3\alpha} + \frac{1}{3.5}\left(\frac{dN_4}{dt}\right)_{^{12}C+\alpha}\right]. \tag{17.4.3}$$

To the extent that, over a thermal pulse cycle, the average number of alpha particles destroyed is equal to the average number of alpha particles created, one has

$$\left\langle \left(\frac{dN_4}{dt}\right)_{CN}\right\rangle = -\left[\left\langle \left(\frac{dN_4}{dt}\right)_{3\alpha}\right\rangle + \left\langle \left(\frac{dN_4}{dt}\right)_{12C+\alpha}\right\rangle\right], \qquad (17.4.4)$$

where angle brackets denote averages over an entire pulse cycle. Then,

$$\frac{\langle L_H\rangle}{\langle L_{He}\rangle} = 10.3 \frac{\langle (dN_4/dt)_{3\alpha}\rangle + \langle (dN_4/dt)_{12C+\alpha}\rangle}{\langle (dN_4/dt)_{3\alpha}\rangle + 3\langle (dN_4/dt)_{12C+\alpha}\rangle}. \qquad (17.4.5)$$

From eq. (17.4.5) it is evident that, if temperatures in the helium burning region are such that ^{12}C is the final product of complete helium burning, the average hydrogen-burning luminosity is 10.3 times the average helium-burning luminosity. If temperatures are such that ^{16}O is the final product of complete helium burning, then, for every oxygen nucleus made, 4 alpha particles have been used, 3 in the triple alpha reaction and 1 in the $^{12}C(\alpha,\gamma)^{16}O$ reaction, with the result that the ratio of average luminosities is $10.3 \times (3+1)/(3+3) = 6.9$.

17.5 The roles of nuclear burning, convective mixing, and gravothermal activity in determining abundance changes and dredge-up during the TPAGB phase

Abundances of the isotopes ^4He, ^{12}C, and ^{16}O at times (1) shortly before the onset of the ninth set of thermal pulses, (2) shortly after the completion of the ninth set, and (3) shortly before the onset of the tenth set of thermal pulses are shown in Figs. 17.5.1–17.5.3. In each figure, compositions at two of the times are compared.

Comparison in Fig. 17.5.1 of compositions at times (1) and (3), shortly before the ninth and shortly before the tenth set of thermal pulses, demonstrates that, just prior to a set of thermal pulses, composition profiles are virtually identical, except for the fact that the profiles for the more evolved model (solid curves and subscripts 10−) are centered at larger masses than the profiles for the less evolved model (dotted curves and subscripts 9−). In both models, the hydrogen-burning shell has a thickness $\Delta M_H \sim 0.9 \times 10^{-4}\ M_\odot$, and the base of the convective envelope lies $\Delta M \sim 1.75 \times 10^{-3}\ M_\odot$ above the center of the hydrogen-burning shell, or about 20 times the thickness of the hydrogen-burning shell.

At any point, the sum of the abundances of the three isotopes is constant and it is evident by inspection that the amount of helium destroyed in an entire pulse cycle is equal to the amount of carbon and oxygen created, as demanded by conservation of alpha particles. It is not as self evident that the amount of helium produced by hydrogen burning during an entire pulse cycle equals the amount of helium destroyed by helium burning, although this is nearly the case. That the amounts are not exactly the same is due to the fact that the average separation in mass between the hydrogen–helium interface and the helium–CO interface decreases as evolution progresses.

The fact that the left hand edge of the region defined by the peak abundance of helium in the more evolved model lies to the left of the right hand edge of the region defined by the peak abundance of helium in the less evolved model demonstrates that the outer edge

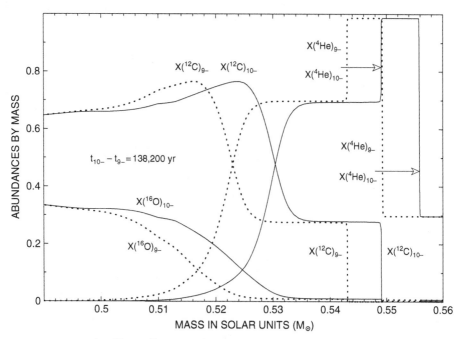

Fig. 17.5.1   Abundances by mass of ^{4}He, ^{12}C, and ^{16}O in a 1 $M_\odot$ model just before the ninth set and just before the tenth set of thermal pulses (Z = 0.015, Y = 0.275)

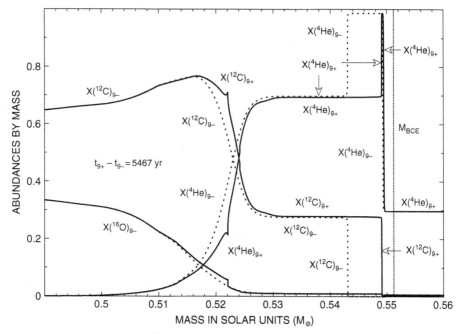

Fig. 17.5.2   Abundances by mass of ^{4}He, ^{12}C, and ^{16}O in a 1 $M_\odot$ model just before and just after the ninth set of thermal pulses (Z = 0.015, Y = 0.275)

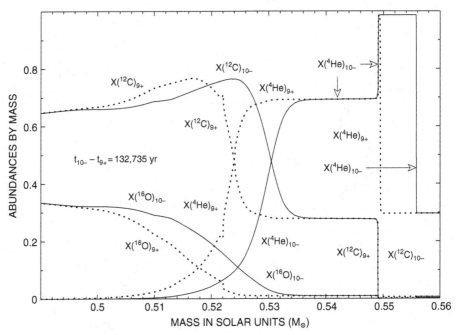

Fig. 17.5.3 Abundances by mass of ^{4}He, ^{12}C, and ^{16}O in a 1 $M_\odot$ model just after the ninth set and just before the tenth set of thermal pulses ($Z = 0.015$, $Y = 0.275$)

of the convective shell formed during the ninth helium shell flash does not reach beyond the hydrogen–helium interface. This is made clearer in Fig. 17.5.2, which compares abundance profiles for the model just after the completion of the ninth set of thermal pulses (solid curves and subscripts 9+) with the abundance profiles for the model just before the onset of the ninth set of pulses (dotted curves and subscripts 9−). The outer edge of the carbon-abundance profile has moved from the vertical portion of the dotted curve at $M \sim 0.543\ M_\odot$ to the vertical portion of the solid curve at $M = 0.549\,079\ M_\odot$, a change brought about by convection during the first of the ninth set of shell flashes. The center of the hydrogen–helium interface has moved outward from $M \sim 0.549\,27\ M_\odot$ to $M \sim 0.549\,455\ M_\odot$, a change which preceeded the flash episodes separating the two models. Thus, in the more evolved model, the outer edge of the carbon-abundance profile and the center of the hydrogen–helium interface are separated by $\Delta M \sim 3.76 \times 10^{-4}\ M_\odot$. This separation, which represents the closest approach between the outer edge of the convective shell during the first flashing episode separating the models and the position of the H-He interface is about four times the thickness of the H-He interface.

The expansion outward of the carbon-abundance profile occurs during the first helium shell flash of a set of thermal pulses and is primarily a consequence of the development of a convective shell in which helium is first mixed inward, converted into carbon by helium burning, and then mixed outward. The amount of fresh carbon appearing in the region between $\sim 0.543\ M_\odot$ and $\sim 0.549\ M_\odot$ in Fig. 17.5.2 is very nearly the same as the mass of helium which disappears in this same region. Thus, although a modest increase in the mass of ^{12}C occurs during the flash in a region of intense helium burning below the base of the convective shell (that is, between $\sim 0.517\ M_\odot$ and $\sim 0.529\ M_\odot$), the most significant change in the ^{12}C abundance distribution during the flash is the creation of carbon at the expense of helium ingested by the convective shell and transferred outward through the shell.

From the perspective of galactic nucleosynthesis, it is of significance that helium burning in the convective shell is incomplete, with hardly any oxygen being produced. Essentially only carbon is carried outward to be in close proximity to the hydrogen–helium interface, a position from which it can be dredged into the envelope, when helium flashes become strong enough, to produce the carbon–star phenomenon and, ultimately, contribute to an increase of the abundance of carbon in the interstellar medium. A final observation is that both before thermal pulses begin and after they have subsided, the base of the convective envelope, defined in Fig. 17.5.2 by the location of the vertical line labeled M_{BCE}, is quite far above the hydrogen–helium interface. The dredge–up phenomenon, characterized by the motion of the base of the convective envelope inward toward the hydrogen–helium interface, occurs during the powerdown phase of the first of every set of helium shell flashes.

Changes in abundances after the end of the ninth set of thermal pulses and before the beginning of the tenth set of thermal pulses are shown by the abundance profiles in Fig. 17.5.3. Profiles for the model just after the completion of the ninth set of thermal pulses are the dotted curves labeled by an abundance with subscript 9+ and profiles for the model just before the onset of the tenth set of pulses are the solid curves labeled by an abundance with subscript 10−. On comparison with the profiles in Fig. 17.5.2, it is evident that most of the conversion of helium into carbon and oxygen occurs during the quiescent helium-burning phase following the thermally pulsing phase and that this conversion takes place in a helium-burning shell under radiative conditions, well below the hydrogen–helium interface. However, both the helium-flash phase and the interflash phase of helium burning contribute to the eventual occurrence of the carbon star phenomenon and to the increase in the abundance of carbon in the interstellar medium.

Although it is true that the creation and diffusion outward of carbon in a convective shell during the first of a set of thermal pulses sets the stage for the eventual dredge-up of carbon into the convective envelope from which it is expelled by a wind into the interstellar medium, it is the increase in the mass of the helium-exhausted core, most of which occurs during the interflash phase and results in an increase in the binding energy of the core, that is responsible for the increase in pulse strength with each successive thermal pulse episode. The larger the binding energy, the more nuclear energy must be converted into thermal energy to counter the binding. The increase in strength contributes not only to a decrease during the first pulse of a set of pulses in the minimum separation in mass between the outer edge of the convective shell and the hydrogen–helium interface but also to a decrease during pulse powerdown in the minimum separation in mass between the inner edge of the convective envelope and the hydrogen–helium interface. Ultimately, the base of the convective envelope reaches and extends inward beyond where the hydrogen–helium interface is located at the beginning of a thermal pulse episode and reaches and extends into the region where fresh carbon has been deposited by shell convection during the first of the set of thermal pulses.

The fact that, following the first helium shell flash of each successive set of thermal pulses, the base of the convective envelope approaches ever closer in mass to the hydrogen–helium interface is due to the fact that, the stronger the flash, the more gravothermal energy is stored in the flashing region. As the stored energy leaks outward, the luminosity near the base of the convective envelope increases, causing the region near the base where the radiative gradient exceeds the adiabatic gradient to extend further inward in mass. The larger the stored energy, the larger is the maximum luminosity achieved at the base of the convective envelope and the further inward does the base extend.

The change in luminosity at and beyond the hydrogen–helium interface have already been described in Fig. 17.4.5 during and between the first two of the fifth set of helium shell flashes. In this figure, the hydrogen–helium interface coincides with the luminosity profile of model 1. Although its location is not shown explicitly in the figure, the base of the convective envelope never extends inward as far as the H-He interface. It is evident that the luminosity at the interface increases as gravothermal energy stored during the first helium shell flash leaks outward (models 2–6) and then declines (model 7), and, even though a second helium shell flash occurs (model 8), the gravothermal energy which is stored and thereafter leaks outward never produces luminosities at the H-He interface which approach the maximum luminosity achieved there after the first flash (models 8 and 9).

Given the observational fact that TPAGB stars of the appropriate composition and mass can become carbon stars, it is worth belaboring the mechanism which is, without doubt, responsible for the observed phenomenon: when a helium flash becomes sufficiently strong, during the powerdown phase after the flash, the base of the convective envelope extends inward beyond the hydrogen–helium interface until it reaches and extends into the region where the convective shell formed during the peak of the flash has deposited carbon-rich material.

The theoretical justification for this interpretation is elaborated again in Figs. 17.5.4–17.5.8 which show nuclear and gravothermal energy-generation rates, composition and luminosity profiles, and the location of the base of the convective envelope associated with the eleventh set of helium shell flashes in the 1 $M_{\odot}$ population I model under consideration. In these figures, the scale for luminosity given along the right hand vertical axis (-2000 $L_{\odot}$ to 9000 $L_{\odot}$) remains the same and the scale for number (not mass) abundances given along the left hand vertical axis (0 to 1) remains the same. However, the scale for energy-generation rates is a 1000 times larger in Fig. 17.5.4 (units of 10^9 erg g^{-1} s^{-1}) than it is in the other three figures (units of 10^6 erg g^{-1} s^{-1}).

Figure 17.5.4 describes conditions in the model near the peak of the first of the eleventh set of thermal pulses which begins approximately 1.292×10^5 yr after the peak of the first flash of the tenth set. Essentially all of the energy produced by helium burning is converted into gravothermal energy ($L_{\mathrm{He}} = 7.064 \times 10^6 \, L_{\odot} \sim - L_{\mathrm{grav}} = 7.203 \times 10^6 \, L_{\odot}$). At its maximum of $\sim 5 \times 10^6 \, L_{\odot}$, the interior luminosity is 1000 times larger than the surface luminosity. The huge fluxes produced by the helium-burning shell sustain a convective region which extends from $M \sim 0.537\,74 \, M_{\odot}$ to $M \sim 0.562\,08 \, M_{\odot}$. Whereas the injection of nuclear energy due to helium burning is confined to a narrow region above the base of the convective shell, the rate at which energy from the ambient flow is converted into gravothermal energy remains at a high level throughout the convective region. Absorption continues for a distance $\Delta M = 0.000\,38 \, M_{\odot}$ beyond the outer edge of the convective region, to a point identified by the location of a relative minimum in luminosity, where $L = - 467.8 \, L_{\odot}$, at $M = 0.562\,46 \, M_{\odot}$.

The positions where the hydrogen abundance is approximately half of the surface abundance of hydrogen and where the nuclear energy-generation rate due to hydrogen

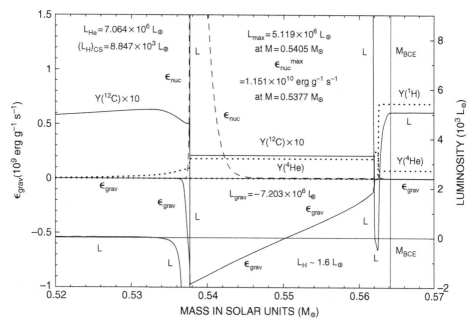

Fig. 17.5.4 Energy-generation rates and composition in a 1 $M_\odot$ model near the peak of the first of the eleventh set of thermal pulses ($Z = 0.015$, $Y = 0.275$)

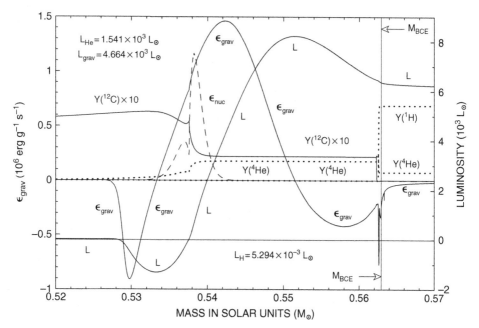

Fig. 17.5.5 Energy-generation rates and composition in a 1 $M_\odot$ model 237 yr after the peak of the first of the eleventh set of thermal pulses ($Z = 0.015$, $Y = 0.275$)

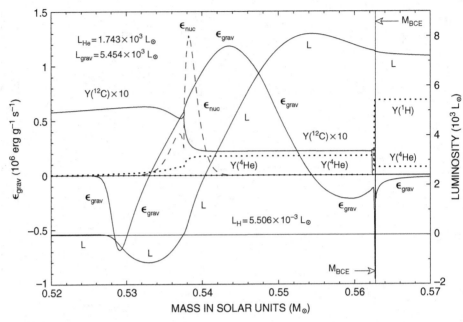

Fig. 17.5.6 Energy-generation rates and composition in a 1 $M_\odot$ model 291 yr after the peak of the first of the eleventh set of thermal pulses ($Z = 0.015, Y = 0.275$)

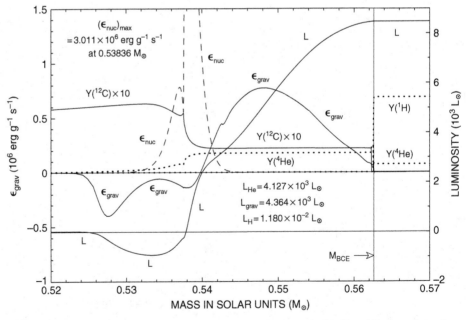

Fig. 17.5.7 Energy-generation rates and composition in a 1 $M_\odot$ model 354 yr after the peak of the first of the eleventh set of thermal pulses ($Z = 0.015, Y = 0.275$)

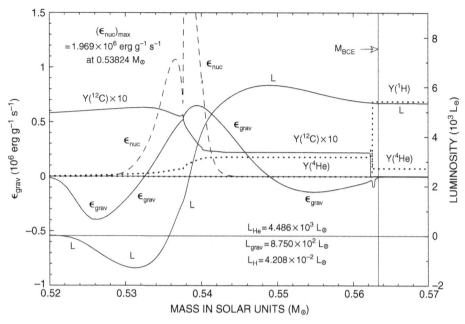

Fig. 17.5.8 Energy-generation rates and composition in a 1 $M_\odot$ model 383 yr after the peak of the second of the eleventh set of thermal pulses ($Z = 0.015$, $Y = 0.275$)

burning is near a relative maximum of 2.45×10^4 erg g^{-1} s^{-1} nearly coincide at a mass $M \sim 0.562\,57\ M_\odot$, approximately $\Delta M \sim 0.000\,11\ M_\odot$ above the luminosity minimum, and the base of the convective envelope lies above the center of the hydrogen-burning shell by $\Delta M \sim 0.001\,62\ M_\odot$. Expansion and cooling in the region above the relative luminosity minimum have occurred to the extent that gravothermal energy is released at a rate that completely overwhelms the rate at which nuclear energy is released. The hydrogen-burning shell contributes only $L_H \sim 1.60\ L_\odot$ to the surface luminosity of $L_s \sim 5.1 \times 10^3\ L_\odot$.

Conditions in the model approximately 237 yr after the peak of the helium shell flash described in Fig. 17.5.4 are shown in Fig. 17.5.5. The expansion and cooling caused by the conversion of energy from the ambient flow into gravothermal energy has decreased the strength of the hydrogen-burning shell centered at the hydrogen–helium interface by an additional factor of ~ 300 ($L_H = 5.294 \times 10^{-3}\ L_\odot$) and hydrogen burning near the base of the helium-burning region is no longer a factor. However, the situation which prevailed during the first of the eleventh set of helium shell flashes, when most of the nuclear energy released was converted into gravothermal energy, has been replaced by a situation in which energy stored in gravothermal form is released at a rate substantially larger than the rate at which nuclear energy is released ($L_{grav} = 4.664 \times 10^3\ L_\odot$ versus $L_{He} = 1.541 \times 10^3\ L_\odot$).

The luminosity between the hydrogen–helium interface and the surface has become larger than during pulse peak. The expansion outward and cooling of the region between the hydrogen–helium interface and the base of the convective envelope has led to an increase in the opacity in this region and, coupled with the increase in the luminosity in the region, the increase in the ratio of the radiative temperature–pressure gradient to the adiabatic temperature–pressure gradient has led to a migration inward in the mass at the base of the convective envelope. The center of the hydrogen–helium interface and the base of the convective envelope are now separated by a mass $\Delta M \sim 0.000\,3663\ M_\odot$, which is approximately four times larger than the thickness of the hydrogen–helium interface.

The trend continues. Figure 17.5.6 shows conditions in the model after an additional 54 yr of evolution and Fig. 17.5.7 shows conditions after a still further 63 yr of evolution. In each case, the luminosity near the base of the convective envelope has increased and the base has in consequence moved closer in mass to the hydrogen–helium interface, actually reaching inward beyond the outer edge of this interface in the second of the two models. In the first of the two models, the two fiducial masses are separated by $\Delta M \sim 0.000\,1798$ $M_\odot$, or about twice the thickness of the hydrogen–helium interface. In the second model, the fiducial masses are separated by $\Delta M \sim 0.000\,077\,50\ M_\odot$, so the base of the convective envelope has actually penetrated beyond the center of the interface prevailing during shell hydrogen burning. This marks the closest approach in mass between the H-He interface and the base of the CE during the eleventh set of helium shell flashes.

The second of the eleventh set of helium shell flashes is underway in the models described in Figs. 17.5.6 and 17.5.7, with the the maximum helium-burning energy-generation rate increasing by a factor of about 2.4 between the two models and the helium-burning luminosity increasing by about the same factor from $L_{He} = 1.743 \times 10^3\ L_\odot$ to $L_{He} = 4.127 \times 10^3\ L_\odot$. The second flash produces a peak luminosity of $L_{He} = 1.665 \times 10^4$ $L_\odot$ approximately 18 yr after the model described in Fig. 17.5.7, or 372 yr after the peak of the first of the eleventh set of flashes. Being much weaker than the first flash, this second flash has no effect on the distribution of number abundances in the region centered at the hydrogen–helium interface. In fact, during the powerdown phase after the second flash, the base of the convective envelope moves outward in mass. Conditions in the model 385 yr after the peak of the second flash are shown in Fig. 17.5.8. The surface luminosity has decreased by more than 50% from that prevailing in the model described in Fig. 17.5.7 and the base of the convective envelope has receded outward to $M_{BCE} \sim 0.563\,354$, where it is separated by $\sim 7.9 \times 10^{-4}\ M_\odot$ from the center of the hydrogen–helium interface, or by approximately nine times the thickness of the interface.

It is clear that, as additional sets of helium-burning shell flashes occur, the increase in the strength of the first flash of each set will result in a penetration of the base of the convective envelope still further into the hydrogen–helium interface prevailing at the start of the flash and it will result ultimately in a penetration into the region where fresh carbon has been deposited by mixing outward in the convective shell produced by the flash.

17.6 Neutron production and neutron capture in helium-burning regions

Given the fact that, when thermal pulses become strong enough, dredge-up carries products of incomplete helium burning into the convective envelope where they can be potentially

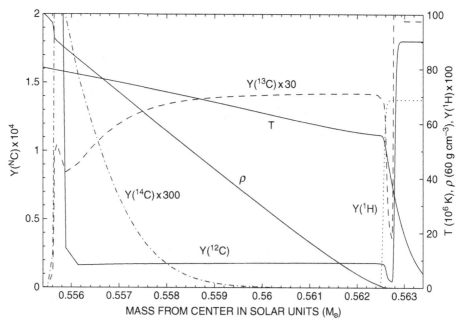

Fig. 17.6.1 Abundances of carbon isotopes below the H-burning shell at the end of the tenth interpulse phase in a 1 $M_\odot$ TPAGB model ($Z = 0.015, Y = 0.275$)

observed and ejected in a wind, and given the fact that s-process elements are observed in some low-mass AGB stars, it is pertinent to examine the potential for producing neutrons in helium-burning regions of the 1 $M_\odot$ model. As elaborated in Chapter 19, both the reactions ^{13}C$(\alpha, n)^{16}$O and ^{22}Ne$(\alpha, n)^{25}$Mg are activated at interesting rates in a 5 $M_\odot$ TPAGB model. ^{13}C is produced in the hydrogen-burning shell during the interpulse phase and ^{22}Ne is made in the convective shell during a helium shell flash in consequence of the sequence of reactions ^{14}N$(\alpha, \gamma)^{18}$F, ^{18}F$(e^+ \nu_e)^{18}$O, and ^{18}O$(\alpha, \gamma)^{22}$Ne.

In Fig. 17.6.1 are shown the abundances of carbon isotopes in the region of mass $\Delta M_H \sim$ 0.0066 $M_\odot$ traversed by the hydrogen-burning shell between the tenth and eleventh sets of helium shell flashes of the 1 $M_\odot$ model. The scale is defined along the left hand vertical axis. The hydrogen abundance is shown by the dotted curve, temperature and density are shown by solid curves labeled T and ρ, respectively, and the scale for all three variables is defined along the right hand vertical axis. In the outer half of the region below the hydrogen profile, all isotopes involved in CN-cycle burning have their equilibrium abundances achieved at the end of hydrogen burning and the abundance of ^{13}C is $Y(^{13}$C$) \sim 4.7 \times 10^{-6}$, which is approximately one-quarter of the abundance of ^{12}C. In the inner half of the region, where temperatures exceed $\sim 70 \times 10^6$ K, the abundance of ^{13}C decreases inward in step with increasing temperature in consequence of the operation of the ^{13}C$(\alpha, n)^{16}$O reaction.

The abundance by number of ^{14}N throughout the region traversed by the hydrogen-burning shell during the interpulse phase is $Y(^{14}$N$) \sim 6.38 \times 10^{-4}$, or over 130 times the abundance of ^{13}C. Most of the neutrons released in the helium-rich region below the

Table 17.6.1 Abundances in the convective shell during the first of the eleventh set of helium shell flashes

M_{CS}	T_{BCS}	^{4}He	^{12}C	^{13}C	^{14}C	^{14}N	^{16}O	^{18}O	^{22}Ne	^{25}Mg
$0.01 M_\odot$	10^6 K	0.1	0.01	10^{-8}	10^{-8}	10^{-5}	10^{-4}	10^{-5}	10^{-4}	10^{-6}
1.465	153.5	1.650	2.576	5.9/−4	3.3/−4	9.6/−3	7.692	4.327	6.299	3.189
1.597	157.8	1.652	2.571	3.6/−4	4.6/−4	1.03/−2	7.652	4.556	6.297	3.189
1.867	167.5	1.663	2.537	2.321	5.86	1.319	7.542	5.081	6.16	3.184
2.432	244.3	1.807	2.090	5.293	67.91	4.586	5.797	6.360	5.242	3.159
2.443	240.4	1.792	2.131	1.308	56.42	1.539	5.768	5.471	5.641	3.204
0.790	218.3	1.773	2.192	2.3/−3	3.114	1.08/−2	5.829	2.097	6.135	3.235

hydrogen-burning shell during the interpulse phase disappear via the ^{14}N$(n, p)^{14}$C reaction. The half life of ^{14}C is 5730 yr, much shorter than the roughly 1.3×10^5 yr between thermal pulse episodes, and this accounts for the fact that the abundance of ^{14}C at any point is much smaller than the amount by which the abundance of ^{13}C has been decreased from the value achieved by complete hydrogen burning.

The total amount of ^{13}C produced during the tenth interpulse phase, as measured by an integral of abundance over mass, is

$$\int Y(^{13}C)\, dM \sim 3.3 \times 10^{-8}\, M_\odot , \qquad (17.6.1)$$

and less than 10% of the total amount is used up in producing neutrons. Thus, most of the ^{13}C produced during the interpulse phase is ingested by the convective shell during the next shell flash and, as it is swept inward by convection to larger and larger temperatures, it experiences the ^{13}C$(\alpha, n)^{16}$O reaction. The total amount of ^{14}N produced during the tenth interpulse phase is, by the same measure,

$$\int Y(^{14}N)\, dM \sim 4.3 \times 10^{-6}\, M_\odot , \qquad (17.6.2)$$

Abundances in the convective shell of several of the isotopes involved in helium burning, neutron production, and neutron capture are presented in Table 17.6.1 at several stages during the first helium shell flash of the eleventh set of thermal pulses. The mass of the convective shell and the temperature at the base of the shell are given, respectively, in the first two columns of the table. All other numerical entries in the table are abundances by number. The fact that ^{13}C and ^{14}N are ingested and burned is demonstrated very clearly by the variation in the abundances in columns 5 and 7 of the table. The fact that neutrons are captured by ^{14}N with the emission of a proton is demonstrated by the variation in the abundance of ^{14}C in column 6. That ^{14}N is converted into ^{18}O and ^{18}O is converted into ^{22}Ne is demonstrated by the variations in abundances in columns 7 and 9. The comparatively large and relatively constant abundance of ^{22}Ne in the convective shell is mostly the consequence of the ^{14}N $\rightarrow$ ^{18}O $\rightarrow$ ^{22}Ne reactions occurring during previous helium-burning phases.

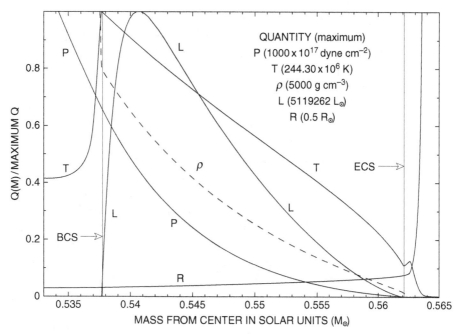

Fig. 17.6.2 Structure in the convective shell in a 1 $M_\odot$ model just after maximum L_{He} during the eleventh set of thermal pulses
($Z = 0.015, Y = 0.275, \Delta t = 10.41518 \times 10^5$ yr)

At its maximimum extent, the convective shell has a mass $M_{CS}^{max} \sim 0.0244\ M_\odot$. If the
ingested ^{13}C were not burned, its abundance in the convective shell would reach
$Y(^{13}C)^{max} \sim 135 \times 10^{-8}$, a number obtained by dividing the quantity in eq. (17.6.1) by
M_{CS}^{max}. But ^{13}C is essentially completely burned up and comparison with the maximum
variation in the number abundance of ^{25}Mg shows that the number of neutrons produced by
the ^{13}C$(\alpha, n)^{16}$O reaction is at least 25 times larger than the number of neutrons produced
by the ^{22}Ne$(\alpha, n)^{25}$Mg reaction. This estimate is actually a lower limit due to an interesting set of reactions involving neutrons and protons which occur with greatest frequency
near the base of the convective shell where temperatures are the highest. ^{13}C$(\alpha, n)^{16}$O
reactions produce neutrons, ^{12}C$(n, \gamma)^{13}$C reactions reconstitute ^{13}C, ^{14}N$(n, p)^{14}$C reactions produce protons, and ^{12}C$(p, \gamma)^{13}$N$(e^+, \nu_e)^{13}$C reactions make still more ^{13}C. At the
base of the convective shell in the model with abundances described in the fourth row of
Table 17.6.1, the first three reactions occur at the rates 2.69×10^{13} g^{-1} s^{-1}, 5.59×10^{12}
g^{-1} s^{-1}, and 1.88×10^{12} g^{-1} s^{-1}, respectively. Although $\langle \sigma v \rangle$ for neutron capture by ^{12}C
is approximately 100 times smaller than $\langle \sigma v \rangle$ for neutron capture by ^{14}N, because ^{12}C
is approximately 4500 times more abundant than ^{14}N, approximately five times as many
neutrons are captured by ^{12}C than are captured by ^{14}N.

Nuclear and gravothermal energy-generation rates in the nuclear active region of the
model with abundances given in the fourth row of Table 17.6.1 have already been described
in Fig. 17.5.4. Structure variables in the model are presented in Fig. 17.6.2 for a region
containing the convective shell. The mixing time in the convective shell is approximately

7000 seconds and the abundances by number of ^{4}He, ^{12}C, ^{13}C, ^{14}C, ^{14}N, and ^{16}O are constant throughout the shell. Protons, neutrons, and ^{13}N are everywhere in local equilibrium in the convective shell; at the base of the shell, their abundances by number are 4.421×10^{-13}, 6.610×10^{-17}, and 1.445×10^{-9}, respectively.

The rate of energy generation by helium-burning reactions at the base of the convective shell is $\epsilon_{He} = 1.13 \times 10^{10}$ erg g^{-1} s^{-1} and the total rate of energy generation by traditional CN-cycle reactions at the base is $\epsilon_{CN} = 1.51 \times 10^7$ erg g^{-1} s^{-1}. The total rate of energy generation by CN-cycle burning in the convective shell is $L_H = 8.847 \times 10^3$ $L_\odot$. This is coincidentally sufficiently close to the $\sim 5.6 \times 10^3$ $L_\odot$ increase in luminosity which occurs through the region of the hydrogen–helium interface between the luminosity minimum at $M \sim 0.5622$ $M_\odot$ and the base of the convective envelope that close inspection is required to recognize that the luminosity increase through the interface is not the consequence of hydrogen burning, offset somewhat by gravothermal absorption, but is due essentially entirely to the release of gravothermal energy. The increase in luminosity across the interface due to hydrogen burning is a paltry $L_H \sim 1.6$ $L_\odot$. Thus, amazingly, CN-cycle reactions occurring at and near the base of the helium-burning shell produce energy at a rate over 5500 times greater than the rate at which they produce energy through the hydrogen–helium interface.

The total number of neutrons released in the convective shell is approximately 10 times smaller than the number of ^{56}Fe nuclei in the shell, so, although some s-process nucleosynthesis occurs (see Chapter 19), it is very minor. As succeeding flashes become stronger, this result will be considerably altered due to the formation of a ^{13}C abundance peak in the region where, after a sufficiently strong flash, the base of the convective envelope penetrates into the ^{12}C layer formed during the flash. That this is bound to come about is made clear by the discussion in Section 17.5 based on the information in Figs. 17.5.5–17.5.8.

At the beginning of the interpulse phase following the penetration of the base of the convective envelope into the carbon-rich region predicted in Section 17.5, (p, γ) reactions on ^{12}C produce a spike in the ^{13}C abundance centered on the interpenetration zone, and, during the next helium flash phase, ingestion and burning of this ^{13}C spike can lead to the production of s-process elements in interesting amounts. The manner in which the spike is formed in a 5 $M_\odot$ TPAGB model, in which flashes are stronger and less complicated than in the 1 $M_\odot$ model, is described in Chapter 19.

Bibliography and references

A. Buzzoni, F. Fusi-Pecci, R. Buononno, & C. E. Corsi, *A&Ap*, **128**, 94, 1938.

Keith H. Despain, *ApJ*, **253**, 811, 1982.

Icko Iben, Jr., *Nature*, **220**, 143, 1968.

Icko Iben, Jr., *ApJ*, **260**, 821, 1982.

Icko Iben, Jr. & Alvio Renzini, *Phys. Rep.*, **105**, 330, 1984.

J. G. Mengel & P. G. Gross, *Ap. Space Sci.*, **41**, 407, 1976.

J. G. Mengel & A. V. Sweigart, *Astrophysical Parameters for Globular Clusters*, Ed. A. G. D. Philip (Dordrecht: Reidel), 1981, p. 277.

J. G. Mengel, A. V. Sweigart, P. Demarque, & P. G. Gross, *ApJS*, **40**, 733, 1969.

Bohdan Paczyñski, *Acta Astron.*, **20**, 47 & 287, 1970.

R. G. Spulak, *ApJ*, **235**, 565, 1980.

Martin Schwarzschild and Richard Harm, *ApJ*, **142**, 855, 1965; **150**, 961, 1967.

Richard Stothers & Chao-Wen Chin, *ApJ*, **292**, 222, 1985.

Alan V. Sweigart & P. G. Gross, *ApJS*, **32**, 367, 1976; **36**, 405, 1978.

Alan V. Sweigart & J. G. Mengel, *ApJ*, **229**, 624, 1979.

U. Uus, *Nauch. Inform. Akad. Nauk.*, **17**, 3, 1970.

This chapter traces the evolution of a 5 $M_\odot$ model of initial composition $(Z, Y) =$ $(0.015, 0.275)$ through the core helium-burning phase (Section 18.1), through the early asymptotic giant branch (EAGB) phase (Section 18.2), and well into the thermally pulsing asymptotic giant branch (TPAGB) phase (Section 18.3).

In contrast with the situation during the main sequence phase, when the convective core shrinks in mass as evolution progresses, during most of the core helium-burning phase, the average mass of the convective core increases as evolution progresses. The basic reason for the difference in behavior is related to a difference in the manner in which the opacity varies across the outer edge of the convective core in the two cases. In both cases, the dominant sources of opacity near the edge of the convective core are electron scattering and free–free absorption. The cross section for electron scattering is proportional to the number abundance of free electrons $\left(Y_e = \sum_i Y_i Z_i = \sum_i (X_i/A_i) Z_i\right)$ and the cross section for free–free absorption is proportional to the sum over all isotopes of the products of the isotopic number abundance and the square of the isotopic electric charge: $\left(\sum_i Y_i Z_i^2 = \sum_i (X_i/A_i) Z_i^2\right)$.

In the main sequence case, the number of free electrons per gram in the convective core becomes smaller with time as hydrogen is converted into helium, but the sum of the products of isotopic number abundance and the square of isotopic electric charge remains the same. Thus, at the outer edge of the convective core, the contribution of electron scattering to the opacity decreases discontinuously inward but the contribution of free–free absorption changes only because of a change in density associated with a change in molecular weight. At every point in the core, the opacity decreases more rapidly with time than the luminosity increases, and this is the principle reason for the decrease with time in the mass of the core.

In the core helium-burning case, the situation is complicated by the fact that, near the outer edge of the convective core, the radiative gradient V_{rad} can pass through a minimum which is very close to the adiabatic gradient V_{ad}. The minimum occurs because V_{rad} is proportional to the opacity κ, which increases outward, and to L/M and P/P_{rad}, both of which decrease outward. When the minimum occurs near the edge of the convective core, where V_{rad} is in any case very close to V_{ad}, in any given time step, the minimum can fall below V_{ad} in a region which was within the convective core during the previous time step. Consequently, the convective core breaks semiperiodically into two convective regions – a smaller convective core and a convective shell the outer edge of which more or less coincides with the outer edge of the original convective core – separated by a radiative region.

The number of free electrons per gram in both of the two convective regions remains constant with time as helium is converted into carbon and oxygen, but the product of the isotopic number abundance and the square of isotopic electric charge increases with time. Thus, across the outer edge of the outermost of the two interior convective zones (convective core or convective shell), the contribution to the opacity of electron scattering does not vary with mass, but the contribution of free–free absorption increases discontinuously inward; the location of the discontinuity defines the outer edge of the outermost convective region. Because the primary source of opacity in the two convective zones is electron scattering and since this contribution does not change with either mass or time, the primary reason for the increase with time in the mass at the outer edge of the outermost interior convective zone is the increase in the value of L/M at this outer edge.

Eventually, as the abundance of helium in central regions decreases below a critical value, the declining outward flux of energy due to core helium burning is no longer capable of sustaining a large convective core. The convective core decreases in mass and ultimately vanishes, with helium burning now occurring in a shell in which temperatures and densities are at first significantly smaller than at model center during most of the preceding core helium-burning phase. The global helium-burning luminosity drops and, in response, the helium-depleted core (including the helium-burning shell) contracts and heats until the global helium-burning luminosity exceeds its value during the core helium-burning phase. At the same time, the increase in the flux of energy at the base of the helium-rich region between the helium-burning shell and the location of the hydrogen–helium interface leads to absorption of energy from the ambient energy flow in this region. Matter at and near the hydrogen–helium interface is forced outward to smaller densities and temperatures until the contibution to the surface luminosity by the hydrogen-burning shell becomes inconsequential and remains so as the helium-burning shell works its way outward in mass.

During the ensuing EAGB phase, the model surface luminosity, which is essentially the helium-burning luminosity, increases by over an order of magnitude and the model radius increases by almost an order of magnitude. In response to an increasing luminosity, the base of the convective envelope moves inward in mass, eventually reaching the hydrogen–helium interface. For a while thereafter, the base of the convective envelope and the hydrogen–helium interface coincide and, as the base of the convective envelope continues to move inward in mass, matter which has experienced complete hydrogen burning is mixed into the convective envelope. The process whereby products of complete hydrogen burning are dredged-up into (incorporated by) the convective envelope is called the second dredge-up process to distinguish it from the first dredge-up process which occurs as the model ascends the red giant branch for the first time after developing a hydrogen-exhausted core (see Section 11.2 in Volume 1).

As the helium-burning shell and the hydrogen–helium interface approach one another in mass, temperatures and densities in and near the interface increase until, ultimately, hydrogen burning is resurrected. This terminates the EAGB phase and begins the TPAGB phase during which the helium-burning shell and the hydrogen-burning shell alternate as major contributors to the surface luminosity.

As described in Section 18.3, each phase of helium-burning dominance begins with a helium-burning runaway during which the nuclear energy liberated by helium burning is stored first as gravothermal energy at the base of the region between the two burning shells and then transferred outward in what is called a thermal pulse episode (even though the expansion associated with the outward transfer of energy is every bit as important as the changes in temperature that occur). As the outward transfer of stored energy continues, matter at the helium–hydrogen interface is forced outward to lower temperatures and densities and hydrogen burning is effectively extinguished. As matter in the helium-burning shell expands, the rate of helium burning declines and temperatures in the shell drop, at first precipitously and then more gradually. Helium burning continues thereafter quiescently, even when hydrogen burning is again resurrected and becomes the dominant source of surface luminosity. During the period of hydrogen-burning dominance, the second most potent contributor to the surface luminosity is gravothermal energy released as fresh helium is added to the contracting helium-rich region between the two burning shells. Helium burning makes a distant third contribution to the surface luminosity until, once the helium-rich zone attains a critical mass, another thermal pulse is initiated.

In the course of a TPAGB cycle, the mass M_{BCE} at the base of the convective envelope alternately increases and decreases. During the quiescent hydrogen-burning portion of the cycle, M_{BCE} increases synchronously with the mass M_H at the center of the hydrogen-burning shell, with $M_{BCE} - M_H$ remaining essentially constant. When the hydrogen-burning shell abruptly becomes extinct, a concomitant abrupt decrease in the luminous flux at the base of the convective envelope causes M_{BCE} to increase abruptly. However, as the gravothermal energy stored between the helium- and hydrogen-burning shells during the peak of a helium shell flash is transferred outward, a luminosity wave reaches the base of the convective envelope and the increase in luminosity there causes the base of the convective envelope to move inward in mass until $M_{BCE} \sim M_H$.

The primary source of opacity at the hydrogen–helium interface is electron scattering so that, when composition variables and logarithmic temperature–pressure gradients are defined at zone centers, as in this book, the radiative gradient V_{rad} is discontinuous across the interface, with the opacity on the helium-rich side of the interface being substantially smaller than on the hydrogen-rich side. The quantity $V_{rad} - V_{ad}$ is formally large and positive on the hydrogen-rich side of the interface and large and negative on the helium-rich side. Since convective velocities are formally large on the hydrogen-rich side, convective overshoot must occur, and in a more realistic treatment, mixing between hydrogen-rich and helium-rich material will produce abundance profiles of every isotope extending in both directions about the formal hydrogen–helium interface by some fraction of a pressure scale height. In the calculations reported here, if overshoot is not explicitly invoked, on reaching the H-He interface for the first time, the base of the convective envelope proceeds no further inward in mass.

However, by insisting that complete mixing of composition variables extends exactly one mass zone inward beyond the base of the region formally unstable against convection into the region formally stable against convection, the hydrogen–helium interface moves gradually inward, eventually encountering and extending into the region where fresh carbon has been deposited during the preceding helium shell flash. Carbon is then injected into the convective envelope in what is called the third dredge-up process. The precise manner in which the process takes place has yet to be satisfactorily modeled, but the fact that carbon stars exist is incontrovertible evidence that the process occurs and is undoubtedly responsible for most of the carbon found in the Universe outside of stars.

During the TPAGB phase, neutrons are produced in helium-burning regions by the $^{13}C(\alpha, n)^{16}O$ and $^{22}Ne(\alpha, n)^{25}Mg$ reactions. A discussion of these reactions and the consequent neutron-capture nucleosynthesis are the subject of Chapter 19.

18.1   Evolution during the core helium-burning phase

The starting point for the calculations which produce the results just described is the model discussed at the end of Section 11.2 in Volume 1. This model has just begun the core helium-burning phase and is at the tip of the red giant branch in the HR diagram of Fig. 11.2.42.

Its structure characteristics are shown in Fig. 11.2.57 and its energy-generation characteristics are shown in Fig. 11.2.58. Evolution in the HR diagram during the core helium-burning phase is shown in Fig. 18.1.1, along with the evolutionary track for the last part of the shell hydrogen-burning phase which precedes core helium burning. A number next to an open circle or square along the evolutionary track gives the model age in units of 10^7 yr at the location of the symbol. The glitch in the track between the times $t_7 = 8.33$ and

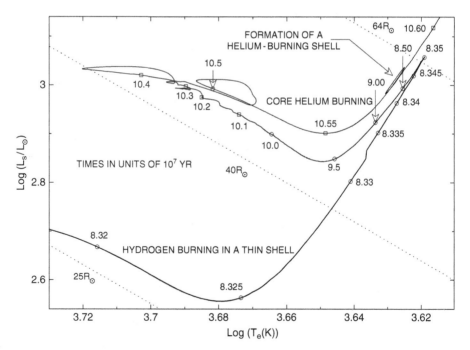

Fig. 18.1.1   Evolution in the HR-diagram of a 5 $M_\odot$ model ($Z = 0.015, Y = 0.275$) just prior to and during core helium burning – first approximation

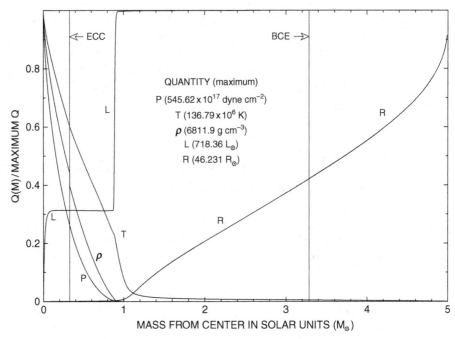

Fig. 18.1.2   Structure of a 5 $M_\odot$ model approximately half way through the core helium-burning phase ($Z = 0.015$, $Y = 0.275$, Age $= 9.421 \times 10^7$ yr)

$t_7 = 8.335$ is a consequence of increasing the mass of the static envelope near a boundary in the grid for static envelopes.

Structure and composition variables are shown in Figs. 18.1.2 and 18.1.3, respectively, for a model of age $t_7 = 9.42$, approximately half way through the core helium-burning phase. From the luminosity profile in Fig. 18.1.2, one sees that hydrogen burning in a shell contributes approximately twice as much to the surface luminosity as does helium burning at the base of a convective core. Comparing Fig. 18.1.2 with Fig. 11.2.57, it is evident that, during evolution from the red giant branch tip, the mass of the convective core has increased by approximately a factor of 2, from ∼0.16 $M_\odot$ to ∼0.32 $M_\odot$, and the base of the convective envelope has receded from a position approximately 1 $M_\odot$ beyond the center to over 3 $M_\odot$ beyond the center. From the luminosity profiles in the two figures, it is evident that the center of the hydrogen-burning shell has moved from ∼0.6 $M_\odot$ to ∼0.9 $M_\odot$, approximately 0.1 $M_\odot$ below the position of the base of the convective envelope at its maximum extent inward during the ascent of the red giant branch prior to the ignition of helium at the center. Figure 18.1.3 shows that approximately 0.15 $M_\odot$ of ^{4}He has been converted into ^{12}C and ^{16}O in the convective core. Comparing the hydrogen-abundance profiles in Figs. 11.2.41 and 18.1.3, one infers that a comparable mass of hydrogen has been converted into helium during the first half of the core helium-burning phase.

Gravothermal energy-generation rates for portions of the model which is almost half-way through the core helium-burning phase are described in Figs. 18.1.4 and 18.1.5. Gravitational acceleration and velocity are also shown in Fig. 18.1.4. In both figures,

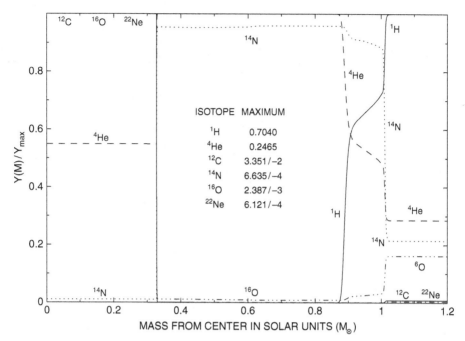

Number abundances of major constituents in a 5 $M_\odot$ model approximately half way through the core helium-burning phase ($Z = 0.015, Y = 0.275$, Age $= 9.421 \times 10^7$ yr)

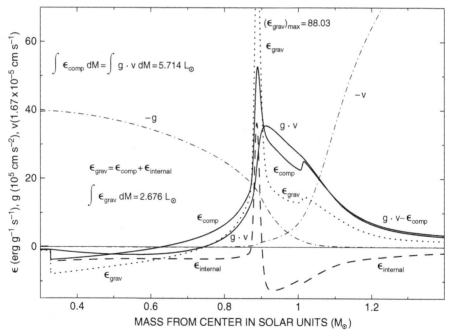

Gravothermal energy-generation rates in a 5 $M_\odot$ model approximately half way through the core helium-burning phase ($Z = 0.015, Y = 0.275$, Age $= 9.421 \times 10^7$ yr)

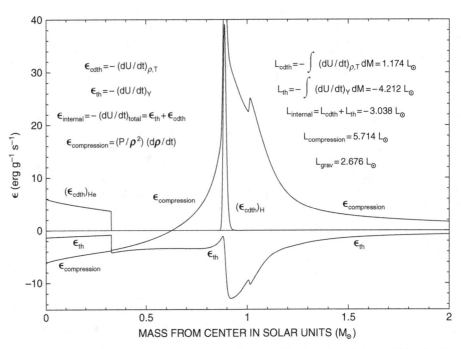

Fig. 18.1.5   Components of the gravothermal energy-generation rate in a 5 $M_\odot$ model midway through the core helium-burning phase ($Z = 0.015$, $Y = 0.275$, Age $= 9.421 \times 10^7$ yr)

discontinuities along several of the curves occur at the outer edge of the convective core. Near discontinuities occur at the position reached by the base of the convective envelope at its maximum extent inward. As always, the overall rate of compressional energy release is equal to the overall rate at which work is done by gravity ($L_{\text{compressional}} = L_{\text{work}} = 5.714\ L_\odot$). In the convective envelope, which is everywhere contracting, the local rate of compressional energy release $\epsilon_{\text{compression}}$ and the local rate at which gravity does work $\epsilon_{\text{work}} = \mathbf{g} \cdot \mathbf{v}$ are essentially identical. Among the most interesting results described in Figs. 18.1.4 and 18.1.5 are that (1) the total rate of energy generation associated with the reduction in the particle number density due to the conversion of helium into carbon and oxygen in the convective core ($L_{\text{cdth}}^{\text{He}} \sim 0.79\ L_\odot$) and the total rate of energy generation associated with the reduction in the particle number density due to the conversion of hydrogen into helium in the hydrogen-burning shell ($L_{\text{cdth}}^{\text{H}} \sim 0.38\ L_\odot$) are of comparable magnitude and that (2) the total rate of energy generation due to the reduction in the particle number density ($L_{\text{cdth}}^{\text{tot}} = 1.174\ L_\odot$) is a subtantial fraction of the total rate of gravothermal energy generation ($L_{\text{grav}} = 2.676\ L_\odot$).

Figure 18.1.6 shows the variations with mass of the radiative and adiabatic logarithmic temperature–pressure gradients and of the ingredients of the radiative gradient over most of the interior. The discontinuity in the radiative gradient across the outer edge of the convective core is due to the discontinuity in composition variables across the edge which produces a discontinuity in the free–free contribution to the opacity. As discussed in the introduction of this chapter and in Section 17.2 in connection with Fig. 17.2.11,

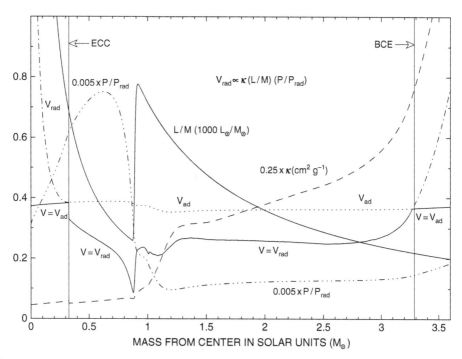

Fig. 18.1.6 Logarithmic gradients in a 5 $M_\odot$ model approximately half way through the core helium-burning phase ($Z = 0.015$, $Y = 0.275$, Age = 9.421×10^7 yr)

the free–free contribution to opacity is proportional to $S = \sum_i Y_i Z_i^2 = \sum_i X_i Z_i^2 / A_i$, where the sum is over all isotopes. In the convective core of the model under discussion, $S = 0.55 \times 0.2465 \times 2^2 + 0.0335 \times 6^2 + 0.002\,387 \times 8^2 + \cdots \sim 1.90$. Just outside the convective core, $S = 0.2465 \times 2^2 + 0.000\,6635 \times 7^2 + \cdots \sim 1.02$. In the 1 $M_\odot$ model described in Section 17.2 which is also approximately halfway through the core helium-burning phase, the discontinuity in composition which leads to a large discontinuity in the opacity and in the radiative gradient occurs at the outer edge of a convective shell outside of the convective core (see Fig. 17.2.11). The comparable 5 $M_\odot$ model does not have a convective shell.

The apparent difference in the locations of convective zones in the two models of different mass is not a fundamental difference. Both core helium-burning models behave very similarly with regard to the fluctuation in mass of the convective core and with regard to the appearance and disappearance of a convective shell above the convective core. It is instructive to compare the profile of the mass of the convective core of the 1 $M_\odot$ core helium-burning model shown in Fig. 17.2.6 with the same profile in the 5 $M_\odot$ core helium-burning model shown in Fig. 18.1.7. In both cases, over the first 25% or so of the core helium-burning phase, the mass of the convective core increases essentially monotonically, with fluctuations in mass being inconsequential. In the 1 $M_\odot$ model, subsequent abrupt decreases in the mass of the convective core are accompanied by the appearance of a convective shell which, as the mass of the convective core grows, is ultimately incorporated into the convective core. The process repeats in a semiperiodic fashion. In the

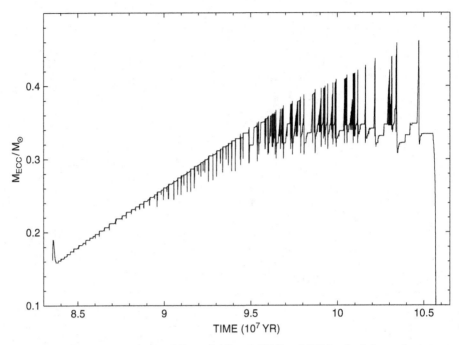

Fig. 18.1.7 Mass of the convective core versus time in a 5 $M_\odot$ model ($Z = 0.015$, $Y = 0.275$) burning helium at the center – first approximation

5 $M_\odot$ model, the appearance of convective shells is delayed to the second half of the core helium-burning phase.

In both cases, the variations in the location and mass of convective zones are, to some extent, a manifestation of numerical noise associated with frequent changes in zoning in mass. To demonstrate this, the 9.42×10^7 yr old 5 $M_\odot$ model has been rezoned so that $\delta q/q \gtrsim 0.01$ over the mass range 0.25–0.5 $M_\odot$. Zoning in this region has been thereafter frozen. In an effort to minimize variations in surface characteristics due to rezoning in the envelope, zoning has also been frozen over the mass range $3.75-5.00$ $M_\odot$. The consequent variations with time in the mass of the convective core are shown in Fig. 18.1.8. Comparing with the variations in Fig. 18.1.7, it is evident that, although the variations with time in the second calculation become more periodic than those in the first calculation, the variation with time of the maximum mass of the convective core and the total duration of the helium-burning phase are very similar.

Although some of the erratic behavior of convective zones in core helium-burning models can be ascribed to numerical noise, the fact that semiperiodic decreases in mass of the convective core occur and are accompanied by the appearance of a convective shell followed by a merger of the two convective zones can be understood as a consequence of the parabolic nature of the radiative temperature–pressure gradient V_{rad} in the region where core and shell convection occur. This is made clear by inspection of the curves for V_{rad} in Figs. 18.1.9, 18.1.10, and 18.1.11 for models in the second calculation of age $t_7 = 9.87$, 10.05, and 10.23, respectively. In each case, between the center and the outer

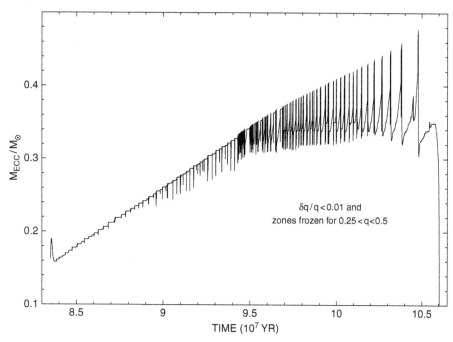

Fig. 18.1.8 Mass of the convective core versus time in a 5 $M_\odot$ model ($Z = 0.015$, $Y = 0.275$) burning helium at the center – second approximation

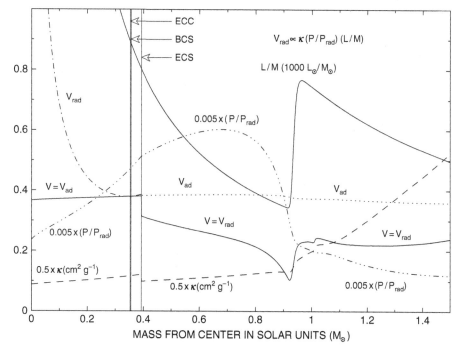

Fig. 18.1.9 Logarithmic gradients in a 5 $M_\odot$ model approximately 67% of the way through the core helium-burning phase ($Z = 0.015$, $Y = 0.275$, Age $= 9.866 \times 10^7$ yr)

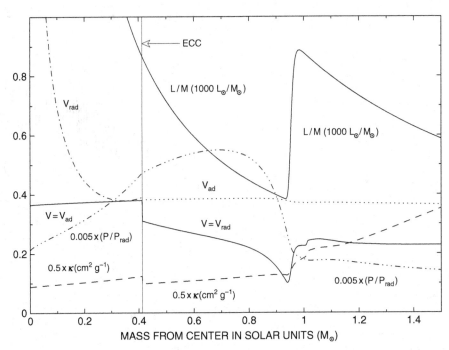

Fig. 18.1.10   Logarithmic gradients in a 5 $M_\odot$ model approximately 75% of the way through the core helium-burning phase ($Z = 0.015$, $Y = 0.275$, Age $= 10.046 \times 10^7$ yr)

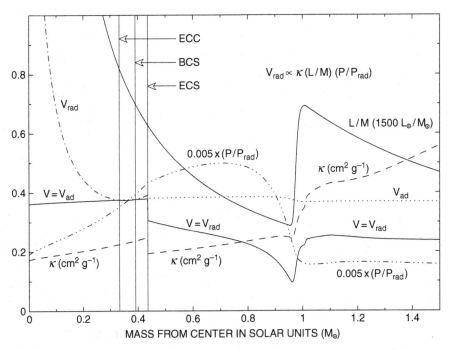

Fig. 18.1.11   Logarithmic gradients in a 5 $M_\odot$ model approximately 84% of the way through the core helium-burning phase ($Z = 0.015$, $Y = 0.275$, Age $= 10.226 \times 10^7$ yr)

edge of the outermost interior convective zone, V_{rad} has a parabolic form, with the base of the parabola being extremely close to the curve for the adiabatic gradient V_{ad}. In the $t_7 = 9.87$ model, the base of the V_{rad} parabola extends ever so slightly below V_{ad} and the base of the convective shell almost touches the outer edge of the convective core. The outer edge of the convective shell coincides with the composition discontinuity which is responsible for the precipitous drop in V_{rad} below V_{ad}. In the $t_7 = 10.05$ model, the base of the V_{rad} parabola remains slightly above V_{ad} and the outer edge of the convective core coincides with the composition discontinuity where V_{rad} drops discontinuously below V_{ad}. Finally, in the $t_7 = 10.23$ model, the base of the V_{rad} parabola falls below the curve for V_{ad} and the parabola remains smaller than V_{ad} for a considerable distance in mass on either side of the base of the parabola. As in the youngest of the three models, the outer edge of the convective core coincides with the composition discontinuity where V_{rad} again drops abruptly below V_{ad}.

The parabolic shape of the V_{rad} curve is the consequence of the fact that, of the three factors whose product make up V_{rad} in the region of nearly homogeneous composition where the convective zones prevail, two (κ and P/P_{rad}) increase outward, and the third (L/M) decreases outward. The point where a minimum in the triple product is reached defines where a radiative zone can appear between a convective core of abruptly decreased mass and a convective shell whose outer edge coincides with the outer edge of the convective core before shrinkage. That the composition remains nearly homogeneous over the entire region from the center to the outer edge of the outermost interior convective zone is demonstrated explicitly by the distribution shown in Fig. 18.1.12 of composition variables in the third of the three models.

Comparing the three quantities which make up V_{rad} in the three models, it is apparent that, at any given mass point within the region where convection occurs, the value of P/P_{rad} decreases with time, the value of κ remains essentially constant, and the value of L/M increases. That the maximum outward extent in mass of the region in which convective motions occur increases with time is due to the circumstance that L/M increases more rapidly at any mass point than P/P_{rad} decreases, a fact which could not have been predicted, but has required an explicit calculation to demonstrate.

That L/M increases with time at any mass point is a consequence of the facts that (1) the average temperature in central regions where helium burns increases with time and (2) the temperature dependent factor in the rate of energy generation (proportional to approximately the 40th power of the temperature) increases with time more rapidly than the square of the helium abundance in the convective core decreases with time. The variation with time in the central temperature is shown in Fig. 18.1.13, along with the variation in central density and pressure. The solid curves are for the first calculation and the dotted curves are for the second calculation. Variations with time of abundances at the center of the model in the second calculation are shown in Fig. 18.1.14 and the global helium-burning luminosity for the model is shown as a function of time in Fig. 18.1.15.

Toward the end of the core helium-burning phase in the second calculation, the evolutionary track in the HR diagram shown in Fig. 18.1.16 extends further to the blue by almost $\Delta \log T_e = 0.02$ than does the evolutionary track shown in Fig. 18.1.1 produced by the first calculation. In both cases, there are wild gyrations in the evolutionary track

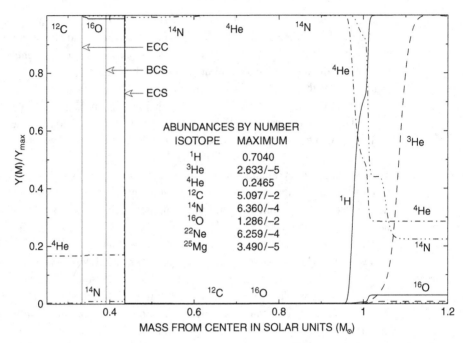

Fig. 18.1.12 Abundances of major constituents in a 5 $M_\odot$ model approximately 84% of the way through the core helium-burning phase ($Z = 0.015$, $Y = 0.275$, Age $= 10.226 \times 10^7$ yr)

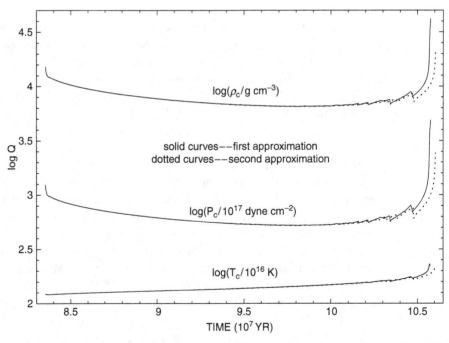

Fig. 18.1.13 Density, temperature, and pressure at the center of a 5 $M_\odot$ model burning helium in a convective core ($Z = 0.015$, $Y = 0.275$)

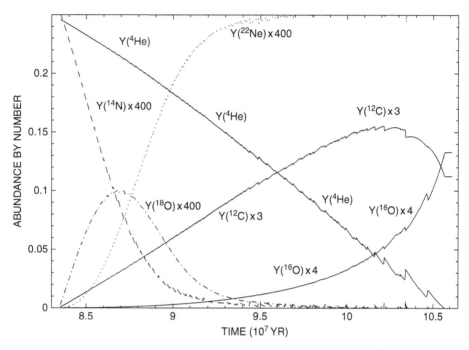

Fig. 18.1.14   Abundances at the center of a 5 M$_\odot$ model burning helium in a convective core ($Z = 0.015$, $Y = 0.275$)

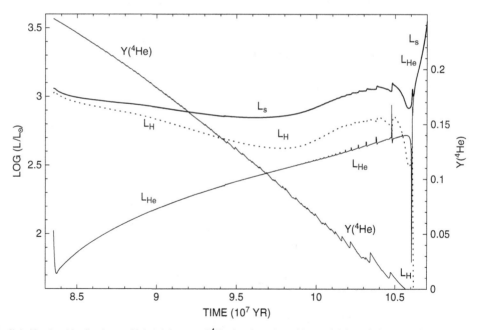

Fig. 18.1.15   Global luminosities (L_H, L_{He}, and L_s) and the central ^{4}He abundance in a 5 M$_\odot$ model through the core helium burning phase and into the EAGB phase ($Z = 0.015$, $Y = 0.275$)

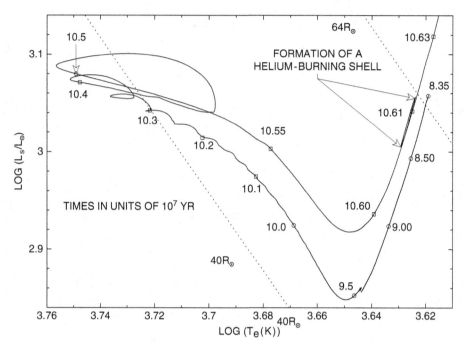

Fig. 18.1.16   Evolution in the HR-diagram of a 5 M$_\odot$ model ($Z = 0.015$, $Y = 0.275$) during core helium burning - second approximation

toward the end of the core helium-burning phase and no attempt is made to analyze these gyrations which are, in any case, of a very short duration relative to the total duration of the core helium-burning phase. In both cases the furthest blueward extent of the core helium-burning track falls short of the Cepheid instability strip for the chosen composition, which, at a luminosity in the range corresponding to $\log(L/L_\odot) = 3.0$–3.1, lies in the range $\log(T_e) = 3.77$–3.82.

Where in the HR diagram a core helium-burning model spends most of its time vis-à-vis the red giant branch and the Cepheid instability strip is sensitive to opacities in the hydrogen-rich envelope of the model and so is sensitive to the choice of metallicity and to the extent to which all sources of opacity are taken into account. The location in the HR diagram is also sensitive to whether or not convective overshoot at the outer edge of the convective core is taken into account and to the ratio of the cross section for the ^{12}C(α, γ)^{16}O reaction to the effective cross section for the triple-alpha reaction. The ratio of cross sections determines the extent to which helium burning produces oxygen and thus influences the duration of the core helium-burning phase.

Several early calculations of the evolution of intermediate mass models of population I composition utilized opacities which did not take bound–bound transitions into account and did not take convective overshoot into account; resultant evolutionary tracks during core helium-burning for 5 $M_\odot$ models of metallicity larger than employed here extend further to the blue and to larger luminosities than found here, in particular crossing and extending to the blue of the Cepheid instability strip as, for example, in the calculations of Iben (1966) for a composition $(Z, Y) = (0.02, 0.27)$. Using, instead, a first generation of opacities including bound–bound absorption produced by Art Cox and his collaborators at Los Alamos Laboratory, but using the same choice for the relevant nuclear cross section factors as used in the Iben (1966) study, the evolutionary track during core helium burning for a 5 $M_\odot$ model of composition $(Z, Y) = (0.02, 0.27)$ does not extend as far to the blue as the Cepheid instability strip, remaining very close to the red giant branch (Iben, 1972). However, reducing the metallicity by a factor of 2, the evolutionary track during the core helium-burning phase again extends to the blue of the Cepheid instability strip (S. A. Becker, I. Iben, Jr., and R. S. Tuggle, 1977) and achieves luminosity levels comparable to those achieved by models made with twice the metallicity but without bound–bound transitions included in the opacity.

When, finally, convective overshoot at the outer edge of the convective core is taken into account, the core helium-burning track of a 5 $M_\odot$ model made with Los Alamos opacities and with standard cross section factors and composition $(Z, Y) = (0.02, 0.27)$ also extends considerably blueward of the Cepheid instability strip (Iben, 1986). As an aside, the three panels in Fig. 6 of this last reference may be profitably compared with Figs. 18.1.9–18.1.11 here as an elucidation of the parabolic nature of the radiative gradient near the outer edge of the convective core.

From the results of many calculations by many authors, it has been established that, whatever the choice of input physics, the larger the mass of the model, the further does the evolutionary track in the HR diagram during the core helium-burning phase extend to the blue and that, for any choice of composition, there exists a mass such that the time spent in the Cepheid instability strip is a maximum, demonstrating that, for an aggregate of stars of the same composition and age, there is a peak in the distribution of Cepheids in number versus luminosity. The location of the maximum in this distribution can be used to estimate the metallicity of the aggregate, as described in Section 20.1.

18.2   Transition to, evolution along, and transition from the early asymptotic giant branch

As the helium abundance near the center of the model decreases below a critical value, the contribution of central regions to the global helium-burning luminosity L_{He} declines and the major contibution to L_{He} shifts to a double shell. Because densities and temperatures in the shell are initially smaller than at the center during most of the preceding core helium-burning phase, L_{He} decreases. The decrease in L_{He} is initially offset by an increase in the rate of gravothermal energy production by the helium-depleted core, which contracts and heats, and by an increase in the rate of energy production by the hydrogen-burning shell in which matter also contracts and heats. Thus, the surface luminosity L_s continues to increase even as L_{He} decreases. These changes are evident in Fig. 18.2.1 where global nuclear burning, gravothermal, and surface luminosities are shown as functions of time over a time period centered on the transition from core to shell helium burning.

Matter in the helium-burning shell also contracts and heats, so, after reaching a minimum, L_{He} begins to increase, ultimately exceeding the value prevailing during most of

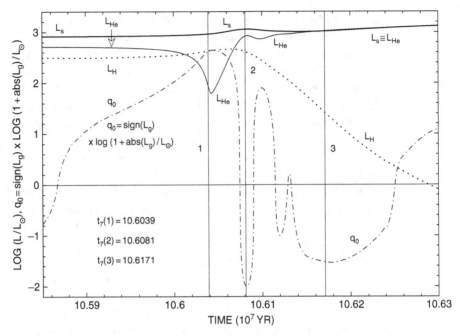

Fig. 18.2.1 Nuclear-burning, gravothermal, and surface luminosities of a 5 $M_\odot$ model transitioning between the core helium-burning and EAGB phases ($Z = 0.015$, $Y = 0.275$)

the core helium-burning phase. Absorption from the ambient energy flow in the region between the helium- and hydrogen-burning shells and beyond leads to expansion and cooling, with the result that not only does the hydrogen-burning luminosity L_H decrease, but the rate of gravothermal energy production L_{grav} becomes temporarily negative. The net result is a temporary decrease in L_s which is reversed as L_{He} continues to increase. As hydrogen burning and gravothermal energy production become minor contributors to the surface luminosity, the model enters the EAGB phase proper and $L_s \sim L_{He}$ during the remainder of the phase.

For the first model calculation discussed in Section 18.1, the location in the HR diagram where the transition from core helium burning to shell helium burning takes place is identified explicitly in the upper right hand corner of Fig. 18.1.1 by a slanted arrow pointing to a portion of the evolutionary track which is bluer than and asymptotic to the first red giant branch track which terminates at age $t_7 = 8.35$. Along this portion of the AGB track, the double reversal in direction is centered at $\log(L/L_\odot) \sim 3.1$ and is better characterized as a thickening of the track than as a loop. For the second model calculation, the location along the AGB track where the double reversal in direction takes place is identified in Fig. 18.1.16 by two arrows pointing to the boundaries of the thickened portion of the track containing the open square at time $t_7 = 10.61$.

Luminosity profiles for three models of ages $t_7 = 10.6039$, 10.6081, and 10.617, respectively, are shown in Fig. 18.2.2. The locations of the three models are identified in Fig. 18.2.1 by vertical lines labeled 1, 2, and 3 and the intersections of these lines with

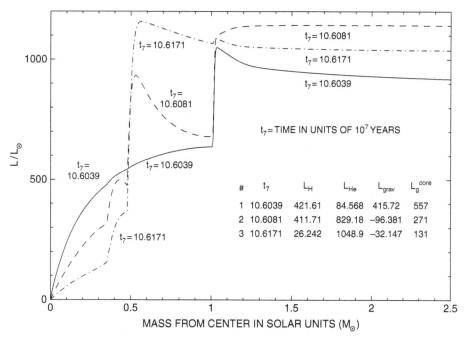

Fig. 18.2.2 Luminosity versus mass in a 5 $M_\odot$ model at three times near the beginning of the EAGB phase ($Z = 0.015$, $Y = 0.275$)

the luminosity–time curves may be used to obtain the global luminosities listed in units of $L_\odot$ in the lower right hand portion of Fig. 18.2.2.

Structure variables for model 1 are shown in Fig. 18.2.3, structure and composition variables for the inner portion of this model are shown in Fig. 18.2.4, and gravothermal and nuclear energy-generation rates for the model are shown as functions of mass in Fig. 18.2.5. Gravothermal and nuclear energy-generation rates for models 2 and 3 are shown as functions of mass in Figs. 18.2.6 and 18.2.7, respectively. The estimates in Fig. 18.2.2 of the gravothermal luminosity L_g^{core} emanating from the helium-depleted core have been obtained by integrating over mass the local gravothermal energy-generation rates for the three models in Figs. 18.2.5–18.2.7.

In model 1, L_{grav} is made up of two components, a positive component, $L_g^{core} \sim 557\, L_\odot$, due to the release of gravothermal energy from the core and a negative component, $L_g^{env} \sim -141\, L_\odot$, due to absorption in the envelope above the hydrogen-burning shell. The contribution of gravothermal and hydrogen-burning energy, $L_g^{core} + L_H \sim 979\, L_\odot$, to the luminosity at the outer edge of the hydrogen-burning shell is almost a dozen times larger than the contribution of energy from helium burning, $L_{He} \sim 84.6\, L_\odot$. As follows from the profile for ϵ_{He} in Fig. 18.2.5, the helium-burning luminosity is made up of three components, $L_{He}^{core} \sim 72\, L_\odot$ produced by the core and $L_{He}^{shell\ 1} \sim 11\, L_\odot$ and $L_{He}^{shell\ 2} \sim 2\, L_\odot$ produced, respectively, by each of the two shells. The occurrence of two helium-burning shells is a consequence of the helium-abundance distribution created during the late stages of core helium burning.

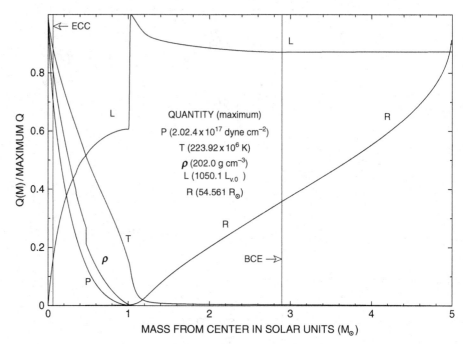

Fig. 18.2.3   Structure of a 5 $M_\odot$ model transitioning from core to shell helium burning ($Z = 0.015$, $Y = 0.275$, AGE $= 10.6039 \times 10^7$ yr)

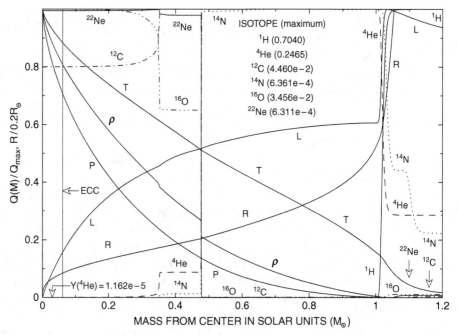

Fig. 18.2.4   Structure and composition variables in a 5 $M_\odot$ model transitioning from core to shell helium burning ($Z = 0.015$, $Y = 0.275$, AGE $= 10.6039 \times 10^7$ yr)

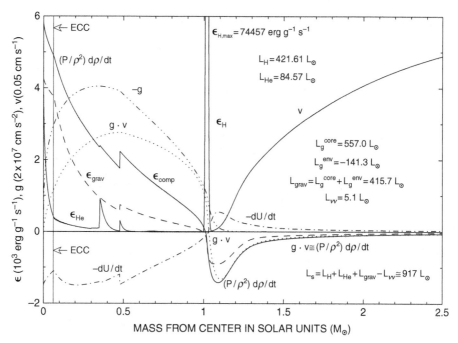

Fig. 18.2.5 Gravothermal and nuclear energy-generation rates in a 5 $M_\odot$ model transitioning from core to shell helium burning ($Z = 0.015$, $Y = 0.275$, AGE $= 10.6039 \times 10^7$ yr)

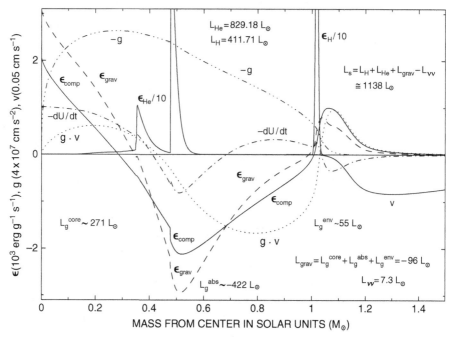

Fig. 18.2.6 Gravothermal and nuclear energy-generation rates in a 5 $M_\odot$ incipient EAGB model burning helium and hydrogen in shells ($Z = 0.015$, $Y = 0.275$, AGE $= 10.6081 \times 10^7$ yr)

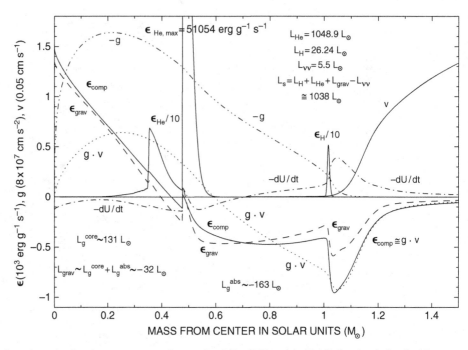

Fig. 18.2.7 Gravothermal and nuclear energy-generation rates in a 5 $M_\odot$ EAGB model with a He-burning shell and a dying H-burning shell ($Z = 0.015, Y = 0.275$, AGE $= 10.6171 \times 10^7$ yr)

From Fig. 18.1.8 it is evident that the convective core attains a maximum mass of $M_{CC} \sim 0.476\ M_\odot$ at $t_7 \sim 10.4$ and from Fig. 18.1.14 it is evident that the helium number abundance at this time is ~ 0.06 times the maximum helium number abundance. Thereafter, M_{CC} drops precipitously to $M_{CC} \sim 0.3\ M_\odot$ and then climbs slowly to $M_{CC} \sim 0.35$ before again decreasing precipitously. Before this last precipitous drop, the helium number abundance in the convective core is just barely resolvable in Fig. 18.1.14. Thus, if no convective shell were to arise during the remainder of the core helium-burning phase, the helium number-abundance profile would consist of a segment extending from the center to $\sim 0.35\ M_\odot$ in which the number abundance is only slightly larger than zero and a segment extending between $\sim 0.35\ M_\odot$ and $\sim 0.476\ M_\odot$ in which the helium number abundance is ~ 0.06 times the maximum number abundance in the model.

The distribution of the helium number abundance in Fig. 18.2.4 corroborates the predicted distribution for $M < 0.35\ M_\odot$ and the prediction of a segment extending from $\sim 0.35\ M_\odot$ to $\sim 0.476\ M_\odot$ in which the helium number abundance is of the order of 0.06 times the maximum number abundance. However, the actual number abundance in this region is ~ 0.08 times the maximum number abundance. Thus, convective mixing in a shell has occurred during the remainder of the core helium-burning phase and injected fresh helium into the region centered at $M \sim 0.415\ M_\odot$. From the difference between the abundance of 0.06 predicted without additional convective mixing in a shell and the abundance 0.08 actually found, one ascertains that the outer edge of the final convective shell reaches a maximum mass which is larger than the maximum mass of the convective core by the modest amount of $\sim 0.003\ M_\odot$.

The abundance of helium in central regions is extremely small (only $Y_{He} = 1.162 \times 10^{-5}$ in the convective core), and nuclear energy production between the center and the first nuclear energy shell source is overwhelmingly by the $^{12}C(\alpha, \gamma)^{16}O$ reaction. In the first shell source (where $Y_{He} = 2.163 \times 10^{-2}$), the rate of energy production by the triple-alpha reaction varies from 3% to 30% of that by the $^{12}C(\alpha, \gamma)^{16}O$ reaction. In the second shell source (where $Y_{He} = 0.2465$), the energy-production rate by the triple-alpha reaction exceeds by one to two orders of magnitude that by the $^{12}C(\alpha, \gamma)^{16}O$ reaction.

It is remarkable that, although the helium number abundance in central regions is three to four orders of magnitude smaller than in the two shell helium-burning regions, the contribution of helium burning by central regions to L_{He} is over five times larger than the rate at which helium burning in the two shells contributes. The fact that the extremely helium-depleted core liberates nuclear energy so much more efficiently than the helium-rich shells is one more demonstration of the very large temperature sensitivity of helium-burning reactions.

In model 2, the helium-burning luminosity has increased by almost a factor of 10 over that of model 1 and the rate at which gravothermal energy is liberated below the two helium-burning shells has been reduced by a factor of only about 2 from the rate in model 1. The consequent large increase in the rate at which energy is injected into the region between the base of the first helium-burning shell and just below the base of the hydrogen-burning shell has led to absorption of energy in this region at the rate $L_g^{abs} \sim -422\ L_\odot$, an estimate which has been obtained by integrating over mass under the negative portion of the curve for ϵ_{grav} in Fig. 18.2.6 between masses $M \sim 0.34\ M_\odot$ and $M \sim 0.98\ M_\odot$. The expansion has not quite reached the hydrogen-burning shell, which continues to produce energy at nearly the same rate as in model 1 and, in fact, instead of expanding outward as in model 1, matter outward from the base of the hydrogen-burning shell is contracting inward in model 2 and matter outward from the outer edge of the burning shell is heating. Gravothermal energy is being liberated above the base of the hydrogen-burning shell at the rate $L_g^{env} \sim 55\ L_\odot$.

During evolution between models 2 and 3, the expansion and cooling wave that has nearly reached the base of the hydrogen-burning shell in model 2 continues to move outward in mass, causing the luminosity of the hydrogen-burning shell to decline steadily. In model 2, L_H is about half of L_{He}, but, in model 3, L_H is about 40 times smaller than L_{He}. The curves for $\mathbf{g} \cdot \mathbf{v}$ and v in Fig. 18.2.7 show that, in model 3, matter between the outer edge of the second helium-burning shell and well into the hydrogen-rich envelope is moving outward, the curve ϵ_{comp} shows that matter outward from the base of the second helium-burning shell is expanding, and the curve for $-dU/dt$ shows that, although matter in the two helium-burning shells is heating, cooling prevails outward from midway between the helium- and hydrogen-burning shells well into the hydrogen-rich envelope.

Thermodynamic characteristics at the center of the helium-burning region in the model and the location in mass and radius of the center of the burning region are shown as functions of time in Fig. 18.2.8. Scale limits are indicated along the left hand vertical axis. Shown also is the helium-burning luminosity, for which scale limits are indicated along the right hand vertical axis. The times associated with models 1–3 are identified by the three vertical lines close to the left hand edge of the figure.

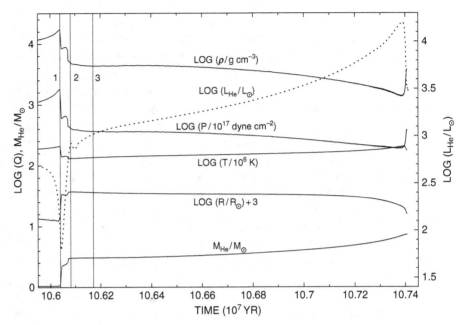

Conditions at the center of the helium-burning shell of a 5 $M_\odot$ model and the helium-burning luminosity during the EAGB phase ($Z = 0.015, Y = 0.275$)

It is evident that model 1 is at a time just prior to the first abrupt drop in all thermodynamic variables at the center of the burning region and the first abrupt increase in the radius and mass marking the spatial location of the burning center. Prior to the first abrupt changes in all variables, helium burning is concentrated in central regions which, as shown by the density and temperature curves, are contracting and heating. The first abrupt decreases in density and temperature identify the transition from central to shell helium burning.

The slight increase in temperature and density at the burning center that occurs during the first part of the evolution between models 1 and 2 shows that shell helium burning is initially concentrated in the shell closer to the center and the second abrupt change in all characteristics that occurs about three quarters of the way between models 1 and 2 marks the transition from the inner to the outer of the two helium-burning shells as the stronger of the two nuclear energy sources.

In model 2, shell helium-burning is well underway, but since hydrogen burning continues to make a significant contribution to the surface luminosity, the model is not yet a bona fide EAGB model. However, model 3, for which $L_s \sim L_{He}$, is well into the EAGB phase proper. For the next $\sim 1.3 \times 10^6$ yr, the model has a structure analogous to that of a low mass model ascending the red giant branch for the first time with a hydrogen-burning shell above an inert core composed of helium, the difference being that the nuclear energy shell source in the EAGB model burns helium and the inert core is composed of carbon and oxygen.

From Fig. 18.2.8, it can be seen that, as the mass of the carbon–oxygen (CO) core grows from $\sim 0.5\ M_\odot$ to $\sim 0.9\ M_\odot$, the helium-burning luminosity increases by over an order of magnitude, from $\sim 1000\ L_\odot$ to $\sim 16\,000\ L_\odot$, but the radius at the center of the

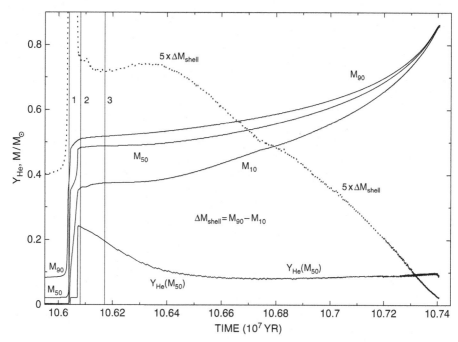

Fig. 18.2.9 Helium abundance and mass at the center, mass boundaries, and total mass of the helium-burning shell in a 5 $M_\odot$ model during the EAGB phase ($Z = 0.015$, $Y = 0.275$)

helium-burning shell decreases by only a modest amount, from $\sim 0.04\ R_\odot$ to $\sim 0.03\ R_\odot$. Thus, as the model progresses through the EAGB phase, the inert CO core maintains linear dimensions significantly larger than that of a cold white dwarf of the same mass, demonstrating that thermal pressure makes an important contribution to the total pressure over a substantial fraction of the core.

In Fig. 18.2.9, the masses below which 10%, 50%, and 90% of the helium-burning luminosity is produced are shown as functions of time by the curves labeled M_{10}, M_{50}, and M_{90}, respectively. Also shown as a function of time are $\Delta M_{shell} = M_{90} - M_{10}$, which is a measure of the mass width of the shell, and $Y_{He}(M_{50})$, the helium number abundance at the center of the shell.

From Figs. 18.2.8 and 18.2.9 it is evident that, as evolution progresses through the EAGB phase, the density at the center of the helium-burning shell decreases by a factor of the order of 3 and the width of the shell decreases by a factor of the order of 40. The number abundance of helium at the center of the shell decreases by a factor of the order of 2, most of the decrease occurring in the first quarter of the EAGB phase. The helium-burning luminosity increases by a factor of about 16.

Given that most of the helium-burning luminosity is produced by the triple-alpha reaction, one may write

$$L_{He} \overset{\sim}{\propto} Y_{He}^2\ \rho^2\ T^{40}\ \Delta M_{shell}, \tag{18.2.1}$$

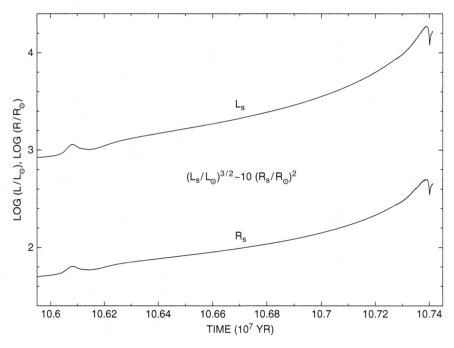

Fig. 18.2.10 Surface luminosity and radius of a 5 $M_\odot$ model during the EAGB phase ($Z = 0.015$, $Y = 0.275$)

where the first three quantities on the right hand side of the equation are to be evaluated at shell center. Inserting the estimates just made, it follows that

$$16 \sim \left(\frac{1}{2}\right)^2 \left(\frac{1}{3}\right)^2 \left(\frac{T_{\text{end}}}{T_{\text{start}}}\right)^{40} \frac{1}{40}, \tag{18.2.2}$$

or

$$\frac{T_{\text{end}}}{T_{\text{start}}} \sim 1.29, \tag{18.2.3}$$

an estimate which is not outrageously different from the change along the temperature curve in Fig. 18.2.8. The change is any case very modest and highlights once again the extreme sensitivity of the triple-alpha reaction rate to the temperature.

Surface luminosity and radius are shown as functions of time during the EAGB phase in Fig. 18.2.10. The important lesson from the curves is that the two surface quantities vary in step with one another. To a surprisingly good approximation,

$$\left(\frac{L_s}{L_\odot}\right)^{3/2} \cong 10 \left(\frac{R_s}{R_\odot}\right)^2. \tag{18.2.4}$$

Using eq. (18.2.4) in conjunction with the relationship $T_e^4 \propto L_s/R_s^2$, one has that, along the EAGB,

$$\left(\frac{d \log L_s}{d \log T_e}\right)_{\text{EAGB}} \cong -8. \tag{18.2.5}$$

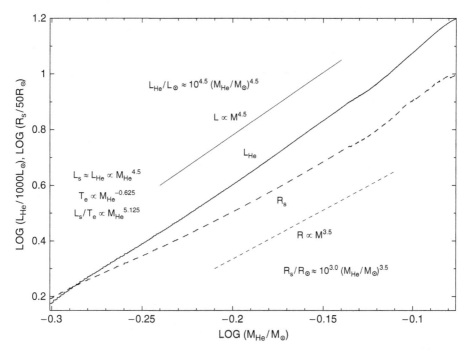

Fig. 18.2.11 He-burning luminosity and surface radius versus mass at the He-burning shell during the EAGB phase of a 5 $M_\odot$ model ($Z = 0.015, Y = 0.275$)

For comparison, a straight line fit to the evolutionary track along the red giant branch in Fig. 11.2.42 in Volume 1 (or Fig. 18.1.1) gives

$$\left(\frac{d \log L_s}{d \log T_e}\right)_{RGB} \stackrel{\sim}{=} -11. \tag{18.2.6}$$

The fact that an extension of the red giant branch to luminosities similar to those along the EAGB would produce a curve which, with increasing luminosity, departs further and further to the blue of the EAGB demonstrates that, for intermediate mass models, the designation AGB is a matter of tradition rather than a matter of topology.

Since there is a simple power law relationship between the surface luminosity and the core mass of a low mass red giant (see Fig. 11.1.33 in Volume 1), it is reasonable to expect that a similarly simple relationship exists between the surface luminosity and core mass of an EAGB model of intermediate mass. Keeping in mind that, during most of the EAGB phase, L_{He} serves as a proxy for L_s, this expectation is confirmed in Fig. 18.2.11 where $\log(L_{He})$ is plotted against $\log(M_{He})$. In rough approximation,

$$\frac{L_{He}}{L_\odot} \sim 10^{4.5} \left(\frac{M_{He}}{M_\odot}\right)^{4.5}, \tag{18.2.7}$$

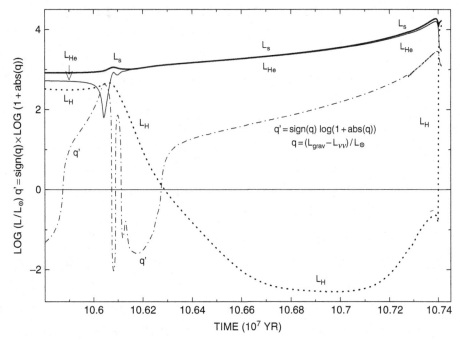

Fig. 18.2.12 Nuclear-burning, gravothermal minus neutrino, and surface luminosities of a 5 $M_\odot$ model during the EAGB phase ($Z = 0.015$, $Y = 0.275$)

which is to be compared with the relationship

$$\frac{L_H}{L_\odot} \sim 10^{5.6} \left(\frac{M_H}{M_\odot}\right)^{6.5}, \tag{18.2.8}$$

which holds for a low mass red giant branch model.

The relationship shown in Fig. 18.2.11 between surface radius and the mass of the CO core of the EAGB model can be approximated by

$$\frac{R_s}{R_\odot} \sim 10^{3.0} \left(\frac{M_{CO}}{M_\odot}\right)^{3.5}. \tag{18.2.9}$$

This relationship is actually useful in estimating the mass of the CO white dwarf formed in a close binary as a function of the separation of the components in the initial binary (see, e.g., Iben & Alexander V. Tutukov, 1993). For comparison, the relationship between the surface radius and the mass of the helium core of a low mass red giant can be approximated by

$$\frac{R_s}{R_\odot} \sim 10^{3.38} \left(\frac{M_{He}}{M_\odot}\right)^{4.0}. \tag{18.2.10}$$

For all intents and purposes, M_{He} in eq. (18.2.7) is the same as M_{CO} in eq. (18.2.9).

Quantities shown in Fig. 18.2.1 for a period of time focussed on the beginning of the EAGB phase are presented again in Fig. 18.2.12, but for a time period which extends

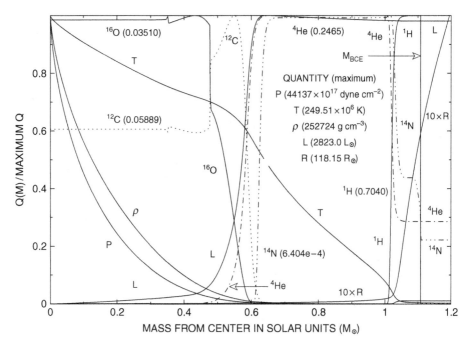

Structure and composition variables in a 5 $M_\odot$ model near minimum L_H during the EAGB phase ($Z = 0.015$, $Y = 0.275$, AGE $= 10.6875 \times 10^7$ yr)

over the entire EAGB phase. In Fig. 18.2.1, the rate of gravothermal energy generation is measured by the quantity

$$q_0 = \text{sign}\,(L_{\text{grav}}) \times \log(1 + \text{abs}(L_{\text{grav}}/L_\odot)), \qquad (18.2.11)$$

and, to first approximation, the rate at which gravothermal energy contributes to the surface luminosity is close to the global value of L_{grav}. However, as densities and temperatures in the helium-exhausted core continue to increase, the rate of neutrino energy loss from the core increases. Since this energy loss is supplied by the release of gravothermal energy in the core, the appropriate global gravothermal quantity to compare with nuclear burning luminosities and the surface luminosity is

$$q = (L_{\text{grav}} - L_{\nu\bar{\nu}})/L_\odot, \qquad (18.2.12)$$

so, in Fig. 18.2.12, the quantity

$$q' = \text{sign}\,(q) \times \log(1 + \text{abs}(q)) \qquad (18.2.13)$$

is shown instead of q_0.

Over most of the EAGB phase, hydrogen burns at an insignificant rate, the smallest hydrogen-burning luminosity, $L_H^{\text{min}} \sim 0.002\,80\ L_\odot$, occurring at $t_7 \sim 10.698$. Several characteristics of a model of age $t_7 = 10.6875$, shortly before minimum L_H is reached, are shown in Figs. 18.2.13–18.2.16. Structure and composition variables in a region extending

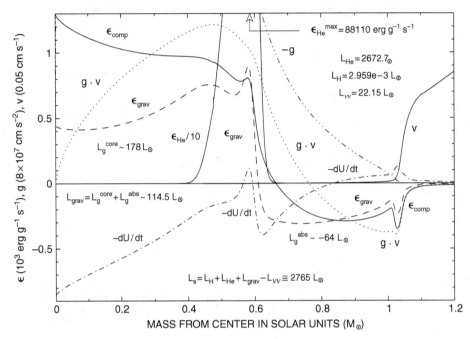

Fig. 18.2.14 Gravothermal and nuclear energy-generation rates in a 5 $M_\odot$ EAGB model near minimum L_H (Z = 0.015, Y = 0.275, AGE = 10.6875×10^7 yr)

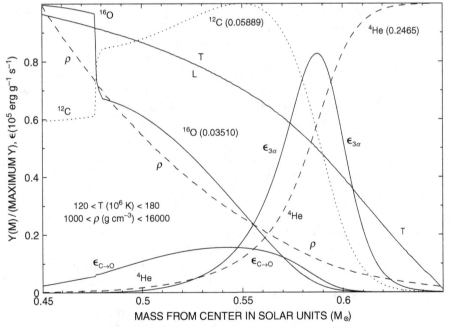

Fig. 18.2.15 Helium-burning rates, composition variables, and ρ and T in a 5 $M_\odot$ model near minimum L_H during the EAGB phase (Z = 0.015, Y = 0.275, AGE = 10.6875×10^7 yr)

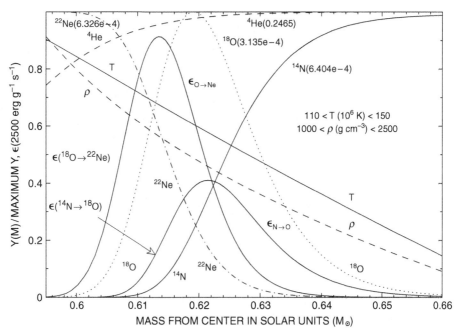

Characteristics in a 5 $M_\odot$ model near minimum L_H during the EAGB phase in the region where $^{14}N \rightarrow {}^{22}Ne$
($Z = 0.015, Y = 0.275, AGE = 10.6875 \times 10^7$ yr)

from the center of the model to slightly beyond the base of the convective envelope are
shown in Fig. 18.2.13 and gravothermal and nuclear energy-generation characteristics over
the same region are shown in Fig. 18.2.14. Although the model is approximately two-
thirds of the way in time through the EAGB phase, the center of the helium-burning shell,
at $M_{He} = 0.574 \ M_\odot$, has moved outward by only a quarter of the way it moves over the
entire EAGB phase, which is from $M_{He} \sim 0.474 \ M_\odot$ to $M_{He} \sim 0.861 \ M_\odot$.

Comparing the abundance profiles for ^{12}C and ^{16}O between the center and $M_{He} \sim$
$0.474 \ M_\odot$ with the profiles for the same isotopes in the mass range $M_{He} \sim 0.474 \ M_\odot$ to
$M_{He} \sim 0.574$, it is evident that the extent to which helium is converted into oxygen is considerably smaller during shell helium burning than during core helium burning, indicative
of a smaller mean temperature in the burning shell than in the convective core during core
helium burning. At $M \sim 1.02 \ M_\odot$, where the hydrogen abundance is at half maximum,
$T_6 \sim 11.2$, $\rho \sim 0.43$ g cm^{-3}, and $\epsilon_H \sim 0.0041$ erg g^{-1} s^{-1}, compared with $T_6 \sim 15$,
$\rho \sim 150$ g cm^{-3}, and $\epsilon_H \sim 100$ erg g^{-1} s^{-1} at the center of the Sun.

The distributions of temperature and density in the helium-exhausted core shown in
Fig. 18.2.13 reflect the fact that electrons in the core are only mildly degenerate. At the
center, where the degree of electron degeneracy is largest, $\epsilon_F/kT = 0.278\,35$, indicating
that matter throughout the core behaves nearly as a perfect gas and that electron conductivity is small. The steepness of the temperature profile is a further indication that the electron
conductivity is small. The opacity is, in fact, primarily due to scattering by non-degenerate
electrons.

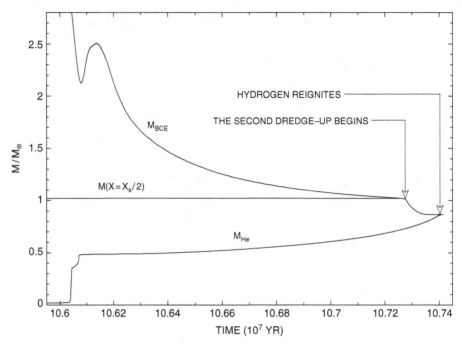

Fig. 18.2.17 Masses at the base of the convective envelope, at the point where $X_H = X_s/2$ and at the center of the helium-burning shell in a 5 $M_\odot$ EAGB model ($Z = 0.015, Y = 0.275$)

From the profiles of the gravothermal energy-generation rates in Fig. 18.2.14 one sees that matter everywhere below the outer edge of the helium-burning shell is being compressed and that, with the exception of a small region near the peak in the nuclear energy-generation rate, matter from the center to midway between the helium-burning shell and the hydrogen–helium interface is heating. The fact that $dU/dt < 0$ over the small region near the peak in the nuclear energy-generation rate is a consequence of the reduction in the number of particles per gram. From the outer edge of the helium-burning shell to well into the convective envelope, matter is expanding, and, outward beyond $M \sim 0.76\ M_\odot$, where the curve for $\mathbf{g} \cdot \mathbf{v}$ passes through zero, it is moving outward as well.

Conditions in the model where the triple-alpha and the $^{12}C(\alpha, \gamma)^{16}C$ reactions are concentrated are presented in Fig. 18.2.15 and conditions in the model where the reactions $^{14}N(\alpha, \gamma)^{18}F(\beta^+, \nu_e)^{18}O$, and $^{18}O(\alpha, \gamma)^{22}Ne$ are concentrated are presented in Fig. 18.2.16. In both figures, the maximum number abundance of an isotope is listed in parenthesis beside the symbol for the isotope. In Fig. 18.2.15, the scale limits for temperature and density are listed in the lower left hand portion of the figure and energy-generation rates are in units of 10^5 erg g^{-1} s^{-1}. In Fig. 18.2.16, the scale limits for T and ρ are given to the right in the figure and energy-generation rates are in units of 2500 erg g^{-1} s^{-1}.

In Fig. 18.2.17, three critical masses are shown as functions of time: the mass M_{He} at the center of the helium-burning shell, the mass $M(X = X_s/2)$ at the hydrogen–helium interface (defined as the point where the hydrogen number abundance is at half maximum), and the mass M_{BCE} at the base of the convective envelope. The first mass increases steadily as the helium-burning shell deposits its ashes into the CO core, the second mass remains constant because hydrogen-burning is ineffective, and the third mass becomes smaller because the increasing luminosity through the envelope causes the radiative gradient to become larger than the adiabatic gradient over a larger and larger fraction of the envelope mass.

Eventually the base of the convective envelope reaches the hydrogen–helium interface and, as convection continues to extend further inward in mass, helium-rich matter is mixed into the envelope. The hydrogen–helium interface and the base of the convective envelope become one and the same. The process whereby matter processed through complete hydrogen burning is carried into (or dredged up into) the convective envelope has come to be known as the second dredge-up process to distinguish it from the first dredge-up process which occurs during the first ascent of the red giant branch after the exhaustion of central hydrogen.

Because the dominant source of opacity at the hydrogen–helium interface is electron scattering, the radiative gradient V_{rad} is discontinuous at the interface, being larger on the side with the greater hydrogen abundance in such a way that $V_{rad} - V_{ad}$ is finite and positive on the hydrogen-rich side and finite and negative on the hydrogen-poor side. This circumstance implies that convective overshoot must occur, leading to a variation in number abundances over some fraction of a pressure scale height and to a profile of the quantity $V_{rad} - V_{ad}$ which passes smoothly through zero. However, in the calculations reported here, no attempt has been made to smooth out abundance profiles and the inward march of the discontinuous profile comes to an artificially premature end.

Nevertheless, the occurrence of a second dredge-up episode has been demonstrated and the surface number-abundance changes obtained for some isotopes are of the order of 5–10%. The changes found are recorded in Table 18.2.1. The first column in the table identifies the isotope and the next two columns give, respectively, number abundances of the isotope before and after the first dredge-up episode, as carried over from Table 11.2.2 in Volume 1. The fourth column gives the ratio of number abundances after and before the first dredge-up episode. The fifth and sixth columns give number abundances before and after the second dredge-up episode and the seventh column gives the ratio of number abundances after and before the second dredge-up episode.

Of particular interest are the facts that the percentage changes in the number abundances of ^{4}He and ^{1}H are approximately four times larger during the second dredge-up episode than during the first dredge-up episode and that, during the second dredge-up episode, the increase in the number abundance of ^{14}N due to the conversion of ^{16}O into ^{14}N is twice as large as the increase due to the conversion of ^{12}C into ^{14}N.

The eighth column in Table 18.2.1 gives the ratio of the final number abundance in column six to the initial number abundance in column 2. Since the real analogue of the 5 $M_\odot$ model will expel into the interstellar medium most of the material that is at this point in the convective envelope, the number abundance changes recorded in column 8 are indicative of the contribution of a 5 $M_\odot$ star to galactic nucleosynthesis. It is evident that such stars are basically destroyers of the light isotopes ^{6}Li, ^{7}Li, ^{9}Be, and ^{10}B and that they significantly deplete ^{11}B, ^{15}N, and ^{18}O. On the other hand, they significantly enhance ^{13}C, ^{14}N, and ^{17}O. And, most importantly, the fact that they eject material in which the ratio of helium to hydrogen is at least 18% larger than in the material out of which they are born accounts readily for the fact that the Y_{He}/Y_H abundance ratio now in interstellar matter

Table 18.2.1 Surface abundances prior to and after the first and second dredge-up episodes

Isotope	start1	dgup1	ratio1	start2	dgup2	ratio2	ratio3
^{1}H	0.7105	0.7040	0.9909	0.7040	0.6765	0.9609	0.9521
^{3}He	2.578/−5	2.610/−5	1.0123	2.587/−5	2.434/−5	0.9409	0.9441
^{4}He	0.06880	0.07039	1.0231	0.07039	0.07726	1.0976	1.1230
^{6}Li	1.167/−10	8.126/−13	0.006962	4.643/−13	4.360/−13	0.9391	0.003747
^{7}Li	1.439/−9	1.934/−11	0.01344	1.304/−11	1.225/−11	0.9391	0.008514
^{9}Be	1.386/−11	3.960/−13	0.02858	3.044/−13	2.858/−13	0.9391	0.02060
^{10}B	8.739/−11	5.117/−12	0.05856	4.334/−12	4.070/−12	0.9391	0.04657
^{11}B	3.517/−10	1.056/−10	0.3004	1.024/−10	9.614/−11	0.9390	0.2734
^{12}C	2.132/−4	1.338/−4	0.6275	1.335/−4	1.258/−4	0.9423	0.5901
^{13}C	2.372/−6	6.235/−6	2.629	6.361/−6	6.117/−6	0.9617	2.579
^{14}N	5.275/−5	1.423/−4	2.697	1.424/−4	1.656/−4	1.1629	3.139
^{15}N	1.940/−7	9.103/−8	0.4693	8.963/−8	8.528/−8	0.9515	0.4396
^{16}O	3.997/−4	3.854/−4	0.9644	3.854/−4	3.701/−4	0.9603	0.9259
^{17}O	1.524/−7	7.371/−7	4.837	7.372/−7	7.605/−7	1.0316	4.990
^{18}O	8.050/−7	5.770/−7	0.7168	5.767/−7	5.433/−7	0.9421	0.6749
^{19}F	1.609/−8	1.519/−8	0.9438	1.519/−8	1.449/−8	0.9538	0.9001
^{22}Ne	5.329/−6	4.796/−6	0.9001	4.796/−6	4.558/−6	0.9504	0.8553

of high metallicity is substantially larger than that prevailing in interstellar matter shortly after the occurrence of the big bang.

At the end of the EAGB phase, the number abundance of ^{12}C at the surface of the 5 $M_\odot$ model is 40% smaller than at the start of the main sequence phase. However, as described in Section 18.3, during the subsequent TPAGB phase, the surface abundance of ^{12}C is enhanced in a series of third dredge-up episodes so that the real analogue of the 5 $M_\odot$ model is a net positive contributor of ^{12}C to the interstellar medium.

Density, temperature, and pressure at the center of the model are shown as functions of time during the EAGB phase in Fig. 18.2.18, along with the neutrino luminosity $L_{\nu\bar{\nu}}$ due to square terms in the weak interaction Hamiltonian. Comparing with the helium-burning luminosity L_{He} in Fig. 18.2.8, it is evident that, between model 3 and when the maxima in both L_{He} and $L_{\nu\bar{\nu}}$ occur near $t_7 = 10.74$, $L_{\nu\bar{\nu}}$ increases by a factor of about 460, while L_{He} increases by only a factor of about 15. Specifically, in model 3, $L_{\nu\bar{\nu}} \sim 5.5\ L_\odot$, compared with $L_{He} \sim 1050\ L_\odot$, but at the maxima in the two luminosities, $L_{\nu\bar{\nu}} \sim 2500$ $L_\odot$, compared with $L_{He} \sim 16\,000\ L_\odot$.

Primarily responsible for the increase in $L_{\nu\bar{\nu}}$ relative to L_{He} over the course of the EAGB phase are the increase in the degree of electron degeneracy and the increase in average temperatures in the helium-exhausted core. Structure and composition variables in the interior of a model which is basically at the end of the EAGB phase are shown in Fig. 18.2.19 and several relevant thermodynamic characteristics are shown in Fig. 18.2.20. The model is of age $t_7 = 10.74$ and the hydrogen- and helium-burning luminosities are almost identical at $L_H \sim 584\ L_\odot$ and $L_{He} \sim 551\ L_\odot$, respectively.

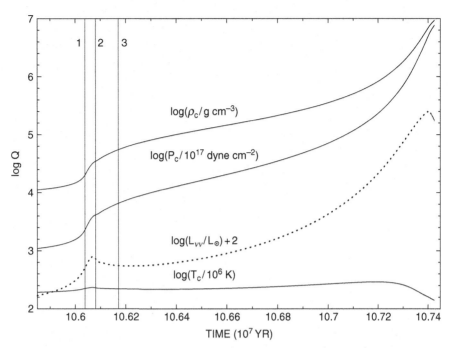

Fig. 18.2.18 Density, temperature, and pressure at the center and neutrino luminosity of a 5 M$_\odot$ model during the EAGB phase ($Z = 0.015$, $Y = 0.275$)

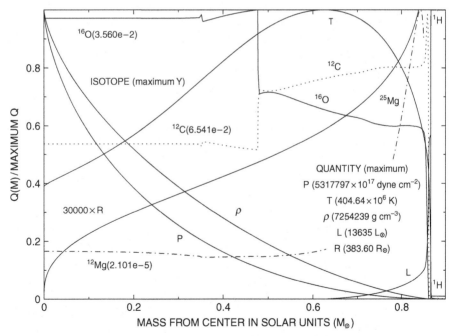

Fig. 18.2.19 Interior structure and composition of a 5 M$_\odot$ model transitioning between the EAGB and TPAGB phases ($Z = 0.015$, $Y = 0.275$, AGE = 10.7403×10^7 yr)

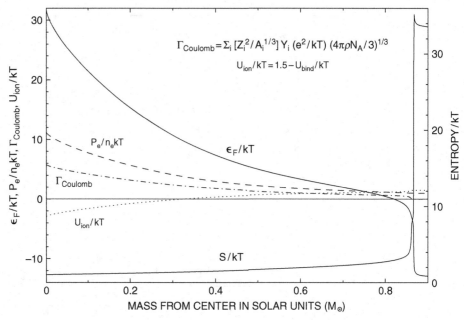

Thermodynamic characteristics of matter in the CO core of a 5 $M_\odot$ model transitioning between the EAGB and TPAGB phases ($Z = 0.015$, $Y = 0.275$, AGE $= 10.7403 \times 10^7$ yr)

At the center of the model, where the density is $\rho_c \sim 7.25 \times 10^6$ g cm^{-3} and temperature is $T_c \sim 160 \times 10^6$ K, the electron Fermi energy is over 30 times larger than kT, the electron pressure is over ten times what it would be if electrons were not degenerate, and the Coulomb binding energy between an average nucleus and electrons is approximately three times larger than $\frac{3}{2} kT$. The quantity $\Gamma_{Coulomb}$, which is another measure of the binding energy between electrons and an average nucleus relative to kT, is sufficiently small ($\Gamma \sim 6$ at the center) compared with a value of the order of 200 required for solidification (see Section 21.4) that one may safely assume that nuclei move randomly as in a gas.

The number-abundance profiles for ^{12}C, ^{16}O, and ^{25}Mg in the mass range $0.5 < M/M_\odot < 0.85$ that are shown in Fig. 18.2.19 demonstrate that, as the helium-burning shell moves outward in mass, the degree to which helium burning progresses beyond the formation of ^{12}C and ^{22}Ne increases, consistent with that fact that, as shown in Fig. 18.2.8, the average temperature in the shell increases with time. Particularly interesting is the fact that, over the last half of the EAGB phase, the abundance of ^{25}Mg left behind by the advancing shell increases by over a factor of 5. The significance of this fact is that, since neutrons are released in the reaction ^{22}Ne(α, n)^{25}Mg and since much larger temperatures arise in the helium-burning shell during the subsequent TPAGB phase, substantial neutron-capture nucleosynthesis may be anticipated during the TPAGB phase.

Energetics in the core of the model below where nuclear energy generation occurs are described in Figs. 18.2.21–18.2.23. In Fig. 18.2.21 are shown gravothermal energy-generation rates and the ingredients (velocity and gravitational acceleration) of the rate at which gravity does work ($\mathbf{g} \cdot \mathbf{v}$). It is evident by inspection that, in Fig. 18.2.21, over

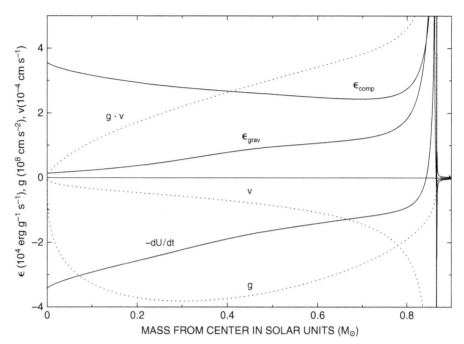

Fig. 18.2.21 Gravothermal energy-generation rates in a 5 M$_\odot$ model transitioning between the EAGB and TPAGB phases (Z = 0.015, Y = 0.275, AGE = 10.7403 × 10^7 yr)

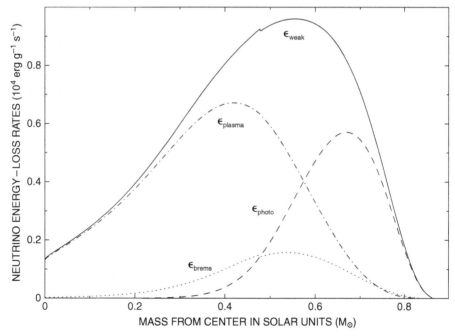

Fig. 18.2.22 Neutrino energy-loss rates in the CO core of a 5 M$_\odot$ model transitioning between the EAGB and TPAGB phases (Z = 0.015, Y = 0.275, AGE = 10.7403 × 10^7 yr)

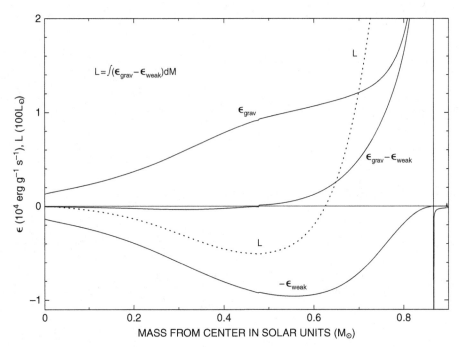

Fig. 18.2.23 Gravothermal and neutrino energy-generation rates in a 5 $M_\odot$ model transitioning between the EAGB and TPAGB phases ($Z = 0.015$, $Y = 0.275$, AGE $= 10.7403 \times 10^7$ yr)

the region between the center and where $dU/dt = 0$ at $M \sim 0.844\ M_\odot$, $\int \mathbf{g} \cdot \mathbf{v}\, dM \sim \int \epsilon_{\text{compression}}\, dM$. In the region beyond $\sim 0.844\ M_\odot$, events in the helium-burning shell play the dominant role in influencing gravothermal activity.

Rates of energy loss from the model associated with several weak-interaction processes are shown as functions of mass in Fig. 18.2.22. In order of importance are the plasma, photo, and bremsstrahlung processes. As expected in a highly electron-degenerate environment which inhibits the formation of real electron–positron pairs, pair annihilation into neutrinos and antineutrinos plays no role. Most fascinating is the comparison in Fig. 18.2.23 between the net gravothermal energy-generation rate ϵ_{grav} from Fig. 18.2.21 and the total weak interaction energy-loss rate ϵ_{weak} from Fig. 18.2.22. Over the inner 0.62 $M_\odot$ of the core, where $L < 0$, the model has cleverly managed to almost balance the two rates locally. That is, the total contribution of gravothermal energy to the inward flow of energy between where $L = 0$ and where $dL/dM = 0$ and the total amount of energy converted from the ambient flow of energy into neutrino energy loss in the region between where $dL/dM = 0$ and the center are both small compared with the amount of gravothermal energy converted locally into neutrino energy loss. Outward from where $L = 0$, gravothermal energy production is more than adequate to supply neutrino energy loss locally, with plenty to spare to contribute to the outward flow of radiant energy.

Note from Fig. 18.2.23 that ϵ_{weak} is essentially zero beyond $M \sim 0.85\ M_\odot$, and that, at $M \sim 0.82 M_\odot$, well below where nuclear reactions are important, $\epsilon_{\text{weak}} \ll \epsilon_{\text{grav}}$. This means that neutrino energy loss due to weak interaction processes is basically confined to the helium-exhausted core and that the neutrino energy loss is supplied almost entirely by

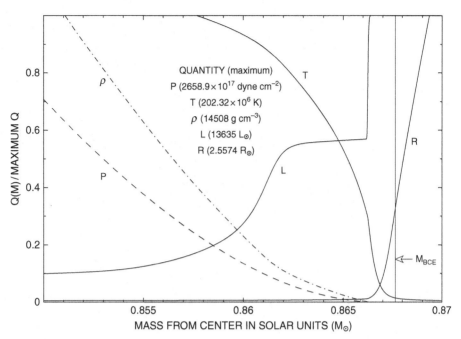

Structure in the nuclear burning region of a 5 $M_\odot$ model transitioning between the EAGB and TPAGB phases
($Z = 0.015, Y = 0.275$, AGE $= 10.7403 \times 10^7$ yr)

gravothermal energy released in the core. Thus, the contribution of gravothermal energy, whether negative or positive, to energy flowing diffusively through the model is given by $L_{\mathrm{grav}}^{\mathrm{diff}} = L_{\mathrm{grav}} - L_{\mathrm{weak}}$. It is for this reason that the quantity q' defined by eqs. (18.2.12)–(18.2.13) has been chosen as the best representation of the contribution of gravothermal energy to the surface luminosity. In the model currently under discussion, $L_{\mathrm{grav}} = 4716.8$ $L_\odot$, $L_{\mathrm{weak}} = 2521.4$ $L_\odot$, and the net contribution of gravothermal energy production to energy diffusing through the model rather than streaming unimpeded out of it in the form of non-interacting neutrinos is $L_{\mathrm{grav}}^{\mathrm{diff}} = 2195.4$ $L_\odot$. An example of a net negative contribution to the surface luminosity is given by model 3 at $t_7 = 10.6171$, for which $L_{\mathrm{grav}} = -32.147$ $L_\odot$, $L_{\mathrm{weak}} = 5.518$ $L_\odot$, and $L_{\mathrm{grav}}^{\mathrm{diff}} = -37.665$ $L_\odot$.

Structure variables and number-abundance distributions in the nuclear burning region of the model which is transitioning between the EAGB and TPAGB phases are shown, respectively, in Figs. 18.2.24 and 18.2.25. Already noted, but demonstrated again by the luminosity profile in Fig. 18.2.24, is the fact that the global hydrogen- and helium-burning luminosities are almost identical. The luminosity profile also demonstrates that gravothermal energy liberated by the contracting core contributes approximately 10% of the surface luminosity.

The circumstance that the two nuclear burning shells are in such close proximity simplifies analysis of the differences in character of the two sources. Among the differences which stand out most clearly is the difference in the mass width of the two regions; in the helium-burning shell, composition profiles and the temperature profile extend over a mass interval which is approximately two orders of magnitude larger than they do in the

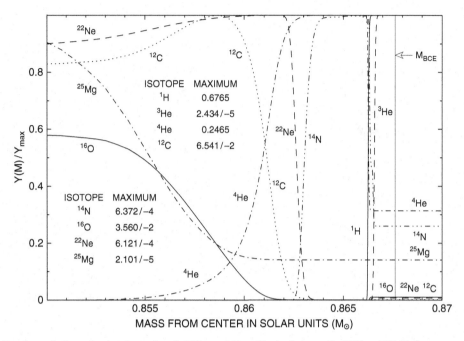

Fig. 18.2.25 Abundances in the nuclear burning region of a 5 $M_\odot$ model transitioning between the EAGB and TPAGB phases ($Z = 0.015$, $Y = 0.275$, AGE $= 10.7403 \times 10^7$ yr)

hydrogen-burning shell. The difference is related to the facts that, as hydrogen is converted into helium, eight particles are replaced by three, whereas, as helium is converted into carbon, nine particles are replaced by seven, the relative particle-abundance change in the hydrogen-burning case being about a factor of 2 larger than in the helium-burning case. The equation of state in the nuclear burning regions is essentially that of a perfect gas, so that $P \propto nkT$, and since pressure decreases continuously with increasing mass, fractional temperature changes through the hydrogen-abundance profile are larger by about a factor of 2 than through the helium-abundance profile.

Since the rate at which any nuclear species changes is proportional to a very high power of temperature, the steeper temperature gradient through the hydrogen-burning shell translates into a hydrogen-burning shell much narrower in mass width than the helium-burning shell. The difference in width is quantified in Figs. 18.2.26 and 18.2.27, which show, respectively, nuclear and gravothermal energy-generation rates in the two burning regions as functions of mass. Defining mass width as the distance in mass between the two edges of each nuclear energy generation profile near the $y = 0$ axis, one has that $\Delta M_{He} \sim 1.2 \times 10^{-2} \, M_\odot$ and $\Delta M_H \sim 2.6 \times 10^{-4} \, M_\odot$.

In both cases, the peaks in all of the gravothermal energy-generation rates coincide with the peaks in the nuclear energy-generation rates. Most fascinating is the fact that the gravothermal energy-generation rates, which are responses to nuclear transformations and are therefore facilitators of evolutionary changes in structure, produce profiles which extend smoothly over much larger regions than do the profiles of the nuclear energy-generation rates, which are by far the major contributors to the ambient flow of energy

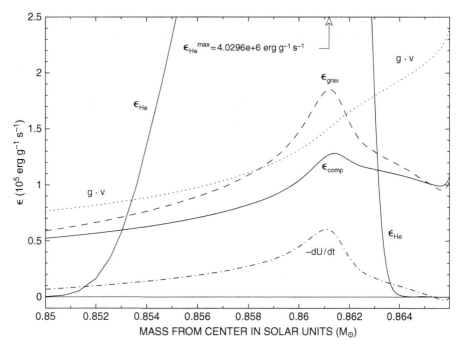

Fig. 18.2.26 Energy-generation rates near the He-burning shell in a 5 $M_\odot$ model transitioning between the EAGB and TPAGB phases ($Z = 0.015$, $Y = 0.275$, AGE $= 10.7403 \times 10^7$ yr)

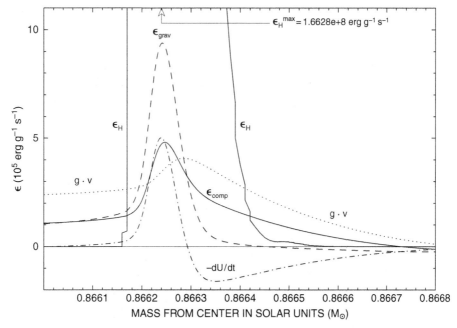

Fig. 18.2.27 Energy-generation rates near the H-burning shell in a 5 $M_\odot$ model transitioning between the EAGB and TPAGB phases ($Z = 0.015$, $Y = 0.275$, AGE $= 10.7403 \times 10^7$ yr)

which ultimately diffuses to the surface to account for the surface luminosity. Thus, although coupled in subtle ways, the two sources – gravothermal and nuclear – have different agendas, or independent lives of their own.

Integrating under the profile for ϵ_{grav} in Fig. 18.2.26, one finds that compression and cooling in the region between $M = 0.850\ M_\odot$ and $M = 0.865\ M_\odot$ contribute approximately $\Delta L_{grav}^{He\ shell} \sim 800\ L_\odot$ to the outward diffusion of energy. Coupling this estimate with the observation from Fig. 18.2.24 that the flow of energy of gravothermal origin from the model core at the point $M = 0.850\ M_\odot$ is $L(0.850 M_\odot) \sim 1363.5\ L_\odot$, one has that gravothermal energy production in the core and the helium-burning shell together contribute approximately $2164\ L_\odot$ to the ambient flow of energy, or approximately 16% of the surface luminosity. An additional contribution to the surface luminosity of $\Delta L_{grav}^{H\ shell} \sim 30\ L_\odot$ is made by the release of gravothermal energy in, and in the vicinity of, the hydrogen-burning shell.

18.3 The thermally pulsing asymptotic giant branch phase and the third dredge-up phenomenon

Surface luminosity and global nuclear burning luminosities are shown in Fig. 18.3.1 over a time interval extending from the end of the EAGB phase through the first four thermal pulse cycles during the TPAGB phase. The intersection of the L_H profile (heavy dotted curve) with the L_{He} profile (solid curve) at the left hand side of the figure occurs almost precisely where the model described by Figs. 18.2.19–18.2.27 is located in time and marks the transition from the EAGB phase to the TPAGB phase. Time Δt is measured in units of 10^3 yr from the location of the most prominent relative maximum in L_{He} that occurs between the first intersection of the L_H and L_{He} profiles (at $\Delta t = -10.752$) and the first of three thermal pulses which develop sharp spikes in L_{He} marking the occurrence of helium shell flashes. The first pulse, identified by the relative maximum in L_{He} at $\Delta t = 0$, is quite mild and does not evolve into a helium shell flash. The spikes in L_{He} produced during helium shell flashes reach far above the surface luminosity L_s and increase in amplitude with each successive pulse.

Except over very narrow time intervals centered on where helium shell flashes occur and where profiles for the various quantities are difficult to disentangle, it can be determined from Fig. 18.3.1 that the sum of the nuclear burning luminosities remains relatively constant relative to the surface luminosity and accounts for approximately 90 ± 5 % of the surface luminosity. To verify this, note first that, at $\Delta t = -10.752$, $L_{He} = L_H \sim 0.42\ L_s$, so that $L_H + L_{He} \sim 0.84\ L_s$. After each of the first three helium shell flashes, at the position where curves for L_H and L_{He} intersect, $L_H + L_{He} \sim 0.86\ L_s$. Finally, it is evident by inspection that, between each intersection point and the next helium shell flash, the total nuclear burning luminosity approaches the surface luminosity ever more closely.

The curve in Fig. 18.3.1 labeled q' describes the variation with time of the gravothermal contribution to energy diffusing through the model as defined by eqs. (18.2.12) and

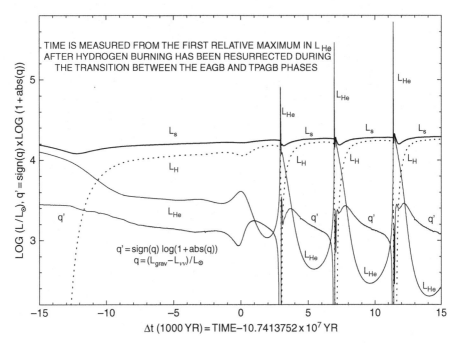

Fig. 18.3.1 Global luminosities of a 5 $M_\odot$ model beginning the thermally pulsing asymptotic giant branch phase ($Z = 0.015$, $Y = 0.275$)

(18.2.13). In very rough approximation, the average value of q' is 3.2, compared with the average value of $\log(L_s/L_\odot) \sim 4.25$, so that, between helium shell flashes, gravothermal energy contributes on average roughly 9% of the surface luminosity.

As described in Sections 17.3–17.4, during each helium shell flash in a 1 $M_\odot$ TPAGB model, the energy produced by helium burning during the peak of a helium-shell flash is converted locally into stored gravothermal energy which subsequently leaks out by diffusion. Exactly the same thing occurs in the 5 $M_\odot$ model, as is demonstrated in Fig. 18.3.2 by the fact that, at the peak of each flash, the global helium-burning luminosity is virtually identical with the global rate of gravothermal energy absorption corrected for the amount of gravothermal energy used in supplying neutrino losses due to weak interaction processes in the helium-exhausted core.

Variations in global luminosities during the third thermal pulse are shown in Fig. 18.3.3 in such a way as to resolve the behavior of the gravothermal source immediately following flash peak. At the luminosities shown, to a very good approximation, the plotted quantity q' is given by

$$q' \stackrel{\sim}{=} \log \frac{L_{grav} - L_{\nu\bar\nu}}{L_\odot} \tag{18.3.1}$$

and thus measures directly the rate at which gravothermal energy contributes to the surface luminosity. The first relative peak in q' occurs at $\Delta t = 7.00 \times 10^3$ yr and reflects the fact that, in response to the abrupt decline in L_H, the model envelope contracts and releases gravothermal energy to partially compensate for the decrease in energy of nuclear origin

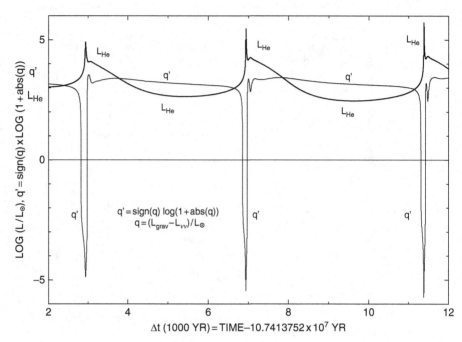

Fig. 18.3.2 Helium burning and gravothermal minus neutrino luminosities of a 5 $M_\odot$ model during the second through fourth thermal pulses of the TPAGB phase ($Z = 0.015, Y = 0.275$)

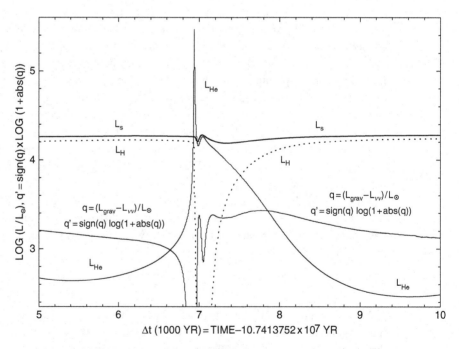

Fig. 18.3.3 Global luminosities of a 5 $M_\odot$ TPAGB model during the third thermal pulse ($Z = 0.015, Y = 0.275$)

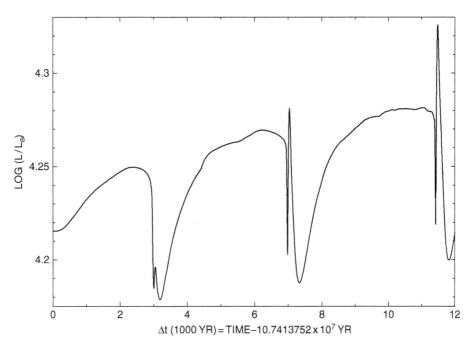

Fig. 18.3.4 Surface luminosity of a 5 M$_\odot$ model beginning the thermally pulsing asymptotic giant branch phase ($Z = 0.015$, $Y = 0.275$)

flowing through the envelope. The second peak in q' occurs at $\Delta t \sim 7.17 \times 10^3$ yr and the difference between this time and the time $\Delta t \sim 6.95 \times 10^3$ yr at which the peak in L_{He} occurs is a measure of the time required for the gravothermal energy stored in the vicinity of the helium-burning shell during the helium shell flash to make its way by diffusion to the model surface ($\sim$200 years).

Since variations in the surface luminosity are difficult to resolve on the scale of Fig. 18.3.1, which highlights the large variations in individual nuclear burning luminosities, surface luminosity is plotted by itself in Fig. 18.3.4 as a function of time. The sharp drop in surface luminosity which occurs after each broad relative maximum in luminosity is due to the precipitous demise of hydrogen burning occasioned by the helium shell flash which forces expansion of immediately overlying layers. The magnitude of the luminosity drop is mitigated by the release of gravitational potential energy in the contracting hydrogen-rich envelope. The sharp rise in surface luminosity which immediately follows the sharp drop is due in part to the release of gravothermal energy by the contracting envelope and in part to the diffusion outward of gravothermal energy stored initially in the helium-burning region during the helium-shell flash. Once this conversion and outward transfer has been completed, the primary source of surface luminosity becomes, first, the conversion of nuclear energy into outward flowing energy by the helium-burning shell, which burns quiescently and gradually declines in strength, and, then, the conversion of nuclear energy into outward flowing energy by the hydrogen-burning shell, which burns quiescently and gradually grows in strength.

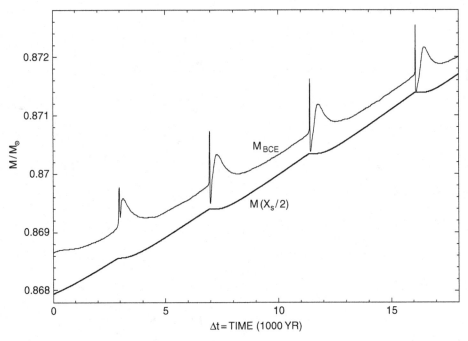

Fig. 18.3.5   Mass at the base of the convective envelope compared with the mass at the hydrogen-abundance half maximum in a
5 $M_\odot$ TPAGB model ($Z = 0.015$, $Y = 0.275$)

During the course of a thermal pulse cycle, the location of the base of the convective
envelope varies in response to changes in conditions near the base, including changes in
the luminosity there. Of major interest is the location in mass of the base of the convective envelope relative to the location in mass of the hydrogen–helium interface, defined
as the position where the hydrogen abundance is half of the surface abundance of hydrogen. These two masses, M_{BCE} and $M(X = X_s/2)$, are shown in Fig. 18.3.5 as functions of
time over the course of four pulse cycles. Between helium shell flashes, as the hydrogen–
helium interface moves outward in mass in consequence of active hydrogen burning, the
base of the convective envelope moves outward as well, with the difference in the two
masses remaining essentially constant. During a helium shell flash, as the flux of energy
through the envelope drops abruptly in response to an abrupt decline in the rate of hydrogen burning, M_{BCE} at first moves rapidly outward. However, as gravothermal energy stored
in the flashing region during the flash diffuses beyond the hydrogen–helium interface and
reaches the base of the convective envelope, the local increase in luminosity causes M_{BCE}
to decrease as rapidly as it has previously increased.

The entire episode of rapid outward and inward motion of the base of the convective
envelope is very brief and is not transparently correlated with changes in surface luminosity
as is the case for the 1 $M_\odot$ TPAGB model discussed in Section 17.3 (see Figs. 17.3.8
and 17.3.9). The difference is due to the extra time it takes energy to diffuse through the
much more massive envelope of the 5 $M_\odot$ model.

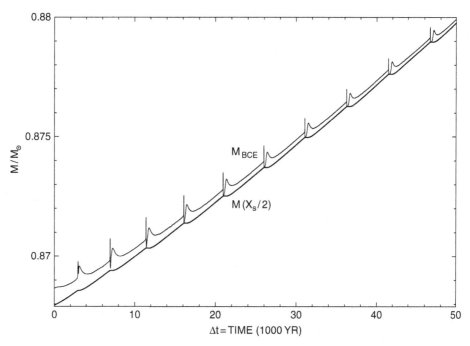

Fig. 18.3.6 Mass at the base of the convective envelope compared with the mass at the hydrogen-abundance half maximum in a 5 $M_\odot$ TPAGB model ($Z = 0.015, Y = 0.275$)

As seen in Fig. 18.3.5, the minimum value of M_{BCE} achieved after each helium shell flash is smaller than the value of M_{BCE} prevailing immediately prior to the flash and, with each successive flash, this minimum value approaches the hydrogen–helium interface more closely, with near contact being achieved after the fourth helium shell flash. The fact that, after reaching the hydrogen–helium interface near the end of the EAGB phase, the base of the convective envelope continues to move inward into the hydrogen-exhausted interior (see Fig. 18.2.17) leads one to expect that the same phenomenon should occur during the TPAGB phase. Surprisingly, as shown in Fig. 18.3.6, this is not the case. Over the next six thermal pulse cycles, M_{BCE} reaches, but does not extend below, $M(X = X_s/2)$, the mass where the hydrogen abundance is half the surface hydrogen abundance.

This surprising behavior is also at variance with experience acquired in investigations which first reported mixing of freshly made carbon into the convective envelope following helium shell flashes in an intermediate mass model (Iben, 1975; 1976). The difference in behavior hinges on whether or not convective overshoot at the base of the convective zone is taken into account in some fashion. In the earlier work, although V_{rad} and V_{ad} were defined at mass zone centers, both composition and structure variables were defined at mass zone boundaries. This meant that, at the beginning of each time step, at the center of the first mass zone below the base of the region which was formally unstable against convection during the preceding time step, every abundance variable was intermediate between its value at the inner edge of this mass zone and its value at the inner edge of the first mass zone in the formally defined convective envelope during the preceding time step. The new

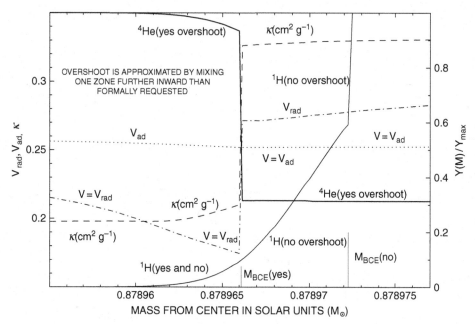

Fig. 18.3.7 V_{rad}, V_{ad}, κ, and hydrogen and helium abundances in a 5 $M_\odot$ model after the tenth helium shell flash ($Z = 0.015$, $Y = 0.275$, $\Delta t = 46704$ yr)

value of V_{rad} calculated at the center of the zone was therefore intermediate between its value in zones on either side, with the possibility that the zone could now be formally unstable against convection, allowing overshoot to occur automatically, with the base of the formally defined convective envelope extending one zone further inward beyond the location of the hydrogen–helium interface at the end of the previous time step.

With each successive pulse in these earlier calculations, the maximum helium-burning luminosity achieved during a pulse increased and the maximum inward extent of the base of the convective envelope into the carbon-rich region produced during the helium shell flash increased in basically a linear fashion with respect to pulse number. Replacing the step-wise variation in composition variables by a smooth distribution over a pressure scale height about the formal boundary between convective and radiative regions did not notice-ably alter the degree of dredge-up achieved, suggesting that dredge-up was taken into account in a fairly realistic fashion. The process was dubbed the third dredge-up process.

Not recognizing that the serendipidous discovery of the third dredge-up process was due to what might be thought of as a technical detail, it came as a total surprise in the present calculations to find that the third dredge-up process did not occur automatically. The realization came slowly that the difference in behavior is due to that fact that, in the calculations reported here, composition variables are defined at the centers of mass zones rather than at the edges of mass zones, as in the earlier calculations.

Relevant to understanding the reasons for the difference in behavior are characteristics shown in Fig. 18.3.7 of two models which have experienced the tenth helium shell flash. The hydrogen-abundance profile shown in the figure is that in the last model produced in the calculations thus far reported at a time when the base of the convective envelope has reached the hydrogen–helium interface and has ceased to move further inward. The helium-abundance profile and all other profiles in the figure are those in a model produced by repeating the calculation which led to the first model but with the modification that, after determining the formal boundaries of the convective envelope at the beginning of a time step, the base of the convective envelope was extended one mass zone further inward beyond the formally estimated boundary and mixing was assumed to occur over the forcibly enlarged convective envelope mass.

The result is that, in the second model, the formal base of the convective envelope extends further into the interior than in the first model, reaching into the helium-rich region of the initial hydrogen–helium interface. The minimum masses achieved at the base of the convective envelope in the first and second calculations are indicated by short vertical lines extending upward from the $y = 0$ axis and labeled $M_{BCE}(no)$ and $M_{BCE}(yes)$, respectively, where yes and no indicate whether or not overshoot has been taken into account. The hydrogen abundance in the second model in the region $M_{BCE}(yes) < M < M_{BCE}(no)$ is the same as in the convective envelope of the model (namely, $Y_H = 0.6765$), and in the region $M < M_{BCE}(yes)$, it is the same as that in the first model (e.g., $Y_H = 0.06765$ just to the left of $M_{BCE}(yes)$).

Of the three quantities to which the radiative gradient is proportional (P/P_{rad}, L/M, and κ), only κ varies discontinuously across the base of the convective envelope in either model. From Fig. 18.3.7, one may determine that, in the second model,

$$\frac{\kappa^+}{\kappa^-} \cong \frac{V_{rad}^+}{V_{rad}^-} \sim 1.55, \tag{18.3.2}$$

where a $+$ superscript denotes that the quantity is evaluated just to the right of $M_{BCE}(yes)$ and a $-$ superscript denotes that the quantity is evaluated just to the left of $M_{BCE}(yes)$. The fact that

$$\frac{1 + Y_H^+}{1 + Y_H^-} = \frac{1 + 0.6765}{1.067\,65} = 1.57 \tag{18.3.3}$$

is effectively the same as the ratios reported in eq. (18.3.2) plus the fact that, when only electron scattering is important, the opacity is proportional to $1 + Y_H$ demonstrate that electron scattering is basically the only source of opacity near the base of the convective envelope. A less convoluted way of arriving at the same conclusion is to begin with the fact that, in the absence of any opacity source other than electron scattering, the opacity is given approximately by $\kappa_e = 0.2(1 + Y_H)$ g cm^2. In the region of the model where $Y_H = 0$, $\kappa_e = 0.2$ g cm^2 differs by less than 1% from the model opacity. In the region of the model where $Y_H = 0.6765$, $\kappa_e = 0.335$ g cm^2 differs on average by only 2% from the model opacity.

The large discontinuity in V_{rad} across the boundary between the convective envelope and the radiative interior associated with the discontinuous change in the abundances of hydrogen and helium across the boundary is the consequence of an inadequate treatment of convection. The fact that V_{rad} is substantially larger than V_{ad} on the hydrogen-rich side of the discontinuity implies large convective velocities at the boundary and hence extensive mixing of hydrogen-richer material with hydrogen-poorer material across the discontinuity, with the result that, in a realistic treatment of convection, the discontinuity in abundances is replaced by a continuous variation in abundances across the formal boundary. Ideally, one would like to devise an algorithm which would lead automatically to a distribution of abundances such that the quantity $\Delta V = V_{rad} - V_{ad}$ decreases smoothly and continuously through the lower portions of the convective envelope and into the radiative zone below it in such a way that ΔV passes through 0, with a negative but finite slope, at what one might accept as a physically meaningful boundary between the envelope convective zone, where fully developed convection occurs, and the radiative interior, at the outer edge of which convective overshoot occurs.

Given the nuisance of artificially spreading out abundances over an arbitrarily chosen mass range, an alternative is to accept a discontinuity in abundances, to insist on some degree of mixing across the formal base of the convective envelope whenever

$$V_{rad}^+ > V_{ad}, \tag{18.3.4}$$

where V_{rad}^+ is the value of V_{rad} in the zone the lower boundary of which defines the formal base, and to hope that

$$V_{rad}^+ - V_{ad}$$

is small compared with

$$V_{ad} - V_{rad}^-,$$

where V_{rad}^- is the value of V_{rad} on the hydrogen-poor side of the abundance discontinuity. These goals have been modestly achieved in calculations continued with the same algorithm used after the tenth helium shell flash to obtain the second of the two models just described.

Resultant variations in M_{BCE} and $M(X = X_s/2)$ are shown as functions of time in Fig. 18.3.8 for two pulse cycles after the tenth helium shell flash. Variations in the two quantities are also shown for the model in which the overshoot algorithm has been implemented after the tenth helium shell flash, just before the base of the convective envelope reaches the position where $X = X_s/2$. After both the eleventh and twelfth helium shell flashes, substantial dredge-up occurs. The degree of dredge-up may be defined by the parameter

$$\lambda = \frac{\Delta M_{dg}}{\Delta M_H}, \tag{18.3.5}$$

where ΔM_{dg} is the amount of mass dredged up in a given dredge-up episode and ΔM_H is the mass which the hydrogen-burning shell traverses during the time between the previous two helium shell flashes. The major reason why the degree of dredge-up is much more substantial after the eleventh than after the tenth helium shell flash is that the algorithm for overshoot was invoked after the peak of the tenth helium shell flash only near the end of the time that conditions near the base of the convective envelope were favorable for overshoot. That the degree of dredge-up is larger after the twelfth helium shell flash ($\lambda \sim 0.284$) than after the eleventh flash ($\lambda \sim 0.160$) is due to the fact that the strength of a helium

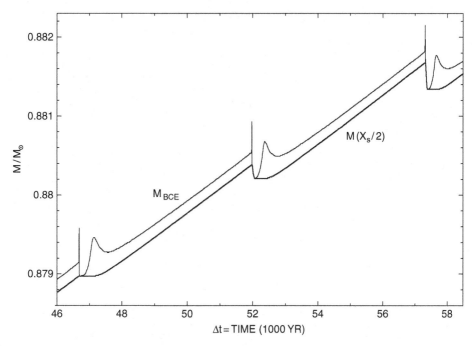

Fig. 18.3.8 Mass at base of convective envelope and mass at the hydrogen-abundance half maximum in a 5 $M_\odot$ TPAGB model with overshoot ($Z = 0.015$, $Y = 0.275$)

flash increases with each pulse cycle, causing the relevant factor L/M in the expression for V_{rad} to remain large enough to require overshoot for a longer time with each successive pulse cycle.

Quantities shown in Figs. 18.3.9–18.3.13 are helpful for understanding the causes and the consequences of the third dredge-up episode which follows the eleventh helium shell flash during the twelfth thermal pulse cycle. Logarithmic temperature–pressure gradients V_{rad} and V_{ad} and the three ingredients of V_{rad} are shown in Fig. 18.3.9 for a model which is just 7.8 days shy of achieving maximum L_{He} during the eleventh helium shell flash. The huge outward flux of energy due to the release of nuclear energy by helium burning produces a convective shell which extends throughout most of the region between the helium-exhausted interior and the hydrogen-rich envelope, from $M \sim 0.877\,58\ M_\odot$ to $M \sim 0.880\,28\ M_\odot$. The boundaries of the convective shell are indicated by vertical lines labeled M_{BCS} and M_{ECS}, respectively. At each end of the convective shell, the luminosity becomes negative for a short distance beyond the shell boundary, suggesting that extensive convective overshoot does not occur at the boundaries.

From the perspective of surface abundances in TPAGB stars, the abundance profiles of most importance in the 5 $M_\odot$ TPAGB model are those of ^{16}O and ^{12}C. It is evident from Fig. 18.3.10 that the abundance of ^{16}O in the convective shell of the model during the peak of the helium shell flash is smaller than the initial abundance in the model. Thus, as thermal pulses continue and matter near the outer edge of the convective shell during pulse peak is dredged into the convective envelope, the ratio of ^{12}C to ^{16}O in the convective

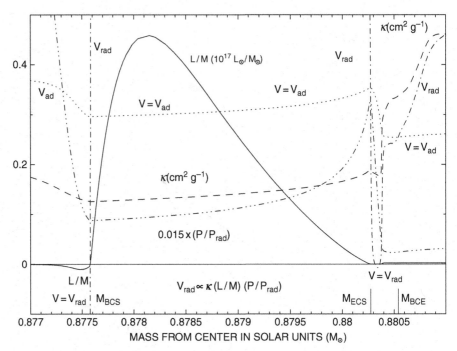

Fig. 18.3.9   Logarithmic gradients in a 5 $M_\odot$ model 7.8 days prior to maximum L_{He} during the eleventh helium shell flash ($Z = 0.015$, $Y = 0.275$, $\Delta t = 51970$ yr)

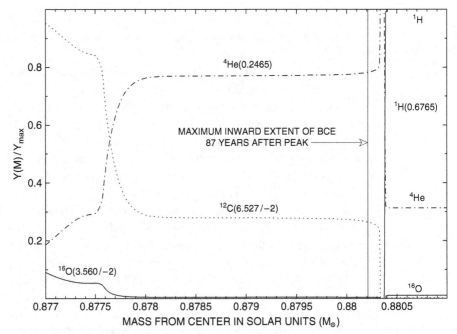

Fig. 18.3.10   Number abundances in a 5 $M_\odot$ TPAGB model 11 yr after maximum L_{He} during the eleventh helium shell flash ($Z = 0.015$, $Y = 0.275$, $\Delta t = 51981$ yr)

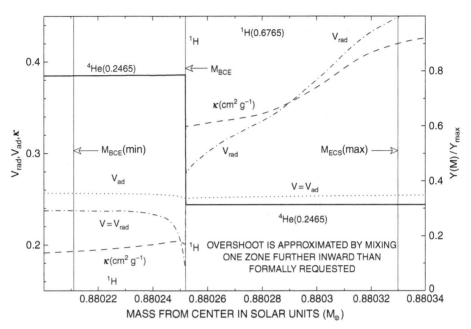

Fig. 18.3.11 V_{rad}, V_{ad}, κ, and ^{1}H and ^{4}He abundances in a 5 $M_\odot$ model during dredge-up after the eleventh helium shell flash ($Z = 0.015$, $Y = 0.275$, $\Delta t = 52033$ yr)

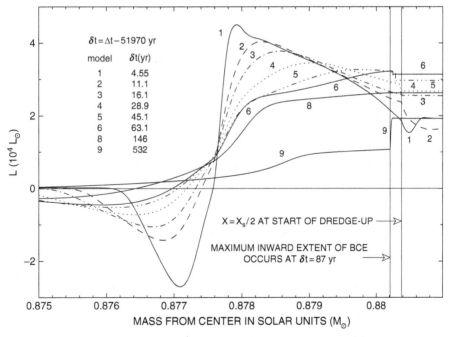

Fig. 18.3.12   Luminosity versus mass during dredge-up (overshoot invoked) after the eleventh helium shell flash in a 5 $M_\odot$ TPAGB model ($Z = 0.015$, $Y = 0.275$)

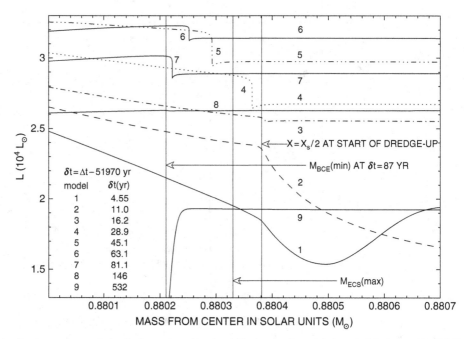

The figure contains the following labels:

$\delta t = \Delta t - 51970$ yr

model	δt(yr)
1	4.55
2	11.0
3	16.2
4	28.9
5	45.1
6	63.1
7	81.1
8	146
9	532

$X = X_s/2$ AT START OF DREDGE-UP

M_{BCE}(min) AT $\delta t = 87$ YR

M_{ECS}(max)

L (10^4 L$_\odot$)

MASS FROM CENTER IN SOLAR UNITS (M$_\odot$)

Fig. 18.3.13 Luminosity versus mass during dredge-up (overshoot invoked) after the eleventh helium shell flash in a 5 M$_\odot$ TPAGB model ($Z = 0.015, Y = 0.275$)

envelope will increase. Unless carbon is converted into nitrogen in the convective envelope between pulses faster than it is dredged up, and if mass loss from the surface of the real counterpart of the model is not excessive, it is conceivable that the abundance of ^{12}C to ^{16}O in the envelope could become larger than unity, causing the real counterpart of the model to appear as a carbon star.

In Fig. 18.3.11 are shown logarithmic temperature–pressure gradients, the opacity, and the hydrogen and helium number abundances for the 5 $M_\odot$ model 63 years after the peak of the eleventh helium shell flash. The vertical line labeled M_{ECS}(max) indicates the maximum mass reached by the outer edge of the convective shell sustained by helium burning during the preceding helium flash and its position coincides with the position of the outer edge of the ^{12}C number-abundance profile in Fig. 18.3.10. The vertical line labeled M_{BCE}(min) indicates the furthest inward extent of the convective envelope during the dredge-up episode. It is evident that, in the model 63 years past flash peak, the base of the convective envelope, now at the position indicated by the vertical line labeled M_{BCE}, has traversed approximately two-thirds of the distance between M_{ECS}(max) and M_{BCE}(min). At the hydrogen–helium discontinuity, $V_{rad}^+ - V_{ad} \sim 0.34 \, (V_{ad} - V_{rad}^-)$.

The primary factor that drives the base of the convective envelope inward in mass is the increase in luminosity near the base due to the diffusion outward of gravothermal energy stored originally in the helium-flashing region. The outward diffusion of gravothermal energy is demonstrated in Fig. 18.3.12 by luminosity profiles for eight models after the peak of the eleventh helium shell flash at times δt(yr) after flash peak indicated in the upper left hand corner of the figure. The profiles extend from below the helium-burning region to beyond the region where dredge-up occurs. Model 6 is the same model for which some characteristics have been described in Fig. 18.3.11.

That dredge-up occurs when the luminosity near the base of the convective envelope is larger than some threshold can be deduced from Fig. 18.3.12, but it is much easier to do so from Fig. 18.3.13 which provides a magnified view of the luminosity profiles centered on the dredge-up region. The figure with magnified profiles features a profile for a model (model 7) not shown in the figure with unmagnified profiles. In models 4–7, which are actively experiencing dredge-up, the location of the inward moving composition discontinuity coincides with the discontinuity in luminosity along the luminosity profile. Overshoot begins between models 3 and 4 and ceases between models 7 and 8. Thus, a luminosity level close to those of models 3 and 8 defines in this instance the threshold for overshooting to occur.

During the dredge-up episode, which lasts approximately 64 years, $\sim$632 models have been calculated, whereas typically only $\sim$165 mass zones are contained in the entire dredge-up region. This means that the formal base of the convective envelope reaches the composition discontinuity on average only every fourth model and that, therefore, the rate at which the base of the convective envelope moves inward in mass is not being over-estimated by the adopted overshoot algorithm. It means further that, just as in earlier calculations in which the assignment of composition variables at mass zone boundaries led automatically to overshoot, the degree of dredge-up achieved with the algorithm here employed will not be altered appreciably by spreading composition variables out over a pressure scale height. The circumstance that the luminosity drop occurs exactly where the composition discontinuity occurs appears to be the model solution to the fact that energy is required to expand matter in the region where the number abundance of particles is or has been increasing, even though the increase does not occur in every time step.

To summarize the results thus far, the third dredge-up phenomenon occurs because (1) the luminosity at the base of the convective envelope increases as gravothermal energy stored in the helium-flashing region during flash peak diffuses outward, and because (2) a discontinuity in $V_{\rm rad} - V_{\rm ad}$ at the hydrogen–helium interface (caused by a discontinuity in the opacity related to the abundance discontinuity) requires convective overshooting to occur at the interface. A discontinuity in the luminosity at the composition discontinuity is the model solution to the complicated set of physical phenomena occurring.

With each successive thermal pulse, the strength of the helium flash and the degree of dredge-up increases. For eight models during the dredge-up episode after the thirteenth shell flash, the three ingredients of the radiative temperature–pressure gradient are shown in Figs. 18.3.14–18.3.16, respectively, and the gradient $V_{\rm rad}$ itself is shown in Fig. 18.3.17. Opacity profiles are shown in Fig. 18.3.14, $P/P_{\rm rad}$ profiles are shown in Fig. 18.3.15, and luminosity profiles are shown in Fig. 18.3.16. Strictly speaking, it is L/M to which $V_{\rm rad}$ is proportional, but the variation in mass over the region shown is so small (less than 0.1%) that L provides essentially the same information as L/M.

Each model is the last one of a calculational run of between 90 and 360 models and each model profile is labeled with the number of the calculational run, numbering beginning with the calculational run which produced the first core helium-burning model described in Section 18.1. The time δt listed for each model in the left hand portion of each figure is

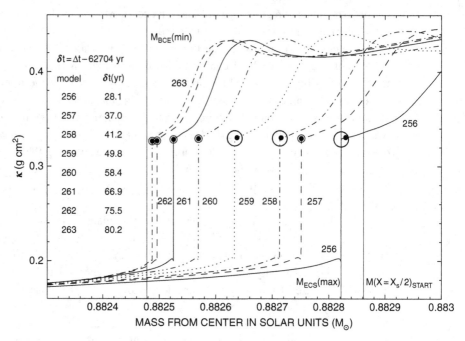

Fig. 18.3.14   Opacity in a 5 $M_\odot$ TPAGB model during dredgeup after the thirteenth helium shell flash ($Z = 0.015$, $Y = 0.275$)

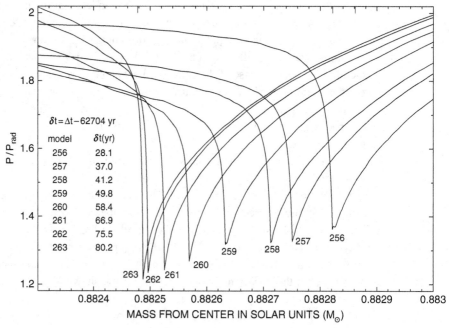

Fig. 18.3.15 P/P_{rad} versus mass in a 5 $M_\odot$ TPAGB model during the dredgeup episode after the thirteenth helium shell flash ($Z = 0.015$, $Y = 0.275$)

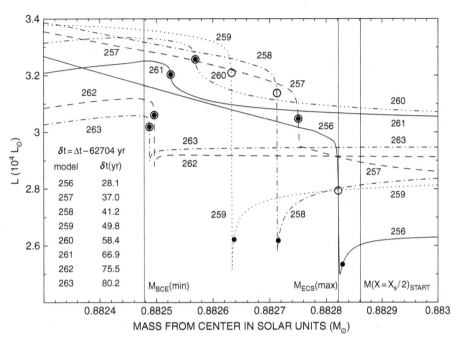

Fig. 18.3.16 Luminosity versus mass in a 5 $M_\odot$ TPAGB model during the dredgeup episode after the thirteenth helium shell flash ($Z = 0.015, Y = 0.275$)

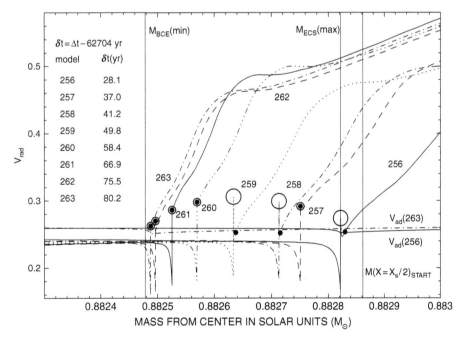

Fig. 18.3.17 VRAD in a 5 $M_\odot$ TPAGB model during dredgeup after the thirteenth helium shell flash ($Z = 0.015, Y = 0.275$)

the time which has elapsed following maximum L_{He} achieved during the thirteenth helium shell flash. A total of 2070 time steps and a total of 52 years separate the last of the eight models from the first, for an average of 0.025 years (9 days) per time step. In any given time step there are typically 200 mass zones in the dredge-up region, so, with overshoot being invoked only once every ten or so time steps, the degree of dredge-up is, on average, being neither overestimated nor underestimated, as found for the dredge-up episode after the eleventh helium shell flash.

Along each opacity profile in Fig. 18.3.14, the filled circle identifies the location of the base of the convective envelope in the model. The larger, open circle identifies the location of the hydrogen–helium discontinuity. In the case of models 256, 257, and 259, the size of the open circle has been exaggerated to make it crystal clear that the formal base of the convective envelope is at a larger mass than is the location of the hydrogen–helium discontinuity. What this mass difference demonstrates is that the overshoot algorithm does not force an inexorable inward movement of the base of the convective envelope. A number of competing physical processes are at work.

It is evident from the P/P_{rad} profiles in Fig. 18.3.15 that over a considerable distance about the composition discontinuity, which coincides with the minimum in each profile, radiation pressure is the dominant contributor to the total pressure. At the discontinuity itself, in going from model 256 to model 261, P_{rad} contributes from ~74% to ~82% of the total pressure.

In Fig. 18.3.16, the curve defined by the intersections of the luminosity profiles with the locations of the composition discontinuity (i.e., by the open circles) is quite smooth, approximating the arc of a parabola. In the first four models (256–259), the luminosity is increasing with time both at the composition discontinuity and over a considerable distance in mass below the discontinuity as the rate at which gravothermal energy stored during the helium shell flash diffuses outward increases. The reverse is the case for the last four models (260–263), in which the luminosity at the composition discontinuity declines with each successive model as the rate at which gravothermal energy diffuses outward decreases. Outward from the composition discontinuity in these four models, the luminosity either decreases monotonically (models 260 and 261) or climbs very modestly outward and then remains nearly constant (models 262 and 263).

In Fig. 18.3.17 which shows the V_{rad} profiles for the eight models, the numbering scheme, the meaning of filled and open circles, and the time δt designations are the same as in the three previous figures. V_{ad} profiles are shown for only two models (256 and 263), but these are sufficient to demonstrate that, on passing inward in mass, V_{ad} for any given model decreases to $V_{\text{ad}} \sim 0.25$ as the composition discontinuity is approached and then increases to a common value of $V_{\text{ad}} \sim 0.26$. These values reflect the facts that radiation pressure is the dominant contributor to the total pressure and that the maximum contribution occurs at the composition discontinuity where P/P_{rad} is at a minimum.

It is simple to verify the behavior of V_{ad} analytically. A change in heat content per unit mass can be written as

$$\delta Q = \delta U - \frac{P}{\rho} \frac{\delta \rho}{\rho}. \qquad (18.3.6)$$

In the current context, the internal energy per unit mass is to a good approximation given by

$$U = \frac{1}{\rho} \left\{ \frac{3}{2} P_{gas} + 3 P_{rad} \right\} = \frac{3}{2} \frac{kT}{\mu} + \frac{aT^4}{\rho}, \tag{18.3.7}$$

so that

$$\delta U = \frac{1}{\rho} \left\{ \left(\frac{3}{2} P_{gas} + 12 \, P_{rad} \right) \frac{\delta T}{T} - 3 \, P_{rad} \frac{\delta \rho}{\rho} \right\} \tag{18.3.8}$$

$$= \frac{P_{gas}}{\rho} \left\{ \left(\frac{3}{2} + 12 \frac{P_{rad}}{P_{gas}} \right) \frac{\delta T}{T} - 3 \frac{P_{rad}}{P_{gas}} \frac{\delta \rho}{\rho} \right\}. \tag{18.3.9}$$

Inserting in eq. (18.3.6) the expression for δU given by eq. (18.3.8), one has

$$\delta Q = \left\{ \left(\frac{3}{2} P_{gas} + 12 P_{rad} \right) \frac{\delta T}{T} - (3 P_{rad} + P) \frac{\delta \rho}{\rho} \right\}. \tag{18.3.10}$$

For an adiabatic change, $\delta Q = 0$, so

$$\left(\frac{\delta \rho}{\rho} \right)_{ad} = \frac{1.5 P_{gas} + 12 P_{rad}}{3 P_{rad} + P} \left(\frac{\delta T}{T} \right)_{ad}. \tag{18.3.11}$$

By inspection, an incremental change in the pressure can be written as

$$\frac{\delta P}{P} = \frac{P_{gas}}{P} \left(\frac{\delta \rho}{\rho} + \frac{\delta T}{T} \right) + \frac{P_{rad}}{P} 4 \frac{\delta T}{T}, \tag{18.3.12}$$

or, rearranging, as

$$\frac{\delta P}{P} = \left(\frac{P_{gas} + 4 P_{rad}}{P} \right) \frac{\delta T}{T} + \frac{P_{gas}}{P} \frac{\delta \rho}{\rho}. \tag{18.3.13}$$

Then, making use of eq. (18.3.11), an adiabatic change in pressure is related to an adiabatic change in temperature by

$$\left(\frac{\delta P}{P} \right)_{ad} = \left(\frac{P_{gas} + 4 P_{rad}}{P} + \frac{P_{gas}}{P} \frac{1.5 P_{gas} + 12 P_{rad}}{3 P_{rad} + P} \right) \left(\frac{\delta T}{T} \right)_{ad}. \tag{18.3.14}$$

Setting

$$\beta = \frac{P_{gas}}{P} \quad \text{and} \quad 1 - \beta = \frac{P_{rad}}{P}, \tag{18.3.15}$$

one obtains, after a bit of manipulation:

$$V_{ad} = \left(\frac{\delta T}{T} \right)_{ad} \div \left(\frac{\delta P}{P} \right)_{ad} = \frac{(4 - 3\beta)}{4(4 - 3\beta) - 1.5 \beta^2} = \frac{1}{4 - 1.5 \beta^2 / (4 - 3\beta)}, \tag{18.3.16}$$

from which the familiar results emerge that $V_{ad} = 0.4$ when radiation pressure is negligible and $V_{ad} = 0.25$ when gas pressure is negligible. At the composition discontinuity in model 256, where $1 - \beta \sim 0.74$, $V_{ad} = 0.251$.

It is instructive to ascertain how each of the three factors to which V_{rad} is proportional influences the V_{rad} profiles. In all eight models, the fact that V_{rad} is discontinuous at the composition discontinuity is primarily a consequence of a discontinuity in the electron-scattering contribution to the opacity. However, the steep decrease with increasing mass in V_{rad} just below the discontinuity is due to a comparably steep decrease with increasing mass in P/P_{rad} and most of the increase with increasing mass in V_{rad} in the convective envelope is due to an increase with increasing mass in P/P_{rad}. It is the luminosity level maintained at the inner edge of the composition discontinuity by the outward diffusion of stored gravothermal energy that provides the critical threshold for intermittently achieving $V_{rad} > V_{ad}$ at the composition discontinuity, despite the drop in luminosity there occasioned by the need to supply the energy necessary for expansion as the number abundance per unit mass increases inward due to overshoot.

From the perspective of whether or not the adopted dredge-up algorithm produces a reasonable approximation to the dredge-up process that occurs in the real counterpart of the model, the most encouraging aspect of the V_{rad} profiles in Fig. 18.3.17 is that the curve defined by the open circles is essentially a parabola which crosses the mean value of V_{ad} at either end of the dredge-up episode. In three of the eight models (256, 258, and 259), the formal base of the convective envelope is at a mass larger than the mass at which the composition discontinuity occurs, demonstrating again that the overshoot algorithm does not inexorably force the base of the convective envelope to move inward in mass. In each of these three models, $V_{rad} > V_{ad}$ in the zone on the hydrogen-rich side of the composition discontinuity and is separated from the formal base of the convective envelope by between eight and 25 mass zones in which $V_{rad} < V_{ad}$. The fact that, in these models, the near discontinuity in the luminosity profile (Fig. 18.3.16) passes through the location of the composition discontinuity rather than through the base of the convective envelope means that the rate of absorption from the ambient energy flow peaks at the composition discontinuity and is therefore influenced primarily by the existence of the composition discontinuity rather than by a transition between energy flow by radiation and energy flow by convection.

Radius and temperature profiles for the eight models are shown in Figs. 18.3.18 and 18.3.19, respectively. Closed and open circles along the radius and temperature profiles have the same meaning as in Figs. 18.3.14–18.3.17. It is clear from these profiles that, at the composition discontinuity and for some distance on either side of the discontinuity, matter is moving outward and cooling. At the discontinuity itself, at first the radius increases and the temperature drops (models 256–258) and then the radius decreases and the temperature increases (models 259–263).

Gravothermal energy-generation rates and the location of convective and radiative regions for model 256 are shown in Fig. 18.3.20. Open circles define the locations of mass zone boundaries, with the vertical coordinate of each circle designating whether the zone is convective (-1) or radiative (-3). The leftmost of the two vertical lines in the figure defines the maximum mass $M_{ECS}(max)$ reached by the convective shell formed during the thirteenth helium-burning flash. Serendipitously, it also defines a relative minimum in the mass at the base of the convective envelope reached three time steps prior to model 256. It thus marks a discontinuity in three major isotopes: ^{12}C, ^4He, and ^1H. The rightmost of the two vertical lines defines the outer edge of the zone for which the leftmost line is the inner boundary. The relatively large mass of this zone is a consequence of the fact that the

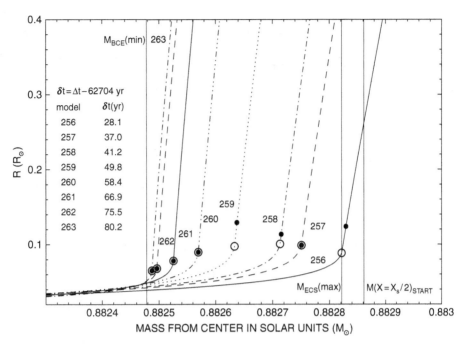

Fig. 18.3.18 Radius versus mass in a 5 M$_\odot$ TPAGB model during dredgeup after the thirteenth helium shell flash ($Z = 0.015$, $Y = 0.275$)

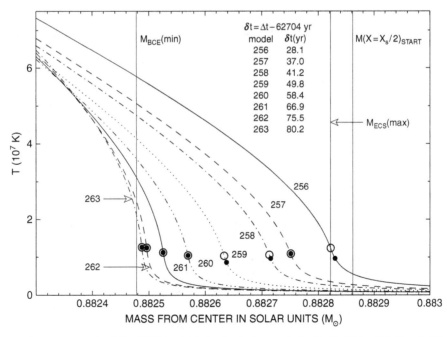

Fig. 18.3.19 Temperature versus mass in a 5 M$_\odot$ TPAGB model during the dredgeup episode after the thirteenth helium shell flash ($Z = 0.015$, $Y = 0.275$)

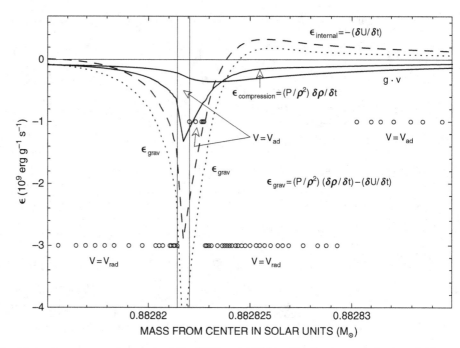

Fig. 18.3.20 Gravothermal energy-generation rates in a 5 $M_\odot$ TPAGB model (256) in which M_{BCE} has reached and receded from $M_{ECS}(\text{max})$ ($Z = 0.015$, $Y = 0.275$, $\delta t = 28.1$ yr)

slope of the gravothermal energy generation rate ϵ_{grav} changes sign in the middle of the zone, causing the rezoning algorithm (which is activated every three time steps) to assume that the zone need not be split. In this mass zone and in four adjacent mass zones of larger mass, energy flow is by convection. In the next 25 zones, energy flow is by radiation. The location of the base of the convective envelope in model 256 coincides with the position of the rightmost open circle with vertical coordinate -3.

Figure 18.3.21 shows for model 261 the same quantities shown in Fig. 18.3.20 for model 256. The only difference in notation is that, when energy flow in the associated zone is convective, the vertical coordinate of an open circle is -3, and, when energy flow in the associated zone is radiative, the vertical coordinate is -5. The morphological similarities between the profiles in Figs. 18.3.21 and 18.3.20 are more significant than the differences in quantitative detail. In both instances the gravothermal absorption rate $-\epsilon_{grav}$ reaches a peak in the mass zone on the hydrogen-rich side of the composition discontinuity and, at the peak and over a finite region about the peak, the rate $-\epsilon_{compression}$ at which energy is required for expansion and the rate $-\epsilon_{internal}$ at which energy is required for increasing the internal energy per unit mass contribute comparably to $-\epsilon_{grav}$. In both cases, matter is expanding ($\epsilon_{comp} \propto d\rho/dt < 0$) and moving outward ($\mathbf{g} \cdot \mathbf{v} < 0$) over a substantial distance in mass on either side of the composition discontinuity.

In both models 256 and 261, $dU/dt > 0$ in a small region in the neighborhood of the composition discontinuity, even though $dU/dt < 0$ below and beyond this small region and even though, from Fig. 18.3.19, $dT/dt \leq 0$ everywhere in this region. If the equation

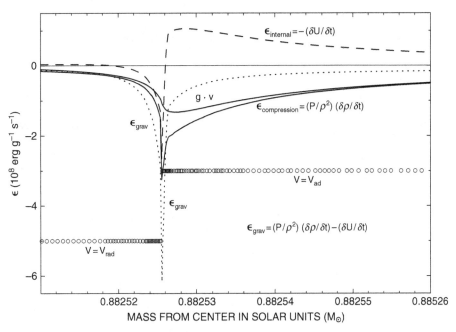

Gravothermal energy-generation rates in a 5 $M_\odot$ TPAGB model (261) in which M_{BCE} is approaching $M_{BCE}(\min)$
($Z = 0.015, Y = 0.275, \delta t = 66.9$ yr)

of state were that of a perfect gas, so that $dU/dt \propto dT/dt$, this circumstance would suggest a contradiction. The contradiction is resolved by recalling from Fig. 18.3.15 that radiation pressure is significantly larger than gas pressure over a fairly large region centered on the composition discontinuity and that the contribution of radiation pressure to the internal energy per gram is inversely proportional the density, which is decreasing. A straightforward qualitative explanation for why δU can be positive even when δT is negative follows from the fact that that the second term in curly brackets in eqs. (18.3.8) and (18.3.9) is positive when density is decreasing. A quantitative explanation is provided in Figs. 18.3.22 and 18.3.23, where logarithmic variations with mass in density, temperature, and radius are shown for models 256 and 261, respectively, along with variations in mass of P_{rad}/P_{gas} and of the two terms in curly brackets in eq. (18.3.9). For both models there is a region centered on the position of the composition discontinuity (where P_{rad}/P peaks) where the dash-dot-dot-dot curve proportional to the second term exceeds the dot-dot-dot-dot curve proportional to the first term.

It is of interest to examine the quantitative changes in surface abundances brought about by the dredge-up episode. During the quiescent hydrogen-burning phase between the twelfth and thirteenth helium shell flashes, the center of the hydrogen-burning shell moves outward by $\Delta M_H \sim 0.00151\ M_\odot$. Over the course of the dredge-up episode after the thirteenth flash, the base of the convective envelope moves a total mass of $\sim 0.00038\ M_\odot$ inward beyond the initial hydrogen–helium interface (between $M_{BCE}(\min)$ and $M(X = X_s/2)$ in, say, Fig. 18.3.17), so the degree of dredge-up associated with the thirteenth flash is, as defined by eq. (18.3.5), $\lambda \sim 0.26$. The first $\sim 0.00004\ M_\odot$ of the dredged-up

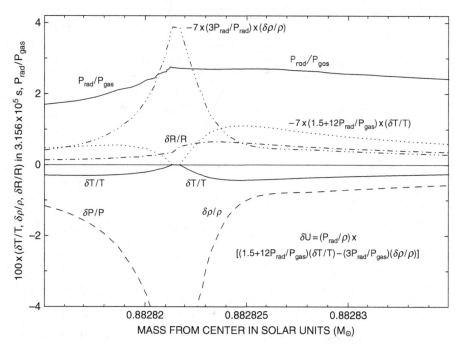

Fig. 18.3.22   Logarithmic increments of T, ρ, and R in a 5 $M_\odot$ TPAGB model (256) beginning dredgeup after the thirteenth helium shell flash (Z = 0.015, Y = .275, δt = 28.1 yr)

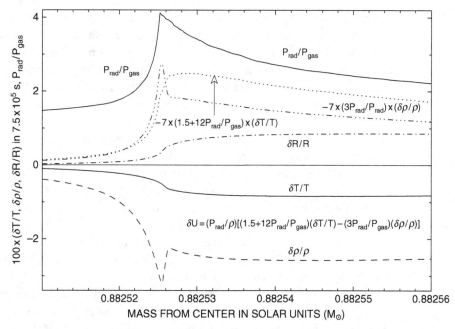

Fig. 18.3.23   Logarithmic increments of T, ρ, and R in a 5 $M_\odot$ TPAGB model (261) near the end of dredgeup after the thirteenth helium shell flash (Z = 0.015, Y = .275, δt = 66.9 yr)

material (between M_{ECS} and $M(X = X_s/2)$ in Fig. 18.3.17) contains fresh products of complete hydrogen burning. The next $\sim 0.000\,34\ M_\odot$ of dredged-up material (between $M(X = X_s/2)$ and $M_{BCE}(\min)$ in Fig. 18.3.17) consists of carbon-rich and helium-rich material left behind by the shrinking convective shell after the peak of the helium shell flash. The number abundance of carbon in this latter region is $Y(^{12}C) \sim 0.0167$, compared with a number abundance of carbon in the convective envelope prior to the dredge-up episode of $Y(^{12}C) \sim 1.275 \times 10^{-4}$. Since the mass of the convective envelope is $\sim 4.12\ M_\odot$, the number abundance of carbon in the envelope at the end of the dredge-up episode is

$$Y(^{12}C) \sim \frac{0.000\,34 \times 0.0167 + 4.12 \times 1.275 \times 10^{-4}}{4.12}$$
$$= 1.378 \times 10^{-6} + 1.275 \times 10^{-4} = 1.289 \times 10^{-4}, \tag{18.3.17}$$

an enhancement of approximately 1.1 %. If this process were to continue in the real analogue of the model, if the same degree of enhancement occurred in each dredge-up episode, if no mass were lost from the surface, and if no carbon were converted into nitrogen in the envelope, after N helium shell flashes, the enhancement of the surface carbon abundance would be

$$\Delta Y(^{12}C) \stackrel{\sim}{=} 1.38 \times 10^{-6}\ N. \tag{18.3.18}$$

The number abundance of oxygen in the convective envelope prior to dredge-up after the 13th helium shell flash is $Y(^{16}O) = 3.701 \times 10^{-4}$, compared with a number abundance before dredge-up of $Y(^{16}O) \sim 1.5 \times 10^{-4}$ in the same region where $Y(^{12}C) \sim 1.67 \times 10^{-2}$. After the dredge-up episode, the surface abundance of ^{16}O is smaller by $\sim 2.2 \times 10^{-8}$, and after N identical dredge-up episodes, the number abundance would be changed by

$$\Delta Y(^{16}O) \stackrel{\sim}{=} -2.2 \times 10^{-8}\ N. \tag{18.3.19}$$

Similar calculations show that the surface helium number abundance would be enhanced by

$$\Delta Y(^4He) \stackrel{\sim}{=} 1.9 \times 10^{-6}\ N \tag{18.3.20}$$

and the surface hydrogen abundance would be decreased by four times this. After approximately 90 dredge-up episodes, the carbon abundance at the surface would double and after approximately 180 episodes, the carbon abundance at the surface would exceed the oxygen abundance, which would be essentially unchanged at $Y(^{16}O) \sim 3.66 \times 10^{-4}$; the helium to hydrogen abundance ratio would be increased by about 2.4%.

However, the strength of a helium shell flash increases with each flash and the degree of dredge-up increases with flash strength. Several characteristics resulting from the last five pulse cycles calculated are detailed in Table 18.3.1. Given in columns from left to right are: cycle number, the maximum helium-burning luminosity L_{He}, the duration Δt_{flash} of a flash, the mass ΔM_H through which the hydrogen-burning shell moves during the interpulse phase, the ratio λ of mass ΔM_{dg} dredged-up after a flash to ΔM_H, and $Y_s(^{12}C)$, the number abundance of ^{12}C at the surface after the third dredge-up. Although $L_{He}^{\max}$ during a flash and

Table 18.3.1 Degree of dredge-up and carbon-abundance enhancements versus strength of the helium shell flash

flash	$\frac{L_{He}^{max}}{L_\odot}$	$\Delta t_{flash}(s)$	$\frac{\Delta M_H}{M_\odot}$	$\lambda = \frac{\Delta M_{dg}}{\Delta M_H}$	$Y_s(^{12}C)$
11	7.2895/6	2/5	0.00142	0.12	1.275/−4
12	1.1305/7	2/5	0.00146	0.22	1.275/−4
13	1.3715/7	2/5	0.00151	0.26	1.289/−4
14	1.6432/7	1/5	0.00155	0.48	1.319/−4
15	2.2208/7	3.156/4	0.00167	0.51	1.353/−4

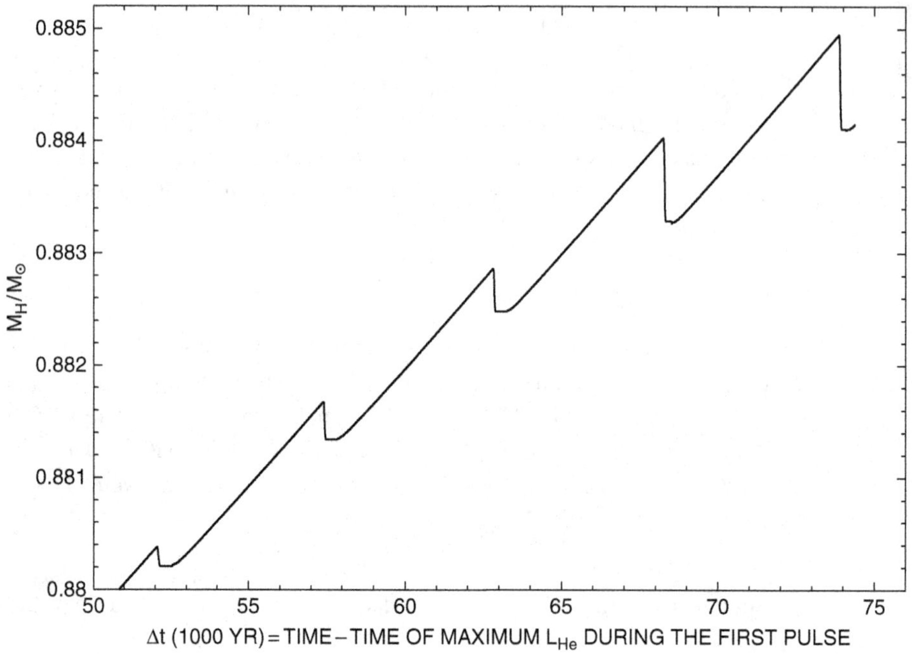

Fig. 18.3.24   Mass at half maximum X_H in a 5 $M_\odot$ model during the 11th through 15th thermal pulses triggered by shell helium-burning flashes (Z = 0.015, Y = 0.275)

the degree of dredge-up λ following the flash are correlated, the relationship is far from smooth, demonstrating that all numerical results are afflicted by numerical noise as well as by human intervention in the form of changes in spatial- and time-zoning criteria. In this particular set of calculations, the time steps near maximum L_{He} differ from one flash calculation to the next and, prior to the dredge-up episodes after the fourteenth and fifteenth flashes, several hundred mass zones have been introduced in the region where dredge-up subsequently occurs and spatial zoning in this region has then been frozen during the dredge-up episode.

The mass at the hydrogen–helium interface during the final five pulse cycles calculated are shown in Fig. 18.3.24. It is evident that finer zoning in both space and time contribute

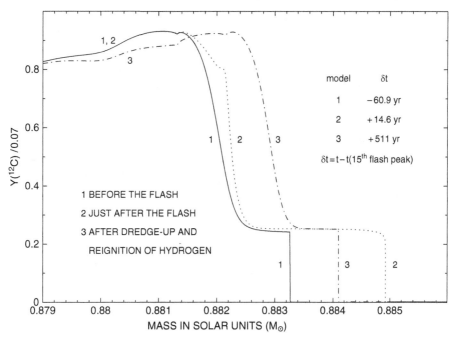

Carbon-abundance profiles before and after the fifteenth helium shell flash in a 5 $M_\odot$ model ($Z = 0.015$, $Y = 0.275$)

to enhancements in the strength of the flash and in the degree of dredge-up. Further calculations to explore this relationship would be worthwhile.

Carbon abundance profiles at three different times before, during, and after the fifteenth helium shell flash are shown in Fig. 18.3.25. During the shell flash, convection carries carbon out to just below the hydrogen–helium interface, so the distance between the outer edges of the carbon profiles for models 1 and 2 is effectively the distance through which the hydrogen-burning shell has moved during the fourteenth interpulse phase, namely $\Delta M_H \sim 0.00167\ M_\odot$. By the time model 3 is reached, dredge-up has been completed and quiescent helium burning has essentially run its course. Note that helium burning continues during the dredge-up episode and that, in fact, the amount of carbon produced during and after this episode is comparable to the carbon produced during the helium shell flash. Note also that carbon-abundance profiles for models 1 and 3 are nearly identical, with the profile for the more evolved model simply displaced to a slightly larger mass than the profile for the less evolved model. The outer edge of the profile for model 3 is approximately half-way between those for models 1 and 2, making it easy to see that the degree of dredge-up is $\lambda \sim 0.5$.

The internal global luminosity characteristics L_{He}, L_H, and L_{grav} of the fifteen pulse cycles punctuated by helium shell flashes that have been calculated are shown as functions of time in Figs. 18.3.26–18.3.28, respectively. In order to compare it directly with the global gravothermal luminosity, the global helium-burning luminosity is shown again by the dashed curve in Fig. 18.3.28. The fact that, during flash peak, $\log L_{He}/L_\odot$ and

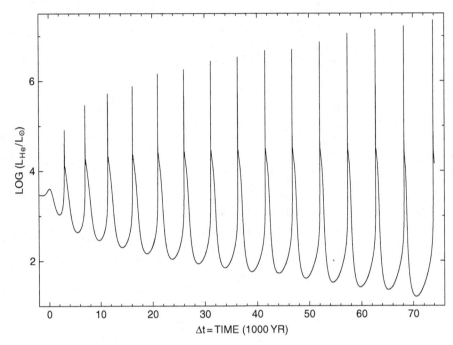

Fig. 18.3.26 Helium-burning luminosity of a 5 $M_\odot$ model ($Z = 0.015, Y = 0.275$) experiencing shell helium-burning flashes during the TPAGB phase

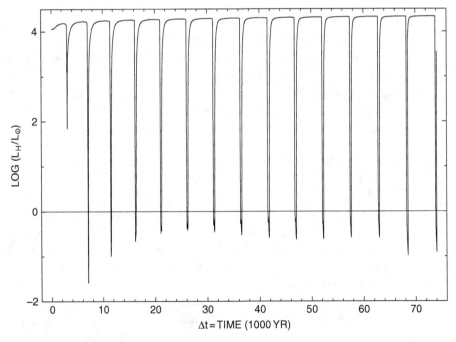

Fig. 18.3.27 Hydrogen-burning luminosity of a 5 $M_\odot$ TPAGB model ($Z = 0.015, Y = 0.275$) experiencing multiple thermal pulses triggered by shell helium-burning flashes

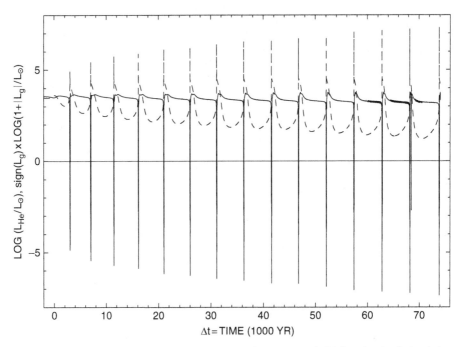

Fig. 18.3.28 Gravothermal and helium-burning luminosities of a 5 $M_\odot$ model experiencing shell helium-burning flashes during the TPAGB phase ($Z = 0.015, Y = 0.275$)

$\log(1 + |L_{\text{grav}}/L_\odot|)$ are identical once again emphasizes the fact that, during each flash, most of the nuclear energy released by helium burning is stored as potential and thermal energy in the region where the rate of nuclear energy production is at its highest.

In Fig. 18.3.29, surface luminosity L_s is compared with the luminosity L_{PU} obtained by inserting in eq. (17.4.1) the values of M_H described in Figs. 18.3.5, 18.3.6, and 18.3.24. As shown in Fig. 17.4.15, L_{PU} gives a fair representation of the mean luminosity of a 1 $M_\odot$ TPAGB model. In contrast, for the 5 $M_\odot$ TPAGB model, L_{PU} appears to act as an asymptote which L_s approaches at maximum luminosity during the quiescent hydrogen-burning phase just prior to the ignition of a helium shell flash.

Irregular low frequency oscillations of variable amplitude appear in L_s during the quiescent hydrogen-burning phases prior to the first twelve helium shell flashes. The oscillations increase in frequency and amplitude during the quiescent hydrogen-burning phases prior to the thirteenth and fourteenth helium shell flashes. These high frequency oscillations are related to similar oscillations in the gravothermal luminosity that occur during the same two time frames and that, because of the vertical scaling, are barely detectible in Fig. 18.3.28. The oscillations tend to be correlated with the zoning changes that are activated every third time step and can be attributed to the numerical noise introduced by such changes into the local gravothermal energy-generation rate.

A simple reason has not been identified for why the high frequency oscillations appear during the quiescent hydrogen-burning phases prior to the thirteenth and fourteenth helium shell flashes but not during previous interflash phases. In an effort to minimize the

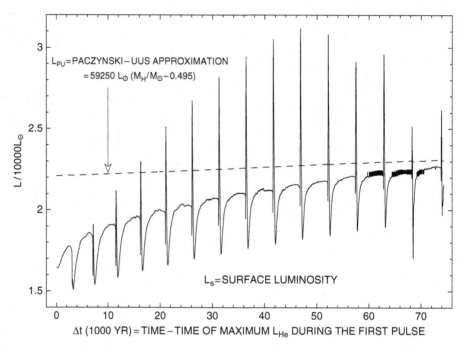

Fig. 18.3.29 Surface luminosity of a 5 $M_\odot$ model experiencing thermal pulses triggered by shell helium-burning flashes ($Z = 0.015, Y = 0.275$)

amplitude and frequency of the oscillations during the fifteenth interflash period, spatial zoning in the entire region below the advancing hydrogen-burning shell has been frozen, with rezoning allowed only in and beyond the shell to insure that the hydrogen profile is properly resolved. As follows from the curves for L_s and L_{grav} for times beyond $\Delta t \sim 72$, this strategem reduces the frequency of the oscillations but does not eliminate oscillations of the sort that occur during earlier interflash phases and are presumably related to noise generated by rezoning in the region of variable hydrogen abundance.

Except for a tiny region just below the helium-flashing zone where energy flows inward during a flash, the behavior of the helium-exhausted core of the model has up to this point been ignored. This neglect is partially remedied in Fig. 18.3.30 where central density and temperature are shown as functions of time. It is evident that the expansion that occurs in and above the helium-burning zone during a flash extends all the way to the center and that some of the energy required for expansion comes locally from the thermal energy of non-degenerate electrons near the top of the electron Fermi sea. Between the short lived episodes of expansion, central density increases as matter containing the fresh carbon and oxygen created during each major helium-burning phase contracts as it is added to the hydrogen-exhausted core during the subsequent hydrogen-burning phase. Cooling continues in consequence of neutrino losses due to the plasma process. The general trend of increasing density and decreasing temperature that occurs at the center prior to each helium

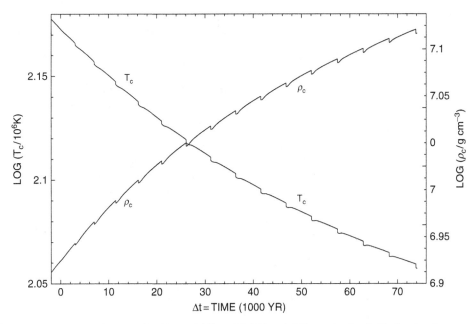

Fig. 18.3.30 Central temperature and density of a 5 $M_\odot$ model ($Z = 0.015$, $Y = 0.275$) experiencing shell helium-burning flashes during the TPAGB phase

flash occurs also in the helium-rich zone where the next flash will begin, and this accounts for the fact that successive flashes grow stronger.

Even in this day of personal computers which exceed the capacity, speed, and power of computers available to entire institutions many decades ago, the calculation of TPAGB evolution is very demanding of time and attention. Carrying the calculation of thermal pulses further to demonstrate that pulse strength increases with each pulse until a local asymptotic strength is reached is left as a project for an interested reader.

In summary, it follows from the discussion here and from the results of many calculations similar to those presented, carbon production during helium shell flashes followed by the third dredge-up can account for the existence of carbon stars. That the two processes can also account for much of the carbon in the Universe follows from the observational fact that TPAGB stars lose mass at rates large compared with the rates at which they process matter by nuclear burning and end their nuclear burning lives as the compact central stars of the planetary nebula formed by the ejected envelope which fluoresces in response to radiation absorbed from the central star. The large observed rates of mass loss are in part a consequence of radiation pressure on carbon- and silicon-rich grains which can be formed because of carbon production during flashes followed by dredge-up and they limit the lifetime of real TPAGB stars to less than $\sim 10^6$ yr. The white dwarf into which a star initially of intermediate mass eventually becomes after evolving into an AGB star is therefore not much more massive than the helium-exhausted core at the beginning of the TPAGB phase. These matters are elaborated in Chapter 21 in the context of the evolution of a low mass star which, after becoming a TPAGB star, evolves in a manner which is qualitatively similar to the evolution of a star initially of intermediate mass.

Bibliography and references

Stephen Allan Becker, *ApJS*, **45**, 475, 1981.

Stephen Allan Becker & Icko Iben, Jr., *ApJ*, **232**, 831, 1979; **237**, 111, 1980.

Stephen A. Becker, Icko Iben, Jr., and Roy S. Tuggle, *ApJ*, **218**, 633, 1977.

E. Hofmeister, R. Kippenhahn, & A. Weigart, *Z. für Ap*, **59**, 215, 1964; **59**, 242, 1964; **60**, 57, 1964.

Icko Iben, Jr., *ApJ*, **143**, 483, 1966; **178**, 433, 1972; **196**, 525, 1975; **208**, 165, 1976; **304**, 201, 1986.

Icko Iben, Jr. & Alexander V. Tutukov, *ApJ*, **418**, 343, 1993.

Icko Iben, Jr. & Alvio Renzini, *ARAA*, **21**, 271, 1983.

Icko Iben, Jr. & Alvio Renzini, *Phys. Rep.*, **105**, No. 6, 330, 1984.

Icko Iben, Jr. & Jim Truran, *ApJ*, **220**, 740, 1978.

Rudolph Kippenhahn, H.-C. Thomas, & A. Weigart, *Z. für Ap*, **61**, 241, 1965; **64**, 373, 1966.

A. Weigart, *Z. für Ap*, **64**, 395, 1966.

Neutron production and neutron capture in a TPAGB model star of intermediate mass

19.1 History of *s*-process nucleosynthesis and outline

In the middle of the twentieth century, Paul W. Merrill (1952) detected the beta-unstable element technetium in the spectra of red supergiants known as S stars, which have long been recognized as being AGB stars. ^{99}Tc is the longest lived isotope of technetium, and its half life of 211 000 yr is several orders of magnitude shorter than the lifetime of a core helium-burning precursor of an AGB star. When observed mass-loss rates are taken into account, the lifetime of a star in the AGB phase is roughly an order of magnitude larger than the half life of ^{99}Tc. It is therefore incontrovertible that the ^{99}Tc observed has been formed in AGB stars. On the other hand, given the temperatures and densities in AGB models, it is clear that an element of atomic charge as large as that of Tc cannot be a consequence of reactions between charged particles and it has become accepted that neutron capture must be involved.

The earliest theoretical studies of neutron-capture nucleosynthesis were carried out by A. G. W. Cameron (1955) and by E. Margaret Burbidge, Geoffrey R. Burbidge, William A. Fowler, and Fred Hoyle (1957) and, over the years, numerous studies have been devoted to examining how the distribution of abundances in an initial set of heavy elements is altered when this set is placed in a bath of neutrons at various assumed number densities for various exposure times. Two regimes have received the most intense study, according to whether the intervals between neutron captures tend to be long compared with beta-decay half lives (the so-called slow or s-process) or short compared with beta-decay half lives (the so-called rapid or r-process).

In the s-process, successively heavier elements are built up along the band-defined by beta-stable isotopes in the proton number versus neutron number (Z-N) plane. In the r-process, heavier elements are built up along bands defined by beta-unstable isotopes which decay by the emission of electrons to the neutron-rich side of the beta-stable band after the termination of the neutron-exposure episode. The two regimes produce two distinct distributions of isotopes and, remarkably, heavy neutron-rich isotopes in the observed solar-system distribution can for the most part be assigned to one or the other of the two theoretical distributions (see, e.g., P. A. Seeger, W. A., Fowler, & D. D. Clayton, 1965).

Although it decays on a time scale long compared with the duration of a neutron-production episode of the s-process variety and is thus technically an r-process isotope, ^{99}Tc is produced preferentially by the s-process and one infers that it is the s-process which operates in TPAGB stars. Cameron identified what have turned out to be the two main sources of neutrons for s-process nucleosynthesis: the exothermic reaction $^{13}C(\alpha, n)^{16}O$ (Cameron, 1955) and the endothermic reaction $^{22}Ne(\alpha, n)^{25}Mg$ (Cameron, 1960). These reactions have since been studied extensively experimentally and theoretically, and it has become evident that the two reactions occur at interesting rates when temperatures approach 10^8 K and 3×10^8 K, respectively.

Some of the uncertainties involved in estimating the rates at which the two major neutron-producing reactions occur have been presented in Section 16.5. In Section 19.2 of this chapter, guesses as to the temperature dependences of the two reactions are reviewed and neutron-capture cross sections of relevance are described. In subsequent sections, the activation of the two main neutron sources and the consequences of their activation are examined during the 14th interpulse phase and during the 15th helium shell flash experienced by the 5 $M_\odot$ TPAGB model described in Chapter 18.

During the dredge-up episode which follows a helium shell flash, H-rich matter meets C-rich matter at an interface produced by the inward movement of the base of the convective envelope. It is to be expected that, during and after dredge-up, diffusion across the interface mixes H-rich matter into the C-rich region and C-rich matter into the H-rich region over a finite region centered on the interface. In Section 19.3, an example of anticipated particle distributions is presented. During subsequent evolution, complete hydrogen burning of matter in the mixed region produces a peak in the ^{13}C abundance where the abundances of ^{12}C and ^{14}N are comparable and are an order of magnitude larger than the maximum ^{13}C abundance. As temperatures in the ^{13}C-peak region approach 10^8 K, an episode of neutron production and capture occurs. By far the major absorber of neutrons is ^{14}N, with most of the neutrons released experiencing the reaction $^{14}N(n, p)^{14}C$. Most of the protons released initiate the $^{12}C(p, \gamma)^{13}N(e^+ \nu_e)^{13}C$ reaction chain. The additional ^{13}C produced experiences (α, n) reactions, so the total number of released neutrons exceeds the initial number of ^{13}C nuclei in the ^{13}C abundance peak. Nevertheless, altogether in the ^{13}C abundance peak, less than one heavy s-process nucleus is produced for every 25 initial ^{13}C nuclei.

In Section 19.4, a description is given of nucleosynthesis in matter outside of the ^{13}C-peak region as the hydrogen-burning shell passes through it and it then experiences quiescent helium burning during the interpulse phase. In approximately two-thirds of the matter processed by hydrogen burning during the interpulse phase, the $^{13}C(\alpha, n)^{16}O$ reaction destroys ^{13}C completely, with ^{14}N again being the dominant absorber of released neutrons. However, the circumstance that the initial abundance of ^{12}C is much smaller than the abundance of ^{14}N (rather than larger, as in the ^{13}C-peak region) alters nucleosynthesis patterns in such a way that approximately one heavy s-process nucleus is produced for every 10 initial ^{13}C nuclei outside of the ^{13}C abundance peak. The net result is that twice as many heavy s-process elements are produced in the region outside of the initial ^{13}C abundance peak as are produced in the region containing the peak.

In Section 19.5, a description is given of the two episodes of neutron-capture nucleosynthesis occurring in the convective shell which forms during the 15th helium shell flash. The first episode occurs when matter which has been processed by hydrogen burning during the interpulse phase but in which ^{13}C has not been affected by (α, n) reactions is ingested by the convective shell. Alpha-particle captures on the ingested ^{13}C nuclei produce neutrons in an environment which differs from the environments prevailing during the two inter-pulse phase s-process episodes in that (1) temperatures at which burning occurs are larger and (2) the ^{14}N present in the convective shell prior to the ingestion of fresh ^{13}C has been destroyed by the ^{14}N$(\alpha, \gamma)^{18}$F reaction. Approximately as many heavy s-process nuclei are produced in the convective shell by the capture of neutrons released by the burning of ingested ^{13}C as are produced outside of the convective shell and then ingested.

The second episode of neutron nucleosynthesis in the convective shell occurs when temperatures near the base of the shell become large enough for the ^{22}Ne$(\alpha, n)^{25}$Mg reaction to occur. In addition to temperatures being larger, the isotope mix is different from the mixes prevailing during preceding episodes. In particular, the dominant neutron absorber among elements lighter than Fe is ^{12}C and heavy s-process elements capture approximately one neutron for every three neutrons captured by lighter elements, with the result that the second episode of neutron-capture nucleosynthesis in the convective shell produces approximately five times as many heavy s-process elements as do the other three neutron-capture nucleosynthesis episodes combined.

19.2 Neutron-production and neutron-capture reaction rates

Some of the physics involved in estimating cross sections for the two main neutron-production mechanisms has been discussed in Section 16.5. Cross sections for the ^{13}C$(\alpha, n)^{16}$O reaction at astrophysically interesting temperatures (5–30 keV) are theoretical extrapolations of experimental data at energies above 0.26 MeV. For $T \sim 10^8$ K, the leading term in the reaction rate vector prepared by G. R. Caughlan and W. A. Fowler (1988) is

$$N_A \langle \sigma\, v \rangle = \left(1 + 0.0060\, T_8^{1/3}\right) \frac{1}{T_8^{2/3}} \exp\left(37.986 - 32.329/T_9^{1/3}\right). \qquad (19.2.1)$$

The form of the analytic approximation is that expected for a non-resonant reaction. The leading term in the extrapolation by C. Angulo $et\ al.$ (1999) is

$$N_A \langle \sigma\, v \rangle = \left(1 + 4.68\, T_8 - 2.92\, T_8^2 + 0.738\, T_8^3\right) \frac{1}{T_8^2} \exp\left(38.171 - 32.333/T_9^{1/3}\right),$$

$$(19.2.2)$$

where the coefficient of the exponential term is to be viewed as a fit to a complicated function rather than as an expression of the physics involved. At $T_8 = 1$, the rate vector given by eq. (19.2.2) is 5.3 times larger than the rate vector given by eq. (19.2.1). Fortunately, the major temperature dependances in the rate vectors are expressed in the exponential terms and these dependances are identical in the two analytical fits. For example, at $T_8 = 3$, the exponential term is $\sim 2 \times 10^9$ larger than at $T_8 = 1$, but the rate vector given by eq. (19.2.2) is only about three times as large as the rate vector given by eq. (19.2.1), a change by less than a factor of 2 in the ratio of the two analytic representations. Thus, the temperatures at which the reaction is expected to be important are nearly independent of which representation is adopted.

In the neighborhood of $T_8 = 3$, the Caughlan and Fowler estimate of the cross section for the $^{22}\text{Ne}(\alpha, n)^{25}\text{Mg}$ reaction is

$$N_A \langle \sigma\, v \rangle = \frac{1}{T_9^{2/3}} \exp\left(45.161 - 47.260/T_9^{1/3} - (0.197/T_9)^{4.82}\right)$$

$$+ \exp\left(-8.846 - 5.577/T_9\right). \tag{19.2.3}$$

The form of the approximation indicates that it has been assumed that the reaction has both a non-resonant and a resonant component. The estimate by Angulo *et al.* is

$$N_A \langle \sigma\, v \rangle = \exp\left(2.001 - 7.79/T_9\right) + T_9^{0.83} \exp\left(-8.948 - 5.52/T_9\right)$$

$$+ T_9^{2.78} \exp\left(9.150 - 11.7/T_9\right) + T_9^{0.892} \exp\left(15.966 - 24.4/T_9\right), \tag{19.2.4}$$

the form of the approximation indicating that it has been assumed that the reaction proceeds through four different resonances. At $T_9 = 0.3$, the Angulo *et al.* estimate is 2.22 times larger than the Caughlan and Fowler estimate.

Because there is no Coulomb barrier to overcome and because neutron-capture cross sections tend to increase with decreasing neutron energy, there exists a wealth of experimental information regarding neutron-capture cross sections, and the uncertainty associated with extrapolation to temperatures of astrophysical interest is significantly smaller than in the case of charged-particle reactions. Results of experimental studies are sometimes displayed in terms of the quantity

$$\sigma_{kT} = \frac{\langle \sigma\, v \rangle}{v_T}, \tag{19.2.5}$$

where

$$\langle \sigma\, v \rangle = \frac{\int \sigma(E) v \, dn}{\int dn} \tag{19.2.6}$$

is the average of σv over dn, the neutron-target distribution in thermal equilibrium at temperature T, and v_T is an average thermal velocity. In regions in TPAGB stars where neutron production is important, the relevant distribution functions are non-relativistic and one can write

$$dn = \frac{4\pi}{h^3} g e^{-E/kT} p^2 dp, \tag{19.2.7}$$

where g is an appropriate statistical weight, $E = p^2/2m$, and m is the reduced mass of the system which, for heavy enough target nuclei, can be taken as the mass of a neutron.

Although the actual mean thermal velocity is

$$\langle v \rangle = \frac{\int v \, dn}{\int dn} = \sqrt{\frac{8}{\pi} \frac{kT}{m}} \tag{19.2.8}$$

and the root mean square thermal velocity is

$$\sqrt{\langle v^2 \rangle} = \sqrt{\frac{\int v^2 \, dn}{\int dn}} = \sqrt{\frac{3kT}{m}}, \tag{19.2.9}$$

the tradition in the field is to define

$$v_{\mathrm{T}} = \sqrt{\frac{2kT}{m}} = \sqrt{\frac{\pi}{4}} \langle v \rangle = \sqrt{\frac{2}{3}} \sqrt{\langle v^2 \rangle}. \tag{19.2.10}$$

From eqs. (19.2.7), (19.2.8), and (19.2.10),

$$\sigma_{kT} = \frac{\langle \sigma v \rangle}{v_T} = \sqrt{\frac{4}{\pi}} \frac{\langle \sigma v \rangle}{\langle v \rangle} = \sqrt{\frac{4}{\pi}} \frac{\int \sigma \, E \exp\left(-E/kT\right) dE}{\int E \, \exp\left(-E/kT\right) dE}. \tag{19.2.11}$$

When m is approximated by the mass of the neutron and σ_{kT} is expressed in millibarns,

$$N_{\mathrm{A}} \langle \sigma v \rangle = N_{\mathrm{A}} \, v_T \, \sigma_{kT} = 2.4557 \times 10^5 \sqrt{T_9} \, \sigma_{kT} (\mathrm{mb}) \ \mathrm{cm}^3 \ \mathrm{g}^{-1} \ \mathrm{s}^{-1}. \tag{19.2.12}$$

Multiplying the rightmost expression by $\sqrt{1 + m_n/M}$, where m_n and M are, respectively, the mass of the neutron and the mass of the target nucleus, corrects for the difference between the reduced mass and the neutron mass.

The values of σ_{kT} in Tables 19.2.1a and 19.2.1b for (n, γ) reactions on various isotopes are taken from Z. Y. Bao, H. Beer, F. Käppeler, F. Voss, K. Wisshak, and T. Rauscher (2000). Temperatures in keV are related to temperatures on other scales by $T(\mathrm{keV}) = 8.65 \, T(10^8 \ \mathrm{K}) = 8.65 \, T_8$ and σ_{kT} is in millibarns ($10^{-27} \ \mathrm{cm}^2$).

During the interpulse phase of an intermediate mass TPAGB model, a substantial fraction of the neutrons produced by the $^{13}\mathrm{C}(\alpha, n)^{16}\mathrm{O}$ reaction are captured by $^{14}\mathrm{N}$, leaving a much smaller fraction for the production of heavy s-process elements. Responsibility for this result is twofold: (1) in regions where neutrons are produced at a high rate, the number abundance of $^{14}\mathrm{N}$ is over an order of magnitude larger than the number abundances of heavy elements in the iron group which act as seeds for the synthesis of heavy s-process elements and (2) (n, p) reactions on $^{14}\mathrm{N}$ are an order of magnitude more frequent than (n, γ) reactions on $^{14}\mathrm{N}$. Thus, for quantitative studies of s-process nucleosynthesis in regions where (α, n) reactions on $^{13}\mathrm{C}$ are the primary source of neutrons, a reliable estimate of the rate of the $^{14}\mathrm{N}(n, p)^{14}\mathrm{C}$ reaction is crucial.

An experiment by P. E. Koehler & H. A. O'Brien (1989) provides estimates of the cross section for the $^{14}\mathrm{N}(n, p)^{14}\mathrm{C}$ reaction over a range of energies ideally suited for determining σ_{kT} and $\langle \sigma v \rangle$ relevant to neutron-capture nucleosynthesis in TPAGB stars. An eye fit to the Koehler & O'Brien data set is presented in Fig. 19.2.1. Results provided by earlier experiments quoted by Koehler & O'Brien and two points provided by T. Kii, T. Shima, H. Sato, T. Baba, & Y. Nagai (1999), $\sigma \sqrt{E} = 0.316(\pm 33\%)$ b $\sqrt{\mathrm{eV}}$ at $E = 35.8$ keV and $\sigma \sqrt{E} = 0.308(\pm 14\%)$ b $\sqrt{\mathrm{eV}}$ at $E = 67.1$ keV, have been taken into account.

The rest energy of the $^{14}\mathrm{N}$ + neutron system is approximately 10.835 MeV above the rest energy of $^{15}\mathrm{N}$ in the ground state and the product system, $^{14}\mathrm{C}$ + proton, has a rest energy approximately 10.208 MeV above the rest energy of $^{15}\mathrm{N}$ in the ground state, so ~ 0.627 MeV is liberated in the reaction. At neutron kinetic energies up to ~ 10 keV, the cross section for the (n, p) reaction is almost exactly inversely proportional to the velocity.

Table 19.2.1a Estimates of $\sigma_{kT} = \langle \sigma v \rangle / v_T$ in millibarns for radiative neutron-capture reactions

T(KeV)	5	10	15	20	25	30	40	50	60	80	100
Proton	0.87	0.53	0.40	0.33	0.29	0.25	0.22	0.19	0.18	0.16	0.15
^{12}C	0.0126	0.0125	0.0130	0.0139	0.0147	0.0154	0.0169	0.0183	0.0196	0.022	0.024
^{13}C	0.0073	0.0081	0.0093	0.012	0.016	0.021	0.032	0.041	0.046	0.050	0.050
^{14}C	0.0037	0.0026	0.0021	0.0019	0.0017	0.0015	0.0013	0.0012	0.0011	0.00093	0.00083
^{14}N	0.183	0.101	0.071	0.057	0.046	0.041	0.033	0.027	0.022	0.019	0.016
^{15}N	0.0025	0.0035	0.0042	0.0048	0.0053	0.0058	0.0065	0.0072	0.0078	0.00087	0.00097
^{16}O	0.016	0.022	0.027	0.031	0.035	0.038	0.044	0.049	0.054	0.062	0.069
^{18}O	0.0037	0.0051	0.0061	0.0070	0.00793	0.0089	0.0108	0.0130	0.0157	0.0223	0.0293
^{19}F	2.0	6.7	7.8	7.4	6.6	5.8	4.6	3.8	3.2	2.5	2.2
^{20}Ne	0.088	0.062	0.053	0.060	0.084	0.119	0.191	0.242	0.259	0.272	0.253
^{21}Ne				1.7	1.6	1.5	1.3	1.2			
^{22}Ne	0.221	0.133	0.103	0.088	0.074	0.059	0.059	0.044	0.044	0.044	0.04
^{23}Na	1.4	5.2	3.4	2.7	2.2	2.1	1.7	1.5	1.4	1.3	1.2
^{24}Mg	0.11	0.48	1.3	2.3	2.9	3.3	3.6	3.4	3.1	2.7	2.1
^{25}Mg	4.8	5.0	5.5	6.0	6.2	6.4	6.2	5.7	5.3	4.4	3.6
^{26}Mg	0.103	0.091	0.098	0.110	0.119	0.126	0.143	0.161	0.165	0.226	0.265

Table 19.2.1b Estimates of $\sigma_{kT} = \langle \sigma v \rangle / v_T$ in millibarns for radiative neutron-capture reactions

T(KeV)	5	10	15	20	25	30	40	50	60	80	100
27Al	11.2	6.8	5.4	4.6	4.1	3.74	3.3	3.0	2.8	2.5	2.3
28Si	0.29	0.86	1.9	2.5	2.8	2.9	2.8	2.7	2.5	2.2	1.9
29Si	10.3	14.4	13.3	11.3	9.5	7.9	5.8	4.4	3.4	2.3	1.7
30Si	124	43	22	13	8.8	6.5	3.8	2.6	1.9	1.3	0.96
31P	0.8	1.8	2.0	1.9	1.8	1.74	1.7	1.7	1.9	1.9	1.9
32S	1.3	3.0	3.7	4.0	4.1	4.1	4.0	3.8	3.6	3.3	2.9
33S	6.3	11	10	9.3	8.2	7.4	6.2	5.3	4.7	3.9	3.4
34S	0.367	0.256	0.213	0.204	0.212	0.226	0.250	0.262	0.267	0.269	0.265
35Cl	26	22.3	17.6	14.1	11.7	10.0	7.9	6.6	5.8	4.8	4.1
40Ar	2.9	2.7	2.5	2.5	2.6	2.6	2.6	2.6	2.5		
39K	24.2	20.2	17.1	14.8	13.1	11.8	9.8	8.5	7.5	6.1	5.1
40Ca	11.3	12.1	10.3	8.6	7.5	6.7	5.8	5.4	5.2	4.9	4.9
56Fe	11.8	9.73	10.8	11.5	11.7	11.7	11.3	10.6	9.97	8.63	7.42

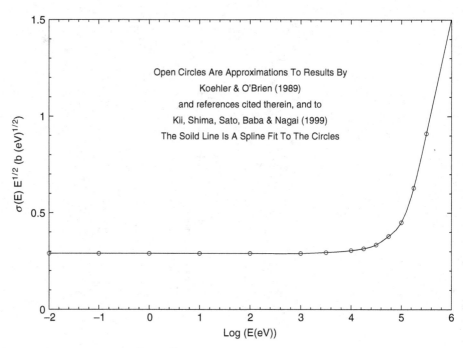

Fig. 19.2.1   Cross section times velocity for the ^{14}N(n,p)^{14}C reaction as estimated by fitting to results of direct neutron-capture experiments

At larger energies, as the reaction proceeds through an energy level in the ^{15}N compound nucleus that is approximately 11.1 MeV above the ground state of ^{15}N, the cross section decreases less rapidly than inversely with the velocity.

Reaction rates at stellar energies follow from the experimental results by use of eq. (19.2.11):

$$\frac{\langle \sigma v \rangle}{\langle v \rangle} = \frac{\int \sigma(E)\, E \exp(-E/kT)\, dE}{\int E \, \exp(-E/kT)\, dE} = \frac{\int \left[\sigma(E)\sqrt{E}\right] \sqrt{E} \exp(-E/kT)\, dE}{\int E \, \exp(-E/kT)\, dE}.$$

$$(19.2.13)$$

Setting $x = (E/kT)$,

$$\frac{\langle \sigma v \rangle}{\langle v \rangle} = \frac{1}{\sqrt{kT}} \frac{\int \left[\sigma(E)\sqrt{E}\right] \sqrt{x}\, e^{-x}\, dx}{\int x\, e^{-x}\, dx} = \frac{1}{\sqrt{kT}} \int \left[\sigma(E)\sqrt{E}\right]\, \sqrt{x}\, e^{-x}\, dx,$$

$$(19.2.14)$$

where the the second equality follows from the fact that $\int_0^\infty x\, e^{-x}\, dx = 1$. Making use of eq. (19.2.8) in eq. (19.2.14), one can write

$$\langle \sigma v \rangle = \sqrt{\frac{8}{\pi}\frac{1}{m}} \int \left[\sigma(E)\sqrt{E}\right]\, \sqrt{x}\, e^{-x}\, dx,$$

$$(19.2.15)$$

and, using eqs. (19.2.8) and (19.2.11) in eq. (19.2.15), one obtains

$$\sigma_{kT} = \frac{2}{\sqrt{\pi}} \frac{1}{\sqrt{kT}} \int \left[\sigma(E)\sqrt{E}\right] \sqrt{x}\, e^{-x}\, dx. \tag{19.2.16}$$

Energy E in the center of mass coordinate system is related to energy E_{lab} in the laboratory system by

$$r = \frac{E}{E_{\text{lab}}} = \frac{M}{m_{\text{n}} + M}, \tag{19.2.17}$$

where M and m_{n} are, respectively, the mass of the target and the mass of the neutron, so

$$\sigma(E)\sqrt{E} = \sigma_{\text{lab}}(E/r)\sqrt{E/r}. \tag{19.2.18}$$

If the product $\sigma(E)\sqrt{E}$ is constant, one can move it outside of the integral signs and, since

$$\int_0^\infty \sqrt{x}\, e^{-x}\, dx = 2\int_0^\infty y^2\, e^{-y^2} dy = \frac{\sqrt{\pi}}{2}, \tag{19.2.19}$$

it follows that

$$\langle \sigma v \rangle = \sqrt{\frac{2}{m}}\, \sigma(E)\sqrt{E} \sim \sqrt{\frac{2 N_{\text{A}}}{r}}\, \sigma(E)\sqrt{E}, \tag{19.2.20}$$

where the last approximate equality follows from the facts that $m = m_{\text{n}} r$ and $m_{\text{n}} \sim 1/N_{\text{A}}$. Using eqs. (19.2.5) and (19.2.10) in eq. (19.2.20), one has

$$\sigma_{kT} = \frac{\langle \sigma v \rangle}{v_{\text{T}}} = \frac{\sigma(E)\sqrt{E}}{\sqrt{kT}}. \tag{19.2.21}$$

If experimental results are given in units of barn $\sqrt{\text{eV}}$, then, in cgs units,

$$\sigma(E)\sqrt{E}\ (\text{cgs}) = \sigma(E)\sqrt{E}\left(\text{b}\sqrt{\text{eV}}\right) \times 10^{-24}\ \text{cm}^2 \sqrt{1.6022 \times 10^{-12}\ \text{erg}}$$

$$= \sigma(E)\sqrt{E}\left(\text{b}\sqrt{\text{eV}}\right) \times 1.2658 \times 10^{-30} \text{cm}^2 \sqrt{\text{erg}}. \tag{19.2.22}$$

In the general case,

$$\langle \sigma v \rangle = \frac{1.2384 \times 10^{12}}{\sqrt{r}} 1.2658 \times 10^{-30} \text{cm}^3\ \text{s}^{-1} \int_0^\infty \sigma(E)\sqrt{E}\left(\text{b}\sqrt{\text{eV}}\right) \sqrt{x}\, e^{-x}\, dx$$

$$= \frac{1.5676 \times 10^{-18}}{\sqrt{r}}\ \text{cm}^3\ \text{s}^{-1} \int_0^\infty \sigma(E)\sqrt{E}\left(\text{b}\sqrt{\text{eV}}\right) \sqrt{x}\, e^{-x}\, dx, \tag{19.2.23}$$

and

$$N_A \langle \sigma v \rangle = \frac{9.4418 \times 10^5}{\sqrt{r}} \ \text{cm}^3 \ \text{g}^{-1} \text{s}^{-1} \int_0^\infty \sigma(E)\sqrt{E} \left(b\sqrt{\text{eV}} \right) \sqrt{x} \ e^{-x} \ dx. \quad (19.2.24)$$

Further,

$$\sigma_{kT} = \frac{1.2658 \times 10^{-30} \ \text{cm}^2 \sqrt{\text{erg}}}{4.002\,75 \times 10^{-5} \sqrt{T(\text{keV})} \ \text{erg}} \int_0^\infty \sigma(E)\sqrt{E} \left(b\sqrt{\text{eV}} \right) \sqrt{x} \ e^{-x} \ dx$$

$$= \frac{0.316\,23 \times 10^{-25} \ \text{cm}^2}{\sqrt{T(\text{keV})}} \int_0^\infty \sigma(E)\sqrt{E} \left(b\sqrt{\text{eV}} \right) \sqrt{x} \ e^{-x} \ dx$$

$$= \frac{10^{-26} \ \text{cm}^2}{\sqrt{kT/10 \ \text{keV}}} \int_0^\infty \sigma(E)\sqrt{E} \left(b\sqrt{\text{eV}} \right) \sqrt{x} \ e^{-x} \ dx. \quad (19.2.25)$$

If $\sigma(E)\sqrt{E}$ were constant,

$$\sigma_{kT} = \frac{\sqrt{\pi}}{2} \frac{10^{-26} \ \text{cm}^2}{\sqrt{kT/10 \ \text{keV}}} \left(\sigma(E)\sqrt{E} \right) \left(b\sqrt{\text{eV}} \right) = \frac{8.8623 \ \text{mb}}{\sqrt{kT/10 \ \text{keV}}} \ \sigma(E)\sqrt{E} \left(b\sqrt{\text{eV}} \right). \quad (19.2.26)$$

From Fig. 19.2.1, at energies less than 1 keV, $\sigma(E)\sqrt{E} \sim 0.29 \ b\sqrt{\text{eV}}$. With this choice for $\sigma(E)\sqrt{E}$ and $kT = 10$ keV, eq. (19.2.26) gives $\sigma_{kT} = 2.57$ mb, compared with a cross section for the (n, γ) reaction on ^{14}N at this temperature of $\sim$0.101 mb, as given in Table 19.2.1a. Note that, at $T = 293$ K $= 0.0253$ eV, the choice of 0.29 b $\sqrt{\text{eV}}$ gives $\sigma(n, p) = 1.82$ b, which is essentially equal to the measured (n, p) cross section at room temperature (1.83 b), and that the ratio $2.57/0.101 \sim 26$ is approximately the same as the ratio $1.83/0.078 \sim 23$ of (n, p) and (n, γ) cross sections at room temperature.

The existence of a stationary state in the ^{15}N compound nucleus at an energy which is slightly above the rest energy of a neutron plus the rest energy of the ^{14}N nucleus in the ground state is responsible for the fact that the cross section for the (n, p) reaction on ^{14}N decreases less rapidly with energy than $1/\sqrt{E}$, leading to a value of $\langle \sigma v \rangle$ which increases with neutron kinetic energy. In Fig. 19.2.2, the solid curves are the results of carrying out the integrations in eqs. (19.2.24) and (19.2.25) with $\sigma(E)\sqrt{E}$ from Fig. 19.2.1. The dotted curves are the corresponding results of assuming that $\sigma(E)\sqrt{E}$ remains constant at 0.29 b$\sqrt{\text{eV}}$. At $kT = 10$ keV, σ_{kT} along the solid curve is $\sim$2.8 mb, only marginally (9%) larger than if $\sigma(E)\sqrt{E}$ remained constant. At $kT = 30$ keV, the two estimates differ by 27%.

Ramifications for s-process nucleosynthesis are straightforward. In typical circumstances in radiative zones where ^{13}C acts as a neutron source, ^{14}N is approximately 25 times as abundant as ^{56}Fe. At a typical temperature of $T_8 = 1$, the cross section for neutron capture on ^{56}Fe, as given by Table 19.2.1a, is $\sigma_{kT} \sim 9.6$ mb compared with $\sigma_{kT} \sim 2.8$ mb for neutron capture on ^{14}N. So, over seven times as many of the neutrons released by the

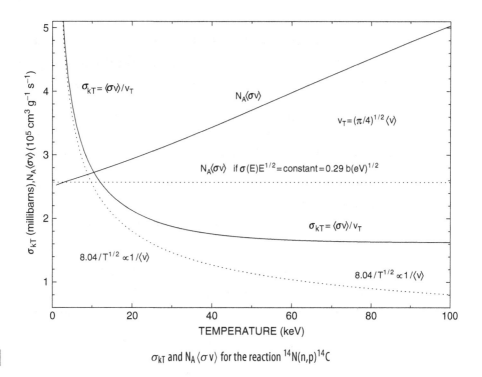

Fig. 19.2.2

σ_{kT} and $N_A \langle \sigma v \rangle$ for the reaction $^{14}N(n,p)^{14}C$

$^{13}C(\alpha, n)^{16}O$ reaction are captured by ^{14}N as are captured by ^{56}Fe. Given the presence of many elements lighter than iron with a total sum of abundances times (n, γ) cross sections comparable with the abundance of Fe times its (n, γ) cross section, less than 7% of the neutrons released by the $^{13}C(\alpha, n)^{16}O$ reaction are captured by ^{56}Fe.

It is amusing that ^{14}C, the main product of this type of s-processing in an intermediate mass TPAGB star, has a half life (5730 yr) which is not only comparable with the time between thermal pulses but is, ironically, large compared with the timescale over which significant neutron-capture nucleosynthesis occurs at any given point in the star.

The following sections describe neutron production and capture during the 14th inter-pulse phase and during the 15th helium shell flash in the $5M_\odot$ TPAGB model discussed in Chapter 18. In order to better assess the effects of neutron capture on light elements in reducing the flow of neutrons to heavy s-process isotopes, the set of isotopes has been aug-mented by the addition of several (mostly neutron-rich) isotopes (^{20}F, ^{23}Ne, ^{24}Na, ^{27}Mg, ^{27}Al, ^{28}Al, ^{29}Si, ^{30}Si, ^{31}Si, ^{31}P, ^{32}P, ^{33}S, and ^{34}S) at number abundances, respectively, of $(0, 0, 0, 0, 2.319 \times 10^{-6}, 0, 1.434 \times 10^{-6}, 9.460 \times 10^{-7}, 0, 1.483 \times 10^{-7}, 0, 9.032 \times 10^{-8},$ and 5.234×10^{-7}). The isotopes ^{21}Na and ^{22}Na have been dropped. In order to moni-tor the extent of neutron capture on heavy s-process elements, the isotope ^{57}Fe has been added at an initial abundance of zero. By insisting that its abundance remain constant, ^{56}Fe acts as a proxy for all heavy s-process elements and ^{57}Fe acts as a neutron counter which monitors the number of neutrons captured by heavy s-process elements. The evolutionary calculations have been repeated with these modifications.

19.3   Formation of a ^{13}C abundance peak and neutron production and neutron capture in the peak

At the end of the dredge-up episode following the 14th helium shell flash in the 5 $M_\odot$ model described in Chapter 18, and before the re-ignition of hydrogen burning following this flash, the abundances of all isotopes vary discontinuously across the formal base of the convective envelope. In Fig. 19.3.1, the number abundances of ^{1}H and ^{12}C in four mass zones about the discontinuity at the end of the dredge-up episode following the shell flash are shown by large open squares and open stars, respectively. The model is the last model of run 296, 269 yr after flash peak, abundances are shown at zone centers, and the locations of three adjacent zone boundaries are indicated by vertical lines. The discontinuity in abundances occurs across the zone boundary at mass $M = 0.883\,269\,331\ M_\odot$. In order to distinguish symbols for isotopes with number abundances which are very small, the vertical (y) axis has been extended to negative values.

In an effort to reduce gravothermal noise generated by spatial rezoning, rezoning was suspended during the dredge-up phase in the region between the center and $M = 0.8844$ $M_\odot$, with the result that, during run 296, the number of zones in the region shown in Fig. 19.3.1 is approximately ten times smaller than had the usual criteria for rezoning been invoked. In runs 297 and after, in order to ensure adequate resolution in the

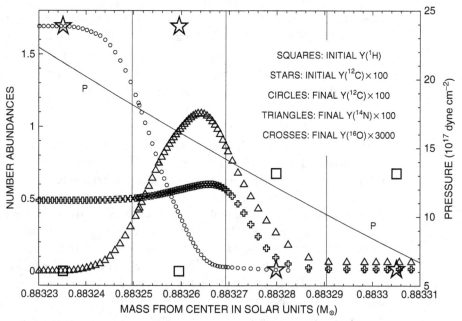

Fig. 19.3.1   Abundances after mixing across the H–C discontinuity and burning hydrogen following the 14th helium shell flash in a 5 $M_\odot$ TPAGB model ($Z = 0.015$, $Y = 0.275$)

hydrogen-burning region as hydrogen re-ignites, rezoning was allowed to occur again beyond $M = M_{freeze} = 0.8832\ M_\odot$. The rezoning algorithm used allows a zone to be split only if the mass at the base of the zone is larger than M_{freeze}. Assignment of composition variables to each half of a split zone makes use of composition variables in immediately adjacent zones, with the result that a degree of numerical mixing takes place.

During rezoning in the present instance, a modest amount of matter which has experienced partial helium burning leaks from the carbon-rich zone into the hydrogen-rich zone and a modest amount of hydrogen leaks into the carbon-rich zone. The outward migration of carbon and inward migration of hydrogen produce a region in which the product of the hydrogen and carbon abundances forms a bell-shaped curve. Temperatures in this region soon become large enough to burn hydrogen completely, with several resulting consequences described in Fig. 19.3.1. In this figure, number abundances at zone centers for ^{12}C, ^{14}N, and ^{16}O in the last model of run 298a, 626 yr after flash peak, are given by open circles, triangles, and crosses, respectively. The bell-shaped distribution for the ^{14}N number abundance reflects the bell-shaped distribution defined by the product of the ^{12}C and ^{1}H abundances after mixing across the H-C interface but before hydrogen burning. At the base of the curve, where the initial hydrogen abundance is much smaller than the initial ^{12}C abundance, the maximum amount of ^{14}N which can be produced is smaller than the intial proton abundance. At the outer edge of the curve, where the initial hydrogen abundance is much larger than the ^{12}C abundance, the maximum amount of ^{14}N which can be made is smaller than the initial ^{12}C abundance.

Comparing the profile of the pressure shown in Fig. 19.3.1 with the ^{14}N number abundance profile in the same figure, it is evident that, over the region in which mixing has taken place across the H-C interface during and after dredge-up, pressure in the region after hydrogen-burning has taken place varies by roughly a factor of 2, or slightly less than one pressure scale height. Thus, although the mixed region has been constructed somewhat artificially, the fact that its extent is of the order of a pressure scale height is consistent with one of the basic assumptions of the mixing length theory of convective diffusion.

In Fig. 19.3.2, the number abundance of ^{13}C produced by complete hydrogen burning in the mixed region is compared with the ^{12}C and ^{14}N number abundances. The total mass of ^{13}C nuclei formed in the abundance peak, as measured by the integral of $Y(^{13}$C$)$ over mass, is

$$\int Y(^{13}\text{C})\ \mathrm{d}M \sim 10^{-8} M_\odot. \tag{19.3.1}$$

The number abundance of ^{13}C is shown in Fig. 19.3.3 by open circles at zone centers over a region which extends from just below the abundance discontinuity at the end of the preceding dredge-up episode to beyond the hydrogen-burning shell, the location of which is manifested by the high density of circles centered at $M \sim 0.883\,337\ M_\odot$. The solid line in Fig. 19.3.3 gives the number abundance of ^{12}C divided by 4. It intersects the curve for the ^{13}C number abundance defined by the open circles at $M \sim 0.883\,375\ M_\odot$. To the left of the intersection, over a region which extends just beyond the base of the ^{13}C abundance

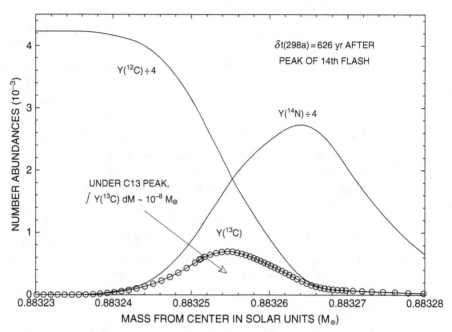

Fig. 19.3.2 Results of hydrogen burning of matter mixed across the H–C discontinuity following the 14th helium shell flash in a 5 $M_\odot$ TPAGB model (Z = 0.015, Y = 0.275)

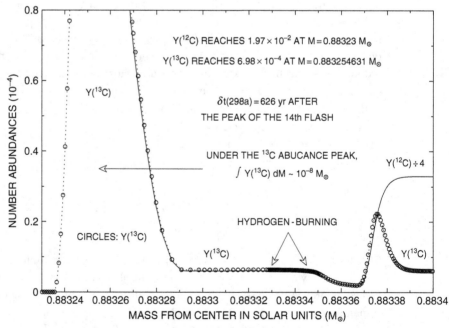

Fig. 19.3.3 Results of hydrogen burning in a 5 $M_\odot$ TPAGB model during the first part of the 14th interpulse phase (Z = 0.015, Y = 0.275)

peak at $M \sim 0.833\,266\ M_\odot$ (see also Fig. 19.3.2), the two isotopes of carbon are in exact equilibrium, signalled by the fact that the ratio of the isotopes is everywhere essentially constant. To the right of the intersection, equilibrium has yet to be achieved.

In Fig. 19.3.4, number abundances of ^{1}H, ^{12}C, ^{13}C, ^{14}N, and ^{16}O are shown for matter near the hydrogen-burning shell in the last model of run 298a, along with the temperature and density. Note that temperatures in the hydrogen-burning shell, at $T_6 \sim 75$, are approximately twice as high as temperatures in the hydrogen-burning shell in the 5 $M_\odot$ model near the tip of the first red giant branch (see Fig. 11.2.52 in Volume 1). The relative minima in the ^{12}C and ^{13}C abundances which occur ahead of the center of the hydrogen-burning shell, at $M \sim 0.883\,365\ M_\odot$, are a consequence of the conversion of ^{12}C into ^{14}N. The occurrence of plateaus in the abundance profiles of the two carbon isotopes within and below the hydrogen-burning shell (at $M \sim 0.883\,345\ M_\odot$) are a consequence of the essentially complete conversion of initial ^{16}O into CN-cycle isotopes.

As temperatures continue to increase in the region defined by the ^{13}C number abundance peak, the rate at which ^{13}C$(\alpha, n)^{16}$O reactions occur increases. Some aspects of the history of the subsequent neutron-production and neutron-capture episode are detailed in Figs. 19.3.5–19.3.12. In four of the figures, profiles of the rate dn/dt at which neutrons are released, the neutron number density n_n, the density ρ and temperature T, and the number abundance of ^{13}C are shown; in three of the figures the number abundance of the heavy element neutron-capture counter ^{57}Fe is shown. In four figures, neutron-absorption rates are compared with the neutron-production rate. In the last figure of the set, the number of neutrons captured by ^{56}Fe, as revealed by the number abundance of the

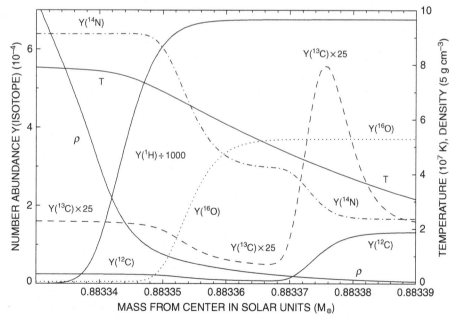

Fig. 19.3.4   Structure and abundance variables near the H-burning shell in a 5 $M_\odot$ TPAGB model 626 yr after the peak of the 14th helium shell flash ($Z = 0.015, Y = 0.275$)

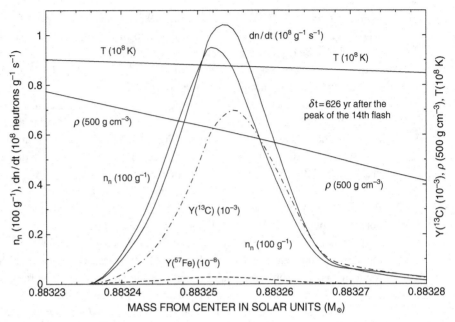

Fig. 19.3.5   Neutron production in the ^{13}C abundance peak in a 5 $M_\odot$ TPAGB model one tenth of the way through the 14th interpulse phase ($Z = 0.015, Y = 0.275$)

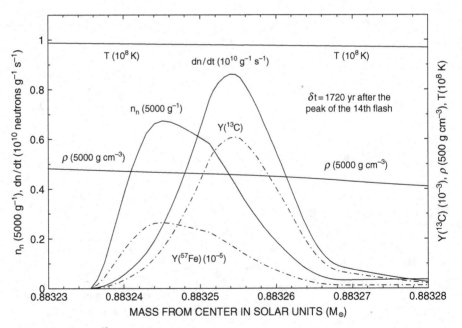

Fig. 19.3.6   Neutron production in the ^{13}C abundance peak in a 5 $M_\odot$ TPAGB model one third of the way through the 14th interpulse phase ($Z = 0.015, Y = 0.275$)

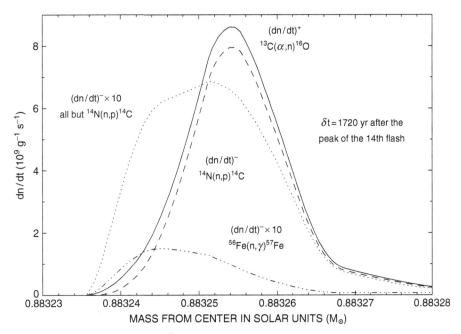

Fig. 19.3.7   Neutron production and absorption in the ^{13}C abundance peak in a 5 $M_\odot$ TPAGB model one third of the way through the 14th interpulse phase ($Z = 0.015, Y = 0.275$)

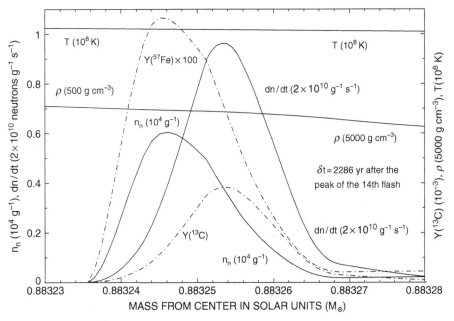

Fig. 19.3.8   Neutron production in the ^{13}C abundance peak in a 5 $M_\odot$ TPAGB model four tenths of the way through the 14th interpulse phase ($Z = 0.015, Y = 0.275$)

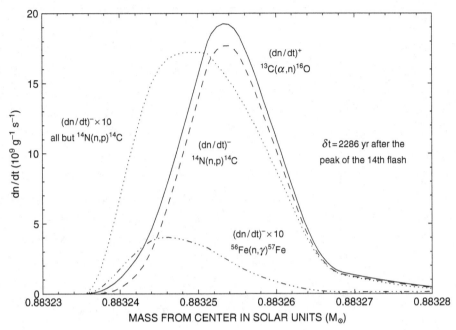

Fig. 19.3.9　Neutron production and absorption in the ^{13}C abundance peak in a 5 $M_\odot$ TPAGB model four tenths of the way through the 14th interpulse phase ($Z = 0.015$, $Y = 0.275$)

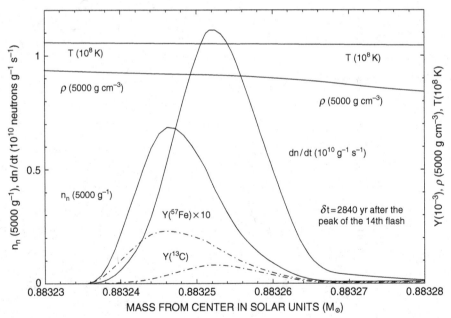

Fig. 19.3.10　Neutron production in the ^{13}C abundance peak in a 5 $M_\odot$ TPAGB model approximately halfway through the 14th interpulse phase ($Z = 0.015$, $Y = 0.275$)

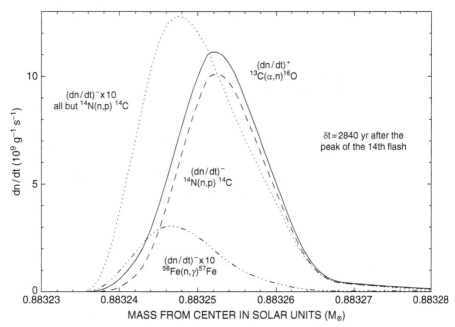

Fig. 19.3.11 Neutron production and absorption in the ^{13}C abundance peak in a 5 M$_\odot$ TPAGB model approximately halfway through the 14th interpulse phase ($Z = 0.015, Y = 0.275$)

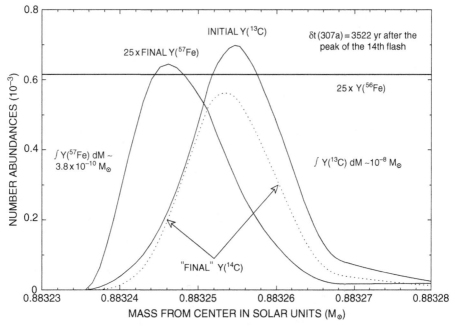

Fig. 19.3.12 Before and after neutron capture in the ^{13}C abundance peak following the 14th helium shell flash in a 5 M$_\odot$ TPAGB model ($Z = 0.015, Y = 0.275$)

neutron-capture counter ^{57}Fe, and the number abundance of ^{14}C, the primary product of the s-process episode, are compared with the initial number abundance of ^{13}C in the ^{13}C abundance peak.

Over the 1094 yr which elapses between the models described in Figs. 19.3.5 and 19.3.6, the mean density in the ^{13}C abundance peak increases by a factor of 10, the temperature increases to almost 10^8 K, and the rate of neutron production increases by a factor of 100, resulting in the capture of a finite number of neutrons by ^{56}Fe, as demonstrated by the profile of $Y(^{57}$Fe$)$ shown in Fig. 19.3.6. In Fig. 19.3.7, the rate of neutron production in the more evolved model is compared with the rates at which ^{14}N, ^{56}Fe, and all isotopes other than ^{14}N absorb neutrons. Comparing the areas under the rate profiles, it is evident that almost 90% of the neutrons released in the ^{13}C abundance peak are absorbed by ^{14}N, that approximately 8% of the neutrons released are absorbed by light elements other than ^{14}N, and that only about one out of every 50 neutrons released is absorbed by the heavy s-process proxy ^{56}Fe.

Figures 19.3.8 and 19.3.9, which describe conditions in the ^{13}C-peak region in a model some 566 yr older than the model described in the previous two figures, tell a similar story. Mean densities have increased by roughly 50%, mean temperatures have increased by approximately 4%, and all neutron production and capture rates have increased by slightly more than a factor of 2, even though the number of ^{13}C nuclei has decreased by approximately 40%. Comparing Figs. 19.3.9 and 19.3.7, it is evident that ratios of the three different neutron-absorption rates have not changed appreciably.

Figures 19.3.10 and 19.3.11 describe conditions in the ^{13}C-peak region in a model some 554 yr older than the model described in the previous two figures, and the story remains similar. Mean densities have increased by roughly 30%, mean temperatures have increased by approximately 3%, and all neutron production and capture rates remain large, though smaller by approximately a factor of 2, as the number of ^{13}C nuclei has decreased by approximately a factor of 4. Comparing Figs. 19.3.11 and 19.3.9, even the ratios of the three different neutron-absorption rates remain nearly the same.

In Fig. 19.3.12, the number of neutrons captured by ^{56}Fe nuclei during the s-process episode is compared with the number of ^{13}C nuclei in the initial ^{13}C-peak region. The model is 682 yr more evolved than the model described in the previous two figures and ^{13}C has been completely destroyed. By inspection, the total number of ^{57}Fe nuclei that have been produced is approximately 25 times smaller than the total number of ^{13}C nuclei in the initial ^{13}C peak and one might infer that one out of every 25 neutrons produced is captured by an ^{56}Fe nucleus. However, the just completed comparison of absorption rates in four different models during the s-process episode shows that ^{56}Fe actually captures approximately one out of every 50 neutrons produced. The resolution of the apparent paradox is that many of the protons produced by the ^{14}N$(n, p)^{14}$C reaction initiate the ^{12}C$(p, \gamma)^{13}$N$(e^+ \nu_e)^{13}$C reactions, adding more ^{13}C to experience the ^{13}C$(\alpha, n)^{16}$O reaction. Thus, the total number of neutrons produced exceeds the number of initial ^{13}C nuclei. A similar conclusion can be drawn from a comparison of the "final" number abundance profile for ^{14}C with the initial number abundance profile for ^{13}C. There is a region where the two profiles almost touch and, since only about 90% of the neutrons released are used up in making ^{14}C and since ^{14}C decays during the roughly thousand year long s-process episode, it is clear that the number of ^{14}C nuclei that are made exceeds the number of initial ^{13}C nuclei.

It is possible to understand analytically how the factor of 2 given by the evolutionary calculations comes about. Let x, z, n, and p stand, respectively, for the number abundances of ^{13}C, ^{57}Fe, neutrons, and protons, and assume that, in first approximation, the abundances of ^{12}C and of ^{14}N remain constant during the s-process episode. Assuming still further that temperatures and densities remain constant, one may write

$$\frac{dx}{dt} = -\frac{x}{\tau} + b_{12}\, p, \tag{19.3.2}$$

$$\frac{dz}{dt} = n\, a_{56}, \tag{19.3.3}$$

$$\frac{dp}{dt} = n\, a_{14} - b\, p, \tag{19.3.4}$$

and

$$\frac{dn}{dt} = \frac{x}{\tau} - n\, a, \tag{19.3.5}$$

where all of the coefficients of the variables are constants. Further,

$$a = a_{14} + a_l + a_{56}, \tag{19.3.6}$$

and

$$b = b_{12} + b_l, \tag{19.3.7}$$

where a_l and b_l are average coefficients for reactions on all light elements in the adopted isotope set, the sum of the abundances being constant.

Assuming that neutron and proton abundances are at their local equilibrium values, one has that

$$n = \frac{x}{\tau\, a} \tag{19.3.8}$$

and

$$p = \frac{a_{14}}{b}\, n = \frac{a_{14}}{b}\, \frac{x}{\tau\, a}. \tag{19.3.9}$$

Combining eq. (19.3.9) with eq. (19.3.2) gives

$$\frac{dx}{dt} = -\frac{x}{\tau}\left(1 - \frac{b_{12}}{b}\, \frac{a_{14}}{a}\right), \tag{19.3.10}$$

the solution of which is

$$x = x_0\, exp\left(-\frac{t}{\tau'}\right), \tag{19.3.11}$$

where x_0 is the initial abundance of ^{13}C and

$$\tau' = \tau\left(1 - \frac{b_{12}}{b}\, \frac{a_{14}}{a}\right)^{-1}. \tag{19.3.12}$$

The final number abundance of ^{57}Fe is given by

$$z_{\text{final}} = a_{56} \int_0^\infty n \, dt = \frac{a_{56}}{\tau a} \int_0^\infty x \, dt = \frac{a_{56}}{\tau a} x_0 \int_0^\infty e^{-t/\tau'} dt = \frac{a_{56}}{a} \frac{\tau'}{\tau} x_0 \quad (19.3.13)$$

$$= \frac{a_{56}}{a} \left(1 - \frac{b_{12}}{b} \frac{a_{14}}{a} \right)^{-1} x_0. \quad (19.3.14)$$

Examination of the absorption profiles in Figs. 19.3.7, 19.3.9, and 19.3.11 suggests mean ratios of

$$\frac{a_{56}}{a} \sim 0.02 \text{ and } \frac{a_{14}}{a} \sim 0.9. \quad (19.3.15)$$

The choice

$$\frac{b_{12}}{b} \sim 0.5 \quad (19.3.16)$$

would give

$$z_{\text{final}} \sim 0.036 \, x_0, \quad (19.3.17)$$

which is consistent with the overall result. However, the fact that the distributions of the initial ^{13}C and the final ^{57}Fe number abundances are displaced relative to one another demonstrate that this explanation of the overall result is spurious. The composition and neutron production and neutron absorption characteristics vary so greatly over the s-processing region that the analytic expressions can only be used to elucidate local results.

Local results differ dramatically across the s-processing region. For example, at the peak in the ^{57}Fe profile in Fig. 9.3.12, $x_{\text{final}}/x_0 \sim 0.12$, whereas, at the peak in the ^{13}C profile, $x_{\text{final}}/x_0 \sim 0.019$. To explore these differences further it is necessary to examine quantitatively the ratio b_{12}/b, which is the fraction of all protons released that are captured by ^{12}C. For the model described in Figs. 19.3.8 and 19.3.9, the ratio can be obtained by examination of the curves in Fig. 19.3.13 which gives the rates of proton captures for various processes as functions of mass. The rate at which protons are produced by the ^{14}N$(n, p)^{14}$C reaction is given by the curve labeled R(^{14}N $+ p$) and, for comparison, the rate at which ^{12}C captures neutrons is given by the curve R(^{12}C $+ n$).

At the position corresponding to the peak in the ^{57}Fe profile in Fig. 9.3.12, Fig. 19.3.13 gives $b_{12}/b \sim 0.79$. For the model described in Fig. 19.3.9, ratios of absorption rates at the same position are $a_{14}/a \sim 0.78$ and $a_{56}/a \sim 0.053$. Assuming these ratios to be typical for the entire s-process episode, eq. (19.3.14) gives $z_{\text{final}}/x_0 \sim 0.14$, reasonably close to the value of 0.12 found from Fig. 19.3.12. At the position corresponding to the peak in the ^{13}C profile in Fig. 9.3.12, Fig. 19.3.13 gives $b_{12}/b \sim 0.61$. Ratios of absorption rates at the same position given by Fig. 19.3.9 are $a_{14}/a \sim 0.925$ and $a_{56}/a \sim 0.011$. Again, assuming these values to be typical for the entire s-process episode, eq. 19.3.14 gives $z_{\text{final}}/x_0 \sim 0.025$, compared with the value of 0.019 given directly by Fig. 19.3.12.

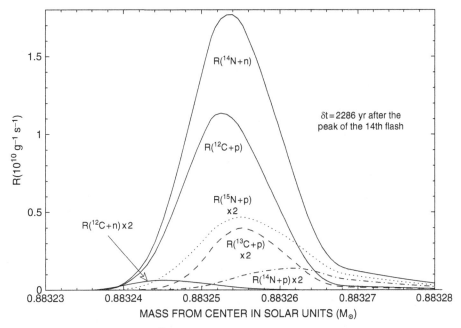

Fig. 19.3.13 Neutron and proton capture rates in the ^{13}C abundance peak in a 5 M$_\odot$ TPAGB model four tenths of the way through the 14th interpulse phase ($Z = 0.015, Y = 0.275$)

19.4 Neutron production and capture during the interpulse phase in matter processed by hydrogen burning

Carbon isotope abundances in three models during the hydrogen-burning phase between the 14th and 15th helium shell flashes are shown in Fig. 19.4.1. The number after the colon specifying an isotopic number abundance Y(isotope) describes the calculational run producing the model to which the abundance applies and δt (run) gives the time which has elapsed since the peak of the 14th flash. As demonstrated in the figure, the region over which ^{13}C has been depleted increases with each successive model, indicating that temperatures at a sufficiently large distance below the hydrogen-burning shell become high enough for the ^{13}C$(\alpha, n)^{16}$O reaction to operate.

In the youngest of the three models, at $\delta t = 1720$ yr, the maximum abundance of ^{13}C in the abundance peak at the left hand edge of the figure is only $\sim 15\%$ smaller than in the model 626 yr past the peak of the 14th flash (see Fig. 19.3.5), so the neutron-production process has barely begun. In the model at $\delta t = 3522$ yr, the ^{13}C abundance peak has disappeared and most of the ^{13}C below the hydrogen-burning shell in the model at $\delta t = 1720$ yr has been destroyed. Finally, at $\delta t = 5603$ yr, most of the ^{13}C below the hydrogen-burning shell in the model at $\delta t = 3522$ yr has been destroyed. As the 15th helium shell flash begins,

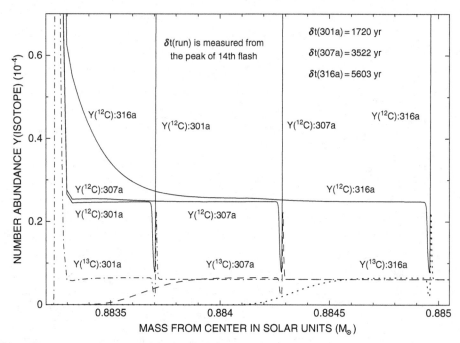

Fig. 19.4.1 ^{12}C and ^{13}C abundances during the hydrogen-burning phase following the 14th helium shell flash in a 5 M$_\odot$ TPAGB model ($Z = 0.015, Y = 0.275$)

the total mass of neutrons which has been released outside of the erstwhile ^{13}C abundance peak during the 14th interpulse phase is

$$-\int_{0.8833M_\odot}^{0.88445M_\odot} \Delta Y(^{13}\text{C}) \, \mathrm{d}M \sim 8 \times 10^{-9} M_\odot, \tag{19.4.1}$$

which is almost the same as the integral with respect to mass of the ^{13}C number abundance in the ^{13}C peak produced during run 279a (see eq. (19.3.1)). Thus, the total mass of neutrons produced below the hydrogen-burning shell by the ^{13}C$(\alpha, n)^{16}$O reaction is of the order of $-\int \Delta Y_{13} \, \mathrm{d}M \sim 18 \times 10^{-9} \ M_\odot$, to which the ^{13}C peak and the region outside of the peak make comparable contributions. The mass of neutrons that can be released by the ^{13}C that remains below the hydrogen-burning shell as the 15th helium shell flash gets underway and is incorporated by the helium-flash driven convective shell during the 15th helium shell flash is of the order of $\int Y_{13} \, \mathrm{d}M \sim 4 \times 10^{-9} \ M_\odot$. As the convective shell formed during the 15th helium shell flash approaches the hydrogen–helium interface, matter which has been exposed to neutrons of mass of the order of $18 \times 10^{-9} \ M_\odot$ is injected into the convective shell and an amount of ^{13}C which will release a mass of neutrons of the order of $4 \times 10^{-9} \ M_\odot$ is also injected into the convective shell, where the neutrons are finally released.

That the triple-alpha reaction is taking place behind the advancing hydrogen-burning shell is evidenced by the ^{12}C abundance profiles in Fig. 19.4.1. In the region below the hydrogen-burning shell in the model at $\delta t = 1720$ yr, the ^{12}C abundance in the model at

$\delta t = 3522$ is slightly larger than in the model at $\delta t = 1720$ yr and the ^{12}C abundance in the model at $\delta t = 5603$ yr is much larger than in either of the two younger models, reflecting the fact that the 15th helium shell flash has already begun in the oldest of the three models. It is worth remarking at this point that the meanings of the words young and old are context dependent. In one context, youngest may refer to the most recent model produced, whereas in another, it may refer to the earliest model produced. The reader must exercise judgement in determining which meaning is being used.

In Figs. 19.4.2 and 19.4.3, number abundance profiles for ^{16}O, ^{14}N, ^{14}C, ^{13}C, ^{12}C, and ^{1}H, and density and temperature profiles are shown for models at the ends of runs 307a ($\delta t = 3522$ yr) and 316a ($\delta t = 5603$ yr), respectively. In both cases, the ratios of abundances of CNO isotopes over the entire region between the initial ^{13}C peak and the hydrogen-burning shell are more or less constant. At $M \sim 0.883\,66\ M_\odot$ in the less evolved (older or younger, according to taste) model, where the ^{13}C abundance has been reduced by helium burning to approximately half of the ^{13}C abundance left behind by the advancing hydrogen-burning shell, $T \sim 10^8$ K and $\rho \sim 3300$ g cm^{-3}. At the corresponding position ($M \sim 0.884\,35\ M_\odot$) in the more evolved (younger or older, according to taste), $T \sim 10^8$ K and $\rho \sim 3000$ g cm^{-3}, almost the same as in the less evolved model. These similarities demonstrate that the neutron production and capture nucleosynthesis taking place in the region between the initial ^{13}C peak and the hydrogen-burning shell is more or less independent of where in this region the nucleosynthesis is occurring.

In both cases, the relative maximum in the ^{14}C abundance profile is approximately 60% of the ^{13}C abundance deposited by the H-burning shell, reinforcing the inference of

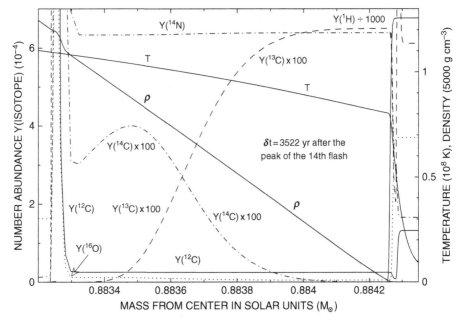

Fig. 19.4.2 Structure and abundance variables below the H-burning shell in a 5 $M_\odot$ TPAGB model six tenths of the way through the 14th interpulse phase ($Z = 0.015$, $Y = 0.275$)

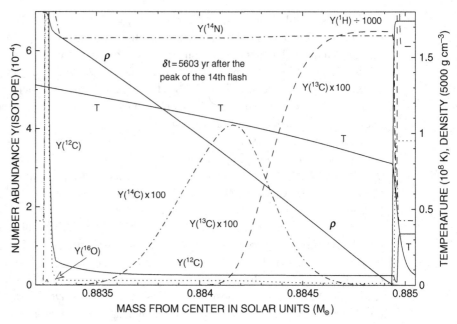

Fig. 19.4.3   Structure and abundance variables below the H-burning shell in a 5 $M_\odot$ TPAGB model beginning the 15th thermal pulse ($Z = 0.015, Y = 0.275$)

position-independent neutron-capture nucleosynthesis and, at the same time, demonstrating that most neutrons disappear by way of the $^{14}N(n, p)^{14}C$ reaction. Since ^{14}C is beta-unstable, the fact that the ^{14}C abundance reaches a maximum and then decreases inward is not surprising. However, the rate of decrease is much steeper than if beta decay were solely responsible for the decline. The ^{13}C and ^{14}C profiles move outward in mass at approximately 3.36×10^{-7} $M_\odot$ yr^{-1}. The distance in mass between the relative maximum and the relative minimum along the ^{14}C profile in the less evolved model (Fig. 19.4.2) is $\sim 2 \times 10^{-4}$ $M_\odot$, so the 25% decrease in the ^{14}C abundance between the two extrema occurs in $\sim$595 yr, short compared with the 8276 yr lifetime of ^{14}C. If beta decay were the only way in which ^{14}C were destroyed, one would expect $Y_{rel.Min} = Y_{rel.Max}$ $\exp(-595/8267) = 0.93$ $Y_{rel.Max}$. Since the decrease in the ^{14}C number abundance is actually over three times greater than this, another process is at work. The process is the reaction $^{14}C(\alpha, \gamma)^{18}O$. The destruction of ^{14}C primarily by alpha capture is even more apparent from a comparison of ^{14}C abundance profiles in the two models; at M = 0.8835 $M_\odot$, where its abundance is at a relative maximum in the less evolved model, ^{14}C is almost completely gone in the more evolved model. Had beta decay been the only destruction mechanism, the abundance of ^{14}C would have decreased by only 22% over the 2081 yr separating the two models.

The history of ^{14}C demonstrates dramatically that, in a region of neutron production initiated by (α, n) reactions, there is an intricate interplay between neutron-capture and alpha-capture reactions. In the present instance, the production of ^{18}O by (α, γ) reactions on ^{14}C is made possible by the prior occurrence of (n, p) reactions on ^{14}N, which reactions are in turn made possible by the occurrence of (α, n) reactions on ^{13}C. In addition,

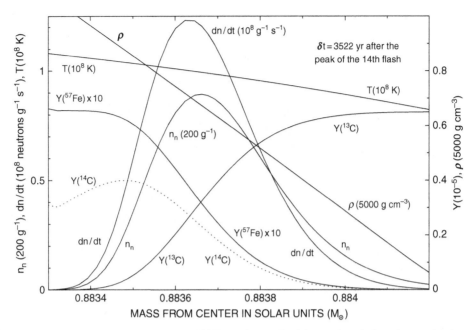

Fig. 19.4.4 Neutron production and s-processing in a 5 $M_\odot$ TPAGB model six tenths of the way through the 14th interpulse phase ($Z = 0.015, Y = 0.275$)

both (α, γ) reactions on ^{14}N and (n, γ) reactions on ^{17}O make further contributions to the production of ^{18}O. The actual change in the abundance of ^{18}O is actually quite minor, but this does not diminish the pleasure one can take in recognizing the intricacies involved.

In Fig. 19.4.4 the rate at which neutrons are produced and the neutron number density in the last model of run 307a are shown as functions of mass by curves labeled, respectively, dn/dt and n_n. The same quantities are shown in Fig. 19.4.5 for the last model of run 316a. The neutron production rate is in units of g^{-1} s^{-1} and the neutron number density is in units of g^{-1}. Shown also in the two figures are temperature, density, and number-abundance profiles for ^{57}Fe, ^{14}C, and ^{13}C.

The neutron number density is the consequence of a balance between the neutron-production rate, the neutron-decay rate, and the sum of all neutron-capture rates. In Fig. 19.4.4, the maximum neutron-production rate is

$$\left(\frac{dn}{dt}\right)^{max} \sim 1.23 \times 10^8 \text{ g}^{-1} \text{ s}^{-1} \tag{19.4.2}$$

and occurs at $\sim 0.883\,62 \ M_\odot$, where the abundance of ^{13}C is approximately four tenths of its maximum abundance. The neutron number density has essentially the same form as the neutron-production rate and its maximum, at

$$n_n^{max} \sim 180 \text{ g}^{-1}, \tag{19.4.3}$$

occurs at nearly the same place as the maximum in the neutron-production rate. In units of cm^{-3}, the maximum neutron number density is

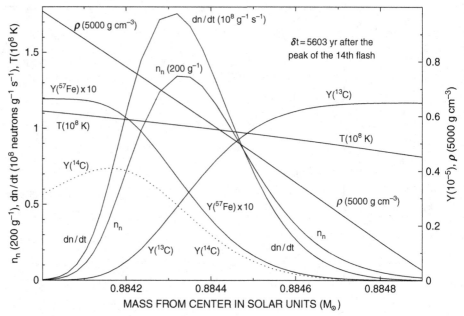

Fig. 19.4.5 Neutron production by the $^{13}C(\alpha,n)^{16}O$ reaction and s-processing in a 5 $M_\odot$ TPAGB model ending the 14th interpulse phase ($Z = 0.015, Y = 0.275$)

$$n'_n = (\rho \, n_n)^{max} \sim 3500 \text{ g cm}^{-3} \, n_n^{max} = 6.3 \times 10^5 \text{ cm}^{-3}. \quad (19.4.4)$$

Apart from small differences in scale and displacement to larger masses, the situation is almost the same in the more evolved model. In Fig. 19.4.5, the maximum neutron-production rate is

$$\left(\frac{dn}{dt}\right)^{max} \sim 1.75 \times 10^8 \text{ g}^{-1} \text{ s}^{-1} \quad (19.4.5)$$

and occurs at $\sim 0.884\,32 \, M_\odot$, where the abundance of ^{13}C is again approximately 40% of its maximum abundance. The neutron number density again has essentially the same form as the neutron-production rate and its maximum, at

$$n_n^{max} \sim 270 \text{ g}^{-1}, \quad (19.4.6)$$

occurs at nearly the same place as the maximum in the neutron-production rate. In units of cm^{-3}, the maximum neutron number density is

$$n'_n = (\rho \, n_n)^{max} \sim 3250 \text{ g cm}^{-3} \, n_n^{max} = 8.8 \times 10^5 \text{ cm}^{-3}. \quad (19.4.7)$$

Comparing the ^{57}Fe and ^{13}C number-abundance profiles in Figs. 19.4.4 and 19.4.5, it is evident that, once ^{13}C has been completely destroyed, approximately one neutron has been captured by an ^{56}Fe nucleus (read: by heavy s-process nuclei) for every ten initial ^{13}C nuclei. By contrast, as demonstrated in Fig. 19.3.12, in the region defined by the ^{13}C abundance peak formed by mixing at the C-H interface followed by complete hydrogen burning, approximately one neutron is captured by an ^{56}Fe nucleus (by heavy s-process nuclei) for every 25 initial ^{13}C nuclei. Thus, the efficiency with which heavy s-process elements are formed is 2.5 times greater in the region between the initial ^{13}C abundance peak and the H-burning shell than it is in the ^{13}C abundance peak.

The basic reason for the difference in efficiency in the two environments is the ratio of ^{14}N to ^{56}Fe prior to the onset of neutron production. In the ^{13}C abundance-peak region, at the location where the final abundance of ^{57}Fe is at a maximum, the initial ratio is

$$\frac{Y(^{14}\text{N})}{Y(^{56}\text{Fe})} = \frac{1.4 \times 10^{-3}}{2.46 \times 10^{-5}} = 56.9 \tag{19.4.8}$$

and, at the location where the initial abundance of ^{13}C is at a maximum, the initial ratio is

$$\frac{Y(^{14}\text{N})}{Y(^{56}\text{Fe})} = \frac{7.6 \times 10^{-3}}{2.46 \times 10^{-5}} = 309. \tag{19.4.9}$$

By contrast, in the region between the location of the initial peak in the ^{13}C abundance distribution and the location of the hydrogen-burning shell, the initial number abundance ratio is constant at

$$\frac{Y(^{14}\text{N})}{Y(^{56}\text{Fe})} = \frac{6.3 \times 10^{-4}}{2.46 \times 10^{-5}} = 25.6. \tag{19.4.10}$$

Another reason why the outcome of neutron-capture nucleosynthesis in the region defined by the initial ^{13}C abundance peak differs from the outcome in the region between the initial ^{13}C peak and the base of the hydrogen-burning shell is that the ratio of the ^{12}C abundance to the ^{14}N abundance differs dramatically between the two regions. Over most of the first region, the abundance of ^{12}C either exceeds or is comparable with the abundance of ^{14}N, ensuring that the majority of protons produced by the ^{14}N$(n, p)^{14}$C reaction participate in the ^{12}C$(p, \gamma)^{13}$N$(e^+ \nu_e)^{13}$C$(\alpha, n)^{16}$O reaction chain, producing additional neutrons. In the second region, the ^{14}N abundance exceeds the ^{12}C abundance by a nearly constant factor of 25, with the consequence that the total number of neutrons produced is not significantly larger than the initial number of ^{13}C nuclei.

This inference is quantified in Fig. 19.4.6, where reaction rates for proton-capture reactions on the four beta-stable CN-cycle isotopes are compared with the rates of several neutron-capture processes in the region between the initial ^{13}C abundance peak and the base of the hydrogen-burning shell. Comparing the sum of the proton-capture rates with the proton-production rate as given approximately by R(^{14}N+n), it is evident that over 80% of the protons produced are captured by CN-cycle isotopes. In the outer part of the region defined by the rate profiles, where neutron production is just beginning, the four beta-stable CN isotopes are at abundances close to the equilibrium abundances prevailing at the end of pure hydrogen burning, which means that newly produced protons are captured at nearly the same rate by each of the four beta-stable isotopes and that, therefore, only about a quarter of the protons produced by the ^{14}N$(n, p)^{14}$C reaction participate in the ^{12}C$(p, \gamma)^{13}$N$(e^+ \nu_e)^{13}$C$(\alpha, n)^{16}$O reaction chain. Approaching the base of the neutron-production region, R(^{13}C+p) becomes small relative to the other proton-capture rates as the ^{13}C abundance tends to zero, and R(^{15}N+p) increases relative to the other two proton-capture rates because (n, γ) reactions on ^{14}N increase the abundance of ^{15}N. The net result

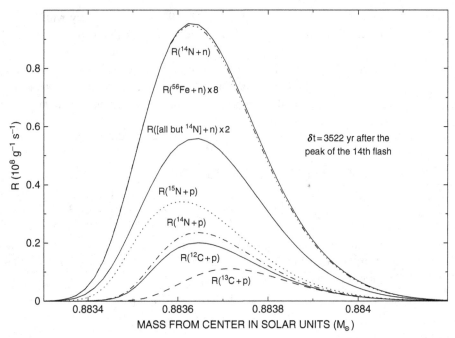

Fig. 19.4.6 Neutron and proton capture rates below the H-burning shell in a 5 $M_\odot$ TPAGB model six tenths of the way through the 14th interpulse phase ($Z = 0.015, Y = 0.275$)

is a decrease in the rate at which ^{12}C captures protons relative to the total rate at which CN-cycle isotopes capture protons.

Another perspective on nucleosynthesis in the region between the helium-burning shell and the hydrogen-burning shell during the interpulse phase is given in Figs. 19.4.7 and 19.4.8, where number abundances of a selection of isotopes in the last models of runs 307a and 316a are plotted logarithmically. It is particularly interesting that the number abundances of protons and neutrons remain relatively constant over a considerable distance below the hydrogen-burning shell, with the ratio of the proton abundance to the neutron abundance being of the order of $10^{5\pm1}$, and that, despite their extremely small abundances, these particles play central roles in ongoing nucleosynthesis.

It is curious that processes in and in the very near vicinity of the hydrogen-burning shell dramatically deplete relatively high Z isotopes such as ^{18}O, ^{22}Ne, and ^{23}Na and that these isotopes are then partially reconstituted in regions far below the hydrogen-burning shell. As seen in Figs. 19.4.7 and 19.4.8, the abundances of these isotopes in the convective envelope are orders of magnitude larger their abundances just below the hydrogen-burning shell but, after reaching relative minima, their abundances increase inward by several orders of magnitude. Figure 19.4.9, which is an enlargement of the crowded region near the hydrogen-burning shell in Fig. 19.4.7, allows one to see some of the details of the dramatic changes from the base of the convective shell to the location of minimum abundances. At mass $M = 0.884\,275\ M_\odot$, where density and temperature are, respectively, $\rho = 2.98$ g cm^{-3} and $T_6 = 62.2$, the $^{22}Ne(p, \gamma)^{23}Na$ reaction occurs at the rate 3.8×10^8 g^{-1} s^{-1} and the

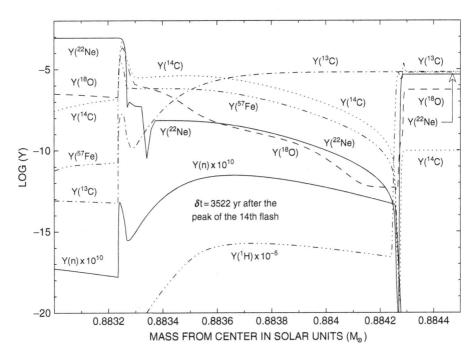

Fig. 19.4.7   Abundances between C–rich and H–rich regions in a 5 $M_\odot$ TPAGB model six tenths of the way through the 14th interpulse phase ($Z = 0.015$, $Y = 0.275$)

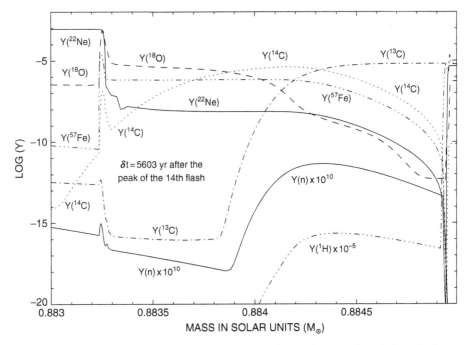

Fig. 19.4.8   Abundances between C–rich and H–rich regions in a 5 $M_\odot$ model TPAGB star beginning the 15th thermal pulse ($Z = 0.015$, $Y = 0.275$)

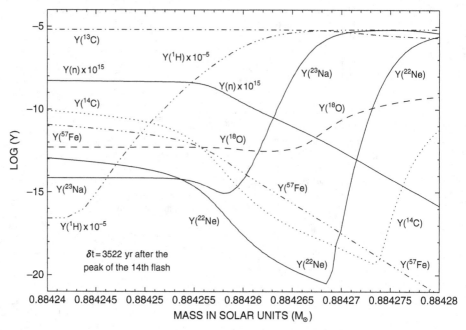

Fig. 19.4.9 Abundances in the vicinity of the H-burning shell in a 5 $M_\odot$ TPAGB model six tenths of the way through the 14th interpulse phase ($Z = 0.015, Y = 0.275$)

reactions $^{23}\mathrm{Na}(p, \gamma)^{24}\mathrm{Mg}$ and $^{23}\mathrm{Na}(p, \alpha)^{20}\mathrm{Ne}$ occur at about the same rate of 2.75×10^8 $\mathrm{g}^{-1}\,\mathrm{s}^{-1}$. At mass $M = 0.884\,265\ M_\odot$, where density and temperature are, respectively, $\rho = 10$ g cm^{-3} and $T_6 = 77.65$, the $^{23}\mathrm{Na}(p, \gamma)^{24}\mathrm{Mg}$ and $^{23}\mathrm{Na}(p, \alpha)^{20}\mathrm{Ne}$ reactions occur at the same rate of 5.4×10^7 g^{-1} s^{-1}. The partial reconstitution of the two isotopes through alpha-particle capture and neutron-capture reactions below hydrogen-rich region is evident in Fig. 19.4.7.

19.5   Neutron-capture nucleosynthesis in the convective shell during the fifteenth helium shell flash

Profiles of luminosity, density, temperature, and number abundances of $^4\mathrm{He}$ and $^{12}\mathrm{C}$ are shown in Figs. 19.5.1–19.5.4 for models at four different stages of the 15th helium shell flash over a region which encompasses where hydrogen and helium burning are most active. The first figure describes conditions in the last model of run 316a, which is also described in Figs. 19.4.3, 19.4.5, and 19.4.8. The shell flash is just beginning and a convective shell driven by helium burning has not yet formed.

The model described in Fig. 19.5.2 is the last model of run 317a. Hydrogen burning is still by far the major source of surface luminosity ($L_\mathrm{H} = 2.1377 \times 10^4\ L_\odot$), but the

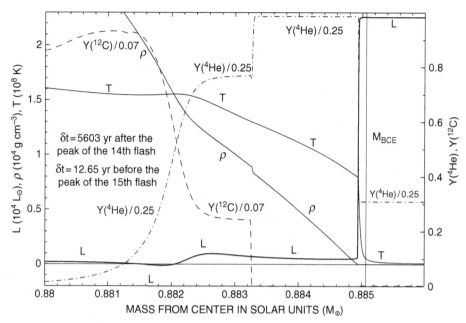

Fig. 19.5.1 Structure variables and number abundances in a 5 $M_\odot$ TPAGB model beginning the 15th thermal pulse ($Z = 0.015$, $Y = 0.275$)

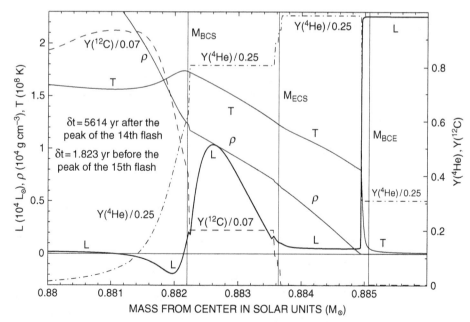

Fig. 19.5.2 Structure variables and number abundances in a 5 $M_\odot$ TPAGB model as a convective shell grows during the 15th helium shell flash ($Z = 0.015$, $Y = 0.275$)

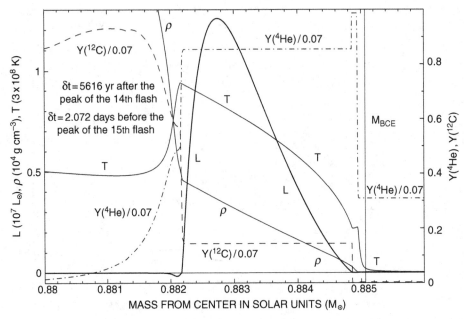

Fig. 19.5.3   Structure variables and number abundances in a 5 $M_\odot$ TPAGB model just before maximum L_{He} during the 15th helium shell flash ($Z = 0.015$, $Y = 0.275$)

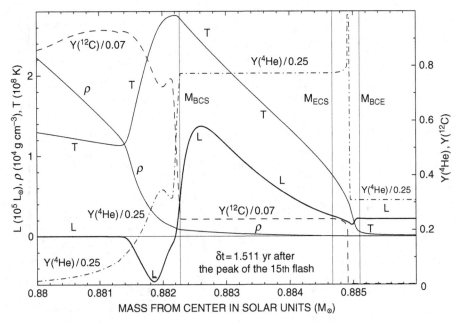

Fig. 19.5.4   Structure variables and number abundances in a 5 $M_\odot$ TPAGB model as the convective shell recedes during the 15th helium shell flash ($Z = 0.015$, $Y = 0.275$)

rate of helium burning has become large enough ($L_{He} = 9390\ L_\odot$) to support a convective shell which extends over approximately half of the distance in mass from the base of the helium-burning region to the hydrogen-burning shell. Most of the energy produced by helium burning is absorbed in heating and expanding matter in and for a short distance above the convective shell ($L_{grav} = -8158\ L_\odot$), and the luminosity above the shell drops to a minimum of $\sim$500 $L_\odot$.

The outer edge of the convective shell extends formally almost 0.0004 $M_\odot$ beyond the location of the outer edge of the ^{12}C profile in Fig. 19.5.1, causing carbon to mix outward and helium to mix inward. The base and outer edge of the convective shell driven by helium burning are indicated in Fig. 19.5.2 by vertical lines labeled M_{BCS} and M_{ECS}, respectively. The circumstance that the abundances of hydrogen and helium are not constant throughout the region between the two shell boundaries demonstrates that the boundaries are growing in opposite directions in response to the general increase in amplitude of the luminosity profile generated by helium burning. That is, the region over which each of the abundances is constant defines the region encompassed by the convective zone at the beginning of the time step which produces the slightly expanded region prevailing at the end of the time step. Mixing over the expanded region occurs at the beginning of the next time step.

The third model, characteristics of which are described in Fig. 19.5.3, is the last model produced during run 320a. The global helium-burning luminosity, at $L_{He} = 1.877 \times 10^7$ $L_\odot$, is near the peak value achieved during the 15th flash and the mass of the convective shell driven by helium burning is approaching its maximum value of $\Delta M_{CS}^{max} \sim 0.0027$ $M_\odot$. The global gravothermal energy-generation rate is $L_{grav} = -1.879 \times 10^7\ L_\odot$. The fact that the absolute value of L_{grav} is almost identical to L_{He}, plus the fact that the local luminosity is slightly negative near the outer edge of the convective shell, demonstrate that all of the helium-burning energy generated is converted locally into heat and gravitational potential energy. Expansion and cooling extend into the hydrogen-burning region and the global hydrogen-burning luminosity has decreased to $L_H = 1.262 \times 10^3\ L_\odot$.

From the temperature and density profiles in Figs. 19.5.2 and 19.5.3, one may deduce that, at the base of the convective shell, the temperature has increased from 164×10^6 K to 296×10^6 K, while the density has decreased from 12 400 g cm^{-3} to 3419 g cm^{-3}. Since the rates of alpha-burning processes are much more sensitive to temperature than to density, these changes mean that the rates of all processes involving alpha capture have accelerating during evolution between the two models represented in the figures. This helps one to understand why, for example, the mass of ^{12}C in the convective shell has increased even though the mass of the shell has almost doubled during evolution between the two models.

Figure 19.5.4 depicts characteristics of the last model of run 323a. The model is near the end of the 15th helium shell flash. The global helium-burning luminosity has decreased to $L_{He} = 2.794 \times 10^5\ L_\odot$ and the hydrogen-burning luminosity has decreased to $L_H = 8.922$ $L_\odot$. The convective shell has begun to shrink and the ^{12}C abundance in the shell has nearly reached its maximum value which is essentially the same as the ^{12}C abundance just below the outer edge of the carbon-abundance profile at the beginning of the flash, as given in Fig. 19.5.1. Thus, during the 15th helium shell flash, fresh carbon is produced and carried outward, resurrecting a ^{12}C abundance profile which is almost identical with that produced during the 14th helium shell flash. Most importantly, the profile almost reaches the hydrogen–helium interface, setting the stage for dredge-up to the surface and mixing between H-rich and C-rich matter necessary for the formation of another ^{13}C abundance peak.

Number-abundance profiles of various isotopes relevant to the production and absorption of neutrons in the convective shell during the 15th helium shell flash are shown in Figs. 19.5.5–19.5.7 for the same models described in Figs. 19.5.2–19.5.4. Included are three isotopes (^{14}N, ^{18}O, and ^{22}Ne) which are involved in the production of neutrons by the ^{22}Ne$(\alpha, n)^{25}$Mg reaction. Also included are the neutron counter ^{57}Fe and the prospective neutron source ^{13}C, both of which are ingested by the convective shell as it grows in mass.

In Fig. 19.5.5, the slight decreases in the abundances of ^{22}Ne and ^{25}Mg in the convective shell relative to their values just below the base of the shell are due to the ingestion by the shell of matter from above the outer edge of the shell where ^{22}Ne is essentially non-existent and the number abundance of ^{25}Mg is approximately five times smaller than in the convective shell. The absence of ^{14}N below the base of the convective shell is due to the conversion of ^{14}N into ^{22}Ne by the ^{14}N$(\alpha, \gamma)^{18}$F$(\beta^{+}\nu_e)^{18}$O and ^{18}O$(\alpha, \gamma)^{22}$Ne reactions during the 14th helium shell flash, whereas the abundance of ^{14}N in the shell is a consequence of the ingestion of ^{14}N-rich matter from above the shell. Temperatures and

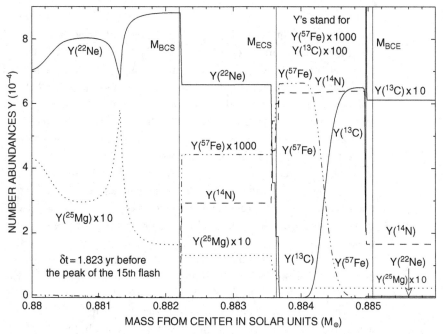

Fig. 19.5.5 Number abundances in a 5 $M_{\odot}$ TPAGB model as a convective shell grows during the 15th helium shell flash ($Z = 0.015, Y = 0.275$)

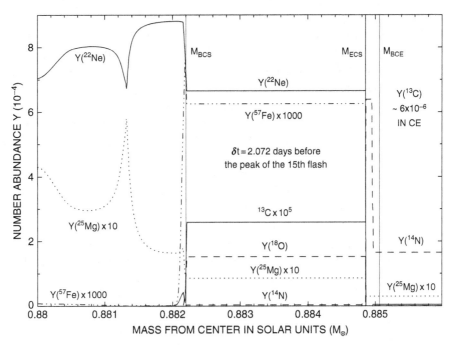

Fig. 19.5.6   Number abundances in nuclear active regions of a 5 $M_\odot$ TPAGB model near the peak of the 15th helium shell flash ($Z = 0.015, Y = 0.275$)

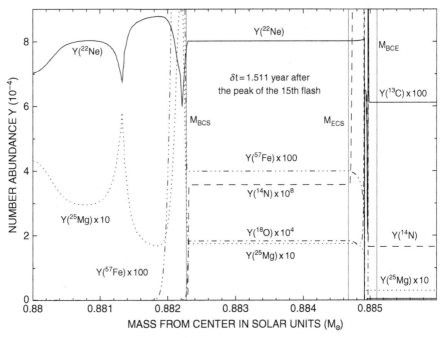

Fig. 19.5.7   Abundances in and in the vicinity of the shrinking convective shell of a 5 $M_\odot$ TPAGB model near the end of the 15th helium shell flash ($Z = 0.015, Y = 0.275$)

densities in the shell ($T_6 \sim 164$ and $\rho \sim 1.24 \times 10^4$ g cm^{-3} at the base of the shell) are not yet large enough to convert much ^{18}O into ^{22}Ne. The fact that the number abundance of ^{25}Mg over most of the convective shell is larger than its number abundance above the shell is a consequence of the activation of the ^{22}Ne$(\alpha, n)^{25}$Mg neutron source during the preceding helium shell flash.

Comparison of number abundances in Fig. 19.5.6 for the model near the peak of the 15th helium shell flash with abundances in Fig. 19.5.5 for the model just beginning the flash shows that essentially all of the ^{14}N that has entered the convective shell has been converted into ^{18}O and that most of the ^{18}O made in the shell has been converted into ^{22}Ne. Had the fresh ^{13}C ingested by the convective shell not burned, the number abundance of ^{13}C in the shell in the more evolved model would be $Y(^{13}$C$) \sim 1.34 \times 10^{-6}$, compared with the actual $Y(^{13}$C$) \sim 2.59 \times 10^{-9}$. Hence most of the ^{13}C that has been ingested by the convective shell has been burned,

Integration under the curves for $Y(^{57}$Fe$)$ in the two figures yields $\int Y(^{57}$Fe$)\, dM \sim 1.08 \times 10^{-9}$ for the less evolved model and $\int Y(^{57}$Fe$)\, dM \sim 1.64 \times 10^{-9}\ M_\odot$ for the more evolved model, which means that the total number of neutrons absorbed by heavy s-process elements has increased by about 50% between the beginning and shortly before the peak of the 15th flash. For the same two models, the total change in the number of ^{25}Mg nuclei in the mass interval defined by the convective shell in the more evolved model is less than 1%, with the integral $\int Y(^{25}$Mg$)\, dM$ increasing from $\sim 2.275 \times 10^{-8} M_\odot$ to $\sim 2.289 \times 10^{-8}\ M_\odot$, a change of only $\sim 1.4 \times 10^{-10}\ M_\odot$. Thus, the total number of neutrons produced by the burning of ^{22}Ne accounts for very little of the increase in the abundance of ^{57}Fe, demonstrating that it is absorption by ^{56}Fe of neutrons produced by the burning of ingested ^{13}C which is responsible for most of the increase in the ^{57}Fe abundance.

In summary, approximately one third of the heavy s-process nuclei created by absorption of neutrons produced by the ^{13}C$(\alpha, n)^{16}$O reaction are created in the convective shell prior to the peak of the helium shell flash. Of the two thirds created outside of the convective shell during the interpulse phase, one third are created in the region of the initial ^{13}C abundance peak and two thirds are created between the abundance peak and the hydrogen-burning shell. Stated in another way, the contributions to heavy s-process nucleosynthesis of the three different ^{13}C$(\alpha, n)^{16}$O-driven sources are in the ratios peak : interpulse : convective shell = 2 : 4 : 3. Near the end of the contribution of the ^{13}C$(\alpha, n)^{16}$O neutron source, the abundance of the neutron counter $Y(^{57}$Fe$)$ in the convective shell reaches $\sim 6.3 \times 10^{-7}$, approximately 2.6% of the abundance of the heavy element s-process proxy ^{56}Fe.

Abundances in the last model of run 323a are shown in Fig. 19.5.7. Over the mass interval $\sim 0.8822 \rightarrow 0.8849\ M_\odot$, $\int Y(^{25}$Mg$)\, dM \sim 4.63 \times 10^{-8}\ M_\odot$. The difference between this mass and the mass given by the same integral for the last model of run 320a is $\Delta M_{\text{neutrons}} \sim 2.34 \times 10^{-8}\ M_\odot$. This difference is approximately equal to the total mass of neutrons produced by the ^{22}Ne$(\alpha, n)^{25}$Mg reaction during the 15th helium shell flash. Integration over the ^{56}Fe abundance gives $\int Y(^{57}$Fe$)\, dM \sim 1.04 \times 10^{-8}\ M_\odot$.

The number abundance of ^{57}Fe in the convective shell in Fig. 19.5.7 is $Y(^{57}$Fe$) \sim 4 \times 10^{-6}$, about 16% of the abundance of ^{56}Fe, compared with the fraction of 2.6% due to the burning of ^{13}C. Hence, approximately five times as many neutrons are captured by heavy s-process elements in consequence of neutron production by the ^{22}Ne$(\alpha, n)^{25}$Mg reaction in the convective shell during the helium shell flash as are produced in consequence of neutron production by the ^{13}C$(\alpha, n)^{16}$O reaction in the region between C-rich and H-rich matter during the interpulse phase and in the convective shell during the early stages of flash-driven convection.

The next six figures describe the neutron density and the rates at which neutrons are produced by the burning of ^{13}C and ^{22}Ne in the convective shell and for a short distance below the convective shell in models at six different times during the 15th helium shell flash. Number abundances of ^{13}C, ^{14}N, ^{18}O, ^{22}Ne, ^{25}Mg, and ^{57}Fe in the convective shell are presented in figure legends, as are the temperature and density at the base of the convective shell. The location of the base of the convective shell is identified in each figure by a thin vertical line labeled M_{BCS} and the mass at the outer edge of the shell, which lies beyond each figure boundary, is presented in each figure legend by the entry labeled M_{ECS}.

Figures 19.5.8–19.5.10 detail characteristics as ^{13}C is being ingested by the convective shell at, respectively, approximately 124 days, 29 days, and 2 days before flash peak. Comparison of M_{ECS} given in the legends on the right hand sides of the figures with the dotted curve for $Y(^{13}$C) in Fig. 19.4.1 and with the curves for $Y(^{13}$C) and $Y(^{57}$Fe) in Fig. 19.5.5 allows one to follow the ingestion process. In the case of the first of the three figures, Fig. 19.5.8, where $M_{\mathrm{BCS}} \sim 0.8842 \, M_{\odot}$, most of the ^{57}Fe created during the

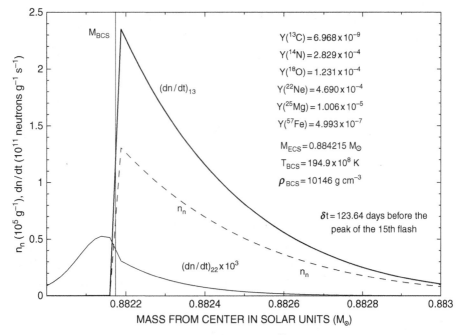

Fig. 19.5.8   Neutron production in the convective shell of a 5 $M_{\odot}$ TPAGB model as ^{13}C ingested by the shell burns during the 15th helium shell flash ($Z = 0.015, Y = 0.275$)

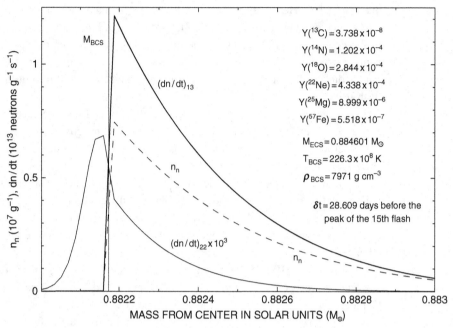

Fig. 19.5.9 Neutron production in the convective shell of a 5 $M_\odot$ TPAGB model as ^{13}C ingested by the shell during the 15th helium shell flash burns ($Z = 0.015$, $Y = 0.275$)

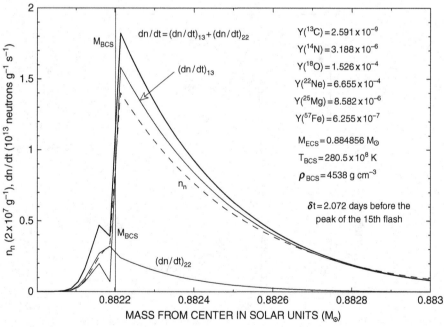

Fig. 19.5.10 Neutron production in the convective shell of a 5 $M_\odot$ TPAGB model as ^{13}C continues to be ingested and burn and ^{22}Ne begins to burn ($Z = 0.015$, $Y = 0.275$)

interpulse phase has been ingested, and fresh ^{13}C is just beginning to be ingested. Even though the abundance of ^{13}C in the convective shell is almost three orders of magnitude smaller than its abundance in the region of neutron production during the interpulse phase (see Figs. 19.4.4 and 19.4.5), the factor of 2 larger temperature at which peak ^{13}C burning occurs at the base of the convective shell results in a neutron production rate which is three orders of magnitude larger than during the interpulse phase and to a neutron density which is also three orders of magnitude larger.

In the model at 29 days before flash peak, with characteristics given in Fig. 19.5.9, the outer edge of the convective shell extends almost to the outer edge of the ^{57}Fe abundance profile built up during the interpulse phase and approximately 40% of the fresh ^{13}C that is ultimately incorporated by the convective shell has been ingested by the shell and partially burned. At the base of the convective shell, the ^{13}C abundance is approximately five times larger than in the model 124 days prior to pulse peak, the temperature is 16% higher, the density is 21% smaller, and the neutron-production rate and the neutron density are approximately 50 times larger than in the less evolved model.

The model 2 days younger than at pulse peak, with characteristics described in Fig. 19.5.10, is the same model as is described in Figs. 19.5.3 and 19.5.6. Most of the fresh ^{13}C ultimately incorporated by the convective shell has been incorporated and burned, but neutron production and absorption continue at rates comparable with those in the model 29 days prior to flash peak. At the base of the convective shell, the ^{13}C abundance is approximately 15 times smaller than in the model 29 days prior to pulse peak, but the temperature is 24% higher, and the density is 57% smaller. The neutron-production rate in the older model is approximately 1.5 times larger, and the neutron density is approximately 3.7 times larger than in the younger model.

Particularly significant with regard to the nature of neutron-capture nucleosynthesis are the facts that the number abundance of ^{14}N has been reduced by a factor of nearly 100, effectively eliminating ^{14}N as a neutron poison which inhibits the production of heavy s-process elements and that temperatures have become high enough for the ^{22}Ne$(\alpha, n)^{25}$Mg neutron source to be activated. Note in Fig. 19.5.10 that, at the base of the convective shell, the ^{22}Ne source produces neutrons at approximately 16% of the rate at which the ^{13}C source produces neutrons.

The peak of the helium shell flash has been arbitrarily defined as the point at which the global helium-burning luminosity reaches a maximum. However, temperatures in the convective shell continue to increase for a while after the peak defined in this way has been reached and, from the point of view of neutron-capture nucleosynthesis, a definition more closely tied with the temperature at the base of the convective shell is perhaps more meaningful. Be that as it may, the traditional definition of flash peak will be retained. Figures 19.5.11 and 19.5.12 show neutron density and neutron production rates in models, respectively, approximately 10 days and 1 month after the flash peak as defined in the traditional way. In both models, although the major source of neutrons is the ^{22}Ne$(\alpha, n)^{25}$Mg reaction, approximately one third of the neutrons so produced in the convective shell are absorbed by the ^{12}C$(n, \gamma)^{13}$C reaction, with the result that, even though ^{13}C is no longer being ingested by the convective shell, the ^{13}C$(\alpha, n)^{16}$O reaction continues to produce neutrons at roughly one third of the rate at which the ^{22}Ne source produces neutrons.

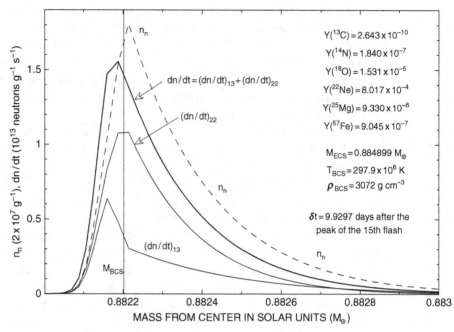

Fig. 19.5.11   Neutron production as ^{22}Ne burns in the convective shell of a 5 $M_\odot$ TPAGB model 10 days after the peak of the 15th helium shell flash ($Z = 0.015$, $Y = 0.275$)

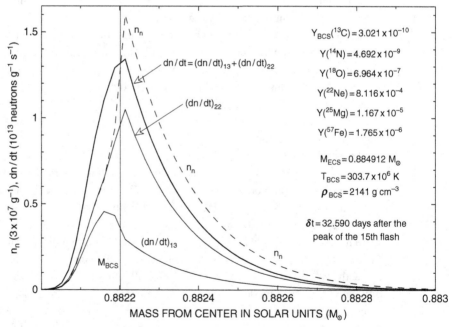

Fig. 19.5.12   Neutron production by ^{22}Ne burning in the convective shell of a 5 $M_\odot$ TPAGB model as the 15th helium shell flash continues ($Z = 0.015$, $Y = 0.275$)

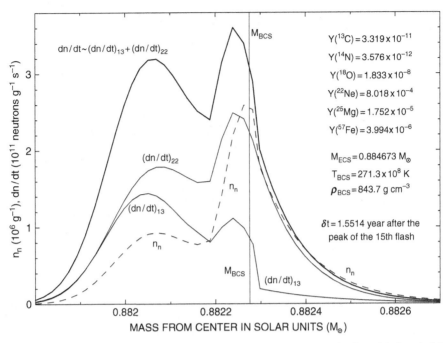

Fig. 19.5.13 Neutron production in and below the shrinking convective shell of a 5 $M_\odot$ TPAGB model as the 15th helium shell flash winds down ($Z = 0.015$, $Y = 0.275$)

Below the convective shell, where the ratio of the number abundance of ^{12}C to the number abundance of ^{22}Ne is much larger, the recycling of neutrons into ^{13}C is correspondingly larger, with the result that the rates of neutron production by ^{13}C and ^{22}Ne are more nearly comparable.

Characteristics and consequences of neutron production in a model 1.55 years past the formal peak of the 15th helium shell flash are shown in Fig. 19.5.13. In the convective shell, temperatures and densities have decreased significantly from those prevailing at flash peak, and neutron-production rates have decreased to levels comparable to those prevailing near the beginning of the flash (see Fig. 19.5.8). The fact that neutron densities are still substantially larger (factor of 20) than those near the beginning of the flash is attributable to the fact that, in the younger model, ^{14}N is the dominant absorber of neutrons, whereas, in the older model, ^{14}N is effectively non-existent.

The efficiency with which neutrons are captured by ^{56}Fe when absorption of neutrons by ^{14}N does not play a role is demonstrated in Figs. 19.5.14 and 19.5.15 where neutron-capture rates, ordered according to their magnitudes, are compared with the overall neutron-production rate in models which are, respectively, roughly 10 days and 1 month past flash peak. In the convective shell of these models, remarkably over half of the neutrons released are captured by the s-process element proxy ^{56}Fe.

This high efficiency of capture helps explain why approximately five times more heavy s-process elements are produced in consequence of ^{22}Ne burning in the convective shell than are produced by all previously occurring processes involving the ^{13}C neutron source

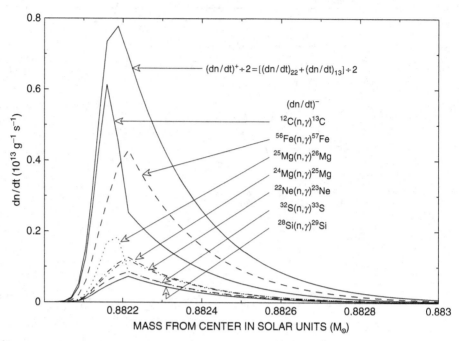

Fig. 19.5.14   Neutron production and absorption in and below the convective shell of a 5 $M_\odot$ TPAGB model 10 days after the peak of the 15th helium shell flash ($Z = 0.015$, $Y = 0.275$)

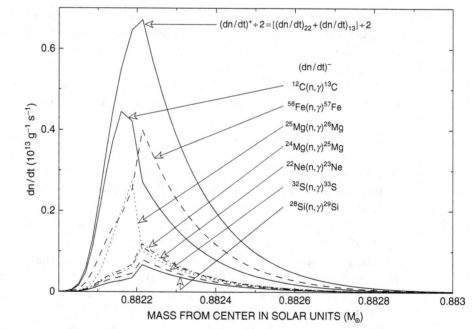

Fig. 19.5.15   Neutron production and absorption in and below the convective shell of a 5 $M_\odot$ TPAGB model a month after the peak of the 15th helium shell flash ($Z = 0.015$, $Y = 0.275$)

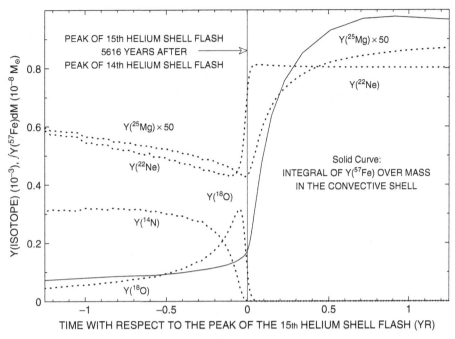

Fig. 19.5.16
Conversion of ^{14}N into ^{18}O, ^{18}O into ^{22}Ne, and ^{22}Ne into ^{25}Mg in the convective shell of a 5 M$_\odot$ TPAGB model during the 15th helium shell flash

when, for example, only one out of ten neutrons released between the ^{13}C abundance peak and the H-burning shell during the interpulse phase is captured by ^{56}Fe. The factor of 5 is demonstrated again in Fig. 19.5.16 where the variation with time of the total mass of ^{57}Fe in the convective shell is compared with the variations of abundances of isotopes involved in the activation of the ^{22}Ne neutron source. Just before flash peak, the acceleration of the rate at which the mass of ^{57}Fe increases is clearly correlated with the decrease of the ^{14}N abundance. After flash peak, ^{14}N is absent, and, as long as the mass of the convective shell continues to increase, the increase in the mass of ^{57}Fe in the shell is clearly correlated with the increase in the ^{25}Mg abundance, which fairly accurately reflects the increase in the number abundance of neutrons released by alpha captures on ^{22}Ne. Ultimately, the mass of the convective shell begins to decrease and, although the number abundance of ^{25}Mg in the shell continues to increase and the total mass of ^{57}Fe in the model continues to increase, the total mass of ^{57}Fe in the shell begins to decrease, no longer serving as a barometer of the progress of heavy element s-process nucleosynthesis. This explains why the integral $\int Y(^{57}\text{Fe})\, dM$ over the entire ^{57}Fe profile in Fig. 19.5.7 yields $\sim 1.04 \times 10^{-8}\ M_\odot$, whereas the maximum of the same integral over just the convective shell yields only $\sim 0.97 \times 10^{-8}$ $M_\odot$, as shown in Fig. 19.5.16.

A final accounting of the contribution of the various episodes of neutron production and neutron capture by the heavy s-process proxy ^{56}Fe is given in Fig. 19.5.17, where the integral $\int Y(^{57}\text{Fe})\, dM$ over the convective shell is plotted as a function of the mass at the edge of the convective shell.

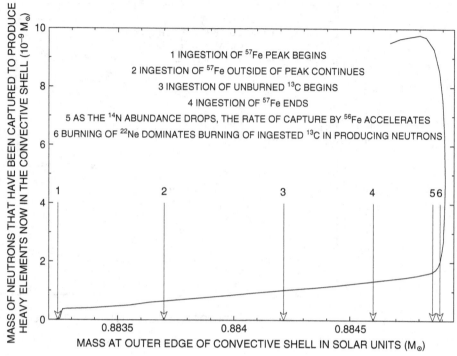

Fig. 19.5.17 Heavy s-process elements in the convective shell of a 5 $M_\odot$ TPAGB model during the 15th helium shell flash ($Z = 0.015, Y = 0.275$)

19.6 Neutron-capture nucleosynthesis in TPAGB stars and heavy s-process element production in the Universe

To summarize the situation with regard to neutron production in a TPAGB model of intermediate mass, there are four distinct environments in which neutrons are produced: (1) a region that (a) is the result of mixing across the H-C discontinuity during and after dredge-up following a helium shell flash, in which (b) ^{13}C is formed by complete hydrogen burning of H-rich and C-rich matter into a peaked distribution; and in which (c) ^{13}C then experiences (α, n) reactions early on during the following interpulse phase; (2) a region in which ^{13}C is produced in equilibrium with respect to other CNO isotopes by the advancing hydrogen-burning shell during the rest of the interpulse phase and is then processed by (α, n) reactions as temperatures in this region reach $\sim 10^8$ K prior to the onset of the next helium shell flash; (3) the convective shell formed during the next helium shell flash, in which the ^{13}C which has been left behind by the hydrogen-burning shell during the interpulse phase but which has not experienced (α, n) reactions is ingested by the convective zone and, as it is swept downward to temperatures of about 10^8 K, experiences (α, n) reactions; (4) the convective shell near the peak of the helium shell flash when the ^{22}Ne formed by the ^{14}N$(\alpha, \gamma)^{18}$F$(\beta^+ \nu_e)^{18}$O and ^{18}O$(\alpha, \gamma)^{22}$Ne reactions experiences ^{12}Ne$(\alpha, n)^{25}$Mg reactions.

Each episode of neutron production occurs under different temperature and density conditions and acts on a different mix of isotopes. The large abundance of ^{14}N present during the first three episodes inhibits the production of heavy s-process elements while the absence of ^{14}N makes the fourth episode, at least in the particular model examined here, much more productive of heavy s-process elements than the other three episodes combined. To the extent that the initial ratio of CNO elements to metals is independent of metallicity, during two of the three episodes relying on ^{13}C as a neutron source and during the episode relying on ^{22}Ne as a neutron source, the ratio of neutrons produced to the abundance of iron seed nuclei is independent of the metallicity. In the case of the first episode which relies on ^{13}C as a neutron source, the amount of ^{13}C that can produce neutrons is determined by the degree of mixing between initial hydrogen-rich material and ^{12}C-rich material produced in the star by helium burning and is thus to a large extent independent of the metallicity.

The exploration of s-process nucleosynthesis in TPAGB stars is a very rich and entertaining enterprise that can be meaningfully explored in many ways. In the present instance, it is evident that, since the amplitude of each thermal pulse increases with each pulse, enough additional pulses should be followed to find the locally asymptotic character of pulses, to determine the asymptotic heavy s-process nucleosynthesis yield per pulse, and to obtain an estimate of the degree of dredge-up in the asymptotic limit. With the added information one can then determine the effect of mixing of nucleosynthesis yields from successive pulses and ascertain the character of the resulting exponential distribution of neutron exposures in the s-process material in the matter eventually ejected by the model in the superwind (see Chapter 21) which terminates the TPAGB phase.

The distribution of s-process elements produced and injected into the interstellar medium by a realistic model TPAGB star depends strongly on the initial mass and metallicity of the model, on the number of thermal pulses the model has experienced, and on the rate at which it loses mass from its surface. As a consequence, most attempts to reproduce theoretically the observed Solar System distribution of s-process elements have suffered from a woefully incomplete data base.

Neutron fluxes occurring in realistic stellar models not being available, the earliest attempts to understand s-process nucleosynthesis in stars were conducted by adopting various imposed neutron fluxes and calculating the distribution of neutron-rich isotopes formed by captures of neutrons on iron-group seed nuclei and on their progeny. These exercises led to an understanding of the basic features of s-process nucleosynthesis and to the identification of what have proved to be the two major sources of neutrons for s-process nucleosynthesis. They did not lead to a quantitative understanding of s-process nucleosynthetic yields.

With the advent of realistic stellar models and the identification of the most likely neutron sources, it has became apparent that elements lighter than iron absorb most of the liberated neutrons and that a determination of the distribution of neutron-rich isotopes produced by stars is a very complex problem which requires that the characteristics of realistic stellar models of different initial masses and compositions be carefully taken into account. It is to be hoped that, ultimately, a careful assessment of the contributions of stars of different initial masses and initial metallicities, taking neutron-production rates as parameters, will produce an understanding of the true dependences on temperature of the rates of the two primary neutron-production processes as well as an understanding of the relative contributions of the two processes to the creation of heavy s-process elements in the Universe.

Bibliography and references

C. Angulo, M. Arnould, M. Rayet, *et al., Nucl. Phys. A*, **656**, 3, 1999.

Z. Y. Bao, H. Beer, F. Käppeler, F. Voss, K. Wisshak, & T. Rauscher, *Atomic Data and Nuclear Data Tables*, **76**, Issue 1, 70, 2000.

E. Margaret Burbidge, Geoffrey R. Burbidge, William A. Fowler, and Fred Hoyle, *Rev. Mod. Phys.*, **29**, 547, 1957.

Alastair G. W. Cameron, *ApJ*, **121**, 144, 1955.

Alastair G. W. Cameron, *AJ*, **65**, 485, 1960.

Georgeanne R. Caughlan and William A. Fowler, *Atomic Data & Nuclear Data Tables*, **40**, 283, 1988.

David Hollowell & Icko Iben, Jr., *ApJL*, **333**, L25, 1988; *ApJ*, **349**, 208, 1990.

Fred Hoyle, *Rev. Mod. Phys.*, **29**, 547, 1957.

Icko Iben, Jr., *ApJ*, **196**, 525, 549, 1975.

Icko Iben, Jr. & Alvio Renzini, *ApJL*, **263**, L23, 1982.

T. Kii, T. Shima, H. Sato, T. Baba, & Y. Nagai, *Phys. Rev. C*, **59**, no. 6, 3397, 1999.

P. E. Koehler & H. A. O'Brien, *Phys. Rev. C*, **39**, no. 4, 1655, 1989.

Maria Lugaro, Falk Herwig, John C. Lattanzio, Roberto Gallino, & Oscar Straniero, *ApJ*, **586**, 1305, 2003.

Paul W. Merrill, *Science*, **115**, 484, 1952.

P. A. Seeger, W. A. Fowler, & D. D. Clayton, *ApJS*, **11**, 121, 1965.

Oscar Straniero, Roberto Gallino, M. Busso, Amedeo Chieffi, & C. M. Raiteri, *ApJL*, **440**, L85, 1995.

James W. Truran & Icko Iben, Jr., *ApJ*, **216**, 797, 1977.

20 Evolution of a massive population I model during helium- and carbon-burning stages

This chapter describes the evolution of a 25 $M_\odot$ population I model during core and shell helium-burning phases and during core and early shell carbon-burning phases. The initial model, described in Volume 1 (Section 11.3), is burning hydrogen in a shell and has just begun to burn helium in central regions. In Section 20.1 of this chapter, the evolution of central and surface characteristics of the model during the bulk of its quiescent nuclear burning lifetime is compared with the evolution of the same characteristics in 1 $M_\odot$ and 5 $M_\odot$ models during quiescent nuclear burning phases up to the TPAGB phase. The location in the HR diagram of a theoretical pulsational instability strip is compared with the location of a band defined by where core helium burning takes place on a long time scale in models of different mass. The fact that the strip and the band have slopes of opposite sign and intersect makes it possible to understand why there exists a peak in the distribution of Cepheids in an aggregate of stars of the same composition, but of different masses and ages. The peak occurs at the intersection of the strip and the band.

In Section 20.2, the evolution of internal structure and composition characteristics of the 25 $M_\odot$ model during the core helium-burning phase is described in some detail, with particular attention paid to the neutron-capture nucleosynthesis in the convective core occasioned by the activation of the ^{22}Ne$(\alpha, n)^{25}$Mg neutron source. Helium ultimately becomes exhausted over a central region of mass ~ 6 $M_\odot$ in which the dominant isotopes are ^{12}C and ^{16}O at number abundances of $Y_{12} = 0.021$ ($X_{12} = 0.252$) and $Y_{16} = 0.045$ ($X_{16} = 0.72$), respectively. Neutron-capture nucleosynthesis in this central region has resulted in the capture of approximately 1.5 neutrons for every ^{56}Fe seed nucleus.

The immediately subsequent evolution is discussed in Section 20.3. Temporarily devoid of a nuclear energy source, the helium-exhausted core contracts and heats, with contraction and heating rates being controlled by neutrino energy losses occasioned by the conversion, due to the weak interaction, of electron–positron pairs into neutrino–antineutrino pairs which escape freely into space. Ultimately, carbon burning begins in central regions. Carbon burning results primarily in the conversion of a mixture of ^{12}C and ^{16}O isotopes into a mixture of ^{16}O and ^{20}Ne isotopes, at number abundances of $Y_{16} = 0.038$ ($X_{16} = 0.61$) and $Y_{20} = 0.0176$ ($X_{20} = 0.352$), respectively. The calculations are terminated when carbon is nearly exhausted over the inner 0.5 $M_\odot$ of the core and carbon burning takes place in a convective shell of mass $M_{CS} \sim 1$ $M_\odot$ outside of the carbon-depleted core. Essentially all of the energy liberated by carbon-burning reactions is lost to neutrino–antineutrino pairs produced by the annihilation of electron–positron pairs. Helium-burning in a shell at the outer edge of the helium-exhausted core is responsible for the surface luminosity of the model.

Numerically following subsequent evolution, which terminates in the formation of a neutron star or black hole remnant and the ejection of the envelope in a type II supernova explosion, is beyond the scope of this book, particularly since evolution shortly becomes dynamic and memory of much of the nucleosynthesis which has occurred in central regions is totally lost during the collapse of the core. However, prior to core collapse and envelope ejection, additional quasistatic phases do occur during which ^{16}O, ^{20}Ne, and ^{28}Si burn, and some of the interesting physics that takes place during these phases is sketched in Sections 20.4 and 20.5.

20.1 Evolution of surface and central characteristics of a 25 $M_\odot$ model during quiescent nuclear burning stages and comparison of characteristics of models of mass 1 $M_\odot$, 5 $M_\odot$, and 25 $M_\odot$

The variation of surface characteristics of the 25 $M_\odot$ model during evolution from the main sequence to the beginning of the core carbon-burning phase is shown in the HR diagram of Fig. 20.1.1. The evolutionary path for temperatures larger than 20 000 K is a repetition of the path in Fig. 11.3.1 in Volume 1 which ends at the beginning of the core helium-burning phase. In both figures, the open circles along the tracks occur at intervals of

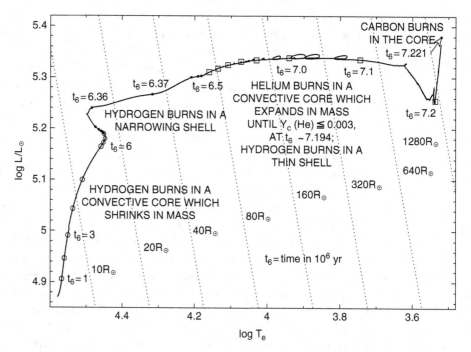

Fig. 20.1.1   Evolutionary path of a 25 $M_\odot$ model through core hydrogen and core helium burning to the onset of core carbon burning ($Z = 0.015$, $Y = 0.275$)

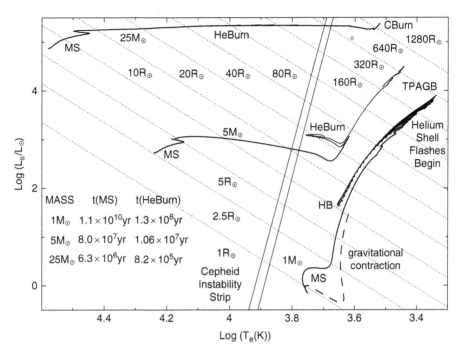

Fig. 20.1.2 Evolution in the HR-diagram of 1 $M_\odot$, 5 $M_\odot$, and 25 $M_\odot$ models during nuclear burning phases ($Z = 0.015$, $Y = 0.275$)

10^6 yr ($\Delta t_6 = 1$) and the solid dots occur at intervals of 10^4 yr ($\Delta t_6 = 0.01$). The open boxes along the track in Fig. 20.1.1 occur at intervals of 10^5 yr ($\Delta t_6 = 0.1$) and define the region in which core helium burning is important. Core helium burning begins in earnest at $t_6 \sim 6.4$ and helium does not vanish at the center until $t_6 \sim 7.21$, so the core helium-burning phase lasts for approximately 8.1×10^5 yr, or about 13% of the 6.36×10^6 yr duration of the core hydrogen-burning phase. Over the major portion of the core helium-burning phase the model resides in the HR diagram approximately halfway between the main sequence and the giant branch.

Evolutionary tracks in the HR diagram of the three main population I models constructed for this monograph are compared in Fig. 20.1.2. Lines of constant radius are provided to permit a more complete appreciation of quantitative differences in global characteristics and qualitative similarities and differences in evolutionary behavior. For example, although the three models differ greatly in luminosity and radius during the main sequence phase, each model experiences approximately the same factor of 2 increase in radius during core hydrogen burning. In each case, the major portion of the quiescent core helium-burning phase occurs at a radius which is from four to eight times larger than at the end of the core hydrogen-burning phase. Finally, the last quiescent nuclear burning phase takes place in a giant configuration at a radius substantially larger than the Earth's orbital radius.

Just as the evolutionary paths traversed by the models during the core hydrogen-burning phase define a band in the HR diagram which may be identified with a population of real (main sequence) stars, so the evolutionary paths traversed by the models during the core helium-burning phase define a band in the HR diagram which may identified with a population of real stars. In Fig. 20.1.2, the core helium-burning band extends from the location on the 1 $M_\odot$ track labeled HB (for horizontal branch) through the locations on the 5 $M_\odot$ and 25 $M_\odot$ tracks labeled HeBurn. Comparing with the locations of real stars in the HR diagram of Fig. 2.2.1 in Volume 1, it is evident that the theoretical core helium-burning band is more or less coincident with the band of real stars extending from Capella to Rigel and Deneb.

A rough approximation to the location of the Cepheid instability strip for a population I composition is provided in Fig. 20.1.2. The left hand (blue) edge of the strip is adapted from Iben & R. S. Tuggle (1975). The red edge is parallel to the blue edge and, at any luminosity, is separated arbitrarily from the blue edge by $\Delta \log T_e = -0.03$. The theoretical band joining the word HeBurn next to the 5 $M_\odot$ track and the same word next to the 25 $M_\odot$ track intersects the instability strip at $\log(L/L_\odot) \sim 3.7 \pm 0.25$, $\log(T_e) \sim 3.76 \pm 0.05$, and $(R/R_\odot) \sim 70 \pm 20$. Interpolating logarithmically in the luminosity, one may infer that, in an aggregate consisting of model stars all of the chosen initial composition, distributed smoothly in initial mass, and produced at a time-independent rate, the distribution of Cepheids will cluster about models of mass $\sim 8\ M_\odot$.

A rough estimate of the relationship between the period, mass, and radius of a model star pulsating in the fundamental mode follows from consideration of a simple pendulum for which

$$P = 2\pi \sqrt{\frac{l}{g}}, \tag{20.1.1}$$

where P and l are, respectively, the period and length of the pendulum, and g is the acceleration due to gravity. Setting

$$l = \frac{R}{2} \text{ and } g = \frac{GM/2}{(R/2)^2}, \tag{20.1.2}$$

one obtains

$$P \sim \pi \sqrt{\frac{R^3}{GM}} = 0.0578 \sqrt{\frac{(R/R_\odot)^3}{M/M_\odot}} \text{ days.} \tag{20.1.3}$$

In the present instance, this translates into a Cepheid distribution which peaks at $P \sim 12$ days.

Much more rigorous estimates of the relationship between period and model properties can of course be constructed. The estimates of Iben & Tuggle (1972a and 1972b) may be approximated by

$$\log P(\text{days}) \sim 0.65 + 0.83 (\log(L/L_\odot) - 3.25)$$

$$- 0.63 (\log(M/M_\odot) - 0.7) - 3.4 (\log(T_e) - 3.77). \tag{20.1.4}$$

Note that, very roughly, $P \stackrel{\sim}{\propto} \left(L/T_e^4\right)^{0.85} M^{-.63} \stackrel{\sim}{\propto} (R^2)^{0.9} M^{-0.6} \stackrel{\sim}{\propto} (R^3/M)^{0.6}$. Inserting $M = 8\ M_\odot$ and $\log T_e = 3.76$ in eq. (20.1.4), one obtains $\log P \sim 0.93$, or $P \sim 8.5$ days.

With decreasing metallicity, the location of a theoretical helium-burning band shifts to higher surface temperature at any given luminosity. Although the instability strip also shifts, the change is much less pronounced. The net result is that, with decreasing metallicity, the intersection between the instability strip and the helium-burning band decreases in luminosity and mass and increases in surface temperature, with the mean period at the intersection also decreasing.

Cepheid distributions in Andromeda and the Milky Way culled from the literature are exhibited in Fig. 7 of S. A. Becker, Iben, & R. S. Tuggle (1977). The distribution of Cepheids in Andromeda extends from $\log P \sim 0.4$ to $\log P \sim 1.7$, with a centroid at $\log P \sim 0.9$, or $P \sim 7.9$ days. The distribution of Cepheids in the Milky Way extends from $\log P \sim 0.2$ to $\log P \sim 1.6$, with a centroid at $\log P \sim 0.8$, or $P \sim 6.3$ days. Given the theoretical dependence on metallicity of Cepheid properties at the intersection of the helium-burning band and the pulsational instability strip, one infers that the average metallicity of Andromeda Cepheids is larger than the average metallicity of Milky Way Cepheids, and that the average metallicity of Cepheids in each of the two observed galaxies is smaller than the metallicity of the models described in Fig. 20.1.2.

It is interesting to compare the tracks in Fig. 20.1.2 with tracks for models of the same mass shown in Fig 2.4.1 in Volume 1. The latter models were constructed in the mid-1970s and differ from the models constructed for this book primarily by the choice of opacity – an analytic approximation to Los Alamos opacities for $Z = 0.02$ rather than tabular Livermore opacities for $Z = 0.015$ used here. By far the most satisfying and encouraging feature of the comparison between the two sets of evolutionary tracks is that, despite the differences in input physics, the qualitative characteristics of the tracks have not changed. The major difference is that the more recently constructed models produce tracks which, in any given evolutionary phase, are slightly less luminous and redder than the tracks produced by the earlier models, a change which is entirely consistent with the fact that the opacities in the envelopes of the more recently constructed models are in general larger than those in the earlier models, thanks to a more complete accounting of bound–bound radiative transitions.

The manner in which density and temperature at the center of the 25 $M_\odot$ model are related during evolution is shown in Fig. 20.1.3. Large circles along the track in the lower left hand portion of the figure indicate where core hydrogen burning takes place and large open boxes along the track in the middle of the figure indicate where core helium burning takes place. Core carbon burning begins along the track in the upper right hand portion of the figure. The time indicators show that the transition time from the core hydrogen-burning phase to the core helium-burning phase is approximately 1.5% of the core hydrogen-burning lifetime and that the transition time from the core helium-burning phase to the core carbon-burning phase is approximately 2.5% of the core helium-burning lifetime.

The thin solid line of constant slope in Fig. 20.1.3 shows where the radiation pressure for a gas composed of equal masses of hydrogen and helium is equal to one fourth of the gas pressure. A comparable line for a pure hydrogen gas lies above the given line by $\Delta \log T_6 = 0.0542$ and a comparable line for a pure helium gas lies below the given line by $\Delta \log T_6 = -0.0877$. It is evident that the ratio of radiation pressure to gas pressure at

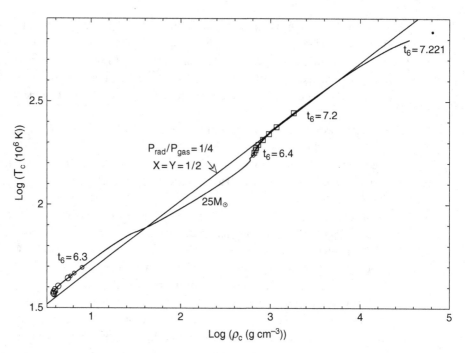

Fig. 20.1.3 Central temperature vs central density in a 25 $M_\odot$ model through core H- and He-burning phases to the onset of core carbon burning ($Z = 0.015, Y = 0.275$)

the center remains fairly constant during the entire evolution and this means that the core contracts and heats more or less adiabatically.

In Fig. 20.1.4, the evolution of temperature with respect to density at the centers of the 1 $M_\odot$ and 5 $M_\odot$ standard models is compared with the evolution of temperature with respect to density at the center of the 25 $M_\odot$ standard model. The slopes of the $\log T_c - \log \rho_c$ curves for all three models during core hydrogen burning are quite similar. In the 5 $M_\odot$ model, central temperature drops briefly as convection and the hydrogen abundance disappear together over central regions, but, as the model evolves toward the core helium-burning stage, the center contracts and heats nearly adiabatically, just as does the center of the 25 $M_\odot$ model. The beginning of core helium burning in the 5 $M_\odot$ model is signalled by an abrupt change in the slope of the $\log T_c - \log \rho_c$ curve, as a rapid injection of nuclear fuel leads to expansion. A change in slope associated with helium ignition is also evident in the $\log T_c - \log \rho_c$ curve for the 25 $M_\odot$ model, but this change is almost entirely due to an increase in the rate of heating, with an almost imperceptible expansion occurring at the beginning of the helium-burning phase.

The lines in Fig. 20.1.4 labeled by the ratio of the electron Fermi kinetic energy ϵ'_F to kT have been constructed by approximating the Fermi energy by its value at zero temperature (ϵ_{F0} as given in Section 4.8 in Volume 1 is a function only of the density and the electron molecular weight). The procedure is justified by the fact that the two curves are approximately parallel over the entire range shown. Comparing the location of the abrupt change in slope along the $\log T_c - \log \rho_c$ curve for the 5 $M_\odot$ model with the location of

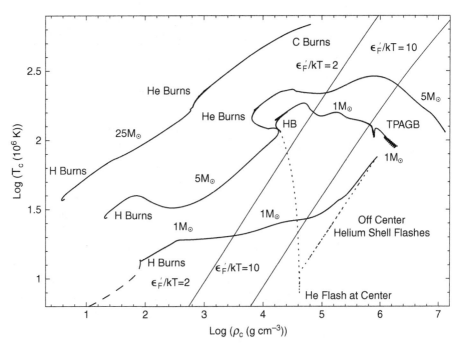

Fig. 20.1.4 Central temperature versus central density during nuclear-burning stages in 1 $M_\odot$, 5 $M_\odot$, and 25 $M_\odot$ models ($Z = 0.015$, $Y = 0.275$)

the curve labeled $\epsilon'_F/kT = 2$, it is clear that electrons in the helium-burning core of the 5 $M_\odot$ model are not very degenerate, but that electron degeneracy sets in after helium has been exhausted in the core. In the 25 $M_\odot$ model, electron degeneracy does not become significant until carbon burning begins at the center.

In contrast with the two more massive models, the 1 $M_\odot$ model does not have a convective core during core hydrogen burning and experiences an extended phase burning hydrogen in a thick shell before the mass of the hydrogen-exhausted core exceeds the Schönberg-Chandrasekhar mass; the core then contracts nearly isothermally and electrons there become degenerate before helium-ignition temperatures are reached. Due to plasma neutrino losses, the maximum temperature in the core of the 1 $M_\odot$ model occurs off center. Once this maximum temperature becomes large enough, a series of off-center helium shell flashes lifts electron degeneracy in regions successively closer in mass to the center, as described in detail in Section 17.1. The evolution of central density and temperature during this phase is traced by the dotted curve of positive slope in Fig. 20.1.4. Until helium is ignited at the center, matter at the center of the model cools and expands nearly adiabatically, with the degree of electron degeneracy, as measured by ϵ'_F/kT, remaining essentially constant, until, after six helium shell flashes, helium is ignited at the center.

In response to the rapid release of nuclear energy in central regions, the core of the 1 $M_\odot$ model heats and expands rapidly, until a quasistatic equilibrium configuration is established in which most of the nuclear energy liberated by helium burning reaches the surface instead of heating and expanding matter against gravity. Evolution in the $\log T_c - \log \rho_c$

diagram from the onset of helium burning at the center to the horizontal branch (clump) configuration where quiescent core helium burning has been established is traced by that portion of the dotted curve in Fig. 20.1.4 which is of negative slope. In summary, off-center helium shell flashes occur along that portion of the dotted curve which has a positive slope and evolution after helium ignition at the center occurs along that portion which has a negative slope.

In the HR diagram of Fig. 20.1.2, the evolutionary track during the helium shell-flash phase is described by the dotted curve which terminates at the clump position labeled HB (for horizontal branch). In contrast with the behavior at the center, which cools and expands adiabatically until helium is ignited at the center and then abruptly heats as it continues to expand, the envelope of the model contracts and heats monotonically during the entire transition from the tip of the first red giant branch to the start of the horizontal branch phase.

Interestingly, the density and temperature at the center of the 1 $M_\odot$ model at the beginning of the HB phase are, respectively, almost identical with the density and temperature at the center of the 5 $M_\odot$ model at the beginning of the core helium-burning phase. This proximity helps explain why, after core helium burning is complete, both models evolve into similar configurations on the TPAGB, both with electron-degenerate CO cores and both experiencing thermal pulses, as quiescent hydrogen burning in a shell is periodically interrupted by a helium shell flash followed by a period of quiescent helium burning in a shell.

20.2   Evolution of internal characteristics and production of light s-process elements in a 25 $M_\odot$ model during core helium burning

Figs. 11.3.17 and 11.3.18 in Volume 1 of this book are the first pair of a set of four pairs of figures which describe various internal characteristics of four 25 $M_\odot$ models during the core helium-burning phase. Figs. 20.2.1–20.2.6 in this section form the additional three sets. The first of each pair of figures describes structure variables and the hydrogen and helium abundance distributions (by number in the first set and by mass in the last three sets). The second of each pair of figures describes the ratio of radiation pressure to gas pressure, the opacity, and various temperature–pressure gradients. The locations of all radiative-convective interfaces are shown in each figure.

The figure sets describe the 25 $M_\odot$ model: (1) at the beginning of core helium burning; (2) approximately half way through the core helium-burning phase; (3) near the end of the core helium-burning phase, but while the mass of the convective core is still increasing, and (4) at the end of the core helium-burning phase, as the helium abundance in and the mass of the convective core begin to decrease precipitously in tandem.

A distinctive feature of the 25 $M_\odot$ model during core helium burning relative to the less massive core helium-burning models described in Chapters 17 and 18 is that convective zones are maintained by energy fluxes from two nuclear sources. The core convective

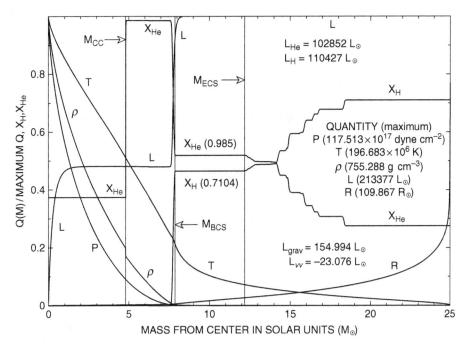

Fig. 20.2.1 Structure of a 25 $M_\odot$ core helium-burning model about halfway through the core helium-burning phase ($Z = 0.015$, $Y = 0.275$, and AGE $= 6.7892 \times 10^6$ yr)

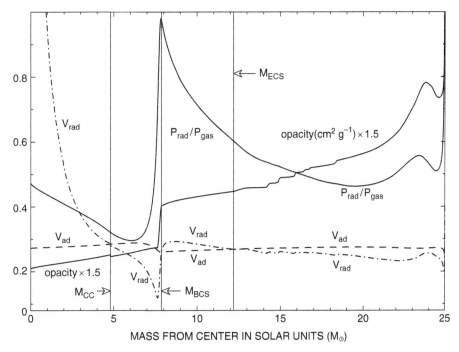

Fig. 20.2.2 Interesting variables in a 25 $M_\odot$ about halfway through the core helium-burning phase ($Z = 0.015$, $Y = 0.275$, and AGE $= 6.7892 \times 10^6$ yr)

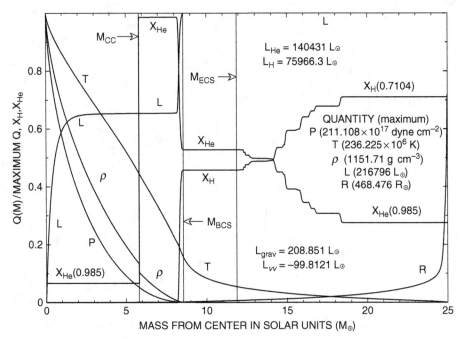

Fig. 20.2.3 Structure of a 25 $M_\odot$ model nearing the end of the core helium-burning phase ($Z = 0.015$, $Y = 0.275$, and AGE $= 7.0969 \times 10^6$ yr)

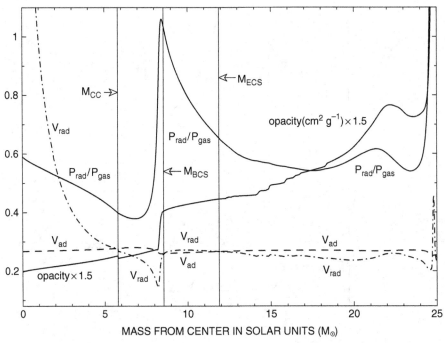

Fig. 20.2.4 Interesting variables in a 25 $M_\odot$ model nearing the end of the core helium-burning phase ($Z = 0.015$, $Y = 0.275$, and AGE $= 7.0969 \times 10^6$ yr)

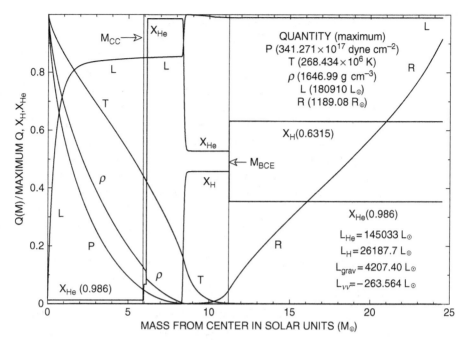

Fig. 20.2.5 Structure of a 25 $M_\odot$ model with a shrinking convective core near the end of the core helium-burning phase ($Z = 0.015$, $Y = 0.275$, and AGE = 7.1933×10^6 yr)

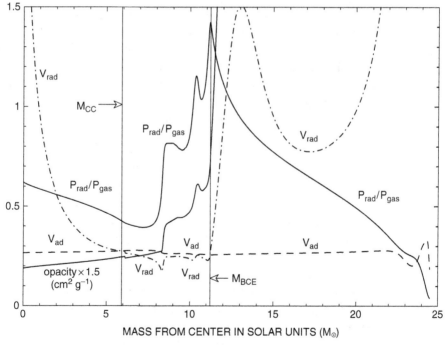

Fig. 20.2.6 Interesting variables in a 25 $M_\odot$ model with a shrinking convective core near the end of the core helium-burning phase ($Z = 0.015$, $Y = 0.275$, and AGE = 7.19330×10^6 yr)

zone, in which convection is maintained by the flux due to helium burning, increases in mass from $M_{CC} \sim 3.6 \ M_\odot$ in the first model to $M_{CC} \sim 6 \ M_\odot$ in the fourth model. A convective shell, in which convection is maintained by fluxes due to both helium and hydrogen burning, decreases in mass from $M_{CS} \sim 5.8 \ M_\odot$ in the first model to $M_{CS} \sim 3.2 \ M_\odot$ in the third model, before disappearing entirely. The variations in mass of the convective core and the convective shell are related to the variations in the fractions of the total nuclear burning energy which the two nuclear energy sources provide; in the four models, the ratio $L_{He}/(L_H + L_{He})$ increases through the values 0.30, 0.38, 0.65, and 0.85.

In the third model, a convective envelope has begun to develop. As the model grows to red giant dimensions in response to a contracting helium-burning core, the base of the convective envelope extends steadily inward in mass. In the fourth model, the base of the convective envelope has extended inward through the region where semiconvection during the main sequence phase has led to an enhancement of the helium to hydrogen ratio and then even further inward to beyond the maximum mass achieved at the outer edge of the erstwhile convective shell previously supported in part by hydrogen burning. The accompanying dredge-up of helium increases the surface ratio of helium to hydrogen from an initial value of $Y_{He}/Y_H \sim 0.10$ on the main sequence to $Y_{He}/Y_H \sim 0.14$. Since the matter in the convective envelope is ultimately returned to the interstellar medium, this enhancement in the helium–hydrogen ratio demonstrates that massive stars contribute to the gradual increase in the abundance of helium in the interstellar medium.

Inspection of the curves for radiation pressure over gas pressure in the second of each pair of figures shows that, although the ratio P_{rad}/P_{gas} varies significantly with respect to position in any given model and the ratio at corresponding positions in different models varies considerably from one model to the next, the mass-averaged value of the ratio varies much less dramatically. For example, in the the inner half of the four models considered, the maximum and minimum ratios differ by factors of approximately 4.5, 3.3, 2.8, and 3.5, respectively. At the centers of the four models, the ratio increases from ~ 0.33 to ~ 0.62 and, at the hydrogen–helium interface, the ratio increases from ~ 0.9 to ~ 1.4. Yet, the mass-averaged value of the ratio varies only from ~ 0.45 in the first model to ~ 0.6 in the fourth model.

In Fig. 20.2.7, gravothermal energy-production rates are shown as functions of mass for the inner half of the second model, which is approximately halfway through the core helium-burning phase. In Fig. 20.2.7a, the gravothermal rates are shown more clearly over a region centered on the hydrogen-burning shell. Over the region shown in Fig. 20.2.7, $\mathbf{g} \cdot \mathbf{v} > 0$, demonstrating that, everywhere in the region, matter is moving inward. Below $M \sim 11 \ M_\odot$, $\epsilon_{compression} > 0$, so that essentially everywhere in the inner half of the model, matter is both moving inward and contracting.

Particularly interesting are relationships between energy-generation rates of the various gravothermal energy sources identified and discussed in Section 18.1 for a 5 $M_\odot$ model which is at a comparable point in the core helium-burning phase, namely, about half way through the phase. The figures for the 5 $M_\odot$ model which are to be compared with Figs. 20.2.7 and 20.2.7a for the 25 $M_\odot$ model are Figs. 18.1.5 and 18.1.4. The rate ϵ_{th} shown in Fig. 18.1.5 is not shown explicitly in Fig. 20.2.7, but, since it is related to the total rate $\epsilon_{internal}$ at which the internal energy changes and to the rate ϵ_{cdth} at which the internal energy changes in response to the creation and destruction of particles by

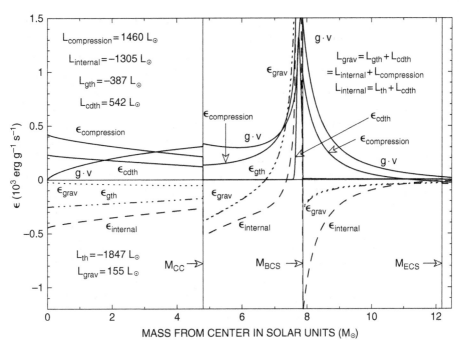

Fig. 20.2.7 Gravothermal energy generation in a 25 $M_\odot$ model approximately halfway through the core helium-burning phase ($Z = 0.015, Y = 0.275, \text{AGE} = 6.7892 \times 10^6$ yr)

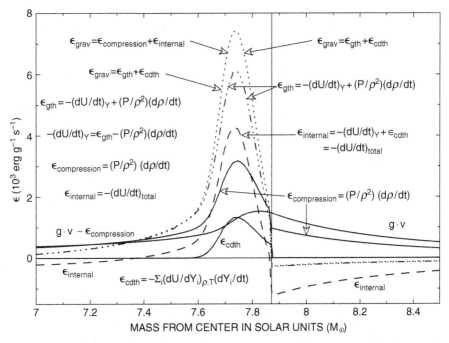

Fig. 20.2.7a Gravothermal energy generation in the H-burning region of a 25 $M_\odot$ model midway in the core helium-burning phase ($Z = 0.015, Y = 0.275, \text{AGE} = 6.7892 \times 10^6$ yr)

$\epsilon_{internal} = \epsilon_{th} + \epsilon_{cdth}$, ϵ_{th} can be visualized in Fig. 20.2.7 by subtracting the curve labeled ϵ_{cdth} from the curve labeled $\epsilon_{internal}$. Thus, for example, in the convective core, a curve for ϵ_{th} would lie below the curve for $\epsilon_{internal}$ by the same amount that the curve for ϵ_{cdth} lies above the x axis. And, since the curve for $\epsilon_{internal}$ lies approximately twice as far below the x axis as the curve for ϵ_{cdth} lies above the x axis, the rate at which the internal energy decreases in consequence of the reduction of particles brought about by the conversion of three alpha particles plus six electrons into one ^{12}C nucleus and six electrons is approximately three times smaller than the rate at which the internal energy increases due to changes in density and temperature.

This ratio persists on average throughout the model; integrated over the entire model, $L_{cdth} = 542\ L_\odot$ is approximately three (actually 3.4) times smaller than $-L_{th} = 1847\ L_\odot$. Evaluating the integral $\int \epsilon_{cdth}\ dM$ over the convective core in Fig. 20.2.7 gives L_{cdth} (core) $\sim 424\ L_\odot$ and the same integration over the region centered on the hydrogen-burning shell in Fig. 20.2.7a gives L_{cdth}(shell) $\sim 118\ L_\odot$, so L_{cdth}(core)$/L_{cdth}$(shell) ~ 3.6. This ratio can be understood in terms of physical processes by following the same steps outlined in Section 18.1 for the 5 $M_\odot$ core helium-burning model.

From the temperature profile in Fig. 20.2.1, it is evident that the mean temperature in the convective core of the 25 $M_\odot$ model is approximately three times larger than the mean temperature in the hydrogen-burning shell. Assume first that ^{12}C is the terminal product of helium burning. Since, as seen in Fig. 20.2.1, the helium-burning luminosity is essentially the same as the hydrogen-burning luminosity, the rate at which ^{12}C nuclei are made in the core relative to the rate at which ^4He nuclei are made in the shell is equal to the inverse of the ratio of energies released in the two burning processes, namely 3.45. Finally, since two particles disappear for every reaction chain which produces ^{12}C and five disappear for every reaction chain which produces ^4He, one has that L_{cdth}(core)$/L_{cdth}$(shell) $\sim 3 \times 3.45 \times 2/5 \sim 4.1$. Suppose next that the terminal result of helium burning is always ^{16}O. Then, ^{16}O nuclei are made $\sim 25\mathrm{MeV}/14.44\mathrm{MeV} = 1.73$ times more frequently in the convective core than ^4He nuclei are made in the shell and three particles disappear for every ^{16}O made in the core compared with the five which disappear for every ^4He nucleus made in the shell. Altogether, then, L_{cdth}(core)$/L_{cdth}$(shell) $\sim 3 \times 1.73 \times 3/5 \sim 3.12$, which is 24% less than the estimate made on the assumption that helium burning terminates with ^{12}C. At the center of the 25 $M_\odot$ model, ^{12}C nuclei are being produced ~ 3.8 times more rapidly than ^{16}O nuclei and this translates into $(L_{cdth}$(core)$/L_{cdth}$(shell)$) \sim (4.1 + 0.26 \times 3.12)/1.26 = 3.90$ which is the ratio actually obtained in the calculated model.

The net result of the competition between the various gravothermal energy-generation sources is that very little gravothermally generated energy makes its way to the surface. For example, in the convective core, the net gravothermal energy-generation rate, which can be alternativly described as $\epsilon_{grav} = \epsilon_{th} + \epsilon_{cdth} + \epsilon_{compression}$, $\epsilon_{grav} = \epsilon_{internal} + \epsilon_{compression}$, $\epsilon_{grav} = \epsilon_{gth} + \epsilon_{cdth}$, and in yet other ways, the net result is quite small compared with the absolute value of components. Integrated over the entire model, $L_{grav} = 155\ L_\odot$ is only of the order of 10% of the separate components $L_{compression} = 1460\ L_\odot$ and $-L_{internal} = 1305\ L_\odot$.

In most of the region between the outer edge of the convective core and the inner edge of the convective shell which forms because of the contribution of hydrogen burning to

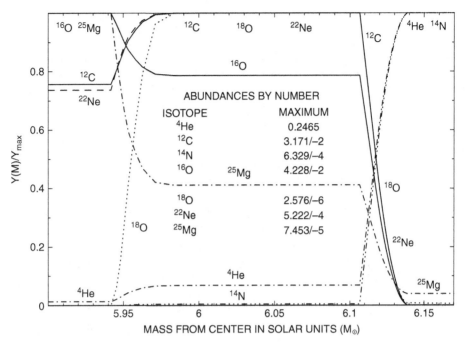

Fig. 20.2.8 Abundances of major constituents in a 25 $M_\odot$ model near the end of the core helium-burning phase ($Z = 0.015$, $Y = 0.275$, AGE $= 7.1933 \times 10^6$ yr)

the energy flux, ϵ_{cdth} is essentially zero due to the absence of nuclear transformations and the consequences thereof, so $\epsilon_{\mathrm{internal}} = \epsilon_{\mathrm{th}}$. The competition in this region is between $\epsilon_{\mathrm{compression}}$ and ϵ_{th}, and the result of the competition is expressed by $\epsilon_{\mathrm{grav}} = \epsilon_{\mathrm{th}}$. Over the inner almost two-thirds of the region, the rate of release of compressional energy is less than the rate of absorption of energy by heating, so $\epsilon_{\mathrm{grav}} < 0$. In the outer one-third of the region, the reverse is the case.

Number abundances of several major element constituents in the neighborhood of the outer edge of the convective core in the last of the four models are shown in Fig. 20.2.8. Note that the abundances listed in the figure legend are maximum abundances in the model and each abundance profile is normalized to the maximum in the model. In the convective core, of mass $M_{\mathrm{CC}} \sim 5.94\ M_\odot$, the number abundances of ^{4}He, ^{12}C, and ^{16}O are, respectively, $Y(^4\mathrm{He}) = 3.1922 \times 10^{-3}$, $Y(^{12}\mathrm{C}) = 2.3981 \times 10^{-2}$, and $Y(^{16}\mathrm{O}) = 4.2282 \times 10^{-2}$. The maximum mass achieved by the convective core is $M_{\mathrm{CC}}^{\mathrm{max}} \sim 6.105\ M_\odot$, and, in the region vacated by convection, between $M = 5.97\ M_\odot$ and $M = 6.105\ M_\odot$, the number abundances of the three most abundant isotopes are, respectively, $Y(^4\mathrm{He}) = 1.6836 \times 10^{-2}$, $Y(^{12}\mathrm{C}) = 3.1713 \times 10^{-2}$, and $Y(^{16}\mathrm{O}) = 3.3249 \times 10^{-2}$. Thus, as the convective core approaches its maximum mass and then shrinks in mass, the ^{12}C/^{16}O ratio in the core changes from being roughly unity to being about 0.6.

When the convective core reaches its maximum outward extent in mass, the abundance of ^{25}Mg exceeds its initial main sequence abundance by a factor of 10, a consequence of the reaction $^{22}\mathrm{Ne}(\alpha, n)^{25}\mathrm{Mg}$ moderated by the reaction $^{25}\mathrm{Mg}(n, \gamma)^{26}\mathrm{Mg}$. As the

Table 20.2.1 Number abundances in a 25 $M_\odot$ model with a convective core that has begun to shrink in mass

isotope	$M = 5.921$	$M = 6.033\ M_\odot$	$M = 6.162\ M_\odot$	$M = 24.5\ M_\odot$
^{1}H	~0.0000	~0.0000	~0.0000	0.63151
^{4}He	3.1922/−3	1.6836/−2	0.24648	8.8511/−2
^{12}C	2.3981/−2	3.1713/−2	1.7781/−5	1.3118/−4
^{14}N	1.9068/−13	2.8046/−6	6.3292/−4	2.2127/−4
^{16}O	4.2282/−2	3.3249/−2	1.8232/−5	3.1084/−4
^{17}O	2.0533/−7	1.0913/−7	1.7475/−6	4.8999/−6
^{18}O	3.5624/−12	2.5750/−6	1.8070/−8	5.0586/−7
^{20}Ne	1.4313/−4	7.9002/−5	5.2692/−5	5.3219/−5
^{21}Ne	1.1364/−5	4.0011/−6	2.1438/−8	1.2975/−6
^{22}Ne	3.8458/−4	5.2216/−4	1.3807/−8	3.7418/−6
^{23}Na	1.0083/−5	9.4479/−6	8.4658/−6	3.0279/−6
^{24}Mg	2.2119/−5	2.2705/−5	2.3470/−5	2.3393/−5
^{25}Mg	7.4535/−5	3.0714/−5	2.9478/−6	2.9623/−6
^{26}Mg	1.7851/−4	7.9145/−5	3.2874/−6	3.2602/−6
^{27}Al	2.3711/−6	2.1194/−6	2.2834/−6	2.3190/−6
^{28}Si	2.3582/−5	2.6301/−5	2.8297/−5	2.8294/−5
^{29}Si	4.0986/−6	2.8572/−6	1.4292/−6	1.4338/−6
^{30}Si	4.2482/−6	2.4518/−6	9.2971/−7	9.4585/−7
^{31}P	2.8624/−6	1.0238/−6	2.6819/−7	1.4828/−7
^{32}S	8.7408/−6	1.0481/−5	1.1985/−5	1.2041/−5
^{33}S	1.2956/−7	1.4492/−7	7.6910/−8	9.0306/−8
^{34}S	6.4316/−7	5.8095/−7	5.3382/−7	5.3232/−7
^{57}Fe	2.4034/−5	1.0109/−5	4.7750/−7	0.0000

convective core shrinks in mass, the mean temperatures in the core increase, accelerating the ^{22}Ne$(\alpha, n)^{25}$Mg and ^{22}Ne$(\alpha, \gamma)^{26}$Mg reactions. This accounts for the fact that, in the convective core of the fourth model, the ^{22}Ne number abundance is 25% smaller than in the convective core at its maximum mass and the number abundance of ^{25}Mg has increased by a factor of 2.5. The decrease in the number abundance of ^{22}Ne is significantly (three times) larger than the increase in the number abundance of ^{25}Mg, this being due in part to the operation of the ^{25}Mg$(n, \gamma)^{26}$Mg reaction, but primarily due to the operation of the ^{22}Ne$(\alpha, \gamma)^{26}$Mg reaction. Because adopted cross sections for both the (α, n) and (α, γ) reactions on ^{22}Ne are in large part guesses, only the sense of the number abundance changes are to be taken seriously.

Number abundances of 24 of the 51 isotopes in the nuclear-reaction network employed in the calculations are listed Table 20.2.1 at four different positions in the fourth of the models being considered. Only those isotopes are listed for which the number abundance of at least one of the chosen positions is larger than 10^{-7}. The four positions are: (1) just below the outer edge of the convective core; (2) the middle of the region just vacated by the shrinking convective core; (3) just outside of the outer edge of the convective core before it began shrinking in mass; and (4) the outer edge of the model interior. The outer edge of the model interior does not coincide with the surface, since the mass of the static envelope has been increased to 0.5 $M_\odot$ in order to avoid an excessive number of interior shells. In the first location, at the outer edge of the convective core, the only isotope not in convective equilibrium is hydrogen.

It is of interest to compare the distributions of isotopes at the different locations. The sum of the principal CNO isotopes in the convective envelope is 6.6329×10^{-4}. At $M = 6.162$ $M_\odot$, where most of the original ^{12}C and ^{16}O isotopes have been converted by hydrogen burning into ^{14}N, the sum is 6.6893×10^{-4}, the slight increase in the sum being due to a very modest degree of helium burning. At $M = 6.033$ $M_\odot$, most of the ^{14}N present at $M = 6.162$ $M_\odot$ has been converted into ^{22}Ne, of which almost 20% has experienced the ^{22}Ne$(\alpha, n)^{25}$Mg or the ^{22}Ne$(\alpha, \gamma)^{26}$Mg reaction. Extensive helium burning has produced large abundances of fresh ^{12}C and ^{16}O, and the neutrons released in the ^{22}Ne$(\alpha, n)^{25}$Mg reaction have been captured by many isotopes, leading to a reshuffling of abundances in the network. In particular, neutron captures on ^{25}Mg have contributed to the increase in the abundance of ^{26}Mg and created an abundance of the neutron counter ^{57}Fe which is about 40% of the abundance of ^{56}Fe, the designated heavy element s-process proxy the abundance of which has been held fixed at $Y(^{56}\text{Fe}) = 2.4614 \times 10^{-5}$.

Number abundances in the convective core, as given by abundances at the mass point $M = 5.921$ $M_\odot$, show that (α, γ) reactions have converted a considerable amount of ^{12}C into ^{16}O, some ^{16}O into ^{20}Ne, and some ^{22}Ne into ^{26}Mg. At the center of the model, the ^{22}Ne$(\alpha, n)^{25}$Mg and ^{22}Ne$(\alpha, \gamma)^{26}$Mg reactions occur at approximately the same rate, 2.48×10^9 g^{-1} s^{-1} and 2.25×10^9 g^{-1} s^{-1}, respectively. At mass point $M = 6.033$ $M_\odot$, the sum of the abundances of isotopes with atomic mass between $A = 22$ and $A = 34$ is 7.101×10^{-4}. At mass point $M = 5.921$ $M_\odot$, the sum is 7.143×10^{-4}. The slight increase in the sum is due to neutron captures on ^{20}Ne and then on ^{21}Ne to produce ^{22}Ne.

An upper limit to the average number abundance of neutrons emitted in the convective core up to the time of the fourth model is given by the difference in the number abundance of ^{14}N at $M = 6.162$ $M_\odot$ and the number abundance of ^{22}Ne at $M = 5.921$ $M_\odot$. This difference is 2.48×10^{-4}, which is approximately ten times the abundance of the heavy element s-process proxy ^{56}Fe at any point. This means that, since the number abundance of ^{57}Fe at any point in the convective core of the fourth model is 2.4034×10^{-4}, or almost equal to the number abundance of ^{56}Fe, approximately one out of every ten neutrons released has been captured by heavy s-process isotopes, the other nine having gone into reshuffling the light isotope distribution.

Another perspective on neutron-capture nucleosynthesis is provided by Table 20.2.2, where the most prominent neutron-capture rates on isotopes in the adopted network are shown at two positions in the convective core of the fourth model under consideration. The total neutron-absorption rate is given in the last row of the table. The capture rate on ^{56}Fe is approximately 10% of the sum of all capture rates on all included isotopes of atomic mass equal or greater to that of ^{12}C. Thus, despite the caveats presented, all results indicate that elements lighter than the heavy s-process elements capture of the order of 90% of the neutrons emitted.

In less than 17 600 yr after the occurrence of the fourth model, helium effectively vanishes over the inner $\sim$6 $M_\odot$ of the model and a convective core ceases to exist. Structure

Table 20.2.2 Neutron-capture rates at two locations in the shrinking convective core of a 25 $M_\odot$ model

isotope	abundance	M = 0.0 $M_\odot$	M = 0.5 $M_\odot$
^{12}C	2.3981/−2	4.121/8	3.546/7
^{16}O	4.2282/−2	1.692/9	1.418/8
^{20}Ne	1.4313/−4	1.250/7	9.649/5
^{21}Ne	1.1364/−5	2.217/7	1.996/6
^{22}Ne	3.8458/−4	3.644/7	3.415/6
^{23}Na	1.0083/−5	2.861/7	2.946/6
^{24}Mg	2.2119/−5	7.159/7	5.756/6
^{25}Mg	7.4535/−5	5.452/8	4.740/7
^{26}Mg	1.7851/−4	2.467/7	2.105/6
^{27}Al	2.3711/−6	1.202/7	1.106/6
^{28}Si	2.3582/−5	7.620/7	6.403/6
^{29}Si	4.0986/−6	4.960/7	4.663/6
^{30}Si	4.2482/−6	4.993/7	5.118/6
^{31}P	2.8624/−6	6.251/6	5.618/5
^{32}S	8.7408/−6	4.250/7	3.701/6
^{33}S	1.2956/−7	4.122/7	3.791/6
^{56}Fe	2.4614/−5	3.417/8	2.985/7
$(dn/dt)_-$		3.465/9	2.969/8

variables in the model at this point are shown in Fig. 20.2.9. Energy production by helium burning is not very important ($L_{He} \sim 800\ L_\odot$ versus a surface luminosity of $L_s \sim 182\,000\ L_\odot$), and, deprived of a potent source of nuclear energy, the entire region between the center and the hydrogen-burning shell contracts, releasing gravothermal energy at the rate $L_{grav}^{core} \sim 167\,400\ L_\odot$. Hydrogen burning continues unabated ($L_H \sim 32\,360\ L_\odot$), but absorption in the expanding envelope reduces the outward flow of energy by $L_{grav}^{envelope} \sim -16\,660\ L_\odot$. The increase in surface luminosity has promoted the continued motion inward in mass of the base of the convective envelope, and dredge-up has increased the surface ratio of helium to hydrogen to $Y(^4He)/Y(^1H) \sim 0.15$.

Gravothermal and nuclear energy-production rates are shown as functions of mass in Fig. 20.2.10. As always, the rate at which gravity does work locally, as given by $\mathbf{g} \cdot \mathbf{v}$, when integrated over mass, is equal to the rate at which compressional energy is released locally, when integrated over mass, despite the fact that the rate at which gravity does work has a maximum near the middle of the contracting region while the maximum in the rate at which compressional energy is released occurs at the center of the model.

Number abundances of a selection of isotopes in the model with a helium-exhausted core of mass $M \gtrsim 6\ M_\odot$ are shown as functions of mass in Fig. 20.2.11. In first approximation, the sum of the number abundances of ^{22}Ne, ^{25}Mg, and ^{26}Mg in the helium-exhausted core is constant and the changes ΔY in several number abundances which have occurred during

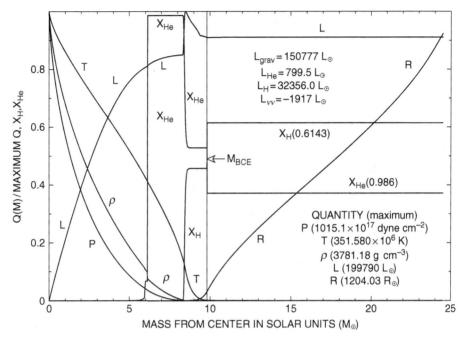

Fig. 20.2.9 Structure of a 25 $M_\odot$ model with a gravitationally contracting helium-exhausted core and a hydrogen-burning shell ($Z = 0.015$, $Y = 0.275$, and AGE $= 7.2109 \times 10^6$ yr)

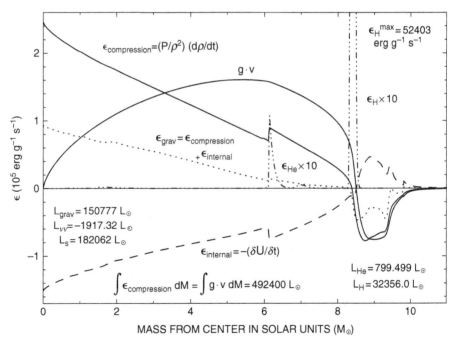

Fig. 20.2.10 Energy-generation rates in a 25 $M_\odot$ model which has just exhausted helium over central regions ($Z = 0.015$, $Y = 0.275$, AGE $= 7.2109 \times 10^6$ yr)

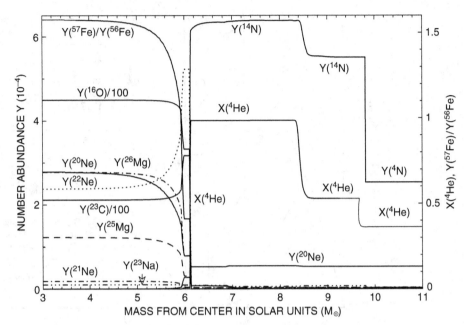

Composition in the gravitationally contracting interior of a 25 $M_\odot$ model with a helium-exhausted core ($Z = 0.015$, $Y = 0.275$, AGE = 7.2109×10^6)

helium burning in the core are related by

$$a \, |\Delta Y(^{22}\text{Ne})| = \Delta Y(^{25}\text{Mg}) + \delta Y_{25} = \Delta Y(n) \tag{20.2.1}$$

and

$$b \, |\Delta Y(^{22}\text{Ne})| = \Delta Y(^{26}\text{Mg}) - \delta Y_{25}, \tag{20.2.2}$$

where a and b are the average fractions of ^{22}Ne nuclei which experience (α, n) and (α, γ) reactions, respectively, during the helium-burning episode, the ΔYs are the changes in number abundances during the episode, with $\Delta Y(n)$ being the total number abundance of neutrons released, and δY_{25} is the number abundance of ^{25}Mg nuclei which experience (n, γ) reactions. Solving eqs. (20.2.1) and (20.2.2),

$$2 \, \delta Y_{25} = (a - b) \, |\Delta Y(^{22}\text{Ne})| + \left(\Delta Y(^{26}\text{Mg}) - \Delta Y(^{26}\text{Mg})\right). \tag{20.2.3}$$

In the region with $M \gtrsim 5 \, M_\odot$, $|\Delta Y(^{22}\text{Ne})| \sim 4.0 \times 10^{-4}$, $\Delta Y(^{25}\text{Mg}) \sim 1.2 \times 10^{-4}$, and $\Delta Y(^{26}\text{Mg}) \sim 2.8 \times 10^{-4}$, so

$$\delta Y_{25} \sim [2.0 \, (a - b) + 0.8] \times 10^{-4}. \tag{20.2.4}$$

It follows from eq. (20.2.1) and the estimate of $\delta Y(^{25}\text{Mg})$ that

$$\Delta Y(n) = \Delta Y(^{25}\text{Mg}) + \delta Y_{25} \sim (a - b + 1) \times 2.0 \times 10^{-4}. \tag{20.2.5}$$

The efficiency with which ^{56}Fe has captured neutrons is given by

$$\text{eff} = \frac{Y(^{57}\text{Fe})}{Y(^{56}\text{Fe})} \frac{2.4614 \times 10^{-5}}{Y(n)} \sim 1.55 \frac{2.4614 \times 10^{-5}}{Y(n)} \sim \frac{0.191}{a - b + 1}. \tag{20.2.6}$$

Coupling this result with the requirement that $(a+b) = 1$, an estimate of the efficiency then determines a and b separately. For an efficiency of 10%, one has $a \sim 0.85$ and $b \sim 0.15$. For an efficiency of 20%, $a \sim 0.48$ and $b \sim 0.52$. At the center of the model from which Table 20.2.2 is derived, the (α, n) and (α, γ) rates for ^{22}Ne are nearly the same, suggesting an efficiency close to 20%. On the other hand, the average efficiency in the convective core is actually closer to 10%, indicating that, during helium burning in the present instance, when averaged over the helium-burning region, the (α, n) rate is several times larger than the (α, γ) rate.

Another perspective follows from the observation that, in the inner ~ 5 $M_\odot$ of the model, the number abundance of ^{22}Ne is approximately 40% of the maximum number abundance achieved prior to helium burning. In the same region, the ratio of ^{57}Fe to ^{56}Fe is ~ 1.55, so the degree to which neutron capture has enhanced the abundances of heavy s-process elements is of the order of 60% of the maximum possible enhancement, regardless of the uncertainties in charged-particle nuclear reaction cross sections involved in the calculations. In other words, the filtering action of light isotopes is such that the maximum number of neutrons available for heavy s-process nucleosynthesis, even if every original ^{22}Ne isotope were to produce a neutron, is ~ 2.5 neutrons per iron seed nucleus.

In closing this section, it is to be remarked that only part of the ~ 6 $M_\odot$ of matter with enhanced s-process isotopes will survive further evolution with its current abundance distribution and contribute to the galactic abundance of s-process elements. First of all, the inner ~ 1.5–2 $M_\odot$ of the region ultimately becomes part of a compact remnant. Second, during the operation of the processes whereby most of the star is ejected, memory of much of the nucleosynthesis which occurred during earlier evolution is lost. Nevertheless, the evidence for s-process nucleosynthesis in and ejection from stars is ubiquitous in the Universe and it is reasonable to suppose that massive stars make a contribution which must be taken into account. This justifies the effort expended in understanding where and how in massive stars the relevant nucleosynthesis occurs.

20.3 Core and shell carbon-burning phases

The rapid contraction and heating of the helium-exhausted core described in the previous section has several consequences. First, the increase in density and temperature in the ~ 6 $M_\odot$ core leads to an increase in the rate at which neutrino energy losses due to weak interactions occur in the core. To make up for this increased energy-loss rate, the rate of

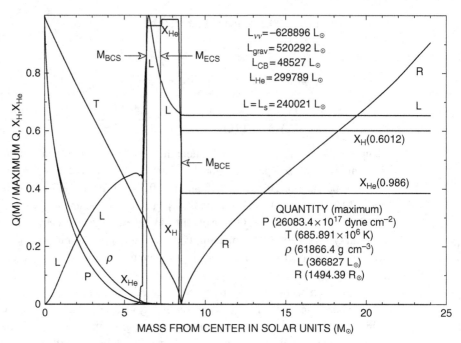

Fig. 20.3.1   Structure of a 25 $M_\odot$ model burning helium in a shell and carbon at the base of a contracting core ($Z = 0.015$, $Y = 0.275$, and AGE $= 7.220963 \times 10^6$ yr)

contraction and heating of the core accelerates. Second, the contraction of the core is accompanied by contraction and heating at the carbon–helium interface and the helium-burning shell is resurrected. The increased flux of energy from the core and the helium-burning shell forces matter above the helium-burning shell to expand and cool, with the result that the hydrogen-burning shell is essentially extinguished.

These developments are reflected in Figs. 20.3.1 and 20.3.2 which describe characteristics of a model which is only $\sim 10^4$ yr older than the model with characteristics described by the corresponding Figs. 20.2.9 and 20.2.10. Structural characteristics of the more evolved model are shown in Fig. 20.3.1 and gravothermal, weak interaction, and nuclear reaction energy-generation characteristics are shown in Fig. 20.3.2. In the 10^4 yr of evolution separating the two models, central density has increased by a factor slightly larger than 16 and central temperature has almost doubled.

The close proximity in Fig. 20.3.2 of the curves for ϵ_{grav} and $\epsilon_{\nu\bar{\nu}}$ in the helium-exhausted core demonstrate that neutrino losses are the basic driving force for the contraction and heating of the core. Nuclear burning at and near the center due to nuclear reactions connected with carbon burning produces a luminosity of $L_{\text{CB}} \sim 48\,500\ L_\odot$, compared with the almost 13 times larger rate at which neutrinos and antineutrinos generated by weak interactions carry off energy. The luminosity at the base of the helium-burning shell, $L_{\text{base}} \sim 1.66 \times 10^5\ L_\odot$, is equal to the rate of gravothermal energy release plus the rate of nuclear energy release due to carbon burning in the helium-exhausted core minus the rate $L_{\nu\bar{\nu}} \sim 6.29 \times 10^5\ L_\odot$ at which neutrinos and antineutrinos generated by weak interactions

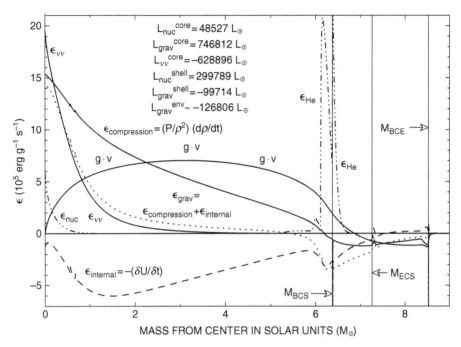

Energy-generation and loss rates in a 25 $M_\odot$ model burning helium in a shell and carbon at the base of a contracting core ($Z = 0.015, Y = 0.275$, AGE $= 7.22096 \times 10^6$ yr)

carry off energy. Thus, the rate at which gravothermal energy is released in the core is $L_{\mathrm{grav}}^{\mathrm{core}} \sim 7.47 \times 10^5\, L_\odot$. Integrated over the entire model, $L_{\mathrm{grav}} \sim 5.20 \times 10^5\, L_\odot$. From the luminosity profile in Fig. 20.3.1, it is evident that the rate at which gravothermal energy is absorbed by matter above the helium-burning shell is $-L_{\mathrm{grav}}^{\mathrm{env}} \sim 1.27 \times 10^5\, L_\odot$. Since the total helium-burning luminosity is $L_{\mathrm{He}} \sim 3.00 \times 10^5\, L_\odot$ and the surface luminosity is $L_{\mathrm{s}} = 2.40 \times 10^5\, L_\odot$, gravothermal energy is being absorbed in the helium-burning shell at the rate $-L_{\mathrm{grav}}^{\mathrm{He\ shell}} \sim 0.997 \times 10^5\, L_\odot$.

Energy-loss rates associated with neutrinos produced by the four neutrino-production processes predicted by square terms in the weak interaction Hamiltonian (Chapter 15) are shown in Fig. 20.3.3. It is evident that the largest energy-loss rates are associated with neutrinos produced by the annihilation of real electron–positron pairs and by the photoneutrino process, with losses associated with neutrinos produced by the plasma and bremsstrahlung processes playing very minor roles. At temperatures in the model here being considered, eq. (15.3.60), when divided by the density, is a relevant first approximation to the $e^+ e^-$ pair-annihilation neutrino–antineutrino energy-loss rate in units of erg g^{-1} s^{-1}:

$$\epsilon_{\nu_e \bar{\nu}_e} (\mathrm{erg\ g}^{-1}\ \mathrm{s}^{-1}) \sim 4.99 \times 10^{18}\, T_9^3\, (1 + 1.10\, T_9)\, \exp\left(-11.86/T_9\right)\, \frac{1}{\rho_0}, \qquad (20.3.1)$$

where ρ_0 is the density in units of g cm^{-3}.

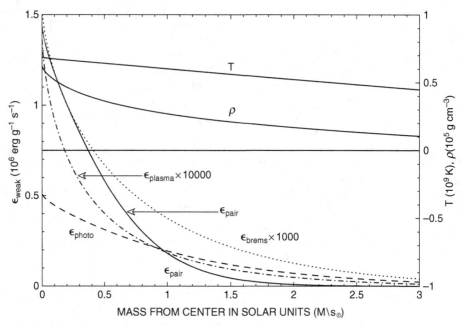

Fig. 20.3.3 Weak interaction neutrino-loss rates in the contracting core of a 25 $M_\odot$ model with a helium-burning shell ($Z = 0.015$, $Y = 0.275$, and AGE $= 7.22096 \times 10^6$ yr)

At the center of the model being described, $T_9 \sim 0.687$ and $\rho_0 \sim 6.24 \times 10^4$. With these values, eq. (20.3.1) gives $\epsilon_{\nu_e \bar{\nu}_e} \sim 1.45 \times 10^6$ erg g^{-1} s^{-1}. As shown in Fig. 20.3.3, the calculated value at the model center is $\epsilon_{\nu_e \bar{\nu}_e} \sim 1.44 \times 10^6$ erg g^{-1} s^{-1}, slightly smaller than the estimate given by eq. (20.3.1). The contribution of neutral currents is not taken into account in the construction of eq. (20.3.1) whereas it is taken into account in the approximations prepared by Itoh and his coworkers and used in the model calculations (Itoh, Hayashi, Nishikawa, & Kohyama, 1996 and their coworkers, who used the formulation of weak interaction theory given by S. Weinberg, 1967, and A. Salam, 1968). The near equality of the two estimates may simply reflect the fact that analytical approximations are, indeed, approximations and not precisely accurate reflections of exact results.

Reactions between heavy ions produce one or more types of ion, each of mass comparable with the sum of the masses of the interacting ion pair. Typically, a proton, alpha particle, or neutron is produced as a byproduct. The combined number of types of hydrogen-burning, alpha-burning, and neutron-capture reactions which follow can be as large as or larger than the number of heavy ion reactions which occur and much of the energy release associated with any given heavy ion reaction comes from reactions other than the initiating reaction. Thus, although the nuclear burning stages involving heavy ion reactions are conventionally labeled with the names of the heavy ions involved, e.g., carbon burning, carbon–oxygen burning, and oxygen burning, in reality a good fraction of the energy generated during the named stages is due to hydrogen-capture, alpha-capture, and neutron-capture reactions.

Two consistent sets of cross sections for reactions between two ^{12}C nuclei at center of mass energies in the range 2.6–6.0 MeV have been obtained by J. R. Patterson, H. Winkler, & C. S. Zaidens (1969) and by H. W. Becker, K. U. Kettner, C. Rolfs, & H. P. Trautvetter (1981). The energy dependences of the cross sections indicate the existence of resonances, but, when averaged over a large enough energy interval, variations about the mean cross section are of the order of 50% and, assuming an overall smooth behavior, one can define a cross section factor that can be extrapolated with some confidence to energies of relevance in the stellar interior. Assuming that the average cross section factor varies monotonically, the maximum in the thermal average of $\langle \sigma\, v \rangle$ occurs at an energy near the energy given by eq. (6.7.17) in Volume 1 as

$$E_0 = 1.22 \left(Z_1^2\, Z_2^2\, A_{12}\, T_6^2 \right)^{1/3} \text{ keV}, \tag{20.3.2}$$

which, in the present instance, is

$$E_0 = 1.22 \left(6^2\, 6^2\, (12/2)\, 10^6\, T_9^2 \right)^{1/3} \text{ keV} = 2.42\, T_9^{2/3} \text{ MeV}. \tag{20.3.3}$$

The energy width over which major contributions to the average occurs is given by eq. (6.7.18) in Volume 1 which, in the present instance, is

$$\Delta E_0 = 0.614 \left(\frac{T_6}{Z_1^2\, Z_2^2\, A_{12}} \right)^{1/6} E_0$$

$$= 236.9 \text{ keV} \left(Z_1^2\, Z_2^2\, A_{12} \right)^{1/6} T_9^{5/6} = 1.054\, T_9^{5/6} \text{ MeV}. \tag{20.3.4}$$

For temperatures in the range $T_9 = 0.7$–1.0, where carbon burning occurs, $E_0 = 2.15$–2.42 MeV and $\Delta E_0 = 0.78$–1.05 MeV, so the estimate of the reaction rate at temperatures of relevance relies on cross sections at energies that extend into the energy range where experimental cross sections are reasonably well established. However, because of the known fluctuations about an averaged cross section, which may extend into the region where no data exists, there is an unavoidable uncertainty of the order of $\pm 50\%$ in the averaged cross section factor at any temperature.

The Patterson *et al.* (1969) data suggests an average center of mass cross section factor of $\bar{S} \sim (3 \pm 1) \times 10^{16}$ MeV b over the energy range studied experimentally and Becker *et al.* (1981) suggest $\bar{S} \sim 2 \times 10^{16}$ MeV b at energies of relevance in stellar interiors. The overall reaction rate, branching ratios, and Q values that are used here are taken from G. R. Caughlan and W. A. Fowler (1988). In particular, the adopted overall reaction rate is

$$N_A\, \langle \sigma\, v \rangle_{12,12} = \exp(61.319 - 84.165/T_9^{1/3} - 0.002\,12\, T_9^3)/T_9^{2/3}. \tag{20.3.5}$$

From the temperature dependence of the rate, an uncertainty of a factor of two in the adopted cross section factor translates into a 7% uncertainty in the temperature at which carbon burning occurs. The branching ratios adopted by Caughlan and Fowler for the reactions which yield neutrons, protons, and alpha particles are 0.0 (-2.598 MeV), 0.44 (2.242 MeV), and 0.56 (4.621 MeV), where the energies in parentheses are the Q values for the reactions.

Table 20.3.1 Nuclear energy generation rates at the center of a $25\,M_\odot$ model at the beginning of the core carbon-burning phase

reaction	ϵ (erg/g/s)	reaction	ϵ (erg/g/s)
$^{12}\mathrm{C}(^{12}\mathrm{C},\alpha)^{20}\mathrm{Ne}$	1.518/5	$^{22}\mathrm{Ne}(\alpha,n)^{25}\mathrm{Mg}$	$-1.883/2$
$^{12}\mathrm{C}(^{12}\mathrm{C},p)^{23}\mathrm{Na}$	5.785/4	$^{22}\mathrm{Ne}(n,\gamma)^{23}\mathrm{Ne}$	2.809/2
$^{12}\mathrm{C}(p,\gamma)^{13}\mathrm{N}$	1.172/5	$^{23}\mathrm{Ne}(e^-\bar{\nu}_e)^{23}\mathrm{Na}$	1.013/2
$^{12}\mathrm{C}(n,\gamma)^{13}\mathrm{C}$	1.034/4	$^{23}\mathrm{Na}(p,\gamma)^{24}\mathrm{Mg}$	6.438/3
$^{13}\mathrm{C}(\alpha,n)^{16}\mathrm{O}$	3.319/4	$^{23}\mathrm{Na}(p,\alpha)^{20}\mathrm{Ne}$	6.775/3
$^{13}\mathrm{N}(\gamma,p)^{12}\mathrm{C}$	$-9.845/4$	$^{23}\mathrm{Na}(n,\gamma)^{24}\mathrm{Na}$	8.077/3
$^{13}\mathrm{N}(e^+\nu_e)^{13}\mathrm{C}$	1.903/4	$^{24}\mathrm{Na}(e^-\bar{\nu}_e)^{24}\mathrm{Mg}$	6.144/3
$^{14}\mathrm{N}(\alpha,\gamma)^{18}\mathrm{O}$	1.320/3	$^{24}\mathrm{Mg}(n,\gamma)^{25}\mathrm{Mg}$	2.795/3
$^{15}\mathrm{N}(p,\alpha)^{12}\mathrm{C}$	1.022/3	$^{25}\mathrm{Mg}(n,\gamma)^{26}\mathrm{Mg}$	3.274/3
$^{16}\mathrm{O}(\alpha,\gamma)^{20}\mathrm{Ne}$	7.528/4	$^{26}\mathrm{Mg}(p,\gamma)^{27}\mathrm{Al}$	3.687/4
$^{16}\mathrm{O}(n,\gamma)^{17}\mathrm{O}$	5.262/4	$^{26}\mathrm{Mg}(n,\gamma)^{27}\mathrm{Mg}$	1.559/3
$^{17}\mathrm{O}(p,\alpha)^{18}\mathrm{F}$	4.673/3	$^{27}\mathrm{Mg}(e^-\bar{\nu}_e)^{27}\mathrm{Al}$	3.865/2
$^{17}\mathrm{O}(\alpha,\gamma)^{21}\mathrm{Ne}$	1.638/3	$^{27}\mathrm{Al}(p,\gamma)^{28}\mathrm{Si}$	94.90
$^{17}\mathrm{O}(\alpha,n)^{20}\mathrm{Ne}$	4.599/3	$^{27}\mathrm{Al}(n,\gamma)^{28}\mathrm{Al}$	2.692/3
$^{18}\mathrm{O}(p,\alpha)^{15}\mathrm{N}$	7.909/2	$^{28}\mathrm{Al}(e^-\bar{\nu}_e)^{28}\mathrm{Si}$	1.019/3
$^{17}\mathrm{F}(\gamma,p)^{16}\mathrm{O}$	$-2.903/2$	$^{28}\mathrm{Si}(n,\gamma)^{29}\mathrm{Si}$	2.290/3
$^{18}\mathrm{F}(e^+\nu_e)^{18}\mathrm{O}$	3.852/2	$^{29}\mathrm{Si}(n,\gamma)^{30}\mathrm{Si}$	1.184/3
$^{20}\mathrm{Ne}(\alpha,\gamma)^{24}\mathrm{Mg}$	8.877/2	$^{30}\mathrm{Si}(n,\gamma)^{31}\mathrm{Si}$	4.665/2
$^{20}\mathrm{Ne}(n,\gamma)^{21}\mathrm{Ne}$	5.273/3	$^{31}\mathrm{P}(n,\gamma)^{32}\mathrm{P}$	4.117/2
$^{21}\mathrm{Ne}(n,\gamma)^{22}\mathrm{Ne}$	1.351/3	$^{32}\mathrm{S}(n,\gamma)^{33}\mathrm{S}$	1.122/3
$^{22}\mathrm{Ne}(p,\gamma)^{23}\mathrm{Na}$	4.434/3	$^{33}\mathrm{S}(n,\alpha)^{30}\mathrm{Si}$	4.452/2
$^{22}\mathrm{Ne}(\alpha,\gamma)^{26}\mathrm{Mg}$	2.456/2	$^{56}\mathrm{Fe}(n,\gamma)^{57}\mathrm{Fe}$	9.968/3

At the center of the model with characteristics described in Figs. 20.3.1 and 20.3.2, the reactions $^{12}\mathrm{C}(^{12}\mathrm{C},\alpha)^{20}\mathrm{Ne}$ and $^{12}\mathrm{C}(^{12}\mathrm{C},p)^{23}\mathrm{Na}$ together release energy at the rate $\epsilon_{\mathrm{C+C}} \sim 2.10 \times 10^5$ erg g^{-1} s^{-1}, but reactions involving proton capture, alpha capture, and neutron capture on isotopes in the nuclear reaction network produce energy at the rates $\sim 2.1 \times 10^5$ erg g^{-1} s^{-1}, $\sim 1.32 \times 10^5$ erg g^{-1} s^{-1}, $\sim 1.1 \times 10^5$ erg g^{-1} s^{-1}, respectively, for a combined energy-production rate of 4.52×10^5 erg g^{-1} s^{-1} over and above the rate of energy release by the initiating heavy ion reactions. In addition to the contribution of nuclear fusion reactions, there are an assortment of beta decay reactions which contribute $\sim 2.7 \times 10^4$ erg g^{-1} s^{-1} to the overall nuclear energy-production rate and, most notably, $^{13}\mathrm{N}(\gamma,p)^{12}\mathrm{C}$ reactions absorb energy at the rate $\sim 9.8 \times 10^4$ erg g^{-1} s^{-1}. Details are given in Table 20.3.1. The selection criterion for inclusion in the table is, with one exception, an energy-generation or energy-absorption rate in excess of ~ 100 erg g^{-1} s^{-1}.

Abundances at the model center of a selection of the heavy isotopes in the isotope network used are given in Table 20.3.2.

Contributions of various nuclear energy sources are shown as functions of mass in Fig. 20.3.4. There, $\epsilon_{\mathrm{C+C}}$ is the rate at which $^{12}\mathrm{C} + ^{12}\mathrm{C}$ reactions produce energy and

Table 20.3.2 Abundances at the center of a 25 $M_\odot$ model at the beginning of the core carbon-burning phase			
isotope	abundance	isotope	abundance
^{12}C	2.030/−2	^{23}Ne	1.594/−18
^{13}C	3.644/−7	^{23}Na	1.579/−4
^{13}N	1.202/−11	^{24}Na	1.692/−12
^{14}N	1.223/−5	^{24}Mg	2.310/−5
^{15}N	1.364/−8	^{25}Mg	1.049/−4
^{16}O	4.488/−2	^{26}Mg	2.772/−4
^{17}O	4.759/−5	^{27}Mg	2.185/−13
^{18}O	2.592/−7	^{27}Al	2.504/−5
^{17}F	1.371/−20	^{28}Al	8.077/−14
^{18}F	3.562/−12	^{28}Si	2.040/−5
^{20}Ne	5.821/−4	^{29}Si	6.081/−6
^{21}Ne	2.239/−5	^{30}Si	6.868/−6
^{21}Ne	2.239/−5	^{31}P	5.2034/−6
^{22}Ne	2.342/−4	^{32}S	6.802/−6
^{56}Fe	2.461/−5	^{33}S	1.498/−7
^{57}Fe	4.790/−5		

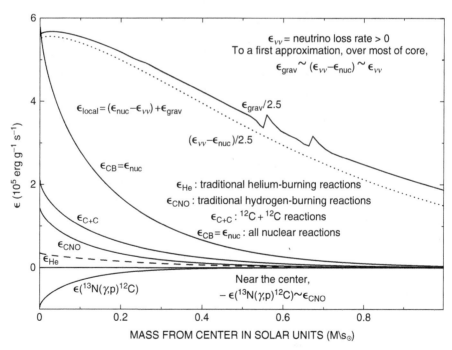

Fig. 20.3.4 Nuclear energy-generation rates in the core of a 25 $M_\odot$ model balancing neutrino losses with gravothermal energy ($Z = 0.015$, $Y = 0.275$, and AGE $= 7.22096 \times 10^6$ yr)

ϵ_{CB} is the total rate at which nuclear reactions produce energy. Energy-production rates by classical hydrogen-burning and helium-burning reactions are given by ϵ_{CNO} and ϵ_{He}, respectively. The lesson from Table 20.3.1 and Fig. 20.3.4 is that, during carbon burning, approximately one third of the nuclear energy liberated comes from the initiating heavy ion reactions and two thirds comes from reactions involving capture of particles liberated by the initiating reactions.

At the model center, two beta-unstable isotopes in the adopted nuclear reaction network, ^{13}N and ^{17}F, have lifetimes against photodisintegration which are shorter than their lifetimes against beta decay. The most conspicuous example of the phenomenon is ^{13}N. From the energetics in Table 20.3.1, it follows that for every 100 ^{13}N nuclei formed by ^{12}C$(p, \gamma)^{13}$N reactions, 83 experience a ^{13}N$(\gamma, p)^{12}$C reaction, while only 17 experience a ^{13}N$(e^+ \nu_e)^{13}$C reaction. As a result, most of the energy liberated by ^{12}C$(p, \gamma)^{13}$N reactions is absorbed in the inverse photodisintegration process. Furthermore, since only a fraction of the ^{13}N nuclei produced decay to ^{13}C, not only is the operation of the rest of the traditional CNO cycle curtailed, but the production of neutrons by the ^{13}C$(\alpha, n)^{16}$O reaction is severely inhibited.

A demonstration of how the phenomenon comes about requires the construction of a theoretical photodisintegration rate. Suppose that a nucleus $I_{Z,N}$ and the nucleus $I_{Z-1,N}$ resulting from the reaction

$$I_{Z,N} + \gamma \rightarrow I_{Z-1,N} + p \tag{20.3.6}$$

are in thermal equilibrium. The relationship between the number abundances of the two isotopes and the number abundance of protons follows from a straightforward generalization of the Saha equation relating the abundances of atoms in two stages of ionization with the abundance of free electrons. In the nuclear physics case, the heavy particles are isotopes which differ by one unit of charge and the free particles are protons. Thus, a straightforward application of eq. (4.15.10) in Volume 1 gives

$$\frac{n_p \, n_{Z-1,N}}{n_{Z,N}} = \left(\frac{2\pi \mu kT}{h^2}\right)^{3/2} \frac{2 \, g_{Z-1,N}}{g_{Z,N}} \exp\left(-Q/kT\right), \tag{20.3.7}$$

where the ns are number densities in units of cm^{-3}, $\mu = \left(M_{Z-1,N} M_p / M_{Z,N}\right)$ is the reduced mass of the system $I_{Z-1,N}$ plus a proton, the factor 2 is the statistical weight of the spin 1/2 proton, $g_{Z-1,N}$ and $g_{Z,N}$ are the statistical weights of the isotopes $I_{Z-1,N}$ and $I_{Z,N}$, and Q is the difference in energy between the two isotopes.

The photodisintegration rate can be written as

$$R_{\gamma p} = \lambda_{\gamma p} \, n_{Z,N} \tag{20.3.8}$$

and the rate of the fusion reaction

$$I_{Z-1,N} + p \rightarrow I_{Z,N} + \gamma \tag{20.3.9}$$

can be written as

$$R_{p\gamma} = n_p \, n_{Z-1,N} \, \langle \sigma_{p\gamma} \, v \rangle. \tag{20.3.10}$$

Equating the two rates gives

$$\lambda_{\gamma p}\, n_{Z,N} = n_p\, n_{Z-1,N} \langle \sigma_{p\gamma}\, v \rangle. \tag{20.3.11}$$

Combining eq. (20.3.6) with eq. (20.3.11) gives

$$\lambda_{\gamma p} = \frac{n_p\, n_{Z-1,N}}{n_{Z,N}}\, \langle \sigma_{p\gamma}\, v \rangle = \left(\frac{2\pi \mu k T}{h^2} \right)^{3/2} \frac{2\, g_{Z-1,N}}{g_{Z,N}}\, \exp\left(-Q/kT \right) \langle \sigma_{p\gamma}\, v \rangle. \tag{20.3.12}$$

Inserting values for fundamental constants,

$$\lambda_{\gamma p} = 5.9417 \times 10^{33}\, (\bar\mu T_9)^{3/2}\, \frac{2\, g_{Z-1,N}}{g_{Z,N}}\, \exp(-11.605\, Q(\mathrm{MeV})/T_9)\, \langle \sigma_{p\gamma}\, v \rangle, \tag{20.3.13}$$

where $\bar\mu = (Z + N - 1)/(Z + N)$ is the reduced mass of the two particle system in units of the proton mass, Q is in MeV, and temperature is in units of $10^9\,$K. In the nuclear astrophysics literature, $\langle \sigma v \rangle$ is normally multiplied by Avogadro's number N_A, so it is useful to rewrite
eq. (20.3.13) as

$$\lambda_{\gamma p} = 0.986\,495 \times 10^{10}\, \frac{2\, g_{Z-1,N}}{g_{Z,N}}\, (\bar\mu T_9)^{3/2}\, \exp(-11.605\, Q(\mathrm{MeV})/T_9)\, \left[N_A \langle \sigma_{p\gamma}\, v \rangle \right]$$

$$= \frac{2\, g_{Z-1,N}}{g_{Z,N}}\, (\bar\mu T_9)^{3/2}\, \exp(23.012 - 11.605\, Q(\mathrm{MeV})/T_9)\, \left[N_A \langle \sigma_{p\gamma}\, v \rangle \right]. \tag{20.3.14}$$

The statistical weights of ^{12}C and ^{13}N in their ground states are 1 and 2, respectively, their ground state energies differ by $Q = 1.943$ MeV, and $\bar\mu = 12/13$. Thus, for ^{13}N, the coefficient of $\left[N_A \langle \sigma_{p\gamma}\, v \rangle \right]$ in eq. (20.3.14) is

$$T_9^{3/2}\, \exp(22.892 - 22.549/T_9). \tag{20.3.15}$$

The lifetime of ^{13}N against positron decay is 9.965 minutes $= 597.9$ s and its lifetime against photodisintegration with the emission of a proton is

$$t_{\gamma p} = \frac{1}{\lambda_{\gamma p}}. \tag{20.3.16}$$

Adopting the Angulo et al. (1999) rate estimates and neglecting the effect of electron screening, at $T_9 = 0.6$, $N_A \langle \sigma_{p\gamma}\, v \rangle = 79.9\ \mathrm{g}^{-1}\ \mathrm{cm}^3\ \mathrm{s}^{-1}$, and, from eqs. (20.3.14)–(20.3.16), $t_{\gamma p} = 2990$ s. At $T_9 = 0.7$, $N_A \langle \sigma_{p\gamma}\, v \rangle = 181.6\ \mathrm{g}^{-1}\ \mathrm{cm}^3\ \mathrm{s}^{-1}$, and $t_{\gamma p} = 105$ s. So, over an interval of $\Delta T_9 = 0.1$, the lifetime of ^{13}N against photodisintegration with the ejection of a proton goes from being five times the positron decay lifetime to being over five times smaller than the positron decay lifetime.

Variations with respect to mass of radiation pressure over gas pressure, of various temperature–pressure gradients, and of the opacity are shown in Fig. 20.3.5 for the incipiently core carbon-burning model. Comparing with the model described in Fig. 20.2.6, it is apparent that, during the approximately 28 000 yr of evolution between the two models, the

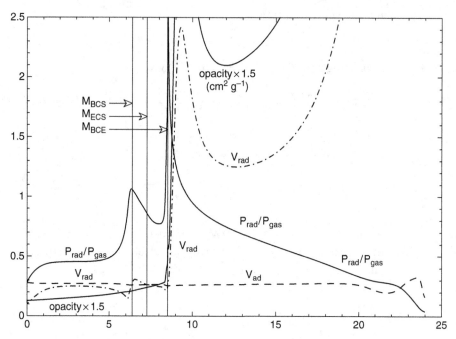

Fig. 20.3.5 Interesting variables in a 25 $M_\odot$ model burning helium in a shell and carbon at the base of a contracting core ($Z = 0.015$, $Y = 0.275$, and AGE $= 7.22096 \times 10^6$ yr)

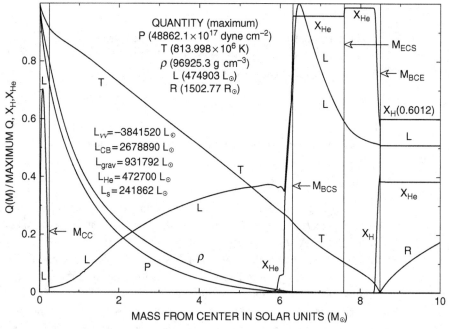

Fig. 20.3.6 Structure of a 25 $M_\odot$ model burning helium in a shell and carbon in a small convective core ($Z = 0.015$, $Y = 0.275$, and AGE $= 7.2213433 \times 10^6$ yr)

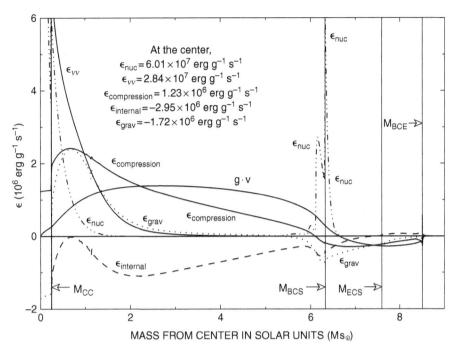

At the center,
$\epsilon_{nuc} = 6.01 \times 10^7$ erg g^{-1} s^{-1}
$\epsilon_{\nu\nu} = 2.84 \times 10^7$ erg g^{-1} s^{-1}
$\epsilon_{compression} = 1.23 \times 10^6$ erg g^{-1} s^{-1}
$\epsilon_{internal} = -2.95 \times 10^6$ erg g^{-1} s^{-1}
$\epsilon_{grav} = -1.72 \times 10^6$ erg g^{-1} s^{-1}

Fig. 20.3.7 Energy-generation and loss rates in a 25 $M_\odot$ model burning helium in a shell and carbon in a small convective core ($Z = 0.015, Y = 0.275, \text{AGE} = 7.221343 \times 10^6$ yr)

ratio P_{rad}/P_{gas} at the base of the convective envelope has nearly doubled and the steepness of the gradients of the ratio on either side of the base have increased even more.

During carbon-burning phases, thanks to the high temperature sensitivities of both nuclear reaction rates and neutrino-loss rates, evolutionary changes in the helium-exhausted core of the model occur at an ever accelerating rate. After only 380 yr of evolution, the model with characteristics depicted in Figs. 20.3.1–20.3.5 is transformed into a model with characteristics depicted in Figs. 20.3.6–20.3.9. Structure variables at the center have changed by $\Delta T/T \sim 0.19$, $\Delta\rho/\rho \sim 0.57$, and $\Delta P/P \sim 0.87$. The nuclear energy-generation rate and the neutrino-loss rate at the center have changed by $\Delta\epsilon_{nuc}/\epsilon_{nuc} \sim 102$ and $\Delta\epsilon_{\nu\bar\nu}/\epsilon_{\nu\bar\nu} \sim 14.6$, respectively, the components of the gravothermal energy-generation rate have changed by $\Delta\epsilon_{compression}/\epsilon_{compression} \sim -0.25$ and $\Delta\epsilon_{internal}/\epsilon_{internal} \sim -23.2$, respectively, and the gravothermal energy-generation rate has changed by $\Delta\epsilon_{grav}/\epsilon_{grav} \sim -0.22$.

These changes show that, at and near the center, energy generation by carbon-burning reactions has replaced grovothermal energy generation as the major contributor to neutrino energy losses. At the center, ϵ_{nuc} exceeds $\epsilon_{\nu\bar\nu}$ by slightly more than a factor of 2 and since $\epsilon_{nuc} > \epsilon_{\nu\bar\nu}$ over the next ~ 0.08 $M_\odot$, the luminosity increases outward, reaching a relative maximum of $L \sim 333\,611$ $L_\odot$ at $M \sim 0.08$ $M_\odot$, as shown in Fig. 20.3.6. Beyond this point, $\epsilon_{\nu\bar\nu} > \epsilon_{nuc}$, and the luminosity decreases outward until almost all of the energy generated by nuclear burning has been converted into neutrino energy losses. However, the large outward flux of energy created in the region of increasing luminosity results in a radiative

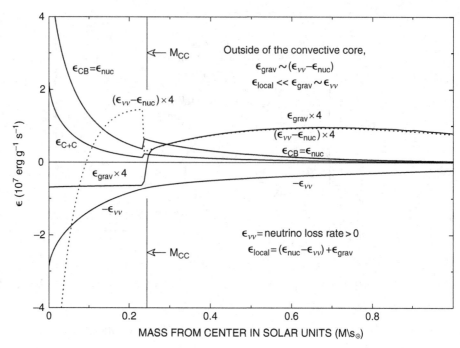

Fig. 20.3.8 Energy-generation and loss rates in a 25 $M_\odot$ model burning carbon in a small convective core ($Z = 0.015$, $Y = 0.275$, and AGE $= 7.221343 \times 10^6$ yr)

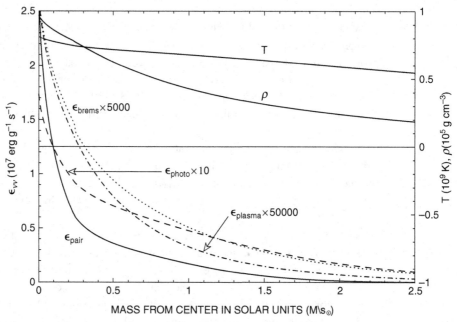

Fig. 20.3.9 Weak interaction neutrino-loss rates in a 25 $M_\odot$ model burning carbon in a convective core ($Z = 0.015$, $Y = 0.275$, and AGE $= 7.22134 \times 10^6$ yr)

temperature–pressure gradient which exceeds the adiabatic temperature–pressure gradient and this situation prevails over the next $\sim$0.16 $M_\odot$, with the result that a small convective core extends over the inner 0.24 $M_\odot$ of the model.

Contraction and heating extend from the center to the helium-burning shell with a consequent change of $\Delta L_{He}/L_{He} \sim 0.56$ in the helium-burning luminosity. In contrast, surface characteristics have changed by only $\Delta L/L \sim 0.0077$ and $\Delta R/R \sim 0.0056$ and the location of the base of the convective envelope is essentially unchanged.

The luminosities generated by the various sources are, by previous standards, huge, certainly far exceeding the luminosities of all but the very largest globular cluster in our galaxy, ω Cen. For example, the luminosity generated by reactions associated with carbon burning is $L_{CB} = 2.678\,8890 \times 10^6\ L_\odot$, and the luminosity generated by weak interaction induced neutrino energy losses is $L_{\nu\bar{\nu}} = -3.841\,520 \times 10^6\ L_\odot$. The luminosities due to gravothermal energy generation and helium-burning reactions are $L_{grav} = 931\,792\ L_\odot$ and $L_{He} = 472\,700\ L_\odot$, respectively.

As shown in Fig. 20.3.7, beyond the outer edge of the convective core, the rate of release of gravothermal energy gradually increases outward with respect to the rate of release of nuclear energy as the primary source of energy feeding neutrino losses, with equality of the two positive energy-generation rates being achieved at $M \sim 0.5\ M_\odot$. Outside of the convective core, as demonstrated in Fig. 20.3.8 by the near identity of the curves for ϵ_{grav} and $(\epsilon_{\nu\bar{\nu}} - \epsilon_{nuc})$ in this region, an almost perfect balance prevails between the neutrino-loss rate and the sum of the rates of nuclear and gravothermal energy generation. The balance is almost perfect in the sense that $\epsilon_{net} = \epsilon_{grav} + \epsilon_{nuc} - \epsilon_{\nu\bar{\nu}} \ll \epsilon_{grav}$, but, nevertheless, $\epsilon_{net} > 0$ and the luminosity continues to increase outward, as demonstrated in Fig. 20.3.6.

Neutrino energy-loss rates due to the weak interaction are shown as functions of mass in Fig. 20.3.9 and are to be compared with the loss rates in Fig. 20.3.3. In a mere 380 yr, the neutrino energy-loss rate due to electron–positron annihilation has increased everywhere by more than an order of magnitude. Energy-loss rates due to the other three mechanisms have also increased, but their relative contributions to the total energy-loss rate by neutrinos and antineutrinos have decreased.

The convective core of the model lasts for only about 270 yr, at its maximum extent having a mass $M_{CC}^{max} \sim 0.37\ M_\odot$. It disappears before carbon is exhausted at the center and is replaced by a convective shell of much larger mass, maintained by fluxes generated by carbon burning at the base of the shell. As convection vanishes at the center, the number abundances of the most abundant isotopes at the model center are $Y(^{16}O) = 3.833 \times 10^{-2}$, $Y(^{20}Ne) = 1.601 \times 10^{-2}$, $Y(^{12}C) = 1.925 \times 10^{-3}$, $Y(^{23}Na) = 0.815 \times 10^{-3}$, $Y(^{27}Al) = 0.246 \times 10^{-3}$, and $Y(^{25}Mg) = 0.200 \times 10^{-3}$. The next most abundant isotopes have number abundances $Y(^{21}Ne) = 7.15 \times 10^{-5}$, $Y(^{57}Fe) = 6.73 \times 10^{-5}$, $Y(^{28}Si) = 6.25 \times 10^{-5}$, $Y(^{26}Mg) = 4.60 \times 10^{-5}$, $Y(^{22}Ne) = 4.23 \times 10^{-5}$, $Y(^{56}Fe) = 2.46 \times 10^{-5}$, and $Y(^{17}F) = 2.37 \times 10^{-5}$, followed by isotopes with number abundances of $Y(^{29}Si) = 1.07 \times 10^{-5}$, $Y(^{30}Si) = 0.836 \times 10^{-5}$, $Y(^{32}S) = 0.621 \times 10^{-5}$, and $Y(^{31}P) = 0.572 \times 10^{-5}$.

Characteristics of the last model calculated are depicted in Figs. 20.3.10–20.3.17. Only 190 yr of evolution separates this last model from the model with characteristics depicted in Figs. 20.3.6–20.3.9, but significant changes have occurred in the distribution of variables in the carbon-burning region. Comparing structure variables in Figs. 20.3.10 and 20.3.11

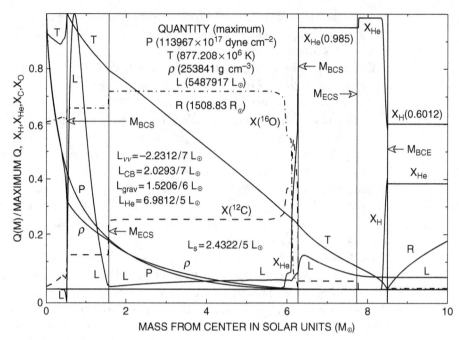

Fig. 20.3.10   Structure of a 25 $M_\odot$ model burning carbon and helium in shells in a contracting core ($Z = 0.015$, $Y = 0.275$, and AGE $= 7.221533 \times 10^6$ yr)

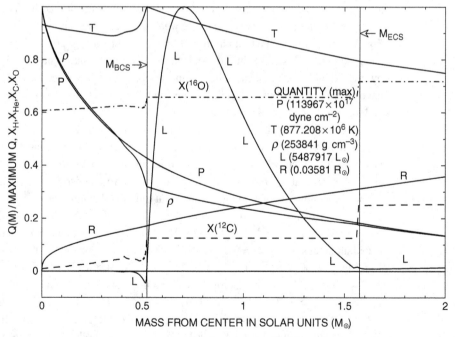

Fig. 20.3.11   Structure of the contracting core of a 25 $M_\odot$ model containing a carbon-burning shell ($Z = 0.015$, $Y = 0.275$, and AGE $= 7.221533 \times 10^6$ yr)

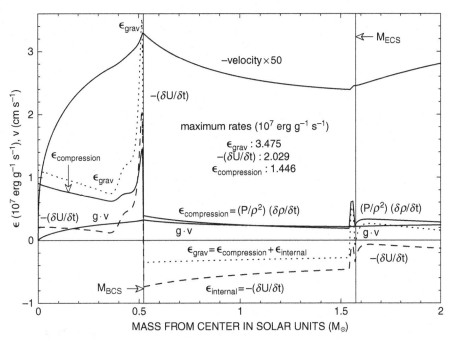

Fig. 20.3.12 Gravothermal energy-generation rates in and near the carbon-burning shell of a 25 $M_\odot$ model ($Z = 0.015$, $Y = 0.275$, AGE = 7.2215334×10^6 yr)

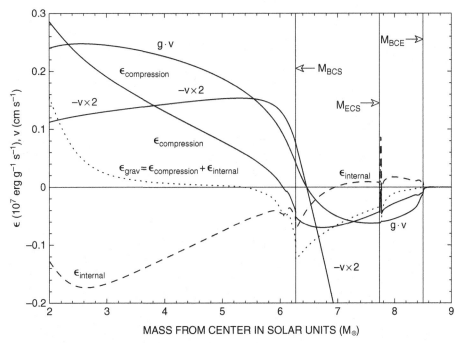

Fig. 20.3.13 Gravothermal energy-generation rates in the outer core of a 25 $M_\odot$ model burning helium and carbon in shells ($Z = 0.015$, $Y = 0.275$, AGE = 7.2215334×10^6 yr)

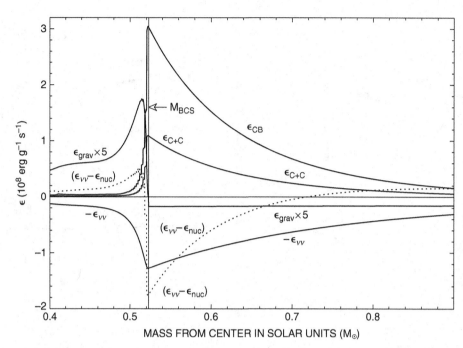

Fig. 20.3.14 Energy-generation and loss rates in and near the carbon-burning shell of a 25 $M_\odot$ model ($Z = 0.015$, $Y = 0.275$, and AGE = 7.2215334×10^6 yr)

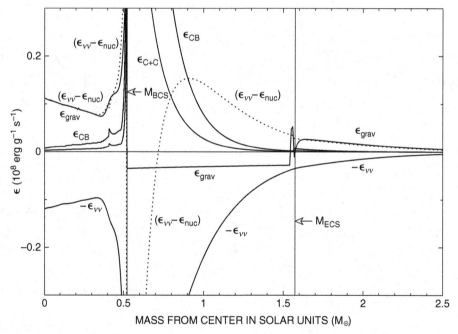

Fig. 20.3.15 Energy-generation and loss rates in the deep interior of a 25 $M_\odot$ model burning carbon in a shell ($Z = 0.015$, $Y = 0.275$, and AGE = 7.2215334×10^6 yr)

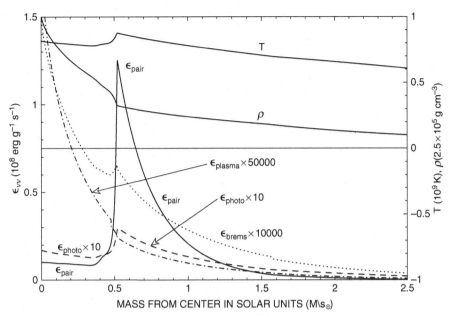

Fig. 20.3.16   Weak interaction neutrino-loss rates in the core of a 25 $M_\odot$ model burning carbon in a shell ($Z = 0.015$, $Y = 0.275$, and AGE $= 7.2215334 \times 10^6$ yr)

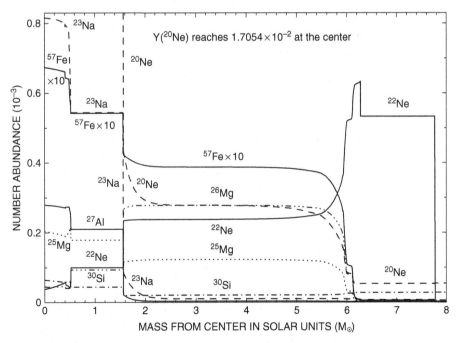

Fig. 20.3.17   Isotopic number abundances in a 25 $M_\odot$ model burning carbon and helium in shells ($Z = 0.015$, $Y = 0.275$, and AGE $= 7.2215334 \times 10^6$ yr)

Table 20.3.3 Global luminosities in a 25 $M_\odot$ model contracting and heating during carbon-burning phases

Δt	L_{CB}	L_{He}	L_H	$L_{\nu_e \bar{\nu}_e}$	L_{grav}	L_s
0.0	0.000	7.995/2	3.2356/4	1.9170/3	1.5078/5	1.8202/5
1000 yr	4.8527/4	2.9979/5	8.6098/−1	6.2890/5	5.2029/5	2.4002/5
380 yr	2.6789/6	4.7270/5	2.1794/−1	3.8415/6	9.3179/5	2.4186/5
190 yr	2.0293/7	6.9812/5	6.9638/−2	2.2312/7	1.5606/6	2.4322/5

with those in Fig. 20.3.6, and making use of the information in legends in the three figures, it is evident that the maximum luminosity in the model has increased by almost a factor of 12 to $L_{max} = 5.49 \times 10^6\ L_\odot$, compared with a surface luminosity which has remained essentially fixed at $L_s \sim 2.43 \times 10^5\ L_\odot$. The mass of the convective shell supported by fluxes produced by the carbon-burning shell, at $M_{CS} \sim 1.05\ M_\odot$, is approximately three times larger than M_{CC}^{max}. These changes are related to increases in both the carbon-burning luminosity L_{CB} and the total neutrino-loss rate $L_{\nu\bar{\nu}}$ which have increased, respectively, by factors of 7.5 and 5.8. The increases in L_{CB} and $L_{\nu\bar{\nu}}$ are, in turn, due to increases in temperature and density in the contracting core. The density at the center has increased by a factor of 2.6 and, while the temperature at the center has increased by only about 0.6%, the maximum in temperature, which has shifted outward to the base of the convective shell, driven by the flux generated by carbon burning in a shell, is about 7% larger than at the centers of the two models.

As in the model with a convective core, contraction and heating extend throughout the helium-exhausted core and into the helium-burning shell, with a consequent increase in the helium-burning luminosity by almost 50% and an increase in the mass of the convective shell supported by fluxes due to helium burning. Again, surface characteristics have changed by only fractions of a percent and the location of the base of the convective envelope is essentially unchanged.

Global luminosities in the models described in Figs. 20.2.9, 20.3.1, 20.3.6, and 20.3.10 are compared in Table 20.3.3. The quantity Δt in the first column of the table is the time elapsing between successive models in the table.

Gravothermal energy-generation rates for the last model are shown as functions of mass in Figs. 20.3.12 and 20.3.13, the first figure describing rates over a region centered on the convective shell associated with shell carbon burning and the second figure describing rates over a region extending from the outer edge of the first region to somewhat beyond the base of the convective envelope. It is evident in Fig. 20.3.12 that, below the base of the inner convective shell driven by carbon burning, temperature is everywhere decreasing and the integral over mass of the rate at which compression does work is approximately three times larger than the integral over mass of the rate at which gravity does work. In the region in the figure beyond the base of the convective shell, the integrals over mass of the two rates are roughly the same. The overall balance between the rate at which compression does work and the rate at which gravity does work is restored by integrating the two rates over mass in Fig. 20.3.13. Shown also in Figs. 20.3.12 and 20.3.13 is the variation with mass of the matter velocity in the final model. Over the entire region between the center and $M \sim 6.43\ M_\odot$, just outside of the outer edge of the helium-burning shell, matter is moving inward. Over the rest of the model, matter is everywhere moving outward.

In Fig. 20.3.14, energy-generation and energy-loss rates are shown as functions of mass in a region centered on the carbon-burning shell, and, in Fig. 20.3.15, they are shown over a larger region in a way which highlights variations outside of the carbon-burning region. As seen most clearly in Fig. 20.3.15, in the region below the base of the convective shell, just short of where the nuclear energy-generation rate rises steeply with increasing mass, gravothermal energy provides most of the energy which balances neutrino losses due to weak interactions. Between $M \sim 0.52\ M_\odot$ and $M \sim 0.71\ M_\odot$, neutrino losses are attended to primarily by the release of nuclear energy, with enough to spare to contribute to an outwardly increasing luminosity (see the maximum in the luminosity profile at $M \sim 0.71\ M_\odot$ in Fig. 20.3.11). Beyond $M \sim 0.71\ M_\odot$, the neutrino energy-loss rate exceeds the rate of nuclear energy generation and, coupled with the fact that, due to heating, the rate of gravothermal energy generation throughout most of the convective shell is negative, almost none of the nuclear energy generated by carbon burning makes its way beyond the outer edge of the convective shell. The outward drop in luminosity is finally reversed in a small region centered on the outer edge of the convective shell, where nuclear energy production is no longer of consequence and gravothermal energy production finally shoulders the responsibility of slightly overbalancing the rate of neutrino energy losses.

Neutrino energy-loss rates in the final model are shown as functions of mass in Fig. 20.3.16. Comparing with these same rates in Fig. 20.3.9 for the model with a small convective core, it is curious to note that the pair neutrino-loss rate at the center has been reduced by a factor of $\sim$2.5 even though the temperature at the center is essentially unchanged. The potential discrepancy disappears when one recognizes that the loss rate in units of erg g^{-1} s^{-1} is the loss rate in units of erg cm^{-3} s^{-1} divided by the density and notices that the density at the center has increased by roughly the same factor of $\sim$2.5; thus, the energy loss rate in units of erg cm^{-3} s^{-1} has not changed, consistent with the fact that, under non-electron-degenerate conditions, the loss rate in these units is a function only of the temperature (see eq. (15.3.26)). At the temperature maximum, where $T_9 = 0.8772\ \epsilon_{\nu\bar{\nu}} = 1.25 \times 10^8$ erg g^{-1} s^{-1}, almost 90 times the loss rate at the center of the model beginning carbon burning at the model center (see Fig. 20.3.3) and about five times larger than the loss rate at the center of the model burning carbon in a convective core (see Fig. 20.3.9). The first order relativistic approximation to the e^+e^- pair-annihilation neutrino energy-loss rate, quoted in eq. (20.3.1), gives $\epsilon_{\nu\bar{\nu}} \sim 1.10 \times 10^8$ erg cm^{-3} s^{-1}, approximately 12% smaller than the value calculated in the model.

It is interesting that, despite the huge energy-loss rates which arise due to electron–positron annihilations into neutrino–antineutrino pairs, the number abundance of positrons is quite modest. Adopting the approximation given by eqs. (4.10.62) and (4.10.64) in Volume 1, the ratio of positrons to electrons contributed by heavy ions is of the order of

$$\frac{n_+}{n_Z} \sim 25.913\ \frac{T_9^3}{\rho_5^2}\ \exp\left(-\frac{11.860}{T_9}\right)(1 + 0.3162\ T_9\ (1 + 0.073\,77\ T_9))^2. \quad (20.3.17)$$

At the temperature maximum in the model, where $T_9 = 0.8772$ and $\rho_5 = 0.8115$, eq. (20.3.17) gives $n_+/n_Z \sim 6.0 \times 10^{-5}$. A check on this estimate is given by comparing in Fig. 4.10.4 in Volume 1 the position defined by $\log T_9 = -0.0569$ and $\log \rho_6 = -1.091$ with the curves of constant n_+/n_e. The position lies slightly below the curve for $n_+/n_e = 0.0001$ by an amount which is of the order of one tenth of the way between curves of constant n_+/n_Z which differ by a factor of 100 in the ratio. In short, the analytical approximation and the exact result are essentially identical. Thus, one may assert that, although the neutrino-loss rate attains a value of over 10^8 erg g^{-1} s^{-1}, there are only about six positrons for every 100 000 electrons, demonstrating that a very large neutrino energy-loss rate due to e^+e^- annihilation does not necessarily imply a positron number abundance large enough to have a noticeable effect on the equation of state.

Number abundances of several heavy isotopes less abundant than the most abundant isotopes, ^{20}Ne, ^{16}O, and ^{12}C, are shown as functions of mass in Fig. 20.3.17. Abundances of 28 of the heavy isotopes in the adopted nuclear reaction network are listed in Table 20.3.4 at three different positions: at the center, in the middle of the carbon-burning shell, and near the base of the region which has experienced complete helium burning, but not carbon burning. Also listed in the table are the number abundances of protons, neutrons, and alpha particles, which are in local equilibrium with respect to creating and destroying reactions.

An interesting lesson from the table is that by far the major product of complete carbon burning is ^{20}Ne, at a number abundance very close to the number abundance of ^{12}C at the end of complete helium burning. At the end of complete helium burning (column 4 in the table), $Y(^{12}$C$) = 0.021\,11$ and $Y(^{20}$Ne$) = 0.000\,29$, whereas, at nearly the end of complete carbon burning (column 2), $Y(^{20}$Ne$) = 0.017\,05$ and $Y(^{12}$C$) = 0.000\,72$, with the sum of the ^{20}Ne and ^{12}C abundances being approximately 84% of the ^{12}C abundance at the end of complete helium burning.

This result suggests that, of the alpha particles emitted in ^{12}C$+^{12}$C reactions, most experience an ^{16}O$(\alpha, \gamma)^{20}$Ne reaction and that, of the protons emitted in ^{12}C $+ ^{12}$C reactions, most experience a ^{23}Na$(p, \alpha)^{20}$Ne reaction. To verify this inference, note that, with the adopted branching ratios, 56% of all ^{12}C $+ ^{12}$C reactions produce a ^{20}Ne nucleus directly. This results in a contribution to the number abundance of ^{20}Ne of $\Delta Y_{20} = 0.56 \times 0.5 \times 0.0211 = 0.0059$, where the factor of 0.5 comes from the fact that two ^{12}C nuclei are destroyed to make one ^{20}Ne nucleus. Similarly, if ^{23}Na were not destroyed by (p, α) and (p, γ) reactions, its number abundance would increase by $\Delta Y_{23} = 0.44 \times 0.5 \times 0.0211 = 0.0046$. However, in consequence primarily of ^{23}Na$(p, \alpha)^{20}$Ne reactions, which occur approximately ten times more frequently than ^{23}Na$(p, \gamma)^{24}$Mg reactions, the actual increase in the abundance of ^{23}Na is only $\Delta Y_{23} = 0.000\,82$, so, instead, the abundance of ^{20}Ne receives another contribution of $\Delta Y_{20}{}' \sim 0.0038$. Finally, due primarily to (α, γ) reactions, the ^{16}O number abundance is decreased by $-\Delta Y_{16} = 0.044\,92 - 0.037\,92 = 0.0070$, resulting in another contribution to the abundance of ^{20}Ne of $\Delta Y_{20}{}'' = 0.0070$. Altogether, one has that the abundance of ^{20}Ne is increased by $\Delta Y_{20} \sim 0.0059 + 0.0038 + 0.0070 = 0.0167$, substantially in agreement with the increase given by the detailed model calculation.

At the center, the most abundant isotopes shown in Fig. 20.3.17 are ^{23}Na and ^{20}Ne and one might expect that the reactions ^{23}Na$(p, \gamma)^{24}$Mg and ^{20}Ne$(\alpha, \gamma)^{24}$Mg would have led

isotope/mass	0.0 $M_\odot$	1.003 $M_\odot$	2.000 $M_\odot$
Table 20.3.4 Number abundances in a 25 $M_\odot$ model with a carbon-burning shell			
proton	3.4825/−20	3.0780/−18	6.0396/−20
neutron	2.1535/−24	3.3449/−22	1.4285/−22
^{4}He	1.0267/−14	8.4699/−13	3.2760/−14
^{12}C	7.1702/−04	1.0425/−02	2.1110/−02
^{13}C	2.5928/−11	8.8674/−09	1.0579/−06
^{14}C	2.5970/−08	1.0954/−07	1.0683/−09
^{13}N	9.4112/−16	1.1206/−12	3.0266/−12
^{14}N	2.5914/−06	7.5926/−06	3.7003/−08
^{15}N	5.4476/−09	1.1777/−08	9.5127/−12
^{16}O	3.7923/−02	4.1106/−02	4.4924/−02
^{17}O	4.5404/−07	1.6450/−06	4.8411/−06
^{18}O	5.8660/−08	1.5065/−07	2.6592/−10
^{17}F	7.9451/−22	2.8076/−20	2.3323/−21
^{18}F	1.0400/−12	3.7478/−11	1.3904/−15
^{19}F	4.8200/−10	8.5614/−10	4.4488/−14
^{20}Ne	1.7054/−02	9.0783/−03	2.9308/−04
^{21}Ne	7.2029/−05	5.1867/−05	1.8668/−05
^{22}Ne	3.7772/−05	9.9901/−05	2.3788/−04
^{23}Na	8.1552/−04	5.4433/−04	1.8792/−05
^{24}Mg	2.3731/−05	2.3348/−05	2.3106/−05
^{25}Mg	2.0407/−04	1.7861/−04	1.2193/−04
^{26}Mg	4.5343/−05	9.3556/−05	2.7836/−04
^{27}Al	2.7971/−04	2.0985/−04	4.0978/−06
^{28}Si	6.4693/−05	4.4746/−05	2.0891/−05
^{29}Si	1.0909/−05	8.2719/−06	4.9603/−06
^{30}Si	8.3218/−06	7.0686/−06	5.7894/−06
^{31}P	5.7275/−06	5.2860/−06	5.0868/−06
^{32}S	6.2390/−06	6.6829/−06	7.3657/−06
^{33}S	1.4202/−07	1.5101/−07	1.3135/−07
^{34}S	7.9330/−07	7.4902/−07	6.9997/−07
^{57}Fe	6.7447/−05	5.4242/−05	3.9257/−05

to a significant increase in the abundance of ^{24}Mg, as is the case in some earlier calculations (e.g., Lamb, Iben, Howard, 1976) using charged particle reaction rates from W. A. Fowler, G. R. Caughlan, and B. A. Zimmerman (1975). The fact that this has not occurred in the present calculations is a mistake in the choice of the rate of the ^{23}Na$(p, \gamma)^{24}$Mg reaction.

In summary, the primary nucleosynthesis result of complete carbon burning is to convert a mixture dominated by ^{12}C and ^{16}O into a mixture dominated by ^{16}O and ^{20}Ne, with a modest reduction in the total number abundance of the dominant isotopes. In the present instance, the sum of the number abundances of ^{12}C, ^{16}O, and ^{20}Ne decreases through the values 0.0663, 0.0606, and 0.0557 in going from the abundance mix prevailing after complete helium burning, through a region of incomplete carbon burning, into a region in which carbon burning is almost completed (columns 4 through 2 in Table 20.3.3). Another way of describing this result is that, for every 2.1 ^{12}C nuclei and 4.5 ^{16}O nuclei present at the end of complete helium burning, there are $\sim$3.8 ^{16}O nuclei and $\sim$1.76 ^{20}Ne nuclei present at the end of complete carbon burning, a reduction by approximately 16% in the number density of heavy ions.

20.4 Comments on neon-, oxygen-, and silicon-burning phases

The last model described in the previous section has become too cumbersome to justify following it further. It has almost 3000 mass zones and the range in values of number abundances has led to difficulties in convergence that would require extensive recoding to overcome. The model should be broken up into several parts in each of which different nuclear reaction networks are employed. In particular, in regions which have experienced complete helium burning, it is foolish to retain and continue to calculate transformations of isotopes of elements below carbon in the periodic table (apart from protons, alpha particles, and neutrons, of course).

More to the point, excellent treatments of the complexities occurring during these advanced phases of evolution already exist, two examples being *Principles of Stellar Evolution and Nucleosynthesis* by Donald D. Clayton (1968) and *Supernovae and Nucleosynthesis* by David Arnett (1996). Particularly relevant extensions of the discussion in this section are Chapter 7 in Clayton's book and Sections 10.2 and 10.3 in Arnett's book.

Neon burning is initiated and driven by the photodisintegration of ^{20}Ne. When temperatures and densities increase into an appropriate regime, the reaction ^{20}Ne$(\gamma, \alpha)^{16}$O occurs more rapidly than the inverse reaction ^{16}O$(\alpha, \gamma)^{20}$Ne. The released alpha particles are captured by ^{20}Ne and ^{24}Mg to make ^{24}Mg and ^{28}Si, respectively. Additional isotopes are made at modest abundances. Results of a neon-burning calculation from Arnett's book are given in Table 20.4.1. Note that two of the isotopes in the final isotope mix contain more neutrons than protons.

It is possible to make a rough estimate of the density–temperature regime where neon burning occurs by examining the conditions for equilibrium between alpha-capture formation and photodisintegration of ^{20}Ne. By making the appropriate modifications, the equilibrium condition follows from eq. (20.3.7):

$$\frac{n_\alpha \, n_{Z-2,N-2}}{n_{Z,N}} = \left(\frac{2\pi \mu kT}{h^2}\right)^{3/2} \frac{g_\alpha \, g_{Z-2,N-2}}{g_{Z,N}} \exp(-Q/kT), \qquad (20.4.1)$$

where $Z = N = 10$, $\mu = 3.2\, M_{\mathrm{H}}$, the statistical weights of all three particles are unity, and $Q = 4.73$ MeV. Inserting numbers, the equilibrium condition becomes

$$\rho_0 \frac{Y_{16}}{Y_{20}} Y_\alpha = T_9^{3/2} \exp(24.757 - 54.89/T_9), \qquad (20.4.2)$$

Table 20.4.1 Number abundances of products of neon burning (Arnett, 1996)						
isotope	^{20}Ne	^{16}O	^{24}Mg	^{27}Al	^{28}Si	^{29}Si
Y_{start}	0.0162	0.0370	1.179/−3	2.22/−4	7.1/−5	———
Y_{end}	5/−5	0.0474	4.33/−3	5.93/−4	2.79/−3	4.83/−4

where the Ys are the number abundances of the three particles. For any specified set of conditions, if the quantity on the left hand side of eq. (20.4.2) is less than the quantity on the right hand side, creation of ^{20}Ne occurs less rapidly than photodisintegration, and vice versa if the quantity on the left hand side is larger than the quantity on the right hand side.

Suppose that density and temperature are related roughly as

$$\rho_0 \sim 2.5 \times 10^5 \left(\frac{T_9}{0.85} \right)^3 , \qquad (20.4.3)$$

as is suggested by the fact that $\rho_c \propto T_c^3$ during the core hydrogen- and helium-burning phases (see Fig. 20.1.3), and that the number abundances of ^{16}O and ^{20}Ne have the starting values listed in Table 20.4.1. Then the equilibrium condition becomes

$$Y_\alpha = T_9^{-3/2} \exp(11.014 - 54.89/T_9). \qquad (20.4.4)$$

At $T_9 = 2$, corresponding to $\rho_0 = 3.3 \times 10^6$ as given by eq. (20.4.3), the right hand side of eq. (20.4.4) is 2.59×10^{-8}. But, from Table 20.3.4, typical alpha particle abundances under burning conditions are many orders of magnitude smaller than this, so photodisintegration prevails. At $T_9 = 1.5$, corresponding to $\rho_0 = 1.4 \times 10^6$, the right hand side of eq. (20.4.4) is 4.24×10^{-12}, similar to the alpha particle abundance in the third column of Table 20.3.4. From Fig. 10.3 in Arnett's book, neon burning occurs in a model with an 8 $M_\odot$ hydrogen-exhausted core at temperatures and densities in the neighborhood of $T_9 \sim 1.6$ and $\rho_0 \sim 6 \times 10^6$, respectively. The fact that the central density in Arnett's model core increases with roughly the 4.5th power of the central temperature accounts for the result that the density at burning is larger than estimated by eq. (20.4.3). In summary, neon burning occurs at temperatures of the order of $T_9 = 1.6$ and at densities of several million g cm^{-3} and the net result is the conversion of a mixture of ^{16}O and ^{20}Ne into a mixture which is roughly 85% ^{16}O, 8% ^{24}Mg, and 5% ^{28}Si.

The next important burning process is initiated by reactions between ^{16}O nuclei. Relevant experiments include those by H. Spinka & H. Winkler (1974), by C.-S. Wu & C. A. Barnes (1984), and by J. Thomas, Y. T. Chen, S. Hinds, D. Meredith, & M. Olson (1986). The channels which are open at the lowest energies reached in the experiments are listed in Table 20.4.2, along with the nuclear energies liberated in each channel, and cross sections at $E_{cm} = 7$ MeV, as estimated, respectively, by Wu & Barnes (1984) and by Thomas $et\ al.$ (1986). The cross sections from Wu and Barnes are direct quotes and the cross sections from Thomas $et\ al.$ are estimates from their Fig. 6. The difficulties presented by the large Coulomb barrier at the lowest energies explored in the experiments are reflected in the large differences in estimated cross sections in Table 20.4.2.

product	Q(MeV)	σ (Wu and Barnes)	σ (Thomas *et al.*)
Table 20.4.2 Cross sections for $^{16}O + ^{16}O$ reactions at $E_{cm} = 7$ MeV			
$^{31}P + p$	7.677	0.00357 mb	0.0025mb
$^{31}S^* + n$	1.453	0.000381 mb	0.00056 mb
$^{28}Si + \alpha$	9.593	0.00309 mb	0.002 mb
$^{32}S + \gamma$	16.54	0.0 mb	0.0 mb
$^{30}Si + 2p$	0.380	0.000743 mb	0.010 mb
$^{30}P^* + p + n$	4.630	0.00243 mb	0.004 mb
all		0.0102 mb	0.0191 mb

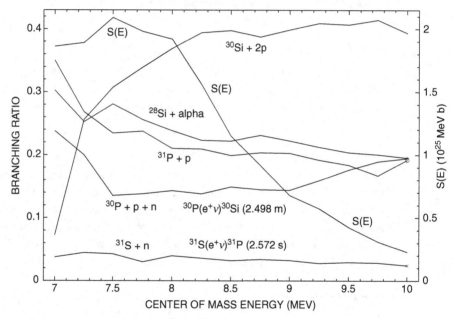

Fig. 20.4.1   Branching ratios and cross section factor for $^{16}O + ^{16}O$ reactions From S.-C. Wu & C. A. Barnes, Nuclear Physics A, 422, 373, 1984

In two channels, the product nuclei are unstable to positron decay: $^{31}S^* \rightarrow ^{31}P + e^+ + \nu_e + 5.44$ MeV, with a 2.572 s half life, and $^{30}P^* \rightarrow ^{31}Si + e^+ + \nu_e + 4.23$ MeV, with a 2.496 min half life. On average, in the first decay, an energy of 2.788 MeV remains in the star and 2.652 MeV is carried off by the neutrino. In the second decay, an energy of 2.304 MeV remains in the star and 1.926 MeV is carried off by the neutrino.

Branching ratios for the various reactions obtained by dividing individual cross sections by the total cross section are shown in Fig. 20.4.1 for the Wu and Barnes data over the range $E = 7$–10 MeV. Shown also is the cross section factor $S(E)$ obtained by dividing the total cross section by the Coulomb penetration factor. Over energies where the experiments give consistent results, the reactions do not exhibit the fluctuations with respect to energy

Table 20.4.3 Branching ratios for $^{16}O + {}^{16}O$ reactions at $E_{cm} = 7.5$ MeV (Wu & Barnes, 1984)

product	$^{30}Si + 2p$	$^{28}Si + \alpha$	$^{31}P + p$	$^{30}P + p + n$	$^{31}S + n$	p	α	n
ratio	0.31	0.29	0.24	0.14	0.04	0.99	0.28	0.18

that would suggest the occurrence of resonances. On the other hand, the fact that the Wu & Barnes cross section factor exhibits a maximum at $E \sim 7.5$ MeV might be taken as evidence for a giant resonance centered at $E \sim 7.5$ MeV, were it not for the fact that, for $E \gtrsim 8$ MeV, Thomas *et al.* estimates of $S(E)$ are consistently larger than Wu & Barnes estimates and continue to increase with decreasing energy over the entire range of energies explored.

The mean of the cross section factors estimated experimentally varies with energy slowly enough that it is reasonable to assume that the major contribution to the overall reaction rate comes from reactions at energies near the energy given by eq. (6.7.17) in Volume 1 as

$$E_0 = 122 \left(8^2\, 8^2\, (16/2)\, T_9^2 \right)^{1/3} \text{ keV} = 3.9\, T_9^{2/3} \text{ MeV}, \tag{20.4.5}$$

and that the energy range over which the major contribution to the reaction rate occurs is given by eq. (6.7.18) in Volume 1 which, in the present instance, is

$$\Delta E_0 = 251.9 \text{ keV} \left(Z_1^2\, Z_2^2\, A_{12} \right)^{1/6} T_9^{5/6} = 1.425\, T_9^{5/6} \text{ MeV}. \tag{20.4.6}$$

For temperatures of the order of $T_9 = 2.0$, where oxygen burning occurs, $E_0 = 6.2$ MeV and $\Delta E_0 = 2.54$ MeV, so, once again, the estimate of the reaction rate at relevant temperatures relies on cross sections at energies that extend into the energy range where experimental estimates of cross sections exist. The overall reaction rate suggested by Caughlan and Fowler (1988) is

$$N_A \langle \sigma v \rangle = \exp \left(84.853 - 135.93/T_9^{1/3} - 0.629\, T_9^{2/3} - 0.445\, T_9^{4/3} + 0.0103\, T_9^2 \right)/T_9^{2/3}, \tag{20.4.7}$$

with a factor of 2 uncertainty in the adopted cross section factor translating into an uncertainty of less than 5% in the temperature at oxygen burning. The appropriate branching ratios to adopt are somewhat more problematic. Given the large difference in the experimental estimates at the lowest energies, it seems prudent to adopt branching ratios given by the experiments at energies which are above where discrepancies are large but which are as close as possible to energies where they are needed in the stellar interior. At $E = 7.5$ MeV, the Wu and Barnes data gives the branching ratios shown in Table 20.4.3. Note that the total number of protons (0.99), alpha particles (0.28), and neutrons (0.18) emitted per average $^{16}O + {}^{16}O$ reaction is 1.45 times larger than the number of $^{16}O + {}^{16}O$ reactions.

Table 20.4.4 Number abundances of products of oxygen burning (Arnett, 1996)					
isotope	^{16}O	^{24}Mg	^{28}Si	^{32}S	^{34}S	^{40}Ca
Y_{start}	4.73/−2	4.29/−3	2.82/−3	2.19/−4		
Y_{end}	6.25/−5	0.0	1.96/−2	8.44/−3	1.15/−3	6.5/−4

Table 20.4.5 Number abundances of products of silicon burning (Arnett, 1996)					
isotope	^{28}Si	^{32}S	^{34}S	^{54}Fe	^{56}Fe	^{58}Ni
Y_{start}	1.96/−2	8.44/−3	1.15/−3	1.11/−4	8.93/−5	
Y_{end}	3.6/−5	3.1/−5		1.20/−2	7.32/−4	2.17/−3

Caughlan & Fowler (1988) recommend similar numbers, 0.85, 0.30, and 0.25, respectively, for protons, alphas, and neutrons, for a total of 1.40 per average $^{16}O + {}^{16}O$ reaction.

The major consequence of reactions which occur during oxygen burning is the conversion of ^{16}O into ^{28}Si and ^{32}S. Number abundances of the most common isotopes at the end of oxygen burning in Arnett's 1996 illustration are given in Table 20.4.4.

The final burning process, which goes by the name silicon burning, involves many photodisintegration reactions and alpha particle captures which convert ^{28}Si and the other products of oxygen burning into iron-peak elements. An early discussion of relevant processes is given by E. M. Burbidge, G. Burbidge, W. A. Fowler, & F. Hoyle (B^2FH) (1957), followed by elaboration and clarification by Fowler and Hoyle (1964), by James W. Truran, A. G. W. Cameron, and A. A. Gilbert (1966), and by D. Bodansky, D. D. Clayton, and W. A. Fowler (1968).

The Bodansky *et al.* exploration emphasizes that, at $T_9 \sim 3$–4, a quasiequilibrium process takes place in which ^{28}Si is almost in equilibrium with respect to ^{24}Mg via the reactions $^{28}Si + \gamma \longleftrightarrow {}^{24}Mg + \alpha$, maintaining the number abundances of ^{24}Mg and ^{28}Si in the ratio $Y(^{24}Mg)/Y(^{28}Si) \sim 10^{-3}$. Other quasiequilibrium reactions which take place are $^{28}Si + \alpha \longleftrightarrow {}^{32}S + \gamma$, $^{32}S + \alpha \longleftrightarrow {}^{36}A + \gamma$ and so forth. ^{28}Si slowly "melts", with iron group nuclei being built up at the expense of isotopes with smaller binding energies per nucleon. When all is said and done, the final result of the process is the formation of isotopes with binding energies per nucleon ((B/A)s) which are near the maximum in the (B/A) vs A curve.

Under the conditions adoped in Section 10.2 of Arnett's book, the most abundant species remaining after silicon burning are given in Table 20.4.5. The results are sensitive to many different factors, including the adopted initial neutron to proton ratio in the isotope mix with which the silicon-burning calculation is begun. The fact that the light curves of type II supernovae indicate the presence of large quantities of ^{56}Ni in the initial supernova ejectum can be taken as a boundary condition for the most likely isotope mix which precedes silicon burning in regions of real massive stars which escape collapse into a condensed core and are ejected.

20.5 More on the relationship between direct and inverse tranformations

The fact that, during the last nuclear burning stages, many isotopes are in near equilibrium with regard to creating and destroying reactions prompts a further exploration of the manner in which relationships between number abundances are established when direct and inverse reaction rates occur at nearly the same rates. In the case of radiative capture and photodisintegration, the relationship follows most easily from a simple modification of an appropriate Saha equation applicable to an atomic situation, as accomplished in Sections (20.3) and (20.4) for (p, γ) and (α, γ) reactions and their inverses (see eqs. (20.3.7) and (20.4.1)).

There are also situations in which a non-radiative reaction between an ion and a nucleon or an alpha particle occur at about the same rate as the inverse reaction between the product particles. For concreteness, consider the reaction

$$I_{Z,N} + \alpha \rightarrow I_{Z+1,N+2} + p \tag{20.5.1}$$

and its inverse. To obtain the relationship between the number abundances of all particles involved when the direct and inverse reactions proceed at the same rate, the probability of a given configuration is maximized subject to constraints that the energy is constant and that the total number of protons and the total number of neutrons are constant. Proceeding as in Section 4.13 in Volume 1 on the assumption that Maxwell–Boltzmann statistics is adequate, the probability of a given configuration is

$$P(n_A, n_\alpha, n_B, n_p) \propto \prod_{i=1}^{\infty} \frac{g_{Ai}^{n_{Ai}}}{n_{Ai}!} \prod_{j=1}^{\infty} \frac{g_{\alpha j}^{n_{\alpha j}}}{n_{\alpha j}!} \prod_{k=1}^{\infty} \frac{g_{Bk}^{n_{Bk}}}{n_{Bk}!} \prod_{l=1}^{\infty} \frac{g_{pl}^{n_{pl}}}{n_{pl}!}, \tag{20.5.2}$$

where A, B, α, and p stand, respectively, for nuclei $I_{Z,N}$ and $I_{Z+1,N+2}$ and alpha particles and protons.

The number conservation constraints are

$$N_{\text{protons}} = Z \sum_i n_{Ai} + 2 \sum_j n_{\alpha j} + (Z+1) \sum_k n_{Bk} + \sum_l n_{pl} = \text{constant}, \tag{20.5.3}$$

$$N_{\text{neutrons}} = N \sum_i n_{Ai} + 2 \sum_j n_{\alpha j} + (N+2) \sum_k n_{Bk} = \text{constant}, \tag{20.5.4}$$

and the energy constraint is

$$E = \sum_i n_{Ai} (\epsilon_{Ai} + Q) + \sum_j n_{\alpha j} \epsilon_{\alpha j} + \sum_k n_{Bk} \epsilon_{Bk} + \sum_l n_{pl} \epsilon_{pl} = \text{constant}.$$

$$\tag{20.5.5}$$

Multiplying eqs. (20.5.3)–(20.5.5) by constants α_1, α_2, and β, respectively, differentiating the modified equations, adding to the result of differentiating eq. (20.5.2), and setting to zero yields

$$\sum_i \delta n_{Ai} \left(\log_e \frac{g_{Ai}}{n_{Ai}} + \alpha_1 Z + \alpha_2 N + \beta \left(\epsilon_{Ai} + Q \right) \right)$$

$$+ \sum_i \delta n_{\alpha j} \left(\log_e \frac{g_{\alpha j}}{n_{\alpha j}} + \alpha_1 + \alpha_2 + \beta \epsilon_{\alpha j} \right)$$

$$+ \sum_k \delta n_{Bk} \left(\log_e \frac{g_{Bk}}{n_{Bk}} + \alpha_1 \left(Z + 1 \right) + \alpha_2 \left(N + 2 \right) + \beta \epsilon_{Bk} \right)$$

$$+ \sum_l \delta n_{pl} \left(\log_e \frac{g_{pl}}{n_{pl}} + \alpha_1 + \beta \epsilon_{\alpha j} \right) = 0. \tag{20.5.6}$$

Setting the coefficients of the differentials in this equation equal to zero results in four equations, the first of which is

$$n_{Ai} = g_{Ai} \, e^{-\beta \epsilon_{Ai}} \, e^{-\alpha_1 Z - \alpha_2 N - \beta Q}. \tag{20.5.7}$$

Setting $\beta = 1/kT$ and summing over all kinetic energies gives

$$n_A = g_A \int \frac{4\pi p^2 \, dp}{h^3} \, e^{-p^2/2M_A kT} \, e^{-\alpha_1 Z - \alpha_2 N - Q/kT}$$

$$= g_A \left(\frac{2\pi M_A \, kT}{h^2} \right)^{3/2} e^{-\alpha_1 Z - \alpha_2 N - Q/kT}. \tag{20.5.8}$$

Similarly,

$$n_\alpha = g_\alpha \left(\frac{2\pi M_\alpha \, kT}{h^2} \right)^{3/2} e^{-(\alpha_1 + \alpha_2) \, 2}, \tag{20.5.9}$$

$$n_B = g_B \left(\frac{2\pi M_B \, kT}{h^2} \right)^{3/2} e^{-\alpha_1 (Z+1) - \alpha_2 (N+2)}, \tag{20.5.10}$$

and

$$n_p = g_p \left(\frac{2\pi M_p \, kT}{h^2} \right)^{3/2} e^{-\alpha_1}. \tag{20.5.11}$$

The unique solution of eqs. (20.5.8)–(20.5.11) is

$$\frac{n_A \, n_\alpha}{n_B \, n_p} = \frac{g_A g_\alpha}{g_B g_p} \left(\frac{M_A M_\alpha}{M_B M_p} \right)^{3/2} \exp\left(-\frac{Q}{kT} \right). \tag{20.5.12}$$

Since the reaction rates for the forward and backward reactions are, respectively, $n_A n_\alpha \langle \sigma_{\alpha,p} \, v \rangle$ and $n_B n_p \langle \sigma_{p,\alpha} \, v \rangle$, one has that

$$\langle \sigma_{\alpha,p} \, v \rangle = \frac{g_A g_\alpha}{g_B g_p} \left(\frac{M_A M_\alpha}{M_B M_p} \right)^{3/2} \langle \sigma_{p,\alpha} \, v \rangle \exp\left(-\frac{Q}{kT} \right). \tag{20.5.13}$$

It is clear that, when screening is taken into account, the screening factor which multiplies the unscreened $\sigma_{p,\alpha}$ must be the same as the screening factor which multiplies the unscreened $\sigma_{\alpha,p}$. The simplest way to achieve this is to assume that the screening potentials $S(\alpha, p)$ and $S(p, \alpha)$ are given by

$$S(\alpha, p) = S(\alpha, \gamma) + S(p, \gamma) = S(p, \alpha), \qquad (20.5.14)$$

where $S(\alpha, \gamma)$ and $S(p, \gamma)$ are screening potentials estimated in the standard way for radiative transitions, and that, e.g.,

$$\langle \sigma_{p,\alpha} \, v \rangle_{\text{screened}} = e^{S(\alpha,p)/kT} \, \langle \sigma_{p,\alpha} \, v \rangle_{\text{unscreened}}, \qquad (20.5.15)$$

and similarly for $\langle \sigma_{\alpha,p} \, v \rangle_{\text{unscreened}}$.

The major lesson from eqs. (20.5.12) and (20.5.13) is that, since the product of the terms on the right hand side of each equation involving statistical weights and particle masses is of the order of unity, once the typical kinetic energies of the participating particles is of the order of or exceeds Q, the abundances of the particles satisfy, in order of magnitude,

$$n_A \, n_\alpha \sim n_B \, n_p. \qquad (20.5.16)$$

It is interesting to apply the relationship between particle numbers in equilibrium given by eq. (20.3.7) to the case of radiative transitions on the supposition that photons can be treated as if they possess a mass. In Section 4.11 of Volume 1, an average number density n_{ph} and an average energy density U_{ph} are defined, respectively, by

$$n_{\text{ph}} = 16\pi \left(\frac{kT}{ch} \right)^3 \times 1.201\,975 \qquad (20.5.17)$$

and

$$U_{\text{ph}} = 48\pi \left(\frac{kT}{ch} \right)^3 kT \times 1.082\,323, \qquad (20.5.18)$$

giving an average energy of

$$E_{\text{ph}} = \frac{U_{\text{ph}}}{n_{\text{ph}}} = 2.701\,362 \, kT. \qquad (20.5.19)$$

Now consider the equilibrium established by (p, γ) and (γ, p) reactions on isotopes, related as in eq. (20.3.6). Having ascribed a mass to the photon, eq. (20.5.12) is applicable, with the modification that the quantities A, α, B and p stand for $I_{Z-1,N}$, P, $I_{Z,N}$, and γ. Then, one has that

$$\frac{n_p \, n_{Z-1,N}}{n_{\text{ph}} \, n_{Z,N}} = \left(\frac{M_p \, M_{Z-1,N}}{M_{\text{ph}} \, M_{Z,N}} \right)^{3/2} \frac{2 \, g_{Z-1,N}}{g_{\text{ph}} \, g_{Z,N}} \exp\left(-Q/kT\right), \qquad (20.5.20)$$

where M_{ph} is the fictitious mass assigned to an average photon and the photon statistical weight is $g_{\text{ph}} = 2$. Dividing the left hand side of eq. (20.5.20) by the left hand side of eq. (20.3.7) and dividing the right hand sides of both equations produces an equation for n_{ph} which, when identified with n_{ph} given by eq. (20.5.17), leads to

$$M_{\mathrm{ph}} \, c^2 = \frac{(8\pi \; 1.201\,975)^{2/3}}{2\pi} \, kT = 1.543\,750 \, kT, \tag{20.5.21}$$

slightly larger than half of the energy of an average photon.

Recall that, before formulating the concepts of general relativity, Einstein suggested that photons passing close to the Sun might be deflected as if they possessed a mass $M_{\mathrm{ph}} \, c^2 = h\nu$. Observations showed a deflection of half the predicted value, in agreement with the deflection predicted later by general relativity. The result here is an interesting addendum to the story, and is not intended to be anything more than that. Photons are neither Maxwell–Boltzmann nor Fermi–Dirac particles and they do not have mass.

Neverthless, the formulation can provide insight into the relationship beween a photodisintegration cross section and a radiative capture cross section. Defining a quantity $\langle \sigma_{\gamma,p} \, v \rangle$ by

$$\langle \sigma_{\gamma,p} \, v \rangle \, n_{\mathrm{ph}} \, n_{Z,N} = \langle \sigma_{p,\gamma} \, v \rangle \, n_p \, n_{Z-1,N}, \tag{20.5.22}$$

and using eqs. (20.5.17) and (20.3.7), one has that

$$\langle \sigma_{\gamma,p} \, v \rangle = \left(0.408\,072 \, \frac{\mu c^2}{kT} \right)^{3/2} \langle \sigma_{p,\gamma} \, v \rangle \frac{2 \, g_{Z-1,N}}{g_{Z,N}} \, e^{-Q/kT}, \tag{20.5.23}$$

where $\mu = M_p M_{Z-1,N} / M_{Z,N}$ and $\langle \sigma_{\gamma,p} \, v \rangle = c \, \langle \sigma_{\gamma,p} \rangle$. Noting that a mean center of mass velocity is proportional to $\sqrt{kT/\mu}$, one has further that

$$\langle \sigma_{\gamma,p} \rangle \propto \frac{e^{-Q/kT}}{\langle (v/c)^2 \rangle} \frac{\langle \sigma_{p,\gamma} v \rangle}{\langle |v| \rangle} \propto \frac{e^{-Q/kT}}{(kT/\mu c^2)} \frac{\langle \sigma_{p,\gamma} v \rangle}{\langle |v| \rangle}. \tag{20.5.24}$$

From the explicitly displayed temperature dependence in eq. (20.5.24), it follows that, for temperatures $kT < Q$, the ratio of the photodisintegration cross section for a typical photon to the velocity-averaged radiative capture cross section increases with increasing temperature. For $kT > Q$, the reverse is the case, but this is more than offset by the fact that the number density of photons increases with the third power of the temperature.

The net effect on the equilibrium ratio of two heavy isotopes can be seen most easily by supposing that the only particles in the system are protons, photons, and the two isotopes $I_{Z-1,N}$ and $I_{Z,N}$. Setting $n = n_{Z-1,N} + n_{Z,N}$ and $n_p = n_{Z-1,N} = cn$, so that $n_{Z,N} = (1-c)n$, one has that

$$\frac{c^2}{1-c} \propto \frac{1}{n} \, e^{-Q/kT} \, (kT)^{3/2}, \tag{20.5.25}$$

a relationship which is also obvious from the Saha-like eq. (20.3.7). With increasing temperature, the neutron-rich isotope increases in abundance. It is this effect which accounts for the fact that, during each successive burning process described in Sections 20.3–20.4, the neutron to proton ratio in the prevailing isotope mix increases.

20.6 Concluding remarks on massive star evolution

In consequence of continued neutrino losses and in the absence of additional nuclear energy sources, central regions in which burning has resulted in the formation of iron peak elements continue to contract and heat. Multiple photodisintegration reactions cause the iron peak elements which have been so elaborately constructed to be decomposed first into alpha particles and then into protons and neutrons, with the energy loss associated with photodisintegration accelerating the rate of core contraction until it proceeds on a dynamic time scale. On the other hand, from the observed properties of type II supernovae, it is evident that most of the mass of the star with an imploding core is expelled, and the fact that the light curve can be understood in terms of the decay of ^{56}Ni into ^{56}Co and of ^{56}Co into ^{56}Fe and of the fact that ^{56}Fe is the most abundant of the heavy isotopes in the Universe demonstrates that all of the reaction processes discussed in previous sections occur in matter which is outside of the collapsing core and is expelled. How all of this comes about is a fascinating story beyond the scope of this book. It is, nevertheless, satisfying to realize that much of the physics that occurs in the expelled matter has been encountered and understood during the quasistatic evolution of the matter that ultimately disappears from view into the interior of a neutron star or black hole.

Bibliography and references

C. Angulo, M. Arnould, M. Rayet, *et al., Nucl. Phys. A*, **656**, 3, 1999.

David Arnett, *Supernovae and Nucleosynthesis*, (Princeton: Princeton University Press, 1996.

H. W. Becker, K. U. Kettner, C. Rolfs, & H. P. Trautvetter, *Zeits. für Phys, A*, **303**, 305, 1981.

Steven A. Becker, Icko Iben, Jr., & R. S. Tuggle, *ApJ*, **218**, 633, 1977.

D. Bodansky, Donald D. Clayton, & William A. Fowler, *ApJS*, **16**, 299, 1968.

E. M. Burbidge, G. Burbidge, W. A. Fowler, & F. Hoyle (B^{2}FH), *Rev. Mod. Phys.*, **29**, No.4, 547, 1957.

Georgeanne R. Caughlan and William A. Fowler, *Atomic Data and Nuclear Data Tables*, **40**, 283, 1988.

Donald D. Clayton, *Principles of Stellar Evolution and Nucleosynthesis*, (New York: McGraw-Hill), 1968.

William A. Fowler, Georgeanne R. Caughlan, and Barbara A. Zimmerman, *ARAA*, **13**, 69, 1975.

William A. Fowler & Fred Hoyle, *ApJS*, **9**, 201, 1964.

N. Itoh, H. Hayashi, A. Nishikawa, & Y. Kohyama, *ApJS*, **102**, 411, 1996.

Icko Iben, Jr., & Roy S. Tuggle, *ApJ*, **197**, 39, 1975; **173**, 135, 1972; **178**, 441, 1972b.

Susan A. Lamb, Icko Iben, Jr., & W. Michael Howard, *ApJ*, **207**, 209, 1976.

J. R. Patterson, H. Winkler, & C. S. Zaidens, *ApJ*, **157**, 367, 1969.

Abdus Salam, *Proceedings of the 8*[th] *Nobel Symposium on Elementary Particle Theory* and Relativistic Groups, 367, 1968.

H. Spinka & H. Winkler, *Nucl. Phys. A*, **233**, 456, 1974.

J. Thomas, Y. T. Chen, S. Hinds, D. Meredith, & M. Olson, *Phys. Rev. C*, **33**, No. 5., 1679, 1986.

James W. Truran, A. G. W. Cameron, & A. A. Gilbert, *Can. J. Phys.*, **44**, 576, 1966.

Steven Weinberg, *Phys. Rev. Lett.*, **19**, 1264, 1967.

C.-S. Wu & C. A. Barnes, *Nucl. Phys. A*, **422**, 373, 1984.

PART VI

TERMINAL EVOLUTION OF LOW AND INTERMEDIATE MASS STARS

21 Wind mass loss on the TPAGB and evolution as a PN central star and as a white dwarf

21.1 Introduction

Single stars and stars in wide binaries which become AGB stars end their lives as white dwarfs. After leaving the main sequence, population I stars of initial mass in the range ~ 1 to ~ 2.25 $M_\odot$ become red giants with an electron-degenerate core composed primarily of helium, experience a set of off-center helium shell flashes before quiescently converting the helium in the core into carbon and oxygen, and reach the TPAGB with a common electron-degenerate CO core of mass ~ 0.5 $M_\odot$. Stars of initial mass in the range ~ 2.25 to ~ 10.5 $M_\odot$ ignite helium quiescently and, after converting helium in central regions into carbon and oxygen become AGB stars with electron-degenerate cores of mass between $\gtrsim 0.5$ $M_\odot$ and ~ 1.37 $M_\odot$. If the mass of the main sequence progenitor lies between ~ 2.25 $M_\odot$ and ~ 8.5 $M_\odot$, the star reaches the TPAGB phase with a CO core of mass between ~ 0.5 $M_\odot$ and 1.1 $M_\odot$. If the mass of the main sequence progenitor lies between ~ 8.5 $M_\odot$ and ~ 10.5 $M_\odot$, nucleosynthesis in the electron-degenerate core continues until the core material is converted into oxygen and neon and the star evolves into a TPAGB star with an ONe core of mass between 1.1 and 1.37 $M_\odot$. The precise limits on progenitor mass depend on the initial composition. Representatives of the first two of the three types of star which become TPAGB stars and then white dwarfs are well represented in this book (Chapters 9 and 11 in Volume 1 and Chapters 17, 18, and 19 in this volume). Evolution of the third type has been addressed in a series of investigations summarized in the fifth paper of a series of papers by Claudio Ritossa, Iben, & Enrique García-Berro (1999).

Thermally pulsing AGB stars become unstable to acoustical pulsations in their envelopes (above the thermally pulsing nuclear burning shells) of the sort exhibited by the variable Mira and, as they evolve to higher luminosities and lower surface temperatures, the amplitude of the acoustical pulsations increases. Mira variables experience rapid mass loss via a stellar wind driven by a complex process involving shocks due to the acoustical pulsations and radiation pressure on grains formed in an extended atmosphere. After ejecting most of their hydrogen-rich envelopes, the stellar remnants of TPAGB stars contract rapidly at nearly constant luminosity to high enough surface temperatures that the rate of emission from the compact remnant of photons with enough energy to ionize hydrogen causes the ejected material to fluoresce as a planetary nebula. Most remnants leave the TPAGB burning hydrogen and, ultimately, as the mass of the hydrogen-rich surface layer decreases below a critical value, the rate of hydrogen burning declines precipitously and the remnant evolves into a white dwarf, radiating energy primarily in consequence of compression and cooling of interior matter.

In Section 21.2, the evidence for the occurrence of a superwind is described, the consequences of this wind for the formation of planetary nebulae are discussed, and references to some of the early theoretical work exploring planetary nebula evolution are given. In Section 21.3, the 1 $M_\odot$ model described in Section 17.5, which is in the process of powering down from the last of 11 sets of thermal pulses as a TPAGB star, is evolved further until it has reached the early part of the 12th quiescent hydrogen-burning phase. It is then subjected to mass loss at a high rate until the mass in its hydrogen-rich envelope has been reduced to approximately 11 times the thickness of the hydrogen-burning shell. At this point the model mass is $\sim$0.565 $M_\odot$ and the model radius is seven times smaller than its radius at the start of the mass-loss episode. In describing the further evolution of the model, particular attention is paid to the relationship between theoretical model properties and the observability of the circumstellar shell created during the superwind phase and shaped by interaction with a fast wind from the central star.

A discussion of the evolution of the model following the planetary nebula phase is preceded by a discussion in Section 21.4 of the nature of the equation of state for liquids and solids and a description in Section 21.5 of algorithms for calculating this EOS that are based on sophisticated treatments in the literature and that have been used in calculating evolutionary models during the white dwarf cooling phase. The evolution of the remnant of the 1 $M_\odot$ model as a white dwarf is discussed in Section 21.6, with a focus on the relationship between gravothermal energy generation, weak interaction induced neutrino losses, and the surface luminosity. During the early part of the white dwarf phase, energy released in consequence of compression due to the work done by gravity acts as the major source of gravothermal energy release. For a period of $\sim$10^7 yr after gravothermal energy release outside of the region of neutrino loss replaces hydrogen burning as the major source of surface luminosity, the neutrino luminosity, also supplied by gravothermal energy release, exceeds the surface luminosity. Over the next $\sim$10^8 yr, cooling by heavy ions transitioning from the gaseous state into the liquid state is primarily responsible for the surface luminosity. After $\sim$2 $\times$ 10^9 yr of evolution as a white dwarf, solidification begins in central regions, being completed over most of the interior after another 7 $\times$ 10^9 yr of evolution. Solid state cooling in the Debye regime reduces the ion-cooling rate in central regions relative to the ion-cooling rate in outer layers as the luminosity of the white dwarf decreases below $\sim$3 $\times$ 10^{-5} $L_\odot$.

The gravitational acceleration in the outer layers of the white dwarf remnant of the 1 $M_\odot$ model is over three orders of magnitude larger than in the outer layers of the 1 $M_\odot$ main sequence precursor. Given that, as shown in Section 12.9, the time scale for diffusion of heavy elements out of the convective envelope of a solar model is of the order of 10^{10} yr, one might anticipate that the time scale for diffusion of heavy elements inward in outer layers of the white dwarf progeny is of the order of 10^7 yr. In Section 21.7, the diffusion time scale in surface layers of the white dwarf is calculated to be $\sim$3 $\times$ 10^7 yr and it is estimated that the surface of the white dwarf remnant of the 1 $M_\odot$ model becomes pure hydrogen after a few times 10^8 yr of evolution. At this point, weak interaction induced neutrino losses are not important and pp-chain reactions have replaced CN-cycle reactions as the dominant hydrogen-burning contributors to surface luminosity.

Although the majority of highly evolved white dwarfs are of the DA variety, exhibiting primarily hydrogen at their surfaces, a substantial minority are non-DA white dwarfs showing primarily helium (DB white dwarfs), primarily carbon (DQ white dwarfs), or even more exotic combinations of elements in their spectra. In Section 21.8, it is argued that a DB progenitor ignites helium either just as the progenitor is leaving the AGB or after it has left the AGB burning hydrogen and before the precipitous decline in the hydrogen-burning rate which precedes the transition to the white dwarf state. If the progenitor ignites helium after departing the AGB as a hydrogen-burning star, it returns to the AGB as a born-again AGB star and loses enough of its hydrogen-rich surface layer that, on leaving the AGB for a second time and developing a high surface temperature, a fast wind can remove all of the remaining hydrogen-rich material. During early evolution as a white dwarf, diffusion then leads to a monoelemental surface abundance which can be either helium or carbon.

The exotic combination of elements found in the spectra of the hot central stars known as PG1159 stars is due to the ignition of a helium shell flash as a hydrogen-burning remnant is transitioning to the white dwarf state. When the convective shell engendered by the flash reaches the hydrogen-rich surface layer, hydrogen is ingested by the convective shell and, as hydrogen diffuses far enough inward, a hydrogen-burning shell flash is ignited, with many interesting consequences for surface abundances. Model tracks in the HR diagram are likewise exotic and can account for the evolutionary tracks and evolutionary time scales of real stars such as FG Sge, a classic born-again AGB star.

21.2 Superwind ejection of the envelope and planetary nebula evolution

A large body of observational evidence demonstrates that the lifetime of a real TPAGB star is much shorter than would be predicted if the star did not lose most of its initial hydrogen-rich envelope by mass loss from its surface and instead continued to burn nuclear fuel until it had consumed all of the hydrogen and helium which it had in its envelope when it first became a TPAGB star. The evidence is both direct and indirect.

One line of indirect evidence involves comparisons between distributions in the HR diagram of AGB stars in galactic and extragalactic clusters of intermediate to old age with the characteristics of theoretical models of stars of initial mass larger than the mass of stars in the clusters which are beginning to leave the main sequence (called turnoff point stars). In the observed clusters, the brightest red giant is considerably less bright than theory predicts if this giant was initially sufficiently more massive than a turnoff point star to have evolved to the TPAGB phase and continued to burn hydrogen and helium alternately in shells until it had exhausted the hydrogen which was in its envelope at the beginning of the AGB phase.

Another line of indirect evidence involves a comparison between masses of observed white dwarfs which are not in close binaries with theoretical relationships between the CO core mass at the start of the TPAGB phase and initial model mass. For example, as shown by J. B. Kaler & G. H. Jacoby (1991), white dwarfs in the Magellanic clouds have masses larger than $\sim$0.55 $M_\odot$ and their distribution in number versus mass peaks at a mass near 0.6 $M_\odot$. A similar result holds for white dwarfs in the solar vicinity (Detlef Koester & Volker Weidemann, 1980). Theoretical models which evolve beyond the main sequence in less than a Hubble time must have an initial mass larger than 0.8–1.0 $M_\odot$ (the precise value being a function of metallicity) and models of initial mass between this lower limit and $\sim$2.3 $M_\odot$ (the precise value again being a function of metallicity) have a CO core mass larger than $\sim$0.53 $M_\odot$ on becoming TPAGB stars. Thus, a typical Magellanic Cloud white dwarf has evolved from a star which has lost at least several tenths of a solar mass after leaving the main sequence.

The direct evidence consists of estimates of mass-loss rates from Mira variables and OH-IR stars which are, in many instances, several orders of magnitude larger than the rates at which theoretical models of the same luminosity consume hydrogen and helium. These mass-loss rates are based on velocities determined by Doppler shift measurements and on densities in surface layers estimated by various means, as described, e.g., by Stuart R. Pottash in the book *Planetary Nebulae* (1984). At any given luminosity, estimated mass-loss rates show a large dispersion but the average rate increases with luminosity. Near the observationally defined tip of the AGB, they can approach of the order of 10^{-4} $M_\odot$ yr^{-1}, which is over a thousand times larger than the average rate at which a theoretical TPAGB model of luminosity equal to that of an observed star consumes hydrogen.

Bogdan Paczyński was the first to exploit this evidence by formally abstracting mass from an AGB model of low mass at a rate much larger than the rate of nuclear fuel consumption and by following the evolution of the remnant star across the HR diagram toward the white dwarf stage (Paczyński, 1971). He identified the ejected material as a planetary nebula and noted that, when its surface became sufficiently hot, the remnant would emit photons with enough energy to ionize hydrogen and trigger physical processes that would cause the ejected nebula to fluoresce for an extended period as it evolved at nearly constant luminosity from the AGB to the beginning of the white dwarf phase.

Alvio Renzini played an important part in establishing the theoretical framework for understanding the relationship between properties of theoretical models and the distribution of observed planetary nebulae, demonstrating, for example, that, during a large fraction of the planetary nebula phase, the remnant is evolving into a hot white dwarf, with its luminosity decreasing and its surface temperature increasing, and that, in a typical theoretical planetary nebula distribution, the remnant of the AGB star causes the ejected nebula to remain visible for approximately 10^4 yr, in agreement with the observed statistics of planetary nebulae. For a summary of this work, see A. Renzini in *Planetary Nebulae* (ed. S. Torres-Peimbert, 1989). Very important ingredients of the early comparison between theory and observation were the theoretical evolutionary tracks by Detlef Schönberner (1981) for model stars of initial mass $M = 0.8$–1.4 $M_\odot$ which, assuming rather modest rates of mass loss compared with what observations now suggest, produced white dwarfs of mass $M_{WD} = 0.546$–0.644 $M_\odot$. Other important sets of early models include those by Peter R. Wood and Donald J. Faulkner (1986).

Mass loss from AGB stars is understood to be a consequence of the action of shocks produced by pulsations (Lee Ann Willson & S. J. Hill, 1979) and radiation pressure on grains (N. C. Wickramasinghe, Bertram Donn, & Ted P. Stecher, 1966, and K. S. Krishna Swamy & Ted P. Stecher, 1969). Incorporating both of these effects in hydrodynamic models, G. H. Bowen (1988) and Bowen and Lee Ann Wilson (1991) produced models in which shocks inflate the atmosphere beyond that found in a hydrostatic model, radiation pressure accelerates grains formed in the cool inflated atmosphere, and collisions between grains and ions accelerate the gas, leading to an outflowing wind. The mechanism for driving pulsations is not well established, but the presence of carbon grains observed in some Mira AGB stars is undoubtedly the consequence of dredge-up during the TPAGB phase which brings fresh carbon to the surface.

Good discussions of the observed properties of planetary nebulae and their central stars and of the physical processes involved in planetary nebula fluorescence are given, respectively, by James B. Kaler in *Stars and their Spectra* (1981) and by Donald E. Osterbrock (1988), and a description of the connection between theoretical models of remnants and the observations is given by Iben (1995).

21.3 Departure of a 1 $M_\odot$ model from the TPAGB, evolution as the central star of a planetary nebula, and the transition from nuclear to gravothermal energy as the primary source of surface luminosity

The last 1 $M_\odot$ model described explicitly in Section 17.5 is powering down after the second of the eleventh set of helium shell flashes on the TPAGB. This model has been evolved at constant mass for another 4.839×10^4 yr, during which time hydrogen re-ignites and the mass at the center of the hydrogen-burning shell increases outward by 1.035×10^{-3} $M_\odot$, or by approximately one sixth of the increase in mass between helium shell flash episodes. At this point, mass is abstracted from the model at the rate 10^{-4} $M_\odot$ yr^{-1} for a period of 4288 yr, at the end of which the model mass is $M_s = 0.571\,24$ $M_\odot$. The center of the hydrogen-burning shell is at $M_H = 0.563\,96$ $M_\odot$, so the mass of the hydrogen-rich envelope is $M_e = M_* - M_H = 7.28 \times 10^{-3}$ $M_\odot$, or approximately 68 times larger than the mass thickness $\Delta M_H = 1.070 \times 10^{-4}$ $M_\odot$ of the hydrogen-burning shell. Global characteristics are $\log(L) = 3.6475$, $\log(T_e) = 3.5226$, $R/R_\odot = 200.90$, $L_H = 4140.6$ $L_\odot$, $L_{He} = 76.803$ $L_\odot$, and $L_{grav} = 236.82$ $L_\odot$. The neutrino–antineutrino luminosity is $L_{\nu\bar\nu} = 15.9$ $L_\odot$.

Over the next 42 yr, mass is abstracted at a variable but declining rate and then terminated when the model mass reaches $M_s = 0.565121$ $M_\odot$. The center of the hydrogen-burning shell is at $M_H = 0.563\,983$ $M_\odot$ and the mass of the hydrogen-rich envelope is $M_e = M_s - M_H = 1.138 \times 10^{-3}$ $M_\odot$, or approximately 11 times larger than the mass thickness $\Delta M_H = 1.034 \times 10^{-4}$ $M_\odot$ of the hydrogen-burning shell. Global characteristics are

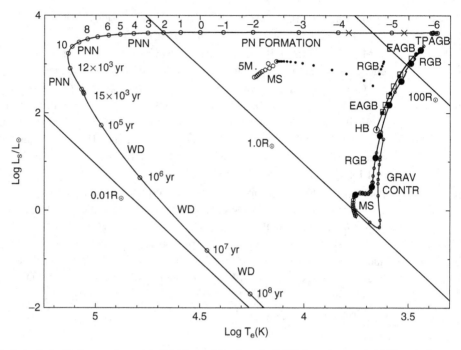

Fig. 21.3.1 Evolution in the HR diagram of a model of initial mass 1 $M_\odot$ from the TPAGB through superwind, PNN, and white dwarf phases

$\log(L) = 3.6476$, $\log(T_e) = 3.7895$, $R/R_\odot = 58.767$, $L_H = 4152.3\ L_\odot$, $L_{He} = 74.504\ L_\odot$, and $L_{grav} = 231.25\ L_\odot$. The neutrino–antineutrino luminosity is still $L_{\nu\bar{\nu}} = 15.9\ L_\odot$.

Evolution in the HR-diagram is shown in Fig. 21.3.1. For orientation, the evolutionary track of the TPAGB precursor is shown on the right hand side of the figure. The track during the pre-main sequence gravitational contraction phase is reproduced from Fig. 11.1.1 in Chapter 11 of Volume 1, the track from the main sequence to the first RGB phase is reproduced from Fig. 11.1.32, and the track from the HB phase through the EAGB phase is reproduced from Fig. 17.2.5 in this volume. The location of the model during the TPAGB phase is identified by the acronym TPAGB. For perspective, the evolutionary track of a 5 $M_\odot$ model evolving from the main sequence to the first red giant branch, as given by Fig. 11.2.42 in Volume 1, is also shown.

The rightmost cross along the nearly horizontal track at the top of Fig. 21.3.1 indicates where the phase of high, constant mass-loss rate terminates and the region between this cross and the leftmost cross indicates where mass loss is phased out. To the left of the leftmost cross, the location along the track is a function only of the mass M_e of the hydrogen-rich envelope, which changes in response to hydrogen burning at the rate

$$\frac{dM_e}{dt} = -\frac{dM_H}{dt} \sim -\frac{1.026 \times 10^{-11}}{X_H} \frac{L_H}{L_\odot}\ M_\odot\ \text{yr}^{-1}, \qquad (21.3.1)$$

where X_H is the abundance by mass of hydrogen in the envelope. In the present instance, $X_H = 0.688$ and $L_H \sim 4.145 \times 10^3 \, M_\odot$, so, along what may be termed the plateau branch of evolution,

$$\frac{dM_e}{dt} = -\frac{dM_H}{dt} \sim -6.0 \times 10^{-8} \, M_\odot \, \text{yr}^{-1}. \tag{21.3.2}$$

The single digit numbers along the plateau branch indicate the time in units of 10^3 yr at which the model reaches each number, with zero time being defined as the time when the model reaches $\log T_e = 4.5$. At time zero, the center of the hydrogen-burning shell is at a mass $M_e = M_* - M_H = 8.47 \times 10^{-4} \, M_\odot$ below the surface and the thickness of the shell is $\Delta M_H = 9.821 \times 10^{-5} \, M_\odot$. If evolution at constant luminosity were to continue, hydrogen would be exhausted in less than 1.41×10^4 yr.

Analysis of observational material (e.g., M. Perinotto, 1990) shows that mass loss occurs along the plateau branch, presumably via a radiatively driven wind, and increases in strength as the star becomes hotter, so eq. (21.3.2) is actually an underestimate of the rate of change of M_e in the real analogue of the theoretical model. In the following development, no attempt is made to take this added complication explicitly into account. This does not mean that the wind from the central star is unimportant. But, since the location along the plateau branch is a function only of the mass in the hydrogen-rich envelope, the effect of mass loss on the central star is simply to increase the rate of evolution along the branch, an increase which can be easily taken into account. The time between two adjacent circles along the track in Fig. 21.3.1 is reduced by the ratio of the rate of change in M_e given by eq. (21.3.2) to the rate of change obtained by adding the adopted mass-loss rate (times X_H) to the rate of change given by eq. (21.3.2).

The importance of the wind from the central star lies in its effect on the nebular matter ejected during the superwind phase. The wind from the central star is a fast wind, with speeds of the order of the escape velocity from the compact central star (several times $1000 \, \text{km s}^{-1}$), in contrast with the slow nebular wind which has speeds of the order of the escape velocity from the supergiant TPAGB precursor (10–$20 \, \text{km s}^{-1}$). The interaction between the fast wind and the slow wind leads to a compression at the point where the two winds meet and, therefore, to an enhancement of the observability of the planetary nebula. The physics of the interaction has been described in a series of papers by Sun Kwok and his coworkers (e.g., S. Kwok, C. R. Purton, & P. M. FitzGerald, 1978) and by F. D. Kahn and his coworkers (e.g., F. D. Kahn & K. A. West, 1985).

The choice of $T_e \sim 30\,000$ K as the zero point for the measurement of time along the plateau branch is a matter of convention, based to a large extent on the properties of observed planetary nebulae, which are the ejecta formed during the superwind phase. The nebula is caused to fluoresce in consequence of absorption of radiation from the remnant central star, once its surface temperature has become hot enough that a sufficiently large fraction of the photons it emits are energetic enough to ionize hydrogen in the nebula. Collisions between electrons and heavy ions excite low lying atomic levels in the ions, and radiative transitions between these levels produce visible light, with forbidden lines of O III ($^1S_0$-$^1D_2$ and $^1D_2$-$^3P_1$) being prominent in many nebular spectra. Absorption by

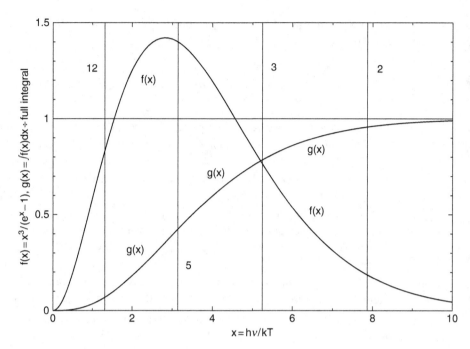

Fig. 21.3.2
Planck distribution of photon energy and integral of the distribution. Vertical lines show where $h\nu \geq 13.61$ eV at temperatures $T_e = 2, 3, 5,$ and 12×10^4 K

atoms of radiation from the central star and recombination of the resulting ions with electrons contribute to the presence of highly excited hydrogen atoms and the Balmer lines of hydrogen are prominent in many nebular spectra.

In first approximation, photons are emitted from the surface of the central star in a distribution given by

$$dN = 8\pi \left(\frac{kT_e}{hc}\right)^3 \frac{x^2 dx}{e^x - 1},\tag{21.3.3}$$

where

$$x = \frac{h\nu}{kT_e}\tag{21.3.4}$$

and dN is the relative number of photons in the interval dx. Energy is radiated by the central star in essentially the Planck distribution given by

$$dE = 8\pi\, h \left(\frac{kT_e}{hc}\right)^4 \frac{x^3 dx}{e^x - 1} = 8\pi\, h \left(\frac{kT_e}{hc}\right)^4 f(x)\, dx.\tag{21.3.5}$$

The x-dependent factor in the distribution given by eq. (21.3.5) is shown in Fig. 21.3.2 by the curve labeled $f(x)$. The integral of $f(x)$ over x up to a given x is proportional to the energy flux of photons up to this value of x. Dividing this integral by the full value of the integral between 0 and ∞ produces the function

$$g(x) = \frac{\int_0^x f(y)\,dy}{\int_0^\infty f(y)\,dy} \tag{21.3.6}$$

which is also plotted in Fig. 21.3.2. The quantity

$$F(x) = 1 - g(x) \tag{21.3.7}$$

gives the fraction of the energy flux from the central star made up of photons with energy equal to or larger than $h\nu = x\,kT$. Thus, for every choice of the surface temperature of the central star there is an x given by

$$x(T_e) = \frac{13.61\ \text{eV}}{kT_e} = \frac{13.61\ \text{eV}}{0.865\ \text{eV}\ T_4} = \frac{15.734}{T_4}, \tag{21.3.8}$$

where T_4 is the surface temperature in units of 10^4 K, such that all photons characterized by values of $x > x(T_e)$ have energies larger than the ionization energy of hydrogen.

Values of $x(T_e)$ are indicated in Fig. 21.3.2 by vertical lines for surface temperatures of $T_4 = 2, 3, 5$, and 12. Values of $F(x)$ at the different temperatures are given by the difference between the horizontal line at unity and the intersection between the appropriate value of $g(x)$ and the vertical line for the chosen temperature. Thus, at $T_e = 20\,000$ K, only about 4% of the energy radiated by the central star involves photons of energy equal to or larger than the ionization energy of hydrogen. At $T_e = 30\,000$ K, slightly more than 20% of the energy radiated involves photons energetic enough to ionize hydrogen. At $T_e = 50\,000$ K and $T_e = 120\,000$ K, the fractions are about 57% and 94%, respectively.

A complete mapping between the quantity $F(x(T_e))$ and $\log(T_e)$ is given in Fig. 21.3.3 by the curve F_{energy}. Exactly the same exercises can be performed with the number distribution given by eq. (21.3.3) as with the energy distribution given by eq. (21.3.5). The end result is shown in Fig. 21.3.3 by the curve F_{number} which gives the number of photons emitted in unit time by the central star with energies large enough to ionize hydrogen relative to the total number of photons radiated in unit of time by the central star. The displacement of the F_{energy} and F_{number} curves is a simple consequence of the fact that low energy photons dominate in the number distribution and high energy photons dominate in the energy distribution.

Both curves in Fig. 21.3.3 demonstrate that the designation of $\log(T_e) = 4.5$ as the point where planetary nebulae first become detectible is rather arbitrary. Nevertheless, a concrete choice allows one to make a first estimate of the duration of the planetary nebula phase. For example, assuming that it is the total energy in ionizing photons that is the most important characteristic of the central star in determining the brightness of a nebula, brightness is proportional to the product of L_{PNN}, the luminosity of the central star (PNN = planetary nebula nucleus), and the fraction $F_{\text{energy}}(x(T_e))$ of the energy emitted in ionizing photons. At $\log(T_e) = 4.5$, $L_{\text{PPN}} \times F_{\text{energy}}(x(T_e)) \sim 4440\ L_\odot \times 0.25 = 1110\ L_\odot$. For $\log(T_e) > 5.1$, $F_{\text{energy}}(x(T_e)) \gtrsim 0.95$, so $L_{\text{PPN}} \sim 1170\ L_\odot$ gives the same value for the product $L_{\text{PPN}} \times F(x(T_e))$ prevailing at $\log(T_e) = 4.5$. From Fig. 21.3.1, this luminosity is

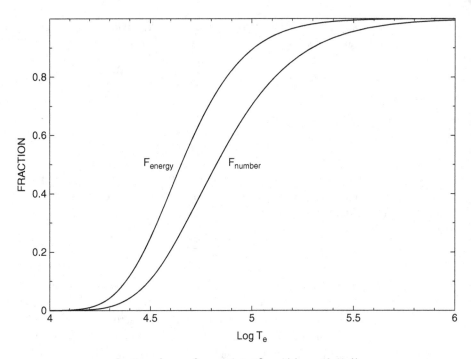

Fig. 21.3.3   Fractions of energy flux and photon flux with $h\nu > 13.61$ eV

reached approximately 1.1×10^4 yr after the time when, by the adopted convention, the planetary first becomes detectible at $\log(T_e) = 4.5$.

On the other hand, it can be argued that the most important characteristic of the central star in determining the observability of the circumstellar nebula is the number, rather than the energy, of ionizing photons which reach the nebula per unit time. The average energy of a photon is $2.71\, kT_e$, so the rate at which ionizing photons reach the circumstellar nebula is

$$\left(\frac{\mathrm{d}N}{\mathrm{d}t}\right)_{h\nu \geq 13.6 \text{ eV}} = \frac{F_{\text{number}}(x(T_e))}{2.7\, kT_e}\, L_{\text{PNN}}. \qquad (21.3.9)$$

At $L_{\text{PNN}} = 4440\, L_{\odot}$ and $\log T_e = 4.5$, this product is $0.014\,05\, L_{\odot}$, and at $L_{\text{PNN}} = 1584$ $L_{\odot}$ and $\log T_e = 5.14$, where $t \sim 10^4$ yr, it is $0.009\,41\, L_{\odot}$.

Many factors other than the rate at which the total energy of ionizing photons and the rate at which the total number of ionizing photons from the central star reach the nebular shell are involved in determining the observability of the nebula, including the amount of mass in and the size of the nebula. Nevertheless, the luminosity of the central star is crucial and this luminosity decreases very rapidly when the mass in the hydrogen-rich envelope decreases below a critical value necessary to sustain CN-cycle hydrogen burning at a high rate. As is evident from Fig. 21.3.1, a rapid drop in luminosity occurs as the model approaches and recedes from a maximum surface temperature of $\log(T_e) \sim 5.14$. The precipitous drop in luminosity begins near $t \sim 11 \times 10^3$ yr and is completed by $t \sim 13 \times 10^3$ yr, when the luminosity has dropped to the extent that $\log(L/L_{\odot}) \sim 2.5$ and the surface

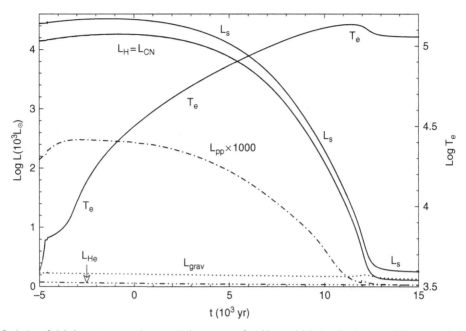

Fig. 21.3.4   Evolution of global energy-generation rates in the remnant of a 1 $M_\odot$ model during the plateau and planetary nebula stages

temperature has dropped to the extent that $\log(T_e) \sim 5.08$. At this point, the product $L_{PPN} \times F_{energy}(x(T_e))$ is about 20% of the value prevailing at $\log(T_e) = 4.5$ and the product $F_{number}(x(T_e))/T_e \, L_{PNN}$ is about 13% of the value prevailing at $\log(T_e) = 4.5$.

In summary, in the adopted scenario, the observable planetary nebula deriving from a 1 $M_\odot$ main sequence star consists of the inner part of an ejectum of mass $\sim 0.43 \, M_\odot$ which is caused to fluoresce by absorption of those photons emitted by the central star with enough energy to ionize hydrogen; the central star, of mass $\sim 0.57 \, M_\odot$, evolves to the blue at a luminosity declining from $\sim 4444 \, L_\odot$ to $\sim 1111 \, L_\odot$, at surface temperatures between $\sim 30\,000$ K and $\sim 140\,000$ K, over a period of the order of 10^4 yr.

Global energy-generation rates in the model during the planetary nebula phase are shown in Fig. 21.3.4. During most of the phase, the primary source of surface luminosity is hydrogen burning by the CN cycle. The rate at which neutrinos abstract energy from the deep interior is not shown in the figure, but it is of the order of 10% of the net rate at which gravothermal energy is produced by the model, so, during the time period shown, most of the gravothermal luminosity contributes to the surface luminosity. The fact that L_{grav} remains nearly constant even as the hydrogen-burning contribution to the surface luminosity declines dramatically demonstrates that most of the energy of gravothermal origin which reaches the surface is not a consequence of compression related to the conversion of hydrogen into helium but is instead a consequence of compression in layers below the hydrogen-burning shell.

Even though it is impossible to define precisely the characteristics of the central star of the first observable planetary nebula in the real Universe, it is nevertheless useful to

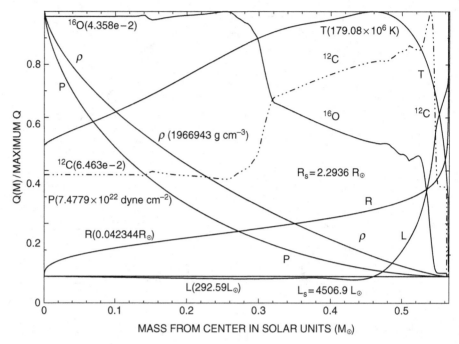

Fig. 21.3.5 Structure and composition in the interior of the remnant of a 1 $M_\odot$ model at the beginning of the observable planetary nebula phase

designate the model which reaches $\log T_e = 4.5$ for the first time as representative of such a central star. Similarly, it is useful to designate the model in which gravothermal energy first replaces nuclear energy as the primary source of surface luminosity as representative of the central star when the real counterpart of the theoretical nebular system is fading from sight. The specific model selected as the representative of the central star during the incipient observable phase has the global luminosity characteristics $L_H = 4255.98\ L_\odot$, $L_{He} = 55.20\ L_\odot$, $L_{grav} = 211.59\ L_\odot$, $L_{\nu\bar\nu} = 16.24\ L_\odot$, and $L_s = 4506.9\ L_\odot$. Its surface characteristics are $\log L_s = 3.6539$, $\log T_e = 4.4954$, and $R_s = 2.2936\ R_\odot$.

The model selected as representative of the central star as the nebula is fading from sight has evolved $\Delta t = 13.16 \times 10^3$ yr beyond the point when $\log T_e = 4.5$. The hydrogen-burning luminosity is slightly smaller than the gravothermal luminosity and the various global luminosities are $L_H = 130.84\ L_\odot$, $L_{He} = 32.06\ L_\odot$, $L_{grav} = 146.61\ L_\odot$, $L_{\nu\bar\nu} = 16.99\ L_\odot$, and $L_s = 292.59\ L_\odot$. Surface characteristics of the model are $\log L_s = 2.4663$, $\log T_e = 5.0654$, and $R_s = 0.042344\ R_\odot$.

The next eight figures compare interior characteristics of the two representative models. Structure and composition characteristics between the centers and the hydrogen-burning regions of the two models are shown in Figs. 21.3.5 and 21.3.6, respectively. In both figures, temperature, density, and pressure variables are normalized to the maximum values of the variables in each model. The distributions of these variables are almost indistinguishable. For example, pressures and densities at the center differ by only 2.2% and 1.5%, respectively, and maximum temperatures differ by only 0.27%. The sense of the

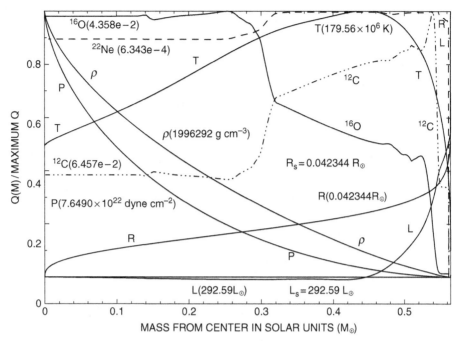

Fig. 21.3.6 Structure and composition in the interior of the remnant of a 1 $M_\odot$ model as gravothermal rivals nuclear energy as the major source of surface luminosity

differences is a reflection of the fact that matter over all of the interior is contracting and, over most of the interior, matter is heating.

The radius and luminosity distributions for both models are normalized to the surface radius and luminosity of the more evolved (older) model. Even though the surface radius of the less evolved (younger) model is 54 times larger than the surface radius of the more evolved model, the radius versus mass distributions below the hydrogen-burning regions remain virtually identical. The same holds true for the luminosity versus mass distributions, except in the outermost regions of the interior, even though the surface luminosity of the more evolved model is 15.4 times smaller than that of the less evolved model. The difference in the luminosity versus mass distributions in the outer 10% of the mass of the models below the hydrogen-burning shells is due to cooling in these regions and to a consequent decrease in the contribution of helium burning to the ambient outward luminosity flux.

In the inner 0.46 $M_\odot$ of the less evolved model and in the inner 0.445 $M_\odot$ of the more evolved model, the luminosity is negative. In both regions of negative luminosity, that fraction of the compressional energy which is released locally and is not converted locally into thermal energy and/or neutrino energy loss contributes to the ambient inward flow of energy which ultimately is used up in supplying the energy carried away by neutrino–antineutrino pairs produced in central regions.

Various sources of energy production in interior regions where the ambient flow of energy is outward and where there is no hydrogen are shown in detail in Figs. 21.3.7 and 21.3.8 for the two models. In both models, the rate at which compressional energy is

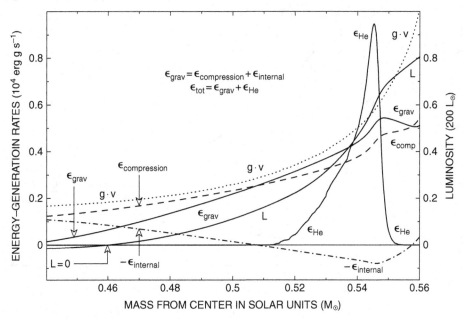

Fig. 21.3.7   Energy-generation rates below the hydrogen-burning shell of the remnant of a 1 $M_\odot$ model at the beginning of the observable planetary nebula phase

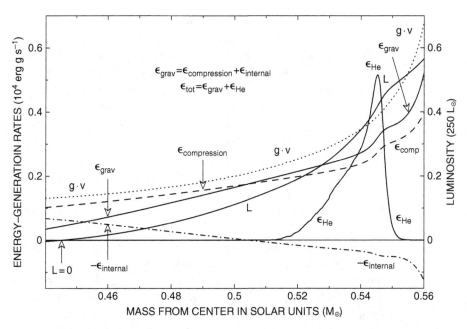

Fig. 21.3.8   Energy-generation rates below the H-burning shell of the remnant of a 1 $M_\odot$ model as gravothermal rivals nuclear energy as the major source of surface luminosity

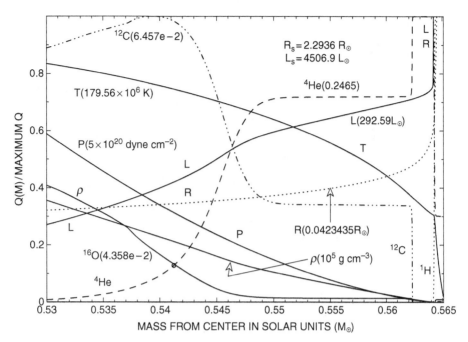

Fig. 21.3.9 Structure and composition in outer layers of the remnant of a 1 $M_\odot$ model at the beginning of the observable planetary nebula phase

released locally dominates the gravothermal energy-generation rate. The contribution of gravothermal energy to the luminosity at the outer edges of the regions, below hydrogen-rich layers, is in both cases nearly the same at $\Delta L_{grav} \sim 105$–$110\ L_\odot$. In regions where helium is burning, temperatures are decreasing and this accounts for the fact the contribution of helium burning to the luminosity at the base of the hydrogen-rich surface layers is smaller by a factor of ~ 1.7 in the older model than in the younger model.

Structure and composition variables in a region centered on the helium-burning shell are shown for the two models in Figs. 21.3.9 and 21.3.10, respectively. Since the scales for all variables are the same in both figures, it is easy to see that, except in the very outermost part of the region shown, temperatures have decreased and densities and pressures have increased with the passage of time.

Structure and composition variables in regions centered on the hydrogen-burning shell are shown for the two models in Figs. 21.3.11 and 21.3.12, respectively. The scale for luminosity is in each case the surface luminosity of the model. For all other variables in the models, the vertical scales are the same. In each case, the center of the hydrogen-burning shell is indicated by a circle along the hydrogen-abundance profile. Although the total mass between end points along the horizontal axis is in each case the same, the end points have been chosen in such a way that the center of the hydrogen-burning shell appears at the same position relative to the low mass and high mass end points. In the case of the younger model, profiles extend only to the outermost mass shell in the quasistatic interior, with

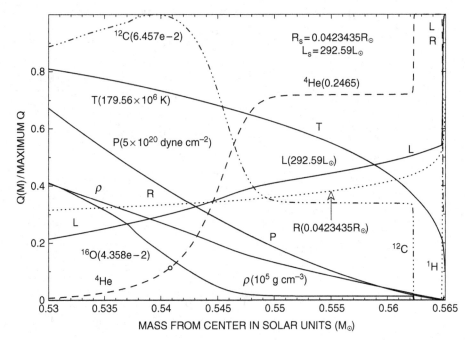

Fig. 21.3.10 Structure and composition in outer layers of the remnant of a 1 $M_\odot$ model as gravothermal rivals nuclear energy as the major source of surface luminosity

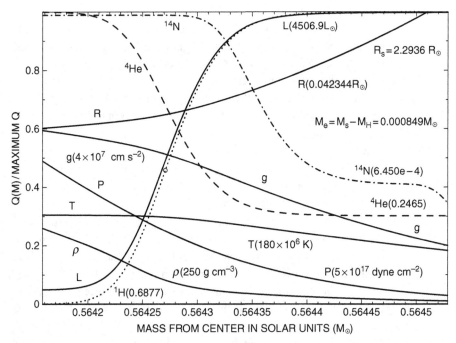

Fig. 21.3.11 Structure and composition in and near the H-burning shell in the remnant of a 1 $M_\odot$ model at the beginning of the observable planetary nebula phase

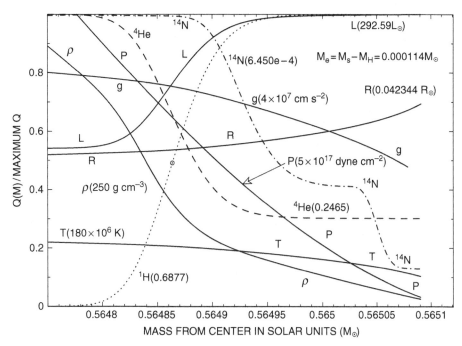

Fig. 21.3.12 Structure and composition in and near the H-burning shell in the remnant of a 1 $M_\odot$ model as gravothermal rivals nuclear energy as the source of surface luminosity

3×10^{-4} $M_\odot$ being contained in a static envelope. In subsequent models, the mass of the static envelope is gradually decreased to 10^{-8} $M_\odot$.

Without examining structure details in layers between the burning shell and the surface, one might be tempted to suppose that it is the decrease with time in the weight of matter above the burning shell that is somehow responsible for the decrease in the strength of the burning shell. In passing from the less evolved (younger) to the more evolved (older) model, the mass between the center of the burning shell and the surface decreases by a factor of approximately 7.45, whereas, from the curves for gravitational acceleration in Figs. 21.3.11 and 21.3.12, one sees that g at the center of the burning shell increases by only a factor of about 1.5, so one might guess that the force which matter above the burning shell exerts at the location of the shell decreases by a factor of the order of 5. However, given the radial variation of all pertinent quantities, weight is more properly measured by pressure. In going from the younger to the older model, pressure at shell center actually *increases* by a factor of about 2.6, with an increase by a factor of ~4.4 in density at shell center offsetting a decrease by ~50% in temperature. Since the pressure at any point measures the weight of a column of matter of *unit* cross section between the point and the surface, it is obviously not a change in the mass in this column that is responsible for the decrease in the strength of the burning shell. The basic reason for the decrease in the strength of the shell is that, as the burning shell and the surface approach one another, temperatures in the shell and the temperature at the surface become closer. That is, it is not a change in the mass above the burning shell but the decrease in shell temperatures brought about by

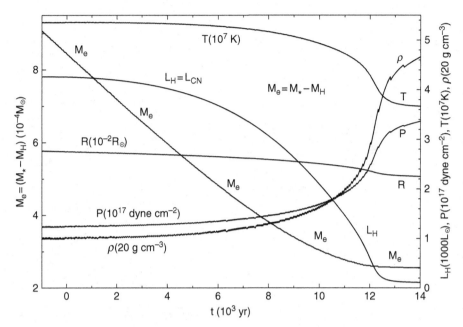

Fig. 21.3.13 Properties at the center of the hydrogen-burning shell in the remnant of a 1 $M_\odot$ model during the planetary nebula stage

the rapprochement of the shell and the surface that is responsible for the decline in shell strength.

Variations with time of several quantities describing conditions at the center of the hydrogen-burning shell are shown in Fig. 21.3.13 over a time period encompassing the two models representing the beginning and end of the planetary nebula phase. The location of the shell center is described in the figure by the curve labeled M_e which is the mass between the center of the shell and the surface of the model of mass $M_* = 0.565\,120\,855$ $M_\odot$. The jaggedness along the curves for pressure and density are due in part to having recorded these quantities with fewer significant figures than warranted by the precision to which they are actually determined.

21.4 Coulomb forces, properties of matter in the solid phase, and a criterion for melting

The approximations to the equation of state described in Volume 1 of this book and used thus far in this volume are adequate over the interiors of most stars during nuclear burning phases. However, in white dwarfs, as the interior cools, matter near the center first lique-fies and then solidifies and the white dwarf becomes a layered structure consisting of a solid core, a liquid mantle, and an electron-degenerate envelope in which the nuclei are in the gaseous state. In the inner two layers, Coulomb forces between electrons and nuclei play a dominant role in controlling the dynamics of nuclear motions. Electrons and nuclei are intimately coupled in the propagation of sound waves. In this section, the focus is on acquiring an intuitive understanding of the physics involved.

In the solid phase, nuclei take up positions in a regular lattice, with each nucleus oscillating about a lattice point. The mean energy of motion is small compared with the electrostatic binding energy between nuclei and electrons and the negative contribution to pressure associated with electrostatic binding dominates the contribution associated with thermal motions. With increasing temperature, beginning at very low temperatures, the thermal contribution increases with the fourth power of the temperature, approaching a maximum of $3kT$ per nucleus: half of this energy is associated with kinetic energy, half with potential energy. The amplitude of oscillations about lattice points increases with increasing temperature and, when the mean vibration amplitude becomes of the order of fifteen percent of the mean distance between adjacent lattice points, corresponding to a ratio of the electrostatic binding energy to kT of the order of 150–200, melting occurs. During the liquid phase, the thermal contribution to the energy remains at $3kT$ per nucleus until, following evaporation, the thermal contribution to the energy reverts to $\frac{3}{2}kT$ per nucleus.

21.4.1 The Wigner–Seitz sphere

In Section 4.17 of Volume 1, the electrostatic binding energy in the gaseous phase when electrons are very degenerate is estimated by imagining that each nucleus is encased in a sphere of uniform electron density and that the nucleus–electron electrostatic binding energy per unit volume is simply the sum of the binding energies of the nuclei and their associated electrons in the unit volume. Each sphere, defined by eq. (4.17.59) in Volume 1 as one which contains Z_i free electrons, where Z_i is the average ion electric charge number, is sometimes called a Wigner–Seitz sphere (E. Wigner & F. Seitz, 1934). For a uniform number density n_e of free electrons and one type of nucleus of charge Z_i and number density n_i, $n_e = Z_i n_i = Z_i(\rho/A_i M_H) = (\rho/\mu_e M_H)$ and the radius $R_{WS,i}$ of the sphere is

$$R_{WS,i} = \left(\frac{3}{4\pi}\frac{1}{n_i}\right)^{1/3} = \left(\frac{3}{4\pi}\frac{A_i}{N_A}\frac{1}{\rho}\right)^{1/3} = 0.734\,60 \times 10^{-8}\ \text{cm}\ \left(\frac{A_i}{\rho_0}\right)^{1/3}. \quad (21.4.1)$$

The fact that the radius of the sphere is independent of the charge Z_i is important when applying expressions tailored to stellar conditions to terrestial metals in which only a small fraction of the electrons are in conduction bands, while most electrons remain bound to nuclei and participate in the oscillatory motion of these nuclei.

A measure of the electrostatic binding energy per nucleus to the classical thermal energy is traditionally expressed by the quantity

$$\Gamma_i = \frac{Z_i^2\, e^2}{R_{WS,i}}\frac{1}{kT} = \frac{Z_i^2}{A_i^{1/3}}\frac{e^2}{kT}\left(\frac{4\pi\rho N_A}{3}\right)^{1/3} = 2.2718 \times 10^5\ \frac{Z_i^2}{A_i^{1/3}}\frac{\rho^{1/3}}{T} \quad (21.4.2)$$

and, as given to first order by eq. (4.17.62) in Volume 1, the energy with which each nucleus is bound to its associated electrons is

$$E_{\text{bind},i} = \frac{9}{10} Z_i^2 \frac{e^2}{R_{\text{ws},i}} = \frac{9}{10} \Gamma_i \, kT. \tag{21.4.3}$$

When the lattice structure of the solid phase is specified, it is possible to find the reduction in the binding energy due to the electrostatic repulsive forces between nuclei. Remarkably, the coefficient of $\Gamma_i \, kT$ in eq. (21.4.3) is decreased by only a very small amount. For a body centered cubic lattice, the coefficient becomes 0.895 929 instead of 0.9, and, for a face centered cubic lattice, it becomes 0.895 874.

It is of interest to compare the electrostatic binding energy between nuclei and electrons in the Wigner–Seitz sphere with the kinetic energy of the degenerate electrons in this sphere. Inserting numbers in eqs. (21.4.2) and (21.4.3),

$$E_{\text{bind},i} = -E_i \sim 2.819\,69 \times 10^{-9} \, Z_i^{5/3} \left(\frac{\rho_6}{A_i} \right)^{1/3} \text{ erg.} \tag{21.4.4}$$

In the zero temperature approximation, the electron Fermi momentum as given by eq. (4.8.7) in Volume 1 is, in units of $m_e c$,

$$p_F = 0.010\,064\,085 \left(\frac{\rho_0}{\mu_e} \right)^{1/3} = 1.006\,4085 \left(\frac{\rho_6}{\mu_e} \right)^{1/3}, \tag{21.4.5}$$

so that the electron Fermi kinetic energy is

$$E_F' = (\epsilon_F - 1) \, m_e c^2 = \left(\sqrt{p_F^2 + 1} - 1 \right) m_e c^2$$

$$= 8.1826 \times 10^{-7} \left(\sqrt{1 + 1.012\,817 \, (\rho_6/\mu_e)^{2/3}} - 1 \right) \text{ erg.} \tag{21.4.6}$$

Altogether,

$$\frac{E_{\text{bind},i}}{Z_i \, E_F'} = 0.003\,4460 \left(Z_i \frac{\rho_6}{\mu_e} \right)^{1/3} \frac{1}{\sqrt{1 + 1.012\,817 \, (\rho_6/\mu_e)^{2/3}} - 1}. \tag{21.4.7}$$

In the limit of very large densities, $(E_{\text{bind},i}/Z_i \, E_F') \to 0.003\,424 \, Z_i^{1/3}$, independent of density. For densities such that electron energies are not relativistic, $(E_{\text{bind},i}/Z_i \, E_F') \to 0.006\,8048 \, (Z_i \, \mu_e/\rho_6)^{1/3} = 0.68048 \, (A_i/\rho_0)^{1/3}$, independent of Z_i. These limiting expressions may be compared with the ratios P_{Coulomb}/P_e given, respectively, by eqs. (4.17.71) and (4.17.72) in Volume 1.

21.4.2 Debye theory and terrestial metals

Although nuclei are not free to move easily through a lattice, they oscillate about their mean positions at amplitudes that increase with increasing temperature. The contribution of lattice vibrations to the bulk thermal energy can be estimated by making use of the formalism of the Debye theory of specific heats (Peter Debye, 1912).

It is fruitful to think in terms of normal modes of weakly coupled simple harmonic oscillators and to call the excitation of any mode a phonon, assuming that the energy of a phonon is quantized in exactly the same way that the energy of a photon is quantized. By analogy with the treatment of photons, the number of normal modes in the frequency range ν to $\nu + d\nu$ is given by

$$dN(\nu) = 3 \frac{4\pi}{h^3} V_i \left(\frac{h\nu}{c_s}\right)^2 d\left(\frac{h\nu}{c_s}\right), \tag{21.4.8}$$

where c_s is defined by

$$\frac{3}{c_s^3} = \frac{2}{v_t^3} + \frac{1}{v_l^3}, \tag{21.4.9}$$

and v_t and v_l are, respectively, the speeds of transverse and longitudinal sound waves. In this picture, all speeds are assumed to be constant. The total number of modes in a mole is given by

$$3N_A = \int_0^{\nu_{max}} dN = \frac{12\pi}{c_s^3} \int_0^{\nu_{max}} \nu^2 \, d\nu \, V_i = \frac{4\pi}{c_s^3} \nu_{max}^3 V_i, \tag{21.4.10}$$

where N_A is Avogadro's number. Equations (21.4.8) and (21.4.10) give

$$dN(\nu) = 9N_A \frac{\nu^2 d\nu}{\nu_{max}^3}. \tag{21.4.11}$$

In these equations,

$$V_i = \frac{N_A M_H \mu_i}{\rho} \sim \frac{N_A M_H A_i}{\rho} \sim \frac{A_i}{\rho} \tag{21.4.12}$$

is the volume occupied by a mole. The nuclear molecular weight μ_i and the atomic number $A_i \sim \mu_i$ are used interchangeably, as are the quantities N_A and $(1/M_H) \sim N_A$. From eqs. (21.4.10), (21.4.12), and (21.4.1), one has

$$\nu_{max} = \left(\frac{3}{4\pi} \frac{N_A}{V_i}\right)^{1/3} c_s = \left(\frac{3}{4\pi} \frac{\rho_0}{A_i M_H}\right)^{1/3} c_s = \left(\frac{3}{4\pi}\right)^{2/3} \frac{c_s}{R_{WS,i}}. \tag{21.4.13}$$

The fact that $R_{WS,i}$ is of the order of half of the mean separation between nearest neighbor nuclei means that the minimum wavelength of acoustical modes is of the order of the mean separation between adjacent nuclei. That is, from the definition $\lambda \nu = c_s$ and eq. (21.4.13), one has

$$\lambda_{min} = \left(\frac{4\pi}{3}\right)^{2/3} R_{WS,i} = 2.5985 \, R_{WS,i}, \tag{21.4.14}$$

which is only 30% larger than the diameter $2R_{WS,i}$ of the Wigner–Seitz sphere. Inserting numbers in eq. (21.4.13),

$$\nu_{max} \, (s^{-1}) = 0.522\,67 \times 10^8 \left(\frac{\rho_0}{A_i}\right)^{1/3} c_s \, (\text{cm s}^{-1}). \tag{21.4.15}$$

Assuming that phonon energies are quantized in integer units of $h\nu$ and that there is no limitation on the amplitude of an oscillator, the average energy of an oscillator may be approximated by

$$\bar{\epsilon}(\nu, T) = \frac{\sum_{n=0}^{\infty} nh\nu \, e^{-nh\nu/kT}}{\sum_{n=0}^{\infty} e^{-nh\nu/kT}} = -\frac{\partial}{\partial(1/kT)} \, \log_e \sum_{n=0}^{\infty} e^{-nh\nu/kT}$$

$$= -\frac{\partial}{\partial(1/kT)} \, \log_e \frac{1}{1 - e^{-h\nu/kT}} = \frac{h\nu}{e^{h\nu/kT} - 1}. \tag{21.4.16}$$

Multiplying this energy by $dN(\nu)$ from eq. (21.4.11) and integrating, one has that the energy in phonons is

$$E = \int_0^{\nu_{max}} \bar{\epsilon}(\nu, T) \, dN(\nu) = 9N_A \frac{1}{\nu_{max}^3} \int_0^{\nu_{max}} \frac{h\nu}{e^{h\nu/kT} - 1} \, \nu^2 d\nu, \tag{21.4.17}$$

or

$$E = 9N_A \, kT \, \frac{1}{x_{max}^3} \int_0^{x_{max}} \frac{x^3 \, dx}{e^x - 1}, \tag{21.4.18}$$

where

$$x_{max} = \frac{h\nu_{max}}{kT} = \frac{\Theta_{Debye}}{T}, \tag{21.4.19}$$

and Θ_{Debye} is the Debye temperature. As $T \to \infty$, $x_{max} \to 0$, and eq. (21.4.18) gives

$$E \to 3N_A \, kT = 6N_A \, \frac{1}{2}kT, \tag{21.4.20}$$

As $T \to 0$, $x_{max} \to \infty$, and eq. (21.4.18) gives

$$E \to 9N_A \, kT \, \frac{1}{x_{max}^3} \, \frac{\pi^4}{15} = \frac{3\pi^4}{5} \, N_A \, kT \, \left(\frac{T}{\Theta_{Debye}}\right)^3. \tag{21.4.21}$$

The derivative with respect to temperature of the energy given by the Debye model reproduces the specific heat of most terrestial solids over a wide range of temperatures. At high temperatures, the derivative of eq. (21.4.20) gives the classical result: $6N_A$ degrees of freedom and a contribution of $1/2 \, k$ per degree of freedom. At low temperatures, the derivative of eq. (21.4.21) agrees with the experimental result that the specific heat varies as the third power of the temperature. The Debye temperature can be chosen in such a way as to give the best fit to experimental results. Values of Θ_{Debye} estimated by fitting to experimentally determined specific heats for several terrestial solids are given in the fifth column of Table 21.4.1 under the heading Θ_{Debye}^{spht}. The first entry in the column is from the 1953 edition of *Introduction to Solid State Physics* by Charles Kittel and the second entry is from the 1996 edition. Here, A is the mean atomic mass, Z is the atomic number, and ρ is the density. Also given in Table 21.4.1 are experimentally determined melting points and the speeds of transverse (v_t) and longitudinal (v_l) sound waves as determined from experimentally determined elastic properties as reported in the 1953 edition of Kittel's book.

				Table 21.4.1 Several properties of a selection of terrestrial metals					
Metal	A	Z	ρ g cm^{-3}	Θ_{Debye}^{spht} °K	T_{melt}^{exp} °K	v_t m s^{-1}	v_l m s^{-1}	c_s m s^{-1}	Θ_{Debye}^{sdsp} °K
Al	27.0	13	2.7	418/428	933	3080	6260	3459	403
Fe	56.9	26	7.9	453/470	1803	3230	5850	3599	470
Ni	58.7	28	8.9	456/450	1725	2960	5630	3310	443
Cu	63.5	29	8.9	315/343	1356	2260	4700	2541	330
Zn	65.4	30	7.1	308/327	693	2410	4170	2675	320
Ag	107.9	47	10.5	215/225	1234	1590	3600	1795	207
Cd	112.4	48	8.6	300/209	594	1500	2780	1674	178
Sn	118.7	50	7.3	200	505	1670	3220	1873	185
W	183.9	74	19.3	379/400	3663	2620	5460	2946	349
Pt	195.0	78	21.4	229/240	2044	1670	3960	1884	226
Au	197.0	79	19.3	180/165	1336	1200	3240	1362	157
Pb	207.2	82	11.4	88/105	600	700	2160	797	76

The Debye temperature for terrestrial solids can also be estimated by using the speeds of transverse and longitudinal sound waves to find c_s as defined by eq. (21.4.9), then using eq. (21.4.15) to determine v_{max}, and finally using eq. (21.4.19) to obtain

$$\Theta_{Debye}^{sdsp} = 0.002\,5084 \left(\frac{\rho_0}{A_i}\right)^{1/3} \left(\frac{c_s}{\text{cm s}^{-1}}\right) \text{°K}. \qquad (21.4.22)$$

Values of Θ_{Debye}^{sdsp} are reported in the last column of Table 21.4.1. There is generally good agreement between Θ_{Debye}^{sdsp} and the most recent estimate of Θ_{Debye}^{spht}, the poorest match being for Pb for which the two estimates differ by $\sim 30\%$.

21.4.3 A characteristic frequency of oscillation in the stellar context

In the stellar environment, one does not have the luxury of directly exploiting experimental properties and dealing with just one element. However, at the large densities of most interest, essentially all electrons are free and, thanks to the Pauli exclusion principle, are at nearly uniform number density, making it feasible to adopt, as the electrical force acting on a nucleus enclosed in its Wigner–Seitz sphere, the derivative of the potential provided by the electrons in this sphere (e.g., E. E. Salpeter, 1961) and to find the characteristic frequency predicted by this force as an estimate of the maximum phonon frequency. Using Newton's result for inverse square law forces (Section 3.1 in Volume 1), one has that, at a distance x from the origin of the sphere, the electrical force per unit charge due to the electrons in the sphere is

$$E = -\frac{e Z_i}{\frac{4\pi}{3} R_{WS,i}^3} \frac{4\pi}{3} x^3 \frac{1}{x^2} = -\frac{4\pi e n_e}{3} x, \qquad (21.4.23)$$

where the second identity follows from eq. (21.4.1) and the fact that $n_e = Z_i n_i$. The equation of motion for the nucleus about the center of the sphere is

$$M_i \ddot{x} = -Z_i e \, \frac{4\pi e n_e}{3} \, x, \qquad (21.4.24)$$

where $M_i = A_i M_H$ is the mass of the nucleus. A relevant solution of this equation is

$$x = a_i \cos{(\omega_i t)}, \qquad (21.4.25)$$

where the amplitude a_i is a constant and, when more than one ion is present,

$$\omega_i = \sqrt{Z_i e \, \frac{4\pi e n_e}{3} \, \frac{1}{M_i}} = \sqrt{N_A \frac{Z_i}{A_i} e^2 \frac{4\pi}{3} N_A \frac{\rho}{\mu_e}} = N_A e \sqrt{\frac{Z_i}{A_i}} \sqrt{\frac{4\pi}{3} \frac{\rho}{\mu_e}} \qquad (21.4.26)$$

is the fundamental oscillation frequency of the ith ion. When only one type of ion is present,

$$\omega_i = N_A e \frac{Z_i}{A_i} \sqrt{\frac{4\pi}{3} \rho}. \qquad (21.4.26a)$$

Note that

$$\omega_i = \frac{1}{\sqrt{3}} \, \Omega_{pi}, \qquad (21.4.27)$$

where

$$\Omega_{pi} = \sqrt{Z_i \frac{4\pi n_e e^2}{M_i}} = \sqrt{\frac{4\pi n_i (Z_i e)^2}{M_i}} \qquad (21.4.28)$$

is the ion plasma frequency when only one type of ion is present. This frequency is not to be confused with

$$\omega_p = \sqrt{\frac{4\pi n_e e^2}{m_e}}, \qquad (21.4.29)$$

the electron plasma frequency, which is the minimum frequency with which an electromagnetic wave can propagate through a medium containing free electrons at the number density n_e.

From the point of view of classical physics, it is counter-intuitive to assume that the distribution of negative charge carried by the light electrons can remain approximately fixed while the heavy nucleus is constrained to move back and forth in this distribution. The justification for the assumption rests with the quantum mechanical exclusion principle which, in this situation, ensures the rigidity of the electron distribution.

The frequency ω_i may be thought of as an upper bound on the frequency of acoustical vibrations. In the real lattice, there are forces which couple the motions of occupants of neighboring lattice sites, leading to a spectrum of lower frequency oscillations. Assuming

Table 21.4.2 Vibration amplitude versus ion separation at the melting point for twelve terrestrial metals

Metal	electron config.	Z_i	A_i	T_{melt} K	ρ g cm^{-3}	c_s km s^{-1}	Θ_{Debye}^{sdsp} K	$\dfrac{R_{WS}^2}{a_{melt}^2}$
Al	$3s^2 5p$	1.57	27.0	933	2.7	3.46	403	41
Fe	$3d^6 4s^2$	2.08	55.9	1803	7.9	3.60	470	47
Ni	$3d^8 4s^2$	1.97	58.7	1725	8.9	3.31	443	45
Cu	$3d^{10} 4s$	1.63	63.5	1356	8.9	2.54	330	36
Zn	$3d^{10} 4s^2$	1.79	65.4	693	7.1	2.68	320	79
Ag	$3s^2 3p$	1.67	107.9	1234	10.5	1.80	207	32
Cd	$4d^{10} 5s^2$	1.78	112.4	594	8.6	1.67	178	62
Sn	$5s^2 5p^2$	1.96	118.7	505	7.3	1.87	185	96
W	$5d^4 6s^2$	3.73	183.9	3663	19.3	2.95	349	51
Pt	$5d^9 6s$	2.25	195.0	2044	21.4	1.88	226	40
Au	$5d^{10} 6s$	1.65	197.0	1336	19.3	1.36	157	32
Pb	$6s^2 6p^2$	1.43	207.2	600	11.4	0.80	95	26

that the effective coupling constant connecting neighboring nuclei is proportional to the spring constant given by the coefficient of x in eq. (21.4.24), one may identify $\omega_i (= 2\pi \nu_i)$ with $2\pi \nu_{max}$, where ν_{max} is given by eq. (21.4.13), and set

$$\Theta_{Debye}^{star} = \frac{h\nu_i}{k} = 4.4909 \times 10^3 \left(\frac{Z_i}{A_i} \sqrt{\rho_0} \right) \text{ K} = 4491 \left(\frac{1}{\mu_e} \sqrt{\rho_0} \right) \text{ K}. \qquad (21.4.30)$$

In cool white dwarfs, with all electrons being free, the typical value of Z_i/A_i is $1/2$, suggesting $\Theta_{Debye} \sim 2245\sqrt{\rho_0}$ K. For $\rho_0 \sim 10$, the value of Θ_{Debye} predicted in this way is over an order of magnitude larger than Θ_{Debye} for typical terrestrial solids.

The major reason for the difference is that, in terrestrial solids, most electrons are not free, but remain bound to the nucleus, with the result that the effective value of Z_i/A_i to use in eq. (21.4.28) is much less than $1/2$. For example, for Fe, insisting that Θ_{Debye}^{star} predicted by eq. (21.4.30) agree with either Θ_{Debye}^{spht} or Θ_{Debye}^{sdsp} gives formally $Z_i \sim 2.09$, suggesting that there are approximately two free electrons for every ion, the rest being in bound orbits, moving with the nucleus. Values of Z_i derived in this way for other metals are given in the second column of Table 21.4.2.

Given that the conditions for the applicability of the model are not even close to being met in terrestrial solids, the formal interpretation presented here is exceedingly naive. Nevertheless, it is a remarkable fact that, in order of magnitude, the prediction of one or two electrons in a conduction band in terrestrial solids is not far from predictions of more sophisticated models. In a magical way, the naive model provides an order of magnitude estimate of the effective force acting on ions. It is interesting to observe that the simple model used here to estimate a fundamental property of metals is just the inverse of the J. J. Thomson plum-pudding model of the atom in vogue prior to the discovery by Rutherford of the nuclear atom in 1910–11. In the plum-pudding model, electrons were thought of as point particles oscillating in a sphere of positive charge having the dimensions of the Wigner–Seitz sphere. In spite of its naivity, the plum-pudding model had its successes.

If one identifies ω_i as given by eq. (21.4.26) with $2\pi \nu_{max}$, with ν_{max} given by eq. (21.4.13), and uses eq. (21.4.1), the equations can be manipulated to yield

$$M_i \, c_s^2 = \left(\frac{1}{2\pi}\right)^2 \left(\frac{4\pi}{3}\right)^{4/3} \frac{(Z_i e)^2}{R_{WS,i}}, \tag{21.4.31}$$

as well as

$$c_s = \frac{1}{2\pi} \left(\frac{4\pi}{3}\right)^{5/6} \frac{e}{M_H^{2/3}} \frac{Z_i}{A_i^{2/3}} \rho^{1/6}$$

$$= 1.7930 \times 10^6 \frac{Z_i}{A_i^{2/3}} \rho_0^{1/6} \text{ cm s}^{-1}. \tag{21.4.32}$$

Setting Θ_{Debye}^{sdsp} given by eq. (21.4.22) equal to Θ_{Debye}^{star} given by eq. (21.4.30) also produces eq. (21.4.32). Belaboring the obvious, values of Z_i obtained by using eq. (21.4.32) and the appropriate experimentally determined properties produces the same values as given by eq. (21.4.30) by setting $\Theta_{Debye}^{star} = \Theta_{Debye}^{sdsp}$. These values differ from the values reported in the table by the factor $\Theta_{Debye}^{sdsp}/\Theta_{Debye}^{spht}$.

21.4.4 Oscillation amplitude and the melting point

The simple model may be explored further to obtain a feeling for the temperature at which the transition from the solid to the liquid state occurs. In the classical approximation, the time-averaged vibrational energy of an individual ion participating in a phonon mode of frequency ω is

$$\langle E_i(\omega) \rangle = 2 \times \left\langle \frac{1}{2} M_i \, a^2(\omega) \, \omega^2 \, \sin^2(\omega t) \right\rangle = \frac{1}{2} M_i \, \omega^2 \, a^2(\omega), \tag{21.4.33}$$

where $a(\omega)$ is the amplitude of the motion. Since a one-dimensional oscillator has two degrees of freedom, thermodynamic equilibrium requires that $\langle E_i(\omega) \rangle = kT$, giving

$$a^2(\omega) = \frac{2 \, kT}{M_i \, \omega^2}. \tag{21.4.34}$$

Combining eqs. (21.4.1), (21.4.34), and (21.4.25) yields

$$\frac{a^2(\omega)}{R_{WS,i}^2} = \frac{2kT}{M_i \, \omega^2} \left(\frac{4\pi}{3} n_i\right)^{2/3} = 2kT \left(\frac{3}{4\pi} \frac{1}{n_i}\right)^{1/3} \frac{1}{Z_i^2 e^2} \frac{\omega_i^2}{\omega^2} = 2kT \frac{R_{WS,i}}{Z_i^2 e^2} \left(\frac{\omega_i}{\omega}\right)^2. \tag{21.4.35}$$

Comparing with eq. (21.4.2), one has also that

$$\frac{a^2(\omega)}{R_{WS,i}^2} = \frac{2}{\Gamma_i} \left(\frac{\omega_i}{\omega}\right)^2. \tag{21.4.36}$$

Assuming that sound speeds are independent of frequency, the average of $1/\omega^2$ over the phonon spectrum defined by eq. (21.4.8) is

$$\left\langle\frac{1}{\omega^2}\right\rangle = \frac{\int_0^{\nu_{max}} \frac{1}{\omega^2}\nu^2\,d\nu}{\int_0^{\nu_{max}} \nu^2\,d\nu} = \frac{3}{\omega_{max}^2}. \tag{21.4.37}$$

Using this result in conjunction with eq. (21.4.33), one has

$$\langle a^2(\omega)\rangle = \frac{6\,kT}{M_i\,\omega_{max}^2}. \tag{21.4.38}$$

Equation (21.4.37) may be used with eqs. (21.4.1) and (21.4.20) to estimate the relationship between the amplitude of vibration of an ion and the average separation between ions at the transition from the solid to the liquid phase. From eqs. (21.4.38) and (21.4.19), it follows that the square of the oscillation amplitude at the melting point $T = T_{melt}$ is

$$a_{melt}^2 = \langle a^2(\omega)\rangle_{melt} = \frac{6\,kT_{melt}}{M_i\,\omega_{max}^2} = \frac{6}{M_i}\frac{h^2}{k}\frac{1}{(2\pi)^2}\frac{T_{melt}}{\Theta_{Debye}^2}. \tag{21.4.39}$$

Dividing by $R_{WS,i}^2$ as given by eq. (21.4.1) results in

$$\left(\frac{a_{melt}}{R_{WS,i}}\right)^2 = \left(\frac{4\pi}{3}\right)^{2/3}\frac{6}{(2\pi)^2}\frac{h^2}{kM_H^{5/3}}\frac{1}{A_i^{5/3}}\frac{T_{melt}}{\Theta_{Debye}^2}\rho^{2/3} = \frac{540.61}{A_i^{5/3}}\frac{T_{melt}}{\Theta_{Debye}^2}\rho_0^{2/3}. \tag{21.4.40}$$

One can also combine eqs. (21.4.13) and (21.4.38) to obtain

$$\left(\frac{a_{melt}}{R_{WS,i}}\right)^2 = \frac{6}{(2\pi)^2}\left(\frac{4\pi}{3}\right)^{4/3}\frac{kT_{melt}}{M_i c_s^2} = 0.008\,4739\left(\frac{km\ s^{-1}}{c_s}\right)^2\frac{T_{melt}}{A_i}. \tag{21.4.41}$$

Equation (21.4.40) with $\Theta_{Debye} = \Theta_{Debye}^{sdsp}$ produces the same result as eq. (21.4.41). The resulting value of R_{ws}^2/a_{melt}^2 is listed in the last column of Table 21.4.2.

In the second column of Table 21.4.2 is the configuration of the electrons in the outer two shells of the atomic form of the element. In the atomic form of five elements (Al, Cu, Ag, Pt, and Au), there is one electron in the outermost occupied shell. The conductivity in the metallic state of four of these elements (all but Pt) is noticeably smaller than that in the metallic state of the other elements in the table (with the exception of Pb). Inferring that the conduction electrons in the metallic form of these five elements are most akin to the free electrons in the stellar context, one may suppose that $(a_{melt}/R_{WS})^2$ at the melting point in the five metals is similar to this ratio at the melting point in a star. For the five selected metals, the average value of $(R_{WS}/a_{melt})^2$ and the inverse of the average of $(a_{melt}/R_{WS})^2$ are essentially identical at 36 ± 0.3. Thus, in the simple model, the transition from solid to liquid occurs when the ratio of the mean vibrational (half) amplitude to the

Wigner–Seitz radius is $\sim 1/6$. For the twelve solids in Table 21.4.2, the nearest neighbor distance is approximately $0.9 \times 2 \, R_{\mathrm{WS,i}}$ (with a deviation of less than 2%), so the full amplitude at the melting point for the five selected solids is $\sim 0.185 \sim 1/5.4$ times the nearest neighbor distance. This may be compared with the value of $\sim 1/4$ which F. A. Lindemann (1910) assigned to the ratio of the full vibration amplitude to the average ionic separation at the melting point. Since two different theories have been used to define a vibration amplitude, the two estimates are not strictly comparable, but it is encouraging that they agree to much better than an order of magnitude.

Using eq. (21.4.36) in conjunction with eq. (21.4.35), one has

$$\left(\frac{\langle a^2(\omega) \rangle}{R_{\mathrm{WS,i}}^2} \right)^2 = \frac{6}{\Gamma_i} \left(\frac{\omega_i}{\omega_{\max}} \right)^2 = \frac{6}{\Gamma_i}, \tag{21.4.42}$$

the final equality following if $\omega_{\max} = \omega_i$. Thus, at the melting point,

$$\Gamma_{\mathrm{melt}} \sim 6 \times 36 = 216. \tag{21.4.43}$$

21.4.5 Application of the Thomas–Fermi model of the atom

In the stellar case, as discussed by L. Mestel and M. A. Ruderman (1967) and by S. L. Shapiro and S. A. Teukolsky (1983), the speed of longitudinal sound waves is not independent of frequency, and this fact must be taken into account in forming the average of $1/\omega^2$ over the phonon spectrum. To understand this point, it is useful to explore the consequences of adopting the statistical model of the many electron atom introduced by L. H. Thomas (1927) and E. Fermi (1928). The two basic assumptions of the Thomas–Fermi model are that, in the electrically neutral (Wigner–Seitz) sphere, (1) the electron Fermi energy plus the electrical potential is constant and (2) all electron states up to the Fermi energy are filled. The normalization of the potential is, of course, arbitrary, but it is useful in the current context to set

$$E_F(r) - e \, \phi(r) = \mathrm{constant} = E_F(R_{\mathrm{WS,i}}) \sim \bar{E}_F, \tag{21.4.44}$$

where $\bar{E}_F$ is the mean electron Fermi energy and $\phi(r)$ is the electrical potential.

Including the electron rest mass in the definiton of E_F, one has $\bar{\epsilon}_F = \bar{E}_F/m_e c^2 = \sqrt{\bar{p}_F^2 + 1}$, and if one assumes that $e\phi(r)/\bar{E}_F \ll 1$, it follows from eq. (21.4.44) that

$$p_F^2 = \bar{\epsilon}_F^2 \left(1 + \frac{e\phi(r)}{\bar{\epsilon}_F m_e c^2} \right)^2 - 1 = \left(\bar{\epsilon}_F^2 - 1 \right) + 2\bar{\epsilon}_F \frac{e\phi(r)}{m_e c^2} + \cdots$$

$$= \bar{p}_F^2 \left(1 + 2 \frac{\bar{\epsilon}_F}{\bar{p}_F^2} \frac{e\phi(r)}{m_e c^2} + \cdots \right). \tag{21.4.45}$$

Invoking the second assumption, the electron density is given as a function of radius by

$$n_e(r) = \frac{8\pi}{3\lambda_C^3}\, p_F^3(r) = \bar{n}_e \left(1 + 3\,\frac{\bar{\epsilon}_F}{\bar{p}_F^2}\,\frac{e\phi(r)}{m_e c^2} + \cdots \right) = \bar{n}_e + n_e'(r), \qquad (21.4.46)$$

where

$$n_e'(r) = \left(3\,\frac{\bar{\epsilon}_F}{\bar{p}_F^2}\,\frac{e\phi(r)}{m_e c^2} + \cdots \right) \bar{n}_e = \left(3\,\frac{\bar{\epsilon}_F}{\bar{p}_F^2}\,\frac{\bar{n}_e}{m_e c^2}\, e \right) \phi(r) + \cdots \qquad (21.4.47)$$

and $\bar{n}_e = Z_i\, n_i$ is the mean number density of electrons in the Wigner–Seitz sphere.

Poisson's equation becomes

$$\nabla^2 \phi(r) = -\,4\pi Z_i e\,\delta(r) + 4\pi e \left[\, \bar{n}_e + n_e{}'(r) \,\right], \qquad (21.4.48)$$

or

$$\left(\nabla^2 - \kappa_{sc}^2 \right) \phi(r) = -\,4\pi Z_i e\,\delta(r) + 4\pi e \bar{n}_e, \qquad (21.4.49)$$

where

$$\kappa_{sc}^2 = 4\pi e \times \left(3\,\frac{\bar{\epsilon}_F}{\bar{p}_F^2}\,\frac{e}{m_e c^2}\,\bar{n}_e \right) = 12\pi\, e^2\,\frac{\bar{\epsilon}_F^2}{\bar{p}_F^2}\,\frac{\bar{n}_e}{\bar{E}_F}. \qquad (21.4.50)$$

When the term on the far right hand side of eq. (21.4.49) is ignored, the general solution is

$$\phi(r) = Z_i e\,\frac{e^{-\kappa_{sc}r}}{r}. \qquad (21.4.51)$$

Thus, $1/\kappa_{sc} = R_{sc}$ is a screening radius. Multiplying and dividing the expression on the far right hand side of eq. (21.4.50) by $V_{WS,i} = (4\pi/3)R_{WS,i}^3$ and using the fact that $\bar{n}_e V_{WS,i} = Z_i$, one has

$$\kappa_{sc}^2 = \frac{1}{R_{sc}^2} = 9\,\frac{Z_i e^2/R_{WS,i}}{\bar{E}_F}\,\frac{\bar{\epsilon}_F^2}{\bar{p}_F^2}\,\frac{1}{R_{WS,i}^2}. \qquad (21.4.52)$$

In the non-relativistic limit,

$$\kappa_{sc}^2 = \frac{1}{R_{sc}^2} = \frac{6\pi Z_i e^2 n_i}{\bar{E}_F'} = \frac{9}{2}\,\frac{Z_i e^2/R_{WS,i}}{\bar{E}_F'}\,\frac{1}{R_{WS,i}^2}, \qquad (21.4.53)$$

where $\bar{E}_F' = \bar{E}_F - m_e c^2 = (\bar{\epsilon}_F - 1)m_e c^2$ is the mean electron Fermi kinetic energy. In the extreme relativistic limit,

$$\kappa_{sc}^2 = \frac{1}{R_{sc}^2} = \frac{12\pi Z_i e^2\, n_i}{\bar{E}_F} = 9\,\frac{Z_i e^2/R_{WS,i}}{\bar{E}_F}\,\frac{1}{R_{WS,i}^2}. \qquad (21.4.54)$$

Table 21.4.3 Several characteristics of solid oxygen in electron-degenerate matter

$\dfrac{\rho}{\mu_e}$ g cm^{-3}	p_F	ϵ_F	$\dfrac{E_{\text{bind}}}{Z E_{\text{Fermi}}}$	Θ_{Debye} 10^6 K	$\dfrac{R_{\text{sc}}}{R_{\text{WS}}}$	Γ_{melt}	T_{melt} 10^6 °K
10^9	10.064	10.114	0.0076	71	9.76	154	81
10^8	4.671	4.777	0.0079	22	9.65	154	38
10^7	2.168	2.388	0.0085	7.1	2.95	171	15
10^6	1.006	1.419	0.0164	2.2	2.60	176	7.1
10^5	0.4671	1.0373	0.0245	0.71	2.01	192	3.0
10^4	0.2168	1.0232	0.0639	0.22	1.42	232	1.1
10^3	0.1006	1.0050	0.1075	0.071	0.98	321	0.39
10^2	0.0467	1.0011	0.2328	0.022	0.67	361	0.16

Using the fact that $e^2/m_e c^2 = \left(e^2/\hbar c\right)^2 a_0$, where a_0 is the Bohr radius of the hydrogen atom in the ground state and $e^2/\hbar c$ is the fine structure constant of electrodynamics, eq. (21.4.52) can be manipulated to read

$$\left(\frac{R_{\text{sc}}}{R_{\text{WS,i}}}\right)^2 = \left(\frac{\hbar c/e^2}{3}\right)^2 \frac{\bar{p}_F^2}{\bar{\epsilon}_F} \frac{R_{\text{WS,i}}}{a_0} = \frac{2086.54}{Z_i} \frac{\bar{p}_F^2}{\bar{\epsilon}_F} \frac{R_{\text{WS,i}}}{Z_i a_0}. \tag{21.4.55}$$

Inserting numbers, eq. (21.4.1) can be written as

$$\frac{R_{\text{WS,i}}}{a_0} = 0.910\,295 \left(\frac{A_i}{\rho_0}\right)^{1/3}. \tag{21.4.56}$$

Combining the last two equations gives

$$\left(\frac{R_{\text{sc}}}{R_{\text{WS,i}}}\right)^2 = \frac{1899.37}{Z_i} \left(\frac{A_i}{\rho_0}\right)^{1/3} \frac{\bar{p}_F^2}{\bar{\epsilon}_F}. \tag{21.4.57}$$

The quantity $R_{\text{sc}}/R_{\text{WS,i}}$ for completely ionized ^{16}O is shown in Table 21.4.3 as a function of density. For densities in excess of 10^4 g cm^{-3}, the screening radius is large compared with the Wigner–Seitz radius. For a density of $10^{3.79}$ g cm^{-3}, $R_{\text{WS,i}} = a_0/8 =$ the radius of the first Bohr orbit for an electron about an isolated oxygen nucleus. As densities are decreased below $10^{3.79}$ g cm^{-3}, the approximation that all electrons are free becomes progressively worse. For example, at $\rho = 10^2$ g cm^{-3}, $R_{\text{WS,i}} = 0.494\, a_0 \sim 4 \times 0.125\, a_0$, and one might expect of the order of two to four electrons to be in bound states about each nucleus. Thus, the entries in the last two rows of Table 21.4.3 are purely formal.

Also given in Table 21.4.3 for completely ionized oxygen are the ratio $(E_{\text{bind,i}}/Z_i\, E_F')$ defined by eq. (21.4.7) and the Debye temperature defined by eq. (21.4.30). In order of magnitude,

$$R_{sc} = \frac{1}{\kappa_{sc}} \sim \sqrt{\frac{E'_{electron}/m_e c^2}{4\pi (e^2/m_e c^2) n_i}} = \sqrt{\frac{E'_{electron}/m_e c^2}{n_i (4\pi/3) r_e^3}} \, r_e$$

$$= \sqrt{\frac{E'_{electron}/m_e c^2}{n_i (4\pi/3) a_0^3}} \left(\frac{\hbar c}{e^2}\right)^2 a_0, \qquad (21.4.58)$$

where $E'_{electron}$ is the average kinetic energy of an electron, $r_e = e^2/m_e c^2$ is the classical radius of an electron, and a_0 is the Bohr radius of the hydrogen atom in the ground state. It is interesting that setting $E'_{electron} \sim kT$ in eq. (21.4.58) gives an approximation to the screening radius appropriate for regions of weak degeneracy, as expressed by eq. (4.17.10) in Volume 1.

When the term $4\pi e \bar{n}_e$ in eq. (21.4.49) is retained, but the delta function is ignored, an approximate solution obtained by quadratures is

$$\phi_1(r) = 4\pi e \bar{n}_e \frac{r^2}{6} \left(1 + \frac{1}{20}(\kappa_{sc} r)^2\right). \qquad (21.4.59)$$

The accuracy of the approximate solution can be gauged by noting that

$$(\nabla^2 - \kappa_{sc}^2)\phi_1(r) = 4\pi e \bar{n}_e \left(1 - \frac{1}{120}(\kappa_{sc} r)^4\right). \qquad (21.4.60)$$

Noting that

$$\frac{4\pi}{3} \bar{n}_e R_{ws,i}^3 = Z_i, \qquad (21.4.61)$$

one can write

$$\phi_1(r) = \frac{Z_i e}{R_{WS,i}} \frac{1}{2} \left(\frac{r}{R_{WS,i}}\right)^2 \left[1 + \frac{1}{20}\left(\frac{r}{R_{sc}}\right)^2\right]. \qquad (21.4.62)$$

Shapiro and Teukolsky argue in their book that the perturbation $\phi'(r)$ of the electrical potential produced by the motion of nuclei participating in longitudinal wave motion can be deduced from the equation

$$\left(\nabla^2 - \kappa_{sc}^2\right)\phi'(r) = -4\pi Z e \, n_i'(r), \qquad (21.4.63)$$

where $n_i'(r)$ is a perturbation of the nuclear number density. Equation (21.4.63) follows from eqs. (21.4.48) and (21.4.49) by adding the source term $-4\pi Z e \, n_i'(r)$ to the right hand sides of these equations and adding $\phi'(r)$ to $\phi(r)$ on the left hand sides. Treating n_i' as the density of a fluid, using eq. (21.4.63) in the equations of fluid dynamics, and inserting a traveling wave solution,

$$\mathbf{x} = \mathbf{a} \, e^{i(\kappa \cdot \mathbf{r} - \omega t)}, \qquad (21.4.64)$$

with $\mathbf{a}$ perpendicular to the velocity $\mathbf{v}$ of the fluid, they find

$$\omega^2 = \frac{\Omega_p^2}{1 + \kappa_{sc}^2/\kappa^2} = \frac{\Omega_p^2}{1 + \lambda^2/R_{sc}^2}, \qquad (21.4.65)$$

where

$$\kappa = \frac{2\pi}{\lambda} \tag{21.4.66}$$

is the wave number, λ is the wave length, ω is the circular frequency, and Ω_p is given by eqs. (21.4.27) and (21.4.28).

For small κ (long wavelength, low frequency),

$$\omega^2 = \frac{4\pi Z_i^2 e^2 \bar{n}_i}{M_i} \left(\frac{\kappa}{\kappa_{sc}}\right)^2 = \frac{1}{3} Z_i \frac{\bar{p}_F^2}{\bar{\epsilon}_F^2} \frac{\bar{E}_F}{M_i} \kappa^2, \tag{21.4.67}$$

where $\bar{n}_i$ is the mean number abundance of nuclei. It follows from eq. (21.4.67) that, for long wavelengths, the velocity of a wave is independent of frequency and equal to

$$v_l = \frac{\omega}{\kappa} \sqrt{\frac{1}{3} Z_i \frac{\bar{p}_F^2}{\bar{\epsilon}_F^2} \frac{\bar{E}_F}{M_i c^2}} \, c = \sqrt{\frac{1}{3} \frac{Z_i}{A_i} \frac{\bar{p}_F^2}{\bar{\epsilon}_F} \frac{m_e}{M_H}} \, c. \tag{21.4.68}$$

Since, in the present context, electrons are the primary contributers to the pressure, and since, in a first approximation, it has been assumed that electron pressure is isotropic and cannot produce shear, the speed of longitudinal sound waves should satisfy

$$c_l \sim \sqrt{\left(\frac{\partial P_e}{\partial \rho}\right)_{ad}}. \tag{21.4.69}$$

Neglecting the temperature dependance of P_e, one has from eqs. (4.8.9) and (4.8.6) in Volume 1 that

$$c_l^2 \sim \frac{\partial P_e}{\partial \rho} = \frac{\partial P_e}{\partial p_F} \frac{\partial p_F}{\partial \rho} = \left(\frac{8\pi}{3} \frac{m_e c^2}{\lambda_C^3} \frac{p_F^2}{\epsilon_F}\right) p_F^2 \frac{1}{3} \frac{p_F}{\rho} = \left(\frac{8\pi}{3} \frac{m_e c^2}{\lambda_C^3} \frac{p_F^2}{\epsilon_F}\right) \frac{1}{3} \frac{p_F^3}{\rho}$$

$$= \left(\frac{8\pi}{3} \frac{m_e c^2}{\lambda_C^3} \frac{p_F^2}{\epsilon_F}\right) \frac{\lambda_C^3}{8\pi} \frac{1}{\mu_e M_H} = \frac{1}{3\mu_e} \frac{m_e}{M_H} \frac{p_F^2}{\epsilon_F} c^2 = \left(\frac{1}{3} \frac{Z_i}{A_i} \frac{p_F^2}{\epsilon_F} \frac{m_e}{M_H}\right) c^2. \tag{21.4.70}$$

showing that c_l is identical with the speed v_l given by eq. (21.4.68). Numerically,

$$c_l = v_l \sim 4.039\,30 \times 10^8 \sqrt{\frac{1}{\mu_e} \frac{\bar{p}_F^2}{\epsilon_F}} \text{ cm s}^{-1}. \tag{21.4.71}$$

Noting the dependence of p_F on density, as given by eq. (21.4.5), and comparing with eq. (21.4.32), it is evident that the speed of longitudinal sound waves is typically over an order of magnitude larger than the speed of transverse sound waves. Equations (21.4.69)–(21.4.71) can also be written as

$$M_i c_l^2 \sim \frac{Z_i}{3} \frac{\epsilon_F + 1}{\epsilon_F} E_F'. \tag{21.4.72}$$

Comparing eq. (21.4.72) with eq. (21.4.31), and using eq. (21.4.3), it is evident that the ratio of squared speeds of transverse and longitudinal waves is roughly equal to the binding energy between a nucleus and the electrons in the Wigner–Seitz sphere about the nucleus divided by the electron Fermi kinetic energy times the number of electrons in the sphere, namely $E_{bind}/(Z E'_F)$. As detailed in the fourth column of Table 21.4.3, as long as electrons are degenerate, this ratio is much smaller than unity.

For large κ (short wavelength, high frequency),

$$\omega_l^2 = \Omega_p^2 \div \left[1 + \left(\frac{\lambda}{\lambda_{min}}\right)^2 \left(\frac{\lambda_{min}}{R_{sc}}\right)^2\right], \tag{21.4.73}$$

where, as given by eq. (21.4.14), $\lambda_{min} \sim 2.6\, R_{WS,i}$. Since, for the longitudinal mode, sound speed varies with frequency, the average of any quantity over the phonon spectrum involves an integration over $\kappa^2 d\kappa$ rather than an integration over $\nu^2 d\nu$. Since there are 2 N_A transverse modes and 1 N_A longitudinal modes in a mole, the normalization is such that $dN_t(\kappa) = 6 (\kappa^2\, d\kappa/\kappa_{max}^3)\, N_A$ and $dN_l(\kappa) = 3 (\kappa^2\, d\kappa/\kappa_{max}^3)\, N_A$, where N_t and N_l are, respectively, the number of transverse and longitudinal modes. Thus,

$$\left\langle\frac{1}{\omega^2}\right\rangle = \left[\int_0^{\kappa_{max}} \frac{1}{\omega_t^2}\, dN_t + \int_0^{\kappa_{max}} \frac{1}{\omega_l^2}\, dN_l\right] \div [N_t + N_l]$$

$$= \left[\int_0^{\kappa_{max}} \left(\frac{6}{\omega_t^2} + \frac{3}{\omega_l^2}\right) \frac{\kappa^2\, d\kappa}{\kappa_{max}^3}\, N_A\right] \div 3\, N_A. \tag{21.4.74}$$

Using eq. (21.4.65) for longitudinal waves and adopting

$$\omega_t = \frac{\kappa}{\kappa_{max}}\, \omega_i = \frac{\kappa}{\kappa_{max}} \frac{1}{\sqrt{3}} \Omega_p \tag{21.4.75}$$

for transverse waves, one has

$$\left\langle\frac{1}{\omega^2}\right\rangle = \left[\int_0^{\kappa_{max}} 2\, \frac{3}{\Omega_p^2} \left(\frac{\kappa_{max}}{\kappa}\right)^2 \kappa^2\, d\kappa + \int_0^{\kappa_{max}} \frac{1}{\Omega_p^2} \left(\frac{\kappa_{sc}^2 + \kappa^2}{\kappa^2}\right) \kappa^2\, d\kappa\right] \frac{1}{\kappa_{max}^3}$$

$$= \frac{1}{\Omega_p^2\, \kappa_{max}^3} \left[6\, \kappa_{max}^3 + \kappa_{sc}^2\, \kappa_{max} + \left(\frac{\kappa_{max}^3}{3}\right)\right] = \frac{6}{\Omega_p^2} \left[\left(1 + \frac{1}{18}\right) + \frac{1}{6} \left(\frac{\kappa_{sc}}{\kappa_{max}}\right)^2\right]$$

$$= \frac{19}{9} \frac{1}{\omega_i^2} \left[1 + \frac{3}{19} \left(\frac{\kappa_{sc}}{\kappa_{max}}\right)^2\right]. \tag{21.4.76}$$

Combining eqs. (21.4.76) and (21.4.36) gives

$$\frac{\langle a^2(\omega)\rangle}{R_{WS,i}^2} = \frac{2}{\Gamma_i} \omega_i^2 \left\langle\frac{1}{\omega^2}\right\rangle = \frac{38}{9} \frac{1}{\Gamma_i} \left[1 + \frac{3}{19} \left(\frac{\kappa_{sc}}{\kappa_{max}}\right)^2\right]. \tag{21.4.77}$$

Neglecting the second term in square brackets in eq. (21.4.77), and adopting the value of $\langle a^2(\omega) \rangle / R_{WS,i}^2 = 1/36$, estimated from the properties of terrestial metals, one has that $\Gamma_{melt} \sim 152$. Keeping the second term gives

$$\Gamma_{melt,i} = 152 \left[1 + \frac{3}{19} \left(\frac{\lambda_{min}}{R_{sc}} \right)^2 \right] = 152 \left[1 + \frac{3}{19} \left(\frac{4\pi}{3} \right)^{4/3} \left(\frac{R_{WS,i}}{R_{sc}} \right)^2 \right]$$

$$= 152 \left[1 + 1.066\,15 \left(\frac{R_{WS,i}}{R_{sc}} \right)^2 \right]. \tag{21.4.78}$$

Equation (21.4.78) produces the values of Γ_{melt} in Table 21.4.3 for pure oxygen. Equations (21.4.1) and (21.4.2) give

$$T_{melt,i} = \left(\frac{4\pi}{3} \frac{N_A}{A_i} \rho \right)^{1/3} \frac{Z^2 e^2}{k} \frac{1}{\Gamma_{melt}}$$

$$= 0.490\,069 \times 10^6 \,^0K \frac{Z_i^2}{A_i^{1/3}} \frac{\rho_0^{1/3}}{\Gamma_{melt,i}}. \tag{21.4.79}$$

Melting temperatures for pure oxygen are given in the last column of Table 21.4.3.

The thermal energy associated with lattice vibrations is given, in the second approximation, by

$$E_{lattice} = E_t + E_l = 6N_A \, kT \, \frac{1}{x_{max}^3} \int_0^{x_{max}} \frac{x^3 \, dx}{e^x - 1} + 3N_A \, kT \, \frac{1}{\kappa_{max}^3} \int_0^{\kappa_{max}} \frac{\bar{x}}{e^{\bar{x}} - 1} \kappa^2 \, d\kappa, \tag{21.4.80}$$

where

$$\bar{x} = \frac{h\nu(\kappa)}{kT} \tag{21.4.81}$$

and

$$\nu(\kappa) = \frac{\Omega_p}{2\pi} \frac{\kappa}{\sqrt{\kappa_{sc}^2 + \kappa^2}}. \tag{21.4.82}$$

21.4.6 The zero-point energy and the Helmholtz free energy

One physical reality that has not yet been taken into account is the fact that a quantum oscillator has a zero-point energy, as required by the Heisenberg uncertainty principle (Werner Heisenberg, 1927). Writing the classical energy of an oscillator as

$$E = \frac{p^2}{2m} + \frac{1}{2}k \, q^2, \tag{21.4.83}$$

where p and q are, respectively, momentum and position coordinates, the average energy of the oscillator is

$$\bar{E} = \frac{1}{2} \left(\frac{p_A^2}{2m} + \frac{1}{2}k \, q_A^2 \right), \tag{21.4.84}$$

where p_A and q_A are the amplitudes, respectively, of the momentum and position coordinates. Replacing p_A by Δp and q_A by Δq and, using the relationship

$$\Delta q \, \Delta p = \hbar, \tag{21.4.85}$$

one has

$$\bar{E} = \frac{1}{2} \left[\frac{\hbar^2}{2m(\Delta q)^2} + \frac{1}{2} k (\Delta q)^2 \right]. \tag{21.4.86}$$

Minimizing the energy by setting $d\bar{E}/d\Delta q = 0$ produces $\Delta q = \left(\frac{\hbar^2}{mk} \right)^{1/2}$, giving

$$\bar{E}_{minimum} = \frac{1}{2} \hbar \left(\frac{k}{m} \right)^{1/2} = \frac{1}{2} \hbar\omega. \tag{21.4.87}$$

Obviously, this is a somewhat contrived result. Nevertheless, it offers insight into the result of a proper quantum mechanical treatment which shows that energy eigenvalues of a simple harmonic oscillator are given by

$$E_n = h\nu \left(n + \frac{1}{2} \right), \tag{21.4.88}$$

where n is an integer.

Doing a sum as in eq. (21.4.16), but including the zero-point energy, the average energy of an oscillator becomes

$$\bar{\epsilon}(\nu, T) = h\nu \left(\frac{1}{e^{h\nu/kT} - 1} + \frac{1}{2} \right) = \frac{h\nu}{e^{h\nu/kT} - 1} \frac{e^{h\nu/kT} + 1}{2} = \frac{h\nu}{2} \coth \left(\frac{h\nu}{2kT} \right). \tag{21.4.89}$$

In a simplified Debye approximation, the appropriately modified energy in phonons is obtained by inserting the quantity $\bar{\epsilon}(\nu, T)$ from eq. (21.4.89) into eq. (21.4.17), giving

$$E = \int_0^{\nu_{max}} \bar{\epsilon}(\nu, T) \, dN(\nu) = 9N_A \frac{1}{\nu_{max}^3} \int_0^{\nu_{max}} \frac{h\nu}{2} \left(\coth \frac{h\nu}{kT} \right) \nu^2 d\nu, \tag{21.4.90}$$

or

$$E = 9N_A \, kT \frac{1}{x_{max}^3} \int_0^{x_{max}} \frac{1}{2} \left(\coth \frac{x}{2} \right) x^3 \, dx, \tag{21.4.91}$$

where x_{max} is given by eq. (21.4.19). Similar considerations can be applied to appropriately modify the second approximation defined by eqs. (21.4.80)–(21.4.82).

In the simplest model, it follows straightforwardly that $P_{phonon} = \frac{1}{2} \frac{E}{V}$. However, in comparing with pertinent theoretical calculations in the literature, it is helpful to understand how phonon pressure can also be derived from the Helmholtz free energy F defined by

$$F = E - TS. \tag{21.4.92}$$

From eq. (21.4.92) and the second law of thermodynamics, it follows that

$$dF = dE - T\,dS - S\,dT = dE - (dE + P\,dV) - S\,dT = -P\,dV - S\,dT. \quad (21.4.93)$$

From eq. (4.13.92), it follows in turn that

$$P = -\left(\frac{\partial F}{\partial V}\right)_T, \quad (21.4.94)$$

$$S = -\left(\frac{\partial F}{\partial T}\right)_V, \quad (21.4.95)$$

and

$$E = F - T\left(\frac{\partial F}{\partial T}\right)_V = -T^2\left(\frac{\partial}{\partial T}\left[\frac{F}{T}\right]\right)_V. \quad (21.4.96)$$

Using in eq. (21.4.90) the fact that

$$-T^2\frac{\partial}{\partial T}\left[\frac{kT\,\log_e\sinh(h\nu/2kT)}{T}\right] = \frac{h\nu}{2}\coth\frac{h\nu}{2kT}, \quad (21.4.97)$$

and comparing with eq. (21.4.96), it is evident that the free energy of the phonons in a mole of a one-component plasma is

$$F = kT\,9N_A\frac{1}{\nu_{max}^3}\int_0^{\nu_{max}}\log_e\left(\sinh\frac{h\nu}{2kT}\right)\nu^2\,d\nu \quad (21.4.98)$$

$$= kT\,9N_A\frac{1}{x_{max}^3}\int_0^{x_{max}}\log_e\left(\sinh\frac{x}{2}\right)x^2\,dx. \quad (21.4.99)$$

From eq. (21.4.25), one has that $\nu_{max} = \omega_i/2\pi = (Z_i e/M_i)\sqrt{2\rho/3\pi} = \text{constant}/\sqrt{V}$ and that, therefore, $x_{max} = \text{constant}'/\sqrt{V}T$. Hence, $(\partial x_{max}/\partial V)_T = -x_{max}/2V$ and $(\partial x_{max}/\partial T)_V = -x_{max}/T$.

From eqs. (21.4.94) and (21.4.99), then,

$$P_{phonons} = -\left(\frac{\partial F}{\partial V}\right)_T = -9N_A\,kT\left(-\frac{3}{x_{max}^4}\right)\left(-\frac{x_{max}}{2V}\right)\int_0^{x_{max}}\log_e\left(\sinh\frac{x}{2}\right)x^2\,dx$$

$$+ \frac{1}{x_{max}^3}\log_e\left(\sinh\frac{x_{max}}{2}\right)x_{max}^2\left(-\frac{x_{max}}{2V}\right)$$

$$= \frac{1}{2}\frac{9N_A\,kT}{V}\left[-\frac{3}{x_{max}^3}\int_0^{x_{max}}\log_e\left(\sinh\frac{x}{2}\right)x^2\,dx + \log_e\left(\sinh\frac{x_{max}}{2}\right)\right]. \quad (21.4.100)$$

Since $\rho V = N_A M_H A$,

$$P_{phonons} = \frac{9}{2}\frac{\rho\,kT}{M_H A}\left[-\frac{3}{x_{max}^3}\int_0^{x_{max}}\log_e\left(\sinh\frac{x}{2}\right)x^2\,dx + \log_e\left(\sinh\frac{x_{max}}{2}\right)\right]. \quad (21.4.101)$$

Similarly, from eqs. (21.4.96) and (21.4.99), one has for the phonon energy density $(U = E/V)$

$$U_{phonons} = 9 \frac{\rho \, kT}{M_H A} \left[-\frac{3}{x_{max}^3} \int_0^{x_{max}} \log_e \left(\sinh \frac{x}{2} \right) x^2 \, dx + \log_e \left(\sinh \frac{x_{max}}{2} \right) \right].$$

$$(21.4.102)$$

Thus,

$$P_{phonons} = \frac{1}{2} U_{phonons}, \qquad (21.4.103)$$

The phonon energy density is also given by

$$U_{phonon} = 9 \frac{\rho \, kT}{M_H A} \frac{1}{x_{max}^3} \int_0^{x_{max}} \frac{1}{2} \left(\coth \frac{x}{2} \right) x^3 \, dx, \qquad (21.4.104)$$

and it is an interesting exercise to demonstrate by direct integration that the two formulations for U_{phonon} agree with one another.

21.5 Algorithms for estimating the energy density and pressure of liquids and solids in stars

The development in Section 21.4 has been given primarily for pedagogical reasons, one of the lessons being that to produce an equation of state for solids using even the simplest of theories requires detailed numerical calculations. In this section are presented expressions which reproduce results of detailed calculations in the literature for both solids and liquids and from which the pressure and energy density can be derived as functions of T, ρ, and composition.

For the solid phase, write

$$F_{solid} = F_{vibration} + F_{Coulomb} \qquad (21.5.1)$$

as the contribution of nuclei to the free energy. Generalizing eq. (21.4.98), one has

$$\frac{F_{vibration}}{kT} = \sum_p \int_0^{\kappa_{max}} \log_e \left(\sinh \frac{\hbar \omega_p(\kappa)}{2kT} \right) \frac{\kappa^2 \, d\kappa}{2\pi^2}, \qquad (21.5.2)$$

where the summation is over the three polarization states. Using data from W. J. Carr (1961), Jim MacDonald has performed the operations indicated in eq. (21.5.2) (inserting a factor of 2 in the argument of the logarithm) and finds that, with a relative error less than 4×10^{-4}, the results are fit by the expression (Appendix A in Iben, Fujimoto, & MacDonald, 1992)

$$\frac{F_{vibration}}{N_A kT} = \alpha \, L \left(\frac{\Lambda}{\Lambda_1} \right) + (1 - \alpha) \, L \left(\frac{\Lambda}{\Lambda_2} \right), \qquad (21.5.3)$$

where

$$L(x) = \frac{9}{8} x + 3 \log_e (1 - e^{-x}) - D(x), \tag{21.5.4}$$

$$D(x) = \frac{3}{x^3} \int_0^x \frac{t^3}{e^t - 1} \, dt, \tag{21.5.5}$$

$\alpha = 0.5711, \Lambda_1 = 1.0643, \Lambda_2 = 2.9438,$

$$\Lambda = \frac{\hbar \Omega_p}{kT}, \tag{21.5.6}$$

and Ω_p is given by eq. (21.4.28).

As estimated by W. L. Slattery, G. D. Doolen, and H. E. DeWitt (1982), the contribution to the free energy by Coulomb interactions between charged particles is, for $\Gamma \geq 160$,

$$\frac{F_{\text{Coulomb}}}{N_A kT} = -0.895\,929\,\Gamma - \frac{3225}{2\Gamma^2}, \tag{21.5.7}$$

where the first term is for a body centered cubic lattice and the second term is a classical anharmonic contribution.

For mixtures of elements, if the solid breaks up into crystal grains of pure elements, one has that

$$F_{\text{Coulomb}} = \frac{\sum_i Y_i \, F_1(\Gamma_{i,s})}{\sum_i Y_i}, \tag{21.5.8}$$

where

$$\Gamma_{i,s} = \frac{Z_i^2}{A_i^{1/3}} \, \Gamma_H, \tag{21.5.9}$$

$$\Gamma_H = \frac{e^2}{kT} \left(\frac{4\pi \rho N_A}{3} \right)^{1/3}, \tag{21.5.10}$$

and $F_1(\Gamma_{i,s})$ is the Coulomb contribution of an individual species. A mean Γ for the solid mixture may be defined as

$$\Gamma = \Gamma_{\text{solid}} = \left(\sum_i Y_i \, \frac{Z_i^2}{A_i^{1/3}} \right) \Gamma_H \div \sum_i Y_i = \left(\sum_i X_i \, \frac{Z_i^2}{A_i^{4/3}} \right) \Gamma_H \div \sum_i \frac{X_i}{A_i}, \tag{21.5.11}$$

and a mean Λ as

$$\Lambda = \Lambda_{\text{solid}} = \left(\sum_i \frac{Z_i^2}{A_i} Y_i \right)^{1/2} \frac{\hbar}{kT} e N_A \, (4\pi \rho)^{1/2}. \tag{21.5.12}$$

The contribution of the solid to pressure and energy density are obtained by taking the derivatives of F_{solid} as given by eqs. (21.4.94) and (21.4.96), respectively.

To describe the liquid phase, one imagines that, in the first approximation, nuclei move similarly to atoms in a gas. However, there are long range correlations between the ions which are difficult to model with a simple theory. In the remainder of this section, fits by

Jim Macdonald to results of model calculations are reported without elaborating on the nature of the models. The free energy in the liquid phase is written as the sum of three terms:

$$F_{liquid} = F_{ideal\ gas} + F_{Coulomb} + F_{quantum}. \tag{21.5.13}$$

The free energy of an ideal gas with no internal energy states (e.g., Landau & Lifschitz, 1958) is given by

$$\frac{F_{ideal\ gas}}{N_A\ kT} = \log_e\left[\frac{N_A}{V}\left(\frac{2\pi\ \hbar^2\ N_A}{A_i\ kT}\right)^{3/2}\right] - 1 = 3\ \log_e \Lambda - \frac{3}{2}\ \log_e \Gamma + \frac{1}{2}\ \log_e \frac{\pi}{6} - 1. \tag{21.5.14}$$

For the Coulomb energy,

$$\frac{U_{Coulomb}}{N_A\ kT} = G(\Gamma), \tag{21.5.15}$$

where

$$G(\Gamma) = \frac{\sqrt{3}}{2}\ G^{3/2} + \Gamma^3\big(-0.141\,642 + 0.516\,331\ \log_e \Gamma - 0.151\,759\ \Gamma^{3/2}\big), 0 \le \Gamma \le 1, \tag{21.5.16}$$

$$G(\Gamma) = 0.897\,744\ \Gamma - 0.950\,43\ \Gamma^{1/4} - 0.189\,56\ \Gamma^{-1/4} + 0.814\,87,\ \ 1 \le \Gamma \le 200, \tag{21.5.17}$$

and

$$G(\Gamma) = 0.895\,929\ \Gamma - 2.964\,673\ \Gamma^{1/4} + \frac{103.5892}{\Gamma},\ \ 200 \le \Gamma. \tag{21.5.18}$$

The first expression for G is based on results of R. Abe (1959), D. L. Bowers & E. E. Salpeter (1960), and Slattery, Doolen, and DeWitt (1980). The second expression is based on results of Monte Carlo simulations by Slattery, Doolen, & DeWitt (1982). The third expression has the limiting behavior of the solid phase as $\Gamma \to \infty$ and the additional terms ensure continuity of G and its first derivative at $\Gamma = 200$.

It follows from eqs. (21.5.15)–(21.5.18) that

$$\frac{F_{Coulomb}}{N_A\ kT} = H(\Gamma), \tag{21.5.19}$$

where

$$H(\Gamma) = \frac{\sqrt{3}}{2}G^{3/2} + \Gamma^3\big(-0.104\,584 + 0.172\,110\ \log_e \Gamma - 0.033\,724\Gamma^{3/2}\big), 0 \le \Gamma \le 1, \tag{21.5.20}$$

$$H(\Gamma) = 0.897\,744\ \Gamma - 3.801\,72\ \Gamma^{1/4} - 0.758\,240\ \Gamma^{-1/4}$$

$$+ 0.814\,87\ \log_e \Gamma + 2.584\,778,\ 1 \le \Gamma \le 200, \tag{21.5.21}$$

and

$$H(\Gamma) = 0.895\,929\,\Gamma - 2.964\,673\,\log_e \Gamma - \frac{103.5892}{\Gamma} + 9.395\,81, \ \ 200 \le \Gamma. \quad (21.5.22)$$

A fit to results by J. P. Hansen & R. Mazighi (1978) and by H. R. Glyde & G. H. Keech (1980), gives, for the quantum contribution to the free energy,

$$\frac{F_{\text{quantum}}}{N_A\,kT} = J(\Lambda) = \frac{1}{36\,a^2}\left(\frac{2}{c} - c - 1\right), \quad (21.5.23)$$

where

$$c = \frac{1}{\sqrt{1 + a^2\,\Lambda^2}} \quad (21.5.24)$$

and

$$a = \frac{1}{18\,\alpha} = 0.051\,931. \quad (21.5.25)$$

Altogether,

$$\frac{F_{\text{liquid}}}{N_A\,kT} = 3\,\log \Lambda - \frac{3}{2}\,\log_e \Gamma - 1.323\,51 - H(\Gamma) + J(\Lambda). \quad (21.5.26)$$

Derivatives of F_{liquid}, as described by eqs. (21.4.94) and (21.4.96), respectively, give the contribution of nuclei in the liquid state to pressure and energy density as functions of temperature, density, and composition.

For a mixture of elements in the liquid phase, the Coulomb contribution to the free energy is (J. P. Hansen, G. M. Torrie, & P. Viellefosse, 1977)

$$F_{\text{Coulomb}} = \frac{\sum_i Y_i\,F_1(\Gamma_{i,1})}{\sum_i Y_i}, \quad (21.5.27)$$

where

$$\Gamma_{i,1} = Z_i^{5/3}\left(\sum_j Y_j Z_j\right)^{1/3}, \quad (21.5.28)$$

Γ_H is given by eq. (21.5.10), and F_1 is given by eqs. (21.5.19)–(21.5.22). The appropriate arguments to use in eq. (21.5.26) are

$$\Gamma = \Gamma_{\text{liquid}} = \sum_i Y_i Z_i^{5/3}\,\frac{(\sum_i Y_i Z_i)^{1/3}}{\sum_i Y_i}\,\Gamma_H \quad (21.5.29)$$

and

$$\Lambda = \Lambda_{\text{liquid}} = \left(\sum_i \frac{Z_i^2}{A_i}\,Y_i\right)^{1/2}\,\frac{\hbar}{kT}\,eN_A\,(4\pi\rho)^{1/2}. \quad (21.5.30)$$

The appropriate equation of state to use is the one for which the free energy is the smaller. Given that the dominant contribution to the pressure is degenerate electrons, the density change across the interface between matter in the liquid phase and matter in the solid phase is small, so the interface can be defined as the position where $F_{solid} = F_{liquid}$. The algorithms presented in this section, when applicable, have been employed in constructing all models discussed in this two volume book.

21.6 White dwarf evolution

Although energy loss due to neutrino–antineutrino pairs escaping from the deep interior is a minor perturbation during most of the planetary nebula stage, the neutrino energy-loss rate increases in relative importance as the model evolves beyond this stage. This is demonstrated in Fig. 21.6.1 where the neutrino luminosity $L_{\nu\bar{\nu}}$ is compared as a function of time with the surface luminosity L_s, the hydrogen-burning luminosity L_H, and the gravothermal luminosity L_{grav}. Also shown is the quantity $L'_{grav} = L_{grav} - L_{\nu\bar{\nu}}$, which is that portion of the gravothermal luminosity that is not used up in balancing the neutrino luminosity and therefore represents the contribution of gravothermal energy to the surface luminosity. In contrast with what might be inferred from the discussion in Section 21.3, after the precipitous decline in the hydrogen-burning luminosity, the relative contributions of gravothermal energy and hydrogen-burning energy to the surface luminosity remain comparable for an additional time approximately as long as the duration of the planetary nebula stage, with L_H being larger than L'_{grav} most of this time. It is not until $t \sim 26\,000$ yr that L_H drops below and remains permanently smaller than L'_{grav}.

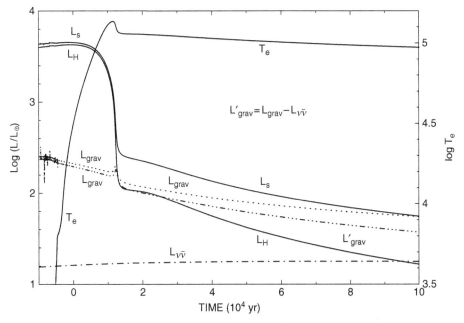

Fig. 21.6.1 Global energy-generation rates in the remnant of a 1 $M_\odot$ model during the planetary nebula phase and the early white dwarf phase

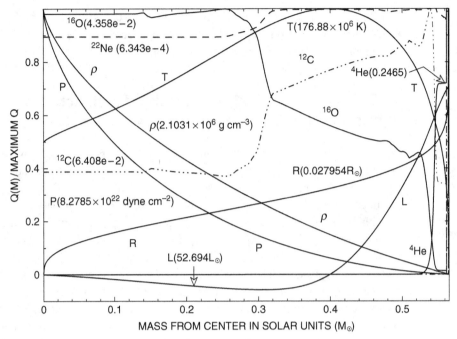

Fig. 21.6.2 Structure and composition in the core of the remnant of a 1 $M_\odot$ model 1.056×10^5 yr after becoming the central star of a planetary nebula

Nevertheless, hydrogen burning continues to be a non-negligible source of surface luminosity long after the planetary nebula phase is over. Characteristics of a model which is the result of 1.056×10^5 yr of evolution after the beginning of the PN phase are shown in Figs. 21.6.2–21.6.5. Comparing quantities in Fig. 21.6.2 with those in Fig. 21.3.6 for the model 13 000 yr beyond the start of the PN phase, one finds that, during the 92 600 yr of evolution separating the two models, changes that have occurred in the deep interior are much more modest than changes in layers near the surface. Central density and pressure have increased by 5.4% and 8.2%, respectively. The maximum temperature has decreased by 1.5% and the mass at which the temperature maximum occurs has moved inward from 0.445 $M_\odot$ to 0.40 $M_\odot$. The radius of the model has decreased by 50% and the surface luminosity has decreased by a factor of 5.6, with the relative contribution of hydrogen burning to the surface luminosity decreasing from $\sim$45% to $\sim$29%.

Comparing profiles in Figs. 21.6.3 and 21.3.10, one sees that temperatures near the base of the helium-abundance profile are typically over 12% smaller in the more evolved model. In the less evolved model, the center of the helium-burning shell is at $M_{He} = 0.541$ $M_\odot$, where $T_6 = 130$, $\rho = 2.59 \times 10^4$ g cm^{-3}, and $R = 1.42 \times 10^{-2}$ $R_\odot$; the helium-burning luminosity is $L_{He} = 32.1$ $L_\odot$. In the more evolved model, $M_{He} = 0.535$ $M_\odot$, where $T_6 = 120$, $\rho = 4.94 \times 10^4$ g cm^{-3}, and $R = 1.29 \times 10^{-2}$ $R_\odot$; the helium-burning luminosity is $L_{He} = 1.55$ $L_\odot$. A comparison of quantities in Figs. 21.6.4 and 21.3.8 shows that, while all gravothermal energy-generation rates have decreased by approximately a factor of 5 during evolution between the two models, the average helium-burning

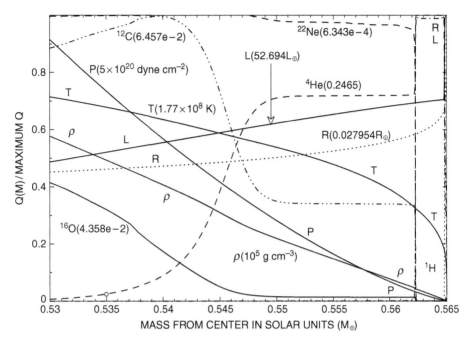

Fig. 21.6.3 Structure and composition below the hydrogen-burning shell in the remnant of a 1 $M_\odot$ model 1.056 × 10^5 yr after becoming a planetary nebula nucleus

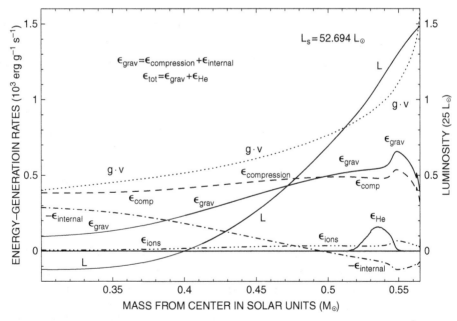

Fig. 21.6.4 Energy-generation rates below the hydrogen-burning shell of the remnant of a 1 $M_\odot$ model 1.056 × 10^5 yr after becoming a planetary nebula nucleus

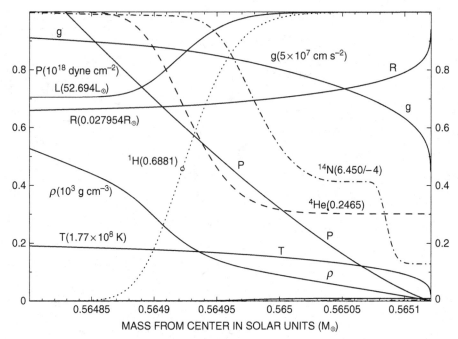

Fig. 21.6.5 Structure and composition in the hydrogen-rich layer of the remnant of a 1 $M_\odot$ model 1.056×10^5 yr after becoming a planetary nebula nucleus

energy-generation rate has decreased by a factor of over 20. Thus, in the 1.06×10^5 year old model, helium burning is not an important energy source relative to the gravothermal source.

The scale for luminosity in Fig. 21.6.4 is such that the maximum luminosity along the luminosity profile is $L_{max} = 37.15$ $L_\odot$ at the very edge of the profile where $M = 0.5648$ $M_\odot$. The remainder of the profile is shown in Fig. 21.6.5, where the luminosity climbs to $L_s = 52.694$ $L_\odot$ through the hydrogen-burning shell. Below the shell, in the region where $L > 0$, between $M = 0.40$ $M_\odot$ and $M = 0.5648$ $M_\odot$, by far the most important contributor to the increase in luminosity outward is the compressional energy released as a consequence of gravitationally induced contraction. The contribution of heavy ion cooling, which is measured by ϵ_{ions}, is quite small relative to the main contributor to the outward flow of energy. Thus it is that, as the hydrogen-burning shell declines in strength, matter below the shell contracts to make up the deficit in the form of the release of compressional energy.

Structure and composition characteristics in a region centered on the location of the hydrogen-burning shell in the 1.06×10^5 year old model are shown in Fig. 21.6.5. Characteristics at the center of the burning shell in this model are given in the third row of Table 21.6.1, while characteristics at the center of the hydrogen-burning shell in the 1.3×10^4 year old model are given in the second row of the table. Pressure is in units of 10^{17} dyne cm^{-2}. During the 92 600 years elapsing between the two models, the center of the hydrogen-burning shell has moved outward in mass by 0.58×10^{-4} $M_\odot$.

time	M_H	T_6	ρ	P_{17}	R	L_H	L_s
yr	$M_\odot$	K	g/cm^3	dyne/cm^2	0.01 $R_\odot$	$L_\odot$	$L_\odot$
0.0	0.564272	53.6	20.5	1.25	2.75	4256	4507
1.30×10^4	0.564864	37.1	88.2	3.24	2.27	131	293
1.06×10^5	0.564922	30.8	212	6.15	1.89	15.4	52.7
2.07×10^6	0.564965	21.0	772	16.3	1.44	0.130	1.66
1.12×10^7	0.564999	14.7	918	18.7	1.32	9.08/−3	0.128
1.81×10^8	0.565012	9.53	1339	20.8	1.26	1.78/−3	7.16/−3
1.50×10^9	0.565020	4.74	1731	20.7	1.24	3.99/−5	4.06/−4
7.50×10^9	0.565020	1.44	2039	21.4	1.23	2.11/−8	3.39/−5

Table 21.6.1 Characteristics at the center of the hydrogen-burning shell during PNN and white dwarf evolution

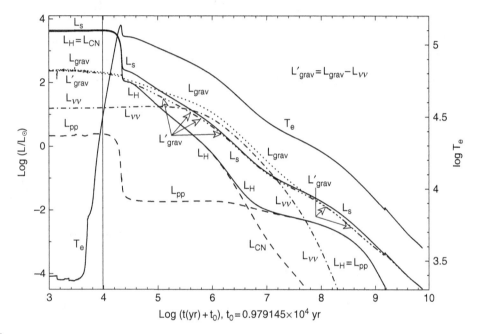

Fig. 21.6.6 Global energy-generation rates in a model of initial mass 1 $M_\odot$ from the onset of the superwind to the fully developed white dwarf stage

Characteristics at the center of the hydrogen-burning shell in the the model at the beginning of the PNN stage are given in the first row of the table.

Global energy-generation rates over a time period extending from near the onset of the superwind to ten billion years past the planetary nebula phase are shown in Fig. 21.6.6. The quantity $t_0 = 0.979\,145 \times 10^4$ yr is the time which elapses between the initiation of the superwind and the time $t = 0$ when the model first reaches $\log T_e = 4.5$, which has

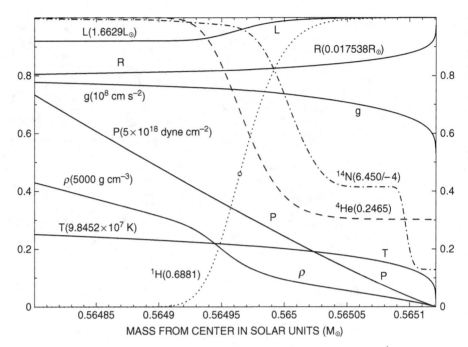

Fig. 21.6.7 Structure and composition in the hydrogen-rich layer of the remnant of a 1 $M_\odot$ model 2.07×10^6 yr after becoming a planetary nebula nucleus

been adopted in Fig. 21.3.4 as the definition of the zero point for time. The zero point is identified in Fig. 21.6.6 by the vertical line at $\log(t + t_0) \sim 4.0$.

For times larger than 10^5 yr, the curves for L_s and L'_{grav} in Fig. 21.6.6 are difficult to distinguish from one another. This is actually a virtue of the presentation, in that it brings home the fact that gravothermal energy remains the primary source of surface luminosity during the entire white dwarf phase, notwithstanding the fact that hydrogen continues to burn in surface layers for over a billion years.

Comparison between the curves for surface luminosity and neutrino luminosity reveals that from $\sim 4 \times 10^5$ yr to $\sim 1.35 \times 10^7$ yr, $L_{\nu\bar{\nu}} > L_s$. Thus, for over thirteen million years, more gravothermal energy is expended in supplying neutrino–antineutrino energy losses than in supplying energy losses by photons from the surface. It would be nice if neutrino detectors eventually become sensitive enough to verify this consequence of weak interaction theory.

Characteristics of a model at $t = 2.066 \times 10^6$ yr, approximately midway (logarithmically) through the period dominated by neutrino–antineutrino energy losses are described in Figs. 21.6.7–21.6.10. In a region centered on the hydrogen-burning shell, structure and composition characteristics are as shown in Fig. 21.6.7. The center of the burning shell is marked by the circle along the hydrogen-abundance profile and characteristics at this point are given in row four of Table 21.6.1. From the luminosity profile in the figure and the last two columns in the table, it is evident that hydrogen burning contributes only about 7% of the surface luminosity.

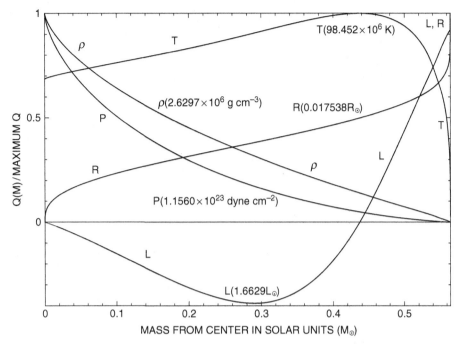

Fig. 21.6.8 Structure variables in the remnant of a 1 $M_\odot$ model 2.07×10^6 yr after becoming the central star of a planetary nebula

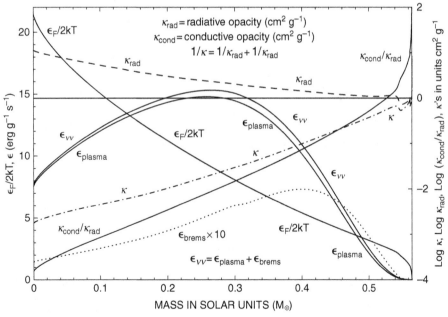

Fig. 21.6.9 Neutrino energy-loss rates, Fermi energy over kT, and opacities in the remnant of a 1 $M_\odot$ model after 2.07×10^6 yr of evolution as a white dwarf

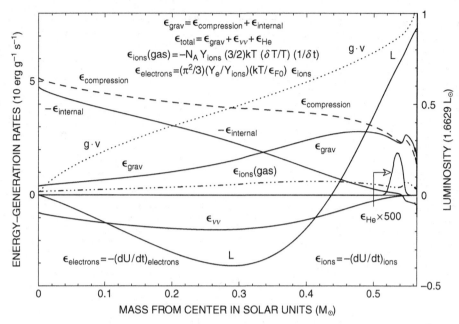

Fig. 21.6.10   Energy-generation rates below the hydrogen-burning shell of the remnant of a 1 $M_\odot$ model after 2.07×10^6 yr of evolution as a white dwarf

The structure variables in the interior of the $\sim 2 \times 10^6$ year old model that are shown in Fig. 21.6.8 may be compared with the same variables for the $\sim 10^5$ year old model shown in Fig. 21.6.2. During evolution between the two models, central density and pressure have increased by 25% and 40%, respectively. The maximum temperature has decreased by 80%, while the temperature at the center has decreased by 30%. Model radius has decreased by 57% and model luminosity has declined by a factor of 31.6.

In Fig. 21.6.9 are shown the variations with mass of quantities associated with several physical processes that are pertinent to understanding the evolution in the interior of the $\sim 2 \times 10^6$ year old model. The quantity $\epsilon_F/2kT$ provides a measure of the degree of electron degeneracy and ϵ_{plasma} and ϵ_{brems} are the rates of neutrino–antineutrino energy loss by the plasma and bremsstrahlung processes, respectively. Everywhere, the total neutrino energy-loss rate $\epsilon_{\nu\bar{\nu}}$ is dominated by the plasma neutrino energy-loss rate.

The curves in Fig. 21.6.9 labeled κ_{rad}, κ, and $\kappa_{cond}/\kappa_{rad}$ give, respectively, the logarithms of the radiative opacity, the total opacity, and the ratio of the conductive to the radiative opacity. Over the inner $\sim 0.525\ M_\odot$ of the model, $\epsilon_F/kT \geq 5$ and the conductive opacity is smaller than the radiative opacity by an amount which approaches a factor of 10^4 at the model center. In the outer $\sim 0.035\ M_\odot$ of the model, the reverse is the case, with the conductive opacity becoming a factor of 10^6 larger than the radiative opacity at the surface.

Gravothermal energy-generation rates for the $\sim 2 \times 10^6$ year old model are shown as functions of mass in Fig. 21.6.10, along with the helium-burning energy-generation rate and the luminosity. It is evident that the helium-burning shell is defunct. Integration under the curves for the rate of compressional energy generation $\left(\epsilon_{compression} = (P/\rho^2)\ (d\rho/dt) \right)$

and the rate at which gravity does work ($\mathbf{g} \cdot \mathbf{v}$) demonstrates that the classical equality of the two integrated rates holds. Comparison of the curves for ϵ_{grav} and $\epsilon_{v\bar{v}}$ with the luminosity profile shows that, in the inner ~ 0.29 $M_\odot$ of the model, where $L < 0$ and $dL/dM < 0$, $|\epsilon_{v\bar{v}}| > \epsilon_{grav}$, so the integrated rate of energy loss by neutrinos up to the point where $dL/dM = 0$ exceeds the integrated rate of gravothermal energy production up to this point. The reverse is the case in the region where $L < 0$ and $dL/dM > 0$. Gravothermal energy released in the region between the luminosity minimum at $M \sim 0.29$ $M_\odot$ and where $L = 0$ at $M \sim 0.44$ $M_\odot$ contributes to an inward flow of energy which remedies the deficiency in the region where $L < 0$ and $dL/dM < 0$.

The fact that, over most of the interior, $-\epsilon_{internal} = dU/dt > 0$ while temperatures are everywhere decreasing shows that, in the present instance, the term gravointernal is a more appropriate description than gravothermal. Under non-electron-degenerate conditions, a decrease in temperature would always correspond to a negative dU/dt and a positive $\epsilon_{internal}$. When electrons are highly degenerate, the internal energy associated with electrons is essentially proportional to the electron Fermi energy and is only weakly dependent on the temperature. Given that over most of the interior there are between six and eight electrons for every heavy ion, a relatively small fraction of the internal energy is in thermal form.

The contribution of heavy ions to the local gravothermal energy-generation rate is

$$\epsilon_{grav}^{ions} = -\frac{dU_{ions}}{dt} + \frac{P_{ions}}{\rho} \frac{1}{\rho} \frac{d\rho}{dt} = -\frac{dU_{ions}}{dt} + \frac{2}{3} U_{ions} \frac{1}{\rho} \frac{d\rho}{dt}, \tag{21.6.1}$$

where

$$\frac{P_{ions}}{\rho} = \frac{2}{3} U_{ions} = N_A Y_{ions} kT, \tag{21.6.2}$$

N_A is Avogadro's number, and Y_{ions} is the sum of the number abundances of the heavy ion components. Differentiation gives

$$\frac{dU_{ions}}{dt} = U_{ions} \frac{1}{T} \frac{dT}{dt} = N_A Y_{ions} \frac{3}{2} kT \frac{1}{T} \frac{dT}{dt}, \tag{21.6.3}$$

and eq. (21.6.1) becomes

$$\epsilon_{grav}^{ions} = U_{ions} \left(-\frac{1}{T} \frac{dT}{dt} + \frac{2}{3} \frac{1}{\rho} \frac{d\rho}{dt} \right) = N_A Y_{ions} \frac{3}{2} kT \left(-\frac{1}{T} \frac{dT}{dt} + \frac{2}{3} \frac{1}{\rho} \frac{d\rho}{dt} \right). \tag{21.6.4}$$

The computational program produces the individual Y_is which make up $Y_{ions} = \sum_i Y_i$ and

$$\frac{1}{T} \frac{dT}{dt} \sim \frac{\delta T}{T} \frac{1}{\delta t} \quad \text{and} \quad \frac{1}{\rho} \frac{d\rho}{dt} \sim \frac{\delta \rho}{\rho} \frac{1}{\delta t}, \tag{21.6.5}$$

where δT and $\delta \rho$ are the changes in T and ρ, respectively, in time step δt. From the abundance distributions in Fig. 21.6.2, one can deduce that, in the region where carbon and oxygen dominate ($M < 0.54$ $M_\odot$), $Y_{ions} \sim 0.072$ to within $\pm 20\%$. In the region where helium is the dominant element, this approximation fails by a factor of 3.4 and in the region where hydrogen is the dominant element, it fails by up to a factor of 11, but the region affected is only the outer 4% of the mass of the model where $\epsilon_{\text{internal}}$ is not overly influenced by the effects of electron degeneracy and the fact that temperatures are decreasing in this region is already reflected in an $\epsilon_{\text{internal}}$ which is positive.

Over the inner 95% of the mass of the models, $(1/\rho)\,(d\rho/dt)$ is more than 100 times smaller than $(1/T)\,(dT/dt)$, so $(P_{\text{ions}}/\rho^2)\,(d\rho/dt)$ can be neglected relative to $-dU_{\text{ions}}/dt$ and, to a good approximation, the rate at which ions contribute to gravothermal energy production is given by the rate of release of kinetic energy per unit mass, namely

$$\epsilon_{\text{grav}}^{\text{ions}} \sim -\frac{dU_{\text{ions}}}{dt} = -U_{\text{ions}}\,\frac{1}{T}\,\frac{\delta T}{\delta t} = -\frac{1.247\,35 \times 10^{14}\,\text{s}}{\delta t}\,Y_{\text{ions}}\,T_6\,\frac{\delta T}{T}\ \text{erg g}^{-1}\,\text{s}^{-1},$$

$$(21.6.6)$$

a quantity that is plotted in both Figs. 21.6.4 and 21.6.10. Comparing the mass dependences of ϵ_{ions} in these figures with the mass dependences of the temperature in Figs. 21.6.2 and 21.6.8, it is evident from eq. (21.6.6) that the local rate of decrease in temperature increases with the temperature. In other words, hotter regions cool faster.

The contribution of electrons to the rate of gravothermal energy generation is given by

$$\epsilon_{\text{grav}}^{\text{electrons}} = \frac{P_e}{\rho}\,\frac{1}{\rho}\,\frac{d\rho}{dt} - \frac{dU_e}{dt}.$$

$$(21.6.7)$$

When electrons are degenerate, to the extent that one can treat the electron pressure as being independent of temperature, one has that

$$U_e = n\,\frac{P_e}{\rho} \propto n\,A(n)\,\rho^{1/n},$$

$$(21.6.8)$$

where $n = 3/2$ if electrons are non-relativistically degenerate and $n = 3$ if electrons are relativistically degenerate. Thus, when electrons are degenerate and regardless of whether or not they have relativistic energies, to zeroth order,

$$\frac{dU_e}{dt} \sim \frac{P_e}{\rho}\,\frac{1}{\rho}\,\frac{d\rho}{dt},$$

$$(21.6.9)$$

and eq. (21.6.7) shows that $\epsilon_{\text{grav}}^{\text{electrons}} \sim 0$.

When electrons are non-relativistically degenerate, the first order correction to the pressure is (see eq. (4.6.41) in Volume 1)

$$\Delta P_e = \frac{\pi^2}{6}\,n_e\,kT\,\frac{kT}{\epsilon_{F0}},$$

$$(21.6.10)$$

where $n_e \propto \rho$ is the electron density and $\epsilon_{F0} \propto \rho^{2/3}$ is the electron Fermi energy at zero temperature (see eq. (4.6.27) in Volume 1). The first order correction to U_e is related to

ΔP_e by $\Delta U_e = (3/2)\,(\Delta P_e/\rho)$. Altogether,

$$\epsilon_{\text{grav}}^{\text{electrons}} = \frac{\Delta P_e}{\rho}\frac{1}{\rho}\frac{d\rho}{dt} - \frac{d}{dt}\,\Delta U_e + \cdots \tag{21.6.11}$$

$$= \frac{\pi^2}{2}\frac{n_e\,kT}{\rho}\frac{kT}{\epsilon_{F0}}\left(-\frac{1}{T}\frac{dT}{dt} + \frac{2}{3}\frac{1}{\rho}\frac{d\rho}{dt}\right) + \cdots$$

$$= \frac{\pi^2}{2}\frac{N_A}{\mu_e}kT\frac{kT}{\epsilon_{F0}}\left(-\frac{1}{T}\frac{dT}{dt} + \frac{2}{3}\frac{1}{\rho}\frac{d\rho}{dt}\right) + \cdots \tag{21.6.12}$$

From eqs. (21.6.12) and (21.6.4) one has that

$$\frac{\epsilon_{\text{grav}}^{\text{electrons}}}{\epsilon_{\text{grav}}^{\text{ions}}} = \frac{\pi^2}{3}\frac{1}{\mu_e Y_{\text{ions}}}\left(\frac{kT}{\epsilon_{F0}}\right) = \frac{\pi^2}{3}\frac{Y_e}{Y_{\text{ions}}}\left(\frac{kT}{\epsilon_{F0}}\right). \tag{21.6.13}$$

At the mass point $M = 0.3\,M_\odot$ in the 2.07×10^6 yr old model, $Y_{\text{ions}} \sim 0.068$, $\mu_e = 2$, and $\epsilon_{F0}/kT = 16$, giving $(\epsilon_{\text{grav}}^{\text{electrons}}/\epsilon_{\text{grav}}^{\text{ions}}) \sim 1.51$, showing that, even though electrons are quite degenerate, electrons at the top of the Fermi sea contribute cooling energy at a larger rate than do heavy ions in the gaseous state. At the center of the model, the ratio is 0.52, and at $M = 0.4\,M_\odot$, the ratio is 2.15.

The rate at which the release of Coulomb binding energy contributes to gravothermal energy production is given by

$$\epsilon_{\text{grav}}^{\text{Coulomb}} = -\frac{dU_{\text{Coulomb}}}{dt} - P_{\text{Coulomb}}\frac{d}{dt}\left(\frac{1}{\rho}\right) = \frac{d|U_{\text{Coulomb}}|}{dt} - \frac{|P_{\text{Coulomb}}|}{\rho}\frac{1}{\rho}\frac{d\rho}{dt}. \tag{21.6.14}$$

At densities and temperatures currently under consideration, the approximations in Section 4.17 of Volume 1 and in Section 21.4 in this volume are appropriate. In Section 4.17, it is argued that $U_{\text{Coulomb}} \propto \rho^{1/3}$ and that $P_{\text{Coulomb}} = (1/3)\,(U_{\text{Coulomb}}/\rho)$. With these prescriptions,

$$\epsilon_{\text{grav}}^{\text{Coulomb}} = 0. \tag{21.6.15}$$

In Fig. 21.6.10, at masses below $M = 0.29\,M_\odot$, where $dL/dM = 0$, most of the compressional energy released locally in consequence of the contraction forced by gravity is used up in increasing the kinetic energy of degenerate electrons, leaving an insufficient residue to balance neutrino energy losses locally. Moving outward from the point where the slope of the luminosity profile changes sign, ϵ_F/kT continues to decrease, as detailed in Fig. 21.6.9, so the relative fraction of the internal energy due to thermal motions increases outward. The net result is that, moving outward, a smaller fraction of the compressional energy released locally is used up in increasing the internal energy of degenerate electrons and more is left to contribute to the ambient flow of energy inward to make up for the locally incompletely compensated neutrino energy loss below the point where $dL/dM = 0$.

Properties of a model that has evolved for $\sim 1.12 \times 10^7$ years beyond the superwind phase are described in Figs. 21.6.11–21.6.14. The curves for L_{pp} and L_{CN} in Fig. 21.6.6

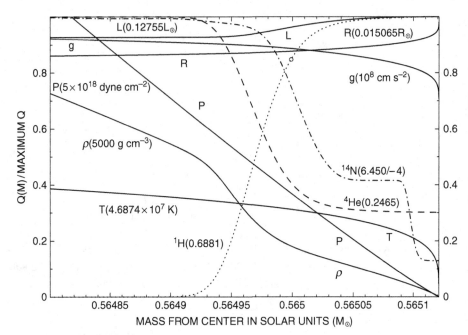

Fig. 21.6.11   Structure and composition in the hydrogen-rich layer of the remnant of a 1 $M_\odot$ model after 1.12×10^7 yr of evolution as a white dwarf

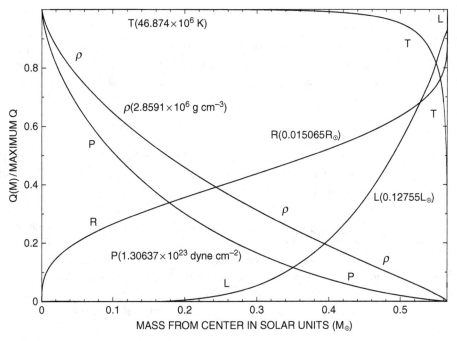

Fig. 21.6.12   Structure variables in the interior of the remnant of a 1 $M_\odot$ model after 1.12×10^7 yr of evolution as a white dwarf

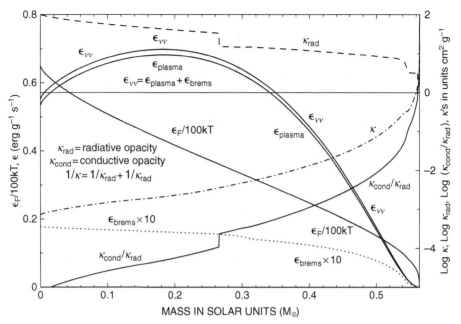

Fig. 21.6.13 Neutrino energy-loss rates, Fermi energy over kT, and opacities in the remnant of a 1 $M_\odot$ model after 1.12×10^7 yr of evolution as a white dwarf

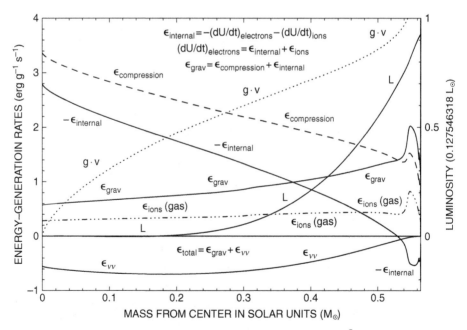

Fig. 21.6.14 Gravothermal energy-generation rates in the remnant of a 1 $M_\odot$ model after 1.12×10^7 yr of evolution as a white dwarf

show that hydrogen burning by the pp chains has replaced CN-cycle burning as the major form of nuclear energy production and the curves for L_s and $L_{\nu\bar{\nu}}$ show that, although the neutrino–antineutrino luminosity is slightly larger than the surface luminosity, this relationship is about to change dramatically.

Structure and composition variables in the hydrogen-burning region of the model are shown in Fig. 21.6.11. The feature which distinguishes the hydrogen-abundance profile in this figure from those in comparable figures for younger models, such as the profile in Fig. 21.6.7 for the $\sim 2 \times 10^6$ year old model, is the location of the center of the hydrogen-burning shell (marked by a circle) along the profile. In the younger CN-cycle burning model, the center of the burning shell is at a hydrogen abundance slightly less than half of the surface abundance of hydrogen. In the pp-chain burning model, the center of the burning shell is at a hydrogen abundance which is approximately 85% of the surface hydrogen abundance. The differences in the relative location of the shell center along the hydrogen profile is of course due to the fact that the pp-chain energy-generation rate is proportional to the square of the hydrogen abundance whereas the CN-cycle energy-generation rate is linearly proportional to the hydrogen abundance.

Conditions at the center of the burning shell in the younger model are described in the fourth row of Table 21.6.1 and conditions at the center of the burning shell in the older model are described in the fifth row of the table. The fraction of the surface luminosity provided by hydrogen burning is approximately the same in both cases, being $\sim 7.8\%$ in the younger model and $\sim 7.1\%$ in the older model.

Structure variables, physical characteristics in which electrons play a prominent role, and gravothermal (gravointernal) energy-generation rates in the 1.12×10^7 year old model are shown as functions of mass in Figs. 21.6.12–21.6.14, respectively. The most dramatic differences between the profiles of structure variables in Fig. 21.6.12 and the profiles of the same variables in Fig. 21.6.8 for the 2.07×10^6 year old model are the temperature and luminosity profiles. In the younger model, temperature decreases inward and the luminosity is negative over $\sim 75\%$ of the mass of the interior of the model. In the older model, temperature is everwhere decreasing outward and the luminosity is everywhere positive. The inner half of the mass of the older model is essentially isothermal at $T \sim 46.9 \times 10^6$ K, compared with a maximum temperature of $T \sim 98.5 \times 10^6$ K and a central temperature of $T \sim 68 \times 10^6$ K in the younger model. In both models, the energy liberated by gravitational contraction continues to help sustain neutrino energy losses from the interior and photon energy losses from the surface, with model radius decreasing by 11.6% and central density increasing by 8.7% during evolution between the two models.

Comparing quantities in Figs. 21.6.13 and 21.6.9, it is evident that the degree of electron degeneracy and the radiative opacity have increased significantly over most of the interior during the evolution between the 2.07×10^6 year old model and the 1.12×10^7 year old model, while the conductive opacity and neutrino energy-loss rates have decreased significantly. At model center, ϵ_F/kT has increased from ~ 44 to ~ 64, the radiative opacity has increased by a factor of ~ 10, and the conductive opacity has decreased by a factor of ~ 2.5. At the mass point $M = 0.3\ M_\odot$, kT/ϵ_{F0} has decreased by a factor of two between the two models, so the contribution to the rate of gravothermal energy release at this mass point by cooling electrons at the top of the Fermi sea relative to the contribution by cooling heavy ions has decreased by a factor of 2 between the two models. Maximum neutrino energy-loss

Table 21.6.2 Central and global characteristics during PNN and white dwarf evolution

time yr	T_c 10^6 K	ρ_c 10^6 g/cm^3	P_c 10^{23} dyne/cm^2	Γ_c	R 0.01 $R_\odot$	L_s $L_\odot$
0.0	88.515	1.9669	0.74779	7.0837	229.36	4506.9
1.30×10^4	88.810	1.9963	0.76490	7.0952	4.2344	292.59
1.06×10^5	88.356	2.1031	0.82785	7.7683	2.7954	52.699
2.07×10^6	67.461	2.6297	1.1560	10.256	1.7538	1.6629
1.12×10^7	46.874	2.8591	1.3064	15.175	1.5065	0.12755
1.81×10^8	14.949	3.0089	1.4029	48.412	1.3763	7.1574/−3
1.50×10^9	5.3778	3.0496	1.4293	135.17	1.3150	4.0590/−4
7.50×10^9	1.5281	3.0679	1.4411	476.57	1.2777	3.3895/−5

rates have decreased by over a factor of 20. The discontinuity in the radiative opacity at 0.265 $M_\odot$ reveals a glitch in the opacity table employed in the model calculations.

The energy-generation profiles in Fig. 21.6.14 demonstrate that, over the inner one third of the mass of the 1.12×10^7 year old model, the neutrino energy loss rate is locally exactly balanced by the rate of release of gravothermal energy. Over the rest of the model, the release of gravothermal energy supplies the energy lost locally to neutrinos, with enough left over that gravothermal energy is by far the major contributor to the energy carried away by photons from the surface. The near constancy of ϵ_{ions} over the inner half of the model mass is a consequence of the fact that this region has become almost isothermal, so that both T_6 and $\delta T/T$ in eq. (21.6.6) are independent of position. The magnitude of ϵ_{ions} relative to the magnitude of ϵ_{grav} in the region where $L > 0$ shows that ion cooling contributes significantly to the surface luminosity. Electrons at the top of the Fermi sea continue to contribute to ϵ_{grav}. At the positions $M = 0.0$, 0.3, and 0.4 $M_\odot$, the quantity $\epsilon_{grav}^{electrons}/\epsilon_{grav}^{ions}$ varies through 0.37, 0.71, and 0.89. The fact that, in the region where $L > 0$, ϵ_{grav} increases outward from being twice as large as ϵ_{ions} to being over three times as large as ϵ_{ions} (gas) shows that the release of compressional energy is still contributing to the surface luminosity.

The global luminosities describing the 1.12×10^7 year old model are $L_{grav} = 0.270\,13$ $L_\odot$, $L_{\nu\bar{\nu}} = 0.151\,67$ $L_\odot$, $L_H = 0.009\,08$ $L_\odot$, and $L_s = 0.127\,55$ $L_\odot$. A check on the accuracy of the calculations is that $L_{grav} + L_H - L_{\nu\bar{\nu}} = L_s$ to five significant figures. Integration under the curve for ϵ_{ions} (gas) in Fig. 21.6.14 gives an ion cooling luminosity of $L_{ions} \sim 0.113$ $L_\odot$. As will become apparent in an upcoming discussion of the quantity Γ in column 5 of Table 21.6.2, ions are gaseous, so L_{ions} is an accurate measure of the contribution of ions to L_{grav}. Table 21.6.2 gives central and global characteristics of the individual models described in this section. Given that the contribution of cooling electrons at the top of the Fermi sea is about 60% of the contribution of cooling ions, it follows that compressional energy contributes about one fourth of the rate of gravothermal energy generation.

Characteristics of a model which has evolved for 1.81×10^8 yr past the planetary nebula phase are shown in Figs. 21.6.15–21.6.18. As is evident from Fig. 21.6.15 and the entries in the sixth row of Table 21.6.1, the contribution of hydrogen burning by the pp chains

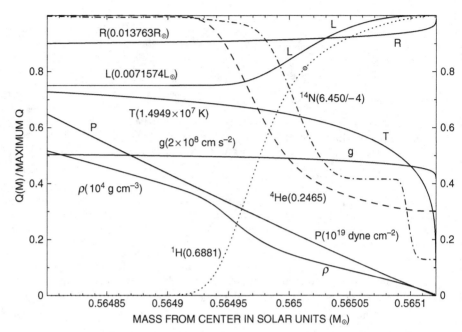

Fig. 21.6.15 Structure and composition in the hydrogen-rich layer of the remnant of a 1 $M_\odot$ model after 1.81×10^8 yr of evolution as a white dwarf

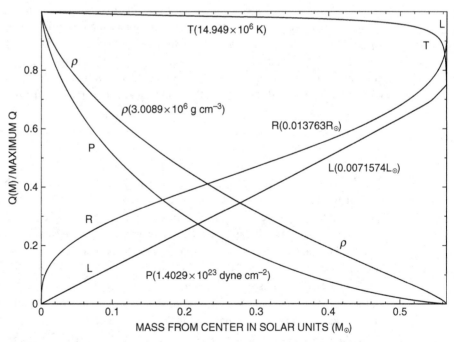

Fig. 21.6.16 Structure variables in the interior of the remnant of a 1 $M_\odot$ model after 1.81×10^8 yr of evolution as a white dwarf

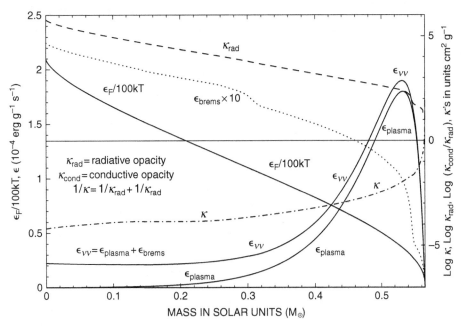

Fig. 21.6.17   Neutrino energy-loss rates, Fermi energy over kT, and opacities in the remnant of a 1 $M_\odot$ model after 1.81×10^8 yr of evolution as a white dwarf

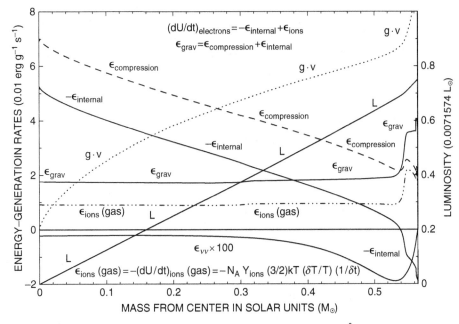

Fig. 21.6.18   Gravothermal Energy-generation rates in the remnant of a 1 $M_\odot$ model after 1.81×10^8 yr of evolution as a white dwarf

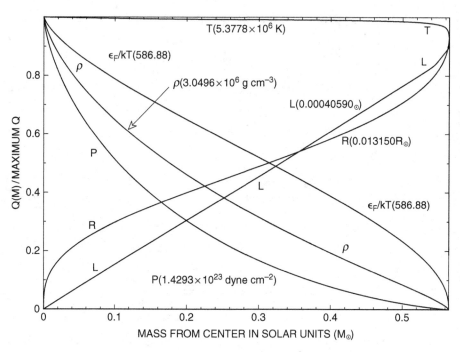

Fig. 21.6.19   Structure variables in the remnant of a 1 $M_\odot$ model after 1.50×10^9 yr of evolution as a white dwarf

to the surface luminosity has increased to ~25% of the surface luminosity, which is a factor of ~3.5 larger than the relative contribution in the 1.12×10^7 year old model. The radius of the model is 9.5% smaller than that of the younger model and the surface luminosity has decreased by a factor of 17.8. Comparing quantities in Fig. 21.6.16 with those in Fig. 21.6.12, one finds that the central density and pressure in the older model have increased relative to those in the younger model by 5.2% and 7.4%, respectively, while the central temperature has decreased by a factor of 3.14.

Comparing neutrino energy-loss rates in Figs. 21.6.17 and 21.6.13, one finds that, while in the younger model the plasma neutrino energy-loss rate is everywhere more than an order of magnitude larger than the bremsstrahlung energy-loss rate, over the inner two-thirds of the mass of the older model, the bremsstrahlung energy-loss rate is larger than the plasma neutrino energy-loss rate. Comparing ϵ_{grav} and $\epsilon_{\nu\bar{\nu}}$ in Figs. 21.6.18 and 21.6.14, it is evident that, whereas the average neutrino energy-loss rate and the gravothermal energy-production rate are of comparable magnitude over much of the interior of the younger model, neutrino-loss rates are over 100 times smaller than gravothermal energy-production rates in the older model. At the center of the older model, $\epsilon_F/kT = 210$, ~ 3.3 times larger than in the younger model, the radiative opacity at the center is five orders of magnitude larger in the older model than in the younger model, and the conductive opacity at the center is an order of magnitude smaller.

Characteristics of a model which has evolved as a white dwarf for ~1.5×10^9 yr are shown in Figs. 21.6.19 and 21.6.20. Comparing structure characteristics in Fig. 21.6.19 with those in Fig. 21.6.16, one sees that during the 1.32×10^9 years of evolution separating

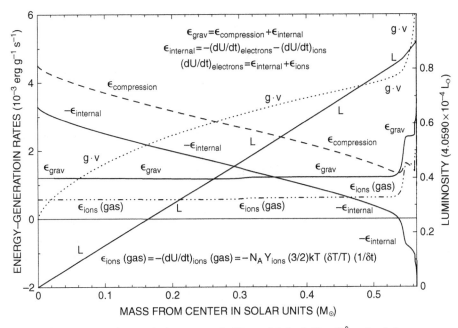

Fig. 21.6.20 Gravothermal energy-generation rates in the remnant of a $1M_\odot$ model after 1.50×10^9 yr of evolution as a white dwarf

the 1.5×10^9 year old model from the 1.81×10^8 year old model, the central temperature has decreased by a factor of 2.8 and the central density has increased by 1.7%. Model radius has decreased by 4.7% and model luminosity has decreased by a factor of 17.6. As detailed in rows 6 and 7 of Table 21.6.1, the contribution of hydrogen burning by the pp chains to the surface luminosity has decreased from 25% to 10%.

More striking than the differences between the two models are the similarities in morphology of the gravothermal energy-generation characteristics of the models as exhibited in Figs. 21.6.20 and 21.6.18. Although the scale for the energy-generation rates in the figure for the older model is ten times smaller than the scale for the younger model (10^{-3} erg g^{-1} s^{-1} for the older model versus 10^{-2} erg g^{-1} s^{-1} for the younger model), the profiles for all quantities are virtually identical in shape. In particular, the net gravothermal energy-generation rate ϵ_{grav} is in both instances almost flat over the large majority of the interior mass where carbon and oxygen are the primary constituents, and this translates into a luminosity profile in both instances which has a constant slope except in the outer few percent of the mass where helium is the primary constituent.

The quantity $\epsilon_{ions}(gas)$ in the figures for both models is the rate of cooling given by eq. (21.6.6), which has been obtained on the assumption that heavy ions are in the gaseous state. The fact that, in both models, $\epsilon_{grav} \sim 2\,\epsilon_{ions}(gas)$ demonstrates that matter over most of the interior of both models is in the liquid or solid state, at temperatures such that the number of degrees of freedom of the heavy ions is exactly twice the number of degrees of freedom in the gaseous state and that the ionic specific heat per ion is essentially the same as that of mercury, of copper, and of some 50 other pure metallic substances at room temperature and atmospheric pressure (Pierre Louis Dulong & Alexis Thèrèse Petit, 1818).

The transition from gas to liquid to solid is related to an increase in the ratio of the Coulomb binding energy between heavy ions and degenerate electrons relative to kT. The ratio is commonly measured by (see eqs. (21.4.2) and (21.4.3))

$$\Gamma = 22.7 \left(\sum_i \frac{Z_i^2}{A_i^{1/3}} Y_i \right) \frac{\rho_6^{1/3}}{T_6} \div \sum_i Y_i, \tag{21.6.16}$$

which is a generalization of eq. (21.4.2) for a multicomponent system. See also eqs. (21.5.11) and (21.5.10). In the younger model, Γ decreases from $\sim$48.4 at the center to $\sim$15.6 at $M = 0.53\ M_\odot$, and in the older model, Γ decreases from $\sim$135 at the center to $\sim$50 at $M = 0.53\ M_\odot$.

Solidification does not occur until Γ exceeds a critical value Γ_{melt} estimated in Section 21.4 to be in the range 172–216. The estimate of $\Gamma_{\text{melt}} = 216$ given by eq. (21.4.42) is obtained on the assumption that, as inferred from the properties of terrestial solids, melting occurs when the half amplitude of thermally-induced oscillations is approximately one-sixth of the radius of the Wigner–Seitz sphere about an average ion. A somewhat more sophisticated treatment making use of the Thomas–Fermi model of the atom produces a Γ_{melt} which depends on the density and, for pure oxygen, varies from 171 at $\rho = 2 \times 10^7$ g cm^{-3} to 192 at $\rho = 2 \times 10^5$ g cm^{-3} (see columns 1 and 7 in Table 21.4.3). On the basis of these estimates, matter over most of the interior of both the 1.81×10^8 and the 1.5×10^9 year old models is in the liquid state, with an atomic specific heat essentially the same as that of mercury at room temperature and atmospheric pressure.

On the supposition that the transition from the gaseous to the liquid state occurs when the half amplitude of vibration is approximately half of the radius of the Wigner–Seitz sphere, one has that the transition occurs when $\Gamma = \Gamma_{\text{condense}} \sim \Gamma_{\text{melt}}/9 \sim 20$–$30$. Central and global characteristics of the models singled out for study in this section are given in Table 21.6.2 and the value of Γ at the center of each model is given in column 5 of the table. It is evident that, with the criterion for liquefaction adopted here, liquefaction at the center occurs at a time somewhere between 10^7 and 10^8 yr.

Characteristics of a model which has evolved as a white dwarf for 7.5×10^9 yr are shown in Figs. 21.6.21 and 21.6.22. During the 6×10^9 years of evolution separating this model from the 1.5×10^9 year old model, the central density has increased by only 0.6% but model radius has decreased by 2.9%. Interior temperatures have decreased by a factor of 3.5 from 5.38×10^6 K to 1.53×10^6 K and ϵ_F/kT has increased everywhere by the same factor.

The differences between the gravothermal energy-generation rate profiles shown in Fig. 21.6.20 for the younger model and in Fig. 21.6.22 for the older model are a consequence of changes in the equation of state for heavy ions that take place during evolution between the two models. In the younger model, ϵ_{grav} is essentially constant over the entire carbon- and oxygen-rich interior, while in the older model, ϵ_{grav} exhibits a discontinuity at mass point $M \sim 0.49\ M_\odot$, and decreases from this point steadily inward, almost vanishing at the center. In the younger model, the curves for $-\epsilon_{\text{internal}}$ and $\epsilon_{\text{compression}}$ are nearly parallel, but in the older model they converge towards one other in the inward direction.

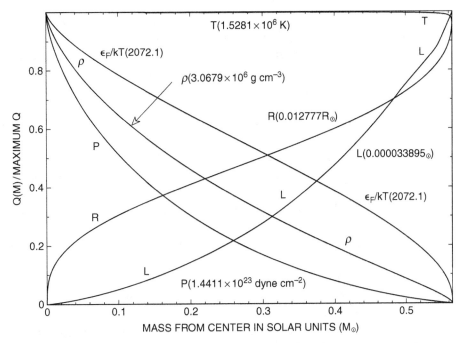

Fig. 21.6.21   Structure variables in the remnant of a $1M_\odot$ model after 7.50×10^9 years of evolution as a white dwarf

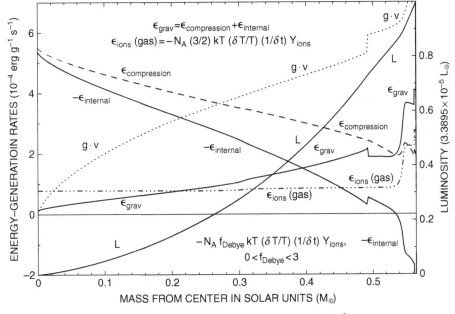

Fig. 21.6.22   Gravothermal energy-generation rates in the remnant of a $1M_\odot$ model after 7.50×10^9 yr of evolution as a white dwarf

On the other hand, between $\sim0.49\ M_\odot$ and the surface, the profiles for each of the three gravothermal energy-generation rate indicators (ϵ_{grav}, $-\epsilon_{internal}$, and $\epsilon_{compression}$) in the older model are morphologically equivalent to the same profiles in the younger model. One infers that the same physics prevails in the outer portions of both models but, over most of the older model, a type of physics prevails which is different from that prevailing in the younger model.

The factor by which $\epsilon_{compression}$ increases inward between the base of the helium-rich layer and the center is nearly the same in the two models, showing that the mechanism for translating work done by gravity into compressional energy has not changed. Since the physics of increasing the internal kinetic energy of electrons also does not change, the fact that $-\epsilon_{internal}$ increases inward more steeply in the older model than in the younger one, coupled with the fact that ϵ_{grav} decreases fairly steeply inward in the older model, while both ϵ_{grav} and ϵ_{ions}(gas) remain nearly constant as functions of mass in the younger model means that, in the older model, the actual rate at which ions cool decreases inward.

Since the interior of each model is essentially isothermal, the local rate of temperature change is the same throughout the interior of each model. In the younger model, the fact that ϵ_{ions}(gas) does not vary appreciably with mass over regions where the number abundance of ions is nearly constant means that the kinetic energy content per unit mass is everywhere linearly proportional to the temperature so that the specific heat is constant, a characteristic of the specific heat of a solid or of a liquid at high temperatures. In a solid at low temperatures, the kinetic energy content per unit mass is proportional to the temperature times the third power of T/Θ_{Debye}, where the Debye temperature Θ_{Debye} is proportional to the square root of the density. Since the temperature is independent of mass and density increases inward, the observed decrease in the specific heat inward suggests that matter over much of the interior of the older model has solidified.

The quantity Γ defined by eq. (21.6.16) is shown as a function of mass for the two models in Fig. 21.6.23. In the model calculations, the properties of the liquid and solid states are taken into account in accordance with algorithms described in Section 21.5. These algorithms lead to a discontinuity in the estimate of internal energy at the formal boundary between solid and liquid and this is reflected by the discontinuities in ϵ_{grav} and $\epsilon_{internal}$ at $M \sim0.49\ M_\odot$ in Fig. 21.6.22. The intersection of a vertical line at this mass and the curve for $\Gamma(t = 7.5 \times 10^9$ yr) in Fig. 21.6.23 produces a Γ equal to

$$\Gamma_{melt} = 185, \tag{21.6.17}$$

which has the significance that solidification occurs when $\Gamma > \Gamma_{melt}$.

Once the transition to the solid phase has occurred, the thermal kinetic energy of ions, as given by Debye theory, behaves according to eqs. (21.4.17)–(21.4.21), declining from $3kT$ per ion at high temperatures and low densities to a value proportional to $kT\ (T/\Theta_{Debye})^3$ at low temperatures and high densities. Below the very outer hydrogen-rich layer, the electron molecular weight is everywhere $\mu_e = 2$, and the Debye temperature, as given by eq. (21.4.30), is

$$\Theta_{Debye} = 2245\ \sqrt{\rho(g/cm^3)}\ K. \tag{21.6.18}$$

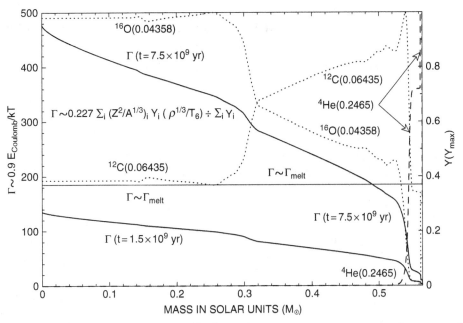

Fig. 21.6.23 Coulomb binding energy over kT for ions (mostly ^{16}O, ^{12}C, and ^{4}He) in white dwarfs that are 1.5×10^9 and 7.5×10^9 years old

The thermal energy-generation rate in the solid state is related to that in the gaseous state by

$$\frac{\epsilon_{\text{ions}}(\text{solid})}{\epsilon_{\text{ions}}(\text{gas})} = \frac{f_{\text{Debye}}}{1.5}, \tag{21.6.19}$$

where, as shown by eqs. (21.4.21) and (21.4.20), f_{Debye} varies from near zero at low temperature and high density to 3 at high temperature and low density. Making use of eq. (21.6.3), one has that

$$\epsilon_{\text{ions}}(\text{solid}) = -\frac{f_{\text{Debye}}}{1.5} \frac{1.247\,35 \times 10^{14}\,s}{\delta t} \, Y_{\text{ions}}\, T_6 \, \frac{\delta T}{T} \, \text{erg g}^{-1}\,\text{s}^{-1}. \tag{21.6.20}$$

In the 7.5×10^9 year old model, the ratio T/Θ_{Debye} varies from 0.385 at the center to 1.23 at $M = 0.49\,M_\odot$. The increase outward in mass from the center to $M = 0.49\,M_\odot$ in ϵ_{grav} reflects the increase outward in $\epsilon_{\text{ions}}(\text{solid})$.

Comparing the profiles of Γ with Γ_{melt} in Fig. 21.6.23, one may estimate that solidification begins at the center of the model at a time which is of the order of 2×10^9 years past the planetary nebula stage and extends to the surface by a time of the order of 9×10^9 years after the planetary nebula stage. Ultimately, temperatures become small enough throughout the model that the ion specific heat becomes everywhere proportional to $(T/\Theta_{\text{Debye}})^3 \propto T^3/\rho^{3/2}$, so that the contribution of ion cooling to the surface luminosity is restricted to a region very near the surface.

21.7 Diffusion and the formation of a pure hydrogen surface abundance

Over the years, it has become clear that white dwarfs for which unambiguous estimates of surface abundances can be made fall into two main categories: those with essentially pure hydrogen surfaces (DA white dwarfs) and those with no hydrogen at the surface (non-DA white dwarfs). Subcategories of non-DA white dwarfs include DB white dwarfs with nearly pure helium surface abundances and DQ white dwarfs with nearly pure carbon surface abundances. Models of the sort described in Sections 21.3 and 21.6 can be used to understand how DA white dwarfs come about.

In the TPAGB precursors of white dwarfs, the abundances of most elements which are not beta-unstable are constant over a large convective region which extends from just above the hydrogen-burning shell to the surface. Between the base of the convective envelope and the surface, the gravitational acceleration is from 1 to 5 orders of magnitude smaller than the gravitational acceleration in the Sun, which means that gravitationally induced diffusion over the lifetime of the TPAGB phase is completely unimportant, even at the base of the convective envelope. However, once most the envelope has been expelled in a superwind and the model begins to contract to white dwarf dimensions, convection in the envelope of the remnant disappears and, at maximum, the gravitational accleration in envelope layers becomes more than three orders of magnitude larger than in the Sun, with the consequence that gravitationally induced diffusion becomes very important.

In the approximation that only hydrogen and helium are present at number abundances $Y_H = 0.68$ and $Y_{He} = 0.075$, respectively, the diffusion velocity of hydrogen, as given by eq. (12.9.2) is

$$\omega_H = 2.021 \, \frac{T_6^{3/2}}{\rho} \, g_7 \, \log_e^{-1} \left(1 + 0.18769 \, x_{st}^{1.2} \right) \text{ cm s}^{-1}, \qquad (21.7.1)$$

where g_7 is the gravitational acceleration in units of 10^7 cm s^{-2} and, as given by eq. (12.6.21),

$$x_{st} = 6.334 \left(\frac{\lambda}{a_0} \right) T_6, \qquad (21.7.2)$$

with λ/a_0 being the larger of the ratios given by eqs. (12.9.4) and (12.9.5), namely,

$$\frac{\lambda}{a_0} = \max \left(\left(\frac{3.571}{\rho_0} \right)^{1/3}, \, 1.34 \left(\frac{T_6}{\rho_0} \right)^{1/2} \right). \qquad (21.7.3)$$

The first entry on the right hand side of eq. (21.7.3) is the ratio of the average separation between ions and the Bohr radius for hydrogen in the ground state and the second entry is the Debye length over the Bohr radius.

Table 21.7.1 Diffusion time scales in hydrogen-rich matter at $M = 0.565\,05\,M_\odot$, 7×10^{-5} $M_\odot$ below the surface

time 10^5 yr	g 10^7cm/s^2	ρ g/cm^3	T 10^6 K	$P/10^{17}$ dyne/cm^2	R $0.01\,R_\odot$	H_P $10^{-4}\,R_\odot$	ω_H cm/yr	τ_{diff} 10^7 yr
1.05	3.68	53.2	23.1	1.69	2.05	12.4	3.72	2.32
20.7	7.06	285	17.6	6.86	1.48	4.90	1.39	2.45
112	8.61	545	13.0	10.6	1.34	3.23	0.790	2.85
1820	9.59	901	8.82	13.1	1.27	2.18	0.443	3.41

A diffusion time scale may be defined by

$$\tau_{\text{diff}} = \frac{H_P}{\omega_H},$$ (21.7.4)

where H_P is given by

$$\frac{1}{H_P} = \left| \frac{1}{P} \frac{dP}{dr} \right| = \frac{g\rho}{P}.$$ (21.7.5)

Using eqs. (21.7.1) and (21.7.5) in eq. (21.7.4) gives

$$\tau_{\text{diff}}(\text{yr}) = 4.929 \times 10^7 \, P_{17} \frac{\log_e \left(1 + 0.187\,69 \, x_{\text{st}}^{1.2}\right)}{T_6^{3/2} \, g_8^2},$$ (21.7.6)

where g_8 is the gravitational acceleration in units of 10^8 cm s^{-2} and P_{17} is the pressure in units of 10^{17} dyne cm^{-2}.

Conditions in the hydrogen-rich region above the hydrogen-burning shell at the mass point $M = 0.565\,05 \, M_\odot$ are given in Table 21.7.1 for four models. The time which the model has evolved past the initiation of the planetary nebula stage is given in column 1. Gravitational acceleration, density, temperature, pressure (in units of 10^{17} dyne cm^{-2}), radius, and pressure scale height H_P are given in the next six columns, respectively. The eighth column gives the diffusion velocity for hydrogen produced by eqs. (21.7.1)–(21.7.3) and the last column gives the diffusion time scale defined by eq. (21.7.6). It is interesting that in all cases the average ionic separation and the Debye length are quite similar. For example, in the third case one finds that $\lambda_{\text{ions}}/a_0 = 0.187$ and $\lambda_D/a_0 = 0.207$. Using the larger value in eq. (21.7.2) gives $x_{\text{st}} = 17.04$ and $\log_e \left(1 + 0.187\,69 \, x_{\text{st}}^{1.2}\right) = 1.894$.

It is evident from the diffusion time scales reported in Table 21.7.1 that, had diffusion been explicitly incorporated in the models, the surface abundance of hydrogen would have begun to increase noticeably by the time evolution had continued for 10^7 yr past the beginning of the planetary nebula stage. Diffusion has not been included due the large number of isotopes employed, but relevant calculations with fewer isotopes and diffusion explicitly included have been conducted by Iben and Jim MacDonald (1985, 1986). The sixth figure in the second of the two cited references is reproduced here as Fig. 21.7.1. The models for which abundances are shown are from an evolutionary sequence calculated for

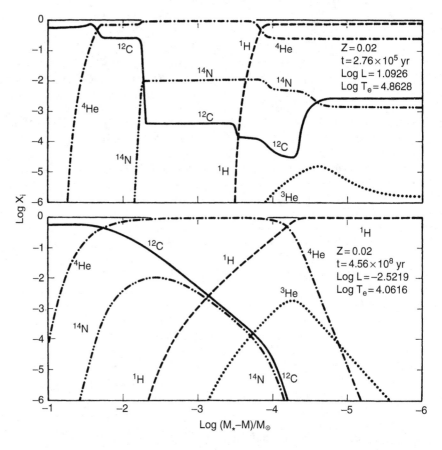

Fig. 21.7.1   Abundance profiles in a model white dwarf experiencing diffusion (Iben & MacDonald, 1986)

the 0.6 $M_\odot$ remnant of a population I model with $Z = 0.02$, using opacities described in Section 7.12 in Volume 1. The models have experienced exactly same kind of evolution as have the models described in Sections 21.3 and 21.6, and evolutionary time scales and the duration of phases in which various physical processes dominate are quite similar.

The distributions of isotope number abundances in the upper panel of Fig. 21.7.1 are those in a model which has evolved for 2.76×10^5 yr past the initiation of the planetary nebula stage and those in the lower panel are for a model which has evolved for 4.56×10^8 yr past the initiation of the planetary nebula stage. It is evident that evolution for only 1% of a diffusion time scale is insufficient to make a noticeable impression on surface abundances, but that evolution for ten diffusion time scales has a dramatic impact on abundances everywhere in the model. In particular, the surface of the model described in the lower panel is free of all isotopes other than hydrogen and all number abundance profiles in the lower panel have been modified relative to those in the upper panel in a way expected if migration of a given isotope is promoted in the direction of a gradient in its distribution. Most noticeably, the innermost edges of the abundance profiles for hydrogen, the helium isotopes, and nitrogen have all been shifted inward. These shifts demonstrate that diffusion

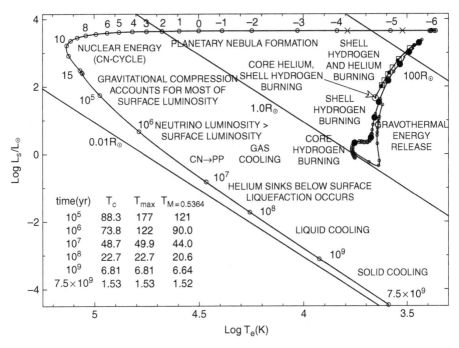

Physical processes in a population 1 model during evolution from before the main sequence to solidification as a white dwarf

induced by concentration gradients is a natural consequence of the use of the algorithms for computing diffusion described in Section 12.11.

It is interesting to record in the HR diagram where along the evolutionary track of the remnant of a low mass TPAGB star gravitational diffusion establishes a monoelemental surface abundance and to compare this position with where various other interesting physical processes occur during the terminal evolution of the remnant. This is done in Fig. 21.7.2. For completeness, interesting physical processes occurring prior to the superwind phase that terminates TPAGB evolution are noted in the upper right hand portion of the figure.

As in Fig. 21.3.1, numbers along the upper, nearly horizontal track in Fig. 21.7.2 denote the time in units of 10^3 yr relative to the time $t = 0$ when the model reaches $\log T_e = 4.5$ where, by decree, the planetary nebula phase officially begins. The crosses along the track close to the numbers -5 and -4 mark where the superwind phase is formally terminated. Exponentiated numbers along the track are in units of years. Eight main physical processes are singled out, beginning with planetary nebula formation and shell hydrogen burning during the planetary nebula stage. During this second stage, hydrogen burning is dominated by CN-cycle reactions and the hydrogen- and helium-exhausted core is kept at high temperatures by the deposition of hydrogen-burning ashes, just as during the prior phase of TPAGB evolution. The gravothermal energy liberated in the deep interior balances neutrino losses occasioned by the weak interaction induced plasma-neutrino process.

The rate of hydrogen-burning drops precipitously between $t \sim 11\,000$ yr and $t \sim 13\,000$ yr and, by $t \sim 30\,000$ yr, gravothermal energy liberated in regions above where gravothermal energy release balances neutrino losses by the plasma process is the dominant source of surface luminosity and this continues until $t \sim 10^7$ yr. The primary gravothermal source is the compressional energy released due to gravitational contraction. Although heavy ion cooling makes a minor direct contribution to the gravothermal energy release rate, it is directly responsible for the contraction which makes the major contribution. From $t \sim 3 \times 10^5$ yr to $\sim 10^7$ yr, the neutrino luminosity exceeds the surface luminosity, and neutrino losses are made up for by the liberation of gravothermal energy in the deep interior, with compressional energy released due to gravitational contraction being at first the major contributor, followed by the release of thermal energy by cooling ions and by cooling electrons near the top of the Fermi sea. Surface luminosity continues to be supplied by the liberation of gravothermal energy in layers of the star outside regions where neutrino losses dominate. At $t \sim 4 \times 10^6$ yr, the rate of hydrogen burning by the pp-chain reactions becomes larger than the rate of hydrogen burning by CN-cycle reactions.

Next, gravitationally induced diffusion becomes important near the surface and liquification begins in central regions. Helium and other heavy elements effectively vanish from the surface by $t \sim 10^8$ yr. From $\sim 10^8$ yr to $\sim 10^9$ yr, the surface luminosity is due primarily to the cooling of heavy ions in the liquid state. Finally, the phase transition to a solid begins at the center and works its way to the surface, until, by $t \sim 9 \times 10^9$ yr, most of the interior has solidified and the surface luminosity is supplied by cooling of heavy ions which are forced to oscillate about fixed sites in a lattice. In regions where temperatures are large compared with the Debye temperature, $\Theta_{\text{Debye}} \propto \sqrt{\rho}$, the cooling rate per heavy ion is the same as in the liquid state, namely, $(d/dt)(6\,kT)$ per ion. In regions where temperatures are such that $T \ll \Theta_{\text{Debye}}$, the cooling rate is proportional to $(T/\Theta_{\text{Debye}})^3$ and the contribution of ion cooling to surface luminosity increases outward in proportion to $T^3/\rho^{3/2}$. This regime begins when the isothermal temperature in the interior drops below $T \sim 5 \times 10^5$ K.

21.8 The relationship between the final white dwarf surface abundance and where in the thermal pulse cycle the precursor first leaves the AGB

The primary factor which determines the final surface abundance of a white dwarf is where in the last thermal pulse cycle its progenitor leaves the TPAGB for the first time. If the progenitor leaves during the first part of a quiescent hydrogen-burning phase, during subsequent evolution to the blue, not enough helium is added to the helium-rich layer below the hydrogen-burning shell for this layer to reach the critical mass necessary for the ignition of a helium shell flash. Evolution procedes in the fashion described in previous sections and the remant evolves into a DA white dwarf with a pure hydrogen surface layer.

If the white dwarf progenitor leaves the TPAGB near the end of a quiescent hydrogen-burning phase, hydrogen burning continues to add to the mass of the helium-rich layer and a final helium shell flash can occur as the progenitor evolves to the blue. What can happen next is shown in Fig. 21.8.1, which is Fig. 10 from Iben (1984). The evolutionary

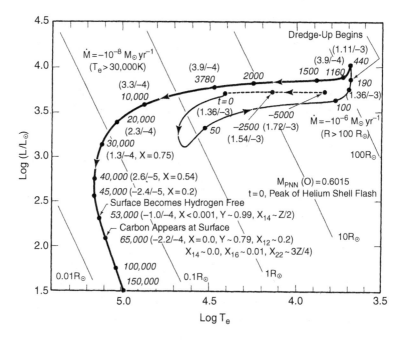

Fig. 21.8.1

Evolutionary track of a model PNN burning helium (Iben, 1984)

track in the figure is that of the remnant a low mass population II TPAGB model which has been subjected to mass loss at a superwind-like rate until the mass of the hydrogen-rich envelope is reduced to $M_e = 1.72 \times 10^{-3} M_\odot$, at which point mass loss has been turned off. The mass of the remnant is $0.6015 M_\odot$. Times along the track are in units of years and the quantities in parentheses give M_e in units of solar masses. As the remnant evolves to the blue along the horizontal dashed track, a helium shell flash is ignited at time $t = 0$ when the mass of the hydrogen-rich surface layer is $M_e = 1.36 \times 10^{-3} M_\odot$. The model at first dims and shrinks, but in less than 50 years it is evolving back toward giant dimensions, reaching a radius of $100 R_\odot$ at $t = 190$ yr.

By this time, hydrogen burning has been extinguished and the density and temperature in the entire hydrogen-rich portion of the model are less even than in matter which was not stripped off during the first superwind phase. Since the luminosity and radius of the model are similar to those of the precursor TPAGB model, and since the density–temperature distributions are nevertheless essentially the same (although the hydrogen profile is shifted to be much closer to the surface), it is reasonable to suppose that a similar type of mass loss will occur during what has come to be known as the born-again AGB phase. In the experiment described in Fig. 21.8.1, mass loss is re-instated at the modest rate of 10^{-6} $M_\odot$ yr^{-1} and continued until the model radius drops below $100 R_\odot$. The whole mass loss episode lasts 970 yr and the mass of the hydrogen-rich surface layer is reduced to $M_e = 3.9 \times 10^{-4} M_\odot$.

The model remnant evolves steadily to the blue burning helium quiescently and, when the surface temperature of the remnant reaches $T_e = 30\,000$ K, a fast, radiatively driven wind is invoked, of strength $\dot{M} = 10^{-8}\ M_\odot\ yr^{-1}$. By time $t = 53\,000$ yr, all hydrogen-rich matter has disappeared from the model, and, in another $12\,000$ yr of evolution, carbon appears at the surface although helium remains the dominant element at the surface. Ultimately, as the model continues to dim, the rate of mass loss decreases and diffusion leads to a pure helium surface which is the characteristic of DB white dwarfs. All of this is, of course, to be interpreted as a qualitative illustration of the sort of evolution that takes place in the real Universe.

The same end result is expected if the white dwarf precursor leaves the TPAGB shortly after a helium shell flash takes place. The only difference is that, instead of losing additional hydrogen-rich matter as a born-again AGB star, the precursor loses this additional hydrogen-rich layer during the helium-flash phase before leaving the TPAGB for the first and only time.

The time scale for evolution as the central star of a planetary nebula differs in two important ways for central stars which are burning helium and for those which are burning hydrogen. The first and most obvious difference is that, for a given PNN mass, the helium-burning central star remains bright and hot longer than its hydrogen-burning counterpart by a factor of roughly 2.5. This happens despite the fact that between ten and seven times more energy is released by the conversion of one gram of hydrogen into helium than by the conversion of one gram of helium into carbon and oxygen. The difference is due to the fact that because all but a small amount of hydrogen-rich matter has been stripped off during the superwind phase, the amount of hydrogen which can be converted into helium in the hydrogen-burning PNN remnant is much smaller than in the precursor during the interpulse phase on the TPAGB, whereas the amount of helium which can be burned in the helium-burning PNN model is essentially the same as during evolution as a TPAGB star. The second difference in time scales between hydrogen-burning PNNi and helium-burning PNNi is the fact that the precipitous decline in the nuclear burning rate that occurs in hydrogen-burning PNNi as the mass in the hydrogen-rich surface layer decreases below a critical value does not occur in helium-burning PNNi.

Since estimates of mass-loss rates from Mira variables show that the average mass-loss rate is correlated with the average luminosity and radius of the mass-losing star, it is reasonable to assume that model mass-loss rates are also correlated with luminosity and/or radius. Figure 17.3.3 demonstrates that the mean surface radius and luminosity during all phases of evolution are tightly coupled, so it is not necessary to decide with which variable the mass-loss rate is most closely tied. Figure 17.4.15 shows that during the quiescent hydrogen-burning phase, the luminosity and therefore the radius of the TPAGB star increase by a factor greater than two by the time a half dozen pulses have occurred. One therefore expects that, if departure from the TPAGB occurs during the quiescent hydrogen-burning phase, the probability of departure increases in proportion to ϕ, the ratio of the time the star has evolved since the beginning of the last pulse to the total time spent between pulses.

During the $\sim 10^3$ yr duration of the first of a set of helium shell flashes and its $\sim 10^4$ yr powerdown, the mean luminosity and radius undergo dramatic changes which must influence the mass-loss rate from the surface significantly. Figure 17.4.6 demonstrates that, after only five sets of pulses, the mean radius and therefore the luminosity reach minima and maxima which differ by over 50% from the radii and luminosities prevailing during the preceding interpulse phase. Figure 17.4.15 shows that the ratio of maxima to pre-pulse luminosities continues to grow with each pulse. Thus, even though the duration of a helium-burning episode is small compared with the interpulse lifetime, the probability of departure from the TPAGB during a helium-burning phase is not proportionately small, but the difference is not known from theory. Given this fact, the fraction of progenitors of white dwarfs which become DB white dwarfs rather than DA white dwarfs is best left to the observations to divulge. Indications are that the fraction is of the order of 10%. The fact that helium-burning central stars of a given mass remain bright and hot roughly 2.5 times as long as hydrogen-burning central stars of the same mass then suggests that perhaps at least 25% percent of all central stars in observed planetary nebulae are burning helium.

Since there is yet another channel for producing central stars of planetary nebulae that are burning helium, this fraction could be a considerable underestimate. This other channel leads to even more exotic surface abundances than have yet been elaborated upon and there are a plethora of observational counterparts which demonstrate that this channel is realized in the real world. Among the counterparts are the class of PG1159 stars that have been extensively analysed by Klaus Werner and his colleagues. For a comprehensive review, see Klaus Werner & Falk Herwig (2006). There are approximately 30 PG1159 stars, with GW Vir being a prominant member. They are all hydrogen-deficient if not completely devoid of hydrogen and typical abundances by number are $(Y(He), Y(C), Y(O)) = (0.08, 0.04, 0.01)$. They are all luminous and hot, with surface temperature in the range given by $4.7 \gtrsim \log T_e \gtrsim 5.2$, and they support wind mass loss at rates of the order of 10^{-8}–10^{-6} $M_\odot \, yr^{-1}$. Approximately half of them are associated with an identified nebula.

The results of processing by both hydrogen burning and helium burning are exhibited in the surface abundances. The most likely explanation of the surface abundances is that the progenitors of the observed remnants left the TPAGB while burning hydrogen with a relatively massive helium-rich layer below the hydrogen-burning shell and that, during the gravitational contraction episode that accompanies the precipitous decline in the hydrogen-burning luminosity which occurs near maximum surface temperature, the helium layer was compressed and heated sufficiently that a helium flash was ignited.

The occurrence of a final helium flash in the compact remnant of a former AGB star as hydrogen burning is being replaced by grovothermal energy production as the major source of surface luminosity was predicted by M. Y. Fujimoto (1977). Modeling the consequences of the flash has progressed through explorations by Iben, J. B. Kaler, J. W. Truran, & A. Renzini (1983), Iben & Jim MacDonald (in *White Dwarfs*, ed. Detlev Koester & Klaus Werner, 1995), F. Herwig, T. Blöcker, N. Langer, & T. Driebe (1999), F. Herwig (2001), and T. M. Lawler & J. MacDonald (2003). The basic physics is this: as the helium shell flash strengthens, the outer edge of the convective shell forced by the energy flux from the helium-burning region reaches the base of the hydrogen profile. Hydrogen is ingested into the convective region and diffuses convectively inward until it reaches high enough temperatures that a hydrogen shell flash is ignited. This flash supports a new convective shell which detaches from the first one and contains products of both complete hydrogen burning and partial helium burning which are now subjected to a second phase of hydrogen burning. Not unexpectedly, the final products are rather exotic.

A most interesting result of the Herwig *et al.* (1999) and Herwig (2001) explorations is that the time scale for evolution in the HR diagram following a helium shell flash which occurs in a central star remant of white dwarf dimensions is very sensitive to the adopted time scale for the inward mixing of hydrogen into the convective shell forced by fluxes from the helium-burning region. By choosing a mixing time scale smaller by a factor of 100 than that predicted by a standard mixing length treatment, Herwig (2001) is able to account for the time scale ($\sim$6 yr) for the evolution in the HR diagram of Sakurai's object, which is commonly acknowledged to have experienced a born-again AGB episode. By choosing a mixing time scale a factor of 100 smaller still, Lawler and MacDonald (2002) are able to account for the very short transit times of V605 Aql ($\sim$2 yr) and Sakurai's object, as well as for the much longer transit time scale ($>$120 yr) of FG Sge, one of the first stars to be recognized as being in the process of experiencing a born-again AGB episode.

21.9 Theoretical and observed white dwarf number–luminosity distributions and the age of the galactic disk

By comparing the number versus magnitude distribution defined by observed central stars of planetary nebulae and observed white dwarfs in the solar vicinity with theoretical distributions constructed from cooling curves given by theoretical calculations, one can estimate both the rate at which stars of low and intermediate mass are born in the galactic disk and the age of the galactic disk as an active star-forming entity.

Since all stars initially in the $\sim$1–2 $M_\odot$ range become AGB stars with a nearly identical hydrogen-exhausted core and since the rate of mass loss during the TPAGB phase is high compared with the rate at which nuclear burning processes matter, these stars become central stars of planetary nebulae with similar luminosities and then evolve on similar time scales into white dwarfs of very similar mass. This accounts for the fact that the distribution of the majority of white dwarfs in the number–mass plane is fairly narrow, peaking at $M_{WD} \sim$0.6 $M_\odot$. A zeroth order approximation to a theoretical distribution in number versus luminosity for white dwarfs and PNNi in the galactic disk can be constructed by assuming that (1) all such stars have the same mass, (2) all follow the same path in luminosity versus time, and (3) their birthrate is constant in time.

With these choices, the number dn of white dwarfs or PNNi per unit volume in a given logarithmic interval in luminosity at luminosity L is proportional to the amount of time spent evolving through this interval and one can write

$$\frac{dn}{d\log L} \propto -\frac{\nu}{V}\left(\frac{d\log L}{dt}\right)^{-1} = -\frac{\nu t}{V}\left(\frac{d\log L}{d\log t}\right)^{-1}, \qquad (21.9.1)$$

where ν is the total birthrate of white dwarfs in the galactic disk, V is the volume of the disk, $d\log L/d\log t$ is the slope of a cooling curve such as the one displayed in Fig. 21.6.6, and t is the time of evolution prior to reaching the luminosity L. Setting $V = 3 \times 10^{11}$ pc, $\nu = 0.5$ yr^{-1}, and adopting a cooling curve which assumes that 75% of all PNNi are

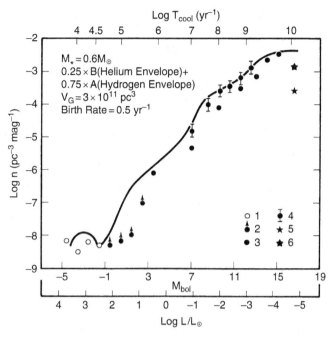

Fig. 21.9.1
Theoretical and observed luminosity functions for white dwarfs in our galaxy (Iben & Tutukov, 1984)

hydrogen burners and 25% are helium burners, all of mass 0.6 $M_\odot$, Iben & Alexander V. Tutukov (1984) constructed the luminosity function shown in Fig. 21.9.1. The points are observational estimates from a variety of sources, including J. B. Kaler (1983) for central stars of planetary nebulae and E. M. Sion & J. Liebert (1977), J. Liebert (1980), and R. F. Green (1980) for white dwarfs.

Comparing the theoretical and observational distributions one sees that, except at the very lowest luminosities, theory and observation are roughly concordant, but, at the lowest luminosities they are quite discordant. PNNi and white dwarfs with large luminosities have been recently formed, their birthrate reflects the current birthrate in the Galaxy, and the concordance between theory and observation can be taken as an indication that a birthrate of the order of one star every 2 years in a volume of 3×10^{11} pc^3 is a fair approximation to the current birthrate in the galactic disk.

The fact that, at the very lowest luminosities, the observed distribution falls far below the theoretical one could be attributed to a faulty equation of state which produces an erroneously large gravothermal energy-generation rate. If the slope of the theoretical cooling curve were constant, i.e., if $d \log L / d \log t = -\alpha =$ constant, eq. (21.9.1) shows that the luminosity function should increase with t and therefore steadily increase with decreasing luminosity. The fact that the luminosity function in Fig. 21.9.1 flattens out at the lowest luminosities is due to the fact that the cooling curves used in constructing the theoretical function steepen ($1/\alpha$ increases) at the lowest luminosities, a fact which might be construed as due to the onset of crystallization. However, it is evident from Fig. 21.6.6 that a cooling curve constructed with a more sophisticated equation of state exhibits an essentially constant slope ($\alpha \sim 0.65$) in the interval $-2 > \log(L/L_\odot) > -4.5$ and a repetition of the exercise that produces Fig. 21.9.1 with $\alpha = 0.65$ leads to a luminosity function which continues to increase with decreasing luminosity at the lowest luminosities, exacerbating the discrepancy between the theoretical and observed functions. Another possibility to consider is that observational selection effects may not have been properly taken into account in constructing the observed luminosity function and that the real space density of white dwarfs at the lowest luminosities has been considerably underestimated.

A much, much simpler interpretation of the dropoff in the luminosity function at the lowest luminosities is that it is primarily the consequence of the finite age of the galactic disk, a reality which has been ignored in the construction of the zeroth order luminosity function. It is obvious that the cooling age of the dimmest white dwarf in the galactic disk must be less than the time between the present and the time when star formation in the galactic disk began by an amount equal to the lifetime of the precursor of this white dwarf and that, therefore, the luminosity function defined by real stars in the disk must fall off sharply to zero at a finite cooling age. The ingredients of a proper theoretical treatment are given by Franca D'Antona & Italo Mazzitelli (1978).

Assuming that the number–mass distribution of white dwarfs at $L = 10^{-2} L_\odot$ is the same as the overall number–mass distribution of white dwarfs, Donald E. Winget et al. (1987) use mass-dependent theoretical cooling curves to construct theoretical luminosity functions which, when compared with the observationally based luminosity function given by J. Liebert, C. C. Donn, & D. G. Monet (1988), lead to an estimate of the age of the galactic disk of $9.3(\pm 2) \times 10^9$ yr, or several billion years younger than estimates of the ages of globular clusters in the galactic halo.

Following the prescriptions of D'Antona and Mazzitelli, Iben & Greg Laughlin (1989) write

$$l\frac{dn(l)}{dl} = -\int_{M_1(l)}^{M_2} \phi(t)\,\frac{dN}{dM}\left(t_{\text{cool}}(l)\frac{d\log t_{\text{cool}}(l)}{d\log l}\right)_{m(M)} dM, \qquad (21.9.2)$$

where $l = L/L_\odot$ is the normalized luminosity of the PNN or white dwarf, $n(l)$ is the space density of compact remnants at luminosity l, $\phi(t)$ is the birth rate of progenitors at time t, dN/dM is the variation with mass of the space density of progenitor stars at time t, $d\log t_{\text{cool}}/d\log l$ is the inverse of the slope of the cooling curve and is, in principle, a function of the mass $m(M)$ of the remnant of the progenitor of mass M. For every value of t, the mass of progenitors at l is given by

$$t + t_{\text{evo}}(M) + t_{\text{cool}}(l, m(M)) = t_{\text{disk}}, \qquad (21.9.3)$$

where $t_{\text{evo}}(M)$ is the pre-white dwarf lifetime of the progenitor of mass M and t_{disk} is the age of the disk as an active producer of stars. Thus, M_1 is defined by

$$t_{\text{evo}}(M_1) + t_{\text{cool}}(l, m(M_1)) = t_{\text{disk}}, \qquad (21.9.4)$$

and M_2 is the maximum mass of a star which can produce a white dwarf.

For simplicity, suppose that the progenitor lifetime prior to formation of a compact remnant is a power law of the form

$$t_{\rm evo}(M) = t_\beta \left(\frac{M_\odot}{M} \right)^\beta, \tag{21.9.5}$$

where t_β and β are constants over extended intervals in M, but can vary from one interval to the next. From Fig. 6 in Iben and Laughlin (1989), which describes results in the literature from a number of studies, for stars less massive than ~ 2 $M_\odot$, $\beta \sim 3.5$ and $t_\beta \sim 10^{10}$ yr, and for more massive stars, $\beta \sim 2.35$ and $t_\beta \sim 5 \times 10^9$ yr. The two approximations match at $M = 1.83$ $M_\odot$. Combining eqs. (21.9.4) and (21.9.5), one has

$$\frac{M_\odot}{M_1} = \left(\frac{t_{\rm disk} - t_{\rm cool}(l, m(M_1))}{t_\beta} \right)^{1/\beta}. \tag{21.9.6}$$

Adopting a Salpeter-like mass function,

$$\frac{{\rm d}N}{{\rm d}M} = 1.3 \, \frac{1}{M_0} \left(\frac{M_0}{M} \right)^{2.3}, \tag{21.9.7}$$

which gives unity when integrated from M_0 to ∞, assuming a constant birthrate function,

$$\phi(t) = \phi_0, \tag{21.9.8}$$

and supposing that the cooling curve is independent of $m(M)$, eq. (21.9.2) becomes

$$l \, \frac{{\rm d}n}{{\rm d}l} = -\phi_0 \left(\frac{M_0}{M_\odot} \right)^{1.3} \frac{{\rm d}\log t_{\rm cool}(l)}{{\rm d}\log l} \, t_{\rm cool}(l) \left\{ \left(\frac{t_{\rm disk} - t_{\rm cool}(l)}{t_\beta} \right)^{1.3/\beta} - \frac{1}{(10.5)^{1.3}} \right\}, \tag{21.9.9}$$

where a maximum mass of $M_2 = 10.5$ $M_\odot$ (Claudio Ritossa, Iben, & Enrique García-Berro, (1999) has been adopted.

Notice that when $t_{\rm cool} \ll t_{\rm disk} \sim 10^{10}$yr, eq. (21.9.9) exhibits essentially the same dependence on cooling curve variables l and $t_{\rm cool}$ as does eq. (21.9.1), the zeroth order approximation to the luminosity function. That is, for cooling times small compared with the age of the disk, the zeroth order approximation is a perfectly adequate formulation and a fit to the luminosity function provides a reasonable estimate of the current birthrate of stars in the galactic disk.

The gradient of a cooling curve is a fairly slowly varying function of $t_{\rm cool}(l)$, as is evident from an inspection of the curve $\log L_s$ versus $\log t$ in Fig. 21.6.6. Over the range $7.25 < \log t < 9.75$, the gradients of mass-dependent cooling curves constructed by Winget et al. (1987) are essentially constant at d $\log t_{\rm cool}(l)/{\rm d} \log l \sim 0.75$, independent of the mass of the white dwarf model. At $\log t \sim 9.75$ the gradients decrease abruply to a value between 0.33 and 0.2 for model masses between 0.8 $M_\odot$ and 1.0 $M_\odot$. Over the range $8.00 < \log t < 9.75$, Iben & McDonald (1986) find a constant slope of 0.62 for an 0.6 $M_\odot$ model.

A straight line approximation to the cooling curve ($\log L_s$ versus $\log t$) in Fig. 21.6.6 which passes through the last point in the figure can be written as

$$t_{cool}(\text{yr}) = 7.5 \times 10^9 \left(\frac{L_0}{L}\right)^\alpha, \tag{21.9.10}$$

where

$$L_0 = 10^{-4.5} L_\odot. \tag{21.9.11}$$

The choice $\alpha = 0.65$ provides a tolerable fit to the cooling curve over the range $8 < \log t < 10$. The choice $\alpha = 0.77$ provides a tolerable fit over the range $4.5 < \log t < 7.5$ and provides a rough fit to the cooling curve over the entire range from $\log t = 4.5$ to $\log t = 10.0$.

Given these results, it is evident that the luminosity function is primarily a function of t_{cool} and that, as t_{cool} approaches t_{disk}, the luminosity function reaches a maximum which can be estimated by taking the derivative of the product of t_{cool} and the first term in curly brackets in eq. (21.9.9) and setting the result equal to zero. This exercise gives

$$t_{cool}^{max} \sim \frac{t_{disk}}{1 + 1.3/\beta} = (0.64 - 0.73)\, t_{disk}, \tag{21.9.12}$$

where the value 0.64 follows when $\beta = 2.35$ and the value 0.73 follows when $\beta = 3.5$. The pre-white dwarf lifetime of the lowest mass stars which contribute to the white dwarf population at the maximum in the luminosity function is $t_{evo}(M_1) = t_{disk} - t_{cool}^{max} \sim 0.36 - 0.27\, t_{disk}$. Thus, if $t_{disk} \sim 10^{10}$ yr, $t_{evo}(M_1) \sim 3.1 \times 10^9$ yr. Choosing $\beta = 3.5$, eq. (21.9.6) gives $M_1 \sim 1.4\ M_\odot$. Stars of such a low mass do not make a major contribution to the luminosity function at its maximum, but their progeny are present in small numbers.

Setting the quantity in curly brackets in eq. (21.9.9) equal to zero determines the relationship between the cutoff luminosity l_{cutoff} and t_{disk} and demonstrates that, at the cutoff, $M_1 = M_2$. Thus, the dimmest white dwarfs are the most massive white dwarfs. Their masses are close to 1.37 $M_\odot$, and they are composed primarily of oxygen and neon. This being the case, the cooling curve for them may differ significantly from that for an 0.6 $M_\odot$ white dwarf composed primarily of carbon and oxygen. However, from a pedagogical perspective, the insight provided by a simple development which ignores this complication is justified.

The term in curly brackets involving $M_2 = 10.5\ M_\odot$ in eq. (21.9.9) is small enough that, for an approximate estimate of the age of the galactic disk, one may set the term in square brackets equal to zero. Thus,

$$t_{disk} \sim 7.5 \times 10^9\ \text{yr} \left(\frac{l_0}{l_{cutoff}}\right)^\alpha. \tag{21.9.13}$$

Precisely because the number of stars in any fixed magnitude interval near cutoff becomes very small, the estimate of the real world cutoff luminosity is not trivial. Among other uncertainties, an estimate of l_{cutoff} depends on how magnitude binning intervals are chosen and the estimate of t_{disk} depends on the choice of α in eq. (21.9.13). For any choice of α, if $l_{cutoff} = l_0$, then $t_{disk} = 7.5 \times 10^9$ yr, a result which is obvious from inspection of Fig. 21.6.6, without going through the lengthy set of steps just concluded. If $l_{cutoff} = 10^{-4.6}$,

$t_{\text{disk}} = 8.7 \times 10^9$ yr for $\alpha = 0.65$ and $t_{\text{disk}} = 9.0 \times 10^9$ yr for $\alpha = 0.77$. If $l_{\text{cutoff}} = 10^{-4.7}$, $t_{\text{disk}} = 10.1 \times 10^9$ yr for $\alpha = 0.65$ and $t_{\text{disk}} = 10.7 \times 10^9$ yr for $\alpha = 0.77$. Although results for two choices for α have been presented, this has been done simply to demonstrate the dependence on α. Given that the smaller choice of α is a better representation near the cutoff luminosity, the smaller ages are to be preferred.

In principle, another way of estimating disk age is to determine the luminosity $l = l_{\text{max}}$ at the maximum in an observed luminosity function, find the corresponding t_{cool} either by inspection of an appropriate cooling curve or by use eq. (21.9.10), and then use eq. (21.9.12) to determine t_{disk}. By insisting that the two methods give the same results, one has from eqs. (21.9.10) and (21.9.13) and from eq. (21.9.12), with the choice 0.64, that

$$\log\left(\frac{l_{\text{max}}}{l_{\text{cutoff}}}\right) = \frac{0.194}{\alpha} \sim 0.30. \tag{21.9.14}$$

The number on the far right of eq. (21.9.14) is the consequence of choosing $\alpha = 0.65$.

The rather clearly defined factor of 2 ratio of l_{max} to l_{cutoff} that simple theory predicts on the assumption that the stellar birthrate is constant is not reflected in most published observationally based luminosity functions which show a ratio lying bieween 3 and 5. The most straightforward explanation of the apparent discrepancy is that the rate of star formation in the disk of the galaxy and the vertical dimensions of the disk have changed with time. A stellar birthrate which passes through a maximum shifts the peak in the theoretical white dwarf luminosity function to a larger luminosity (shorter cooling time) and a gradual increase with time in the thickness of the disk translates into an underestimate of the current number density of the oldest white dwarfs and, thus, to an underestimate of the cooling time corresponding to the age of the disk. It is evident that, because of the many uncertainities involved, in the physics of cooling white dwarfs, in the physics of galactic formation and evolution, and in the establishment of a reliable observational luminosity function for white dwarfs, a definitive determination of the age of the galactic disk has yet to be achieved. This should be viewed as a positive result, in that future generations inherit a challenge to confront rather than a fait accomplis.

This concludes a long journey exploring physical processes occurring in stars and examining how these processes influence the evolution of stars during quasistatic and dynamic phases. The processes and the evolution are entertaining in their own right. Just as important, contemplation of the processes and of the evolution enhance our enjoyment of (1) the individual stars we view in the night sky, (2) the interstellar medium whose characteristics are both responsible for the formation of stars and dramatically affected by the heavy-element-enhanced matter which stars return to the medium and the kinetic energy which they inject into the medium, (3) life itself, which would be impossible without the nucleosynthesis that occurs in stars and the radiant energy which they emit, (4) galaxies, whose distances and properties can be deep searched by examining the luminosities and kinematics of the stars which they contain, and (5) the Universe, understanding the characteristics of which has been enormously aided by the fact that evolving stars of various types act as standard candles, laying the foundation for the cosmological distance scale.

Bibliography and references

R. Abe, *Prog. Theor. Phys.*, **22**, 213, 1959.

Franca D'Antona & Italo Mazzitelli, *A&A*, **66**, 453, 1978.

G. H. Bowen, *ApJ*, **329**, 299, 1988.

G. H. Bowen & Lee Ann Wilson, *ApJL*, **375**, L53, 1991.

D. L. Bowers & Edwin E. Salpeter, *Phys. Rev.*, **119**, 1180, 1960.

W. J. Carr, *Phys. Rev.*, **122**, 1437, 1961.

Peter Debye, *Ann. Physik*, **39**, 789, 1912.

Pierre Louis Dulong & Alexis Thérèse Petit, *Annalen der Physik*, **58**, 254D, 1818.

Enrico Fermi, *Zeits. für Physik*, **48**, 73, 1928.

Masayuki Y. Fujimoto, *PASJ*, **29**. 331, 1977.

H. R. Glyde & G. H. Keech, *Ann. Phys.*, **127**, 330, 1980.

R. F. Green, *ApJ*, **238**, 685, 1980.

J. P. Hansen & R. Mazighi, *Phys. Rev. A.*, **18**, 1282, 1978.

J. P. Hansen, G. M. Torrie, & P. Viellefosse, *Phys. Rev. A.*, **16**, 2153, 1977.

Werner Heisenberg, *Zeits. für Physik*, **43**, 172, 1927.

Falk Herwig, *ApJ*, **554**, L71, 2001.

Falk Herwig, Thomas Blöcker, Norbert Langer, & T. Driebe, *A&A*, **349**, L5, 1999.

Icko Iben, Jr., *ApJ*, **277**, 333, 1984.

Icko Iben, Jr., *Phys. Rep.*, **250**, 1, 1995.

Icko Iben, Jr., Masayuki Y. Fujimoto, & Jim MacDonald, *ApJ*, **388**, 521, 1992.

Icko Iben, Jr., James B. Kaler, James W. Truran, & Alvio Renzini, *ApJ*, **264**, 605, 1983.

Icko Iben, Jr. & Greg Laughlin, *ApJ*, **341**, 312, 1989.

Icko Iben, Jr. & Jim MacDonald, *ApJ*, **296**, 540, 1985; **301**, 164, 1986.

Icko Iben, Jr. & Jim MacDonald, in *White Dwarfs*, eds. Detlev Koester & Klaus Werner, (Berlin: Springer), 48, 1995.

Icko Iben, Jr. & Alexander V. Tutukov, *ApJ*, **282**, 615, 1984; *ApJ*, **418**, 343, 1993.

F. D. Kahn & K. A. West, *MNRAS*, **212**, 837, 1985.

James B. Kaler, *Stars and their Spectra*, (Cambridge: Cambridge University Press), 1981.

James B. Kaler, *ApJ*, **217**, 188, 1983.

James B. Kaler & G. H. Jacoby, *ApJ*, **382**, 134, 1991.

Charles Kittel, *Introduction to Solid State Physics*, (New York: Wiley), 1953, second edition, 1957, fourth edition, 1996.

Detlef Koester & Volker Weidemann, *A&A*, **81**, 145, 1980.

Sun Kwok, C. R. Purton, & P. M. FitzGerald, *ApJL*, **219**, L127, 1978.

Lev D. Landau & E. M. Lifschitz, *Statistical Physics*, (London: Pergamon), 1958.

T. M. Lawler & Jim MacDonald, *ApJ*, **583**, 913, 2003.

James Liebert, *ARAA*, **18**, 363, 1980.

James Liebert, C. C. Donn, & D. G. Monet, *ApJ*, **332**, 1988.

F. A. Lindemann, *Phys. Zeits.*, **11**, 609, 1910.

Leon Mestel and Mal A. Ruderman, *MNRAS*, **136**, 27, 1967.

Donald E. Osterbrock, *PASP*, **100**, 412, 1988.

Bogdan Paczyǹski, *Acta Astron.*, **21**, 417, 1971.

M. Perinotto, *Rep. Prog. Phys.*, **53**, 1559, 1990.

Stuart R. Pottash, *Planetary Nebulae*, (Dordrecht: Kluwer), 1984.

Alvio Renzini, in *Planetary Nebulae*, ed. Sylvia Torres-Peimbert, (Dordrecht: Kluwer), 391, 1989.

Claudio Ritossa, Icko Iben, Jr., & Enrique García-Berro, *ApJ*, **515**, 382, 1999.

Edwin E. Salpeter, *ApJ*, **134**, 669, 1961.

Detlef Schönberner, *A&A*, **103**, 119, 1981.

Stuart L. Shapiro and Saul A. Teukolsky, *Black Holes, White Dwarfs, and Neutron Stars*, (New York: Wiley), 1983.

Edward M. Sion & Jim Liebert, *ApJ*, **213**, 468, 1977.

W. L. Slattery, G. D. Doolen, & H. E. DeWitt, *Phys. Rev. A.*, **21**, 2087, 1980; **26**, 2255, 1982.

K. S. Krishna Swamy & Ted P. Stecher, *PASP*, **81**, 873K, 1969.

L. H. Thomas, *Proc. Cambridge. Phil. Soc.*, **23**, 542, 1927.

Klaus Werner & Falk Herwig, *PASP*, **118**, 183, 2006.

N. C. Wickramasinghe, Bertram Donn, & Ted P. Stecher, *ApJ*, **146**, 590, 1966.

Eugene Wigner & Fred Seitz, *Phys. Rev.*, **46**, 523, 1934.

Lee Ann Willson & S. J. Hill, *ApJ*, **228**, 854, 1979.

Donald E. Winget, Carl J. Hansen, James Liebert, *et al., ApJL*, **315**, L77, 1987.

Peter R. Wood and Donald J. Faulkner, *ApJ*, **307**, 659, 1986.

Index

Printed in the United States
by Baker & Taylor Publisher Services